Computational Models for Polydisperse Particulate and Multiphase Systems

Providing a clear description of the theory of polydisperse multiphase flows, with emphasis on the mesoscale modeling approach and its relationship with microscale and macroscale models, this all-inclusive introduction is ideal, whether you are working in industry or academia. Theory is linked to practice through discussions of key real-world cases (particle/droplet/bubble coalescence, breakup, nucleation, advection and diffusion, and physical- and phase-space), providing valuable experience in simulating systems that can be applied to your own applications. Practical cases of QMOM, DQMOM, CQMOM, EQMOM, and ECQMOM are also discussed and compared, as are realizable finite-volume methods. This provides the tools you need to use quadrature-based moment methods, choose from the many available options, and design high-order numerical methods that guarantee realizable moment sets. In addition to the numerous practical examples, MATLAB scripts for several algorithms are also provided, so you can apply the methods described to practical problems straight away.

Daniele L. Marchisio is an Associate Professor at the Politecnico di Torino, Italy, where he received his Ph.D. in 2001. He has held visiting positions at the Laboratoire des Sciences du Génie Chimique, CNRS–ENSIC (Nancy, France), Iowa State University (USA), Eidgenössische Technische Hochschule Zürich (Switzerland), and University College London (UK), and has been an invited professor at Aalborg University (Denmark) and the University of Valladolid (Spain). He acts as a referee for the key international journals in his field of research. He has authored more than 60 scientific papers and 5 book chapters, and co-edited the volume *Multiphase Reacting Flows* (Springer, 2007).

Rodney O. Fox is the Anson Marston Distinguished Professor of Engineering at Iowa State University (USA), an Associate Scientist at the US-DOE Ames Laboratory, and a Senior Research Fellow in the EM2C laboratory at the Ecole Centrale Paris (France). His numerous professional awards include an NSF Presidential Young Investigator Award in 1992 and Fellow of the American Physical Society in 2007. The impact of Fox's work touches every technological area dealing with multiphase flow and chemical reactions. His monograph *Computational Models for Turbulent Reacting Flows* (Cambridge University Press, 2003) offers an authoritative treatment of the field.

Computational Models for Polydisperse Particulate and Multiphase Systems

DANIELE L. MARCHISIO
Politecnico di Torino

RODNEY O. FOX
Iowa State University

CAMBRIDGE
UNIVERSITY PRESS

CAMBRIDGE UNIVERSITY PRESS
Cambridge, New York, Melbourne, Madrid, Cape Town,
Singapore, São Paulo, Delhi, Mexico City

Cambridge University Press
The Edinburgh Building, Cambridge CB2 8RU, UK

Published in the United States of America by Cambridge University Press, New York

www.cambridge.org
Information on this title: www.cambridge.org/9780521858489

First published 2013

Printed and bound in the United Kingdom by the MPG Books Group

A catalog record for this publication is available from the British Library

Library of Congress Cataloging in Publication Data
Marchisio, Daniele L.
Computational models for polydisperse particulate and multiphase systems / Daniele L.
Marchisio, Politecnico di Torino, Rodney O. Fox, Iowa State University.
pages cm. – (Cambridge series in chemical engineering)
ISBN 978-0-521-85848-9
1. Multiphase flow – Mathematical models. 2. Chemical reactions – Mathematical models.
3. Transport theory. 4. Dispersion – Mathematical models. I. Fox, Rodney O., 1959–
II. Title.
TA357.5.M84M37 2013
532′.56–dc23
2012044073

ISBN 978-0-521-85848-9 Hardback

a Giampaolo
à Roberte

Contents

Preface

This book is intended for graduate students in different branches of science and engineering (i.e. chemical, mechanical, environmental, energetics, etc.) interested in the simulation of *polydisperse multiphase flows*, as well as for scientists and engineers already working in this field. The book provides, in fact, a systematic and consistent discussion of the basic theory that governs polydisperse multiphase systems, which is suitable for a neophyte, and presents a particular class of computational methods for their actual simulation, which might interest the more experienced scholar.

As explained throughout the book, disperse multiphase systems are characterized by multiple phases, with one phase continuous and the others dispersed (i.e. in the form of distinct particles, droplets, or bubbles). The term polydisperse is used in this context to specify that the relevant properties characterizing the elements of the disperse phases, such as mass, momentum, or energy, change from element to element, generating what are commonly called *distributions*. Typical distributions, which are often used as characteristic signatures of multiphase systems, are, for example, a crystal-size distribution (CSD), a particle-size distribution (PSD), and a particle-velocity distribution.

The problem of describing the evolution (in space and time) of these distributions has been treated in many ways by different scientific communities, focusing on aspects most relevant to their community. For example, in the field of crystallization and precipitation, the problem is described (often neglecting spatial inhomogeneities) in terms of crystal or particle size, and the resulting governing equation is called a *population-balance equation* (PBE). In the field of evaporating (and non-evaporating) sprays the problem is formulated in terms of the particle surface area and the governing equation is referred to as the *Williams–Boltzmann equation*. In this and other fields great emphasis has been placed on the fact that the investigated systems are spatially inhomogeneous. Aerosols and ultra-fine particles are often described in terms of particle mass, and the final governing equation is called the *particle-dynamics equation*. Particulate systems involved in granular flows have instead been investigated in terms of particle velocity only, and the governing equation is the inelastic extension to multiphase systems of the well-known *Boltzmann equation* (BE) used to describe molecular velocity distributions in gas dynamics.

Although these apparently different theoretical frameworks are referred to by different names, the underlying theory (which has its foundation in classical statistical mechanics) is exactly the same. This has also generated a plethora of numerical methods for the solution of the governing equations, often sharing many common elements, but generally with a specific focus on only part of the problem. For example, in a PBE the distribution representing the elements constituting the multiphase system is often discretized into *classes* or *sections*, generating the so-called discretized population-balance equation (DPBE). Among the many methods developed, one widely used among practitioners in computational fluid dynamics (CFD) is the multiple-size-group (MUSIG) method. This approach resembles, in its basic ideas, the discretization carried out for the BE in the so-called discrete-velocity method (DVM). Analogously, the *method of moments* (MOM) has been used for the solution of both PBE and BE, but the resulting *closure problem* is overcome by following different strategies in the two cases. In the case of the BE the most popular moment closure is the one proposed by Grad, which is based on the solution of a subset of 13 or 26 moments, coupled with a presumed functional form for the velocity distribution. In contrast, in the case of a PBE the closure strategy often involves interpolation among the known moments (as in the method of moments with interpolative closure, MOMIC). Given the plethora of approaches, for the novice it is often impossible to see the connections between the methods employed by the different communities.

This book provides a consistent treatment of these issues that is based on a general theoretical framework. This, in turn, stems from the *generalized population-balance equation* (GPBE), which includes as special cases all the other governing equations previously mentioned (e.g. PBE and BE). After discussing how this equation originates, the different computational models for its numerical solution are presented. The book is structured as follows.

- Chapter 1 introduces key concepts, such as flow regimes and relevant dimensionless numbers, by using two examples: the PBE for fine particles and the KE for gas–particle flow. Subsequently the mesoscale modeling approach used throughout the book is explained in detail, with particular focus on the relation to microscale and macroscale models and the resulting closure problems.

- Chapter 2 provides a brief introduction to the mesoscale description of polydisperse systems. In this chapter the many possible number-density functions (NDF), formulated with different choices for the internal coordinates, are presented, followed by an introduction to the PBE in their various forms. The chapter concludes with a short discussion on the differences between the moment-transport equations associated with the PBE, and those arising due to ensemble averaging in turbulence theory.

- Chapter 3 provides an introduction to Gaussian quadrature and the moment-inversion algorithms used in quadrature-based moment methods (QBMM). In this chapter, the product–difference (PD) and Wheeler algorithms employed for the classical univariate quadrature method of moments (QMOM) are discussed, together with the brute-force, tensor-product, and conditional QMOM developed for multivariate problems. The chapter concludes with a discussion of the extended quadrature method of moments (EQMOM) and the direct quadrature method of moments (DQMOM).

- In Chapter 4 the GPBE is derived, highlighting the closures that must be introduced for the passage from the microscale to the mesoscale model. This chapter also contains an overview of the mathematical steps needed to derive the transport equations for the moments of the NDF from the GPBE. The resulting moment-closure problem is also throughly discussed.

- Chapter 5 focuses on selected mesoscale models from the literature for key physical and chemical processes. The chapter begins with a general discussion of the mesoscale modeling philosophy and its mathematical framework. Since the number of mesoscale models proposed in the literature is enormous, the goal of the chapter is to introduce examples of models for advection and diffusion in real and phase space and zeroth-, first-, and second-order point processes, such as nucleation, breakage, and aggregation.

- Chapter 6 is devoted to the topic of hard-sphere collision models (and related simpler kinetic models) in the context of QBMM. In particular, the exact source terms for integer moments due to collisions are derived in the case of inelastic binary collisions between two particles with different diameters/masses, and the use of QBMM to overcome the closure problem is illustrated.

- Chapter 7 is devoted to solution methods of the spatially homogeneous GPBE, including class and sectional methods, MOM and QBMM, and Monte Carlo methods. The chapter concludes with a few examples comparing solution methods for selected homogeneous PBE.

- Chapter 8 focuses on the use of moment methods for solving a spatially inhomogeneous GPBE. Critical issues with spatially inhomogeneous systems are moment realizability and corruption (due to numerical advection and diffusion operator) and the presence of particle trajectory crossing (PTC). These are discussed after introducing kinetics-based finite-volume methods, by presenting numerical schemes capable of preserving moment realizability and by demonstrating with practical examples that QBMM are ideally suited for capturing PTC. The chapter concludes with a number of spatially one-dimensional numerical examples.

- To complete the book, four appendices are included. Appendix A contains the MATLAB scripts for the most common moment-inversion algorithms presented in Chapter 3. Appendix B discusses in more detail the kinetics-based finite-volume methods introduced in Chapter 8. Finally, the key issues of PTC in phase space, which occurs in systems far from collisional equilibrium, and moment conservation with some QBMM are discussed in Appendix C and Appendix D, respectively.

The authors are greatly indebted to the many people who contributed in different ways to the completion of this work. Central in this book is the pioneering research of Dr. Robert L. McGraw, who was the first to develop QMOM and the Jacobian matrix transformation (which is the basis for DQMOM) for the solution of the PBE, and brought to our attention the importance of moment corruption and realizability when using moment methods. The authors are therefore especially grateful to Professor Daniel E. Rosner, who in 1999 directed their attention to the newly published work of Dr. McGraw on QMOM. They would also like to thank Professor R. Dennis Vigil for recognizing the capability of QMOM for solving aggregation and breakage problems, and Professor Prakash Vedula for providing the mathematical framework used to compute the moment source terms for hard-sphere collisions reported in Chapter 6.

A central theme of the solution methods described in this book is the importance of maintaining the realizability of moment sets in the numerical approximation. On this point, the authors are especially indebted to Professor Marc Massot for enlightening them on the subtleties of kinetics-based methods for hyperbolic systems and the general topic of particle trajectory crossings. Thanks to the excellent numerical analysis skills of Professor Olivier Desjardins and a key suggestion by Dr. Philippe Villedieu during the 2006 Summer Program at the Center for Turbulence Research, Professor Massot's remarks eventually pointed us in the direction of the realizable finite-volume schemes described in Chapter 8. In this regard, we also want to acknowledge the key contributions of Professor Z. J. Wang in the area of high-order finite-volume schemes and Dr. Varun Vikas for the development and implementation of the realizable quasi-high-order schemes described in Appendix B.

The idea of publishing this book with Cambridge University Press is the result of the interest shown in the topic by Professor Massimo Morbidelli. The contribution of many other colleagues is also gratefully acknowledged, among them Antonello A. Barresi, Marco Vanni, Giancarlo Baldi, Miroslav Soos, Jan Sefcik, Christophe Chalons, Frédérique Laurent, Hailiang Liu, Alberto Passalacqua, Venkat Raman, Julien Reveillon, and Shankar Subramaniam. All the graduate students and post-doctoral researchers supervised by the authors in the last ten years who have contributed to the findings reported in this book are gratefully acknowledged and their specific contributions are meticulously cited.

The research work behind this book has been funded by many institutions and among them are worth mentioning the European Commission (DLM), the Italian Ministry of Education, University, and Research (DLM), the ISI Foundation (DLM, ROF), the US National Science Foundation (ROF), the US Department of Energy (ROF), the Ecole Centrale Paris (ROF), and the Center for Turbulence Research at Stanford University (ROF). The constant stimulus and financial support of numerous industrial collaborators (ENI, Italy; BASF, Germany; BP Chemicals, USA; Conoco Phillips, USA; and Univation Technologies, USA) are also deeply appreciated.

Notation

Upper-case Roman

$\mathbb{A}$	generic particle acceleration due to buoyancy, gravity, and drag
$\mathbf{A}$	coefficient matrix in DQMOM and brute-force QMOM for determining the quadrature approximation
$\mathbf{A}$	coefficient matrix constituted by mixed moments for calculating velocity parameters u_n in inhomogeneous systems
A_{D}	particle cross-section surface area
A_{eq}	area of equivalent sphere
A_{H}	Hamaker parameter
A_{p}	particle surface area
$\mathbf{A}_{\mathrm{f}}$	fluid acceleration due to body forces
$\mathbf{A}_{\mathrm{fp}}$	pure particle acceleration due to fluid–particle momentum exchange
$\mathbf{A}_{\mathrm{p}}$	pure particle acceleration due to body forces
$\mathbf{A}_{\mathrm{p},0}$	mean particle-acceleration term
$\mathbf{A}_{\mathrm{pf}}$	pure fluid acceleration due to fluid–particle momentum exchange
$\mathbf{A}^{(n)}$	acceleration acting on nth particle due to body forces (for particles in vacuum)
$\mathbf{A}_{\mathrm{f}}^{(n)}$	acceleration acting on the fluid in the neighborhood of the nth particle due to pressure, body, and viscous forces
$\mathbf{A}_{\mathrm{fp}}^{(n)}$	acceleration acting on the nth particle due to fluid–particle forces
$\mathbf{A}_{\mathrm{p}}^{(n)}$	acceleration acting on the nth particle due to body forces (for particles suspended in a fluid)
$\mathbf{A}_{\mathrm{pf}}^{(n)}$	acceleration acting on the fluid in the neighborhood of the nth particle due to fluid–particle forces
$\mathbf{A}_{\mathrm{fp}}^{*}$	global particle acceleration due to fluid–particle momentum exchange (including diffusion terms)
$\mathbf{A}_{i}^{*}$	coefficient matrix constituted by moments $m_{j,k;i}^{*}$ for calculating velocity parameters u_n in inhomogeneous systems with FVM
A_{kl}^{ij}	collision frequencies between particle-velocity classes in DVM
$\langle \mathbf{A}_{\mathrm{f}} \rangle_1$	multi-particle conditional expected fluid acceleration due to body forces
$\langle \mathbf{A}_{\mathrm{fp}} \rangle_1$	multi-particle conditional expected particle acceleration due to fluid–particle forces

$\langle \mathbf{A}_{\text{p}} \rangle_1$	multi-particle conditional expected particle acceleration due to field forces
$\langle \mathbf{A}_{\text{pf}} \rangle_1$	multi-particle conditional expected fluid acceleration due to fluid–particle forces
$\langle \mathbf{A}_{\text{fp}}^{(n)} \rangle_1$	single-particle conditional expected continuous particle acceleration due to fluid–particle forces
$\langle \mathbf{A}_{\text{p}}^{(n)} \rangle_1$	single-particle conditional expected continuous particle acceleration due to field forces
$[\![\mathbf{A}_{\text{f}}]\!]_{\text{f}}$	total acceleration of the fluid seen by the particles due to forces in the fluid phase
$[\![\mathbf{A}_{\text{fp}}]\!]_{\text{N}}$	total acceleration acting on monodisperse particles (of constant size and mass) due to drag, lift, and pressure forces
$[\![\mathbf{A}_{\text{fp}}]\!]_{\text{p}}$	total acceleration acting on particles due to drag, lift, and pressure forces
$[\![\mathbf{A}_{\text{p}}]\!]_{\text{N}}$	total acceleration acting on monodisperse particles (of constant size and mass) due to body forces
$[\![\mathbf{A}_{\text{p}}]\!]_{\text{p}}$	total acceleration acting on particles due to body forces
$[\![\mathbf{A}_{\text{pf}}]\!]_{\text{f}}$	total acceleration of fluid seen by the particles due to momentum transfer between phases
Ar_{p}	Archimedes number for disperse phase
$B(x, y)$	beta function
$B(\mathbf{g}, \mathbf{x})$	hard-sphere collision kernel
B_{ij}	rate of change of the particle-number density in intervals $I_1^{(i)}$ and $I_2^{(j)}$ due to a generic point process (in CM for bivariate systems)
B_i^{agg}	rate of change of the particle number in interval I_i due to aggregation (in CM for univariate systems)
B_i^{break}	rate of change of the particle number in interval I_i due to breakage (in CM for univariate systems)
$\mathbf{B}$	coefficient matrix constituted by mixed moments for calculating velocity parameters u_n (in inhomogeneous systems)
$\mathbf{B}_{\text{fv}}$	mixed phase-space diffusion tensor for fluid velocity and particle velocity
$\mathbf{B}_{\text{fv}_\text{f}}$	pure phase-space diffusion tensor for fluid velocity
$\mathbf{B}_{\text{f}\xi}$	mixed phase-space diffusion tensor for fluid velocity and particle internal coordinate
$\mathbf{B}_{\text{f}\xi_\text{f}}$	mixed phase-space diffusion tensor for fluid velocity and fluid internal coordinate
$\mathbf{B}_{\text{pv}}$	pure phase-space diffusion tensor for particle velocity
$\mathbf{B}_{\text{pv}_\text{f}}$	mixed phase-space diffusion tensor for particle velocity and fluid velocity
$\mathbf{B}_{\text{p}\xi}$	mixed phase-space diffusion tensor for particle velocity and particle internal coordinate
$\mathbf{B}_{\text{p}\xi_\text{f}}$	mixed phase-space diffusion tensor for particle velocity and fluid internal coordinate
Bo	Bond number (equal to Eötvös number)
C	constant appearing in the parabolic daughter distribution function for particle breakage
$C(\psi)$	collisional source term for monodisperse system

C_1	coefficient appearing in definition of the fluid effective (laminar plus turbulent) viscosity $\mu_{f,eff}$
C_1–C_7	constants appearing in breakage kernel
C_D	particle-drag coefficient
C_D^*	particle-drag coefficient including the Cunningham correction factor for rarefied continuous phase
C_i	concentration of the potential-determining ions
C_L	lift-force coefficient
C_m	momentum exchange coefficient appearing in thermophoretic force
C_s	thermal slip coefficient appearing in thermophoretic force
C_t	thermal exchange coefficient appearing in thermophoretic force
C_{vm}	virtual-mass force coefficient identifying the fraction of fluid volume moving with a particle
C_α	coefficients appearing in the functional expansion of the NDF
$C_{\alpha\beta}(\psi)$	collisional source term for polydisperse system (particles of types α and β)
$C_{\alpha\beta}^{(m)}(\psi)$	terms appearing in the expansion of the collisional source term for polydisperse systems (particles of types α and β)
$C_{ijk}^{*\gamma}$	approximate collision source term for velocity moments of global order γ (monodisperse systems)
$C_{ijk,\alpha}^{*\gamma}$	approximate collision source term for velocity moments of global order γ of particle type α (polydisperse systems)
$\mathbb{C}$	generic collisional source term
$\mathbb{C}_{\alpha\beta}$	collisional source term for particles of types α and β
C_{N_p}	N_p-particle collision operator
$\mathbf{C}_{f\mathbf{v}}$	mixed phase-space diffusion tensor for fluid internal coordinate and particle velocity
$\mathbf{C}_{f\mathbf{v}_f}$	mixed phase-space diffusion tensor for fluid internal coordinate and fluid velocity
$\mathbf{C}_{f\xi}$	mixed phase-space diffusion tensor for fluid internal coordinate and particle internal coordinate
$\mathbf{C}_{f\xi_f}$	pure phase-space diffusion tensor for fluid internal coordinate
$\mathbf{C}_{p\mathbf{v}}$	mixed phase-space diffusion tensor for particle internal coordinate and particle velocity
$\mathbf{C}_{p\mathbf{v}_f}$	mixed phase-space diffusion tensor for particle internal coordinate and fluid velocity
$\mathbf{C}_{p\xi}$	pure phase-space diffusion tensor for particle internal coordinate
$\mathbf{C}_{p\xi_f}$	mixed phase-space diffusion tensor for particle internal coordinate and fluid internal coordinate
C_i^{pzc}	concentration of potential determining ions at point of zero charge
$C_D^{(n)}$	drag coefficient for the nth particle
$\mathbf{C}_{f\xi}^{(n)}$	rate of change of the internal coordinate vector for the fluid surrounding the nth particle due to discontinuous events
$\mathbf{C}_{pU}^{(n)}$	rate of change of velocity for the nth particle due to collisions (particles suspended in fluid)

$\mathbf{C}_{\mathrm{p}\xi}^{(n)}$	rate of change of particle internal coordinate vector for nth particle due to collisions (particles suspended in fluid)
$\mathbf{C}_{U}^{(n)}$	rate of change of velocity for the nth particle due to discontinuous particle collisions (in vacuum)
$\mathbf{C}_{\xi}^{(n)}$	rate of change of particle internal coordinate vector for nth particle due to particle collisions (in vacuum)
$C_{1}^{(n)}$	single-particle collision operator
$C_{l_1 l_2 l_3}^{(m)}$	collision source terms for integer moments of orders l_1, l_2, and l_3 with respect to the three velocity components
D	solute molecular diffusion coefficient
D_{b}	diameter of a stable bridge between two aggregating particles
D_{f}	particle fractal dimension
D_{G}	average size of objects constituting a porous medium
D_0	cut-off distance for calculating the Hamaker parameter
$\mathbf{D}$	symmetric $\mathcal{N} \times \mathcal{N}$ diffusion matrix
$\hat{\mathbf{D}}$	volume-average symmetric $\mathcal{N} \times \mathcal{N}$ diffusion matrix
$\mathbf{D}^*$	dimensionless normalized diffusion matrix
Da_{a}	aggregation Damköhler number
Da_{b}	breakage Damköhler number
E	bubble aspect ratio
$E_{\alpha\beta}$	energy scaling factor in polydisperse systems (particle types α and β)
E_{p}	total particle granular energy
Eo	Eötvös number
F	inter-particle force
$F(\zeta)$	dimensionless normalized NDF
F_{t}	cumulative probability distribution for the quiescence time in MC methods
$F_{i,l_1 l_2 l_3}^{\gamma}$	ith component of spatial flux for moment of global order γ
$F_{i,l_1 l_2 l_3}^{+}$	ith component of spatial flux for moment of global order γ corresponding to positive velocity
$F_{i,l_1 l_2 l_3}^{-}$	ith component of spatial flux for moment of global order γ corresponding to negative velocity
$\mathbf{F}(\mathbf{M})$	generic moment flux function
$\mathbf{F}_{\mathrm{fp}}$	drag and buoyancy fluid–particle force
$\mathbf{F}_{l_1 l_2 l_3}^{\gamma}$	spatial flux for moment of global order γ
$\mathbf{F}_{l_1 l_2 l_3}^{+}$	spatial flux for moment of global order γ corresponding to positive velocity
$\mathbf{F}_{l_1 l_2 l_3}^{-}$	spatial flux for moment of global order γ corresponding to negative velocity
Fl	flow number for particle aggregation
Fr_{g}	Froude number for the continuous phase
$G(n_l, n_r)$	numerical flux function for inhomogeneous systems discretized with FVM
G_{f}	fluid shear rate

G_i	gain rate of particles with velocity $\boldsymbol{\xi}_i$ due to collisions in DVM
$G_{ijk,\alpha\beta}$	Gaussian moments with mean velocity $\mathbf{U}_{\alpha\beta}$ and covariance matrix $\boldsymbol{\sigma}_{\alpha\beta}$
G_{L}	continuous rate of change of particle size (growth rate)
$G_{\mathrm{L},k}$	average particle growth rate for moment of order k
$G_m(\psi)$	component m of the collisional-flux term
$G_{m,l_1l_2l_3}$	collisional-flux term for integer moments of orders l_1, l_2, and l_3 with respect to the three velocity components
G_{p}	pure advection component for the rate of change of crystal size
$G_{\mathrm{f}}^{\mathrm{o}}$	correction for fluid-dynamic interactions between particles
G_n^k	coefficients appearing in the calculation of the velocity parameters u_n for inhomogeneous systems
G_{p}^*	global rate of change of crystal size
$G_{\mathrm{p}1}^*$	mass-transfer rate from fluid to particle
$\mathbf{G}$	vectorial numerical flux function for inhomogeneous systems discretized with FVM
$\mathbf{G}(\psi)$	collisional-flux term for monodisperse systems
$\mathbf{G}_{\mathrm{f}}$	pure mesoscale advection model for the fluid internal coordinate
$\mathbf{G}_{l_1l_2l_3}$	collisional-flux vector for velocity moment of order $l_1l_2l_3$
$\mathbf{G}_{\mathrm{p}}$	pure mesoscale advection model for particle internal coordinate
$\mathbf{G}_{\alpha\beta}(\psi)$	collisional-flux term for polydisperse systems constituted by particle types α and β
$\mathbf{G}^{(n)}$	continuous rate of change of the internal coordinate vector for the nth particle (particles in vacuum)
$\mathbf{G}_{\mathrm{f}}^{(n)}$	continuous rate of change of the internal coordinate vector for the fluid surrounding the nth particle
$\mathbf{G}_{\mathrm{p}}^{(n)}$	continuous rate of change of the internal coordinate vector for the nth particle (particles suspended in fluid)
$\mathbf{G}_{\alpha\beta}^{(m)}(\psi)$	terms appearing in the expansion for the collisional-flux term for polydisperse systems with particle types α and β
$\langle \mathbf{G}_{\mathrm{f}} \rangle_1$	multi-particle conditional expected continuous rate of change of fluid internal coordinate vector
$\langle \mathbf{G}_{\mathrm{f}}^{(n)} \rangle_1$	single-particle conditional expected continuous rate of change of fluid internal coordinate vector
$\langle \mathbf{G}_{\mathrm{p}} \rangle_1$	multi-particle conditional expected continuous rate of change of particle internal coordinate vector
$\langle \mathbf{G}_{\mathrm{p}}^{(n)} \rangle_1$	single-particle conditional expected continuous rate of change of particle internal coordinate vector
$[\![G_{\mathrm{f}}]\!]$	rate of change of fluid-phase mass density due to continuous processes
$[\![G_{\mathrm{p}}]\!]$	rate of change of disperse-phase mass density due to continuous processes
$[\![G_{\mathrm{f}}]\!]_{\mathrm{V}}$	rate of change of fluid-phase volume fraction due to continuous processes
$[\![G_{\mathrm{p}}]\!]_{\mathrm{V}}$	rate of change of disperse-phase volume fraction due to continuous processes
$[\![\mathbf{G}_{\mathrm{f}}]\!]_{\mathrm{f}}$	global fluid momentum rate of change due to mass transfer from fluid phase

$[\![\mathbf{G}_\mathrm{p}]\!]_\mathrm{p}$	global particle momentum rate of change due to mass transfer from liquid phase
$H(\xi)$	Heaviside step function
$\mathcal{H}$	functional appearing in the general definition of the pair distribution function (function of the moments)
H_o	distance between two primary particles within an aggregate
$H_{m,l_1l_2l_3}$	symmetric change of variable involving $I^{(m)}_{l_1l_2l_3}$
I_i	finite ith internal coordinate interval used in CM
I_r	relative turbulence intensity
$I_{[x_1,x_2]}(x)$	indicator function equal to unity if $x \in [x_1, x_2]$ and zero otherwise
$I_j^{(i)}$	finite ith interval for the jth internal coordinate in CM when extended to multivariate problems
$I^{(m)}_{l_1l_2l_3}$	factor appearing in collision term for integer velocity moments of orders l_1, l_2, and l_3
$I^{(p,q)}_{l_1l_2l_3}$	factor appearing in the collision term for polydisperse systems for integer velocity moments of orders l_1, l_2, and l_3
J	molar flux of solute molecules at particle surface
$J(\phi_\mathrm{f})$	rate of particle formation
$J(\tilde{\boldsymbol{\eta}}, \boldsymbol{\eta})$	Jacobian of variable transformation relating phase-space variables before and after collision
J_f	rate of particle formation
$\hat{J}_\mathrm{f}$	volume-average rate of particle formation
$\overline{J}_k$	kth moment of the rate of particle formation in univariate GPBE
$\overline{J}_\mathbf{k}$	**k**th moment of the rate of particle formation in multivariate GPBE
J^*	dimensionless and normalized rate of particle formation
K_B	history-force kernel
$K^{(m)}_{ijk}$	integral over collision angles
$K^{(mn)}_{ijk}$	coefficients appearing in the third integral over collision angles
$\mathbf{K}$	exponent matrix used to build the quadrature approximation
$\mathbf{K}$	moment vector used in the definition of the moment set $\mathbf{M}$
$\mathbf{K}^m_{i,\alpha}$	reconstructed $\mathbf{K}$ in the ith cell at time step m employed in FVM
$\mathbf{K}^\pm_{\alpha,l/r}$	$\mathbf{K}$ evaluated with $v^+_{\alpha,l}$ or $v^-_{\alpha,r}$
Kn	Knudsen number for continuous phase (relative to particle diameter)
Kn*	Knudsen number for continuous phase (relative to particle radius)
Kn$_\mathrm{p}$	Knudsen number for disperse phase
L	characteristic length of the system under investigation
L	particle length
L_i	loss rate of particles with velocity $\boldsymbol{\xi}_i$ due to collisions in DVM
L_v	latent heat of evaporation
L_{10}	number-average mean particle length
L_{32}	area-average mean particle length or Sauter diameter
L_{43}	volume-average mean particle length
$\mathbf{L}$	transformation matrix between laboratory and collision frames
Le	Lewis number

M	number of internal coordinates appearing in the NDF
M	number of sections or classes used in CM and DVM
$M_{\rm f}$	number of fluid internal coordinates appearing in the NDF
$M_{\rm G}$	ratio between particle and collector size
M_k	number of intervals used for the kth internal coordinate in multivariate CM
M_i	mass of the particles in the interval I_i in CM
$M_{\rm p}$	number of particle internal coordinates appearing in the NDF
$M_{\rm p}$	particle mass
$M_{\rm w}$	molecular weight (relative molecular mass) of chemical species
$M_{\rm w1}$	molecular weight of the evaporating component
$M_{\rm w2}$	molecular weight of the stagnant component
$M_i^{\rm agg}$	rate of change of particle mass in interval I_i due to aggregation in CM
$M_i^{\rm break}$	rate of change of particle mass in interval I_i due to breakage in CM
M_{ijk}^{γ}	velocity distribution moment of global order $\gamma = i + j + k$
$M_{ijk}^{\gamma *}$	velocity equilibrium moment of global order $\gamma = i + j + k$
$M_{ijk,\alpha}^{\gamma}$	velocity distribution moment of global order $\gamma = i + j + k$ for particles of type α in polydisperse systems
$M_{i,l_1 l_2 l_3}^{+}$	positive integer moment of the velocity distribution in the ith direction
$M_{i,l_1 l_2 l_3}^{-}$	negative integer moment of the velocity distribution in the ith direction
$\mathbf{M}$	vector defining the tracked moment set
$\mathbf{M}^{+}$	positive half-moment set (integration for positive velocity)
$\mathbf{M}^{-}$	negative half-moment set (integration for negative velocity)
$\mathbf{M}_i^{m}$	volume-averaged moment set in the ith cell at time step m defined in FVM for a 1D grid
$\mathbf{M}_{ijk}^{m}$	volume-averaged moment set at cell Ω_{ijk} and time step m defined in FVM for a structured 3D grid
$(\mathbf{M}_{ijk}^{m})_p$	updated moment set at cell Ω_{ijk} and time step m defined in FVM for a structured 3D grid calculated with permutation p
$\mathbf{M}_{ijk}^{(h)}$	moment set at cell Ω_{ijk} updated after advection in the h direction in FVM for a structured 3D grid
$\mathbf{M}_i^{(1)}$	first-stage moment set in the ith cell at time step m defined in FVM for a 1D grid
$\mathbf{M}_{ijk}^{+}$	positive half-moment set at cell Ω_{ijk} in FVM (3D)
$\mathbf{M}_{ijk}^{-}$	negative half-moment set at cell Ω_{ijk} in FVM (3D)
$\mathbf{M}_i^{*}$	second-stage moment set in the ith cell at time step m defined in FVM for a 1D grid
Ma	Mach number for continuous phase
$\mathrm{Ma}_{\rm p}$	Mach number for disperse phase
Mo	Morton number
N	order of the quadrature approximation
$N^{\pm}$	number of quadrature nodes used in the calculation of the positive and negative fluxes

$N(t, \mathbf{x})$	total particle-number concentration (or density)
$N(V\|V')$	volume-based daughter distribution function
$N(\boldsymbol{\xi}'\|\boldsymbol{\xi})$	daughter-particle conditional NDF
N_{d}	global number of degrees of freedom of the multiphase system
N_i	number of particles belonging to the ith interval in CM and DVM
N_{ij}	number of particles belonging to intervals $I_1^{(i)}$ and $I_2^{(j)}$ for bivariate CM
$N_{\mathbf{i}}$	number density of particles with velocity equal to $\boldsymbol{\xi}_{\mathbf{i}}$ in DVM
N_{p}	number of particles considered in NDF definition
N_{p}	number of primary particles forming a fractal object
N_{t}	total particle-number density
N'	number of weights and abscissas for Laguerre-polynomial recursion coefficients in EQMOM
Nu	Nusselt number
$P(\boldsymbol{\xi}'\|\boldsymbol{\xi})$	daughter-particle conditional PDF
$P_\alpha(\xi)$	polynomial of order α orthogonal to the NDF
$P_{\alpha,\beta}$	elements of the matrix used in the PD algorithm
P_{ij}	components of the second-order pressure tensor
P_{ij}	probability for the encounter of particles i and j in MC simulations
$\mathbf{P}$	matrix used in the PD algorithm
$\mathbf{P}$	total particle stress tensor and second-order pressure tensor
$\mathbf{P}_\alpha$	second-order pressure tensor in polydisperse systems for particles of type α
Pe	Péclet number
Pr	Prandtl number
$\mathbf{Q}^+$	positive moment flux (integration for positive velocity)
$\mathbf{Q}^-$	negative moment flux (integration for negative velocity)
R	ideal-gas constant
R_{g}	radius of gyration of the particle
Re_{g}	continuous (gas)-phase Reynolds number
Re_{p}	disperse (particulate)-phase Reynolds number
$\mathrm{Re}_{\mathrm{p}}^{\mathrm{c}}$	critical particle Reynolds number
$\mathrm{Re}_{\mathrm{p}}^{\mathrm{M}}$	meta-critical particle Reynolds number
$\mathrm{Re}_{\mathrm{p}}^{*}$	modified disperse-phase Reynolds number
S	supersaturation
S	comprehensive source term in the GPBE including drift, diffusion, and point processes
S_{c}	particle collisional cross-sectional area
$S_{k,i}$	reconstructed slope in the ith cell for the solution of the moment transport equation discretized with FVM (1D)
$S^n_{\alpha\beta,i}$	reconstructed slope in the ith cell at time instant n from EQMOM discretized with FVM (1D)
$\mathbf{S}$	moment set source term
$\mathcal{S}$	generic source term due to discontinuous events for the GPBE
$\mathcal{S}_{\mathrm{fV}}$	source term due to discontinuous events for the fluid-phase volume fraction
$\mathcal{S}_{\mathrm{M}}$	source term due to discontinuous events for the disperse-phase mass density

$\mathcal{S}_{\mathrm{N}}$	source term due to discontinuous events for the total disperse-phase number density
$\mathcal{S}_{\mathrm{V}}$	source term due to discontinuous events for the disperse-phase volume fraction
$\mathcal{S}_{1}$	generic source term due to discontinuous events for GPBE
$\mathbf{S}_{\mathrm{f}}$	viscous and pressure stress tensor for fluid phase
$\overline{S}_{k}$	source term for the kth moment of the NDF
$\overline{S}_{\mathbf{k}}$	source term for the moment of order $\mathbf{k}$ of the multivariate NDF
$\overline{S}_{ijk}$	source term of the moments of orders i, j, and k with respect to the particle-velocity components for the NDF
$S^{(m)}_{l_1 l_2 l_3}$	factor appearing in the collision term for velocity moments of the NDF of orders l_1, l_2, and l_3 with respect to the three components
$S^{\gamma}_{l_1 l_2 l_3}$	source terms for moment of global order $\gamma = l_1 + l_2 + l_3$
$[\![\mathbf{S}]\!]_{\mathrm{p}}$	rate of change for particle momentum due to discontinuous events
$\mathbb{S}^{+}$	collision cross section
Sc	Schmidt number
Sh	Sherwood number
St	Stokes number
St_{p}	particle Stokes number
T_{f}	continuous (fluid) phase temperature
T_{p}	disperse (particulate) phase temperature
T_{ref}	reference temperature for liquid boiling
T_{s}	temperature on particle surface
T_{sat}	saturation temperature for the continuous phase
U	characteristic continuous phase velocity
$U(\xi)$	particle velocity conditioned on particle size ξ
$\mathcal{U}$	characteristic particle velocity
U_{g}	continuous phase velocity
U_{p}	mean particle velocity
U_{r}	impact, or relative, velocity for fragmenting particles
U_{∞}	particle terminal velocity
$\mathbf{U}_{\mathrm{f}}$	mean fluid velocity field
$\mathbf{U}_{\mathrm{fV}}$	volume-average fluid velocity
$\mathbf{U}_{\mathrm{M}}$	mass-average particle velocity
$\mathbf{U}_{\mathrm{mix}}$	mass-average mixture velocity
$\mathbf{U}_{\mathrm{N}}$	number-average particle velocity
$\mathbf{U}_{\mathrm{p}}$	mean particle velocity
$\mathbf{U}_{\mathrm{p},2}$	second-order particle velocity-moment tensor
$\mathbf{U}_{\mathrm{p},k}$	conditional particle velocity for $\boldsymbol{\xi}_{\mathrm{p}} = \boldsymbol{\xi}_{\mathrm{p}k}$
$\mathbf{U}_{\mathrm{pM}}$	fluid-mass-average particle velocity
$\mathbf{U}_{\mathrm{V}}$	volume-average particle velocity
$\mathbf{U}_{\alpha\beta}$	mean velocities for polydisperse Gaussian distributions
$U^{\dagger}_{\mathrm{p}}$	characteristic disperse-phase velocity
$\mathbf{U}^{(n)}$	velocity of the nth particle (in vacuum)
$\mathbf{U}^{(n)}_{\mathrm{f}}$	fluid velocity in the neighborhood of the nth particle

$\mathbf{U}_{\mathrm{f}}^{*}$	mesoscale variable describing fluid velocity
$\mathbf{U}_{\mathrm{p}}^{(n)}$	velocity of the nth particle
$\mathbf{U}_{\mathrm{p}}^{*}$	mesoscale variable describing particle velocity
$\langle \mathbf{U}_{\mathrm{f}} \rangle$	Reynolds-average fluid velocity field
$\langle \mathbf{U}_{\mathrm{p}} \vert \boldsymbol{\xi} = \boldsymbol{\zeta} \rangle$	particle conditional velocity for $\boldsymbol{\xi} = \boldsymbol{\zeta}$
$\langle \mathbf{U}_{\mathrm{p}} \vert \boldsymbol{\xi}_{\mathrm{p}} \rangle$	expected mean particle velocity for internal coordinates equal to $\boldsymbol{\xi}_{\mathrm{p}}$
$[\![\mathbf{U}_{\mathrm{f}}]\!]_{\mathrm{p}}$	particle-mass-average fluid velocity
V	particle volume
V_{f}	fluid volume seen by the particle
V_{L}	length-based volume density function
V_{p}	particle volume
V_{W}	sample volume used in the estimation of the NDF
$V^{*}_{\alpha\beta;i}$	particle velocity with EQMOM in the ith cell after the advection step when using FVM (1D) with time splitting
$V^{\dagger}_{\alpha\beta;i}$	particle velocity with EQMOM in the ith cell after the advection step when using FVM (1D) with time splitting
$\mathbf{V}$	dimensionless and normalized particle velocity
$\mathbf{V}_{\mathrm{f}}^{(n)}$	fluid-velocity space for fluid surrounding the nth particle
$\mathbf{V}_{\mathrm{p}}^{(n)}$	particle-velocity space for the nth particle
$W(t)$	generic Wiener process
$W_{\alpha\beta}$	product of w_{α} and $w_{\alpha\beta}$ in EQMOM
$W^{*}_{\alpha\beta}$	product of w^{*}_{α} and $w^{*}_{\alpha\beta}$ in EQMOM and calculated after the advection step with time splitting
$\mathbf{W}(t)$	generic vectorial Wiener process
We	Weber number
$\mathbf{X}$	dimensionless and normalized spatial coordinate
$\mathbf{X}^{\mathrm{T}}$	abscissa (or node) matrix
$\mathbf{X}^{(n)}$	center of mass of the nth particle
$\mathbf{X}_{\mathrm{p}}^{*}$	mesoscale variable describing particle position
Y_{f1}	molar fraction of evaporating component in the gas phase
Y_{s1}	molar fraction of evaporating component on droplet surface
Y_1	gas-phase molar fraction of evaporating component
Y_2	gas-phase molar fraction of stagnant component
$\mathbf{Y}_{\mathrm{f}}$	fluid-phase species mass fractions
$\mathbf{Y}_{\mathrm{p}}$	particle species mass fractions
$\mathbf{Y}_{\alpha} = w_{\alpha}\boldsymbol{\xi}_{\alpha}$	weighted node (or abscissa) of the M-dimensional quadrature approximation
$\mathbf{Y}^{\mathrm{T}}$	weighted-abscissa (or weighted-node) matrix

Lower-case Roman

a	aggregation kernel
a	breakage kernel (Chapter 7)
a_0	constant-breakage kernel (Chapter 7)

a_{ij}	affinity parameter for i–j aggregation (with i = A, B and j = A, B)
a_α	coefficients of recursive formula for orthogonal polynomials
a_α	source term for the evolution equation of weight w_α in DQMOM
a'_α	coefficients of recursive formula appearing in Wheeler algorithm
$\langle a \rangle$	volume-average aggregation kernel
$\mathbf{a}^m_{ijk}$	slope vector employed in second-order spatial reconstructions for FVM (3D)
b	frequency of first-order process (breakage kernel)
$b(\xi'\|\xi)$	daughter distribution function (Chapter 7)
$b_{i,\alpha}$	source term for the evolution equation of weighted node α for the ith internal coordinate in DQMOM
b_α	coefficients of recursive formula for orthogonal polynomials
b_α	source term for the evolution equation of weighted node α in univariate DQMOM
b'_α	coefficients of recursive formula appearing in Wheeler algorithm
b^*	dimensionless and normalized kernel for first-order process
$\hat{b}$	volume-average frequency of first-order process
$\langle b \rangle$	volume-average breakage kernel
$\overline{b}^k_\alpha$	moment transform of order k of the daughter distribution function for ξ_α in univariate problems
$\overline{b}^{\mathbf{k}}_\alpha$	moment transform of order $\mathbf{k}$ of the daughter distribution function for $\boldsymbol{\xi}_\alpha$ in multivariate problems
$\mathbf{b}_{\mathrm{f}\mathbf{v}_\mathrm{f}}$	fluid velocity coefficient for fluid fluctuations
$\mathbf{b}_k$	response vector for third-order differences to a unit increment
$\mathbf{b}_{\mathrm{p}\mathbf{v}_\mathrm{f}}$	particle-velocity coefficient for fluid fluctuations
c_{eq}	equilibrium solute concentration
$c_{\mathrm{p}\xi_\mathrm{p}}$	self-diffusion component of crystal size growth rate
$c_{\alpha\beta}$	constant appearing in the definition of the pair distribution function
$c^\pm$	coefficients appearing in upwind reconstruction schemes
c^p_f	specific heat of fluid phase
c^p_p	particle specific heat
$\mathbf{c}_\alpha$	lattice velocities used in LBM
d	degree of accuracy of the quadrature approximation
d	molecular diameter of a solute molecule
d_o	diameter of primary particles in an aggregate
d_p	particle size
$d_{10\%}$	size corresponding to 10% of the smaller particles
$d_{90\%}$	size corresponding to 10% of the larger particles
d_α	size of the particles of type α taking part in a collision
d_β	size of the particles of type β taking part in a collision
$d_{\alpha\beta}$	arithmetic average of colliding particle size (polydisperse systems)
d^*_p	characteristic particle length
d^{ij}_k	net flux of particles for the kth internal coordinate due to phase-space diffusion in bivariate CM
$\mathbf{d}$	vector containing the moments source terms in DQMOM
$\mathbf{d}_n$	difference vector of order n of natural logarithm of NDF moments

e	coefficient of restitution for particle–particle collisions
e	elementary charge
e_i	particle specific energy for the ith velocity component
e_p	particle specific energy
$e_{\alpha\beta}$	restitution coefficient in polydisperse systems for collisions between particles of types α and β
$f(t, \mathbf{x}, \mathbf{v})$	particle-velocity NDF for monodisperse system
$f(\xi_1, ..., \xi_{M-1})$	marginal NDF used in CQMOM
$f(\xi_M\|...)$	conditional NDF used in CQMOM
$f(\delta)$	shape function for colliding particles
f_f	one-point PDF for the fluid
f_{G_f}	volume distribution of the fluid shear rate
f_{N_p}	multi-particle joint PDF
f_t	probability density function for the quiescence time in MC methods
f_eq	equilibrium velocity NDF
f_α	weights used in reconstructing the NDF in LBM
$f_\alpha(t, \mathbf{x}, \mathbf{v})$	particle velocity NDF for particle type α
$f_\beta(t, \mathbf{x}, \mathbf{v})$	particle velocity NDF for particle type β
f_ε	volume distribution of the fluid turbulent dissipation rate
$f^{(2)}$	pair correlation velocity NDF
$f^*(\mathbf{v})$	equilibrium distribution function in the BGK kinetic model and in Grad's moment closure
$f_1^{(n)}$	single-particle joint PDF
f_k^{ij}	net flux of particles for the kth internal coordinate due to phase-space drift in bivariate CM
f_α^*	weights used in LBM corresponding to the equilibrium NDF
g	gravity acceleration constant
$g(\epsilon)$	Kuwabara function for particle deposition in porous media
$g_0(\alpha_\mathrm{p}/\alpha_\mathrm{p}^*)$	particle radial distribution function
$g_{\alpha\beta}$	pair correlation function for particles of types α and β
$g_{0,\alpha\beta}$	constant appearing in the pair correlation function
$g_n(\xi)$	velocity parameters used in conjunction with EQMOM
$\langle g \rangle$	mean velocity difference used to approximate $\|\mathbf{v}_1 - \mathbf{v}_2\|$
h	size of the regular discretization used in CM
$h(\boldsymbol{\xi})$	function used to model the second-order tensor for mixed advection
$h_i(\xi)$	numerical NDF in the ith cell for FVM (1D)
h_L	discontinuous event term for length-based formulation
$h_{\mathrm{L},k}$	moment of order k of source term for discontinuous event
h_W	constant kernel function used as filter to estimate NDF
$\mathbf{h}_1$	particle collisional acceleration term
h^+	rate of particle formation due to discontinuous events
h^-	rate of particle disappearance due to discontinuous events
$\mathbf{i}$	index vector identifying a discrete particle velocity in DVM
k	order of moment for univariate NDF
k_A	particle surface shape factor

k_B	Boltzmann constant
k_c	corrective growth crowding factor
k_{co}	coordination number for an aggregate
k_d	particle mass-transfer coefficient
k_f	fluid-phase turbulent kinetic energy
k_f	fluid-phase thermal conductivity
k_g	fractal scaling factor of order unity
k_h	particle heat-transfer coefficient
$k_i(\xi)$	polynomial used to represent the NDF in the I_i interval in CM
k_p	thermal conductivity of particle
k_V	particle volumetric shape factor
k_A^*	equivalent particle surface shape factor
k_V^*	equivalent particle volumetric shape factor
$\mathbf{k}$	exponent vector for the order of moment in multivariate NDF
$\mathbf{k}_\xi$	internal coordinate exponent vector
m	particle mass (used in daughter NDF for breakage)
$m(k)$	moment of order k of univariate NDF
$m(\mathbf{k})$	moment of order $\mathbf{k} = (k_1, \ldots, k_M)$ of multivariate NDF
m_c	mass of newly formed particle (nucleus mass)
$\dot{m}_{cp}$	mass of liquid evaporating per unit volume and unit time
m_k	moment of order k of NDF
m_{ijk}	moment of order i, j, and k with respect to the three velocity components of the particle velocity NDF
$m_{j,k;i}$	mixed moment of orders j and k for the ith cell in FVM (1D)
$m_{L,k}$	kth moment of length-based NDF
$m_{M,k}$	kth moment of mass-based NDF
$m_{U,k}$	kth moment of velocity-based NDF
$m_{V,k}$	kth moment of volume-based NDF
$m_{\xi,\mathbf{k}}$	moment of order $\mathbf{k} = (k_1, \ldots, k_M)$ of the multivariate NDF
m_α	mass of particle of type α taking part in collision
m_β	mass of particle of type β taking part in collision
m_k^*	moment definition used in EQMOM
$m_{k,i}^n$	reconstructed moment for the ith cell at time instant n in FVM
$m_{j,k;i}^*$	mixed moment of order j and k for the ith cell in FVM when using time splitting after convection
$m_{j,k;i}^\dagger$	mixed moment of orders j and k for the ith cell in FVM when using time splitting after drag
n	generic NDF appearing in GPBE
n_{eq}	equilibrium NDF
n_L	length-based NDF
n_M	mass-based NDF
n_U	velocity-based NDF
n_V	volume-based NDF
n_ξ	generic NDF
$\langle n_\xi \rangle$	Reynolds-average NDF
$\hat{n}$	volume-average NDF

n^*	reconstructed NDF
n^*	equilibrium Maxwellian NDF
$n^{[13]}$	NDF reconstructed in Grad's 13-moment closure
$n_i^m(\xi)$	NDF in the ith cell at time step m reconstructed in FVM (1D) from $\mathbf{M}_i^m$
$n_\alpha^\pm$	weights of the quadrature approximation calculated from the positive and negative moments of the velocity distribution
$n_{ijk}^{(h)}$	NDF in Ω_{ijk} updated after advection in the h direction for FVM
n_{ijk}^m	NDF in cell Ω_{ijk} at time $t = m\,\Delta t$ for FVM
n_{ijk}^+	contribution for updating the NDF due to advection from positive velocity in cell Ω_{ijk} for FVM
n_{ijk}^-	contribution for updating the NDF due to advection from negative velocity in cell Ω_{ijk} for FVM
p_{f}	pressure of the fluid phase
p_{f1}	partial pressure of evaporating component in gas phase
p_{g}	pressure of the gas phase
p_{ref}	reference pressure for boiling liquid
p_{s1}	partial pressure of evaporating component on droplet surface
p_α	orthogonal polynomials of order α used in functional expansion of NDF
p_α	granular pressure of particles of type α
q	heat flux to surface of particle
q_i	skewness of the NDF with respect to the ith velocity component
$\mathbf{q}$	total particle energy flux
s	specific surface area of the porous medium
s	ratio of geometric grids employed in CM
s	particle surface area
$t_{\alpha\beta}$	abscissa computed from Laguerre-polynomial recursion coefficients used in EQMOM
u	generic known disperse-phase velocity (1D)
$u(\xi)$	known particle velocity conditioned on internal coordinate ξ
u_k	(with $k = 0, 1, 2$) flow-dependent velocity parameters
u_α	velocity node α of the quadrature approximation
$u_{i,l}^+$	positive i-component of velocity evaluated at left face of cell Ω_{ijk} in FVM (1D)
$u_{i,r}^-$	negative i-component of velocity evaluated at right face of cell Ω_{ijk} in FVM (1D)
$[\![\mathbf{u}_{\mathrm{p}}\mathbf{u}_{\mathrm{p}}]\!]_{\mathrm{N}}$	disperse-phase stress tensor
v	disperse-phase velocity (1D)
v_{f}	fluid-phase velocity (1D)
v_i^+	positive i-component of velocity evaluated at left face of cell Ω_{ijk} in FVM
v_i^-	negative i-component of velocity evaluated at right face cell of Ω_{ijk} in FVM

v_α^+	velocity node α calculated from positive half-moment set
v_α^-	velocity node α calculated from negative half-moment set
$\mathbf{v}$	phase-space variable for particle velocity vector (in vacuum)
$\mathbf{v}_\mathrm{f}$	phase-space variable for fluid velocity
$\mathbf{v}_\mathrm{p}$	phase-space variable for particle velocity vector
$\mathbf{v}_\mathrm{r}$	relative velocity of two colliding particles
$\mathbf{v}_{12}$	velocity difference before a direct collision
$\mathbf{v}_\alpha^\pm$	nodes of the quadrature approximation calculated from the positive and negative moments of the velocity distribution
$\mathbf{v}'_{12}$	velocity difference after a direct collision
$\mathbf{v}''_{12}$	velocity difference before an inverse collision
$\mathbf{v}_{ijk,\alpha}^m$	velocity node α at cell Ω_{ijk} and time $m\,\Delta t$ in FVM (3D)
$\mathbf{v}_{ijk,\alpha}^+$	velocity node α at cell Ω_{ijk} calculated from $\mathbf{M}_{ijk}^+$ in FVM for 3D structured grids
$\mathbf{v}_{ijk,\alpha}^-$	velocity node α at cell Ω_{ijk} calculated from $\mathbf{M}_{ijk}^-$ in FVM for 3D structured grids
w	weight function corresponding to a family of orthogonal polynomials in functional expansion of NDF
w_α	generic weight of a quadrature approximation
w_{α_1}	weight in CQMOM
$w_{\alpha_1;\alpha_2}$	conditional weight in CQMOM
$w_{\alpha_1;\alpha_2;\alpha_3}$	conditional weight in CQMOM
$w_{\alpha\beta}$	weight computed from Laguerre polynomial recursion coefficients in EQMOM
$w_{i,\alpha}^m$	reconstructed weight α in the ith cell at time instant m employed in FVM (1D)
$w_{ijk,\alpha}^m$	weight α at cell Ω_{ijk} at time $m\,\Delta t$ employed in FVM for 3D structured grids
w_α^+	weight calculated from the positive half-moment set
w_α^-	weight calculated from the negative half-moment set
$w_{ijk,\alpha}^+$	weight α at cell Ω_{ijk} calculated from $\mathbf{M}_{ijk}^+$ in FVM for 3D structured grids
$w_{ijk,\alpha}^-$	weight α at Ω_{ijk} calculated from $\mathbf{M}_{ijk}^-$ in FVM for 3D structured grids
$\mathbf{w}$	weight vector
x	cosine of collision angle α_x
$\mathbf{x}$	physical point coordinate vector
$\mathbf{x}_{12}$	collision vector in laboratory frame
$\mathbf{x}_{12}^\dagger$	collision vector in collision frame
$\mathbf{x}^{(n)}$	particle-position space for nth particle
y	cosine of collision angle α_y
z	valency of potential-determining ions

Upper-case Greek

Γ	disperse-phase diffusion coefficient
$\Gamma(x)$	gamma function

Γ_{p}	effective disperse-phase diffusion coefficient
Γ_{T}	turbulent disperse-phase diffusion coefficient
$\mathbf{\Gamma}$	second-order tensor for multidimensional diffusion
Δc	mass-transfer driving force
Δt	uniform time step employed in FVM
Δx	uniform spatial grid spacing in x direction employed in FVM
Δy	uniform spatial grid spacing in y direction employed in FVM
Δz	uniform spatial grid spacing in z direction employed in FVM
$\Delta \xi$	size of velocity intervals in DVM
Θ	generic granular temperature
Θ_{p}	disperse-phase granular temperature
Θ_{α}	granular temperature for particles of type α
Θ_{β}	granular temperature for particles of type β
$\Theta_{\alpha\beta}$	granular temperature for polydisperse systems of particle types α and β
$\Pi_{i,j,k}$	moment of order i, j, and k of the collision integral with respect to the three velocity components
$\mathbf{\Sigma}$	generic velocity covariance matrix
$\mathbf{\Sigma}_{\mathrm{p}}$	disperse-phase velocity covariance matrix
Ψ_{s}	particle surface potential
$\Omega_{\mathbf{x}}$	control volume in physical space
Ω_{ijk}	cell for a structured Cartesian grid centered at $(i\,\Delta x, j\,\Delta y, k\,\Delta z)$
$\lvert\Omega_{ijk}\rvert$	volume of the cell Ω_{ijk}
$\partial\Omega_{\mathbf{x}}$	boundaries of physical-space control volume
Ω_{ξ}	integration or support interval
$\Omega_{\boldsymbol{\xi}}$	control volume in phase space
$\partial\Omega_{\boldsymbol{\xi}}$	boundaries of phase-space control volume

Lower-case Greek

α	aggregation efficiency
α_{f}	volume fraction of fluid phase
α_{g}	volume fraction of gas phase
α_{ikl}	rate of gain of particles with velocity ξ_i due to collision between particles with velocities ξ_k and ξ_l in DVM
α_{p}	volume fraction of disperse phase
α_{V}	length-based volume-fraction density function
α_x	angle formed by $\boldsymbol{\xi}$ and $\mathbf{x}$ in the formulation of the isotropic collision operator
α_y	angle formed by $\boldsymbol{\xi}_*$ and $\mathbf{x}$ in the formulation of the isotropic collision operator
α_1	volume fraction of particles of type 1
α_2	volume fraction of particles of type 2
$\alpha_{\mathrm{f}}^{\mathrm{c}}$	thermal conductivity of continuous phase
$\alpha_{\mathrm{p}}^{\mathrm{c}}$	thermal conductivity of disperse phase
α_{p}^{*}	packing-limit volume fraction for disperse phase
$\boldsymbol{\alpha}$	vector with source terms for weights and nodes in DQMOM

β	evaporation frequency
β	kernel for generic second-order process
$\hat{\beta}$	volume-average kernel for generic second-order process
β_x	angle formed by $\boldsymbol{\xi}'$ and $\mathbf{x}$ for the isotropic collision operator
β_y	angle formed by $\boldsymbol{\xi}'_*$ and $\mathbf{x}$ for the isotropic collision operator
β_0	constant aggregation kernel
β^*	dimensionless normalized kernel for second-order process
β^V	collision rate in the computational box in MC simulations
$\gamma = \|\mathbf{k}\|$	global moment order
γ	effective volume coefficient
γ_x	ratio between time step and grid spacing in x direction
γ_y	ratio between time step and grid spacing in y direction
γ_z	ratio between time step and grid spacing in z direction
γ_i^i	reassignment factor for CM to describe discontinuous events
δ	size ratio of colliding particles
$\delta(x)$	Dirac delta function
δ_σ	kernel density function in EQMOM
$\boldsymbol{\delta}(\mathbf{x})$	multidimensional Dirac delta function
ϵ	porosity of porous medium
ε	turbulence-dissipation rate
$\hat{\varepsilon}$	volume-average turbulence-dissipation rate
ε_{f}	fluid-phase turbulence-dissipation rate
ε_{r}	relative permittivity constant
ε_0	absolute permittivity constant
ζ	model constant for BGK collision model (inverse of Pr)
ζ_i	pivot point for interval I_i used in CM
$\overline{\zeta}_i$	adaptive pivot point in interval I_i in CM with cell averaging
ζ_α	coefficients of the continued fraction in PD algorithm
$\zeta_k^{(i)}$	pivot point for interval I_i for the kth internal coordinate in multivariate CM
$\boldsymbol{\zeta}$	dimensionless and normalized internal coordinate vector
$\dot{\boldsymbol{\zeta}}$	dimensionless and normalized continuous rate of change of the internal coordinate vector
η	coalescence efficiency
η_{D}	particle-deposition efficiency due to Brownian motion
η_i	kurtosis of the NDF with respect to the ith velocity component
η_{ij}	reassignment coefficient for $I_1^{(i)}$ and $I_2^{(j)}$ in bivariate CM
η_{I}	particle-deposition efficiency due to interception
$\boldsymbol{\eta}$	phase-space vector for particles involved in second-order processes
$\boldsymbol{\eta}_{\mathrm{f}}^{(n)}$	internal-coordinate space for fluid surrounding nth particle
$\boldsymbol{\eta}_{\mathrm{p}}^{(n)}$	internal-coordinate space for nth particle
θ	collision angle for $\mathbf{x}_{12}$
θ_1	collision angle for $\mathbf{v}_{12}$
κ	Debye–Hückel parameter
$\kappa_{\alpha\beta}$	parameter appearing in the radial distribution function

$\kappa_{\alpha\beta}$	collision-rate parameters for polydisperse BGK-like models
$\boldsymbol{\kappa}$	symmetric second-order tensor constructed from $I_{ijk}^{(0)}$ appearing in approximate collision models
λ	ratio between time step and grid spacing in FVM (1D)
λ_{f}	mean free path of molecules in continuous phase
λ_k	moment of order k of a log-normal distribution
λ_1	diffusion coefficient appearing in the moment-transport equation discretized with FVM
λ_2	convective coefficient appearing in the moment-transport equation discretized with FVM
$\lambda^{\pm}(\xi)$	function appearing in CM with the cell averaging
$\boldsymbol{\lambda}$	second-order velocity tensor in the equilibrium NDF employed in BGK model
$\boldsymbol{\lambda}$	non-negative second-order tensor for mixed advection
μ	mean internal coordinate for Gaussian and log-normal distributions
μ_k	moments in the positive tail of a Gaussian distribution
μ_{f}	dynamic viscosity for continuous phase
$\mu_{\mathrm{f,eff}}$	effective dynamic viscosity (including turbulent viscosity) for continuous phase
μ_{p}	granular viscosity
$\mu_{\alpha\beta}$	reduced mass of particles in polydisperse system
$\nu(\mathbf{v}_{\mathrm{p}}, \boldsymbol{\xi}_{\mathrm{p}})$	number of daughter particles formed by first-order process
ν_{f}	kinematic viscosity of continuous fluid phase
ν_{g}	kinematic viscosity of continuous (gas) phase
ν_k	coefficients calculated from modified moments in Wheeler algorithm
ν_{ij}	rate of loss of particles with velocity ξ_i due to collisions with particles with velocity ξ_j in DVM
$\tilde{\nu}$	solute molecular volume
ξ	internal coordinate
$\dot{\xi}$	continuous rate of change of particle internal coordinate
ξ_α	abscissa or node α of the quadrature approximation
$\xi_{\beta,\alpha}$	component for the β internal coordinate of the abscissa α of the M-dimensional quadrature approximation
$\xi_{\alpha\beta}$	node of the quadrature approximation for EQMOM
$\xi_{\alpha\beta;i}$	node of the quadrature approximation for EQMOM in the ith cell with FVM (1D)
$\xi_{i,\alpha}^{m}$	reconstructed node α in the ith cell at time $m\,\Delta t$ in FVM
$\xi_{\alpha\beta;i}^{*}$	node of the quadrature approximation for EQMOM in the ith cell with FVM and time splitting after the advection step
$\boldsymbol{\xi}$	internal coordinate vector
$\boldsymbol{\xi}$	test particle pre-collision velocity
$\dot{\boldsymbol{\xi}}$	continuous rate of change of particle internal coordinate
$\hat{\dot{\boldsymbol{\xi}}}$	volume-average continuous rate of change of internal coordinate
$\boldsymbol{\xi}'$	test particle post-collision velocity
$\boldsymbol{\xi}_{\mathrm{f}}$	phase-space variable for fluid characteristic properties
$\boldsymbol{\xi}_{\mathrm{fc}}$	fluid internal coordinate seen by newly formed particles
$\boldsymbol{\xi}_i$	discrete particle velocity used in DVM
$\boldsymbol{\xi}_{\mathbf{i}}$	discrete velocity with index $\mathbf{i} = (i_1, i_2, i_3)$ used in DVM

ξ_{p}	phase-space variable for particle internal coordinate
ξ_{pc}	internal coordinate of newly formed particles
ξ_{α}	node or abscissa of M-dimensional quadrature approximation
ξ_{*}	field particle pre-collision velocity
ξ'_{*}	field particle post-collision velocity
$\xi^{(n)}$	internal coordinate vector (of size M) of the nth particle (in vacuum)
$\xi_{\mathrm{f}}^{(n)}$	vector for characteristic properties of fluid surrounding nth particle
ξ_{f}^{*}	mesoscale variable describing fluid internal coordinate
$\xi_{i,\alpha}^{m}$	reconstructed node α in the ith cell at time $m\,\Delta t$ in FVM (1D)
$\xi_{ijk,\alpha}^{m}$	node α at cell Ω_{ijk} at time $m\,\Delta t$ in FVM (3D)
$\xi_{ijk,\alpha}^{+}$	node α at cell Ω_{ijk} calculated from $\mathbf{M}_{ijk}^{+}$ in FVM (3D)
$\xi_{ijk,\alpha}^{-}$	node α at cell Ω_{ijk} calculated from $\mathbf{M}_{ijk}^{-}$ in FVM (3D)
$\xi_{\mathrm{p}}^{(n)}$	internal coordinate vector (of size M) of the nth particle
ξ_{p}^{*}	mesoscale variable describing particle internal coordinate
$\langle \xi^{k} \rangle$	moment of order k of a univariate NDF
$\langle \xi_1^{k_1} \xi_2^{k_2} \dots \xi_M^{k_M} \rangle$	moment of order $\mathbf{k} = (k_1, \dots, k_M)$ of a multivariate NDF
$\left\langle \xi_2^{k_2} \right\rangle$	conditional moment of order k_2
$\left\langle \xi_3^{k_3} \right\rangle$	conditional moment of order k_3
$\pi_{\alpha}(\xi)$	set of basis functions employed in Wheeler algorithm to calculate weights and abscissas
ρ_{f}	material density of fluid phase
ρ_{g}	material density of gas phase
ρ_{p}	material density of particles
ρ_{1}	material density of particles of type 1
ρ_{2}	material density of particles of type 2
ϱ_{f}	fluid-phase mass density
$\varrho_{\mathrm{f}\alpha}$	fluid-phase mass density for component α
ϱ_{mix}	mixture mass density
ϱ_{p}	disperse-phase mass density
$\varrho_{\mathrm{p}\alpha}$	disperse-phase mass density for component α
σ	parameter used in definition of kernel density function in EQMOM
σ	interfacial tension between continuous and disperse phases
$\sigma_{\alpha\beta}$	velocity covariance matrix for polydisperse system
$\sigma_{\alpha,\beta}$	matrix used to calculate Jacobi matrix in Wheeler algorithm
$\boldsymbol{\sigma}$	velocity-covariance tensor
$\boldsymbol{\sigma}_{\mathrm{p}}$	velocity-covariance matrix
$\boldsymbol{\sigma}_{\alpha}$	velocity-covariance matrix for particles of type α in polydisperse system
ς_{α}	weighted node α of quadrature approximation in DQMOM
$\varsigma_{i,\alpha}$	weighted node α of quadrature approximation for the ith internal coordinate in DQMOM
τ_{a}	characteristic time scale for particle aggregation
τ_{b}	characteristic time scale for particle breakage
τ_{c}	characteristic time scale for particle collision

τ_{col}	characteristic collisional time for BGK-like models
τ_d	characteristic time scale for particle–fluid drag force
$\tau_{D,ij}$	characteristic time scale for diffusion in phase space
τ_f	characteristic time scale for fluid phase
τ_J	characteristic time scale for particle formation
τ_M	characteristic particle advection time scale
τ_n	characteristic time scale for growing a connecting neck between two colliding particles
τ_p	particle relaxation time
τ_s	aggregate strength
τ_W	autocorrelation time related to spatial filter used to estimate an NDF
τ_I	characteristic time scale for first-order process
τ_{II}	characteristic time scale for second-order process
$\boldsymbol{\tau}_p$	traceless granular tensor
$\boldsymbol{\tau}_\alpha$	traceless granular tensor for particles of type α in polydisperse system
$\boldsymbol{\tau}_{\dot{\xi}}$	characteristic time scale for continuous change of internal coordinates
$\langle \tau \rangle$	expected interval of quiescence in MC simulation
ϕ	particle composition
ϕ	collision angle for $\mathbf{x}_{12}$
ϕ_c	size-correction factor
ϕ_{max}	energy barrier for particle aggregation
ϕ_p	volume fraction of solid within an aggregate
ϕ_1	spherical angle for $\mathbf{v}_{12}$
ϕ_1	density ratio between disperse and continuous (gaseous) phases
ϕ_2	mass ratio between disperse and continuous (gaseous) phases
ϕ_3	characteristic velocity ratio of disperse and continuous phases
$\boldsymbol{\phi}_f$	fluid composition vector
χ	factor relating velocities for direct and inverse collisions
$\chi_{\alpha\beta}$	factor relating velocities for direct and inverse collisions for polydisperse particles of types α and β
$\psi(\xi)$	dimensionless self-similar PSD
$\psi(\mathbf{v})$	generic property change during a collision
ω	restitution parameter relating velocities for direct and inverse collisions
$\omega_{\alpha\beta}$	mass-average restitution parameter for polydisperse particles of types α and β

Superscripts and angle brackets

$'$	fluctuating component
$'$	property of particle undergoing breakage
$'$	property of *field* particle undergoing aggregation

$'$	property of particle after a direct collision
$''$	property of particle before an inverse collision
$\dagger$	variable in collision frame of reference
$\tilde{}$	property of *test* particle undergoing aggregation
$\bar{}$	time average
$\langle\ \rangle$	ensemble and volume average

Subscripts

$_\mathrm{f}$	property of fluid phase
$_\mathrm{g}$	property of gas phase
$_\mathrm{p}$	property of disperse phase
$_{i-1/2}$	interface between the grid cells $i-1$ and i in FVM
$_{i+1/2}$	interface between the grid cells i and $i+1$ in FVM
$_\mathrm{l}$	limit value approaching the cell interface from the left in FVM
$_\mathrm{r}$	limit value approaching the cell interface from the right in FVM

Operators

$\partial/\partial\mathbf{x}$	gradient operator in physical space
$\nabla_\mathbf{x}$	gradient operator in physical space
$\partial/\partial\boldsymbol{\xi}$	gradient operator in internal-coordinate phase space
∇_{ξ}	gradient operator in internal-coordinate phase space
$\partial/\partial\mathbf{v}$	gradient operator in velocity phase space
$\nabla_\mathbf{v}$	gradient operator in velocity phase space
$\cdot$	scalar product
$\otimes$	dyadic or tensorial product
$:$	total tensor product

Abbreviations

BGK	Bhatnagar–Gross–Krook
CFD	computational fluid dynamics
CFL	Courant–Friedrichs–Lewy
CM	classes methods
CN	Crank–Nicolson
CQMOM	conditional quadrature method of moments
CSD	crystal-size distribution
DEM	discrete-element method
DNS	direct numerical simulation
DPBE	discrete population-balance equation
DQMOM	direct quadrature method of moments
DQMOM-FC	direct quadrature method of moments, fully conservative
DVM	discrete-velocity method
ECQMOM	extended conditional quadrature method of moments
EQMOM	extended quadrature method of moments
ES-BGK	ellipsoidal–statistical Bhatnagar–Gross–Krook
FEM	finite-element method
FVM	finite-volume method

GPBE	generalized population-balance equation
KBFVM	kinetics-based finite-volume method
KE	kinetic equation
LBM	lattice Boltzmann method
LES	large-eddy simulation
MCM	Monte Carlo method
MOM	method of moments
MOMIC	method of moments with interpolative closure
MUSCL	monotone upstream-centered scheme for conservation laws
MUSIG	multiple-size group
MWR	method of weighted residuals
NDF	number-density function
NS	Navier–Stokes equation
ODE	ordinary differential equation
PBE	population-balance equation
PD	product difference
PDF	probability density function
PSD	particle-size distribution
PTC	particle trajectory crossing
QBMM	quadrature-based moment method
QMOM	quadrature method of moments
QUICK	quadratic upstream interpolation for convective kinematics
RANS	Reynolds-average Navier–Stokes equation
RK2SSP	second-order two-stage strong stability-preserving Runge–Kutta

1

Introduction

1.1 Disperse multiphase flows

The majority of the equipment used in the chemical process industry employs multiphase flow. Bubble columns, fluidized beds, flame reactors, and equipment for liquid–liquid extraction, for solid drying, and size enlargement or reduction are common examples. In order to efficiently design, optimize, and scale up industrial systems, computational tools for simulating multiphase flows are very important. Polydisperse multiphase flows are also common in other areas, such as fuel sprays in auto and aircraft engines, brown-out conditions in aerospace vehicles and particulate flows occurring in the environment. Although at first glance the multifarious industrial and environmental multiphase flows appear to be very different from each other, they have a very important common element: it is possible to identify a continuous phase and a disperse phase (usually with a distribution of characteristic "particle sizes").

Historically the development of the theoretical framework and of computational models for disperse multiphase flows has focused on two different aspects: (i) the evolution of the disperse phase (e.g. breakage and coalescence of bubbles or droplets, particle–particle collisions, etc.) and (ii) multiphase fluid dynamics. The first class of models is mainly concerned with the description of the disperse phase, and is based on the solution of the *spatially homogeneous*[1] population balance equation (PBE). A PBE is a continuity statement written in terms of a *number density function* (NDF), which will be described in detail in Chapter 2. The NDF contains information about how the population constituted by the discrete elements of the disperse phase is distributed over certain characteristic properties that determine, for example, product quality. For example, in crystallization the final quality of the crystals very often depends on the crystal-size distribution (CSD). The CSD (or NDF) defines how the population of crystals is distributed over crystal size, and the PBE is a partial integro-differential equation that defines the evolution of the NDF, as described by Ramkrishna (2000) and Randolph & Larson (1971). Generally these models consist of a PBE coupled with *spatially homogeneous* mass and energy balances, and can be profitably used in modeling, design, and scale-up of process equipment, and in the development of

[1] In the more sophisticated treatments, spatial inhomogeneities are modeled by connected "zones" or regions of space that are assumed to be homogeneous. The resulting population-balance equation (PBE) does not explicitly account for *local* variations in the flow field.

control strategies for particulate systems (Chiu & Christofides, 1999; Crowley *et al.*, 2000; Nagy, 2009).

However, in many industrial and environmental applications the fluid-dynamic interactions between the disperse and continuous phases are also very important, and much research effort has focused on these aspects of the problem (e.g. Delnoij *et al.*, 1997; Fox, 2012; Gavi *et al.*, 2007; Gerber & Mousavi, 2007; Lain *et al.*, 2002; Laurent & Massot, 2001; Marchisio & Fox, 2007; Monahan & Fox, 2007; Petitti *et al.*, 2010; Prat & Ducoste, 2006; Rigopoulos, 2010; Sanyal *et al.*, 2005; Venneker *et al.*, 2002). For example, in gas–solid systems the application of the kinetic theory of gases to granular flows and the development of multiphase computational models have led to a deeper understanding of these issues (Gidaspow, 1994; Goldhirsch, 2003; Jenkins & Mancini, 1989). Generally these transport-phenomena-based models consist of *spatially inhomogeneous* mass, momentum, and energy (thermal and granular) balances between the disperse phase and the continuous phase, and spatial dependences are handled by using a finite-volume approach (Leveque, 2002) in the context of computational fluid dynamics (CFD). Important industrial examples of such flows include fluidized beds and riser flows, slurry-flow reactors, and bubble columns. In all of these examples, the coupling due to mass, momentum, and energy exchange between the disperse and continuous phases results in flow dynamics that are distinctly different than that observed in single-phase flows. Thus, the CFD models used for describing the fluid dynamics of disperse multiphase flows usually involve multiple continuity and momentum equations that are tightly coupled through phase-interaction terms (Drew & Passman, 1999; van der Hoef *et al.*, 2008; Ishii, 1975; Portela & Oliemans, 2006). For polydisperse multiphase flows, the situation becomes more complex because it is necessary to describe the "particle-size" distribution of the disperse phase (De Chaisemartin *et al.*, 2009; Fan *et al.*, 2004; Fox, 2007; Riber *et al.*, 2009), as well as the coupling with the continuous phase for particles of different sizes. One of the primary goals of the present work is to present a systematic modeling framework for accomplishing this task.

In general, the first class of models mentioned above is able to describe the evolution of the NDF that characterizes the disperse phase. For example, if this approach is applied to crystallization problems, it is possible to describe the evolution of the CSD in a crystallizer working under certain operating conditions, taking into account all possible physical and chemical processes such as nucleation, molecular growth, aggregation, and breakage. However, these models are unable to take into account spatial gradients of properties and how they relate to the continuous or disperse phases, fluid-dynamic interactions between crystals and the continuous phase, and interactions among the crystals themselves. As a consequence, the first class of models is unable to predict the effect of local flow interactions on the evolution of the CSD. On the other hand, the second class of models is capable of describing detailed fluid-dynamic interactions, but for simplicity often assumes a constant (or even monodisperse) CSD. The principal aim of this book is to present the common underlying theory and, through the introduction of appropriate computational methods, to create a bridge between the two approaches. The resulting CFD-based computational models can then be used to solve a large variety of *polydisperse* particulate and multiphase flow problems.

A working definition of polydisperse multiphase flows

For the purposes of this book, it will be useful to have a clear definition of what is meant by *disperse* and *polydisperse* in the context of multiphase flows. By disperse, we mean that one or more of the phases is composed of clearly identifiable discrete entities such as solid particles, drops, or bubbles. By polydisperse, we mean that the properties of the disperse-phase entities can be different for each entity (e.g. particles with different mass, composition, or temperature). For example, fuel sprays have a region near the nozzle where the liquid jet is *not* disperse, followed by a region after breakup of the primary jet that is composed of individual droplets. The latter region would be considered a polydisperse multiphase flow and could be modeled using the methods described in this book. In contrast, bubble-column flows, for example, can be entirely monodisperse because all of the bubbles have (approximately) the same properties (e.g. bubble diameter). In practice, monodisperse multiphase flows are relatively rare and, hence, it will be important to have a modeling framework that naturally accounts for polydispersity.

Another important manifestation of "polydispersity" is the presence of disperse-phase entities (even those with identical physical properties) with different velocities. The reader familiar with the kinetic theory of molecular gases will recognize this type of polydispersity as leading to the *velocity-distribution function*, which plays an important role in the transport theory of polydisperse multiphase flows. As in the molecular kinetic theory, the *mesoscale* description of a monodisperse multiphase flow can be formulated in terms of a *kinetic equation* for the velocity-distribution function. However, unlike in molecular gases at standard temperatures and pressures, the disperse-phase entities often interact infrequently due to collisions, so the standard hydrodynamic approximations valid in the collision-dominated regime are no longer accurate. In analogy to molecular gases, such multiphase flows behave as *rarefied granular gases* wherein processes besides collisions (e.g. momentum exchange with the continuous phase) are dominant in determining the flow regime. For this reason, it is often necessary to retain the mesoscale description as the starting point for describing the disperse-phase flow dynamics.

In summary, a polydisperse multiphase flow consists of one (or more) disperse phases with entities of possibly different physical properties and velocities. The mesoscale modeling approach (described in more detail in Section 1.3) for describing such flows is the primary focus of this book. For multiphase flows that are not composed of a clearly distinct disperse phase, other modeling approaches must be followed (e.g. methods that resolve the dynamics of the interface separating the phases, or volume-averaging approaches and the concomitant *ad hoc* closures of the phase-interaction and transport terms). As will become clearer to the reader in subsequent chapters, the mesoscale modeling approach allows rigorous derivation of the macroscale transport equations, thereby ensuring that the resulting CFD models will be as accurate as possible when simulating real polydisperse multiphase flows.

1.2 Two example systems

In this section, we provide two examples of polydisperse multiphase systems. In the first example, the particles are assumed to be very small so that their inertia is small enough to be negligible relative to that of the fluid phase. Thus, the particle-size distribution (PSD) is governed by a PBE and undergoes changes due to aggregation and breakage. In the second example, the particles are assumed to be large enough to have finite inertia relative to the surrounding fluid, and we let all particles be identical and consider the distribution of the particle velocity, which is described by a kinetic equation (KE).

1.2.1 The population-balance equation for fine particles

As a first example, we consider a population of solid particles suspended in a liquid. We assume that the density and size of the particles are such that they move together with the fluid. Local velocity gradients in the fluid, which are generally quantified by the shear rate, induce particle aggregation and breakage, changing the PSD, as dictated by the corresponding PBE. In this context we will assume (as is often the case for colloidal systems) that aggregation and breakage are completely reversible, which means that, when particles aggregate, the aggregates formed can later be fragmented by breakage, and that the fragments formed by breakage can, in turn, aggregate together. In what follows we will analyze the PBE and introduce a set of dimensionless numbers and characteristic time scales that will turn out to be useful in the investigation of multiphase systems.

Relevant dimensionless numbers and time scales

As described in Chapter 4, the following PBE governs the evolution of the PSD, representing the state of the solid aggregates:

$$\frac{\partial n}{\partial t} + \frac{\partial v_{\mathrm{f}} n}{\partial x} = \Gamma \frac{\partial^2 n}{\partial x^2} + \mathbb{C}, \tag{1.1}$$

where v_{f} is the known fluid velocity with which the particles are advected and Γ is a diffusion coefficient. For simplicity, in this example, only one spatial coordinate x is considered and the diffusion coefficient Γ is constant. The PSD, denoted by n, is a number-density function (NDF) representing the number concentration of particles with volumes between V and $V + \mathrm{d}V$. The PSD is a function of time t, of the spatial coordinate x, and of the volume of the aggregates V. The aggregates are composed of spherical primary particles with fixed diameter d_{p}. If just one primary particle is included in the aggregate then $V = \pi d_{\mathrm{p}}^3/6$, whereas for an aggregate composed of two primary particles $V = \pi d_{\mathrm{p}}^3/3$, and so on for larger aggregates. As described in Chapter 5, the source term due to aggregation and breakage is

$$\begin{aligned}\mathbb{C} = \frac{1}{2}\int_0^V a(V - V', V')n(V - V')n(V')\mathrm{d}V' - \int_0^\infty a(V, V')n(V)n(V')\mathrm{d}V' \\ + \int_V^\infty b(V')N(V|V')n(V')\mathrm{d}V' - b(V)n(V),\end{aligned} \tag{1.2}$$

where $a(V, V')$ and $b(V)$ are the aggregation and breakage kernels, $N(V|V')$ is the daughter distribution function, and the time and space dependences are omitted for clarity. As will be described in more detail in later chapters, the aggregation and breakage kernels express the tendency of the particles to aggregate and break. Generally speaking, these kernels depend strongly on the spatial coordinate x, since real multiphase systems are characterized by regions where aggregation and breakage occur at very different rates. For example, particles suspended in a stirred tank might experience intense breakage near the impeller and strong aggregation in other regions. The daughter distribution function gives instead the size distribution of particles generated by the breakup of a single aggregate.

In order to highlight the different regimes exhibited by Eq. (1.1) and the different treatments that are more suitable for its solution in each regime, the PBE has to be normalized and made dimensionless. Different quantities can be used to this end, such as the moments of order zero and one of the PSD:

$$\begin{aligned} m_0(t, x) &= \int_0^\infty n(t, x, V)\mathrm{d}V, \\ m_1(t, x) &= \int_0^\infty V n(t, x, V)\mathrm{d}V. \end{aligned} \tag{1.3}$$

The zeroth-order moment identifies the total aggregate number density (i.e. the total number of aggregates per unit volume), whereas the first-order moment is the disperse-phase volume fraction, more frequently denoted by α_p. For this simple two-phase system, if α_f is used to indicate the fluid-phase volume fraction, by definition $\alpha_p + \alpha_f = 1$. By using these moments, one can define a mean particle volume corresponding to the ratio between the moments of order one and zero and, in addition, one can define the dimensionless PSD $\psi(\xi) = n(t, x, V)\, m_1(t, x)/m_0^2(t, x)$ with $\xi = V m_0/m_1$. The dimensionless PSD has first- and second-order moments equal to unity, and is useful when searching for the self-similar solution that characterizes pure aggregation, pure breakage and simultaneous aggregation and breakage problems (Ramkrishna, 2000). The moment of order two of the PSD,

$$m_2(t, x) = \int_0^\infty V^2 n(t, x, V) \mathrm{d}V, \tag{1.4}$$

is instead useful in the definition of the standard deviation of the PSD, which is equal to $m_2 - m_1^2/m_0$. Here the PSD and Eq. (1.1) are made dimensionless by introducing the characteristic length L, velocity U, aggregate volume ζ, and total aggregate number density N_t. For example, for particles suspended in a stirred tank, L could be the tank diameter and U the impeller-tip velocity, whereas ζ and N_t could be extracted by the volume-average PSD in the vessel at steady state. The last two quantities can in turn be used to define the characteristic disperse-phase volume fraction: $\phi_p = \zeta N_t$.

Knowledge of these characteristic quantities suffices for the calculation of the characteristic time scales for aggregation $\tau_a = [a(\zeta, \zeta)N_t]^{-1}$ and breakage $\tau_b = [b(\zeta)]^{-1}$. These two time scales represent the average time interval between two subsequent aggregation or breakage events. When these time scales are very short, aggregation and breakage events are very frequent, resulting in rapid evolution of the PSD. In fact, as particles move and diffuse in the domain, many aggregation and breakage events occur, resulting in a quick adaptation of the PSD to its local "equilibrium" value $n_{eq}(t, x, V)$, which is dictated by the local values of the aggregation and breakage kernels. Here the term equilibrium refers to the steady state reached when aggregation and breakage counterbalance each other.[2] Because of the hypothesis of reversible aggregation and breakage, this equilibrium will be different from point to point, since it depends on the local values of the aggregation and breakage kernels. The equilibrium solution corresponds to the resulting PSD, in the case of particles unable to move away from point x, but evolving in time according to the local aggregation and breakage kernels at point x. In contrast, when the characteristic time scales are very long, aggregation and breakage events are not very frequent, and the evolution of the PSD becomes quite slow. Therefore as particles move and diffuse in the domain, due to the low aggregation and breakage frequencies and the slow changes in the PSD, only the averaged rates are perceived. The evolution of the PSD is then determined by the volume-average kernels, rather than by their local values.

Closer observation of the aggregation time scale shows that it is inversely proportional to the characteristic total aggregate number density N_t, which is related to the characteristic disperse-phase volume fraction ϕ_p. When the system is dilute (i.e. $\phi_p \ll 1$), N_t is

[2] Depending on the functional form of the aggregation and breakage kernels, different scenarios are possible. When aggregation prevails, the system can undergo extensive aggregation. Since when particles aggregate their number is reduced, the *gelling limit*, where the total number of particles tends to zero, can be reached. In contrast, when breakage prevails, the total number of particles can go to infinity, giving the opposite limit known as *shattering*. When aggregation and breakage counterbalance each other, a population of particles can evolve to a steady state. This steady state can be calculated by setting the collision source term in Eq. (1.2) equal to zero. We refer to this steady-state solution as the *equilibrium PSD*.

small and as a consequence the aggregation time scale is large. Under these conditions aggregation events are separated by long time intervals. In fact, for dilute systems with very few particles, longer time intervals are necessary to observe a collision between two particles. In contrast, as the system becomes more concentrated (i.e. as N_t and ϕ_p increase) the time interval between two aggregation events becomes very short, since as more particles are suspended in the fluid more collisions (and aggregation events) will occur. The characteristic disperse-phase volume fraction also affects the breakage time scales; in fact, as aggregation events become more frequent the resulting characteristic aggregate volume ζ increases. Since generally the breakage kernel increases with increasing particle size (i.e. larger aggregates are more likely to break than smaller ones), the breakage time scale also becomes shorter. In summary, dilute systems are characterized by large aggregation and breakage time scales, whereas for dense systems these time scales are very short.

A more quantitative definition of dilute and dense systems is possible only by analyzing the normalized and dimensionless PBE. By letting $\xi = V/\zeta$, $n^* = n\zeta/N_t$, $x^* = x/L$, $v_f^* = v_f/U$, and $t^* = t\Gamma/L^2$ the following dimensionless PBE is obtained:

$$\begin{aligned}\frac{\partial n^*}{\partial t^*} + \mathrm{Pe}\frac{\partial v_f^* n^*}{\partial x^*} - \frac{\partial^2 n^*}{\partial x^{*2}} = \mathrm{Da_a}\bigg[\frac{1}{2}\int_0^{\xi} & a^*(\xi-\xi',\xi')n^*(\xi-\xi')n^*(\xi')\mathrm{d}\xi' \\ & - \int_0^{+\infty} a^*(\xi,\xi')n^*(\xi)n^*(\xi')\mathrm{d}\xi'\bigg] \\ + \mathrm{Da_b}\bigg[\int_{\xi}^{+\infty} & b^*(\xi')N(\xi|\xi')n^*(\xi')\mathrm{d}\xi' - b^*(\xi)n^*(\xi)\bigg],\end{aligned} \tag{1.5}$$

where the normalized aggregation and breakage kernels are given by $a^*(\xi,\xi') = a(V,V')/a(\zeta,\zeta)$ and $b^*(\xi) = b(V)/b(\zeta)$. The Péclet number, defined by

$$\mathrm{Pe} = \frac{UL}{\Gamma} = \frac{L^2/\Gamma}{L/U}, \tag{1.6}$$

is the ratio between the particle-diffusion and particle-advection time scales. The aggregation Damköhler number, defined by

$$\mathrm{Da_a} = \frac{a(\zeta,\zeta)L^2 N_t}{\Gamma} = \frac{L^2/\Gamma}{1/(a(\zeta,\zeta)N_t)}, \tag{1.7}$$

is the ratio between the particle-diffusion and particle-aggregation time scales. The breakage Damköhler number, defined by

$$\mathrm{Da_b} = \frac{b(\zeta)L^2}{\Gamma} = \frac{L^2/\Gamma}{1/b(\zeta)}, \tag{1.8}$$

is the ratio between the particle-diffusion and breakage time scales. Depending on the values of Pe, $\mathrm{Da_a}$, and $\mathrm{Da_b}$ different regimes can be identified. For example, when Pe becomes much smaller than unity, diffusive particle transport becomes more important than advective transport. Under this condition, different scenarios are still possible, depending on the values of $\mathrm{Da_a}$ and $\mathrm{Da_b}$. When both numbers are much smaller than unity, diffusion is much faster than aggregation and breakage, whereas when both are much greater than unity aggregation and breakage are faster than particle diffusion. It is important to recall here that particle diffusion tends to homogenize the PSD in the spatial domain, whereas, since kernels generally contain some form of spatial dependence, aggregation and breakage tend to create gradients in the PSD.

Major operating regimes

From the discussion above, three major operating regimes can be identified from Eq. (1.5).

(i) Under **dilute** conditions such that $\phi_p \ll 1$, aggregation and breakage are slower than particle diffusion, or, in other words, $\mathrm{Da_a} \ll 1$ and $\mathrm{Da_b} \ll 1$. The evolution of the PSD is not controlled by the local values of the aggregation and breakage kernels, but rather by their volume-average values. Under dilute conditions, since diffusion mixes particles faster than the kernels let them aggregate and break, the two-phase system is generally considered as spatially homogeneous or well mixed (i.e. the PSDs at different points of the domain are identical). The evolution of the PSD for a well-mixed system $\hat{n}(t, V)$ is governed by a volume-average PBE:

$$\frac{\partial \hat{n}}{\partial t} = \frac{1}{2}\int_0^V \hat{a}(V - V', V')\hat{n}(V - V')\hat{n}(V')\mathrm{d}V' - \int_0^\infty \hat{a}(V, V')\hat{n}(V)\hat{n}(V')\mathrm{d}V' + \int_V^\infty \hat{b}(V')N(V|V')\hat{n}(V')\mathrm{d}V' - \hat{b}(V)\hat{n}(V), \tag{1.9}$$

where $\hat{a}$ and $\hat{b}$ are the volume-average aggregation and breakage kernels, which, due to their non-linear dependence on the spatial coordinates, are generally quite different from the local kernels a and b. A system is generally considered dilute (and therefore well mixed) when $\mathrm{Da_a}$ and $\mathrm{Da_b}$ are smaller than 10^{-2}. Depending on the type of problem and the functional form of the aggregation and breakage kernels, these conditions result in different constraints for the characteristic disperse-phase volume fraction. For most applications, it is reasonable to consider the operating regime as dilute when $\phi_p < 10^{-3}$.

(ii) Under **moderately dense** conditions, $\mathrm{Da_a}$ and $\mathrm{Da_b}$ are greater than 10^{-2} but smaller than 10^2. The characteristic time scales for aggregation and breakage are then comparable to that of particle diffusion. Under these conditions the system cannot be considered to be well mixed, and the PSD is subjected to large variations across the spatial domain. As a consequence, the movement of particles in physical space and their aggregation and breakage cannot be decoupled, and the evolution of the two-phase system is found by solving Eq. (1.5).

(iii) Under **dense** conditions $\mathrm{Da_a}$ and $\mathrm{Da_b}$ are both greater than 10^2, and hence aggregation and breakage are very fast (or almost instantaneous) compared with particle diffusion. Thanks to the hypothesis of reversibility of aggregation and breakage, as particles diffuse in the domain, they instantaneously adapt to the local equilibrium solution, i.e. n_{eq}. In this case the evolution of the PSD is completely determined by the local aggregation and breakage kernels and the resulting local steady state.

In summary, for fine particles undergoing aggregation and breakage different modeling approaches should be used for each of the regimes described above. Under dilute conditions, the system can be approximated as a well-mixed system, since spatial gradients in the PSD are not relevant. The evolution of the system is governed by Eq. (1.9) using the volume-average aggregation and breakage kernels. Under moderately dense conditions, spatial gradients of the PSD are not negligible and therefore the evolution of the system is governed by Eq. (1.5). Under dense conditions the system evolves according to the equilibrium solution found by setting the right-hand side of Eq. (1.2) equal to zero.

1.2.2 The kinetic equation for gas–particle flow

As a second example, we consider the kinetic equation (KE) for monodisperse, isothermal solid particles suspended in a constant-density gas phase. For clarity, we assume that the particle material density is significantly larger than that of the gas so that only the fluid drag and buoyancy terms are needed to account for momentum exchange between the two phases (Maxey & Riley, 1983). In this example, the particles are large enough to have finite inertia and thus they evolve with a velocity that can be quite different than that of the gas phase.

Relevant dimensionless numbers

As in all mathematical descriptions of transport phenomena, the theory of polydisperse multiphase flows introduces a set of dimensionless numbers that are pertinent in describing the behavior of the flow. Depending on the complexity of the flow (e.g. variations in physical properties due to chemical reactions, collisions, etc.), the set of dimensionless numbers can be quite large. (Details on the physical models for momentum exchange are given in Chapter 5.) As will be described in detail in Chapter 4, a kinetic equation can be derived for the number-density function (NDF) of the velocity of the disperse phase $n(t, x, v)$. Also in this example, for clarity, we will assume that the problem has only one particle velocity component v and is one-dimensional in physical space with coordinate x at time t. Furthermore, we will assume that the NDF has been normalized (by multiplying it by the volume of a particle) such that the first three velocity moments are

$$\begin{aligned} \alpha_{\mathrm{p}} &\equiv \int_{-\infty}^{+\infty} n \, \mathrm{d}v, \\ \alpha_{\mathrm{p}} U_{\mathrm{p}} &\equiv \int_{-\infty}^{+\infty} v n \, \mathrm{d}v, \\ \alpha_{\mathrm{p}} \Theta_{\mathrm{p}} + \alpha_{\mathrm{p}} U_{\mathrm{p}}^2 &\equiv \int_{-\infty}^{+\infty} v^2 n \, \mathrm{d}v, \end{aligned} \tag{1.10}$$

where α_{p}, U_{p}, and Θ_{p} are the volume fraction, mean velocity, and granular temperature of the disperse phase, respectively. The example kinetic equation is then given by

$$\frac{\partial n}{\partial t} + \frac{\partial v n}{\partial x} + \frac{\partial \mathbb{A} n}{\partial v} = \mathbb{C}, \tag{1.11}$$

where the acceleration (due to buoyancy, gravity, and drag) and collision terms are defined, respectively, by

$$\begin{aligned} \mathbb{A} &\equiv -\frac{1}{\rho_{\mathrm{p}}} \frac{\partial p_{\mathrm{g}}}{\partial x} + g_x + \frac{1}{\tau_{\mathrm{p}}} (U_{\mathrm{g}} - v), \\ \mathbb{C} &\equiv \frac{1}{\tau_{\mathrm{c}}} (n_{\mathrm{eq}} - n). \end{aligned} \tag{1.12}$$

In these expressions, g_x is the x-component of the gravity force, ρ_{p} is the particle density, and p_{g} is the gas-phase pressure. τ_{p} and τ_{c} are characteristic time scales for drag and collisions, respectively. U_{g} is the velocity of the gas phase, and n_{eq} is the equilibrium distribution. For simplicity, we have used a linear collision model[3] in which the first two moments of n_{eq} are the same as those of n, and the third is

$$\frac{1+e^2}{2}\alpha_{\text{p}}\Theta_{\text{p}} + \alpha_{\text{p}} U_{\text{p}}^2 = \int_{-\infty}^{+\infty} v^2 n_{\text{eq}}\, \mathrm{d}v, \tag{1.13}$$

and $0 \le e \le 1$ is the coefficient of restitution for particle–particle collisions. For elastic collisions, $e = 1$, whereas $e < 1$ corresponds to inelastic collisions.

The disperse-phase kinetic equation is coupled to the continuity and momentum equations for the continuous phase, which are given, respectively, by

$$\begin{aligned} &\frac{\partial \alpha_{\text{g}}}{\partial t} + \frac{\partial(\alpha_{\text{g}} U_{\text{g}})}{\partial x} = 0, \\ &\frac{\partial(\alpha_{\text{g}} U_{\text{g}})}{\partial t} + \frac{\partial(\alpha_{\text{g}} U_{\text{g}} U_{\text{g}})}{\partial x} = \alpha_{\text{g}} \nu_{\text{g}} \frac{\partial^2 U_{\text{g}}}{\partial x^2} - \frac{\alpha_{\text{g}}}{\rho_{\text{g}}}\frac{\partial p_{\text{g}}}{\partial x} + \alpha_{\text{g}} g_x + \frac{\rho_{\text{p}}\alpha_{\text{p}}}{\rho_{\text{g}}\tau_{\text{p}}}(U_{\text{p}} - U_{\text{g}}), \end{aligned} \tag{1.14}$$

where α_{g}, ρ_{g}, and ν_{g} are the gas-phase volume fraction, density, and kinematic viscosity,[4] respectively. By definition, $\alpha_{\text{p}} + \alpha_{\text{g}} = 1$. In order to make Eqs. (1.14) dimensionless, we introduce a characteristic gas-phase velocity U and characteristic length L. The dimensionless time is then $t^* = tU/L$. On noting that α_{g} is dimensionless and letting $x^* = x/L$, $U_{\text{g}}^* = U_{\text{g}}/U$, and $U_{\text{p}}^* = U_{\text{p}}/U$, we then have

$$\begin{aligned} &\frac{\partial \alpha_{\text{g}}}{\partial t^*} + \frac{\partial \alpha_{\text{g}} U_{\text{g}}^*}{\partial x^*} = 0, \\ &\frac{\partial(\alpha_{\text{g}} U_{\text{g}}^*)}{\partial t^*} + \frac{\partial(\alpha_{\text{g}} U_{\text{g}}^* U_{\text{g}}^*)}{\partial x^*} = \frac{\alpha_{\text{g}}}{\text{Re}_{\text{g}}}\frac{\partial^2 U_{\text{g}}^*}{(\partial x^*)^2} - \alpha_{\text{g}}\frac{\partial p_{\text{g}}^*}{\partial x^*} + \frac{\alpha_{\text{g}}}{\text{Fr}_{\text{g}}^2} g_x^* + \frac{\alpha_{\text{g}}\phi_2}{\text{St}}(U_{\text{p}}^* - U_{\text{g}}^*), \end{aligned} \tag{1.15}$$

where $p_{\text{g}}^* = p/(\rho_{\text{g}} U^2)$ and $g_x^* = g_x/g$. The gas-phase Reynolds and Froude numbers are defined, respectively, by $\text{Re}_{\text{g}} = UL/\nu_{\text{g}}$ and $\text{Fr}_{\text{g}} = U/(Lg)^{1/2}$, where g is the gravitational acceleration constant. The Stokes number is defined by $\text{St} = U\tau_{\text{p}}/L$, and the phase-mass ratio by $\phi_2 = \alpha_{\text{p}}\rho_{\text{p}}/(\alpha_{\text{g}}\rho_{\text{g}})$. The latter is an important parameter insofar as it determines whether the gas phase "sees" the disperse phase (i.e. momentum coupling is negligible when $\phi_2 \ll 1$).

The kinetic equation given by Eq. (1.11) is made dimensionless by defining $v^* = v/U$ and $n^* = Un$:

$$\frac{\partial n^*}{\partial t^*} + \frac{\partial(v^* n^*)}{\partial x^*} + \frac{\partial(\mathbb{A}^* n^*)}{\partial v^*} = \mathbb{C}^*, \tag{1.16}$$

where $n^*(t^*, x^*, v^*)$ is the dimensionless velocity distribution function,[5] and

$$\begin{aligned} \mathbb{A}^* &\equiv -\frac{1}{\phi_1}\frac{\partial p_{\text{g}}^*}{\partial x^*} + \frac{g_x^*}{\text{Fr}_{\text{g}}^2} + \frac{1}{\text{St}}(U_{\text{g}}^* - v^*), \\ \mathbb{C}^* &\equiv \frac{\phi_3}{\text{Kn}_{\text{p}}}(n^* - n). \end{aligned} \tag{1.17}$$

---

[3]See Chapter 6 for more details on collision models.

[4]For a 1D flow the viscosity term would normally be zero. However, we include the viscosity term as a placeholder for the fully 3D case in order to show the Reynolds-number dependence.

[5]Note that n and n_{eq} are made dimensionless in a manner such that $\alpha_{\text{p}} = \int n^*\, \mathrm{d}v^*$. Because the kinetic equation is linear in n, no new dimensionless numbers are generated by this process.

The phase-density ratio is defined by $\phi_1 = \rho_p/\rho_g$ and, since $\phi_1 \gg 1$, the buoyancy term is negligible. The two new dimensionless numbers generated in this process are the phase-velocity ratio $\phi_3 = U_p^\dagger/U$,[6] and the disperse-phase Knudsen number $\text{Kn}_p = U_p^\dagger\tau_c/L$. In addition, the dimensionless form of Eqs. (1.10),

$$\begin{aligned} \alpha_p &= \int_{-\infty}^{+\infty} n^* \, dv^*, \\ \alpha_p U_p^* &= \int_{-\infty}^{+\infty} v^* n^* \, dv^*, \\ \alpha_p \frac{\phi_3^2}{\text{Ma}_p^2} + \alpha_p (U_p^*)^2 &= \int_{-\infty}^{+\infty} (v^*)^2 n^* \, dv^*, \end{aligned} \tag{1.18}$$

introduces the disperse-phase Mach number $\text{Ma}_p = U_p^\dagger/\Theta_p^{1/2}$. By analogy to compressible gas flows, Ma_p is the ratio of the characteristic mean particle velocity $U_p^\dagger$ and the speed of "sound" ($\Theta_p^{1/2}$) in the disperse phase. Thus for $\text{Ma}_p \ll 1$ transport in the disperse phase is predominantely due to velocity fluctuations, whereas for $\text{Ma}_p \gg 1$ it is due to mean advection.

Following the convention used in gas dynamics, we will define the velocity ratio and disperse-phase Knudsen number differently according to whether the disperse phase is subsonic or supersonic:

$$\begin{aligned} &\text{for } \text{Ma}_p \le 1, \quad \phi_3 = \frac{\Theta_p^{1/2}}{U}, \quad \text{Kn}_p = \frac{\Theta_p^{1/2}\tau_c}{L} = \frac{\pi^{1/2} d_p}{12\alpha_p g_0 L}; \\ &\text{for } \text{Ma}_p \ge 1, \quad \phi_3 = \frac{U_p^\dagger}{U}, \quad \text{Kn}_p = \frac{U_p^\dagger \tau_c}{L} = \frac{\pi^{1/2} d_p}{12\alpha_p g_0 L}\text{Ma}_p. \end{aligned} \tag{1.19}$$

Here d_p is the particle diameter,[7] $g_0(\alpha_p/\alpha_p^*)$ is the radial distribution function, and $\alpha_p^* \approx 0.63$ is the maximum volume fraction at close packing. For $\alpha_p \ll \alpha_p^*$, $g_0 \approx 1$. However, as α_p approaches α_p^*, g_0 diverges to infinity.

Note that the ratio ϕ_3 will depend strongly on the Stokes number,

$$\text{St} = \frac{U\tau_p}{L} = \frac{\rho_p d_p^2 U}{18\rho_g \nu_g L}, \tag{1.20}$$

where this expression for τ_p is valid for small particle Reynolds numbers, defined by

$$\text{Re}_p \equiv \frac{d_p |U_p - U_g|}{\nu_g}. \tag{1.21}$$

When $\text{St} \gg 1$ the kinetic equation will be uncoupled from the gas phase and the disperse phase will behave as a granular gas. In the opposite limit where $\text{St} \ll 1$, $\Theta_p \approx 0$ and $U_p^\dagger \approx U$, so that the disperse-phase Mach number will be very large and $\phi_3 \approx 1$. At intermediate values of the Stokes number, a rich variety of flow phenomena depending on all the values of the dimensionless parameters can be observed.

[6] Because ϕ_3 and Kn_p appear together in Eq. (1.17), they could be combined into one dimensionless number. We keep them separate so that Kn_p and Ma_p depend only on disperse-phase variables.

[7] At fixed volume fraction α_p, reducing d_p increases the collision rate because the total number concentration of particles N increases significantly.

Table 1.1. Relevant dimensionless numbers for gas–particle flow with $\mathrm{Re_p} \ll 1$ and $\rho_\mathrm{g} \ll \rho_\mathrm{p}$

Restitution coefficient	e
Maximum volume fraction	α_p^*
Radial distribution function	$g_0(\alpha_\mathrm{p}/\alpha_\mathrm{p}^*)$
Gas-phase Reynolds number	$\mathrm{Re_g} = UL/\nu_\mathrm{g}$
Gas-phase Froude number	$\mathrm{Fr_g} = U/\sqrt{Lg}$
Phase-density ratio	$\phi_1 = \rho_\mathrm{p}/\rho_\mathrm{g}$
Phase-mass ratio	$\phi_2 = \alpha_\mathrm{p}\rho_\mathrm{p}/(\alpha_\mathrm{g}\rho_\mathrm{g})$
Stokes number	$\mathrm{St} = \rho_\mathrm{p} d_\mathrm{p}^2 U/(18\rho_\mathrm{g}\nu_\mathrm{g} L)$
Particle Reynolds number	$\mathrm{Re_p} = d_\mathrm{p}\lvert U_\mathrm{p} - U_\mathrm{g}\rvert/\nu_\mathrm{g}$
Disperse-phase Mach number	$\mathrm{Ma_p} = U_\mathrm{p}^\dagger/\sqrt{\Theta_\mathrm{p}}$
Disperse-phase Knudsen number	$\mathrm{Kn_p} = \sqrt{\pi} d_\mathrm{p}/(12\alpha_\mathrm{p} g_0 L)$ for $\mathrm{Ma_p} \le 1$
	$\mathrm{Kn_p} = \sqrt{\pi} d_\mathrm{p}/(12\alpha_\mathrm{p} g_0 L)\mathrm{Ma_p}$ for $\mathrm{Ma_p} \ge 1$
Phase-velocity ratio	$\phi_3 = \sqrt{\Theta_\mathrm{p}}/U$ for $\mathrm{Ma_p} \le 1$
	$\phi_3 = U_\mathrm{p}^\dagger/U$ for $\mathrm{Ma_p} \ge 1$
Particle Archimedes number	$\mathrm{Ar_p} = \dfrac{g d_\mathrm{p}^3}{\nu_\mathrm{g}^2}\left(\dfrac{\rho_\mathrm{p} - \rho_\mathrm{g}}{\rho_\mathrm{g}}\right) \approx \dfrac{\rho_\mathrm{p} g d_\mathrm{p}^3}{\rho_\mathrm{g}\nu_\mathrm{g}^2}$

For convenience, the relevant dimensionless numbers for gas–particle flow derived in this section are collected in Table 1.1. In practice, one must choose appropriate values for U and L corresponding to a particular problem. For example, they may be determined by the inlet and/or boundary conditions. However, one case of particular interest is particles falling in an unbounded domain for which convenient choices are $L = d_\mathrm{p}$ and $U = U_\mathrm{p}^\dagger = \lvert U_\mathrm{p} - U_\mathrm{g}\rvert = g\tau_\mathrm{p}\lvert g_x^*\rvert$ (i.e. the settling velocity). For this case, there is no source term for Θ_p and so it relaxes to zero at steady state due to the drag.[8] The disperse-phase Mach number thus becomes infinite. For settling problems, the particle Archimedes number (see Table 1.1) is often used in place of the Froude number.

The reader may be surprised not to find a Reynolds number defined specifically for the disperse phase. This is because the disperse-phase viscosity is well defined only for $\mathrm{Kn_p} \ll 1$ (i.e. the collision-dominated or hydrodynamic regime). In this limit, $\nu_\mathrm{p} \propto \Theta_\mathrm{p}\tau_\mathrm{c} \propto \mathrm{Kn_p}/\mathrm{Ma_p}$ so that the disperse-phase Reynolds number would be proportional to $\mathrm{Ma_p}/\mathrm{Kn_p}$ when $\mathrm{Ma_p} \le 1$. However, in many gas–particle flows the disperse-phase Knudsen number will not be small, even for $\alpha_\mathrm{p} \sim 0.1$, because the granular temperature (and hence the collision frequency) will be strongly reduced by drag and inelastic collisions. In comparison, molecular gases at standard temperature and pressure have $\mathrm{Kn_p} \ll 1$ even though the volume fraction occupied by the molecules is on the order of 0.001. This fact can be

[8] Hydrodynamic interactions, which are not included in the kinetic equation for this example, can lead to a finite Θ_p. However, in gas–particle flows the disperse-phase Mach number will usually be very large.

explained by the high root-mean-square (RMS) velocity of molecules. Indeed, if the granular temperature in a typical gas–particle flow were converted to degrees Kelvin, it would be extremely small compared with room temperature.

Major flow regimes

Given the relatively large number of dimensionless parameters in Table 1.1, it should not be surprising to learn that the number of different flow regimes observed in gas–particle flows is also large. The same is true in general for other polydispere multiphase flows. For example, if we consider flows for which the Stokes number is very small ($\mathrm{St} \ll 1$) so that the disperse-phase velocity is nearly the same as the continuous-phase velocity, then we can expect to observe flow regimes similar to those observed for a constant-density single-phase flow when $\phi_1 \approx 1$. However, if the phase densities are very different, and there are important spatial variations in ϕ_2 (e.g. a stratified flow with a large ϕ_2 on top – relative to gravity – of a small ϕ_2), then one would observe flow regimes typical of variable-density single-phase flow. Despite this plethora of possible flow regimes, it is possible to identify a few *major* regimes that are characteristic of disperse multiphase flows. The purpose of this section is thus to introduce the reader to the major flow regimes which commonly appear in the multiphase-flow literature.

In many discussions on disperse multiphase flows, the value of the disperse-phase volume fraction α_p is used to distinguish between major flow regimes (Balachandar & Eaton, 2010). From Table 1.1, we can observe that α_p appears in ϕ_2, Kn_p, and g_0. As noted earlier, ϕ_2 is related to the momentum coupling between the disperse and continuous phases, and, all other parameters being fixed, ϕ_2 can be reduced by lowering α_p. The disperse-phase Knudsen number and the radial distribution function enter the governing kinetic equation through the collision term. As discussed in detail in Chapter 6, the collision term makes two contributions to the mean momentum equation for polydisperse systems: (1) a "pure" pointwise collision term (which is zero for monodisperse systems) and (2) a collisional-flux term due to the finite diameter of the particles. When $g_0 \approx 1$ (i.e. $\alpha_\mathrm{p} \ll \alpha_\mathrm{p}^*$) the second contribution is negligible. Thus, generally speaking, the value of Kn_p indicates the relative importance of collisions (i.e. when $\mathrm{Kn}_\mathrm{p} \ll 1$ the flow is *collision-dominated*, and when $\mathrm{Kn}_\mathrm{p} = \infty$ it is *collisionless*), and increasing α_p will usually increase the rate of collisions by lowering Kn_p. However, it is important to note that, when the restitution coefficient e is significantly less than unity, the granular temperature will also be affected by α_p. In fact, it is possible to drive the granular temperature to zero due to loss of energy during collisions, in which case the disperse-phase Knudsen number would go to infinity.[9] Thus, in general, the value of α_p is not necessarily a reliable indicator of the rate of collisions, since the latter depends also on the granular temperature.

The other parameter often used to distinguish between flow regimes is the Stokes number St. In the limit of very large St, the particle acceleration (Eq. (1.17)) does not depend on the continuous-phase velocity, and the disperse phase behaves as a *granular gas*. In the limit of small St, there are two important regimes separated by a critical Stokes number St_c (Jabin, 2002). For $\mathrm{St} \leq \mathrm{St}_\mathrm{c}$ the trajectories of any two particles never cross each other, which implies that at any point in the flow at a given time only one particle velocity will be observed. Such flows are called *monokinetic* (Massot, 2007) or *equilibrium* (Ferry & Balachandar, 2001), and, since the velocity distribution is a delta function, the granular

[9]The disperse-phase Knudsen number can also be infinity for dense, elastic systems due to initial conditions. For example, if one releases a dense assembly of particles that are initially at rest in a vacuum, the particles will accelerate due to gravity such that they all have the same velocity. For this case, the granular temperature is null and $\mathrm{Kn}_\mathrm{p} = \infty$.

temperature in such flows is zero.[10] For $\mathrm{St} > \mathrm{St_c}$, particle trajectory crossing is possible, so that locally a non-zero granular temperature will be generated even if it was initially zero.

In terms of α_p and St, some of the major flow regimes discussed in the literature are as follows.

(i) **Very dilute.** For $\alpha_\mathrm{p} \ll 1$ and $\phi_2 \ll 1$, neither collisions nor coupling with the continuous phase are important. This regime is often referred to as *one-way coupling*. The very dilute regime can be further subdivided according to the Stokes number.

(a) $\mathrm{St} \ll \mathrm{St_c}$. Particles have the same velocity as the continuous phase. This flow regime is often referred to as a *dusty gas*.

(b) $\mathrm{St} \le \mathrm{St_c}$. Particles have slightly different velocities than the continuous phase, but are locally unique.

(c) $\mathrm{St} > \mathrm{St_c}$. The kinetic regime where in the particle-velocity distribution is multi-valued due to the finite inertia of the particles.

(d) $\mathrm{St} \gg 1$. Collisionless granular gas.

(ii) **Dilute.** For $\alpha_\mathrm{p} \ll 1$ and $\phi_2 \ge 1$, collisions and/or coupling with the continuous phase are non-negligible. When collisions can be neglected ($\mathrm{Kn_p} \gg 1$), this regime is referred to as *two-way coupling*. When collisions are significant ($\mathrm{Kn_p} \le 1$), this regime is referred to as *four-way coupling*. The case in which collisions are dominant ($\mathrm{Kn_p} \ll 1$) is referred to as the *hydrodynamic* regime. The dilute regime can be further subdivided using the Stokes number, and the results are similar to those for the very dilute regime, with one important difference. Because $\phi_2 \ge 1$, the momentum coupling can lead to new flow instabilities that are not observed in the very dilute regime. For example, a dilute flow can become turbulent at Reynolds numbers $\mathrm{Re_g}$ for which a very dilute flow is laminar (Passalacqua *et al.*, 2010).

(iii) **Moderately dense.** When $\alpha_\mathrm{p} < \alpha_\mathrm{p}^*$ so that $g_0 > 1$, the collisional-flux term cannot be neglected. For gas–particle flows such as fluidized beds, this regime occurs when $0.02 < \alpha_\mathrm{p} < \alpha_\mathrm{p}^*$. Liquid–particle flows with $\phi_1 \approx 1$ (e.g. slurries) and $\mathrm{Re_g} \ll 1$ are often in the moderately dense regime, but the particle–particle interactions are more complicated than simple binary collisions due to long-range hydrodynamic interactions. For example, hydrodynamic interactions change the functional form of the two-particle radial distribution function (Ladd, 1993).

(iv) **Dense.** When $\alpha_\mathrm{p} \approx \alpha_\mathrm{p}^*$, particle–particle contact is sustained and cannot be represented by instantaneous collisions. The binary collision model must then be replaced by a model that accounts for long-range interactions between particles occurring in "force chains" (Aarons *et al.*, 2010).

The flow regimes discussed above are often used to describe systems with $\phi_1 \ll 1$. Other flow regimes occur for $\phi_1 \gg 1$ (e.g. bubbly flows) because additional force models (such as added mass) introduce new dimensionless parameters, and because the momentum of the continuous phase becomes dominant. For example, in bubbly flow when $\mathrm{Re_p} > 1$ the turbulent liquid wakes behind bubbles lead to "pseudo-turbulence" (Mudde *et al.*, 2009; Riboux *et al.*, 2010; Sato & Sekoguchi, 1975) that changes the nature of bubble–bubble interactions through the continuous phase.

---

[10]Note that, if the Stokes number is very small due to the particle diameter being sub-micron, velocity fluctuations due to Brownian motion will be important.

Finally, we should also note that disperse multiphase flows can be turbulent. Indeed, many multiphase flows encountered in industrial and environmental applications are turbulent. Although they are not the major emphasis of this book, multiphase turbulence models are a subject of considerable practical importance. Generally speaking, in the very dilute regime described above the turbulence model for the continuous phase will be exactly the same as for single-phase flows. However, it is necessary to modify the fluid velocity appearing in the drag term to account for the effect of unresolved fluid-velocity fluctuations (Buyevich, 1972; Derevich & Zaichik, 1990; Minier & Peirano, 2001; Pandya & Mashayek, 2003a; Reeks, 1977, 1991; Zaichik *et al.*, 2003, 2004, 2008; Zaichik & Alipchenkov, 1998). On the other hand, in the dilute regime the single-phase turbulence model for the continuous phase must be modified to include the effects of momentum exchange (both mean and fluctuating) between phases (Sommerfeld & Ho, 2003). More generally, when $\phi_2 \geq 1$ new turbulent-flow regimes can arise due to volume-fraction gradients, which are generated spontaneously by the dynamics of the flow. These truly multiphase turbulent-flow regimes exhibit turbulence-production mechanisms that are not present in single-phase flow, which makes them more difficult to model. For example, gas–particle flows can become turbulent due to the nonlinear dependence of the drag law on the disperse-phase volume fraction. Because this instability first occurs on relatively small scales and grows to larger scales, it cannot be successfully modeled as a simple turbulent energy cascade from integral to dissipative scales. Thus, at a very minimum, it will be necessary to add new terms in the turbulence models to account for the additional physics present in multiphase flows. In this work, we will focus on the derivation of the fundamental equations for disperse multiphase flows that can subsequently be used as the starting point for deriving turbulence models using the classical Reynolds-average and large-eddy simulation approaches (Fox, 2012; Pope, 2000). As we go along we will, however, point out the critical differences in the forms of the model equations as compared with the Navier–Stokes equation and comment on how these differences would affect the turbulence-modeling process.

1.3 The mesoscale modeling approach

The modeling approach that we will use to derive the fundamental governing equations for polydisperse multiphase flows will be at the *mesoscale* level. A schematic overview of the modeling framework is shown in Figure 1.1. We will use the term *microscale* to describe modeling approaches that completely resolve all relevant length and time scales at the level of individual particles. For example, a microscale model for fluid–particle flows would resolve all flow structures in the continuous phase, including the boundary layers around the surfaces of particles, as well as particle–particle collisions of all particle pairs in the system. In the context of single-phase flows, microscale models are referred to as direct numerical simulations (DNS). For multiphase systems, a microscale model uses a DNS for each phase and completely resolves the inter-phase coupling terms.

The term *macroscale* will be used to denote multiphase models that employ a hydrodynamic description of the disperse phase. Such models are also called multi-fluid models (because the disperse phase is treated as an effective fluid), or Euler–Euler models. The name of the latter comes from the numerical treatment of the disperse phase (i.e. discretization on a fixed grid), as opposed to Euler–Lagrange models wherein the disperse phase is tracked in a Lagrangian framework as discrete entities. We should note that, in the

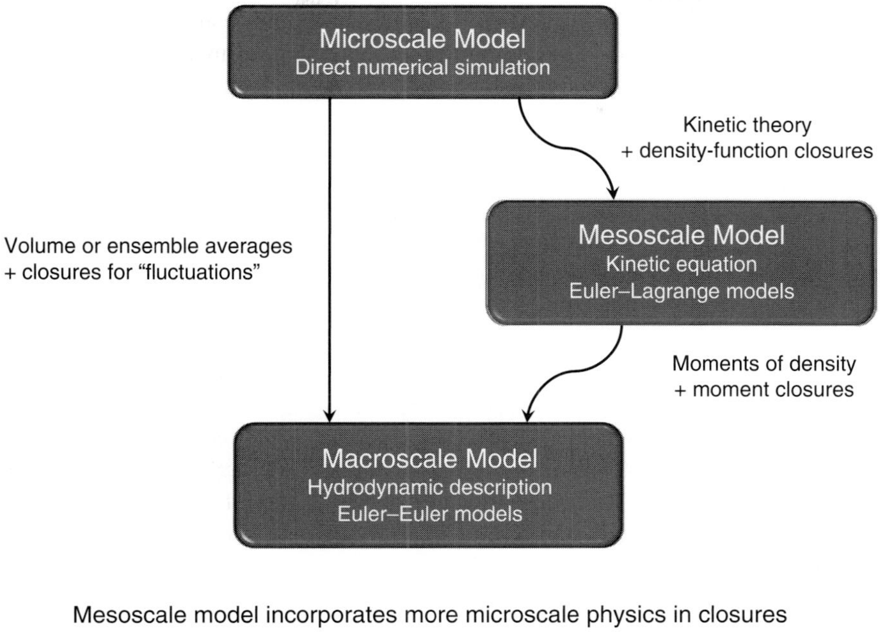

Figure 1.1. Modeling framework for disperse flows.

context of mesoscale models, the Lagrangian treatment of the disperse phase is essentially a stochastic representation of the solution to the kinetic equation (e.g. equivalent to direct-simulation Monte Carlo (DSMC) methods used in kinetic theory (Bird, 1994)). This can be contrasted with microscale models of granular flow that are based on molecular-dynamics (MD) simulations. For the latter, a *closed* kinetic equation is not guaranteed (due to the presence of the two-particle distribution function in the collision term), whereas the former is applied after the kinetic equation has been closed.

In general, a macroscale model can be found from a microscale model using volume- or ensemble-averaging approaches (Drew, 1971; Drew & Passman, 1999), or by first deriving a mesoscale model and then evaluating the moments of the resulting kinetic equation. The first step (deriving the mesoscale model) involves mainly *physical* modeling and physics-based simplifications in order to arrive at a *closed* kinetic equation. The second step (deriving the macroscale model from the mesoscale model) involves *mathematical* approximations to close the moment-transport equations. In other words, a well-defined macroscale model is deemed to be accurate if it reproduces the moments found from the exact solution to the mesoscale model. In contrast, the accuracy of the mesoscale model depends on the fidelity of the physical models derived from the solution to the microscale model. This division of physical and mathematical modeling errors present in the mesoscale modeling approach is not present in volume- and ensemble-averaging approaches. In the latter approaches, one derives directly the moment-transport equations (although they are usually referred to not as moments but as average mass and average momentum, etc.) from the microscale transport equations, and develops closures (or "constitutive" equations) directly for the unknown terms appearing in the moment-transport equations. This lack of separation between the physical and mathematical errors makes it difficult to propose accurate and well-posed macroscale models using averaging approaches. In comparison, by first deriving a mesoscale model it is possible to isolate physical modeling errors by direct comparison with solutions to microscale models.

As we shall see in later chapters, for polydisperse multiphase flows the separation of modeling errors turns out to be extremely valuable for arriving at well-posed macroscale models. A similar situation arises in turbulent reacting flows (Fox, 2003), in particular, and turbulent flows (Pope, 2000), in general. By employing probability density function (PDF) methods, it is possible to write a transport equation for the joint PDF of the velocity and composition variables, which is equivalent to a kinetic equation. Closure of the PDF transport equation involves models for physical processes such as molecular mixing (or micromixing). The PDF transport equation is thus a mesoscale model for turbulent flow, and its moments are the Reynolds-average quantities appearing in classical turbulence models. For turbulent reacting flows, the chemical source term is closed in the mesoscale model but unclosed in the macroscale model. The latter requires a *mathematical* closure, which is different in nature from the *physical* closure used for the micromixing model. Because of the additional physics present in multiphase flows, deriving direct closures for the macroscale equations is even more difficult than for single-phase turbulent flows (Fox, 2012). Another important distinction between PDF methods for turbulent flows and the mesoscale models described in this book is that the former treat ensemble averages arising due to turbulence, while the latter are kinetic-theory models for non-turbulent multiphase systems. In fact, one could derive PDF methods for *turbulent* multiphase flows starting from the fundamental macroscale model equations derived in this book.

1.3.1 Relation to microscale models

As illustrated in Figure 1.2, the key modeling steps in the mesoscale approach are (1) defining the mesoscale variables and (2) deriving a model for how *one* particle's mesoscale variables change due to the microscale physics involving *all* particles. In some cases, the choice of the mesoscale variables is straightforward. For example, due to mass and

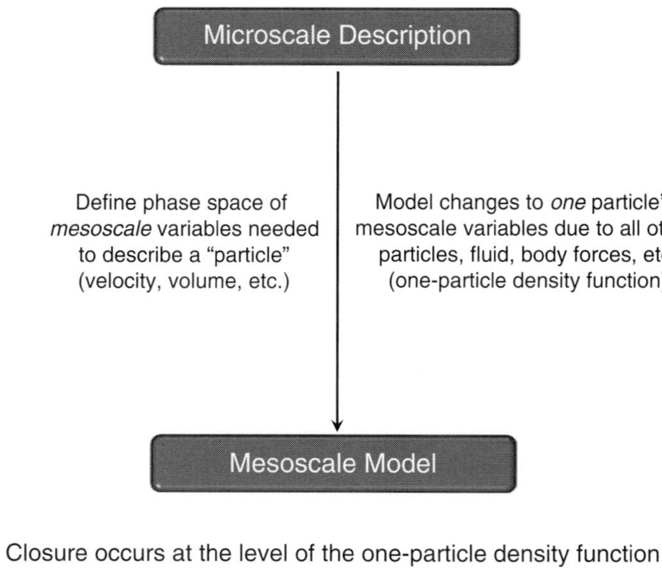

Figure 1.2. From microscale description to mesoscale model.

momentum conservation, the particle mass and velocity are typically used as mesoscale variables, along with particle temperature and chemical composition for systems for mass transfer and/or chemical reactions. In some cases, it may also be necessary to define mesoscale variables accounting for particle properties such as shape, internal structure, etc. In any case, the choice of the mesoscale variables is closely linked to the second step of deriving a model for how these variables change for a single particle. For mass and momentum changes, developing such models is usually straightforward, but this is not always true for other types of mesoscale variables. As discussed in Chapter 5, such models often include a stochastic component to represent the unresolved scales present in the microscale model. A typical example is the random fluctuations in particle velocity generated by the turbulent wakes of neighboring particles in the flow. The principal objective of the mesoscale modeling effort is to capture as much as possible of the microscale physics in the mesoscale model equations for a single particle.

The validation of mesoscale models can be carried out using the numerical solutions to microscale models (Tenneti *et al.*, 2010), in much the same way as that in which DNS is used for model validation in turbulent single-phase flows. A typical mesoscale modeling strategy consists of four steps.

1. Using insights based on the microscale description and a good understanding of the physics, develop a mesoscale model.
2. Design a microscale "numerical experiment" that tests key assumptions made in the development of the mesoscale model.
3. Use the microscale simulation results to validate the predictions of the mesoscale model.
4. Return to step 1 to improve the mesoscale model by applying new insights gained from discrepancies seen during validation.

Note that the order of these steps is important. Indeed, it is very rare for a successful mesoscale model to result from simply studying the results from a microscale simulation that is not specifically designed to test a particular assumption or hypothesis. In general, a microscale simulation will produce an enormous quantity of data if all flow variables are stored at every time step, and it is not realistic to expect that such data sets will "reveal" the correct form of the mesoscale models without some *a priori* understanding of the physics. On the other hand, microscale simulations offer an invaluable and often unique tool for sorting out the validity of a proposed mesoscale model because they contain all of the microscale variables, many of which are impossible to measure in laboratory experiments.

In summary, the microscale description provides two important pieces of information needed for the development of mesoscale models. First, the mathematical formulation of the microscale model, which includes all of the relevant physics needed to completely describe a disperse multiphase flow, provides valuable insights into what mesoscale variables are needed and how these variables interact with each other at the mesoscale. These insights are used to formulate a mesoscale model. Second, the detailed numerical solutions from the microscale model are directly used for validation of a proposed mesoscale model. When significant deviations between the mesoscale model predictions and the microscale simulations are observed, these differences lead to a reformulation of the mesoscale model in order to improve the physical description. Note that it is important to remember that this validation step should be done by comparing *exact* solutions to the mesoscale model with the microscale results, not *approximate* solutions that result from solving the macroscale model. As mentioned earlier, the macroscale model includes additional closure steps not present at the mesoscale level, thus disagreement between the microscale and macroscale models could be simply due to the inadequacy of such closures.

1.3.2 Number-density functions

In the formulation of a mesoscale model, the *number-density function* (NDF) plays a key role. For this reason, we discuss the properties of the NDF in some detail in Chapter 2. In words, the NDF is the number of particles per unit volume with a given set of values for the mesoscale variables. Since at any time instant a microscale particle will have a unique set of microscale variables, the NDF is also referred to as the *one-particle* NDF. In general, the one-particle NDF is nonzero only for *realizable* values of the mesoscale variables. In other words, the realizable mesoscale values are the ones observed in the *ensemble* of all particles appearing in the microscale simulation. In contrast, sets of mesoscale values that are never observed in the microscale simulations are *non-realizable*. Realizability constraints may occur for a variety of reasons, e.g. due to conservation of mass, momentum, energy, etc., and are intrinsic properties of the microscale model. It is also important to note that the mesoscale values are usually strongly correlated. By this we mean that the NDF for any two mesoscale variables cannot be reconstructed from knowledge of the separate NDFs for each variable. Thus, by construction, the one-particle NDF contains all of the underlying correlations between the mesoscale variables for only *one* particle.

As discussed in Chapter 2, the one-particle NDF does not usually provide a complete description of the microscale system. For example, a microscale system containing N particles would be completely described by an N-particle NDF. This is because the mesoscale variable in any one particle can, in principle, be influenced by the mesoscale variables in all N particles. Or, in other words, the N sets of mesoscale variables can be correlated with each other. For example, a system of particles interacting through binary collisions exhibits correlations between the velocities of the two particles before and after a collision. Thus, the *time evolution* of the one-particle NDF for velocity will involve the two-particle NDF due to the collisions. In the mesoscale modeling approach, the primary physical modeling step involves the approximation of the N-particle NDF (i.e. the exact microscale model) by a functional of the one-particle NDF. A typical example is the closure of the collisionterm (see Chapter 6) by approximating the two-particle NDF by the product of two one-particle NDFs.

For polydisperse multiphase flows, the fact that the one-particle NDF contains all correlations between the mesoscale variables makes the mesoscale modeling approach extremely useful for deriving macroscale models. For example, the particle size in a gas–particle flow has a strong influence on the Stokes number, and thus has a strong influence on the acceleration of the particle due to coupling with the gas phase. If, in addition, the particle size is changing due to aggregation or breakage, the coupling between particle size and velocity will be quite complicated. Because the NDF can describe exactly such complex correlations, the mesoscale model will be able to predict the evolution of the mean particle size and mean particle velocity. In contrast, if one tries to derive directly a macroscale model for the mean particle size it will be very difficult to capture the strong coupling with the particle velocity using an *ad hoc* closure.

In summary, the complexity of the correlations between the mesoscale variables describing polydisperse multiphase flows requires us to employ a mathematical formulation that is capable of exactly describing such correlations. The one-particle NDF contains all correlations between the mesoscale variables in one particle and is thus ideally suited for developing a mesoscale modeling framework.

1.3.3 The kinetic equation for the disperse phase

Mathematically speaking, the one-particle NDF is governed by a *kinetic equation*. However, the governing equation goes by other names for specific applications. For example, in kinetic theory the velocity NDF is governed by the Boltzmann equation (Boltzmann, 1872; Cercignani *et al.*, 1994), and in aerosol science the relevant kinetic equation is referred to as the general particle-dynamic equation (Friedlander, 2000; Gelbard & Seinfeld, 1978), while in the field of sprays it is labeled as the Williams–Boltzmann equation (Williams, 1985). In this work, we will refer to kinetic equations that do not include the velocity, but include mesoscale variables such as particle size, as population-balance equations (PBE), and to kinetic equations that include both velocity and "internal" mesoscale variables as generalized population-balance equations (GPBE). While this division is somewhat arbitrary from a mathematical perspective, it has the merit of emphasizing the polydisperse nature of the underlying disperse phase. In examples where the mesoscale variables are limited to the particle velocity, we will follow the nomenclature used in kinetic theory.

At the mesoscale level, the independent variables appearing in the NDF are time, physical location, and the mesoscale variables. In order to distinguish between a physical point and a "mesoscale" point, we introduce the concept of *phase space*. Thus, a physical point takes on a value in real space, and a mesoscale point takes on a value in phase space. However, both types of variables appear in the kinetic equation as independent variables (along with time) so that the NDF is a function of the *independent* variables. In contrast, a macroscale model has only time and space as independent variables. Thus, a macroscale model can be found from the kinetic equation by integrating over the phase space in order to remove the dependence on the mesoscale variables. This process of "integrating over phase space" is related to evaluating the moments of the NDF. For example, in Section 1.2.1 we used the relationship between the first two moments of the particle-size NDF and the macroscale variables representing the total aggregate number density and the disperse-phase volume fraction; in Section 1.2.2 we used instead the relationship of the first three moments of the velocity NDF and the macroscale variables representing disperse-phase volume fraction, mean particle velocity, and granular temperature.

The kinetic equation for the mesoscale variables contains terms for each of the microscale physical processes affecting the disperse phase. For example, in Section 1.2.2, the kinetic equation for gas–particle flow contains terms for acceleration due to buoyancy, gravity, and fluid drag, and changes in velocity due to collisions. In the context of mesoscale modeling, each of these terms is a *model* for the underlying microscale physics, each of which could be validated by comparison with microscale simulations. In general, when a new mesoscale variable is added (e.g. particle size), new terms will be added to the kinetic equation to account for changes in that variable. It is important to recognize that the terms in the kinetic equation are said to be *closed* if they are formulated in terms of the independent variables and are functionals of the one-particle NDF. For example, processes such as gravity that act independently on each particle are linear functions of the NDF. The ultimate objective of the mesoscale modeling approach is to arrive at a closed kinetic equation that accurately describes the effects of the microscale physics on the mesoscale variables.

1.3.4 Closure at the mesoscale level

In most cases, closure of the terms in the kinetic equation will require prior knowledge of how the mesoscale variables are influenced by the underlying physics. Taking the fluid-drag term as an example, the simplest model has the form

$$\mathbb{A}_{\mathrm{d}} = \frac{1}{\tau_{\mathrm{p}}}(U_{\mathrm{g}} - v), \tag{1.22}$$

where τ_{p} is constant (see Eq. (1.20)) for the special case of Stokes flow ($\mathrm{Re}_{\mathrm{p}} = 0$) around a single sphere in an infinite domain ($\alpha_{\mathrm{p}} = 0$). For finite Re_{p}, τ_{p} depends on a drag coefficient, which is a function of $|U_{\mathrm{g}} - v|$. Likewise, for $\alpha_{\mathrm{p}} > 0$ the drag model depends on the disperse-phase volume fraction (i.e. $\mathbb{A}_{\mathrm{d}}$ increases with increasing α_{p}). Changing the particle shape (or other mesoscale variables) will usually also affect the drag model. Thus, using the results from microscale simulations, it is necessary to formulate a drag model similar to Eq. (1.22) that depends on (possibly) all of the mesoscale variables. (See Chapter 5 for a more detailed discussion of drag models.)

The process of formulating mesoscale models from the microscale equations is widely used in transport phenomena (Ferziger & Kaper, 1972). For example, heat transfer between the disperse phase and the fluid depends on the Nusselt number, and mass transfer depends on the Sherwood number. Correlations for how the Nusselt and Sherwood numbers depend on the mesoscale variables and the moments of the NDF (e.g. mean particle temperature and mean particle concentration) are available in the literature. As microscale simulations become more and more sophisticated, modified correlations that are based on the microscale results will become more and more common (Beetstra *et al.*, 2007; Holloway *et al.*, 2010; Tenneti *et al.*, 2010). Note that, because the kinetic equation requires mesoscale models that are valid *locally in phase space* (i.e. for a particular set of mesoscale variables) as opposed to averaged correlations found from macroscale variables, direct numerical simulation of the microscale model is perhaps the only way to obtain the data necessary in order for such models to be thoroughly validated. For example, a macroscale model will depend on the average drag, which is denoted by

$$\langle \mathbb{A}_{\mathrm{d}} \rangle = \left\langle \frac{1}{\tau_{\mathrm{p}}}(U_{\mathrm{g}} - v) \right\rangle \approx \frac{1}{\langle \tau_{\mathrm{p}} \rangle}(U_{\mathrm{g}} - U_{\mathrm{p}}), \tag{1.23}$$

where $\langle \tau_{\mathrm{p}} \rangle$ is a macroscale drag time scale depending only on mean quantities such as $|U_{\mathrm{g}} - U_{\mathrm{p}}|$. In contrast, τ_{p} depends on the mesoscale variables, and thus potentially contains a more complete description of the fluid drag. The same remark applies to the other mesoscale terms appearing in the kinetic equation for the NDF.

1.3.5 Relation to macroscale models

As mentioned above, macroscale models are written in terms of transport equations for the lower-order moments of the NDF. The different types of moments will be discussed in Chapters 2 and 4. However, the lower-order moments that usually appear in macroscale models for monodisperse particles are the disperse-phase volume fraction, the disperse-phase mean velocity, and the disperse-phase granular temperature. When the particles are polydisperse, a description of the PSD requires (at a minimum) the mean and standard deviation of the particle size, or in other words the first three moments of the PSD. However, a more complete description of the PSD will require a larger set of particle-size moments, whose transport equations are often coupled to the velocity moments in a nontrivial way when the Stokes number is nonzero.

The transport equations appearing in macroscale models can be derived from the kinetic equation using the definition of the moment of interest. For example, if the moment of interest is the disperse-phase volume fraction, then it suffices to integrate over the mesoscale variables. (See Section 4.3 for a detailed discussion of this process.) Using the velocity-distribution function from Section 1.2.2 as an example, this process yields

$$\int_{-\infty}^{+\infty} \left(\frac{\partial n(t,x,v)}{\partial t} + \frac{\partial v n(t,x,v)}{\partial x} + \frac{\partial \mathbb{A}(t,x,v) n(t,x,v)}{\partial v} - \mathbb{C}(t,x,v,n(t,x,v)) \right) \mathrm{d}v = 0 \quad (1.24)$$

or, using the definitions in Eq. (1.10),

$$\frac{\partial \alpha_{\mathrm{p}}(t,x)}{\partial t} + \frac{\partial \alpha_{\mathrm{p}}(t,x) U_{\mathrm{p}}(t,x)}{\partial x} + \int_{-\infty}^{+\infty} \frac{\partial \mathbb{A}(t,x,v) n(t,x,v)}{\partial v} \mathrm{d}v = \int_{-\infty}^{+\infty} \mathbb{C}(t,x,v,n(t,x,v)) \mathrm{d}v. \quad (1.25)$$

Owing to conservation of mass during collisions, the integral involving $\mathbb{C}$ is null. The integral involving $\mathbb{A}$ is simplified using the fact that the number-density function is integrable and hence bounded for $|v| \to \infty$:

$$\int_{-\infty}^{+\infty} \frac{\partial \mathbb{A}(t,x,v) n(t,x,v)}{\partial v} \mathrm{d}v = \mathbb{A}(t,x,+\infty) n(t,x,+\infty) - \mathbb{A}(t,x,-\infty) n(t,x,-\infty) = 0. \quad (1.26)$$

Thus, Eq. (1.25) reduces to the continuity equation for the disperse-phase volume fraction. However, the disperse-phase mean velocity U_{p} is unknown, and must be found by solving a separate transport equation.

The disperse-phase mean momentum equation is found from the first-order moment of the velocity distribution function:

$$\int_{-\infty}^{+\infty} v \left(\frac{\partial n(t,x,v)}{\partial t} + \frac{\partial v n(t,x,v)}{\partial x} + \frac{\partial \mathbb{A}(t,x,v) n(t,x,v)}{\partial v} - \mathbb{C}(t,x,v,n(t,x,v)) \right) \mathrm{d}v = 0 \quad (1.27)$$

or, using the definitions in Eq. (1.10),

$$\frac{\partial \alpha_{\mathrm{p}}(t,x) U_{\mathrm{p}}(t,x)}{\partial t} + \frac{\partial}{\partial x} \left(\alpha_{\mathrm{p}}(t,x) U_{\mathrm{p}}^2(t,x) + \alpha_{\mathrm{p}}(t,x) \Theta_{\mathrm{p}}(t,x) \right) + \int_{-\infty}^{+\infty} v \frac{\partial \mathbb{A}(t,x,v) n(t,x,v)}{\partial v} \mathrm{d}v = \int_{-\infty}^{+\infty} v \mathbb{C}(t,x,v,n(t,x,v)) \mathrm{d}v. \quad (1.28)$$

Owing to conservation of momentum during collisions, the integral involving $\mathbb{C}$ is again null. The integral involving $\mathbb{A}$ is simplified using integration by parts:

$$\begin{aligned} \int_{-\infty}^{+\infty} v \frac{\partial \mathbb{A}(t,x,v) n(t,x,v)}{\partial v} \mathrm{d}v &= v \mathbb{A}(t,x,v) n(t,x,v) \Big|_{v=-\infty}^{v=+\infty} - \int_{-\infty}^{+\infty} \mathbb{A}(t,x,v) n(t,x,v) \mathrm{d}v \\ &= -\langle \mathbb{A} \rangle (t,x). \end{aligned} \quad (1.29)$$

Thus, Eq. (1.28) reduces to the transport equation for the disperse-phase mean momentum:

$$\frac{\partial \alpha_p(t,x)U_p(t,x)}{\partial t} + \frac{\partial}{\partial x}\left(\alpha_p(t,x)U_p^2(t,x) + \alpha_p(t,x)\Theta_p(t,x)\right) = \langle \mathbb{A} \rangle(t,x). \tag{1.30}$$

However, the disperse-phase granular temperature Θ_p and the mean acceleration $\langle \mathbb{A} \rangle$ are unknown, and must be modeled using *moment closures*.

This example illustrates the classical problem faced when working with macroscale models (Struchtrup, 2005). No matter how the transport equations for the moments are derived, they will always contain unclosed terms that depend on higher-order moments (e.g. U_p depends on Θ_p, etc.). In comparison, the solution to the kinetic equation for the NDF contains information about *all possible* moments. In other words, if we could compute $n(t, x, v)$ directly, it would not be necessary to work with the macroscale model equations. The obvious question then arises: "Why don't we simply solve the kinetic equation for the mesoscale model instead of working with the macroscale model?"

The answer to this question is mainly driven by the computational cost of solving the kinetic equation due to the large number of independent variables. In the simplest example of a 3D velocity-distribution function $n(t, \mathbf{x}, \mathbf{v})$ the number of independent variables is $1 + 3 + 3 = 7$. However, for polydisperse multiphase flows the number of mesoscale variables can be much larger than three. In comparison, the moment-transport equations involve four independent variables (physical space and time). Furthermore, the form of the moment-transport equations is such that they can be easily integrated into standard computational-fluid-dynamics (CFD) codes. Direct solvers for the kinetic equation are much more difficult to construct and require specialized numerical methods if accurate results are to be obtained (Filbet & Russo, 2003). For example, with a direct solver it is necessary to discretize all of phase space since *a priori* the location of nonzero values of n is unknown, which can be very costly when phase space is not bounded.

In practice, the most commonly used method for simulating the mesoscale model is to resort to a statistical description wherein the NDF is approximated by a finite ensemble of computational "parcels" (Bird, 1994). In the multiphase-flow literature, such methods are known as Euler–Lagrange models because the continuous phase is treated using an Eulerian formulation while one uses a Lagrangian particle-tracking approach for the discrete phase. For cases in which the coupling between phases is weak or nonexistent (e.g. very dilute flows), Euler–Lagrange simulations can be made quite accurate by increasing the number of parcels in order to minimize the statistical errors. However, when the coupling between phases is strong, even small statistical errors in the disperse-phase simulation can lead to qualitatively different flow regimes than would be observed from the direct solution to the kinetic equation (Passalacqua *et al.*, 2010). Also, it is important to realize that, with a fixed number of parcels, the accuracy of the statistics computed from an Euler–Lagrange simulation will decrease as the number of mesoscale variables increases. Thus, even though it is usually straightforward to increase the number of mesoscale variables in an Euler–Lagrange simulation code, this does not imply that the statistical accuracy is not a concern. Another challenge with Euler–Lagrange simulations is the need for *time-accurate* statistics (not just time-average values) for the coupling terms appearing in the continuous-phase transport equations. For all of these reasons, there is strong motivation to find accurate methods to close the moment-transport equations.

1.4 Closure methods for moment-transport equations

As noted above, solving the mesoscale model directly for the NDF will rarely be possible for complicated industrial and environmental flows. Instead, we would like to derive *accurate* macroscale models whose solutions are consistent with the moments found from the solution to the mesoscale model. Because the macroscale models are formulated in terms of transport equations for the moments of the NDF, we will need to find accurate closures for the moment-transport equations, preferably using only the smallest possible moment set. From kinetic theory (Chapman & Cowling, 1961), we know that for cases in which the collision time scale is much smaller than all other processes (i.e. when the disperse-phase Knudsen number is very small), it is possible to formally derive a *hydrodynamic* model that uses only the five moment equations for mass, momentum, and total energy. However, when the collision time scale is not small, it is necessary to develop moment closures that are valid for all values of the disperse-phase Knudsen number (Grad, 1949b; Struchtrup, 2005). In this section, we provide a brief overview of the conceptual foundations behind these two closure methods. More details are presented in later chapters using specific examples. For simplicity, we will restrict our attention in this section to the velocity-distribution function and the corresponding Boltzmann equation.

1.4.1 Hydrodynamic models

Hydrodynamic models are derived from the mesoscale model (e.g. the Boltzmann equation) using a Chapman–Enskog expansion in powers of the Knudsen number (Bardos *et al.*, 1991; Cercignani *et al.*, 1994; Chapman & Cowling, 1961; Ferziger & Kaper, 1972; Jenkins & Mancini, 1989). The basic idea is that the collision term will drive the velocity distribution n towards an equilibrium function n_{eq} (i.e. the solution to $\mathbb{C}(n_{\mathrm{eq}}) = 0$), and thus the deviation from equilibrium can be approximated by $n \approx n_{\mathrm{eq}} + \mathrm{Kn} n_1$. From the linearized Boltzmann equation, it is possible to solve for n_1 and thus to close the moment equations for the conserved quantities (α_{p}, $\mathbf{U}_{\mathrm{p}}$, Θ_{p}) that appear in the definition of n_{eq} and n_1. Hence, by construction, hydrodynamic models are most accurate for $\mathrm{Kn} \ll 1$. In Figure 1.3, the different flow regimes for the Boltzmann equation are presented. The ranges of validity of the two corresponding hydrodynamic models are $\mathrm{Kn} = 0$ for the Euler equation and $\mathrm{Kn} \leq 0.1$ for the Navier–Stokes (NS) equation. In some cases, it is possible to obtain reasonable predictions using the NS equation for slightly larger Knudsen numbers by adjusting the boundary conditions (i.e. partial slip). However, inevitably for larger Knudsen numbers the flow will begin to exhibit non-equilibrium behavior[11] that cannot be captured by using only five moment equations (Grad, 1949b). Recall, however, that the exact solution to the Boltzmann equation (i.e. the mesoscale model) is valid for *all* Knudsen numbers. Thus, the failure of the hydrodynamic models is a result of the *mathematical approximation* used to close the moment equations, rather than being an intrinsic property of the mesoscale modeling approach.

For gas–particle flows, the most obvious manifestation of non-equilibrium behavior is *particle trajectory crossing* (PTC) at $\mathrm{Kn}_{\mathrm{p}} = \infty$ (i.e. no collisions). PTC occurs in the very-dilute-flow regime ($\alpha_{\mathrm{p}} \ll 1$) and is most easily identified when the granular temperature is null ($\mathrm{Ma}_{\mathrm{p}} = \infty$). An example of PTC is shown in Figure 1.4. The panel on the

[11] Equilibrium refers to the form of the velocity-distribution function. When $n = n_{\mathrm{eq}}$, the flow is in equilibrium. When $n \approx n_{\mathrm{eq}} + \mathrm{Kn} n_1$, the flow is near equilibrium. Otherwise, the flow is non-equilibrium and f must be found by solving the mesoscale model.

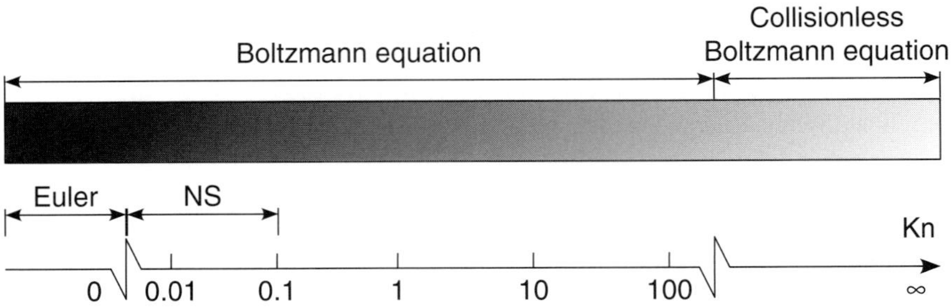

Figure 1.3. Ranges of validity of hydrodynamic models for the velocity kinetic equation in terms of the disperse-phase Knudsen number. NS, Navier–Stokes. Adapted from Bird (1994) by permission of Oxford University Press.

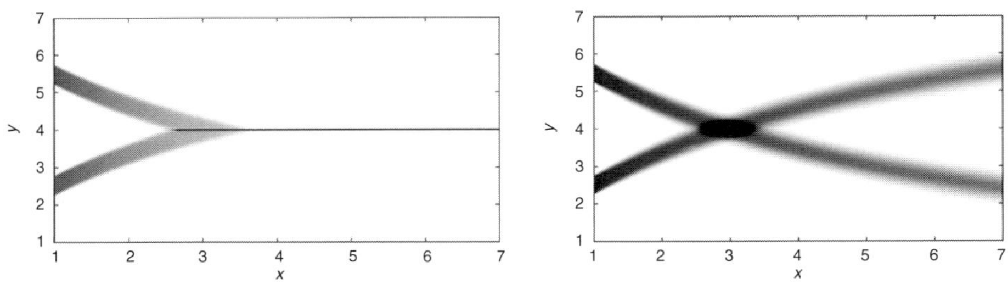

Figure 1.4. An example of the volume-fraction field for a highly non-equilibrium flow with $\mathrm{Kn_p} = \infty$. Left: the hydrodynamic model predicts that particle clouds cannot cross and all particles end up with zero y momentum. Right: the mesoscale model predicts that particle clouds will cross in the absence of collisions. Adapted from Freret *et al.* (2008).

left shows the result from solving the hydrodynamic model (albeit well outside its range of validity!), and the panel on the right shows the solution to the mesoscale model (Kah *et al.*, 2010). The accumulation of mass at $y = 0$ seen in the hydrodynamic model is known in the literature as a *delta shock* (Chen & Liu, 2003). Mathematically, it results from the fact that in the hydrodynamic model there exists only one value of the mesoscale velocity at each spatial location (because the granular temperature and, hence, the velocity fluctuations are zero). When particle clouds try to cross each other, the local velocity is replaced by the mean velocity, which is null. In contrast, the mesoscale model allows multiple velocities at each spatial location and thus PTC can occur unhindered by the mathematical approximation. For finite $\mathrm{Kn_p}$, collisions will cause the velocity to be multi-valued (but not necessarily isotropic), and the particle clouds will scatter. Only in the limit where the collisions are rapid relative to transport will the hydrodynamic model provide accurate predictions for gas–particle flows.

In principle, the range of validity of the hydrodynamic model could be improved by expanding to higher orders (e.g. $n \approx n_{\mathrm{eq}} + \mathrm{Kn}n_1 + \mathrm{Kn}^2 n_2$, etc.). For rarefied gases (e.g. the Boltzmann equation), this procedure has had some success but it does not greatly improve the range of Knudsen numbers that can be accurately simulated (Struchtrup, 2005). For more complicated mesoscale models with multiple dimensionless parameters (e.g. $\mathrm{Kn_p}$, St, etc.), it is not at all obvious that consistent higher-order expansions will provide better predictions than the first-order hydrodynamic models. In fact, in many polydisperse multiphase flows the range of disperse-phase Knudsen numbers observed can vary widely *within the same flow* in an unsteady manner (Passalacqua & Fox, 2011). For these reasons, it is desirable to develop an approximation method that is valid for *all possible values* of the dimensionless parameters appearing in the mesoscale model. At present, only moment methods provide such flexibility.

1.4.2 Moment methods

Moment methods come in many different variations, but the general idea is to increase the number of transported moments (beyond the hydrodynamic variables) in order to improve the description of non-equilibrium behavior. As noted earlier, the moment-transport equations are usually not closed in terms of any finite set of moments. Thus, the first step in any moment method is to apply a closure procedure to the truncated set of moment equations. Broadly speaking, this can be done in one of two ways.

1. Work directly with the unclosed terms in the moment-transport equations to find a functional form to close them (Struchtrup, 2005). For example, a spatial flux involving moment M_{k+1} might be closed using a gradient-diffusion model involving moment M_k.

2. Use the transported moments to reconstruct the unknown density function and then compute the unclosed terms in the moment transport equations. For example, in Grad's 13-moment method (Grad, 1949b) the reconstructed density function uses moments up to third order, and the fourth-order moments are found by integration.

For a general multivariate mesoscale model, closing directly the moment equations can be quite difficult. For example, there is no guarantee that such closures will yield *realizable* moments (i.e. moments that correspond to a nonnegative NDF). Because the space of realizable moments has quite complicated boundaries and, by their very nature, highly non-equilibrium problems have moments close to the boundaries, adding a closure to the moment-transport equations can easily cause the moments to become unrealizable. For this reason, direct closure of the moment equations is usually successful only for problems that are not too far from equilibrium. When used for problems far from equilibrium, the main shortcoming of such methods is not just the fact that they are inaccurate, but rather the fact that the numerical properties of the equations can change, causing the solver to fail catastrophically. To avoid such difficulties, it is then necessary to implement moment-correction algorithms that are not necessarily consistent with the underlying physics of the flow. Hence, from the perspective of both the physics and the numerics, it is highly desirable to use only moment closures that are guaranteed to produce realizable moments. The latter is most easily accomplished by reconstructing the density function.

The essential ideas behind moment methods that are based on reconstruction of the NDF are presented in Figure 1.5. Starting from the kinetic equation, a direct kinetic solver would compute the density $n(t, \mathbf{x}, \mathbf{v})$ using time stepping on a 6D computational mesh, and then integrate over phase space to find the moments $M_k(t, \mathbf{x})$. As discussed earlier, the prohibitive computational cost of a direct solver can be avoided by first integrating over phase space to find the unclosed moment-transport equations. In principle, these equations can be solved using time stepping on a 3D finite-volume mesh to find the moments. However, in order to close the moment-transport equations, the density function must be reconstructed in each mesh cell from the values of the moments in that cell. The reconstructed density function is denoted by $n^*(t, \mathbf{x}, \mathbf{v})$. Once n^* is known, the moment-transport equations are

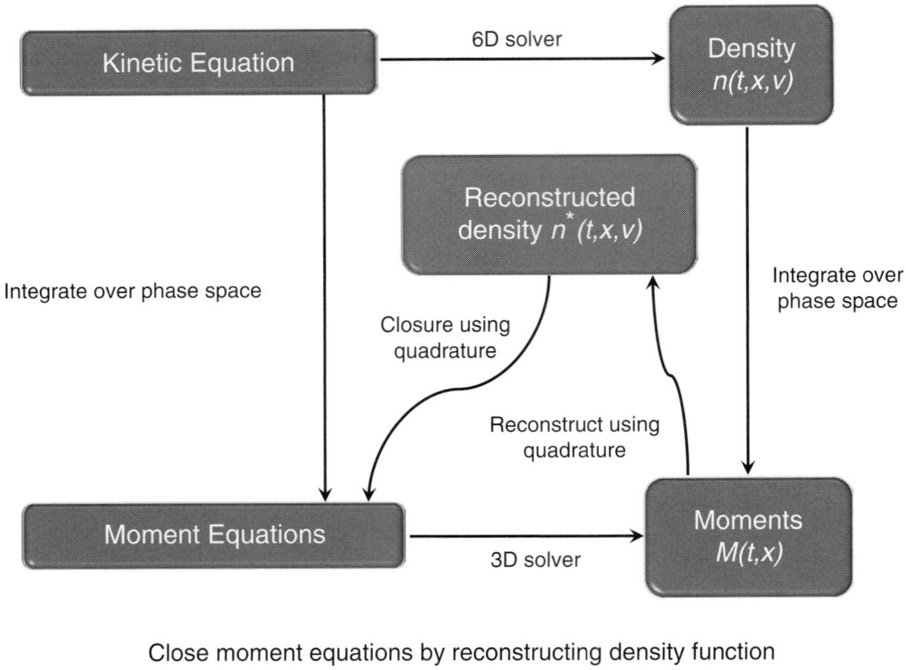

Figure 1.5. From the kinetic equation to macroscale models using moment methods that are based on reconstruction of the NDF.

closed at time t and can be used to compute the moments at time $t + \delta t$. The reconstruction procedure is then repeated to find $n^*(t + \Delta t, \mathbf{v}, \mathbf{x})$, and the moments are updated to the next time. From this description, it should be evident that the reconstruction procedure is the key step in ensuring the consistency of the moment closure. In this book, we focus on quadrature-based moment methods (see Chapter 3). However, it will be useful to have a general overview of the technical issues associated with moment methods that are based on reconstruction, as will be described next.

Four necessary properties of the moment-inversion algorithm are the following.

(i) **Positivity.** By definition, the NDF is nonnegative. Thus, the reconstructed NDF should be nonnegative for all values of its independent variables (i.e. $n^* \geq 0$).

(ii) **Realizability.** The mesoscale variables may take on physically realizable values in a finite subset of phase space. For example, the particle size is always positive and the particle surface area has a lower bound. The reconstructed NDF should be nonzero only for physically realizable values of the mesoscale variables.

(iii) **Extendible to arbitrary order.** In order to increase the accuracy of the moment closure, it is often necessary to increase the order of the moment set used for reconstruction. Thus, the moment-inversion algorithm should be able to increase systematically the order of the moments in such a way that n^* converges to n as the order of the moments goes to infinity.

(iv) **Non-iterative.** Because it is to be applied at every grid cell at every time step, the moment-inversion algorithm must be computationally efficient. In practice, this implies that a non-iterative method (i.e. simple analytical formulas if possible) is preferred. However, if iterations are unavoidable (e.g. a nonlinear solver), the method should guarantee convergence.

In practice, it can be difficult to attain all four properties in a single algorithm. For example, one way to guarantee positivity is to choose a functional form for n^* that depends on a small set of unknown parameters. If the functional form is very simple, then it may be possible to relate the moments to the parameters analytically (e.g. a Gaussian distribution). However, it is difficult to guarantee realizability and to capture complexly shaped NDF with simple functional forms, and hence their extension to higher-order moments cannot be done systematically. With other methods, such as entropy maximization (Massot *et al.*, 2010; Mead & Papanicolaou, 1984; Tagliani, 1999), positivity and realizability can be achieved, but at the cost of using an expensive iterative solver to determine the parameters.

Another option is to use a form for n^* involving a truncated functional expansion (e.g. Grad's method), which can be systematically extended to higher order. The advantage of this option is that the coefficients can be found by solving a linear system constructed from the definition of the moments (Favero & Lage, 2012; Grad, 1949b; Lage, 2011; Strumendo & Arastoopour, 2008). However, since there is no guarantee that the coefficients will be nonnegative, it is highly likely that the reconstructed NDF will be negative (i.e. unrealizable) at some points in phase space. Also, the realizability condition is difficult to guarantee when the nonzero subset of phase space has nonorthogonal boundaries.

In summary, the necessary properties of an ideal moment-inversion algorithm are difficult to attain. Methods that extend to arbitrary order and are guaranteed to be positive almost always require an iterative solver (e.g. entropy-maximization methods). One exception to this rule is quadrature-based moment methods (Desjardins *et al.*, 2008; Fox *et al.*, 2008; Fox & Vedula, 2010; Kah *et al.*, 2010; Marchisio & Fox, 2005; McGraw, 1997; Passalacqua *et al.*, 2011; Vikas *et al.*, 2011b; Yuan & Fox, 2011; Zou *et al.*, 2010) wherein the parameters are determined by an eigenvalue problem (see Chapter 3 for details). Furthermore, although they are not guaranteed to satisfy the realizability property for general multivariate distribution functions, quadrature-based methods offer several numerical advantages when combined with finite-volume solvers for the moment transport equations (Vikas *et al.*, 2011a). The construction of moment closures that are derived from quadrature-based moment methods will be described in detail in later chapters.

1.5 A road map to Chapters 2–8

The remaining chapters in this book are organized as follows. Chapter 2 provides a brief introduction to the mesoscale description of polydisperse systems. There, the mathematical definition of a number-density function (NDF) formulated in terms of different choices for the internal coordinates is described, followed by an introduction to population-balance equations (PBE) in their various forms. Chapter 2 concludes with a short discussion on the differences between the moment-transport equations associated with the PBE and those arising due to ensemble averaging in turbulence theory. This difference is very important, and the reader should keep in mind that at the mesoscale level the microscale turbulence appears in the form of correlations for fluid drag, mass transfer, etc., and thus the mesoscale models can have non-turbulent solutions even when the microscale flow is turbulent (i.e. turbulent wakes behind individual particles). Thus, when dealing with *turbulence models* for mesoscale flows, a separate ensemble-averaging procedure must be applied to the moment-transport equations of the PBE (or to the PBE itself). In this book, we are primarily concerned with the formulation and direct solution of mesoscale models, and hence we do not focus on multiphase-turbulence models. Readers interested in multiphase-turbulence models can find reviews in the literature (Balachandar & Eaton, 2010; Crowe *et al.*, 1996; Fox, 2012; Jiang *et al.*, 2010; Mashayek & Pandya, 2003; Salazar & Collins, 2009).

Chapter 3 provides an introduction to Gaussian quadrature and moment-inversion algorithms used in quadrature-based moment methods (QBMM). Basic knowledge of this material is required in order to formulate the moment closures for the PBE discussed in later chapters. Because the number of internal variables is typically large, a direct solution of the PBE for spatially inhomogeneous problems will usually be intractable. Thus, in this book, we focus on QBMM in order to reduce the dimensionality of the problem. The key step in QBMM is the reconstruction of the NDF starting from its moments. Thus, in Chapter 3, we review quadrature-based moment inversion for univariate and multivariate NDF. Readers already familiar with QBMM may still find this chapter useful since it introduces much of the notation and terminology used in later chapters. Example moment-inversion algorithms are provided in Appendix A.

The detailed derivation of the generalized population-balance equation (GPBE) is the main topic of Chapter 4. This derivation starts from an ensemble of interacting particles (i.e. the disperse phase) in a fluid (i.e. the continuous phase) and arrives at the single-particle GPBE. During this process, closures at the level of the multi-particle joint distribution function are introduced and correspond to the passage from the microscale to the mesoscale model (as discussed in Section 1.3). Chapter 4 is manifestly the most theoretical chapter in the book, nevertheless a clear grasp of the multi-particle closure problem is mandatory before attempting to formulate mesoscale models that correctly capture the microscale physics of interacting-particle systems. Once written in closed form, the GPBE provides the mesoscale (or kinetic) model for the disperse multiphase flow. Moreover, the GPBE is the starting point for deriving the transport equations for the moments of the NDF. Thus, in Chapter 4, we provide a brief overview of the mathematical steps needed to derive the moment-transport equations by integration over the phase-space variables. As discussed in Section 1.4, this procedure leads to a second closure problem because the moment-transport equations are usually not closed and it is this second closure that is the focus of QBMM.

Chapter 5 focuses on selected mesoscale models from the literature for key physical and chemical processes. As the reader might imagine, the number of mesoscale models that has been proposed in the literature is enormous. Our goal in Chapter 5 is thus to provide examples of models for generic phenomena such as advection in real and phase space, diffusion in real and phase space, and point processes such as nucleation of the disperse phase. In real applications, the accuracy of the mesoscale models employed will be critical to the overall accuracy of a simulation. However, because our focus in this book is primarily on how to formulate and solve moment-transport equations in a generic setting, we do not aim at an exhaustive categorization of the available mesoscale models. Nonetheless, the reader should find Chapter 5 to be a useful introduction to mesoscale models and a good starting point for developing new ones. In subsequent chapters, we use mesoscale models from Chapter 5 to illustrate the numerical methods for solving the GPBE.

Many disperse-phase systems involve collisions between particles, and the archetypical example is hard-sphere collisions. Thus, Chapter 6 is devoted to the topic of hard-sphere collision models in the context of QBMM. In particular, because the moment-transport equations for a GBPE with hard-sphere collisions contain a source term for the rate of change of the NDF during a collision, it is necessary to derive analytical expressions for these source terms (Fox & Vedula, 2010). In Chapter 6, the exact source terms are derived for the integer moments for inelastic, binary collisions between two particles with different diameters/masses. Although the technical details of the derivations are rather complex, the resulting source terms have fairly simple forms, which are tabulated in Chapter 6 for future use. A discussion of simpler kinetic models for collisions is given, and selected moment-transport equations are provided as examples. Finally, the chapter concludes with a brief discussion on how to use QBMM with the derived analytical source terms to close the collisional-flux and collision source terms in the moment transport equations.

After our having introduced the basic concepts and models in earlier chapters, Chapters 7 and 8 address the key computational methods needed for simulating polydisperse particulate and multiphase systems. As is often done in simulation codes by employing operator splitting, we consider first the numerical methods for spatially homogeneous systems in Chapter 7, followed by the treatment of spatially inhomogeneous systems in Chapter 8. In the context of homogeneous systems, we review solution methods for directly solving the GPBE by discretizing the phase-space variables, and indirect methods that solve for the moments of the NDF using a moment closure. As is standard practice for homogeneous systems, methods for validating the moment closures by comparing direct and indirect solutions are also presented.

Chapter 8 is devoted to moment methods for solving a spatially inhomogeneous GPBE. A critical issue with spatially inhomogeneous systems is moment realizability of the numerical advection (diffusion) operator, and thus we introduce kinetics-based finite-volume methods (KBFVM) (Perthame, 2003) to deal with this problem. More details on KBFVM applied to moment-transport equations are provided in Appendix B. Other key issues are moment closure in the presence of particle trajectory crossing (PTC) in phase space, which occurs in systems far from collisional equilibrium, and moment conservation with some QBMM. These two issues are discussed in Appendix C and Appendix D, respectively. As shown using an analytical example in Appendix C, PTC corresponds to the crossing of characteristics in the hyperbolic GPBE. Thus, in Chapter 8, we demonstrate that QBMM are ideally suited for capturing PTC at the level of the moment-transport equations.

To conclude the book, spatially 1D numerical examples with different types of GPBE are solved using QBMM combined with KBFVM for spatial transport (Vikas *et al.*, 2011a) in Chapter 8. We provide the exact formulas used in the numerical implementation, since it is the authors' hope that the reader will attempt to reproduce some of these examples and, thereby, gain valuable experience in simulating polydisperse multiphase systems that can be applied to the reader's own applications.

2

Mesoscale description of polydisperse systems

In this chapter, the governing equations needed to describe polydisperse multiphase flows are presented without a rigorous derivation from the microscale model. (See Chapter 4 for a complete derivation.) For clarity, the discussion of the governing equations in this chapter will be limited to particulate systems (e.g. crystallizers, fluidized beds, and aerosol processes). However, the reader familiar with disperse multiphase flow modeling will recognize that our comments hold in a much more general context. Indeed, the extension of the modeling concepts developed in this chapter to many other multiphase systems is straightforward, and will be discussed in later chapters.

The primary purpose of this chapter is to introduce the key concepts and notation needed to develop models for polydisperse multiphase flows. We thus begin with a general discussion of the number-density function (NDF) in its various forms, followed by example transport equations for the NDF with known (PBE) and computed (GPBE) particle velocity. These transport equations are written in terms of "averaged" quantities whose precise definitions will be presented in Chapter 4. We then consider the moment-transport equations that are derived from the NDF transport equation by integration over phase space. Finally, we briefly describe how turbulence modeling can be undertaken starting from the moment-transport equations.

2.1 Number-density functions (NDF)

The disperse phase is constituted by discrete elements. One of the main assumptions of our analysis is that the characteristic length scales of the elements are smaller than the characteristic length scale of the variation of properties of interest (i.e. chemical species concentration, temperature, continuous phase velocities). If this hypothesis holds, the particulate system can be described by a continuum or mean-field theory. Each element of the disperse phase is generally identified by a number of properties known as *coordinates*. Two elements are identical if they have identical values for their coordinates, otherwise elements are indistinguishable. Usually coordinates are classified as *internal* and *external*. External coordinates are spatial coordinates; in fact, the position of the elements in physical space is not an internal property of the elements. Internal coordinates refer to more intimate properties of the elements such as their momenta (or velocities), their enthalpy (or temperatures), or their volumes, surface areas, sizes, or ages. When using an Eulerian approach, the disperse phase is generally described by an NDF. As mentioned in Chapter 1, the NDF contains information about how the population of particles inside an infinitesimal control volume is distributed over the properties of interest.

Let us consider a population of disperse entities such as solid particles or liquid droplets inside an infinitesimal control volume located at the physical point $\mathbf{x} \equiv (x_1, x_2, x_3)$ and of measure $\mathrm{d}\mathbf{x} \equiv \mathrm{d}x_1\,\mathrm{d}x_2\,\mathrm{d}x_3$. Let $\boldsymbol{\xi} \equiv (\xi_1, \xi_2, \ldots, \xi_M)$ be the internal-coordinate vector, namely the vector containing the M internal coordinates of interest. The NDF $n_\xi(t, \mathbf{x}, \boldsymbol{\xi})$ is defined as the expected number[1] of entities in the infinitesimal physical volume $\mathrm{d}\mathbf{x}$ and in the infinitesimal phase-space volume $\mathrm{d}\boldsymbol{\xi}$: $n_\xi\,\mathrm{d}\mathbf{x}\,\mathrm{d}\boldsymbol{\xi}$. Note that n_ξ is a function of time t, space $\mathbf{x}$, and the internal-coordinate vector $\boldsymbol{\xi}$. Henceforth we will include the explicit dependence on these variables only when required for clarity. As shown in detail in Chapter 4, the NDF is an average quantity of the disperse phase. In fact, it has the mathematical characteristics of an averaged function: it is smooth and differentiable[2] with respect to time, physical space, and internal-coordinate (or phase) space.

It is clear that the underlying definition of the NDF is based on a stochastic description of the microscale system. In fact, although the behavior of any single particle in a population is regarded as random, the behavior of a population of particles can be treated as deterministic if the number of particles in the population is large. As discussed in Chapter 4, the NDF identifies the entire population of particles at any instant and at any given point in the computational domain and considers the probability associated with the state of each of these particles. We refer the reader to specialized texts for speculations about the nature of this average and the treatment of very dilute systems where the deterministic description fails because the number of particles in the population is too small; see for example Chapter 7 of Ramkrishna (2000). For the time being, we just highlight here that this random behavior may reflect some chaotic properties of the very intimate nature of the disperse phase or some other chaotic behavior induced, for example, by microscale turbulence in the continuous fluid phase. However, as will become clearer in Section 2.4, when dealing with mesoscale turbulence in solutions to the NDF transport equation, the NDF can be treated just like any other scalar field (albeit infinite-dimensional) in a turbulent flow.

It is straightforward that the quantity $n_\xi(\boldsymbol{\xi})\mathrm{d}\boldsymbol{\xi}$ represents the number density of disperse entities contained in the phase-space volume $\mathrm{d}\boldsymbol{\xi}$ centered at $\boldsymbol{\xi}$ per unit of physical volume. If we integrate the NDF over all possible values of the internal-coordinate vector Ω_ξ we obtain the total number concentration $N(t, \mathbf{x})$:

$$N \equiv m_{\xi,0} \equiv \int_{\Omega_\xi} n_\xi(\boldsymbol{\xi})\mathrm{d}\boldsymbol{\xi}, \tag{2.1}$$

which is the total number of entities per unit volume located at time t and at point $\mathbf{x}$, and corresponds to the zeroth-order moment of the density function. Likewise, an arbitrary moment of the NDF can be defined by

$$m_{\xi,\mathbf{k}} \equiv \int_{\Omega_\xi} \xi_1^{k_1} \cdots \xi_M^{k_M} n_\xi(\boldsymbol{\xi})\mathrm{d}\boldsymbol{\xi} \tag{2.2}$$

---

[1] We shall see in Chapter 7 that stochastic algorithms can be used to simulate the evolution of a population of particles represented by the NDF. However, the NDF is not a random quantity; rather it is the ensemble average of an infinite number of realizations of the stochastic algorithm (assuming that the latter is unbiased).

[2] When the NDF represents the particle velocity, the transport equation is hyperbolic and thus the solution need not be smooth and need not be differentiable at every point in space. In fact, the treatment of shocks in the NDF is a significant challenge.

where $\mathbf{k} \equiv (k_1, \ldots, k_M)$ is a vector containing the order of the moments with respect to each of the components of $\boldsymbol{\xi}$.

2.1.1 Length-based NDF

In order to introduce some particular concepts and properties that will turn out to be very useful in the development of the discussion, we consider a population of particles that can be described by using only one internal coordinate. As already mentioned, particles are indistinguishable except through the chosen internal coordinate. Let us imagine that the particles can be characterized by their size or length L, which in this case constitutes the only component of the internal-coordinate vector. Using this notation, $n_{\mathrm{L}}(t, \mathbf{x}, L)$ is the length-based NDF, and $n_{\mathrm{L}}\, \mathrm{d}L$ represents the expected number density of particles with length between L and $L+\mathrm{d}L$. As reported above in Eq. (2.1), the total number concentration can be calculated as

$$N \equiv m_{\mathrm{L},0} = \int_0^\infty n_{\mathrm{L}}(L)\mathrm{d}L, \tag{2.3}$$

whereas a mean particle size can be defined as

$$L_{10} \equiv \frac{1}{N} \int_0^\infty L n_{\mathrm{L}}(L)\mathrm{d}L = \frac{m_{\mathrm{L},1}}{N}. \tag{2.4}$$

This average property of the distribution is defined through the number density itself and it represents the mean particle size with respect to the number of particles in the system. Of course, other definitions of the mean particle size are possible, as will become clear below.

If we define the kth moment of the length-based NDF as

$$m_{\mathrm{L},k} \equiv \int_0^\infty L^k n_{\mathrm{L}}(L)\mathrm{d}L, \tag{2.5}$$

it is easy to see that a mean particle size can be defined as the ratio $m_{\mathrm{L},k+1}/m_{\mathrm{L},k}$ for any value of k. For example, the Sauter mean diameter is defined by setting $k = 2$, giving $L_{32} \equiv m_{\mathrm{L},3}/m_{\mathrm{L},2}$, and is useful for modeling processes such mass transfer in which the surface area is a key parameter. If volume scales with the third power of length, we can relate the particle volume V and particle length L through a volumetric shape factor k_{V}:[3] $V = k_{\mathrm{V}} L^3$. The resulting *volume-density function* $V_{\mathrm{L}}(t, \mathbf{x}, L)$ represents the volume of particles per unit spatial volume with lengths between L and $L + \mathrm{d}L$, and is defined by $V_{\mathrm{L}} \equiv k_{\mathrm{V}} L^3 n_{\mathrm{L}}$. After normalizing V_{L} to unity, the *volume-fraction density function* $\alpha_{\mathrm{V}}(t, \mathbf{x}, L)$ represents the volume fraction of particles with a specific length over the total particle volume:

$$\alpha_{\mathrm{V}} \equiv \frac{L^3 n_{\mathrm{L}}}{\int_0^\infty L^3 n_{\mathrm{L}}(L)\mathrm{d}L}. \tag{2.6}$$

The mean particle length calculated from this volume-fraction density function is

$$L_{43} = \int_0^\infty \alpha_{\mathrm{V}}(L) L\, \mathrm{d}L = \frac{m_{\mathrm{L},4}}{m_{\mathrm{L},3}}. \tag{2.7}$$

[3]For fractal-shaped particles, this volume cannot be used to compute the mass of the particle. Instead, the mass is related to the length through the *fractal dimension*.

This definition of particle length is widely used for particulate systems. As mentioned above, L_{10} is the mean particle length averaged with respect to particle number density, whereas L_{43} is the mean particle length averaged with respect to particle volume-fraction density function. Naturally, all of these definitions of the particle length can be found starting from the length-based NDF, so it is not necessary to develop separate theories for each new type of NDF.

2.1.2 Volume-based NDF

Another very common internal coordinate for particulate systems is the particle volume V. Using a definition very similar to that of the NDF, we can define the volume-based NDF as the expected number of particles with volume between V and $V + \mathrm{d}V$: $n_\mathrm{V}\, \mathrm{d}\mathbf{x}\, \mathrm{d}V$. If volume and length scale with the third power as $V_\mathrm{L} \equiv k_\mathrm{V} L^3 n_\mathrm{L}$, the relationship between the length-based and the volume-based NDF is straightforward. If fact, $n_\mathrm{L}\, \mathrm{d}L$ quantifies the expected number of particles with length between L and $L + \mathrm{d}L$, which must be equal to the expected number of particles with volume between V and $V + \mathrm{d}V$. Moreover, it is easy to see that $\mathrm{d}V = 3k_\mathrm{V} L^2\, \mathrm{d}L$. Therefore, equality in the expected numbers requires $n_\mathrm{L}(L)\mathrm{d}L = n_\mathrm{V}(V)\mathrm{d}V$ or $n_\mathrm{L}(L)\mathrm{d}L = n_\mathrm{V}(k_\mathrm{V} L^3) 3k_\mathrm{V} L^2\, \mathrm{d}L$, from which we find that $n_\mathrm{L}(L) = 3k_\mathrm{V} L^2 n_\mathrm{V}(k_\mathrm{V} L^3)$, which essentially tells us how to define n_L in terms of n_V. The same procedure can be used to find the NDF in terms of any well-defined function of L. The moments of the volume-based number NDF are defined as

$$m_{\mathrm{V},k} \equiv \int_0^\infty V^k n_\mathrm{V}(V)\mathrm{d}V, \tag{2.8}$$

and can be easily related to the moments of the length-based NDF:

$$m_{\mathrm{V},k} = \int_0^\infty (k_\mathrm{V} L^3)^k n_\mathrm{L}(L)\mathrm{d}L = k_\mathrm{V}^k m_{\mathrm{L},3k}. \tag{2.9}$$

In practical application, the choice of the NDF with which to work often depends on which moments can be measured experimentally.

2.1.3 Mass-based NDF

In the literature on particle aggregation and droplet coalescence, a mass-based NDF $n_\mathrm{M}(t, \mathbf{x}, M)$ is often preferred because mass is a conserved quantity. For these systems, it is possible to relate the mass of a particle to its characteristic length using $M = k_\mathrm{M} L^{D_\mathrm{f}}$, where $1 < D_\mathrm{f} \leq 3$ is the *fractal dimension*.[4] Using the procedure outlined above, we can show that the length-based and mass-based NDF are related by $n_\mathrm{L}(L) = D_\mathrm{f} k_\mathrm{M} L^{D_\mathrm{f}-1} n_\mathrm{M}(k_\mathrm{M} L^{D_\mathrm{f}})$. Likewise, their moments are related by $m_{\mathrm{M},k} = k_\mathrm{M}^k m_{\mathrm{L},D_\mathrm{f}k}$. As we shall see in the following chapters, it is often the case that in closed systems the total particle mass is conserved, i.e. $m_{\mathrm{M},1}$ is constant. In a length-based analysis, this corresponds to conservation of the fractional moment $m_{\mathrm{L},D_\mathrm{f}}$. Ideally, numerical algorithms for solving PBEs should conserve mass regardless of the type of NDF used to express the problem. However, this is not always the case due to the numerical approximation used in the model. We shall return to the subject of choosing the appropriate NDF in Chapter 5.

[4] $D_\mathrm{f} = 3$ for spherical particles and $D_\mathrm{f} \approx 1.8$ for diffusion-limited aggregates.

2.1.4 Velocity-based NDF

A special case of considerable interest occurs when the internal-coordinate vector is the particle-velocity vector, which we will denote by the phase-space variable $\mathbf{v}$.[5] In fact, particle velocity is a *special* internal coordinate since it is related to particle position (i.e. external coordinates) through Newton's law, and therefore a special treatment is necessary. We will come back to this aspect later, but for the time being let us imagine that otherwise identical[6] particles are moving with velocities that may be different from particle to particle (and different from the surrounding fluid velocity). It is therefore possible to define a velocity-based NDF $n_{\mathrm{U}}(t, \mathbf{x}, \mathbf{v})$ that is parameterized by the velocity components $\mathbf{v} = (v_1, v_2, v_3)$. In order to obtain the total number concentration (i.e. number of particles per unit volume) it is sufficient to integrate over all possible values of particle velocity $\Omega_{\mathbf{v}}$:

$$N = m_{\mathrm{U},0} \equiv \int_{\Omega_{\mathbf{v}}} n_{\mathrm{U}}(\mathbf{v})\mathrm{d}\mathbf{v}, \tag{2.10}$$

which is just the zeroth-order moment of the velocity-based NDF. Because we have assumed that particles may have different velocities, we can calculate for each spatial location and at each time instant an average particle velocity $\mathbf{U}_{\mathrm{p}}$ using the first two moments of the velocity-based NDF:

$$\mathbf{U}_{\mathrm{p}} \equiv \frac{1}{N} \int_{\Omega_{\mathbf{v}}} \mathbf{v} n_{\mathrm{U}}(\mathbf{v})\mathrm{d}\mathbf{v} = \frac{m_{\mathrm{U},1}}{m_{\mathrm{U},0}}, \tag{2.11}$$

which is analogous to the definition of the mean particle length reported in Eq. (2.4). The reader should note that this definition of the mean particle velocity is appropriate only when all particles have the same mass. If the particle mass is an internal coordinate, then the mean particle velocity is defined using the ratio of the mean disperse-phase momentum and the mean particle mass.

It is clear that the NDF n_{ξ} and the velocity-based NDF n_{U} have the same mathematical meaning, and can be treated in a similar framework. However, as has already been highlighted, particle velocity is a particular internal coordinate and therefore theories and computational models for the two classes of problems have usually been treated separately. In what follows we will derive the transport equations for these NDF, keeping the treatment for *passive* internal coordinates, such as particle size, volume, area, or temperature, separate from that of *active* internal coordinates such as particle velocity. We will refer to the first transport equation as the population-balance equation (PBE)[7] and to the second one as the generalized population-balance equation (GPBE).[8] In the following chapters we will then develop a general framework to describe and solve both problems numerically. However, the reader should keep in mind that the key distinction between the two models is that for the PBE the velocity of the disperse phase is assumed to be known (e.g. identical to the fluid velocity in the continuous phase), and thus the NDF can be treated as a passive scalar field, whereas the scalar fields appearing in the GPBE are usually active (i.e. momentum

[5]For multiphase flows, it will be necessary to distinguish between the velocities of the various phases. For example, for gas–solid flow we will denote the fluid-velocity vector by $\mathbf{U}_{\mathrm{f}}$. However, for much of the present discussion, we can assume that the particles are traveling in a vacuum.

[6]We will relax this requirement later by introducing particles whose properties (length, volume, mass, etc.) are also different.

[7]The PBE is also known as the general dynamic equation in the aerosol literature.

[8]In the literature the GPBE is known by many different names, depending on the research field. For example, in statistical physics, when written for the velocity NDF, it is called the Boltzmann kinetic equation; and in the combustion literature, when written for droplets, it is known as the spray or Williams equation.

exchange between particles depends on internal coordinates such as length), and thus the velocity of the disperse phase must be computed from a consistent momentum balance.

2.2 The NDF transport equation

We now consider the evolution of the NDF[9] in time, phase space, and physical space. We shall see that the underlying mathematical structure for the PBE and the GPBE is very similar.

2.2.1 The population-balance equation (PBE)

The PBE is a simple continuity statement written in terms of the NDF. It can be derived as a balance for particles in some fixed subregion of phase and physical space (Ramkrishna, 2000). Let us consider a finite control volume in physical space $\Omega_{\mathbf{x}}$ and in phase space $\Omega_{\boldsymbol{\xi}}$ with boundaries defined as $\partial\Omega_{\mathbf{x}}$ and $\partial\Omega_{\boldsymbol{\xi}}$, respectively. In the PBE, the advection velocity $\mathbf{v}$ is assumed to be known (e.g. equal to the local fluid velocity in the continuous phase or directly derivable from this variable). The particle-number-balance equation can be written as

$$\frac{\partial}{\partial t}\left(\int_{\Omega_{\mathbf{x}}} \mathrm{d}\mathbf{x} \int_{\Omega_{\boldsymbol{\xi}}} \mathrm{d}\boldsymbol{\xi}\, n_{\boldsymbol{\xi}}\right) + \int_{\Omega_{\boldsymbol{\xi}}} \mathrm{d}\boldsymbol{\xi} \int_{\partial\Omega_{\mathbf{x}}} \left(n_{\boldsymbol{\xi}}\mathbf{v}\right)\cdot \mathrm{d}\mathbf{A}_{\mathbf{x}} + \int_{\Omega_{\mathbf{x}}} \mathrm{d}\mathbf{x} \int_{\partial\Omega_{\boldsymbol{\xi}}} \left(n_{\boldsymbol{\xi}}\dot{\boldsymbol{\xi}}\right)\cdot \mathrm{d}\mathbf{A}_{\boldsymbol{\xi}} = \int_{\Omega_{\mathbf{x}}} \mathrm{d}\mathbf{x} \int_{\Omega_{\boldsymbol{\xi}}} \mathrm{d}\boldsymbol{\xi}\, h_{\boldsymbol{\xi}}, \tag{2.12}$$

where $\mathbf{v}$ is the velocity vector for the particulate system, $\dot{\boldsymbol{\xi}}$ is the continuous rate of change in phase space, and $h_{\boldsymbol{\xi}}$ is the discontinuous jump function representing discrete events. In addition to the advection velocity, fine particles (less than approximately one micron in size) and low-inertia particles with turbulent wakes (e.g. bubbles) will have a diffusive-flux term due to Brownian and turbulent fluctuations, respectively.

The first term in Eq. (2.12) is an integral over the control volume and represents accumulation. The second term is a surface integral over the boundary of the control volume[10] and represents the net flux of number density due to convection in physical space. The third term represents instead convection in phase space. In fact, particles move in physical space because of the particle velocity $\mathbf{v}$ and in phase space because of an internal-coordinate velocity $\dot{\boldsymbol{\xi}} = \mathrm{d}\boldsymbol{\xi}/\mathrm{d}t$. This vector contains in each component the rate of change of the ith internal coordinate for continuous processes. We generally refer to *continuous* processes when internal coordinates change continuously with a time scale much smaller than the one characterizing solutions to the PBE. Therefore, although all processes (at least at the molecular level) are discontinuous, these processes can be treated as continuous in Eq. (2.12). For example, if particles are growing in size because molecules or groups of molecules are continuously deposited on the particle surface, and this occurs on a time and length scale much smaller than the time and length scale related to the size of the particle, then these processes can be treated as continuous and their mesoscale description[11] is the third term

[9]The same analysis holds for density functions in general, including probability density functions.

[10]Integrals over a control-volume boundary are defined through the infinitesimal surface unit vector $\mathrm{d}\mathbf{A}_{\mathbf{x}}$ with magnitude equal to the measure of the infinitesimal surface and direction defined by the vector normal to the surface itself.

[11]Even at the microscale level, molecular-scale events are described by a continuous process.

of Eq. (2.12). The terms in Eq. (2.12) referring to convection in physical or phase space are usually called drift terms. On the other hand, if particles are growing because of collisions and subsequent aggregation, since the length scale of this process is of the same order of magnitude as the particle size, it cannot be treated as a continuous change in phase space. This process falls in the category of discontinuous jumps in phase space and is represented by the term on the right-hand side of Eq. (2.12). These concepts will become clearer in Chapter 5, where we look at typical models for continuous and discrete processes.

If the Reynolds–Gauss theorem is applied[12] to Eq. (2.12) and the integrals on the boundary of the control volume are written in terms of integrals over the control volume itself, it is straightforward to obtain

$$\frac{\partial}{\partial t}\left(\int_{\Omega_{\mathbf{x}}} \mathrm{d}\mathbf{x} \int_{\Omega_{\xi}} \mathrm{d}\boldsymbol{\xi}\, n_{\xi}\right) + \int_{\Omega_{\mathbf{x}}} \mathrm{d}\mathbf{x} \int_{\Omega_{\xi}} \mathrm{d}\boldsymbol{\xi} \frac{\partial}{\partial \mathbf{x}} \cdot \left(\mathbf{v} n_{\xi}\right) + \int_{\Omega_{\mathbf{x}}} \mathrm{d}\mathbf{x} \int_{\Omega_{\xi}} \mathrm{d}\boldsymbol{\xi} \frac{\partial}{\partial \boldsymbol{\xi}} \cdot \left(\dot{\boldsymbol{\xi}} n_{\xi}\right) = \int_{\Omega_{\mathbf{x}}} \mathrm{d}\mathbf{x} \int_{\Omega_{\xi}} \mathrm{d}\boldsymbol{\xi}\, h_{\xi}, \tag{2.13}$$

where $\partial/\partial\mathbf{x} = \nabla_{\mathbf{x}} = (\partial/\partial x_1, \partial/\partial x_2, \partial/\partial x_3)$ is the gradient operator in physical space and $\partial/\partial\boldsymbol{\xi} = \nabla_{\xi} = (\partial/\partial\xi_1, \ldots, \partial/\partial\xi_m)$ is the gradient operator in phase space. In order for Eq. (2.13) to be satisfied for any arbitrary control volumes $\Omega_{\mathbf{x}}$ and Ω_{ξ}, the integrand must satisfy the relation

$$\frac{\partial n_{\xi}}{\partial t} + \frac{\partial}{\partial \mathbf{x}} \cdot \left(\mathbf{v} n_{\xi}\right) + \frac{\partial}{\partial \boldsymbol{\xi}} \cdot \left(\dot{\boldsymbol{\xi}} n_{\xi}\right) = h_{\xi} \tag{2.14}$$

or, using Einstein notation (i.e. repeated Roman indices imply summation),

$$\frac{\partial n_{\xi}}{\partial t} + \frac{\partial}{\partial x_i}\left(v_i n_{\xi}\right) + \frac{\partial}{\partial \xi_i}\left(\dot{\xi}_i n_{\xi}\right) = h_{\xi}. \tag{2.15}$$

Note that, although we treat $\mathbf{x}$ and $\boldsymbol{\xi}$ in the same manner, they are in fact different types of vectors. The vectors $\mathbf{x}$ and $\mathbf{v}$ are the standard vectors for position and velocity used in continuum mechanics. The internal-coordinate vector $\boldsymbol{\xi}$, on the other hand, is a generalized vector of length N in the sense of linear algebra.

Equation (2.14) must be coupled with initial conditions given for the starting time and with boundaries conditions in physical space $\Omega_{\mathbf{x}}$ and in phase space Ω_{ξ}. Analytical solutions to Eq. (2.14) are available for a few special cases and only under conditions specified by some very simple hypotheses. However, numerical methods can be used to solve this equation and will be presented in Chapters 7 and 8. The numerical solution of Eq. (2.14) provides knowledge of the NDF for each time instant and at every physical point in the computational domain, as well as at every point in phase space. As has already been mentioned, sometimes the population of particles is described by just one internal coordinate, for example particle length (i.e. $\xi = L$), and the PBE is said to be *univariate*. When two internal coordinates are needed, for example particle volume and surface area (i.e. $\boldsymbol{\xi} = (v, a)$), the PBE is said to be *bivariate*. More generally, higher-dimensional cases are referred to as *multivariate* PBEs. Another important case occurs when part of the internal-coordinate vector is equal to the particle-velocity vector (i.e. when the particles are characterized not by a unique velocity field but by their own velocity distribution). In that case, the PBE becomes the GPBE, as described next.

[12] See Aris (1962) for a rigorous derivation of the Reynolds–Gauss transport theorem.

2.2.2 The generalized population-balance equation (GPBE)

By using a very similar approach to the one outlined above for the PBE, it is possible to derive a GPBE for an NDF that includes particle velocity as an internal variable. We will denote this general NDF as $n(t, \mathbf{x}, \mathbf{v}, \boldsymbol{\xi})$ (i.e. without subscripts on n). The simplest GPBE (i.e. velocity without other internal coordinates) is known as the Boltzmann kinetic equation and was first derived in the context of gas theory (Chapman & Cowling, 1961). The final form of the GPBE is

$$\frac{\partial n}{\partial t} + \frac{\partial}{\partial \mathbf{x}} \cdot (\mathbf{v}n) + \frac{\partial}{\partial \mathbf{v}} \cdot \left(\mathbf{A}_{\mathrm{p}}n\right) + \frac{\partial}{\partial \boldsymbol{\xi}} \cdot \left(\dot{\boldsymbol{\xi}}n\right) = h, \tag{2.16}$$

where $\mathbf{A}_{\mathrm{p}}$ is the continuous rate of change of particle velocity (i.e. acceleration), or the force per unit mass acting on particles (e.g. gravity, fluid drag, etc.). The right-hand side h is again the discontinuous jump term, but now including discontinuous changes in particle momentum (e.g. collisions). In the GPBE, there are never terms of diffusion in physical spaces. However, there are often terms for diffusion in velocity phase space due to random (e.g. Brownian) forces, or to turbulent fluctuations in the continuous phases (e.g. turbulent wakes behind inertial particles). The first term on the left-hand side of Eq. (2.16) is the accumulation term, whereas the second one is a drift term that accounts for convection in physical space due to the particle velocity. The third term is the drift term representing convection in particle-velocity phase space. In fact, a force acting on particles produces a uniform acceleration, or in other words a continuous change in particle velocity. The term on the right-hand side represents discontinuous jumps in phase space due to discrete events, mainly collisions between particles, and for this reason it is also known as a collision integral.[13] When two particles collide they exchange momentum and therefore in a single instant they change their momenta (and therefore velocities) from one set of specific values to another one. In this respect the collision process is discrete and produces a finite and discontinuous change in particle-velocity phase space.[14] Moreover, if collisions result in changes in the internal coordinates (e.g. bubble coalescence, particle aggregation, etc.), then the collision term can be quite complicated compared with the hard-sphere collisions used in gas dynamics.

2.2.3 The closure problem

Now that we have introduced the GPBE, it is appropriate to discuss the problem of closure. In general, we say that Eq. (2.16) is *closed* if all of the terms in the equation can be computed from knowledge of $n(t, \mathbf{x}, \mathbf{v}, \boldsymbol{\xi})$ and its moments.[15] In this book, we will in fact assume that the GPBE of interest is closed and focus on methods for computing $n(t, \mathbf{x}, \mathbf{v}, \boldsymbol{\xi})$. As discussed in Chapter 4, the problem of finding closures for multi-particle correlations that occur, for example, due to particle-momentum exchange during collisions or through

[13] As is well known in statistical physics, the collision integral is not closed since it involves velocity correlations between two particles. Thus, a closure hypothesis must be introduced to define h in terms of n. See Chapter 6 for more details.

[14] Collisions can be described as discontinuous only when they occur on time and length scales that are much shorter than those characterizing changes in the NDF. This is indeed the case of particle–particle collision occurring under a hard-sphere potential, but under different potentials, if, for example, particles start perceiving each other's presence at large distances, this might no longer be the case.

[15] Since the latter are computed from n, mention of the moments is redundant. Nevertheless, we mention them anyway to remind the reader that closure at the level of the GPBE includes as a special case closure at the level of moments.

interactions mediated by the continuous fluid phase is highly nontrivial. In order to close the Boltzmann kinetic equation, for example, one usually appeals to the *molecular-chaos* assumption and thereby approximates the two-particle NDF by the product of two one-particle NDF. (See Chapter 6 for details.) As mentioned in Chapter 1, the focus of this book is *not* on methods that can be used to close the GPBE. We will instead refer the reader to the specialized technical literature for closure methods that have been derived in order to close the GPBE. Finally, we want to remind the reader that, even if the GPBE is closed (i.e. amenable to the computation methods discussed in this book), this does not imply that the *moment-transport equations* will be closed. In fact, in most problems of practical interest this will not be the case. Nonetheless, it is often useful to know the form of the (unclosed) moment-transport equations. In the next section, we give a brief overview on how to find the moment-transport equations starting from the PBE and GPBE.

2.3 Moment-transport equations

It has already been mentioned that the PBE (Eq. (2.14)) and the GPBE (Eq. (2.16)) have very similar structures. They both quantify the rate of change of an NDF, representing a population of discrete elements that undergoes some continuous processes (e.g. molecular growth or acceleration due to physical forces) and some discrete processes (e.g. collision and aggregation). It is now interesting to derive from the NDF transport equations some integral quantities of interest (namely the moments), and to define their transport equations.

2.3.1 Moment-transport equations for a PBE

First, for simplicity, let us consider a PBE with only particle length as the internal coordinate, for which the NDF is $n_{\mathrm{L}}(t, \mathbf{x}, L)$. The PBE reported in Eq. (2.14) then becomes

$$\frac{\partial n_{\mathrm{L}}}{\partial t} + \frac{\partial}{\partial \mathbf{x}} \cdot (\mathbf{v} n_{\mathrm{L}}) + \frac{\partial}{\partial L}(G_{\mathrm{L}} n_{\mathrm{L}}) = h_{\mathrm{L}}, \tag{2.17}$$

where the continuous rate of change of particle length $\dot{L}$ is denoted by G_{L} and discrete events are denoted by h_{L}. As with all PBE, we assume that the velocity $\mathbf{v}$ is known. Recall also that for fine particles an additional diffusive flux in physical space would appear on the right-hand side due to Brownian motion of the particles in the surrounding fluid phase.[16] However, because the diffusion coefficient is inversely proportional to the particle mass, in many systems the diffusive-flux contribution will be very small, if not negligible. On applying the moment transform (Eq. (2.5) on page 32) to Eq. (2.17) we obtain[17]

$$\frac{\partial m_{\mathrm{L},k}}{\partial t} + \frac{\partial}{\partial \mathbf{x}} \cdot (\mathbf{v} m_{\mathrm{L},k}) = k \overline{G}_{\mathrm{L},k-1} m_{\mathrm{L},k-1} + \overline{h}_{\mathrm{L},k}, \tag{2.18}$$

where the average particle-growth rate is defined by

$$\overline{G}_{\mathrm{L},k} = \frac{\int_0^{\infty} G_{\mathrm{L}} n_{\mathrm{L}}(L) L^k \, \mathrm{d}L}{\int_0^{\infty} n_{\mathrm{L}}(L) L^k \, \mathrm{d}L}, \tag{2.19}$$

[16]When deriving the PBE, it is implicitly assumed that the average distance between particles is large enough that particle motion is driven by the fluid phase. If this were not the case, then the correct starting point for describing the particles would be the GPBE. Thus, for the PBE, the diffusive-flux term can be approximated by a dilute binary diffusion coefficient for a particle of size L in the continuous fluid phase.

[17]If $G_{\mathrm{L}}(L)$ is negative for $L = 0$, then an additional sink term may be needed in this equation in order to account for particle loss due to disappearance at the origin. This occurs, for example, in systems composed of evaporating droplets when the evaporation rate is proportional to the surface area.

and the moment transform of h_L by

$$h_{L,k} = \int_0^\infty h_L L^k \, dL. \tag{2.20}$$

In the equations below, for simplicity, we will assume that the particle-growth rate G_L is size-independent so that $G_{L,k} = G_L$. However, in general, the functional dependence of G_L on size must be provided by the modeler in order to apply Eq. (2.19).

The zeroth moment $m_{L,0}$ is equivalent to the total number concentration N and its transport equation follows from Eq. (2.18):

$$\frac{\partial N}{\partial t} + \frac{\partial}{\partial \mathbf{x}} \cdot (\mathbf{v}N) = h_{L,0}. \tag{2.21}$$

As it is possible to see, the drift term has disappeared since the continuous growth of particle size does not change the total number concentration (if $G_L > 0$). However, N is influenced by the rate of formation of particles (e.g. nucleation), and the rates of aggregation and breakage, which cause appearance and disappearance of particles. These processes are all contained in the source term $h_{L,0}$. The third-order moment $m_{L,3}$ is related to the fraction of volume occupied by particles with respect to the suspending fluid and can be easily found from Eq. (2.18):

$$\alpha_p = \int_0^\infty k_V L^3 n_L(L) dL = k_V m_{L,3}, \tag{2.22}$$

where α_p indicates the volume fraction for the particle phase. The fluid phase is denoted by α_f, and $\alpha_p + \alpha_f = 1$. The resulting transport equation for the volume fraction of the disperse phase is

$$\frac{\partial \alpha_p}{\partial t} + \frac{\partial}{\partial \mathbf{x}} \cdot \left(\mathbf{v}\alpha_p\right) = 3k_V G_L m_{L,2} + k_V h_{L,3}. \tag{2.23}$$

It is interesting to note that, if the drift term is null (i.e. particles are not growing in size) and if there is no introduction of new particles into the system (i.e. the nucleation rate is null[18]), the right-hand side of Eq. (2.23) is null, resulting in a continuity equation for the particle phase of the form

$$\frac{\partial \alpha_p}{\partial t} + \frac{\partial}{\partial \mathbf{x}} \cdot \left(\mathbf{v}\alpha_p\right) = 0. \tag{2.24}$$

Since $\mathbf{v}$ is known, this closed hyperbolic advection equation for the disperse-phase volume fraction is in conservative form.

The process of finding moment-transport equations starting from the PBE can be continued to arbitrary order. We should note that in most applications the resulting moment-transport equations *will not be closed*. In other words, the moment-transport equation of order k will involve moments of order higher than k. We will discuss moment-closure methods in Chapters 7 and 8 in the context of quadrature-based moment methods.

[18] A very common case for which $h_{L,3}$ is null but the nucleation rate is nonzero occurs when nucleation produces nuclei with zero size. For this case, in fact, only $h_{L,0}$ is nonzero.

2.3.2 Moment-transport equations for a GPBE

The very same approach as described above for the PBE can be applied to the GPBE (Eq. (2.16) on page 37). On applying the moment transform to Eq. (2.16) for the zeroth-order moment, the transport equation for the total number concentration is obtained:

$$\frac{\partial N}{\partial t} + \frac{\partial}{\partial \mathbf{x}} \cdot \left(\mathbf{U}_{\mathrm{p}} N\right) = h_0, \tag{2.25}$$

where the moment transform of the collision term is

$$h_0 \equiv \int_{\Omega_\xi} \int_{\Omega_{\mathrm{v}}} h(\mathbf{v}, \boldsymbol{\xi}) \mathrm{d}\mathbf{v}\, \mathrm{d}\boldsymbol{\xi} \tag{2.26}$$

and (see Eq. (2.11) on page 34) $\mathbf{U}_{\mathrm{p}}$ is the average particle velocity. This transport equation is very similar to Eq. (2.21). We can note that, for the case in which the particle size is constant ($h_0 = 0$), the relation $\alpha_{\mathrm{p}} = k_{\mathrm{V}} L^3 N$ will allow us to rewrite Eq. (2.25) as a balance for the disperse-phase volume fraction (or, by multiplying by the particle density ρ_{p}, the particle mass fraction):

$$\frac{\partial \rho_{\mathrm{p}} \alpha_{\mathrm{p}}}{\partial t} + \frac{\partial}{\partial \mathbf{x}} \cdot \left(\rho_{\mathrm{p}} \alpha_{\mathrm{p}} \mathbf{U}_{\mathrm{p}}\right) = 0. \tag{2.27}$$

This expression is the fundamental statement of the conservation of mass of the disperse phase. If the first-order particle-velocity moment transform is applied, we obtain

$$\frac{\partial \mathbf{U}_{\mathrm{p}} N}{\partial t} + \frac{\partial}{\partial \mathbf{x}} \cdot \left(\mathbf{U}_{\mathrm{p},2} N\right) = \mathbf{A}_{\mathrm{p},0} + \mathbf{h}_1, \tag{2.28}$$

where the second-order velocity-moment tensor is defined by

$$\mathbf{U}_{\mathrm{p},2} \equiv \frac{1}{N} \int_{\Omega_\xi} \int_{\Omega_{\mathrm{v}}} (\mathbf{v} \otimes \mathbf{v}) n(\mathbf{v}, \boldsymbol{\xi}) \mathrm{d}\mathbf{v}\, \mathrm{d}\boldsymbol{\xi}, \tag{2.29}$$

the mean acceleration term by

$$\mathbf{A}_{\mathrm{p},0} \equiv \int_{\Omega_\xi} \int_{\Omega_{\mathrm{v}}} \mathbf{A}_{\mathrm{p}} n(\mathbf{v}, \boldsymbol{\xi}) \mathrm{d}\mathbf{v}\, \mathrm{d}\boldsymbol{\xi}, \tag{2.30}$$

and the collisional acceleration term by

$$\mathbf{h}_1 \equiv \int_{\Omega_\xi} \int_{\Omega_{\mathrm{v}}} \mathbf{v} h(\mathbf{v}, \boldsymbol{\xi}) \mathrm{d}\mathbf{v}\, \mathrm{d}\boldsymbol{\xi}. \tag{2.31}$$

Equation (2.28) is a transport equation for the mean particle velocity $\mathbf{U}_{\mathrm{p}}$ and has a structure very similar to the ones previously reported.[19] The first term on the left-hand side represents accumulation, whereas the second one represents convection of velocity in physical space. The first term on the right-hand side represents the acceleration produced by a force acting on the particles, whereas the second term represents the change of velocity due to discrete events such as particle collisions.

[19] Note that the term $\partial/\partial\mathbf{x} \cdot (\mathbf{U}_{\mathrm{p},2} N)$ is a scalar product of the spatial gradient vector and the second-order tensor produced by the dyadic product of velocity vectors.

We note in passing that nearly all of the terms in Eq. (2.28) are unclosed at the moment level. We can also note that, for the case in which the particle density and size are constant, the relation $\rho_p \alpha_p = \rho_p k_V L^3 N$ will allow us to rewrite Eq. (2.28) as a balance for the disperse-phase momentum ($\rho_p \alpha_p \mathbf{U}_p$):[20]

$$\frac{\partial \rho_p \alpha_p \mathbf{U}_p}{\partial t} + \frac{\partial}{\partial \mathbf{x}} \cdot \left(\rho_p \alpha_p \mathbf{U}_{p,2} \right) = m_p \mathbf{A}_{p,0}, \tag{2.32}$$

where m_p is the (constant) particle mass. This expression is the fundamental statement of the conservation of momentum for the disperse phase. The term on the right-hand side represents the momentum exchange with the continuous fluid phase and body forces (e.g. gravity). The second-order tensor $\mathbf{U}_{p,2}$ is unclosed at the level of the moment-transport equations. The simplest possible closure is $\mathbf{U}_{p,2} = \mathbf{U}_p \otimes \mathbf{U}_p$, which neglects velocity fluctuations in the disperse phase. Using this closure, Eqs. (2.27) and (2.32) describe mass and momentum conservation of the disperse phase written in conservative form.[21] Because the disperse-phase mass and momentum are fundamental conserved quantities, it is often useful to develop numerical schemes starting from the general conservation equations for these quantities. In order to derive them from the GPBE, we must know the functional dependence of the particle mass on the internal coordinates, which we denote by $m(\boldsymbol{\xi})$. By definition, the particle material density and disperse-phase volume fraction (i.e. the disperse-phase mass) are related to the mass by

$$\rho_p \alpha_p = \int_{\Omega_\xi} \int_{\Omega_v} m(\boldsymbol{\xi}) n(\mathbf{v}, \boldsymbol{\xi}) \mathrm{d}\mathbf{v} \, \mathrm{d}\boldsymbol{\xi}. \tag{2.33}$$

Likewise, the disperse-phase momentum is related to the particle mass by

$$\rho_p \alpha_p \mathbf{U}_p = \int_{\Omega_\xi} \int_{\Omega_v} \mathbf{v} m(\boldsymbol{\xi}) n(\mathbf{v}, \boldsymbol{\xi}) \mathrm{d}\mathbf{v} \, \mathrm{d}\boldsymbol{\xi}. \tag{2.34}$$

Thus, in order to derive the disperse-phase mass- and momentum-conservation equations, it suffices to multiply the GPBE by $m(\boldsymbol{\xi})$ and $\mathbf{v}m(\boldsymbol{\xi})$, respectively, and to integrate over the phase-space variables.

When developing models for polydisperse multiphase flows, it is often useful to resort to *conditioning* on particle size. For example, in gas–solid flows the momentum-exchange terms between the gas phase and a solid particle will depend on the particle size. Thus, the conditional particle velocity given that the particle has internal-coordinate vector $\boldsymbol{\xi} = \boldsymbol{\zeta}$ will depend on the value of $\boldsymbol{\zeta}$. Mathematically, we can define a *conditional probability density function*[22] (PDF) $f_{\mathbf{U}|\xi}$ by $n(\mathbf{v}, \boldsymbol{\zeta}) = f_{\mathbf{U}|\xi}(\mathbf{v}|\boldsymbol{\zeta}) n_\xi(\boldsymbol{\zeta})$, where, using the definition of the NDF,

$$n_\xi(\boldsymbol{\zeta}) = \int_{\Omega_v} n(\mathbf{v}, \boldsymbol{\zeta}) \mathrm{d}\mathbf{v}. \tag{2.35}$$

Note that $f_{\mathbf{U}|\xi}$ is indeed a conditional PDF since

$$\int_{\Omega_v} f_{\mathbf{U}|\xi}(\mathbf{v}|\boldsymbol{\zeta}) \mathrm{d}\mathbf{v} = 1 \tag{2.36}$$

[20] Conservation of momentum during collisions for particles of constant mass requires $h_1 = 0$.

[21] The closure for the term $\mathbf{A}_{p,0}$ couples the disperse phase to the continuous phase, and requires a separate model (e.g. Stokes drag).

[22] $f_{\mathbf{U}|\xi}(\mathbf{v}, \boldsymbol{\zeta})$ can also be interpreted as the *fraction* of particles with velocity $\mathbf{v}$ among all particles with internal coordinates $\boldsymbol{\zeta}$.

for all values of ζ. Using the conditional PDF, we can define conditional velocity moments of arbitrary order. For example, the first two moments are

$$\langle \mathbf{U}_\mathrm{p} | \xi = \zeta \rangle \equiv \int_{\Omega_\mathrm{v}} \mathbf{v} f_{\mathrm{U}|\xi}(\mathbf{v}|\zeta) \mathrm{d}\mathbf{v} \tag{2.37}$$

and

$$\langle \mathbf{U}_\mathrm{p} \otimes \mathbf{U}_\mathrm{p} | \xi = \zeta \rangle \equiv \int_{\Omega_\mathrm{v}} (\mathbf{v} \otimes \mathbf{v}) f_{\mathrm{U}|\xi}(\mathbf{v}|\zeta) \mathrm{d}\mathbf{v}. \tag{2.38}$$

Starting from the GPBE (Eq. (2.16)), we can then find a transport equation for $n_\xi(\zeta)$ by substituting $f_{\mathrm{U}|\xi}(\mathbf{v}|\zeta) n_\xi(\zeta)$ for n and integrating over velocity space:

$$\frac{\partial n_\xi}{\partial t} + \frac{\partial}{\partial \mathbf{x}} \cdot \left(\langle \mathbf{U}_\mathrm{p} | \xi = \zeta \rangle n_\xi \right) + \frac{\partial}{\partial \zeta} \cdot \left(\langle \dot{\xi} | \xi = \zeta \rangle n_\xi \right) = \int_{\Omega_\mathrm{v}} h(\mathbf{v}, \zeta) \mathrm{d}\mathbf{v} \tag{2.39}$$

where

$$\langle \dot{\xi} | \xi = \zeta \rangle \equiv \int_{\Omega_\mathrm{v}} \dot{\xi}(\mathbf{v}, \zeta) f_{\mathrm{U}|\xi}(\mathbf{v}|\zeta) \mathrm{d}\mathbf{v}. \tag{2.40}$$

On comparing Eq. (2.39) with the PBE (Eq. (2.14)), we see that they will be equivalent if the velocity is independent of the internal coordinates:[23] $\langle \mathbf{U}_\mathrm{p} | \xi = \zeta \rangle = \mathbf{U}_\mathrm{p}$, if the growth term does not depend on $\mathbf{v}$, $\langle \dot{\xi} | \xi = \zeta \rangle = \dot{\xi}(\zeta)$, and if

$$h_\xi = \int_{\Omega_\mathrm{v}} h(\mathbf{v}, \zeta) \mathrm{d}\mathbf{v}. \tag{2.41}$$

In general, these conditions will not hold and thus Eq. (2.39) should be used to find n_ξ for systems in which the particle velocity differs from particle to particle.

A transport equation for the conditional particle velocity can also be found starting from the GPBE:

$$\begin{aligned} &\frac{\partial \langle \mathbf{U}_\mathrm{p} | \xi = \zeta \rangle n_\xi}{\partial t} + \frac{\partial}{\partial \mathbf{x}} \cdot \left(\langle \mathbf{U}_p \otimes \mathbf{U}_p | \xi = \zeta \rangle n_\xi \right) \\ &\quad = \langle \mathbf{A}_\mathrm{p} | \xi = \zeta \rangle n_\xi - \frac{\partial}{\partial \zeta} \cdot \left(\langle \mathbf{U}_\mathrm{p} \dot{\xi} | \xi = \zeta \rangle n_\xi \right) - \int_{\Omega_\mathrm{v}} \mathbf{v} h(\mathbf{v}, \xi) \mathrm{d}\mathbf{v}, \end{aligned} \tag{2.42}$$

where

$$\langle \mathbf{U}_\mathrm{p} \dot{\xi} | \xi = \zeta \rangle \equiv \int_{\Omega_\mathrm{v}} \mathbf{v} \dot{\xi}(\mathbf{v}, \zeta) f_{\mathrm{U}|\xi}(\mathbf{v}|\zeta) \mathrm{d}\mathbf{v}. \tag{2.43}$$

While this equation contains several unclosed terms, it does offer the possibility of accounting for particle-size effects in the context of multiphase flow models without resorting to the full GPBE. Instead, Eqs. (2.39) and (2.42) could be solved together with appropriate closures. In a typical application, the conditional particle velocity is found by neglecting the left-hand side of Eq. (2.42), leaving only an algebraic expression for $\langle \mathbf{U}_\mathrm{p} | \xi = \zeta \rangle$. This approximation is appropriate for cases in which the acceleration terms (e.g. fluid drag[24]) are much stronger than the effects of spatial transport.

---

[23] Recall that, in the PBE, we denote the known velocity by $\mathbf{v}$, which is equivalent in Eq. (2.39) to $\mathbf{U}_\mathrm{p} = \mathbf{v}$.

[24] The fluid drag term has the form $(\mathbf{v} - \mathbf{U}_\mathrm{f})/\tau(\xi)$, where $\tau(\xi)$ is the drag time scale for particles with internal coordinate ξ. The conditional average is then $(\langle \mathbf{U}_\mathrm{p} | \xi = \zeta \rangle - \mathbf{U}_\mathrm{f})/\tau(\zeta)$. For small $\tau(\zeta)$ (i.e. small Stokes number), the terms on the right-hand side can be combined to find $\langle \mathbf{U}_\mathrm{p} | \xi = \zeta \rangle = \mathbf{U}_\mathrm{f} + \tau(\zeta) \cdot \mathrm{RHS}$. This result can then be used in Eq. (2.39) to solve for n_ξ, where RHS are terms for the deviation of the particle velocity from the fluid velocity found from Eq. (2.42). Thus, up to first order in the Stokes number and in the absence of body forces and collisions, we find $\langle \mathbf{U}_\mathrm{p} | \xi = \zeta \rangle n_\xi = \mathbf{U}_\mathrm{f} n_\xi \left[1 - \tau(\zeta) \partial / \partial \zeta \cdot \left(\langle \dot{\xi} | \xi = \zeta \rangle n_\xi \right) \right]$.

2.4 Flow regimes for the PBE

The PBE in Eq. (2.14) contains a term for transport in physical space and, as written, assumes that the advection velocity $\mathbf{v}(t, \mathbf{x})$ is known exactly. However, in many practical applications the advection velocity will be turbulent and only the lower-order turbulence statistics will be known. If the velocity is laminar, the NDF changes due to advection and, for fine particles, diffusion due to Brownian motion, whereas when the velocity is turbulent there will be turbulent fluctuations in n_ξ due to turbulent mixing, which must also be taken into account. We show how this works for two simple examples below. The treatment of turbulence in the context of the GPBE is much more complicated than with the PBE because with the GPBE the turbulence is often generated by the coupling between the disperse and continuous phases. In other words, the NDF in a PBE is a *passive* scalar that inherits its turbulent attributes from the turbulence in the continuous phase, while the NDF in a GPBE is a *dynamically active* scalar that can generate its own turbulence. In this book we do not consider issues dealing with models for multiphase turbulent flows in the context of the GPBE. Readers interested in the subject can consult the review by Fox (2012). Generally speaking, the current state of the art in multiphase-flow models for device-scale applications is strongly influenced by single-phase turbulence models, such as those described in Pope (2000). For example, most models start from *ad hoc* closures for the disperse-phase mass and momentum equations (i.e. Eqs. (2.27) and (2.32)) and introduce transport equations for the disperse-phase turbulent kinetic equation and energy dissipation. As described in Fox (2012), a more fundamental approach consists of working directly with the GPBE using mesoscale closures that have previously been validated against fully resolved solutions to the GPBE. Our focus in this book is thus on simulation methods to obtain fully resolved solutions to the GPBE.

2.4.1 Laminar PBE

In the case of laminar flow, the NDF is advected with the fluid velocity $\mathbf{v}$. However, it also is interesting to note that a population of particles with different sizes can have different velocities because of the different drag forces acting on each particle. Therefore, as described above for the small-Stokes-number limit of Eq. (2.42), the effective advection velocity $\langle \mathbf{U}_\mathrm{p} | \boldsymbol{\xi} = \boldsymbol{\zeta} \rangle$ can depend on particle size. Likewise, when particles are small enough, they will be subject to Brownian motion, which can be thought of as a random, chaotic component for particle velocity that has to be added to the mean velocity. When describing Brownian motion in a Lagrangian framework, the conditional particle velocity $\langle \mathbf{U}_\mathrm{p} | \boldsymbol{\xi} = \boldsymbol{\zeta} \rangle$ can be written as the sum of the mean component plus a fluctuating one, whereas in an Eulerian framework it can be represented by a flux of particles expressed as a size-dependent diffusion coefficient multiplied by the mean NDF gradient using the very same approach as that used to define molecular properties (such as fluid viscosity) in kinetic theory. In this limit, Eq. (2.39) can be written in closed form as

$$\frac{\partial n_\xi}{\partial t} + \frac{\partial}{\partial \mathbf{x}} \cdot \left(\mathbf{v} n_\xi - \Gamma_\mathrm{p}(\boldsymbol{\xi}) \frac{\partial n_\xi}{\partial \mathbf{x}} \right) + \frac{\partial}{\partial \boldsymbol{\xi}} \cdot \left(\dot{\boldsymbol{\xi}} n_\xi \right) = h_\xi, \tag{2.44}$$

where $\Gamma_\mathrm{p}(\boldsymbol{\xi})$ is the effective diffusion coefficient, which can be estimated, for example, by Nernst–Einstein theory (Bird *et al.*, 1960). Note that, when writing this equation, we implicitly assume that h_ξ can be expressed in terms of n_ξ so that the right-hand side is *closed*. In summary, we should note that the numerical solution of Eq. (2.44) is quite challenging because the effective diffusivity is often many orders of magnitude smaller than the advection velocity. In other words, the Schmidt number associated with Eq. (2.44) (which depends on $\boldsymbol{\xi}$) is very large in most practical applications.

2.4.2 Turbulent PBE

In the case of turbulent advection velocity, the transported quantity in the PBE (i.e. the NDF) fluctuates around its mean value. These fluctuations are due to the nonlinear convection term in the momentum equation of the continuous phase. In turbulent flows usually the Reynolds average is introduced (Pope, 2000).[25] It consists of calculating ensemble-averaged quantities of interest (usually lower-order moments). Given a fluctuating property of a turbulent flow $\phi(t, \mathbf{x})$, its Reynolds average at a fixed point in time and space can be written as

$$\langle \phi \rangle = \int_{-\infty}^{+\infty} \psi f_\phi(\psi) \mathrm{d}\psi, \tag{2.45}$$

where $f_\phi(t, \mathbf{x}, \psi)$ is the one-point PDF of $\phi(t, \mathbf{x})$ defined by the following probability statement: $f_\phi(\psi)\mathrm{d}\psi = P[\psi \leq \phi(t, \mathbf{x}) < \psi + \mathrm{d}\psi]$. The probability on the right-hand side of this expression is computed from the ensemble of all random fields $\phi(t, \mathbf{x})$ observable in a turbulent flow (Fox, 2003; Pope, 2000). The random fields of primary interest in the PBE are the velocity of the fluid phase $\mathbf{U}_\mathrm{f}$ and the NDF $n_{\boldsymbol{\xi}}$ defined by the PBE (Eq. (2.44)). We can define the one-point joint PDF of $\mathbf{U}_\mathrm{f}$ and $n_{\boldsymbol{\xi}}$ as $f_{\mathbf{U}_\mathrm{f},n_{\boldsymbol{\xi}}}(\mathbf{v}, n^*_{\boldsymbol{\xi}})\mathrm{d}\mathbf{v}\,\mathrm{d}n^*_{\boldsymbol{\xi}} = P[(\mathbf{v} \leq \mathbf{U}_\mathrm{f}(t, \mathbf{x}) < \mathbf{v} + \mathrm{d}\mathbf{v}) \cap (n^*_{\boldsymbol{\xi}} \leq n_{\boldsymbol{\xi}}(t, \mathbf{x}) < n^*_{\boldsymbol{\xi}} + \mathrm{d}n^*_{\boldsymbol{\xi}})]$. Note that $\mathrm{d}n^*_{\boldsymbol{\xi}}$ is a functional derivative and represents a positive perturbation of the entire function $n^*_{\boldsymbol{\xi}}$, and $\mathbf{v}$ is the fluid-velocity state space, which includes all the possible values of the fluid velocity $\mathbf{U}_\mathrm{f}$. Note that conceptually $f_{\mathbf{U}_\mathrm{f},n_{\boldsymbol{\xi}}}$ has the same meaning as the joint velocity–composition PDF used in turbulent reacting flows (Fox, 2003). In fact, the function $n^*_{\boldsymbol{\xi}}$ represents an infinite number of scalar fields parameterized by $\boldsymbol{\xi}$. In other words, each value $\boldsymbol{\xi}$ corresponds to a passive scalar with "concentration" $n_{\boldsymbol{\xi}}$ that is transported by the turbulent velocity field $\mathbf{U}_\mathrm{f}$.

We can now define the Reynolds-averaged fluid velocity as

$$\langle \mathbf{U}_\mathrm{f} \rangle \equiv \iint_{-\infty}^{+\infty} \mathbf{v} f_{\mathbf{U}_\mathrm{f},n_{\boldsymbol{\xi}}}(\mathbf{v}, n^*_{\boldsymbol{\xi}})\mathrm{d}\mathbf{v}\,\mathrm{d}n^*_{\boldsymbol{\xi}}. \tag{2.46}$$

Likewise, the Reynolds-averaged NDF is defined by

$$\langle n_{\boldsymbol{\xi}} \rangle \equiv \iint_{-\infty}^{+\infty} n^*_{\boldsymbol{\xi}} f_{\mathbf{U}_\mathrm{f},n_{\boldsymbol{\xi}}}(\mathbf{v}, n^*_{\boldsymbol{\xi}})\mathrm{d}\mathbf{v}\,\mathrm{d}n^*_{\boldsymbol{\xi}}. \tag{2.47}$$

It is also possible to define a fluctuating component for each quantity as the difference between the instantaneous value and the Reynolds-average value: $\mathbf{u}_\mathrm{f} = \mathbf{U}_\mathrm{f} - \langle \mathbf{U}_\mathrm{f} \rangle$ and $n'_{\boldsymbol{\xi}} = n_{\boldsymbol{\xi}} - \langle n_{\boldsymbol{\xi}} \rangle$. These quantities are used for Reynolds decomposition of higher-order terms: $\langle \mathbf{U}_\mathrm{f} n_{\boldsymbol{\xi}} \rangle = \langle \mathbf{U}_\mathrm{f} \rangle \langle n_{\boldsymbol{\xi}} \rangle + \langle \mathbf{u}_\mathrm{f} n'_{\boldsymbol{\xi}} \rangle$, where $\langle \mathbf{u}_\mathrm{f} n'_{\boldsymbol{\xi}} \rangle$ is called the *turbulent flux* of the NDF. Physically, this term denotes the advection of fluid elements with a given $n'_{\boldsymbol{\xi}}$ by fluid velocity fluctuations, and thus enters the transport equation as a *spatial* flux. The Reynolds-average NDF can be found from Eq. (2.44):[26]

$$\frac{\partial \langle n_{\boldsymbol{\xi}} \rangle}{\partial t} + \frac{\partial}{\partial \mathbf{x}} \cdot \left(\langle \mathbf{U}_\mathrm{f} n_{\boldsymbol{\xi}} \rangle - \Gamma_\mathrm{p}(\boldsymbol{\xi}) \frac{\partial \langle n_{\boldsymbol{\xi}} \rangle}{\partial \mathbf{x}} \right) + \frac{\partial}{\partial \boldsymbol{\xi}} \cdot \left(\dot{\boldsymbol{\xi}} \langle n_{\boldsymbol{\xi}} \rangle \right) = \langle h_{\boldsymbol{\xi}} \rangle, \tag{2.48}$$

[25] A similar theory can be developed in the context of large-eddy simulations (LES) (Fox, 2012).

[26] When performing the Reynolds averaging, the internal-coordinate vector $\boldsymbol{\xi}$ is not affected. Thus, the Reynolds average commutes with time, space, and phase-space gradients. See Fox (2003) for a discussion of this topic.

where we have assumed, for simplicity, that $\dot{\xi}$ depends only on ξ and not, for example, on other scalar fields transported by the flow.[27] Using the Reynolds decomposition and a gradient-diffusion model for the turbulent flux (Fox, 2003), we can write $\langle \mathbf{U}_\mathrm{f} n_\xi \rangle = \langle \mathbf{U}_\mathrm{f} \rangle \langle n_\xi \rangle - \Gamma_\mathrm{T} \nabla_\mathbf{x} \langle n_\xi \rangle$, which closes the convection term in Eq. (2.48). The turbulent diffusivity Γ_T is found by solving a separate model for the turbulence kinetic energy (k_f) and turbulence-dissipation rate (ε_f) in the fluid phase. Thus, using this approximation, Eq. (2.48) reduces to

$$\frac{\partial \langle n_\xi \rangle}{\partial t} + \frac{\partial}{\partial \mathbf{x}} \cdot \left(\langle \mathbf{U}_\mathrm{f} \rangle \langle n_\xi \rangle \right) + \frac{\partial}{\partial \xi} \cdot \left(\dot{\xi} \langle n_\xi \rangle \right) = \frac{\partial}{\partial \mathbf{x}} \cdot \left[\left(\Gamma_\mathrm{T} + \Gamma_\mathrm{p}(\xi) \right) \frac{\partial \langle n_\xi \rangle}{\partial \mathbf{x}} \right] + \langle h_\xi \rangle. \tag{2.49}$$

In high-Reynolds-number turbulent flows, the "molecular" diffusion term $\Gamma_\mathrm{p}(\xi)$ will be negligible compared with Γ_T (Fox, 2003). Except for the last term, Eq. (2.49) is closed.

In many practical applications (e.g. particle aggregation), h_ξ will be quadratic in n_ξ. Thus, the closure problem associated with $\langle h_\xi \rangle$ is very similar to that faced when closing the chemical source term for turbulent reacting flows (Fox, 2003). For example, if the time scales of the phenomena accounted for in h_ξ are much slower than the turbulence time scales, then we can close $\langle h_\xi \rangle$ in terms of $\langle n_\xi \rangle$.[28] However, in general, one must resort to the use of a *micromixing model* to capture interactions between the turbulent velocity field and the NDF. Other common unclosed phenomena are fluid-shear-induced particle aggregation and breakage, which depend on the magnitude of the *instantaneous* spatial gradients of $\mathbf{v}$. In a turbulent flow field, $|\nabla_\mathbf{x} \mathbf{v}| \gg |\nabla_\mathbf{x} \langle \mathbf{U}_\mathrm{f} \rangle|$ and, moreover, $|\nabla_\mathbf{x} \mathbf{v}|$ varies strongly in space and time. The Reynolds-average aggregation and breakage rate must therefore be estimated by using values of k_f and ε_f that are based on high-Reynolds-number turbulence scaling laws (Fox, 2003). Finally, the reader should keep clearly in mind the fact that, even when Eq. (2.49) is closed, the NDF $\langle n_\xi \rangle$ still depends on many independent variables (i.e. time, space, and internal coordinates). In practice, the direct solution of Eq. (2.49) by discretizing all of the independent variables is intractable and additional approximations are required. In this book, our primary method for simplifying the PBE is moment closure.

2.5 The moment-closure problem

To conclude this chapter, we look briefly at the moment-closure problem using the NDF transport equation in Eq. (2.49). For clarity, we will drop the angle brackets and assume that the source terms in the PBE have been closed. The PBE then reads

$$\frac{\partial n_\xi}{\partial t} + \frac{\partial}{\partial \mathbf{x}} \cdot \left(\mathbf{U}_\mathrm{f} n_\xi \right) + \frac{\partial}{\partial \xi} \cdot \left(\dot{\xi} n_\xi \right) = \frac{\partial}{\partial \mathbf{x}} \cdot \left[\left(\Gamma_\mathrm{T} + \Gamma_\mathrm{p}(\xi) \right) \frac{\partial n_\xi}{\partial \mathbf{x}} \right] + h_\xi. \tag{2.50}$$

In order to reduce the number of independent variables, we use the definition of the moments in Eq. (2.2) to find the moment-transport equation corresponding to Eq. (2.50):

[27] If, for example, the growth rate depends on the concentration of a chemical species in the fluid phase, this assumption would not hold.

[28] In turbulent reacting flows, this is known as the slow-chemistry limit (Fox, 2003).

$$\frac{\partial m_{\xi,\mathbf{k}}}{\partial t} + \frac{\partial}{\partial \mathbf{x}} \cdot \left(\mathbf{U}_{\mathrm{f}} m_{\xi,\mathbf{k}}\right) - \frac{\partial}{\partial \mathbf{x}} \cdot \left(\Gamma_{\mathrm{T}} \frac{\partial m_{\xi,\mathbf{k}}}{\partial \mathbf{x}}\right) = -\int_{\Omega_\xi} (\xi_1^{k_1} \cdots \xi_M^{k_M}) \frac{\partial}{\partial \boldsymbol{\xi}} \cdot \left(\dot{\boldsymbol{\xi}} n_\xi\right) \mathrm{d}\boldsymbol{\xi}$$
$$+ \frac{\partial}{\partial \mathbf{x}} \cdot \left(\int_{\Omega_\xi} (\xi_1^{k_1} \cdots \xi_M^{k_M}) \Gamma_{\mathrm{p}}(\boldsymbol{\xi}) \frac{\partial n_\xi}{\partial \mathbf{x}} \mathrm{d}\boldsymbol{\xi}\right)$$
$$+ \int_{\Omega_\xi} (\xi_1^{k_1} \cdots \xi_M^{k_M}) h_\xi \, \mathrm{d}\boldsymbol{\xi}. \tag{2.51}$$

In this expression, the closed terms appear on the left-hand side and the unclosed terms on the right-hand side of the equal sign. The moment-closure problem then consists of finding functional approximations for the terms on the right-hand side that depend only on the moments $m_{\xi,\mathbf{k}}$ for a finite set of values $\mathbf{k}$. In practice, the moment closures are recomputed "on the fly" during the numerical simulation of Eq. (2.51). Because the moments depend on time t and space $\mathbf{x}$, the moment closures must be computed many, many times during the course of a single simulation. For this reason, the computational algorithm must be fast and efficient (i.e. an algorithm based on multivariate iterative solutions is unlikely to be tractable). In this book, we focus on quadrature-based moment methods (QBMM) that reconstruct n_ξ from a finite set of moments $m_{\xi,\mathbf{k}}$. In Chapter 3 we provide a detailed introduction to the moment-inversion algorithms used in QBMM. In subsequent chapters we discuss in depth how QBMM are used to close the moment-transport equation, as well as robust numerical algorithms for solving Eq. (2.51) and related moment-transport equations derived from the GPBE.

3

Quadrature-based moment methods

In this chapter we discuss the basic theory of Gaussian quadrature, which is at the heart of quadrature-based moment methods (QBMM). Proofs for most of the results are not included and, for readers requiring more extensive analytical treatments, references to the literature are made. In addition to a summary of the relevant theory, different algorithms to calculate the abscissas (or nodes) and the weights of the quadrature approximation from a known set of moments are presented, and their advantages and disadvantages are critically discussed. It is important to remind readers that most of the theory for quadrature formulas was developed for mono-dimensional integrals, namely integrals of a single independent variable. Therefore the discussion below starts from univariate distributions, for which the Gaussian quadrature theory applies exactly, and subsequently moves to bivariate and multivariate distributions. Although in the latter cases the quadrature is no longer strictly Gaussian, most of its interesting properties are still valid. In the univariate case, the weights and abscissas are used in the quadrature method of moments (QMOM) to solve moment-transport equations. Thus, we will refer to moment-inversion algorithms that use a full set of moments as QMOM, while other methods that use a reduced set will be referred to differently.

3.1 Univariate distributions

As described in earlier chapters, for univariate QMOM the closure problem appears in different parts of the transport equations for the moments of the NDF. However, it can often be reduced to the following integral:

$$I = \int_{\Omega_\xi} n(\xi) g(\xi) \mathrm{d}\xi, \tag{3.1}$$

where $n(\xi)$ is the unknown univariate NDF, $g(\xi)$ is a functional grouping together all the terms appearing in the integral apart from the NDF itself, and Ω_ξ is the integration interval, which is also known as the support interval. As has already been reported, in the case of particle mass (or particle size or surface area) as internal coordinate, the support interval becomes $\Omega_\xi = [0, \infty)$, whereas in the case of particle velocity $\Omega_\xi = (-\infty, +\infty)$. The closure problem can be overcome by any numerical scheme capable of calculating accurately the integral reported in Eq. (3.1). It is important to stress here that, since the NDF is unknown, and only its moments are transported (and therefore known), this numerical scheme must be constructed starting from the transported moments of the NDF. The Gaussian quadrature is such a scheme, but, before discussing the details, let us introduce some preliminary concepts.

In Gaussian quadrature theory the NDF is called the weight function or measure. The weight function must be nonnegative and non-null in the integration interval and all its moments,

$$m_k = \int_{\Omega_\xi} n(\xi)\xi^k \, \mathrm{d}\xi, \quad k = 0, 1, 2, \ldots, \tag{3.2}$$

must exist (Gautschi, 2004; Press *et al.*, 1992). These requirements are clearly met by any NDF representing a real multiphase particulate system. In this chapter a moment of order k will also be indicated as $m(k)$ or as $\langle \xi^k \rangle$ and these three notations will be used interchangeably. Another important definition is that of orthogonal polynomials (Gautschi, 2004). A set of polynomials $\{P_0(\xi), P_1(\xi), \ldots, P_\alpha(\xi), \ldots\}$, with $P_\alpha(\xi) = k_{\alpha,0}x^\alpha + k_{\alpha,1}x^{\alpha-1} + \cdots + k_{\alpha,\alpha}$, is said to be orthogonal in the integration interval Ω_ξ, with respect to the weight function, if

$$\int_{\Omega_\xi} n(\xi)P_\alpha(\xi)P_\beta(\xi)\mathrm{d}\xi \begin{cases} = 0 & \text{for } \alpha \neq \beta, \\ > 0 & \text{for } \alpha = \beta, \end{cases} \tag{3.3}$$

and is orthonormal if

$$\int_{\Omega_\xi} n(\xi)P_\alpha(\xi)P_\beta(\xi)\mathrm{d}\xi = \begin{cases} 0 & \text{for } \alpha \neq \beta, \\ 1 & \text{for } \alpha = \beta. \end{cases} \tag{3.4}$$

The integration domain Ω_ξ and the weight function $n(\xi)$ uniquely define the family of polynomials $\{P_\alpha(\xi)\}$. A polynomial is defined as monic when its leading coefficient (i.e. $k_{\alpha,0}$) is equal to unity. Below two important theorems (without proof) are reported.

Theorem 3.1 *Any set of orthogonal polynomials* $\{P_\alpha(\xi)\}$ *has a recurrence formula relating any three consecutive polynomials in the following sequence:*

$$P_{\alpha+1}(\xi) = (\xi - a_\alpha)P_\alpha(\xi) - b_\alpha P_{\alpha-1}(\xi), \quad \alpha = 0, 1, 2, \ldots \tag{3.5}$$

with $P_{-1}(\xi) \equiv 0$ *and* $P_0(\xi) \equiv 1$, *and where*

$$a_\alpha = \frac{\int_{\Omega_\xi} n(\xi)\xi P_\alpha(\xi)P_\alpha(\xi)\mathrm{d}\xi}{\int_{\Omega_\xi} n(\xi)P_\alpha(\xi)P_\alpha(\xi)\mathrm{d}\xi}, \quad \alpha = 0, 1, 2, \ldots, \tag{3.6}$$

$$b_\alpha = \frac{\int_{\Omega_\xi} n(\xi)P_\alpha(\xi)P_\alpha(\xi)\mathrm{d}\xi}{\int_{\Omega_\xi} n(\xi)P_{\alpha-1}(\xi)P_{\alpha-1}(\xi)\mathrm{d}\xi}, \quad \alpha = 1, 2, \ldots. \tag{3.7}$$

The recursive relation is the most important property for constructive and computational use of orthogonal polynomials. In fact, as will be shown below, knowledge of the recursion coefficients allows the zeros of the orthogonal polynomials to be computed, and with them the quadrature rule. Therefore the calculation of the coefficients of this three-term recurrence relation is of paramount importance. The recursive relationship in Eq. (3.5) generates a sequence of monic polynomials that are orthogonal with respect to the weight function in the considered integration interval. One can calculate a_0 from Eq. (3.6), then $P_1(\xi)$ from Eq. (3.5). Knowing $P_0(\xi)$ and $P_1(\xi)$, one can calculate a_1 and b_1 from Eqs. (3.6) and (3.7), and so on. It is also easy to show that, if the polynomials are explicitly represented and the multiplications are carried out term by term, eventually the coefficients a_α and b_α can be written in terms of the moments of the NDF. Moreover, the coefficients necessary for the construction of a polynomial of order N can be calculated from the first $2N - 1$ moments of the NDF, namely $m_0, m_1, \ldots, m_{2N-2}, m_{2N-1}$. For example, with m_0, m_1, m_2, and m_3, it is possible to calculate the following coefficients:

$$\begin{aligned} a_0 &= \frac{m_1}{m_0}, \\ a_1 &= \frac{m_3 m_0^2 + m_1^3 - 2m_2 m_1 m_0}{m_2 m_0 + m_1^2 - 2m_1^2 m_0}, \\ b_1 &= \frac{m_2 m_0 + m_1^2 - 2m_1^2 m_0}{m_0^2}, \end{aligned} \tag{3.8}$$

which, in turn, suffice for the calculation of the polynomial $P_2(\xi)$ (of order two) orthogonal to the weight function in the integration interval.

Below is another important theorem that ensures that once the integration interval has been specified the zeros of the orthogonal polynomials must be constrained inside this interval. This is an important feature of orthogonal polynomials because, as will become clearer below, it guarantees that the internal coordinates are always sampled within the support of the NDF.

Theorem 3.2 *Each polynomial in an orthogonal sequence has all its roots real, distinct, and strictly inside the integration interval (which is also called the orthogonality interval). The roots of each polynomial lie strictly between the roots of the next higher-order polynomial in the sequence (interlacing roots).*

Why are we interested in orthogonal polynomials? Because, as will be explained in the next section, their roots are the nodes of Gaussian quadrature approximations.

3.1.1 Gaussian quadrature

In univariate QMOM, the closure problem can be overcome by any numerical scheme capable of calculating the integral in Eq. (3.1). A very simple way to do this is to resort to an interpolation formula:

$$\int_{\Omega_\xi} n(\xi) g(\xi) \mathrm{d}\xi \approx \sum_{\alpha=1}^{N} w_\alpha g(\xi_\alpha), \tag{3.9}$$

where w_α and ξ_α are, respectively, the weights and the nodes of the interpolation formula, and N is the number of nodes used for the interpolation (or order of the formula). The accuracy of an interpolation formula is quantified by its degree of accuracy.[1] The degree of accuracy is equal to d if the interpolation formula is exact when the integrand is a polynomial of order less than or equal to d and there exists at least one polynomial of order $d + 1$ that makes the interpolation formula inexact. One of the main requirements of the interpolation formula is to have distinct nodes, and, according to how they are placed in

[1] The degree of accuracy is also known as the degree of exactness.

the integration interval, different degrees of accuracy can be achieved. For example, an interpolation formula with N equally spaced nodes (i.e. Newton–Cotes) has a degree of accuracy of $N - 1$, whereas a Gaussian quadrature has a degree of accuracy of $2N - 1$. It is therefore clear why quadrature approximations are so interesting: with the same number of nodes (and therefore of function evaluations) their degree of accuracy is essentially twice that of other formulas. It is, however, important to remind readers that the definition of the degree of accuracy is based on the performance of the quadrature in handling polynomials as the integrand, therefore high order translates into high accuracy only when the integrand is smooth enough to be approximated by a polynomial. A Gaussian quadrature is an interpolation formula whose N nodes are the roots of the polynomial $P_N(\xi)$ orthogonal to the weight function $n(\xi)$ as reported in the next theorem (Gautschi, 1997, 2004).

Theorem 3.3 *The necessary and sufficient condition for the formula*

$$\int_{\Omega_\xi} n(\xi) g(\xi) \mathrm{d}\xi = \sum_{\alpha=1}^{N} g(\xi_\alpha) w_\alpha + R_N(g) \tag{3.10}$$

to be a Gaussian quadrature approximation or, equivalently, that it have a degree of accuracy of $2N - 1$*, is that its nodes* $\{\xi_\alpha\}$ *coincide with the* N *roots of the polynomial* $P_N(\xi)$ *of order* N *orthogonal in* Ω_ξ *with respect to the weight function* $n(\xi)$.

Theorem 3.3 implies that the quadrature approximation of order[2] N is exact (i.e. the rest function $R_N(g)$ is null) whenever the integrand $g(\xi)$ is a polynomial of order less than or equal to $2N - 1$.

Well-known Gaussian quadratures include those associated with classical orthogonal polynomials, such as Gauss–Legendre, Gauss–Jacobi, Gauss–Laguerre, Gauss–Hermite, and Gauss–Chebyshev polynomials. However, given a specific non-classical weight function, specialized algorithms can be used to calculate the corresponding quadrature. One straightforward way to carry out this calculation is to exploit its high degree of accuracy. Because a Gaussian quadrature of order N has a degree of accuracy of $2N - 1$, it will be the only approximation able to capture the first $2N - 1$ moments of the weight function. Therefore the N weights and N abscissas can be determined by solving the following nonlinear system:

$$\begin{aligned} m_0 &= \sum_{\alpha=1}^{N} w_\alpha, \\ m_1 &= \sum_{\alpha=1}^{N} w_\alpha \xi_\alpha, \\ &\vdots \\ m_{2N-2} &= \sum_{\alpha=1}^{N} w_\alpha \xi_\alpha^{2N-2}, \\ m_{2N-1} &= \sum_{\alpha=1}^{N} w_\alpha \xi_\alpha^{2N-1}. \end{aligned} \tag{3.11}$$

This nonlinear system can be solved by using the Newton–Raphson method, or any other non-linear equation solver. However, this is not the best way since a very good initial guess is needed in order to ensure convergence.

[2] Also referred to as N-point quadrature.

A smarter way to find the roots is to employ the recursive relationship in Eq. (3.5) for the orthogonal polynomials (Dette & Studden, 1997; Golub & Welsch, 1969; Wilf, 1962). The sequence of recursive relationships with $\alpha = 0, 1, 2, \ldots, N$ can be interpreted in matrix form as

$$\xi \begin{bmatrix} P_0(\xi) \\ P_1(\xi) \\ P_2(\xi) \\ P_3(\xi) \\ \vdots \\ P_{N-2}(\xi) \\ P_{N-1}(\xi) \end{bmatrix} = \begin{bmatrix} a_0 & 1 & & & & & \\ b_1 & a_1 & 1 & & & & \\ & b_2 & a_2 & 1 & & & \\ & & b_3 & a_3 & \ddots & & \\ & & & \ddots & \ddots & \ddots & \\ & & & & \ddots & a_{N-2} & 1 \\ & & & & & b_{N-1} & a_{N-1} \end{bmatrix} \begin{bmatrix} P_0(\xi) \\ P_1(\xi) \\ P_2(\xi) \\ P_3(\xi) \\ \vdots \\ P_{N-2}(\xi) \\ P_{N-1}(\xi) \end{bmatrix} + \begin{bmatrix} 0 \\ 0 \\ 0 \\ 0 \\ \vdots \\ 0 \\ P_N(\xi) \end{bmatrix}. \tag{3.12}$$

Equation (3.12) clearly illustrates that the roots of $P_N(\xi)$, namely the nodes of the quadrature approximation $\{\xi_\alpha\}$, are the eigenvalues of the tridiagonal matrix appearing in the equation. This matrix can be made symmetric (preserving the eigenvalues) by a diagonal similarity transformation to give a Jacobi matrix:

$$\mathbf{J} = \begin{bmatrix} a_0 & \sqrt{b_1} & & & & & \\ \sqrt{b_1} & a_1 & \sqrt{b_2} & & & & \\ & \sqrt{b_2} & a_2 & \sqrt{b_3} & & & \\ & & \sqrt{b_3} & a_3 & \ddots & & \\ & & & \ddots & \ddots & \ddots & \\ & & & & \ddots & a_{N-2} & \sqrt{b_{N-1}} \\ & & & & & \sqrt{b_{N-1}} & a_{N-1} \end{bmatrix}. \tag{3.13}$$

This procedure transforms the ill-conditioned problem of finding the roots of a polynomial into the well-conditioned problem of finding the eigenvalues and eigenvectors of a tridiagonal symmetric matrix. As shown by Wilf (1962), the N weights can then be calculated as $w_\alpha = m_0\varphi_{\alpha 1}^2$, where $\varphi_{\alpha 1}$ is the first component of the αth eigenvector $\boldsymbol{\varphi}_\alpha$ of the Jacobi matrix.

Of course, the method described above requires knowledge of the coefficients a_α and b_α appearing in the recursive relationship reported in Eq. (3.5). The coefficients can, in turn, be calculated simply by resorting to the orthogonality condition and by using the moments of the weight function. Below two algorithms for efficiently carrying out these calculations are presented and discussed. The reader can note that, if the NDF corresponds to the weight function of a known family of orthogonal polynomials (e.g. a Gaussian NDF is the weight function for the Hermite polynomials (Gautschi, 2004)), the recursion coefficients are known in advance, so Eq. (3.13) can be applied directly to find the weights and abscissas without using the moments. In Section 3.3, we make use of this fact to develop an extended quadrature. In contrast, for an *arbitrary* NDF the recursion coefficients can be computed from the moments.

3.1.2 The product–difference (PD) algorithm

The product–difference (PD) algorithm was developed by Gordon (1968) and is based on the theory of continued fractions of Stieltjes. The first step is to construct a matrix **P** with components $P_{\alpha,\beta}$ starting from the moments of the weight function. The components in the first column of $\mathbf{P}$ are $P_{\alpha,1} = \delta_{\alpha 1}$ for $\alpha \in 1, \ldots, 2N+1$, where $\delta_{\alpha 1}$ is the Kronecker delta. The components in the second column of $\mathbf{P}$ are $P_{\alpha,2} = (-1)^{\alpha-1} m_{\alpha-1}$ for $\alpha \in 1, \ldots, 2N$. It should be highlighted that, since the final weights can be corrected by multiplying by the true m_0, the calculations can also be done assuming a normalized distribution (i.e. $m_0 = 1$). The remaining components are found by the PD algorithm: $P_{\alpha,\beta} = P_{1,\beta-1} P_{\alpha+1,\beta-2} - P_{1,\beta-2} P_{\alpha+1,\beta-1}$ for $\beta \in 3, \ldots, 2N+1$ and $\alpha \in 1, \ldots, 2N+2-j$. In the case of $N = 2$, for example, the $\mathbf{P}$ matrix is

$$\mathbf{P} = \begin{bmatrix} 1 & m_0 & m_1 & m_0 m_2 - (m_1)^2 & m_0\left(m_3 m_1 - (m_2)^2\right) \\ 0 & -m_1 & -m_2 & -(m_0 m_3 - m_2 m_1) & \\ 0 & m_2 & m_3 & & \\ 0 & -m_3 & & & \\ 0 & & & & \end{bmatrix}. \tag{3.14}$$

The coefficients of the continued fraction $\{\zeta_\alpha\}$ necessary to determine the coefficients of the three-term recursive relation are generated by setting the first element equal to zero ($\zeta_1 = 0$), and then computing the others according to the following recursive relationship:

$$\zeta_\alpha = \frac{P_{1,\alpha+1}}{P_{1,\alpha} P_{1,\alpha-1}}, \quad \alpha \in 2, \ldots, 2N. \tag{3.15}$$

The coefficients of the symmetric tridiagonal Jacobi matrix are then obtained from sums and products of ζ_α: $a_\alpha = \zeta_{2\alpha} + \zeta_{2\alpha-1}$ for $\alpha \in 1, \ldots, N$ and $b_\alpha = -\sqrt{\zeta_{2\alpha+1}\zeta_{2\alpha}}$ for $\alpha \in 1, \ldots, N-1$. A MATLAB script implementing the PD algorithm can be found in Section A.1.1 of Appendix A.

Exercise 3.1 *Consider the following normal (or Gaussian) distribution:*

$$n(\xi) = \frac{1}{\sqrt{2\pi\sigma^2}} \exp\left(-\frac{(\xi-\mu)^2}{2\sigma^2}\right), \tag{3.16}$$

where, as is well known, μ is the mean and σ^2 is the variance of the distribution (Walpole & Myers, 1985). The moments of this distribution can be easily calculated through the following equations (reported only for the first nine moments):

$$\begin{aligned} m_0 &= 1, \\ m_1 &= \mu, \\ m_2 &= \mu^2 + \sigma^2, \\ m_3 &= \mu^3 + 3\mu\sigma^2, \\ m_4 &= \mu^4 + 6\mu^2\sigma^2 + 3\sigma^4, \\ m_5 &= \mu^5 + 10\mu^3\sigma^2 + 15\mu\sigma^4, \\ m_6 &= \mu^6 + 15\mu^4\sigma^2 + 45\mu^2\sigma^4 + 15\sigma^6, \\ m_7 &= \mu^7 + 21\mu^5\sigma^2 + 105\mu^3\sigma^4 + 105\mu\sigma^6, \\ m_8 &= \mu^8 + 28\mu^6\sigma^2 + 210\mu^4\sigma^4 + 420\mu^2\sigma^6 + 105\sigma^8. \end{aligned} \tag{3.17}$$

Use the PD algorithm to calculate the quadrature approximation of order four (i.e. $N = 4$) for $\mu = 5$ and $\sigma = 1$. For this calculation the first eight moments of the distribution are needed:

$$
\begin{aligned}
m_0 &= 1,\\
m_1 &= 5,\\
m_2 &= 26,\\
m_3 &= 140,\\
m_4 &= 778,\\
m_5 &= 4450,\\
m_6 &= 26\,140,\\
m_7 &= 157\,400.
\end{aligned}
$$

After applying the PD algorithm the following Jacobi matrix is obtained:

$$
\mathbf{J} = \begin{bmatrix} 5 & -1 & 0 & 0 \\ -1 & 5 & -\sqrt{2} & 0 \\ 0 & -\sqrt{2} & 5 & -\sqrt{3} \\ 0 & 0 & -\sqrt{3} & 5 \end{bmatrix}, \tag{3.18}
$$

resulting in the weights $w_1 = 0.0459$, $w_2 = 0.4541$, $w_3 = 0.4541$, *and* $w_4 = 0.0459$, *and nodes* $\xi_1 = 2.6656$, $\xi_2 = 4.2580$, $\xi_3 = 5.7420$, *and* $\xi_4 = 7.3344$. *It is left to the reader to verify that with this quadrature approximation the first eight moments can be exactly calculated, whereas an error is committed for* m_8 *(as well as for higher-order moments).*

The PD algorithm is quite efficient in a number of practical cases; however, it generally becomes less stable as N increases. It is difficult to predict *a priori* when this will occur since it depends on the absolute values of the moments, but typically problems can be expected when $N > 10$. Another important issue is related to the fact that for distributions with zero mean[3] (or in other words with $m_1 = 0$) the algorithm blows up, due to a division by zero during the calculation of the coefficients of the continued fraction $\{\zeta_\alpha\}$. The Wheeler algorithm, which will be reported in the next section, does not suffer from these problems.

3.1.3 The Wheeler algorithm

Sack & Donovan (1971) proposed an alternative approach for the calculation of the coefficient of the recursive formula reported in Eq. (3.5) (and appearing also in the Jacobi matrix) that resulted in higher stability. This approach is based on the idea of using a different set of basis functions $\pi_\alpha(\xi)$ to represent the orthogonal polynomials, rather than the usual powers of ξ. The improved stability results from the ability of the new polynomial basis to better sample the integration interval Ω_ξ. The coefficients are calculated from the modified moments defined as follows:

$$
\nu_k = \int_{\Omega_\xi} \pi_k(\xi) n(\xi) \mathrm{d}\xi, \quad k = 0, 1, \ldots, 2N-1, \tag{3.19}
$$

[3]But m_1 is usually not zero when the internal coordinate ξ represents particle mass, surface area, size, etc. In these cases the PD algorithm can be safely used. The case of null m_1 occurs more often when the internal coordinate is a particle velocity that, ranging from negative to positive real values, can result in distributions with zero mean velocity. Another frequent case in which the mean is null is when central moments (moments translated with respect to the mean of the distribution) are used to build the quadrature approximation. These cases will be discussed later on, when describing the algorithms for building multivariate quadratures.

analogously to what is done with standard moments. Moreover, it is assumed that $\pi_k(\xi)$ satisfies the following recursive relation:

$$\begin{aligned}\pi_{-1}(\xi) &\equiv 0,\\ \pi_0(\xi) &\equiv 1,\\ \pi_{\alpha+1}(\xi) &= (\xi - a'_\alpha)\pi_\alpha(\xi) - b'_\alpha\pi_{\alpha-1}(\xi),\end{aligned} \tag{3.20}$$

where, of course, the coefficients a'_α and b'_α have to be known explicitly. From these findings, Wheeler (1974) developed an efficient algorithm to calculate the coefficients of the Jacobi matrix through some intermediate quantities:

$$\sigma_{\alpha,\beta} = \int_{\Omega_\xi} n(\xi)\pi_\alpha(\xi)\pi_\beta(\xi)\mathrm{d}\xi, \quad \alpha,\beta \geq -1. \tag{3.21}$$

These quantities can be calculated by initializing

$$\begin{aligned}\sigma_{-1,\alpha} &= 0, \quad \alpha = 1,2,\ldots,2N-2,\\ \sigma_{0,\alpha} &= \nu_\alpha \quad \alpha = 0,1,\ldots,2N-1',\\ a_0 &= a'_0 + \frac{\nu_1}{\nu_0},\\ b_0 &= 0;\end{aligned} \tag{3.22}$$

and then by computing for $\alpha = 1,2,\ldots,N-1$ (where again N is the order of the quadrature approximation and of the orthogonal polynomial) the following equation is obtained: $\sigma_{\alpha,\beta} = \sigma_{\alpha-1,\beta+1} - \left(a_{k-1} - a'_\beta\right)\sigma_{\alpha-1,\beta} - b_{\beta-1}\sigma_{\alpha-2,\beta} + b'_\beta\sigma_{\alpha-1,\beta-1}$ for $\beta = \alpha, \alpha+1, \ldots, 2N-\alpha-1$. Finally, the coefficients for the Jacobi matrix are computed as follows:

$$\begin{aligned}a_\alpha &= a'_\alpha - \frac{\sigma_{\alpha-1,\alpha}}{\sigma_{\alpha-1,\alpha-1}} + \frac{\sigma_{\alpha,\alpha+1}}{\sigma_{\alpha,\alpha}},\\ b_\alpha &= \frac{\sigma_{\alpha,\alpha}}{\sigma_{\alpha-1,\alpha-1}}.\end{aligned} \tag{3.23}$$

Details on the derivation of this algorithm can be found in Gautschi (1997, 2004). The Wheeler algorithm results in higher stability and can generally be applied for the calculation of higher-order quadrature approximations (as compared with the PD algorithm). Although the choice of the modified moments, and of the basis function $\pi_\alpha(\xi)$, plays a very important role in the performance of the algorithm, it is interesting to note that even with modified moments ν_k identical to standard moments μ_k (or in other words with $a'_\alpha = b'_\alpha = 0$) the Wheeler algorithm is more stable. The Wheeler algorithm presents the additional advantage of being able to calculate the coefficients of the Jacobi matrix in the case of distributions with zero mean (i.e. $m_1 = 0$). A MATLAB script implementing the Wheeler algorithm can be found in Section A.1.2 of Appendix A.

Exercise 3.2 *Consider again the distribution reported in Eq. (3.16) of Exercise 3.1. Let us now calculate the quadrature approximation of order four (i.e. $N = 4$) for $\mu = 0$ and $\sigma = 1$ using the Wheeler algorithm. For this calculation the first eight moments of the distribution are needed:*

$$
\begin{aligned}
m_0 &= 1, \\
m_1 &= 0, \\
m_2 &= 1, \\
m_3 &= 0, \\
m_4 &= 3, \\
m_5 &= 0, \\
m_6 &= 15, \\
m_7 &= 0.
\end{aligned}
$$

Notice that, because the distribution has zero mean and is perfectly symmetric (zero skewness), the odd moments are all null.

After applying the Wheeler algorithm, the following Jacobi matrix is obtained:

$$
\mathbf{J} = \begin{bmatrix} 0 & -1 & 0 & 0 \\ -1 & 0 & -\sqrt{2} & 0 \\ 0 & -\sqrt{2} & 0 & -\sqrt{3} \\ 0 & 0 & -\sqrt{3} & 0 \end{bmatrix}, \tag{3.24}
$$

resulting in the weights $w_1 = 0.0459$, $w_2 = 0.4541$, $w_3 = 0.4541$, *and* $w_4 = 0.0459$, *and nodes* $\xi_1 = -2.3344$, $\xi_2 = -0.7420$, $\xi_3 = 0.7420$, *and* $\xi_4 = 2.3344$. *It is left to the reader to verify that with this quadrature approximation the first eight moments can be exactly calculated, whereas an error is committed for* m_8 *(and higher-order moments). It is also easy to verify that the PD algorithm is unable to calculate the coefficients of the Jacobi matrix in this case.*

In practice, the Wheeler algorithm returns a full set of weights and abscissas for a realizable[4] moment set. However, for unrealizable moment sets it will fail. An adaptive version of the Wheeler algorithm has been proposed by Yuan & Fox (2011) that checks for realizability and returns the largest set of weights and abscissas possible. To accomplish this task, the fact that $0 < \sigma_{\alpha,\alpha}$ for realizable moment sets is used to check, starting from $\alpha = 1$, whether $\sigma_{\alpha,\alpha} \leq 0$ for each value of $1 \leq \alpha \leq N$. If for $\alpha = m$, $\sigma_{m,m} \leq 0$, then only $m - 1$ weights and abscissas can be computed. The adaptive Wheeler algorithm is particularly useful for degenerate cases in which the exact NDF is composed of fewer than N weighted Dirac delta functions. In such cases, the adaptive Wheeler algorithm will return the exact NDF. More details on adaptive quadrature can be found in Yuan & Fox (2011).

3.1.4 Consistency of a moment set

The PD and Wheeler algorithms are quite efficient in calculating univariate quadrature approximations. The resulting nodes are always within the support (and therefore represent realizable values of the internal coordinate), and the weights are always positive, ensuring that, when the quadrature is used to overcome the closure problem, accurate results are obtained. Nevertheless, these properties are respected only if the moment set is realizable, meaning that there exists an NDF resulting in that specific set of moments. If the PD or Wheeler algorithms were fed with invalid moment sets, unrealizable nodes would be calculated, jeopardizing the stability of the simulation. It is therefore clear that algorithms

[4]The concept of realizability or *consistency* of a moment set will be introduced in Section 3.1.4.

capable of determining the realizability and consistency of a moment set, and of correcting an invalid moment set, are particularly useful in this context. Let us therefore analyze the mathematical constraints that link together a set of moments.

The moments of an NDF represent some important physical properties of the underlying population of elements constituting the multiphase system under study. For this reason, they have to satisfy some simple rules. For instance, the positiveness of the density function over its support implies that the moment of order zero must be positive. Additionally, there are other simple, intuitive rules. For example, if the internal coordinate assumes only positive values (or in other words the moment is defined on a positive support) then the moment of order one (as well as all the other moments) must be positive. Another important property of the distribution is its variance (i.e. $\sigma^2 = m_2 - m_1^2/m_0$), which must be zero for a delta-function distribution, while it must be positive for polydisperse distributions. Accordingly, it has to be $m_2 \geq m_1^2/m_0$. For higher-order moments the mathematical constraints are less intuitive and cannot be directly related to specific global properties of the multiphase systems. Fortunately, the theory of moments provides some interesting theorems that turn out to be very useful in determining whether a set of moments is invalid.

Theorem 3.4 *A moment set is said to be realizable if the Hankel–Hadamard determinants (Gautschi, 2004; Shohat & Tamarkin, 1943) are all non-negative:*

$$\Delta_{k,l} = \begin{vmatrix} m_k & m_{k+1} & \cdots & m_{k+l} \\ m_{k+1} & m_{k+2} & \cdots & m_{k+l+1} \\ \vdots & \vdots & \vdots & \vdots \\ m_{k+l} & m_{k+l+1} & \cdots & m_{k+l+l} \end{vmatrix} \geq 0 \tag{3.25}$$

for $k = 0, 1$ *and* $l \geq 0$.

Note that for $k = 0$ and $l = 1$ the Hankel–Hadamard determinant is $m_0 m_2 - m_1^2$; the validity condition thus reduces to $m_2 \geq m_1^2/m_0$. A less stringent condition for validity is defined in the following theorem (Shohat & Tamarkin, 1943).

Theorem 3.5 *A necessary (but not sufficient) condition for validity is that of convexity of the function* $\ln(m_k)$ *with respect to k:*

$$\frac{\ln(m_k) + \ln(m_{k-2})}{2} \geq \ln(m_{k-1}), \quad k = 2, 3, \ldots; \tag{3.26}$$

or, equivalently, $m_k m_{k-2} - m_{k-1}^2 \geq 0$ *for* $k = 2, 3, \ldots$.

It is trivial to show that the convexity condition is equivalent to the positiveness of the Hankel–Hadamard determinants for $k = 0, 1$ and $l = 0, 1, 2$, or in other words for the first four moments. For higher-order moments the equivalence is lost, since more stringent conditions are required by Theorem 3.4. However, it turns out that the convexity condition is useful for reasons that will be discussed below.

The convexity of the function $\ln(m_k)$ with respect to k can be easily verified by building a difference table of $\ln(m_k)$. In the first column of the table ($\mathbf{d}_0$ in Tables 3.1 and 3.2 of Exercise 3.3), the natural logarithms of all moments are reported. Elements in subsequent columns are calculated from the previous ones by subtracting from the element in the same row the element in the row below it. Convexity of $\ln(m_k)$ is ensured by the positiveness of the elements of the column relative to the second-order differences (i.e. $\mathbf{d}_2$). It is also interesting to note that the third-order differences vanish when $\ln(m_k)$ is a quadratic function of k as for the moments of a log-normal distribution. The log-normal is therefore the smoothest distribution that results in realizable moments (i.e. positive second-order differences with smallest third-order differences) when applying the convexity condition. The two validity criteria are illustrated in the following example.

Exercise 3.3 *Let us verify that the set of moments of Exercise 3.1 is realizable. By using the first eight moments,*

$$\begin{aligned} m_0 &= 1, \\ m_1 &= 5, \\ m_2 &= 26, \\ m_3 &= 140, \\ m_4 &= 778, \\ m_5 &= 4450, \\ m_6 &= 26\,140, \\ m_7 &= 157\,400, \end{aligned}$$

the largest Hankel–Hadamard matrices that can be calculated are $\Delta_{0,3}$ and $\Delta_{1,3}$, resulting in

$$\Delta_{0,3} = \begin{bmatrix} 1 & 5 & 26 & 140 \\ 5 & 26 & 140 & 778 \\ 26 & 140 & 778 & 4450 \\ 140 & 778 & 4450 & 26\,140 \end{bmatrix} \tag{3.27}$$

and

$$\Delta_{1,3} = \begin{bmatrix} 5 & 26 & 140 & 778 \\ 26 & 140 & 778 & 4450 \\ 140 & 778 & 4450 & 26\,140 \\ 778 & 4450 & 26\,140 & 157\,400 \end{bmatrix}, \tag{3.28}$$

whose determinants are 12 and 5736, respectively. Because these two values are positive, the moment set is realizable. In fact, as has already been shown in Exercise 3.1, when this moment set is fed to the PD (or Wheeler) algorithm, it results in realizable nodes. As expected, the first seven moments are perfectly reproduced by using a quadrature approximation of order four. The same conclusion can be drawn by calculating the difference table of $\ln(m_k)$ *as shown in Table 3.1. Here the natural logarithms of the set of moments are reported in the first column of the table (i.e.* $\mathbf{d}_0$*) and the other elements are calculated as the difference of the elements in subsequent rows using the previous columns. The elements of the second column (*$\mathbf{d}_2$*), containing the second-order differences, are all positive, confirming that the moment set is realizable.*

It is now sufficient to modify the second-order moment to 25 (instead of 26, corresponding to a difference of only 4%) to make the moment set unrealizable. The Hankel–Hadamard determinants are now equal to $-179\,352$ *and* $-12\,362\,344$*, respectively, whereas the difference table (see Table 3.2) presents negative elements in the column containing the second-order differences (namely* $\mathbf{d}_2$*). If this moment set is fed to the PD algorithm, the resulting quadrature is unable to reproduce the moment set.*

In general, simulations carried out starting from realizable moment sets should result in realizable moment sets. Nevertheless, the moment-transport equations are integrated numerically with some finite discretization errors. As Wright (2007) clearly reports, most of the problems are caused by the approximation of the convective term, in particular with higher-order discretization schemes, which can turn a realizable set of moments into

Table 3.1. Difference table of $\ln(m_k)$ for a realizable moment set

k	$\mathbf{d}_0 = \ln(m_k)$	$\mathbf{d}_1$	$\mathbf{d}_2$	$\mathbf{d}_3$	$\mathbf{d}_4$	$\mathbf{d}_5$	$\mathbf{d}_6$	$\mathbf{d}_7$
0	0	1.609	0.039	−0.0043	0.0010	−0.0003	0.0001	−0.0001
1	1.609	1.648	0.034	−0.0033	0.0007	−0.0002	0.0001	0
2	3.258	1.683	0.031	−0.0027	0.0005	−0.0001	0	0
3	4.941	1.715	0.028	−0.0022	0.0004	0	0	0
4	6.656	1.743	0.026	−0.0019	0	0	0	0
5	8.400	1.770	0.024	0	0	0	0	0
6	10.171	1.795	0	0	0	0	0	0
7	11.966	0	0	0	0	0	0	0

Table 3.2. Difference table of $\ln(m_k)$ for an unrealizable moment set

k	$\mathbf{d}_0 = \ln(m_k)$	$\mathbf{d}_1$	$\mathbf{d}_2$	$\mathbf{d}_3$	$\mathbf{d}_4$	$\mathbf{d}_5$	$\mathbf{d}_6$	$\mathbf{d}_7$
0	0	1.609	0	0.113	−0.2343	0.391	−0.588	0.823
1	1.609	1.609	0.113	−0.121	0.1575	−0.196	0.235	0
2	3.218	1.722	−0.007	0.036	−0.0388	0.039	0	0
3	4.941	1.715	0.028	−0.002	0.0004	0	0	0
4	6.656	1.743	0.026	−0.001	0	0	0	0
5	8.400	1.770	0.024	0	0	0	0	0
6	10.171	1.795	0	0	0	0	0	0
7	11.966	0	0	0	0	0	0	0

an unrealizable one. This phenomenon is called moment corruption and poses a serious threat to quadrature-based moment methods because when an unrealizable set of moments is fed to the PD or Wheeler algorithms it might yield unrealizable nodes and lead to numerical instabilities. In general, the best remedy for moment corruption is to use numerical algorithms that are guaranteed to generate realizable moments (see Appendix B for details). However, if a "black-box" transport code is used (which in the authors' opinion should be avoided!), a consistent strategy to correct corrupted moments must be employed.

A strategy that might overcome the problem of moment corruption is replacing unrealizable moment sets as soon as they appear. An algorithm of this kind was developed by McGraw (2006). The algorithm first verifies whether the moment set is realizable (by looking at the second-order difference vector or by looking at the Hankel–Hadamard determinants). If the moment set is unrealizable it proceeds with the correction. The moment-correction algorithm could be based on changes in the moments that transform negative second-order differences into positive ones. However, this strategy presents two issues. The first one is that positiveness of the second-order differences is only a necessary condition for validity, not a sufficient one. Therefore, it is risky to base the correction on a necessary condition, since it might be possible to have after correction a moment set that is still unrealizable. The second issue is related to the fact that this correction strategy introduces the arbitrary decision of how positive the second-order differences have to be after correction, or in other words, how far from the condition of unrealizability one wants the moment set to stay.

An alternative is to carry out an iterative minimization that identifies and corrects the moment which after adjustment maximizes the smoothness of the function $\ln(m_k)$. This is done by minimizing $\mathbf{d}_3$, the column in the difference table containing the third-order differences of $\ln(m_k)$, which, as has already been reported, vanish for the moments of a log-normal distribution. Maximizing the smoothness of $\ln(m_k)$ (or, equivalently, minimizing the third-order differences) represents a more stringent constraint than ensuring the positiveness of the second-order differences (which is only a necessary condition) and usually results in successful convergence after a few iterations. This is particularly true for sets of moments for which unrealizability is caused by the value of a single moment. In cases in which more than one moment is badly estimated, the algorithm can fail and an alternative approach must be used. The simplest and most effective consists of replacing the moments with those of a log-normal distribution that shares with the original one only some correct moments, as suggested by Wright (2007). Because the corruption usually affects higher-order moments, one can often find the parameters of the log-normal distribution from the lower-order moments m_0, m_1, and m_2, whose physical meaning, as stressed, is particularly important. Then, using the log-normal distribution, one can calculate the other moments. We will now discuss in detail these moment-correction algorithms.

The correction algorithm of McGraw

The realizability of a moment set can be verified by calculating a difference table of the function $\ln(m_k)$. In the first column ($\mathbf{d}_0$) the sequence of moments, in ascending order, is reported. In the next columns the differences between two neighboring elements, sitting in the same precedent column and in two subsequent rows, are reported. The procedure is continued until a zero appears in the first row. The values reported in the different columns represent the second-, third-, and higher-order difference vectors of the function $\ln(m_k)$. If a moment m_k is changed as $(m_k)_1 = c_k(m_k)_0$, its natural logarithm undergoes the following change: $\ln(m_k)_1 = \ln(c_k) + \ln(m_k)_0$, where $(m_k)_0$ and $\ln(m_k)_0$ are the initial value and its logarithm, whereas $(m_k)_1$ and $\ln(m_k)_1$ are the final value and its logarithm. The consequent variation in the third-order difference vector results in $(\mathbf{d}_3)_1 - (\mathbf{d}_3)_0 = \ln(c_k)\mathbf{b}_k$, where $(\mathbf{d}_3)_0$ and $(\mathbf{d}_3)_1$ are, respectively, the initial and final vectors of third-order differences, while $\mathbf{b}_k$ is the response vector describing the change of the vector of the third-order differences to a unit increment in the natural logarithm of the kth moment. Thus the final vector is $(\mathbf{d}_3)_1 = (\mathbf{d}_3)_0 + \ln(c_k)\mathbf{b}_k$. As has already been discussed, an efficient way to correct the moment set is to minimize the third-order difference vector. By simple geometric arguments it is immediately evident that the modified third-order difference vector $(\mathbf{d}_3)_1$ presents a minimum length when it is orthogonal to the vector $\mathbf{b}_k$: $[(\mathbf{d}_3)_0 + \ln(c_k)\mathbf{b}_k] \cdot \mathbf{b}_k = 0$. The value of c_k through which $(\mathbf{d}_3)_1$ is minimized is thus obtained from the following relationship:

$$\ln(c_k) = -\cos((\mathbf{d}_3)_0, \mathbf{b}_k)\frac{|(\mathbf{d}_3)_0|}{|\mathbf{b}_k|} = -\frac{(\mathbf{d}_3)_0 \cdot \mathbf{b}_k}{|\mathbf{b}_k|^2}, \tag{3.29}$$

where $\cos((\mathbf{d}_3)_0, \mathbf{b}_k)$ is the angle formed by the vector $(\mathbf{d}_3)_0$ and the vector $\mathbf{b}_k$. The resulting minimum-squared amplitude results in $|(\mathbf{d}_3)_1|^2 = |(\mathbf{d}_3)_0 + \ln(c_k)\mathbf{b}_k|^2 = |(\mathbf{d}_3)_0|^2 \times \cos^2((\mathbf{d}_3)_0, \mathbf{b}_k)$. In the algorithm, the choice of the index of the moment to correct is made with the objective of achieving the maximum smoothness for the function $\ln(m_k)$. The moment chosen for correction is the one that results in the minimum possible magnitude of the third-order difference vector. The index k^* of the moment that must be corrected is chosen as the one that gives the largest $\cos^2((\mathbf{d}_3)_0, \mathbf{b}_k)$, for any moment of index k. The entity of the correction is given by the orthogonality relationship, namely Eq. (3.29), calculated for $k = k^*$. A MATLAB script implementing this correction algorithm can be found in Section A.2.1 of Appendix A.

Exercise 3.4 *Use the moment-correction algorithm to modify the following moment set:*

$$
\begin{aligned}
m_0 &= 1,\\
m_1 &= 5,\\
m_2 &= 26,\\
m_3 &= 101,\\
m_4 &= 778,\\
m_5 &= 4450,\\
m_6 &= 26\,140,\\
m_7 &= 157\,400.
\end{aligned}
$$

On comparing this set of moments with that of Exercise 3.1, it is clear that the moment that should be modified is m_3, whose original value was 140. The correction algorithm recognizes that the set is invalid, and identifies $k^* = 3$ as the order of the moment that has to be changed to maximize the smoothness of the function $\ln(m_k)$. The correction results in $m_3 = 140.001$, a value astonishingly close to the original one. It is left to the reader to verify that, on changing any of the other moments, the algorithm is capable not only of making the moment set realizable, but also of restoring the moments to their original values.

Exercise 3.5 *Use the moment-correction algorithm to modify the following moment set:*

$$
\begin{aligned}
m_0 &= 1,\\
m_1 &= 5,\\
m_2 &= 12,\\
m_3 &= 101,\\
m_4 &= 778,\\
m_5 &= 4450,\\
m_6 &= 26\,140,\\
m_7 &= 157\,400.
\end{aligned}
$$

In this case the moments that have been changed, which are responsible for the invalidity of the set, are m_2 and m_3. When this moment set is fed to the correction algorithm, the following answer is obtained:

$$
\begin{aligned}
m_0 &= 0.3689,\\
m_1 &= 2.378,\\
m_2 &= 15.65,\\
m_3 &= 104.5,\\
m_4 &= 696.1,\\
m_5 &= 4482,\\
m_6 &= 27\,282,\\
m_7 &= 156\,228.
\end{aligned}
$$

It can be seen that all the moments have been changed and, although the final moment set is realizable, it is not the original one.

In summary, the correction algorithm due to McGraw is successful in most of the investigated cases at correcting the moment set, or in other words transforming an unrealizable moment set into a realizable one. Moreover, when the algorithm is applied with a number of moments in the set greater than four, if only one moment is changed, the algorithm is generally capable of restoring the original moment set. If more moments are changed, the algorithm is capable of transforming the set into a valid one, but typically it does not restore it to the original values. When the moment set contains only four moments the third-order difference vector is constituted by only one component and the orthogonality condition (Eq. (3.29)) becomes meaningless. In this limiting case a single change in any of the four moments constituting the set will result in a null third-order difference; however, only some of them will result in a realizable set (with positive second-order differences). In these cases, as well as in all the other cases for which the algorithm fails (even for sets constituted by more than four moments), it is more convenient to use the procedure described next.

The correction algorithm of Wright

This second correction algorithm consists of calculating the new moment set from those corresponding to log-normal distributions. A log-normal distribution has the following functional form:

$$n(\xi) = \frac{N_t}{\sigma\sqrt{2\pi}} \exp\left(\frac{-(\ln(\xi)-\mu)^2}{2\sigma^2}\right), \tag{3.30}$$

corresponding to the following moment of order k:

$$\lambda_k = N_t \exp\left(k\mu + \frac{k^2\sigma^2}{2}\right), \tag{3.31}$$

where ξ is the internal coordinate, N_t is the total particle number density (identical to the moment of order zero of the distribution λ_0), and μ and σ are two parameters related to the mean and variance of the distribution. It is evident that a log-normal distribution, characterized by three parameters (i.e. N_t, μ, and σ), can accommodate a maximum of three moments belonging to a specific set. Moreover, one of them has to be the moment of order zero, whereas the other two can be any of the other moments. The main idea behind this correction procedure is the following: three of the moments of the original set are chosen and are used to calculate the three parameters of the log-normal distribution (i.e. N_t, μ, and σ); then the entire set of moments is recalculated using Eq. (3.31). Since, as already highlighted, of the three moments one has to be the moment of order zero, N_t is simply taken to be equal to m_0. If the other two chosen moments are those of order i and j, the following relationships can be used to determine

$$\mu = \frac{j}{ij-i^2}\ln\left(\frac{m_i}{m_0}\right) + \frac{i}{ij-j^2}\ln\left(\frac{m_j}{m_0}\right), \tag{3.32}$$

and

$$\sigma^2 = \frac{1}{1-i/j}\left[\frac{2}{j^2}\ln\left(\frac{m_j}{m_0}\right) - \frac{2}{ij}\ln\left(\frac{m_i}{m_0}\right)\right] \tag{3.33}$$

of the log-normal distribution. Depending on the problem under investigation and on the nature of the internal coordinate, different moments can be chosen for this calculation.

Typically two of the lower-order moments m_1, m_2, and m_3 are selected. However, in order to accommodate more moments of the original set, sometimes the final moments are calculated as the arithmetic average of the moments of two log-normal distributions. A typical choice could be to use for the first log-normal distribution m_0, m_1, and m_3, and for the second m_0, m_2, and m_3. In this way, after the correction only m_0 and m_3 will be identical to the original ones, but a certain degree of control is achieved on both m_1 and m_2. A MATLAB script implementing this correction algorithm can be found in Section A.2.2 of Appendix A.

3.2 Multivariate distributions

The theory of Gaussian quadrature applies only to univariate distributions. However, in practical cases, the study of distributions with multiple internal coordinates is often necessary. In these cases the closure problem generally assumes the following form:

$$I = \int_{\Omega_{\xi}} n(\boldsymbol{\xi}) g(\boldsymbol{\xi}) \mathrm{d}\boldsymbol{\xi}, \tag{3.34}$$

where $n(\boldsymbol{\xi})$ is the (generally unknown) NDF, $g(\boldsymbol{\xi})$ is a functional grouping together all the terms appearing in the integral that are not the NDF, $\boldsymbol{\xi} = (\xi_1, \xi_2, \ldots, \xi_M)$ is the internal-coordinate vector (of dimension M), and the compact notation of Eq. (3.34) indicates the following multidimensional integral:

$$I = \int_{\Omega_{\xi_l}} \mathrm{d}\xi_1 \int_{\Omega_{\xi_2}} \mathrm{d}\xi_2 \ldots \int_{\Omega_{\xi_M}} \mathrm{d}\xi_M \, n(\boldsymbol{\xi}) g(\boldsymbol{\xi}) \tag{3.35}$$

over the integration domain $\Omega_{\xi} = \Omega_{\xi_l} \otimes \Omega_{\xi_2} \otimes \cdots \otimes \Omega_{\xi_M}$. A general quadrature of N nodes for M internal coordinates requires knowledge of N weights, w_α, and N vector abscissas (or nodes) of length M, $\boldsymbol{\xi}_\alpha = (\xi_{1,\alpha}, \xi_{2,\alpha}, \ldots, \xi_{M,\alpha})$. The N-point, M-dimensional quadrature (referred to as a multivariate QMOM) overcomes the closure problem by approximating the integral reported in Eq. (3.34) with the following summation:

$$\int_{\Omega_{\xi}} n(\boldsymbol{\xi}) g(\boldsymbol{\xi}) \mathrm{d}\boldsymbol{\xi} \approx \sum_{\alpha=1}^{N} w_\alpha g(\boldsymbol{\xi}_\alpha). \tag{3.36}$$

As has already been mentioned, Eq. (3.36) is not a Gaussian quadrature approximation; its degree of accuracy is not known *a priori* (and strongly depends on the choice of moments on which the formula is constructed) and the algorithms for its derivation, from the moments of the NDF, are not well known (unlike for the univariate case).

Before describing in detail three different algorithms, it is interesting to recall here that Eq. (3.36) corresponds to the following assumption on the NDF:

$$n(\boldsymbol{\xi}) = \sum_{\alpha=1}^{N} w_\alpha \delta(\boldsymbol{\xi} - \boldsymbol{\xi}_\alpha), \tag{3.37}$$

where

$$\delta(\boldsymbol{\xi} - \boldsymbol{\xi}_\alpha) = \prod_{\beta=1}^{M} \delta(\xi_\beta - \xi_{\beta,\alpha}) \tag{3.38}$$

and the first index β identifies the internal coordinate while the second index α identifies the specific node or abscissa. In this chapter we will refer to the quadrature approximation by using the usual formulation reported in Eq. (3.36), or the corresponding assumption on the NDF in Eq. (3.37). It is also useful to remind readers that for this multivariate NDF the mixed moments are defined as follows:

$$m_{k_1,k_2,\ldots,k_M} = m(\mathbf{k}) = \left\langle \xi_1^{k_1} \xi_2^{k_2} \ldots \xi_M^{k_M} \right\rangle = \int_{\Omega_\xi} n(\boldsymbol{\xi}) \xi_1^{k_1} \xi_2^{k_2} \ldots \xi_M^{k_M} \, \mathrm{d}\boldsymbol{\xi}, \tag{3.39}$$

where the three different notations ($m_{k_1,k_2,\ldots,k_M}$, $m(\mathbf{k})$, and $\langle \xi_1^{k_1} \xi_2^{k_2} \ldots \xi_M^{k_M} \rangle$) will be used throughout this book interchangeably, and where $\mathbf{k} = (k_1, k_2, \ldots, k_M)$ is called the exponent vector. The multivariate quadrature approximation is built such that for a certain set of moments the following equivalence is exact:

$$m_{k_1,k_2,\ldots,k_M} = \sum_{\alpha=1}^{N} w_\alpha \prod_{\beta=1}^{M} \xi_{\beta,\alpha}^{k_\beta}, \tag{3.40}$$

or, in other words, the multivariate quadrature approximation is able to exactly reproduce the set of moments used to construct it.

Before introducing three different multivariate moment-inversion algorithms, it is important to highlight here that, although the mathematical theory behind multivariate quadrature approximations is not as articulated as that for univariate quadrature approximations, the ideal moment-inversion algorithm should retain some of the characteristic features of the PD and Wheeler algorithms. First, an ideal algorithm should be non-iterative. In fact, these calculations are typically implemented in finite-volume (or finite-element) codes and are repeated over a large number of computational cells. It is therefore not affordable to have iterative calculations carried out throughout the entire computational domain. Second, it is evident that since the NDF is positive in the integration interval (or support) the moment-inversion algorithm must result in positive weights. Third, the abscissas (or nodes) must remain within the integration interval. This is of crucial importance in many applications, such as in cases where the internal coordinates represent species concentrations or particle volume, just to cite two common examples. It is in fact clear that, if the internal coordinate can assume only positive values, then negative abscissas, representing unrealizable solutions, must be avoided, since they result in instabilities that can indeed jeopardize convergence of the simulation code. A last very important desired property of the moment-inversion algorithm (which is met both by the PD and algorithm and by the Wheeler algorithm) is the ability to exactly represent an N-point distribution.[5] This can be easily verified simply by calculating the moments corresponding to an N-point distribution and by checking that the weights and abscissas obtained reproduce exactly the initial distribution.

3.2.1 Brute-force QMOM

Brute-force QMOM, or the direct-inversion method, has been proposed by Wright *et al.* (2001b) in the context of particle coalescence and sintering, and is referred to as a multivariate QMOM approximation. This method is based on the very simple idea of

[5] An N-point distribution is composed of exactly N distinct delta functions with nonzero weights.

choosing $N(M+1)$ mixed moments and solving the nonlinear system constituted by the corresponding $N(M+1)$ equations

$$m_{k_{i1},k_{i2},\ldots,k_{iM}} = m(\mathbf{k}_i) = \sum_{\alpha=1}^{N} w_\alpha \prod_{\beta=1}^{M} \xi_{\beta,\alpha}^{k_{i\beta}}, \quad 1 \leq i \leq N(M+1) \tag{3.41}$$

obtained with $N(M+1)$ different values of the exponent vector $\mathbf{k}_i = (k_{i1}, k_{i2}, \ldots, k_{iM})$. A convenient way to compactly indicate the chosen moments is to group the different exponent vectors (sorted by the global moment order $\gamma_i = |\mathbf{k}_i| = k_{i1} + k_{i2} + \cdots + k_{iM}$) into the so-called exponent matrix.[6] For example, a two-point quadrature ($N = 2$), for an NDF with two internal coordinates ($M = 2$), can be constructed from six mixed moments. If $m_{0,0}$, $m_{1,0}$, $m_{0,1}$, $m_{1,1}$, $m_{2,0}$, and $m_{0,2}$ are chosen, then the moment-order matrix becomes

$$\mathbf{K} = \begin{bmatrix} \mathbf{k}_1 \\ \mathbf{k}_2 \\ \mathbf{k}_3 \\ \mathbf{k}_4 \\ \mathbf{k}_5 \\ \mathbf{k}_6 \end{bmatrix} = \begin{bmatrix} 0 & 0 \\ 1 & 0 \\ 0 & 1 \\ 2 & 0 \\ 1 & 1 \\ 0 & 2 \end{bmatrix} \begin{matrix} \leftarrow \gamma_1 = 0 \\ \leftarrow \gamma_2 = 1 \\ \leftarrow \gamma_3 = 1 \\ \leftarrow \gamma_4 = 2 \\ \leftarrow \gamma_5 = 2 \\ \leftarrow \gamma_6 = 2 \end{matrix} \tag{3.42}$$

The system of $N(M+1)$ equations is then solved by using standard nonlinear solvers such as the Newton–Raphson method or the conjugate-gradient minimization algorithm, both of which are described in Press *et al.* (1992).

A convenient way to formulate the problem is to write it in terms of weights and weighted abscissas (instead of abscissas). These quantities are usually grouped together in the weight vector, $\mathbf{w}^{\mathrm{T}} = [w_1, w_2, \ldots, w_{N-1}, w_N]$, of dimension N, and in the weighted-abscissa matrix, $\mathbf{Y}^{\mathrm{T}} = [\mathbf{Y}_1^{\mathrm{T}}, \mathbf{Y}_2^{\mathrm{T}}, \ldots, \mathbf{Y}_{N-1}^{\mathrm{T}}, \mathbf{Y}_N^{\mathrm{T}}]$, of dimension NM, where $\mathbf{Y}_\alpha^{\mathrm{T}} = w_\alpha \boldsymbol{\xi}_\alpha = [w_\alpha \xi_{1,\alpha}, w_\alpha \xi_{2,\alpha}, \ldots, w_\alpha \xi_{M,\alpha}]$. The final nonlinear system of $N(M+1)$ equations (corresponding to the different moments) in $N(M+1)$ unknowns can be written as follows:

$$\mathbf{F}(\mathbf{Z}) = \mathbf{0}, \tag{3.43}$$

where $\mathbf{Z}^{\mathrm{T}} = [\mathbf{w}^{\mathrm{T}}\, \mathbf{Y}^{\mathrm{T}}]$ represents the vectors of unknown weights and weighted abscissas. By employing the Newton–Raphson method the following iterative scheme can be used to solve Eq. (3.43): $\mathbf{Z}_{n+1} = \mathbf{Z}_n - \mathbf{A}^{-1}(\mathbf{K}, \mathbf{X}_n)\mathbf{F}(\mathbf{Z}_n)$. The matrix $\mathbf{A}$ corresponds to the Jacobian of the nonlinear system in Eq. (3.43) and is a function of the exponent matrix $\mathbf{K}$ and of the abscissas $\mathbf{X}^{\mathrm{T}} = [\boldsymbol{\xi}_1, \boldsymbol{\xi}_2, \ldots, \boldsymbol{\xi}_{N-1}, \boldsymbol{\xi}_N]$ only. The components of $\mathbf{A}$ can be calculated as

$$a_{ij} = \begin{cases} (1-\gamma_i) \prod_{\alpha=1}^{M} \xi_{\alpha,p}^{k_{i\alpha}} & \text{for } p = j, \text{ if } 1 \leq j \leq N, \\ \left(k_{i1}/\xi_{1,p}\right) \prod_{\alpha=1}^{M} \xi_{\alpha,p}^{k_{i\alpha}} & \text{for } p = j - N, \text{ if } N+1 \leq j \leq 2N, \\ \vdots & \vdots \\ \left(k_{iM}/\xi_{M,p}\right) \prod_{\alpha=1}^{M} \xi_{\alpha,M}^{k_{i\alpha}} & \text{for } p = j - MN, \text{ if } MN+1 \leq j \leq (1+M)N. \end{cases} \tag{3.44}$$

Since the matrix has to be inverted (when using the Newton–Raphson method), it must be non-singular throughout the entire calculation.[7]

---

[6] Throughout the book the global moment order will sometimes be indicated as a superscript of the moment, whereas the components of the exponent vector generally appear at subscript level, i.e. $m^{\gamma}_{k_1,k_2,\ldots,k_M}$.

[7] It is interesting to highlight here that the Jacobian matrix $\mathbf{A}$ is exactly the matrix employed in DQMOM for the calculation of the source terms, see for example Eq. (7.118) for a generic univariate case and Eq. (7.128) for a specific bivariate case.

For $M = 1$ (univariate cases), it can easily be shown that **A** is of full rank if and only if the abscissas are distinct. The latter will always be the case if the corresponding NDF is non-degenerate. Furthermore, for the univariate case, it is possible to choose independent non-integer moments without affecting the rank of **A**. For multivariate cases, having distinct abscissas does not guarantee that **A** will be of full rank for every distinct choice of moments. In fact, it can be shown that, for fixed N and M, certain distinct moments are linearly dependent when $M \geq 1$ for all possible sets of abscissas. It is therefore necessary to identify a moment set for which **A** is always non-singular for all non-degenerate points in phase space for given values of M and N. For this reason, in what follows we define optimal and realizable moment sets and discuss different strategies by which to determine them.

Definition of an optimal moment set

We define an optimal moment set for a given number of internal coordinates M to have the following properties (Fox, 2009b).

1. An optimal moment set consists of $N(M + 1)$ distinct moments.
2. An optimal moment set will result in a full-rank square matrix **A** for all possible sets of N distinct, non-degenerate abscissas.
3. An optimal moment set includes all linearly independent moments of a particular order γ_i before adding moments of higher order in order to result in a perfectly symmetric treatment of the internal coordinates.

The last property ensures that all the lower-order mixed moments are included in the set. This turns out to be very important since generally neglecting mixed moments leads to abscissas that lie on lower-dimensional subspaces of the M-dimensional phase space. Whenever the dynamics of the investigated problem is not confining the abscissas on lower-dimensional supports, it is appropriate to choose a moment set to define **A** that is not restricted to generating such behavior. Therefore, it is suggested for many applications that one should use moment sets that treat all the M internal coordinates equally. If we relax this third condition a valid (but not optimal) moment set is instead obtained.

The methodology proposed for finding optimal moment sets for a given M and N is as follows.

1. The distinct moments of a particular order are dependent on M. Thus, we begin by defining all possible rows of **A** up to a maximum order of $2N$. Note that the matrix **A** constructed in this step will have many more rows than columns.
2. Certain rows generated in the first step will be linearly dependent for any choice of abscissas. Therefore, we generate a set of N optimal abscissas (defined below) and, starting at the lowest order, remove rows from **A** one at a time if they are linearly independent. This procedure terminates when **A** is of full rank and square.
3. The moment choice found in the previous step results in **A** being of full rank for a particular choice of abscissas. For the moment choice to be optimal, it must be shown that **A** is of full rank for all choices of non-degenerate abscissas. This can be done by randomly generating abscissas and checking the condition number of **A**.

4. Test simulations must be carried out to determine whether the weights can become negative with the proposed moment set. If negative weights are observed, the moment set is rejected as non-optimal. An alternative strategy, which can be employed in order to avoid the presence of negative weights, is to define the weight vector $\mathbf{w}$ in terms of w_α^2 instead of w_α.

The procedure outlined above is applied for a given value of $N \geq 2$. If it fails, then no optimal moment set can be found for that value of N, so the procedure must be repeated with the next larger N.

Let us illustrate the procedure for a bivariate case (i.e. $M = 2$). When a quadrature approximation with two nodes is used (i.e. $N = 2$) then six moments are needed to solve the nonlinear system and calculate the nodes and weights. In this case the moment of global order zero, $m_{0,0}$, and the moments of global order one, $m_{1,0}$ and $m_{0,1}$, can indeed be chosen as part of the optimal moment set. Among the moments of global order two only two can be chosen (because the third one will be linearly dependent). In order to treat the two internal coordinates equally, $m_{2,0}$ and $m_{0,2}$ must be chosen. One last moment has to be selected to saturate the degrees of freedom. By selecting any of the available moments, among those of global order three (i.e. $m_{3,0}$, $m_{2,1}$, $m_{1,2}$, and $m_{3,0}$), the symmetry will be lost, implying that the two internal coordinates will not be treated equally. The only way to follow a choice that preserves symmetry is to choose the fourth-order moment $m_{2,2}$. This particular choice would result in a valid, but non-optimal, moment set, since before going to the next global order all independent moments of the previous global order must be selected. The example is illustrated in Table 3.3. It is therefore clear that for $M = 2$ there is no optimal moment set for $N = 2$. Although the moment set reported in Table 3.3 is not optimal, it is still valid, meaning that it can be successfully used to build a two-point bivariate quadrature approximation. If the number of nodes is increased to three, then nine moments are needed in order to solve the nonlinear system. In this case all six moments of global order less than or equal to two can be chosen (i.e. $m_{0,0}$, $m_{1,0}$, $m_{0,1}$, $m_{2,0}$, $m_{1,1}$, and $m_{0,2}$). The remaining three moments can be chosen among the four of global order three. Whatever choice is made, the final moment set will not be symmetric with respect to the two internal coordinates, as illustrated in Table 3.4. Again also for this case this choice is not an optimal moment set, but it is a valid one. Eventually, when four nodes are used

Table 3.3. Moments used to build a bivariate quadrature approximation ($M = 2$) for $N = 2$. In this case $m_{0,3}$ is chosen as the third-order moment in order to saturate the degrees of freedom.

$m(2,0)$			
$m(1,0)$			
$m(0,0)$	$m(0,1)$	$m(0,2)$	$m(0,3)$

Table 3.4. Moments used to build a bivariate quadrature approximation ($M = 2$) for $N = 3$. In this case $m_{2,1}$, $m_{1,2}$, and $m_{0,3}$ are chosen among the third-order moments in order to saturate the degrees of freedom.

$m(2,0)$	$m(2,1)$		
$m(1,0)$	$m(1,1)$	$m(1,2)$	
$m(0,0)$	$m(0,1)$	$m(0,2)$	$m(0,3)$

Table 3.5. The optimal moment set used to build a bivariate quadrature approximation ($M = 2$) for $N = 4$

$m(3,0)$	$m(3,1)$		
$m(2,0)$	$m(2,1)$		
$m(1,0)$	$m(1,1)$	$m(1,2)$	$m(1,3)$
$m(0,0)$	$m(0,1)$	$m(0,2)$	$m(0,3)$

Table 3.6. The optimal moment set used to build a bivariate quadrature approximation ($M = 2$) for $N = 9$

$m(5,0)$	$m(5,1)$	$m(5,2)$			
$m(4,0)$	$m(4,1)$	$m(4,2)$			
$m(3,0)$	$m(3,1)$	$m(3,2)$			
$m(2,0)$	$m(2,1)$	$m(2,2)$	$m(2,3)$	$m(2,4)$	$m(2,5)$
$m(1,0)$	$m(1,1)$	$m(1,2)$	$m(1,3)$	$m(1,4)$	$m(1,5)$
$m(0,0)$	$m(0,1)$	$m(0,2)$	$m(0,3)$	$m(0,4)$	$m(0,5)$

($N = 4$) for building a bivariate quadrature approximation ($M = 2$), the optimal moment set constituted by 12 moments and reported in Table 3.5 is obtained. It can be seen that the moment set is symmetric with respect to the two internal coordinates and all the lower-order mixed moments are included in the set, resulting in a non-singular Jacobian matrix. Another optimal moment set of 27 moments is identified for $M = 2$ when using $N = 9$ as reported in Table 3.6.

Optimal and valid moment sets can be defined also for $M = 3$. When eight nodes are used, a total of 32 moments must be selected, with the components of the exponent matrix taking on integer values in the set $\{0, 1, 2, 3\}$. In this case the first 20 moments in the optimal moment set are the distinct moments of global order three and smaller. The remaining 12 moments are of global order four or five. The complete set of optimal moments (written in terms of the exponent matrix) is given in Table 3.7. It is interesting to note that the higher-order moments appear in symmetric combinations. For example, because $m_{3,2,0}$ is not linearly independent, neither is $m_{0,2,3}$ (or any other permutation of the same exponents.) This fact greatly simplifies the task of finding the linearly independent moments for larger values of M. For $M = 3$ and $N = 27$ in total 108 moments belong to the optimal set. The moment exponents take on integer values in the set $\{0, 1, 2, 3, 4, 5\}$. The first 56 moments in the optimal moment set are the distinct moments of order five and smaller. The order of the remaining 52 moments ranges from six to nine (Fox, 2009b). As noted earlier, all permutations of the moment exponents appear in the set, so it is necessary only to list one such moment.

Once the moment set (either optimal or simply valid) has been defined the search for the final quadrature can be started. As for the univariate distributions, also for the multivariate case a good initial guess is indeed required for convergence. This is often computed by selecting from the mixed moments (when available) the subset of the first $2N - 1$ pure moments with respect to each internal coordinate (for example for ξ_1 by using $m_{0,0,\dots,0}$, $m_{1,0,\dots,0}$, $m_{2,0,\dots,0}$, $\dots$, $m_{2N-2,0,\dots,0}$, and $m_{2N-1,0,\dots,0}$). However, strategies for efficient calculation of the initial guess have been developed only for bivariate distributions (i.e. $M = 2$) and cannot easily be generalized to multivariate distributions. Readers interested in the details are refereed to the specialized literature (McGraw & Wright, 2003; Rosner *et al.*, 2003; Wright *et al.*, 2001b; Yoon & McGraw, 2004a, 2004b). A MATLAB script implementing the bivariate quadrature algorithm can be found in Section A.3.1 of Appendix A. It is left to the

Table 3.7. The optimal moment set used to build a trivariate quadrature approximation ($M = 3$) for $N = 8$

i	1	2	3	4	5	6	7	8	9	10
k_{i1}	0	1	0	0	2	1	1	0	0	0
k_{i2}	0	0	1	0	0	1	0	2	1	0
k_{i3}	0	0	0	1	0	0	1	0	1	2
i	11	12	13	14	15	16	17	18	19	20
k_{i1}	3	2	2	1	1	1	0	0	0	0
k_{i2}	0	1	0	2	1	0	3	2	1	0
k_{i3}	0	0	1	0	1	2	0	1	2	3
i	21	22	23	24	25	26	27	28	29	
k_{i1}	3	3	1	1	0	0	2	1	1	
k_{i2}	0	1	3	0	3	1	1	2	1	
k_{i3}	1	0	0	3	1	3	1	1	2	
i	30	31	32							
k_{i1}	3	1	1							
k_{i2}	1	3	1							
k_{i3}	1	1	3							

reader to verify that the brute-force QMOM described in this section is capable of retrieving an N-point distribution starting from the corresponding moments when the initial guess is close enough to the final exact solution. As the initial guess worsens, a decrease in the under-relaxation factor might help in convergence, but beyond a certain point the method fails. A numerical example that compares the different methods is reported at the end of this section.

3.2.2 Tensor-product QMOM

Tensor-product QMOM was introduced by Yoon & McGraw (2004a, 2004b) in the context of aerosol modeling and by Fox (2008, 2009a) in the context of dilute gas–solid flows. The proposed algorithms differ depending on the selection of the mixed moments used for the inversion and on the type of internal coordinates (e.g. particle size, volume, or velocity). In this section a generalized version of the tensor-product algorithm (which is valid for any type of distribution written in terms of a generic internal-coordinate vector, $\boldsymbol{\xi}$) is discussed. The method constructs a quadrature approximation of order $N = N_1 N_2 \dots N_M$, where M is the number of internal coordinates and $N_1, N_2, \dots, N_M$ are the orders of the univariate quadrature approximations built first in each of the M directions. In fact, the construction of the univariate quadratures (from the first $2N_\beta$ with $1 \leq \beta \leq M$ pure moments with respect to the different M internal coordinates) is the first step of the procedure. This is done by applying the PD or the Wheeler algorithm for the first internal coordinate, $(m_{0,0,\dots,0}, m_{1,0,\dots,0}, \dots, m_{2N_1-2,0,\dots,0}, m_{2N_1-1,0,\dots,0}) \rightarrow (w_1^1, w_2^1, \dots, w_{N_1-1}^1, w_{N_1}^1; \xi_{1,1}^*, \xi_{1,2}^*, \dots, \xi_{1,N_1-1}^*, \xi_{1,N_1}^*)$. The procedure is then repeated for the second internal coordinate, $(m_{0,0,\dots,0},$

$m_{0,1,\ldots,0}, \ldots, m_{0,2N_2-2,\ldots,0}, m_{0,2N_2-1,\ldots,0}) \rightarrow (w_1^2, w_2^2, \ldots, w_{N_2-1}^2, w_{N_2}^2; \xi_{2,1}^*, \xi_{2,2}^*, \ldots, \xi_{2,N_2-1}^*, \xi_{2,N_2}^*)$, and for all the other internal coordinates, including the Mth: $(m_{0,0,\ldots,0}, m_{0,0,\ldots,1}, \ldots, m_{0,0,\ldots,2N_M-2}, m_{0,0,\ldots,2N_M-1}) \rightarrow (w_1^M, w_2^M, \ldots, w_{N_M-1}^M, w_{N_M}^M; \xi_{M,1}^*, \xi_{M,2}^*, \ldots, \xi_{M,N_M-1}^*, \xi_{M,N_M}^*)$. The construction of the N-point quadrature approximation requires the definition of the following abscissa basis vectors:

$$\begin{aligned}
\boldsymbol{\delta}^1 &\equiv \left[\delta\left(\xi_1 - \xi_{1,1}^*\right), \delta\left(\xi_1 - \xi_{1,2}^*\right), \ldots, \delta\left(\xi_1 - \xi_{1,N_1-1}^*\right), \delta\left(\xi_1 - \xi_{1,N_1}^*\right)\right],\\
\boldsymbol{\delta}^2 &\equiv \left[\delta\left(\xi_2 - \xi_{2,1}^*\right), \delta\left(\xi_2 - \xi_{2,2}^*\right), \ldots, \delta\left(\xi_2 - \xi_{2,N_2-1}^*\right), \delta\left(\xi_2 - \xi_{2,N_2}^*\right)\right],\\
&\vdots\\
\boldsymbol{\delta}^M &\equiv \left[\delta\left(\xi_M - \xi_{M,1}^*\right), \delta\left(\xi_M - \xi_{M,2}^*\right), \ldots, \delta\left(\xi_M - \xi_{M,N_M-1}^*\right), \delta\left(\xi_M - \xi_{M,N_M}^*\right)\right].
\end{aligned}$$

There are M different vectors composed of $N_1, N_2, \ldots, N_M$ components, with Dirac delta functions centered on the univariate abscissas (in each of the M directions) corresponding to the basis function set used to approximate the functional form of the multivariate NDF.

The final set of N multivariate abscissas is obtained by using the following tensor product: $\boldsymbol{\Delta} = \left[\boldsymbol{\delta}^1 \otimes \boldsymbol{\delta}^2 \otimes \ldots \otimes \boldsymbol{\delta}^M\right]$. This results in a vector with $N = N_1 N_2 \ldots N_M$ components corresponding to all the possible combinations of abscissas for the M internal coordinates, namely $\boldsymbol{\Delta} = [\delta(\xi_1 - \xi_{1,1}^*)\delta(\xi_2 - \xi_{2,1}^*)\ldots\delta(\xi_M - \xi_{M,1}^*), \delta(\xi_1 - \xi_{1,2}^*)\delta(\xi_2 - \xi_{2,1}^*)\ldots\delta(\xi_M - \xi_{M,1}^*), \delta(\xi_1 - \xi_{1,1}^*)\delta(\xi_2 - \xi_{2,2}^*)\ldots\delta(\xi_M - \xi_{M,1}^*), \ldots, \delta(\xi_1 - \xi_{1,N_1}^*)\delta(\xi_2 - \xi_{2,N_2}^*)\ldots\delta(\xi_M - \xi_{M,N_M}^*)]$, or, with a notation consistent with Eq. (3.36),

$$\begin{aligned}
\boldsymbol{\xi}_1 &= [\xi_{1,1}^*, \xi_{2,1}^*, \ldots, \xi_{M,1}^*],\\
\boldsymbol{\xi}_2 &= [\xi_{1,2}^*, \xi_{2,1}^*, \ldots, \xi_{M,1}^*],\\
\boldsymbol{\xi}_3 &= [\xi_{1,1}^*, \xi_{2,2}^*, \ldots, \xi_{M,1}^*],\\
&\vdots\\
\boldsymbol{\xi}_N &= [\xi_{1,N_1}^*, \xi_{2,N_2}^*, \xi_{M,N_M}^*],
\end{aligned}$$

where $N = N_1 N_2 \ldots N_M$. If the final (unknown) weight vector of the N-point quadrature is defined as $\mathbf{w} \equiv (w_1, w_2, \ldots, w_{N-1}, w_N)$, the final quadrature approximation (written in terms of the assumption for the functional form of the NDF as in Eq. (3.37)) results from the scalar product of these two vectors (i.e. $\mathbf{w} \cdot \boldsymbol{\Delta}$). The unknown weight vector is determined by solving N linear equations, ensuring that the quadrature approximation correctly represents N mixed moments. The first $N_1 + (N_2 - 1) + \cdots + (N_M - 1)$ equations are determined by forcing agreement with the M univariate quadratures,[8] whereas the remaining equations can be determined by fixing some mixed moments. No guarantee concerning the signs of the resulting weights is given, and, depending on the choice, some of the weights might turn out to be negative. In general, the remaining equations are determined by selecting the lowest-order mixed moments as illustrated in the next examples. In order to more clearly illustrate the potentials and limitations of this algorithm a few examples are discussed.

Exercise 3.6 *Consider a bivariate distribution ($M = 2$) with two internal coordinates ξ_1 and ξ_2, and let us construct a four-point quadrature approximation, resulting from univariate quadratures of order $N_1 = N_2 = 2$. Knowledge of the first, $2N_1 = 2N_2 = 4$, pure moments with respect to the first, ξ_1, and second, ξ_2, internal coordinates, suffices for*

[8] As is self-evident, $N_1 + N_2 + \cdots + N_M$ equations can be written by forcing agreement with the univariate quadratures, but only $N_1 + (N_2 - 1) + \cdots + (N_M - 1)$ are linearly independent.

the calculation of the initial univariate quadratures. In fact, using the PD (or Wheeler) algorithm, the following calculations can be easily carried out:

$$(m_{0,0}, m_{1,0}, m_{2,0}, m_{3,0}) \rightarrow \left(w_1^1, w_2^1; \xi_{1,1}^*, \xi_{1,2}^*\right),$$
$$(m_{0,0}, m_{0,1}, m_{0,2}, m_{0,3}) \rightarrow \left(w_1^2, w_2^2; \xi_{2,1}^*, \xi_{2,2}^*\right).$$

Using tensor-product QMOM, the final four-point quadrature approximation is centered on the following bivariate nodes:

$$\begin{aligned} \boldsymbol{\xi}_1 &= \left[\xi_{1,1}^*, \xi_{2,1}^*\right], \\ \boldsymbol{\xi}_2 &= \left[\xi_{1,1}^*, \xi_{2,2}^*\right], \\ \boldsymbol{\xi}_3 &= \left[\xi_{1,2}^*, \xi_{2,1}^*\right], \\ \boldsymbol{\xi}_4 &= \left[\xi_{1,2}^*, \xi_{2,2}^*\right], \end{aligned} \tag{3.45}$$

corresponding to the following assumption on the NDF:

$$n(\boldsymbol{\xi}) = \sum_{\alpha=1}^{4} w_\alpha \delta\left(\boldsymbol{\xi} - \boldsymbol{\xi}_\alpha\right), \tag{3.46}$$

where $\boldsymbol{\xi}_\alpha$ with $\alpha = 1, \dots, 4$ are defined in Eqs. (3.45), whereas $\mathbf{w} = [w_1, w_2, w_3, w_4]$ can be calculated by solving a linear system. This linear system is obtained by forcing agreement with the univariate distributions. Because the two univariate quadratures are constructed on eight pure moments (four with respect to ξ_1 and another four with respect to ξ_2), and because $m_{0,0}$ is identical for the two moment sets, theoretically the following seven equations could be written:

$$\begin{aligned} m_{0,0} &= w_1^1 + w_2^1 = w_1^2 + w_2^2 = w_1 + w_2 + w_3 + w_4, \\ m_{1,0} &= w_1^1\xi_{1,1}^* + w_2^1\xi_{1,2}^* = w_1\xi_{1,1}^* + w_2\xi_{1,1}^* + w_3\xi_{1,1}^* + w_4\xi_{1,1}^*, \\ m_{2,0} &= w_1^1(\xi_{1,1}^*)^2 + w_2^1(\xi_{1,2}^*)^2 = w_1(\xi_{1,1}^*)^2 + w_2(\xi_{1,1}^*)^2 + w_3(\xi_{1,1}^*)^2 + w_4(\xi_{1,1}^*)^2, \\ m_{3,0} &= w_1^1(\xi_{1,1}^*)^3 + w_2^1(\xi_{1,2}^*)^3 = w_1(\xi_{1,1}^*)^3 + w_2(\xi_{1,1}^*)^3 + w_3(\xi_{1,1}^*)^3 + w_4(\xi_{1,1}^*)^3, \\ m_{0,1} &= w_1^2\xi_{2,1}^* + w_2^2\xi_{2,2}^* = w_1\xi_{2,1}^* + w_2\xi_{2,2}^* + w_3\xi_{2,1}^* + w_4\xi_{2,2}^*, \\ m_{0,2} &= w_1^2(\xi_{2,1}^*)^2 + w_2^2(\xi_{2,2}^*)^2 = w_1(\xi_{2,1}^*)^2 + w_2(\xi_{2,2}^*)^2 + w_3(\xi_{2,1}^*)^2 + w_4(\xi_{2,2}^*)^2, \\ m_{0,3} &= w_1^2(\xi_{2,1}^*)^3 + w_2^2(\xi_{2,2}^*)^3 = w_1(\xi_{2,1}^*)^3 + w_2(\xi_{2,2}^*)^3 + w_3(\xi_{2,1}^*)^3 + w_4(\xi_{2,2}^*)^3. \end{aligned} \tag{3.47}$$

However, of these seven, only three are linearly independent (i.e. $N_1 + N_2 - 1 = 3$). It is also interesting to highlight that an alternative way to write these three equations is to force that the summation of the final weights is equal to the original univariate weight in the corresponding direction. For example, the weight w_1 lies on the same line as w_2 (meaning that they all share the same abscissa $\xi_{1,1}^*$),

$$w_1 + w_2 = w_1^1, \tag{3.48}$$

and correspondingly for $\xi_{1,2}^*$,

$$w_3 + w_4 = w_2^1, \tag{3.49}$$

Table 3.8. The moment set used to build a bivariate quadrature approximation ($M = 2$) for $N = 4$ with the tensor-product QMOM

$m(3,0)$			
$m(2,0)$			
$m(1,0)$	$m(1,1)$		
$m(0,0)$	$m(0,1)$	$m(0,2)$	$m(0,3)$

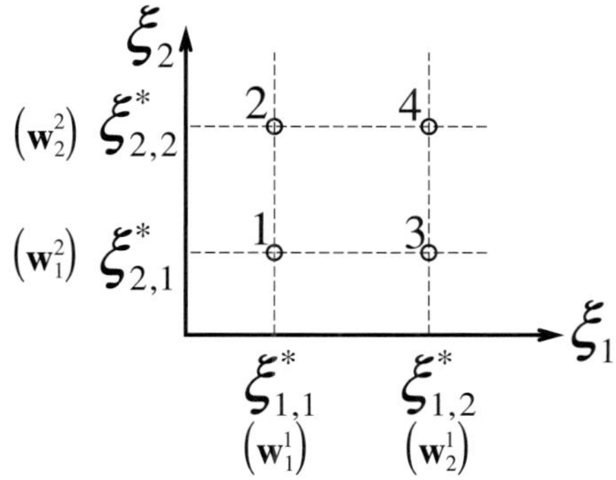

Figure 3.1. Positions in the internal-coordinate plane of the four nodes of the bivariate tensor-product QMOM ($M = 2$) obtained with two-point univariate quadratures ($N_1 = N_2 = 2$).

for the first internal coordinate both equations can be written (i.e. $N_1 = 2$), whereas for the second only one (i.e. $N_2 - 1 = 1$) can be written, resulting in

$$w_1 + w_3 = w_1^2, \tag{3.50}$$

where of course the other is a linear combination of the previous three. The derivation of Eqs. (3.48)–(3.50) can be intuitively understood through the observation of the position of the four bivariate nodes in the phase space (see Figure 3.1). The final missing equation, to solve the linear system of rank four, comes from the enforcement of one extra moment. A usual choice is to enforce one mixed moment, producing an additional equation so that the final linear system is non-singular. If, for example, $m_{1,1}$ is chosen, the following equation is obtained:

$$w_1\xi_{11}\xi_{21} + w_2\xi_{11}\xi_{22} + w_3\xi_{12}\xi_{21} + w_4\xi_{12}\xi_{22} = m_{1,1}. \tag{3.51}$$

The system formed by Eqs. (3.48)–(3.51) can be solved to find the weights. It can be seen that the four-point bivariate quadrature approximation accommodates eight moments, namely the moment of order zero with respect to both the internal coordinates, six pure moments, and one mixed moment. Table 3.8 reports in matrix form these eight moments, and comparison with Table 3.5 clearly shows that this is a subset of the optimal moment set.

Since for velocity distributions (in three spatial dimensions) three internal coordinates ($M = 3$) are needed, we discuss in the following example the construction of a quadrature approximation for a trivariate tensor-product QMOM.

Exercise 3.7 *Consider a trivariate distribution ($M = 3$) of three internal coordinates ξ_1, ξ_2, and ξ_3. Let us construct an eight-point tensor-product QMOM resulting from univariate* *quadratures of order* $N_1 = N_2 = N_3 = 2$. *As in the previous example, the three initial univariate quadratures are built from the first* $2N_1 = 2N_2 = 2N_3 = 4$ *pure moments with respect to three internal coordinates, by using the PD (or Wheeler) algorithm:*

$$\begin{aligned}(m_{0,0,0}, m_{1,0,0}, m_{2,0,0}, m_{3,0,0}) &\rightarrow \left(w_1^1, w_2^1; \xi_{1,1}^*, \xi_{1,2}^*\right),\\ (m_{0,0,0}, m_{0,1,0}, m_{0,2,0}, m_{0,3,0}) &\rightarrow \left(w_1^2, w_2^2; \xi_{2,1}^*, \xi_{2,2}^*\right),\\ (m_{0,0,0}, m_{0,0,1}, m_{0,0,2}, m_{0,0,3}) &\rightarrow \left(w_1^3, w_2^3; \xi_{3,1}^*, \xi_{3,2}^*\right).\end{aligned}$$

Using the tensor-product QMOM, the final eight-point quadrature approximation is centered on the following trivariate nodes:

$$\begin{aligned}\boldsymbol{\xi}_1 &= \left[\xi_{1,1}^*, \xi_{2,1}^*, \xi_{3,1}^*\right],\\ \boldsymbol{\xi}_2 &= \left[\xi_{1,2}^*, \xi_{2,1}^*, \xi_{3,1}^*\right],\\ \boldsymbol{\xi}_3 &= \left[\xi_{1,1}^*, \xi_{2,2}^*, \xi_{3,1}^*\right],\\ \boldsymbol{\xi}_4 &= \left[\xi_{1,2}^*, \xi_{2,2}^*, \xi_{3,1}^*\right],\\ \boldsymbol{\xi}_5 &= \left[\xi_{1,1}^*, \xi_{2,1}^*, \xi_{3,2}^*\right],\\ \boldsymbol{\xi}_6 &= \left[\xi_{1,2}^*, \xi_{2,1}^*, \xi_{3,2}^*\right],\\ \boldsymbol{\xi}_7 &= \left[\xi_{1,1}^*, \xi_{2,2}^*, \xi_{3,2}^*\right],\\ \boldsymbol{\xi}_8 &= \left[\xi_{1,2}^*, \xi_{2,2}^*, \xi_{3,2}^*\right].\end{aligned} \tag{3.52}$$

The weights **w** *can be calculated by solving a linear system, where the first* $N_1 + (N_2 - 1) + (N_3 - 1) = 4$ *equations are obtained by forcing agreement with the univariate quadratures:*

$$\begin{aligned}w_1 + w_3 + w_5 + w_7 &= w_1^1,\\ w_2 + w_4 + w_6 + w_8 &= w_2^1,\\ w_1 + w_2 + w_5 + w_6 &= w_1^2,\\ w_1 + w_2 + w_3 + w_4 &= w_1^3.\end{aligned} \tag{3.53}$$

As in Exercise 3.6, these four equations can easily be found by the observation of the position of the eight nodes in the phase space of the three internal coordinates (see Figure 3.2). The additional four equations are obtained by fixing some mixed moments (i.e. below we show the equations for the most natural choice of fixing the lowest-order moments possible: $m_{1,1,0}$, $m_{1,0,1}$, $m_{0,1,1}$, *and* $m_{1,1,1}$*) resulting in*

$$\begin{aligned}w_1\xi_{1,1}\xi_{2,1} + w_2\xi_{1,2}\xi_{2,1} + w_3\xi_{1,1}\xi_{2,2} + w_4\xi_{1,2}\xi_{2,2} + w_5\xi_{1,1}\xi_{2,1}&\\ + w_6\xi_{1,2}\xi_{2,1} + w_7\xi_{1,1}\xi_{2,2} + w_8\xi_{1,2}\xi_{2,2} &= m_{1,1,0},\end{aligned} \tag{3.54}$$

$$\begin{aligned}w_1\xi_{1,1}\xi_{3,1} + w_2\xi_{1,2}\xi_{3,1} + w_3\xi_{1,1}\xi_{3,1} + w_4\xi_{1,2}\xi_{3,1} + w_5\xi_{1,1}\xi_{3,2}&\\ + w_6\xi_{1,2}\xi_{3,2} + w_7\xi_{1,1}\xi_{3,2} + w_8\xi_{1,2}\xi_{3,2} &= m_{1,0,1},\end{aligned} \tag{3.55}$$

$$\begin{aligned}w_1\xi_{2,1}\xi_{3,1} + w_2\xi_{2,1}\xi_{3,1} + w_3\xi_{2,2}\xi_{3,1} + w_4\xi_{2,2}\xi_{3,1} + w_5\xi_{2,1}\xi_{3,2}&\\ + w_6\xi_{2,1}\xi_{3,2} + w_7\xi_{2,2}\xi_{3,2} + w_8\xi_{2,2}\xi_{3,2} &= m_{0,1,1},\end{aligned} \tag{3.56}$$

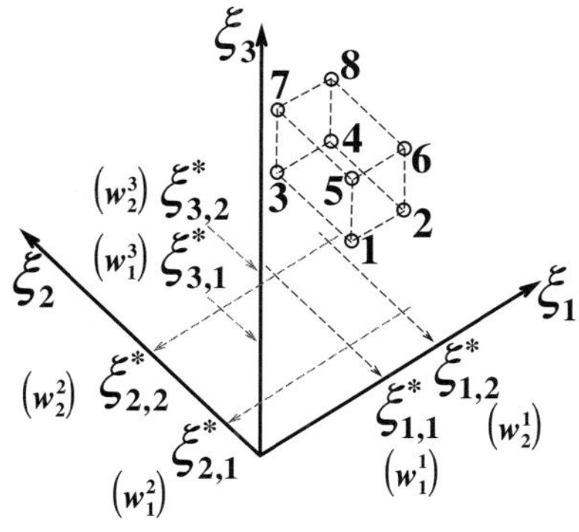

Figure 3.2. Positions in phase space of the eight nodes of the trivariate tensor-product QMOM (i.e. $M = 3$) obtained with two-point univariate quadratures ($N_1 = N_2 = N_3 = 2$).

$$w_1\xi_{1,1}\xi_{2,1}\xi_{3,1} + w_2\xi_{1,2}\xi_{2,1}\xi_{3,1} + w_3\xi_{1,1}\xi_{2,2}\xi_{3,1} + w_4\xi_{1,2}\xi_{2,2}\xi_{3,1} + w_5\xi_{1,1}\xi_{2,1}\xi_{3,2} + w_6\xi_{1,2}\xi_{2,1}\xi_{3,2} + w_7\xi_{1,1}\xi_{2,2}\xi_{3,2} + w_8\xi_{1,2}\xi_{2,2}\xi_{3,2} = m_{1,1,1}. \quad (3.57)$$

This eight-point tensor-product QMOM fixes the moment of order zero with respect to all the internal coordinates, nine pure moments plus four mixed moments, for a total of 14 moments. The highest-order moment accommodated by this particular choice is $m_{1,1,1}$ of global order $\gamma = 3$. Comparison of this moment set with the optimal moment set reported in Table 3.7 for $M = 3$ and $N = 8$ clearly shows that is a subset of the 32 optimal moments.

In summary, the moment-inversion algorithm for a tensor-product QMOM allows us to build (without an iterative procedure as in the case of brute-force QMOM) an M-dimensional, N-point quadrature approximation starting from M univariate quadratures of order $N_1, N_2, \ldots, N_M$ in each of the M directions. As has already been reported, the final number of quadrature nodes is equal to $N_1N_2 \ldots N_M$. The total number of moments controlled (and therefore adopted) by the moment-inversion algorithm is equal to $N_1N_2 \ldots N_M + N_1 + N_2 + \cdots + N_M$. The N nodes are determined by applying the PD (or Wheeler) algorithm M times for each internal coordinate starting from the first $2N_\beta$ with $1 \leq \beta \leq M$ pure moments with respect to the different internal coordinates) and then applying the tensor-product QMOM. The N weights are calculated by solving a linear system. The N linear equations are obtained by forcing agreement with the original univariate quadratures, resulting in $N_1 + (N_2 - 1) \ldots (N_M - 1)$ linearly independent equations. These equations can be found by directly enforcing agreement or by adopting the following procedure. N_1 equations are written for the first internal coordinate and then $N_2 - 1$, $N_3 - 1, \ldots, N_M - 1$ are written in the other remaining directions. As reported in the previous exercises, this is done by forcing the summation of the final weights along one internal coordinate to be equal to the original weight (obtained from the univariate quadratures) in the corresponding direction. The other remaining equations are determined by fixing mixed moments. As has already been mentioned, the selection of the mixed moments cannot be done arbitrarily, because, as illustrated in the examples, for a particular mixed moment the resulting equation must be linearly independent. In general, the choice of mixed moments with indices less than or equal to $N_\beta - 1$ with $1 \leq \beta \leq M$ ensures that the linear system is non-singular and appears to be unique. Implementation for values of N_β larger than the ones discussed in this section is theoretically possible and no specific problems are expected.

To conclude this section on tensor-product QMOM, it is important to highlight that, although it is not necessary, the formulation of the problem in terms of translated (i.e. centered on the mean) and rotated (i.e. with diagonal covariance matrix) internal coordinates can be advantageous. In fact, if a change of variables is implemented[9] so that the distribution is rewritten with respect to its principal coordinates, the calculations for the derivation of the quadrature approximation are simplified. These concepts will be illustrated in Exercise 3.8. A MATLAB script implementing a tensor-product QMOM can be found in Section A.3.2 of Appendix A.

3.2.3 Conditional QMOM

The conditional quadrature method of moments (CQMOM) is based on the concept of a conditional density function (Yuan & Fox, 2011). Conditional density functions represent, in turn, the probability of having one internal coordinate within an infinitesimal limit when one or more of the other internal coordinates are fixed and equal to specific values. For example, in the case of a generic NDF the expression

$$f(\xi_M|\xi_1,\xi_2,\ldots,\xi_{M-2},\xi_{M-1}) = \frac{n(\boldsymbol{\xi})}{f(\xi_1,\xi_2,\ldots,\xi_{M-2},\xi_{M-1})} \tag{3.58}$$

defines the probability of having the Mth internal coordinate within the range from ξ_M to $\xi_M + \mathrm{d}\xi_M$ when all the other internal coordinates (ξ_β with $\beta = 1,\ldots,M-1$) assume fixed constant values. By definition, the conditional density function integrates to unity and the denominator of Eq. (3.58) is defined by integrating out the Mth internal coordinate:

$$f(\xi_1,\xi_2,\ldots,\xi_{M-2},\xi_{M-1}) = \int_{\Omega_{\xi_M}} n(\boldsymbol{\xi})\mathrm{d}\xi_M. \tag{3.59}$$

Analogously we can define the following conditional density function:

$$f(\xi_{M-1}|\xi_1,\xi_2,\ldots,\xi_{M-3},\xi_{M-2}) = \frac{f(\xi_1,\xi_2,\ldots,\xi_{M-2},\xi_{M-1})}{f(\xi_1,\xi_2,\ldots,\xi_{M-3},\xi_{M-2})}, \tag{3.60}$$

defining the probability of having ξ_{M-1} within the infinitesimal interval while ξ_β (with $\beta = 1,\ldots,M-2$) assume constant and specific values. By constructing conditional density functions that are based on smaller and smaller subsets of internal coordinates, the following identity can be obtained:

$$\begin{aligned} n(\boldsymbol{\xi}) = {}& f(\xi_M|\xi_1,\xi_2,\ldots,\xi_{M-2},\xi_{M-1})f(\xi_{M-1}|\xi_1,\xi_2,\ldots,\xi_{M-3},\xi_{M-2}) \\ & \times\cdots\times f(\xi_4|\xi_1,\xi_2,\xi_3)f(\xi_3|\xi_1,\xi_2)f(\xi_2|\xi_1)f(\xi_1). \end{aligned} \tag{3.61}$$

Then it is immediately evident that the following equality holds:

$$m_0 = m(\mathbf{0}) = \int_{\Omega_{\xi_1}} f(\xi_1)\mathrm{d}\xi_1. \tag{3.62}$$

These concepts can be used to construct a multivariate quadrature.

---

[9] Generally the change of variables consists of translating with respect to the mean and rotating into a stress-free (diagonal) coordinate system. This translation ensures that the quadrature remains Galilean invariant and realizable (i.e. nonnegative weights).

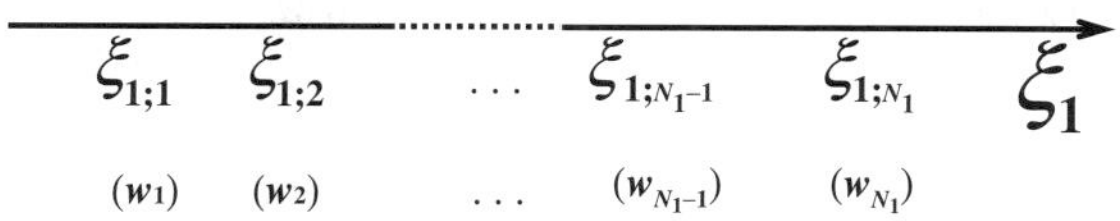

Figure 3.3. The first step in the construction of the multivariate quadrature in the CQMOM. Determination of the N_1 nodes and N_1 weights for the first internal coordinate ξ_1.

The construction of the multivariate quadrature begins with the calculation of the univariate quadrature of order N_1 for the first internal coordinate by using the Wheeler algorithm with the first $2N_1 - 1$ moments:

$$\begin{pmatrix} m_{0,0,\ldots,0,0} \\ m_{1,0,\ldots,0,0} \\ \vdots \\ m_{2N_1-2,0,\ldots,0,0} \\ m_{2N_1-1,0,\ldots,0,0} \end{pmatrix} \xrightarrow{\text{Wheeler}} \begin{pmatrix} w_1 \\ w_2 \\ \vdots \\ w_{N_1-1} \\ w_{N_1} \end{pmatrix} \begin{pmatrix} \xi_{1;1} \\ \xi_{1;1} \\ \vdots \\ \xi_{1;N_1-1} \\ \xi_{1;N_1} \end{pmatrix}.$$

The N_1 weights, w_{α_1}, are identified by a single index, whereas the N_1 nodes, $\xi_{1;\alpha_1}$, are identified by two indices, of which the first one indicates the internal coordinate and $\alpha_1 = 1, \ldots, N_1$. This first step is illustrated in Figure 3.3 and, recalling the equivalence of using quadrature approximations and reconstructing NDFs, the following expression can be obtained:

$$f(\xi_1) = \sum_{\alpha_1=1}^{N_1} w_{\alpha_1} \delta\left(\xi_1 - \xi_{1;\alpha_1}\right). \tag{3.63}$$

By using the definition reported in Eq. (3.61) and exploiting the univariate quadrature over the first internal coordinate, the generic mixed moment of order $\mathbf{k} = [k_1, k_2, \ldots, k_{M-1}, k_M]$ can be calculated as

$$\begin{aligned} m_{k_1,k_2,\ldots,k_{M-1},k_M} = {} & \sum_{\alpha_1=1}^{N_1} w_{\alpha_1} \xi_{1;\alpha_1}^{k_1} \\ & \times \int_{\Omega_{\xi_M}} f(\xi_M | \xi_{1;\alpha_1}, \xi_2, \ldots, \xi_{M-1}) f(\xi_{M-1} | \xi_{1;\alpha_1}, \xi_2, \ldots, \xi_{M-2}) f(\xi_2 | \xi_{1;\alpha_1}) \\ & \times \xi_2^{k_2} \xi_3^{k_3} \ldots \xi_{M-1}^{k_{M-1}} \xi_M^{k_M} \, \mathrm{d}\xi_2 \ldots \mathrm{d}\xi_{M-1} \, \mathrm{d}\xi_M. \end{aligned} \tag{3.64}$$

When $\mathbf{k} = \mathbf{0}$ the quadrature exactly reproduces the moment of order zero with respect to all internal coordinates:

$$m_{0,0,\ldots,0,0} = m(\mathbf{0}) = \sum_{\alpha_1=1}^{N_1} w_{\alpha_1}. \tag{3.65}$$

Likewise, when the mixed moments of order k_1 with respect to the first internal coordinate and k_2 with respect to the second internal are taken, the following equality is obtained:

$$m_{k_1,k_2,0,\ldots,0,0} = \sum_{\alpha_1=1}^{N_1} w_{\alpha_1} \xi_{1;\alpha_1}^{k_1} \int_{\Omega_{\xi_2}} f(\xi_2 | \xi_{1;\alpha_1}) \xi_2^{k_2} \, \mathrm{d}\xi_2, \tag{3.66}$$

since all the other conditional density functions integrate to unity. Equation (3.66) contains the definition of the conditional moment of order k_2 with respect to the second internal coordinate conditioned on the value of the first internal coordinate (i.e. $\xi_1 = \xi_{1;\alpha_1}$),

$$\left\langle \xi_2^{k_2} \right\rangle (\xi_{1;\alpha_1}) = \int_{\Omega_{\xi_2}} f(\xi_2|\xi_{1;\alpha_1}) \xi_2^{k_2} \, d\xi_2, \tag{3.67}$$

which for brevity will be denoted by $\langle \xi_2^{k_2} \rangle_{\alpha_1}$.

For each of the N_1 nodes $\xi_{1;\alpha_1}$, knowledge of the first $2N_2$ conditional moments suffices to calculate N_1 different quadratures by applying the Wheeler algorithm:

$$\begin{pmatrix} 1 \\ \langle \xi_2 \rangle_{\alpha_1} \\ \vdots \\ \left\langle \xi_2^{2N_2-2} \right\rangle_{\alpha_1} \\ \left\langle \xi_2^{2N_2-1} \right\rangle_{\alpha_1} \end{pmatrix} \xrightarrow{\text{Wheeler}} \begin{pmatrix} w_{\alpha_1,1} \\ w_{\alpha_1,2} \\ \vdots \\ w_{\alpha_1,N_2-1} \\ w_{\alpha_1,N_2} \end{pmatrix} \begin{pmatrix} \xi_{2;\alpha_1,1} \\ \xi_{2;\alpha_1,1} \\ \vdots \\ \xi_{2;\alpha_1,N_2-1} \\ \xi_{2;\alpha_1,N_2} \end{pmatrix},$$

where $\alpha_1 = 1, \ldots, N_1$. The necessary conditional moments can be calculated by using Eq. (3.66) for $k_1 = 0, \ldots, N_1$ and for $k_2 = 0, \ldots, 2N_2 - 1$, resulting in the following $2N_2 - 1$ linear systems (of rank N_1) of Vandermonde form:

$$\mathbf{V}_1 \mathbf{R}_1 \begin{bmatrix} \left\langle \xi_2^{k_2} \right\rangle_1 \\ \left\langle \xi_2^{k_2} \right\rangle_2 \\ \vdots \\ \left\langle \xi_2^{k_2} \right\rangle_{N_1} \end{bmatrix} = \begin{bmatrix} m_{0,k_2,0,\ldots,0} \\ m_{1,k_2,0,\ldots,0} \\ \vdots \\ m_{N_1,k_2,0,\ldots,0} \end{bmatrix}, \tag{3.68}$$

where the coefficient matrices are defined by

$$\mathbf{V}_1 \equiv \begin{bmatrix} 1 & \ldots & 1 \\ \xi_{1;1} & \ldots & \xi_{1;N_1} \\ \vdots & \ddots & \vdots \\ (\xi_{1;1})^{N_1-1} & \ldots & (\xi_{1;N_1})^{N_1-1} \end{bmatrix} \tag{3.69}$$

and

$$\mathbf{R}_1 \equiv \begin{bmatrix} w_1 & & \\ & \ddots & \\ & & w_{N_1} \end{bmatrix}, \tag{3.70}$$

and $k_2 = 1, \ldots, 2N_2 - 1$. The Vandermonde linear system of Eq. (3.68) is non-singular as long as the nodes $\xi_{1;\alpha_1}$ are distinct, and can be solved by using the efficient algorithm proposed by Rybicki (Press *et al.*, 1992).

The final quadrature (for the first two internal coordinates) is sketched in Figure 3.4 and corresponds to the following assumption on the NDF:

$$f(\xi_1, \xi_2) = f(\xi_2|\xi_1) f(\xi_1) = \sum_{\alpha_1=1}^{N_1} \sum_{\alpha_2=1}^{N_2} w_{\alpha_1} w_{\alpha_1;\alpha_2} \delta(\xi_1 - \xi_{1;\alpha_1}) \delta(\xi_2 - \xi_{2;\alpha_1,\alpha_2}). \tag{3.71}$$

If now the mixed moments of order $\mathbf{k} = [k_1, k_2, k_3, 0, \ldots, 0]$ are calculated by using Eq. (3.61) and Eq. (3.71) the following expression is obtained:

$$m_{k_1,k_2,k_3,0,\ldots,0} = \sum_{\alpha_1=1}^{N_1} \sum_{\alpha_2=1}^{N_2} w_{\alpha_1} w_{\alpha_1;\alpha_2} (\xi_{1;\alpha_1})^{k_1} (\xi_{2;\alpha_1,\alpha_2})^{k_2} \int_{\Omega_{\xi_3}} f(\xi_3|\xi_{1;\alpha_1}, \xi_{2;\alpha_1,\alpha_2}) \xi_3^{k_3} \, d\xi_3, \tag{3.72}$$

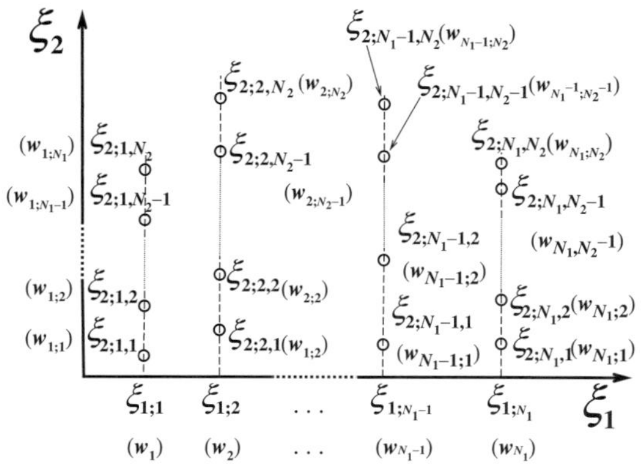

Figure 3.4. The second step in the construction of the multivariate quadrature with the CQMOM. Determination of the N_1N_2 nodes and N_1N_2 weights for the second internal coordinate ξ_2.

where again the definition of the conditional moment of order k_3 with respect to the third internal coordinate, conditioned on the values of the first two internal coordinates (i.e. $\xi_1 = \xi_{1;\alpha_1}$ and $\xi_2 = \xi_{2;\alpha_1,\alpha_2}$), is

$$\left\langle \xi_3^{k_3} \right\rangle (\xi_{1;\alpha_1}, \xi_{2;\alpha_1,\alpha_2}) = \int_{\Omega_{\xi_3}} f(\xi_3 | \xi_{1;\alpha_1}, \xi_{2;\alpha_1,\alpha_2}) \xi_3^{k_3} \, \mathrm{d}\xi_3, \tag{3.73}$$

where $\left\langle \xi_3^{k_3} \right\rangle (\xi_{1;\alpha_1}, \xi_{2;\alpha_1,\alpha_2})$ will be denoted by $\left\langle \xi_3^{k_3} \right\rangle_{\alpha_1,\alpha_2}$. If we now define the quantity

$$\zeta_{\alpha_1}^{k_2,k_3} = \sum_{\alpha_2=1}^{N_2} w_{\alpha_1,\alpha_2} (\xi_{2;\alpha_1,\alpha_2})^{k_2} \left\langle \xi_3^{k_3} \right\rangle_{\alpha_1,\alpha_2}, \tag{3.74}$$

Eq. (3.72) becomes

$$m_{k_1,k_2,k_3,0,\dots,0} = \sum_{\alpha_1=1}^{N_1} w_{\alpha_1} (\xi_{1;\alpha_1})^{k_1} \zeta_{\alpha_1}^{k_2,k_3}, \tag{3.75}$$

and for $k_1 = 0, \dots, N_1 - 1$ the following linear systems are generated:

$$\mathbf{V}_1 \mathbf{R}_1 \begin{bmatrix} \zeta_1^{k_2,k_3} \\ \zeta_2^{k_2,k_3} \\ \vdots \\ \zeta_{N_1-1}^{k_2,k_3} \\ \zeta_{N_1}^{k_2,k_3} \end{bmatrix} = \begin{bmatrix} m_{0,k_2,k_3,0,\dots,0} \\ m_{1,k_2,k_3,0,\dots,0} \\ \vdots \\ m_{N_1-2,k_2,k_3,0,\dots,0} \\ m_{N_1-1,k_2,k_3,0,\dots,0} \end{bmatrix} \tag{3.76}$$

for each pair of $k_2 = 0, \dots, N_2 - 1$ and $k_3 = 1, \dots, 2N_3 - 1$, where N_3 is the number of quadrature nodes employed. Note that these linear systems have the same coefficient matrices as Eq. (3.68), and thus we are guaranteed to find a solution for $\zeta_{\alpha_1}^{k_2,k_3}$.

For each value of α_1, the final step is to solve the linear system generated by Eq. (3.74), which assumes again the Vandermonde form:

$$\mathbf{V}_{2;\alpha_1}\mathbf{R}_{2;\alpha_1}\begin{bmatrix} \left\langle \xi_3^{k_3} \right\rangle_{\alpha_1,1} \\ \left\langle \xi_3^{k_3} \right\rangle_{\alpha_1,2} \\ \vdots \\ \left\langle \xi_3^{k_3} \right\rangle_{\alpha_1,N_2-1} \\ \left\langle \xi_3^{k_3} \right\rangle_{\alpha_1,N_2} \end{bmatrix} = \begin{bmatrix} \zeta_{\alpha_1}^{0,k_3} \\ \zeta_{\alpha_1}^{1,k_3} \\ \vdots \\ \zeta_{\alpha_1}^{N_2-2,k_3} \\ \zeta_{\alpha_1}^{N_2-1,k_3} \end{bmatrix} \tag{3.77}$$

where the coefficient matrices are defined by

$$\mathbf{V}_{2,\alpha_1} \equiv \begin{bmatrix} 1 & \cdots & 1 \\ \xi_{2;\alpha_1,1} & \cdots & \xi_{2;\alpha_1,N_2} \\ \vdots & \ddots & \vdots \\ (\xi_{2;\alpha_1,1})^{N_2-1} & \cdots & (\xi_{2;\alpha_1,N_2})^{N_2-1} \end{bmatrix} \tag{3.78}$$

and

$$\mathbf{R}_{2,\alpha_2} \equiv \begin{bmatrix} w_{\alpha_1,1} & & \\ & \ddots & \\ & & w_{\alpha_1,N_2} \end{bmatrix}. \tag{3.79}$$

In summary, for each pair of $k_2 = 0, \ldots, N_2 - 1$ and $k_3 = 1, \ldots, 2N_3 - 1$, the quantities $\zeta_{\alpha_1}^{k_2,k_3}$ are calculated by solving (with the Rybicki algorithm) the linear system reported in Eq. (3.76). Subsequently these quantities are used to solve for each value of $\alpha_1 = 1, \ldots, N_1$ the linear system reported in Eq. (3.77). Finally, the quadrature nodes for the third internal coordinate are calculated by applying the Wheeler algorithm to the first $2N_3$ conditional moments with respect to the third internal coordinate:

$$\begin{pmatrix} 1 \\ \langle \xi_3 \rangle_{\alpha_1,\alpha_2} \\ \vdots \\ \left\langle \xi_3^{2N_3-2} \right\rangle_{\alpha_1,\alpha_2} \\ \left\langle \xi_3^{2N_3-1} \right\rangle_{\alpha_1,\alpha_2} \end{pmatrix} \xrightarrow{\text{Wheeler}} \begin{pmatrix} w_{\alpha_1,\alpha_2,1} \\ w_{\alpha_1,\alpha_2,2} \\ \vdots \\ w_{\alpha_1,\alpha_2,N_3-1} \\ w_{\alpha_1,\alpha_2,N_3} \end{pmatrix} \begin{pmatrix} \xi_{3;\alpha_1,\alpha_2,1} \\ \xi_{3;\alpha_1,\alpha_2,2} \\ \vdots \\ \xi_{3;\alpha_1,\alpha_2,N_3-1} \\ \xi_{3;\alpha_1,\alpha_2,N_3} \end{pmatrix}.$$

Note that for the third internal coordinate the quadrature nodes are calculated from each of the N_1N_2 nodes of Figure 3.4, along vertical lines perpendicular to the plane generated in phase space by ξ_1 and ξ_2, resulting in $N_1N_2N_3$ nodes and $N_1N_2N_3$ weights. This corresponds to the following functional assumption for the NDF:

$$\begin{aligned} f(\xi_1, \xi_2, \xi_3) &= f(\xi_3|\xi_1, \xi_2) f(\xi_2|\xi_1) f(\xi_1) \\ &= \sum_{\alpha_1=1}^{N_1} \sum_{\alpha_2=1}^{N_2} \sum_{\alpha_3=1}^{N_3} w_{\alpha_1} w_{\alpha_1;\alpha_2} w_{\alpha_1;\alpha_2;\alpha_3} \delta\left(\xi_1 - \xi_{1;\alpha_1}\right) \delta\left(\xi_2 - \xi_{2;\alpha_1,\alpha_2}\right) \delta\left(\xi_3 - \xi_{3;\alpha_1,\alpha_2,\alpha_3}\right). \end{aligned} \tag{3.80}$$

By following the same procedure, quadrature nodes can be calculated for the other internal coordinates.

Table 3.9. Moments used to build a bivariate CQMOM quadrature approximation ($M = 2$) for $N_1 = N_2 = 3$ with ξ_2 conditioned on ξ_1

$m(5, 0)$					
$m(4, 0)$					
$m(3, 0)$					
$m(2, 0)$	$m(2, 1)$	$m(2, 2)$	$m(2, 3)$	$m(2, 4)$	$m(2, 5)$
$m(1, 0)$	$m(1, 1)$	$m(1, 2)$	$m(1, 3)$	$m(1, 4)$	$m(1, 5)$
$m(0, 0)$	$m(0, 1)$	$m(0, 2)$	$m(0, 3)$	$m(0, 4)$	$m(0, 5)$

Table 3.10. Moments used to build a bivariate CQMOM quadrature approximation ($M = 2$) for $N_1 = N_2 = 3$ with ξ_1 conditioned on ξ_2

$m(5, 0)$	$m(5, 1)$	$m(5, 2)$			
$m(4, 0)$	$m(4, 1)$	$m(4, 2)$			
$m(3, 0)$	$m(3, 1)$	$m(3, 2)$			
$m(2, 0)$	$m(2, 1)$	$m(2, 2)$			
$m(1, 0)$	$m(1, 1)$	$m(1, 2)$			
$m(0, 0)$	$m(0, 1)$	$m(0, 2)$	$m(0, 3)$	$m(0, 4)$	$m(0, 5)$

Consider, as an example, a simple bivariate case ($M = 2$), and let us identify the moments used to build the quadrature approximation. If N_1 is taken equal to three, the first six moments with respect to ξ_1 will be used to build the initial univariate quadrature approximation, namely $m(0,0)$, $m(1,0)$, $m(2,0)$, $m(3,0)$, $m(4,0)$, and $m(5,0)$. If we then take also N_2 equal to three, the quadrature approximation on ξ_2 has to be built for each of the three nodes on ξ_1. The moments used to calculate the conditional moments, and to determine the three monovariate quadrature approximations, are $m(0,1)$, $m(1,1)$, $m(2,1)$, $m(0,2)$, $m(1,2)$, $m(2,2)$, $m(0,3)$, $m(1,3)$, $m(2,3)$, $m(0,4)$, $m(1,4)$, $m(2,4)$, $m(0,5)$, $m(1,5)$, and finally $m(2,5)$. The total number of moments employed (for this bivariate case) is equal to 21, as illustrated in Table 3.9. Comparison of Table 3.9 with Table 3.6 shows that these 21 moments are a subset of the 27 optimal moments. It is immediately evident that if the procedure is inverted and the first internal coordinate is conditioned on the second one the 21 moments reported in Table 3.10 will have to be used. As is clear on comparing Table 3.9 and Table 3.10, the moment set obtained by the union of the two sets corresponds to the optimal moment set for $N = 9$. By considering the two possible alternatives, the highest-order moment controlled is of global order 7 (i.e. $m(2,5)$ or $m(5,2)$), whereas the highest pure moments controlled are $m(5,0)$ and $m(5,0)$ of global order 5.

If, for the same bivariate case, N_1 and N_2 are now taken equal to four, more moments can be controlled. For the construction of the first monovariate quadrature with respect to ξ_1, the first $2N_1$ pure moments are used, namely $m(0,0)$, $m(1,0)$, $m(2,0)$, $m(3,0)$, $m(4,0)$, $m(5,0)$, $m(6,0)$, and $m(7,0)$. Then the conditional moments and the resulting N_1 univariate quadratures (of order N_2) are calculated by employing the following moments: $m(0,1)$, $m(0,2)$, $m(0,3)$, $m(0,4)$, $m(0,5)$, $m(0,6)$, $m(0,7)$, and $m(1,1)$, $m(1,2)$, $m(1,3)$, $m(1,4)$, $m(1,5)$, $m(1,6)$, $m(1,7)$, and $m(2,1)$, $m(2,2)$, $m(2,3)$, $m(2,4)$, $m(2,5)$, $m(2,6)$, $m(2,7)$, and finally $m(3,1)$, $m(3,2)$, $m(3,3)$, $m(3,4)$, $m(3,5)$, $m(3,6)$, $m(3,7)$. The moments used in this case are reported in Table 3.11. In this case, 36 moments are controlled, including

Table 3.11. Moments used to build a bivariate CQMOM quadrature approximation ($M = 2$) for $N_1 = N_2 = 4$ with ξ_2 conditioned on ξ_1

$m(7, 0)$							
$m(6, 0)$							
$m(5, 0)$							
$m(4, 0)$							
$m(3, 0)$	$m(3, 1)$	$m(3, 2)$	$m(3, 3)$	$m(3, 4)$	$m(3, 5)$	$m(3, 6)$	$m(3, 7)$
$m(2, 0)$	$m(2, 1)$	$m(2, 2)$	$m(2, 3)$	$m(2, 4)$	$m(2, 5)$	$m(2, 6)$	$m(2, 7)$
$m(1, 0)$	$m(1, 1)$	$m(1, 2)$	$m(1, 3)$	$m(1, 4)$	$m(1, 5)$	$m(1, 6)$	$m(1, 7)$
$m(0, 0)$	$m(0, 1)$	$m(0, 2)$	$m(0, 3)$	$m(0, 4)$	$m(0, 5)$	$m(0, 6)$	$m(0, 7)$

mixed moments of order up to ten and pure moments of order up to seven. It is easy to show that this moment set united with the moment set obtained by using the CQMOM by conditioning ξ_1 on ξ_2 results in the set of 48 optimal moments for $N = 16$. It is left to the reader to verify that this is valid also for trivariate distributions. For example, with $N_1 = N_2 = N_3 = 4$, the CQMOM (with one permutation) controls 168 of the 256 optimal moments up to order 13, while the 64-node tensor-product QMOM controls only 76 moments.

It is interesting to compare the CQMOM with the methods previously introduced. First, an important property of the CQMOM is that it will never have negative weights, since the weights are computed using monovariate quadratures. Thus, as long as the moment set used in the monovariate quadrature is valid,[10] the CQMOM will provide a realizable quadrature for the multivariate moment set. As we saw earlier, this property is not guaranteed for the tensor-product QMOM. Second, the CQMOM controls a larger set of moments than does the tensor-product QMOM, and almost as many as does brute-force QMOM. Third, unlike brute-force QMOM, the CQMOM offers a direct (non-iterative) solution method. A MATLAB script implementing the CQMOM algorithm can be found in Section A.3.3 of Appendix A.

Exercise 3.8 *Determine two quadrature approximations of orders 4 and 9 for the following bivariate Gaussian distribution:*

$$f(\xi_1, \xi_2) = \frac{1}{2\pi\sigma_1\sigma_2\sqrt{1-\rho^2}} \exp\left(-\frac{1}{2(1-\rho^2)}\left[\frac{(\xi_1-\mu_1)^2}{\sigma_1^2} - \frac{2\rho(\xi_1-\mu_1)(\xi_2-\mu_2)}{\sigma_1\sigma_2} + \frac{(\xi_2-\mu_2)^2}{\sigma_2^2}\right]\right)$$

with $\mu_1 = 10$, $\mu_2 = 20$, $\sigma_1 = \sigma_2 = 2$, and different ρ values, by employing the algorithms `BF`, `TP` *and* `CQMOM` *reported in Appendix A. As ρ is increased from zero to one, the distribution is stretched, resulting in the same pure moments, but different mixed moments. It is interesting to evaluate the performance and the results of the three algorithms when $\rho = 0$ and $\rho = 0.5$. Results are reported in Figure 3.5 for $N = 4$ (left) and $N = 9$ (right), and for $\rho = 0$ (top) and $\rho = 0.5$ (bottom). As expected, when $\rho = 0$ the three methods return the very same values. This is due to the fact that the three methods use the same pure moments, and differ only in the number and type of mixed moments employed. It should be highlighted, moreover, that when using brute-force QMOM the iterative procedure converges*

[10] As has previously been explained, it is necessary to determine whether a particular set of monovariate moments is realizable, in order to apply the quadrature algorithm. A moment set is realizable if it corresponds to some nonnegative distribution function.

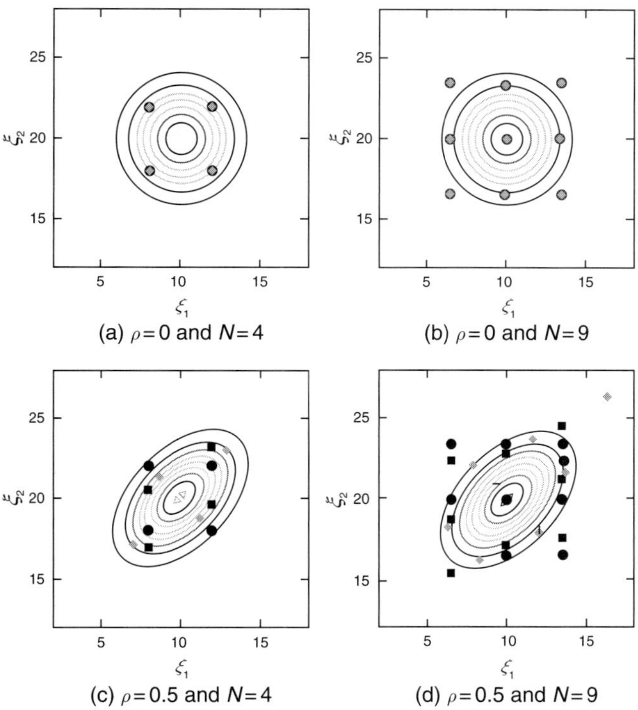

Figure 3.5. Quadrature approximations for bivariate Gaussian distributions with $\mu_1 = 10$, $\mu_2 = 20$, $\sigma_1 = \sigma_2 = 2$, and $\rho = 0$ (top) and $\rho = 0.5$ (bottom) for $N = 4$ (left) and for $N = 9$ (right), namely brute-force QMOM (diamonds), tensor-product QMOM (circles) and CQMOM (squares).

only if initial guesses that are close enough to the final solution are available. Depending on the distance of the initial guess and on the value of the under-relaxation factor as many as 1000 iterations are required in order to calculate the final quadrature. In contrast, as has already been reported, the tensor-product QMOM and CQMOM provide the final solution with a non-iterative calculation.

When $\rho = 0.5$ the results of the different algorithms differ. This is due to the fact that on changing ρ the mixed moments change, and, because the different algorithms use the same pure moments but different mixed moments, each method results in a different final quadrature. It can be seen that the positions in phase space of the nodes of the quadrature approximation calculated by the tensor-product QMOM for $\rho = 0.5$ are the same as those for $\rho = 0$. This is clearly a consequence of the fact that the nodes are calculated with this method from pure moments only, which do not vary when ρ is increased. The final weights are of course different for $\rho = 0$ and $\rho = 0.5$, and, when ρ is increased, the weights of the top-left node and of the right-bottom node are reduced. It is easy to verify that for $\rho = 0.6$ these weights become negative! As anticipated the calculation of the quadrature on the basis of central rotated moments can indeed help in some cases.

When the CQMOM is employed for $\rho = 0.5$ the final nodes automatically move in the internal-coordinate space without rotating the reference frame. It is interesting to highlight that, because the calculation is carried out here by conditioning on the first internal coordinates, when ρ varies the nodes shift only for the second internal coordinate. A shift in the *other direction would be detected if the conditional moments conditioned on the first internal coordinates were used instead. It is also important to recall here that with the CQMOM (as with the tensor-product QMOM) only a subset of the optimal moment set is used. The quadrature calculated with brute-force QMOM more efficiently adapts to the new shape of the distribution, since it is only with brute-force QMOM that all the moments of the optimal moment set are employed. However, the number of iterations required increases significantly and the under-relaxation factor has to be constrained to values much smaller than unity in order to avoid divergence. Test inversions operated on this and other (bivariate and trivariate) distributions have shown that the CQMOM is the more robust and efficient of the three algorithms.*

Our discussion of the CQMOM above assumes that the multivariate moment set corresponds to a continuous distribution. Moreover, we have implicitly assumed that the conditional moment sets generated at each step are all realizable. In practice, situations in which these assumptions do not hold are common and, thus, it is necessary to modify the CQMOM algorithm to handle such cases. As shown in Yuan & Fox (2011), the modifications to the CQMOM algorithm are fairly simple to implement and rely on the adaptive Wheeler algorithm mentioned at the end of Section 3.1.3. The basic idea is that, for each set of conditional moments, the adaptive Wheeler algorithm determines the maximum number of nodes that can be used. Since the CQMOM quadrature is constructed from multiple 1D quadratures, using the adaptive Wheeler algorithm for the latter is a elegant way to ensure that the CQMOM will always return a well-defined multivariate quadrature. More details and examples can be found in Yuan & Fox (2011).

3.3 The extended quadrature method of moments (EQMOM)

The Gaussian quadrature algorithm introduced in Section 3.1.1 is equivalent to approximating the univariate NDF by a sum of Dirac delta functions:

$$n(\xi) \approx \sum_{\alpha=1}^{N} w_\alpha \delta(\xi - \xi_\alpha). \tag{3.81}$$

The order of this quadrature approximation is determined by N, and increasing the value from N to $N + 1$ requires two additional moments (i.e. the QMOM always uses an *even* number of moments). For many applications (Yuan *et al.*, 2012), the form in Eq. (3.81) can be generalized by introducing kernel density functions with a finite (or infinite) support determined by a parameter σ:

$$n(\xi) \approx \sum_{\alpha=1}^{N} w_\alpha \delta_\sigma(\xi, \xi_\alpha). \tag{3.82}$$

The value of σ is determined by fixing one additional moment (a total of $2N + 1$ moments, i.e. an *odd* number of moments). In order to distinguish between moment methods using Eq. (3.81) and those using Eq. (3.82), we will refer to the former as the quadrature moment of moments (QMOM) and the latter as the *extended* quadrature method of moments (EQMOM) (Yuan *et al.*, 2012). The principal advantage of using the EQMOM instead of the QMOM is that with one additional moment it is possible to reconstruct a smooth, non-negative NDF that exactly reproduces the first $2N + 1$ moments. However, there are several practical considerations when using the EQMOM and Eq. (3.82), which we discuss below. The most important property for the function $\delta_\sigma(x, y)$ is that its integer moments with respect to x lead to expressions that are easily inverted to find a set of N weights w_α, N abscissas ξ_α, and σ needed to construct quadrature approximations of integrals involving the NDF. In the limit $\sigma = 0$, Eq. (3.82) should smoothly reduce to Eq. (3.81) (which is not the case for reconstruction methods that are based on entropy minimization!). For all values of σ, the integer moments with respect to ξ of Eq. (3.82) should be easily evaluated in order to facilitate application of the moment-inversion algorithm used in the EQMOM (Yuan *et al.*, 2012). Also, the support of $\delta_\sigma(x, y)$ should not exceed the range of nonzero values of the true NDF (e.g. if x is always nonnegative, $\delta_\sigma(x, y)$ should be null for $x < 0$). When applicable, an excellent choice for δ_σ can often be found from the weight functions corresponding to a family of orthogonal polynomials.

3.3.1 Relationship to orthogonal polynomials

A convenient choice for univariate EQMOM is to define $\delta_\sigma(x, y)$ in terms of the weight function $w(t)$ for a known family of orthogonal polynomials (Gautschi, 2004). For example, on the interval $[0, \infty)$ the associated Laguerre polynomials have the weight function

$$w(t) = t^\alpha \mathrm{e}^{-t}, \tag{3.83}$$

where the parameter $\alpha > -1$ defines a particular family of Laguerre polynomials.[11] An NDF with $\xi \in [0, \infty)$ can then be approximated by using

$$\delta_\sigma(x, y) = \frac{1}{\sigma\Gamma(y/\sigma)} \left(\frac{x}{\sigma}\right)^{y/\sigma - 1} \mathrm{e}^{-x/\sigma}, \tag{3.84}$$

where the right-hand side is a scaled form of Eq. (3.83) with the properties

$$\int_0^\infty \delta_\sigma(x, y)\mathrm{d}x = 1, \quad \int_0^\infty x\delta_\sigma(x, y)\mathrm{d}x = y, \quad \int_0^\infty x^2\delta_\sigma(x, y)\mathrm{d}x = y(y + \sigma). \tag{3.85}$$

Example NDF found using Eq. (3.84) and $N = 2$ for four values of σ are shown in Figure 3.6. The principal advantage of using a known family of orthogonal polynomials is that the recursion coefficients a_α and b_α in Eq. (3.5) are known in advance (Gautschi, 2004) (i.e. we do not need to compute them from the moments). Thus, the Jacobi matrix in Eq. (3.13) can be computed with great accuracy for large values of N (which is not usually the case when the recursion coefficients are computed from the moments).

The fact that the Jacobi matrix is known can be used to evaluate very accurately integrals involving $\delta_\sigma(x, y)$ by applying Gaussian quadrature. For example, Eq. (3.84) leads to

$$\int_0^\infty g(x)\delta_\sigma(x, \xi_\alpha)\mathrm{d}x = \frac{1}{\Gamma(k_\alpha)} \int_0^\infty g(\sigma t)t^{k_\alpha - 1}\mathrm{e}^{-t}\,\mathrm{d}t \approx \frac{1}{\Gamma(k_\alpha)} \sum_{\beta=1}^{N'} w_{\alpha\beta} g(\sigma t_{\alpha\beta}), \tag{3.86}$$

where $k_\alpha = \xi_\alpha/\sigma$ and $\{w_{\alpha\beta}, t_{\alpha\beta}\}$ are the N' weights and N' abscissas computed from the Laguerre polynomial recursion coefficients for a given value of k_α.[12] Here, it is important to recognize that N' in Eq. (3.86) can be chosen independently from the value of N

[11] The zeros of the associated Laguerre polynomials $P_n^{(\alpha)}(t)$ are the n abscissas t_α in the interval $[0, \infty)$ (Gautschi, 2004). By defining $t = x - a$, the interval can be shifted to $[a, \infty)$, and thus the abscissas for the latter will be $x_\alpha = t_\alpha + a$.

[12] Note that a different set of Laguerre polynomials, parameterized by $k_\alpha > 0$, is used for each value of α. Thus, the weights and abscissas must be computed separately for each value of α. As with any Gaussian quadrature, Eq. (3.86) is exact when $g(\xi)$ is a polynomial of order less than $2N'$ (Gautschi, 2004).

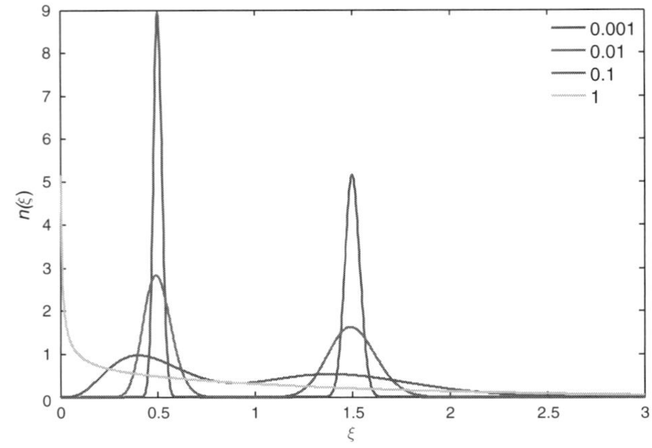

Figure 3.6. Examples of $n(\xi)$ found with Eq. (3.84) for four values of σ. The other parameters are $w_1 = w_2 = 0.5$, $\xi_1 = 0.5$, and $\xi_2 = 1.5$.

used in Eq. (3.82). In other words, the order of the Gaussian quadrature used to evaluate Eq. (3.86) (i.e. $2N'$) is independent of the number of transported moments used in the EQMOM (i.e. $2N + 1$), and we can choose $N' \gg N$ to improve the accuracy of Eq. (3.86). Finally, using the result in Eq. (3.86), we can see that the EQMOM closure defined by Eq. (3.84) approximates integrals with respect to the NDF using a *dual-quadrature representation*:

$$\int_0^{\infty} g(\xi) n(\xi) \mathrm{d}\xi \approx \sum_{\alpha=1}^{N} \frac{w_\alpha}{\Gamma(k_\alpha)} \sum_{\beta=1}^{N'} w_{\alpha\beta} g(\sigma t_{\alpha\beta}), \tag{3.87}$$

where the N weights w_α, N abscissas ξ_α, and σ are found from the first $2N + 1$ integer moments of $n(\xi)$. As we shall see below, the main computational challenge when working with the EQMOM is the definition of an algorithm to find σ. The general approach is to find analytical expressions for the first $2N + 1$ integer moments, and use these to define an iterative procedure to find σ. Note that a robust EQMOM algorithm must also be able to handle degenerate cases such as $\sigma = 0$ and/or when the $2N + 1$ integer moments can be captured by fewer than N quadrature nodes.

3.3.2 Univariate EQMOM

In this section, we discuss three types of univariate EQMOM that are based on weight functions with infinite, semi-infinite, and finite support (Chalons *et al.*, 2010; Yuan *et al.*, 2012). Example numerical algorithms are provided in Section A.4 of Appendix A.

The Gaussian distribution

As an example of univariate EQMOM, we will consider an NDF with $\xi \in (-\infty, +\infty)$ and define kernel density functions using Gaussian distributions (Chalons *et al.*, 2010):

$$\delta_\sigma(x, y) = \frac{1}{\sqrt{2\pi}\sigma} \exp\left(-\frac{(x-y)^2}{2\sigma^2}\right). \tag{3.88}$$

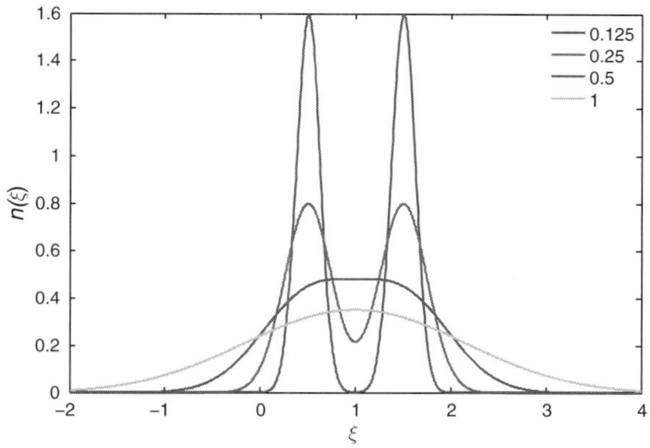

Figure 3.7. Examples of $n(\xi)$ found with Eq. (3.88) for four values of σ. The other parameters are $w_1 = w_2 = 0.5$, $\xi_1 = 0.5$, and $\xi_2 = 1.5$.

Example NDF found using Eq. (3.88) and $N = 2$ for four values of σ are shown in Figure 3.7. Note that, when $y = 0$, the right-hand side of Eq. (3.88) corresponds to the (scaled and shifted) weight function for the Hermite polynomials.[13] With this choice for $\delta_\sigma(x, y)$, it will thus be straightforward to evaluate integrals like that in Eq. (3.87) using the known recursion coefficients for Hermite polynomials (Gautschi, 2004). The integer moments centered at y of Eq. (3.88) are easily computed using their definition:

$$\begin{aligned}\int_{-\infty}^{+\infty} x^k \delta_\sigma(x, y)\mathrm{d}x &= \frac{1}{\sqrt{2\pi}\sigma} \int_{-\infty}^{+\infty} (x+y)^k \exp\left(-\frac{x^2}{2\sigma^2}\right) \mathrm{d}x \\ &= \sum_{i=0}^{[k/2]} \frac{k!}{i!(k-2i)!} \left(\frac{\sigma^2}{2}\right)^i y^{k-2i} \end{aligned} \tag{3.89}$$

where $[k/2] = k/2$ if k is even, and $[k/2] = (k-1)/2$ if k is odd. The first five integer moments centered at $y = \xi_\alpha$ are thus

$$\begin{aligned}\int_{-\infty}^{+\infty} \delta_\sigma(\xi, \xi_\alpha)\mathrm{d}\xi &= 1, \\ \int_{-\infty}^{+\infty} \xi\delta_\sigma(\xi, \xi_\alpha)\mathrm{d}\xi &= \xi_\alpha, \\ \int_{-\infty}^{+\infty} \xi^2\delta_\sigma(\xi, \xi_\alpha)\mathrm{d}\xi &= \sigma^2 + \xi_\alpha^2, \\ \int_{-\infty}^{+\infty} \xi^3\delta_\sigma(\xi, \xi_\alpha)\mathrm{d}\xi &= 3\sigma^2\xi_\alpha + \xi_\alpha^3, \\ \int_{-\infty}^{+\infty} \xi^4\delta_\sigma(\xi, \xi_\alpha)\mathrm{d}\xi &= 3\sigma^4 + 6\sigma^2\xi_\alpha^2 + \xi_\alpha^4. \end{aligned} \tag{3.90}$$

This set of five moments is used for bi-Gaussian EQMOM with $N = 2$ (Chalons *et al.*, 2010).

[13] The Hermite polynomials $P_n^{(\alpha)}(t)$ defined for $t \in (-\infty, \infty)$ have weight functions $w(t) = |t|^{2\alpha}\mathrm{e}^{-t^2}$ (Gautschi, 2004). By shifting and scaling, they can be written for $x = \sqrt{2}\sigma t + y$. Thus, if t_α are the n roots of the Hermite polynomial $P_n^{(0)}(t)$, then $x_\alpha = \sqrt{2}\sigma t_\alpha + y$ are the n abscissas for Eq. (3.88).

From Eq. (3.89), the integer moments of Eq. (3.82) can be expressed as

$$m_k = m_k^* + \sum_{i=1}^{[k/2]} \frac{k!}{i!(k-2i)!} \left(\frac{\sigma^2}{2}\right)^i m_{k-2i}^*, \tag{3.91}$$

where

$$m_k^* = \sum_{\alpha=1}^{N} w_\alpha \xi_\alpha^k \tag{3.92}$$

are the QMOM moments (i.e. when $\sigma = 0$). Note that Eq. (3.91) forms a lower-triangular linear system that can be inverted using forward substitutions:

$$\begin{aligned}
m_0^* &= m_0, \\
m_1^* &= m_1, \\
m_2^* &= m_2 - m_0^*\sigma^2, \\
m_3^* &= m_3 - 3m_1^*\sigma^2, \\
m_4^* &= m_4 - 6m_2^*\sigma^2 - 3m_0^*\sigma^4, \\
&\vdots \\
m_k^* &= m_k - \sum_{i=1}^{[k/2]} \frac{k!}{i!(k-2i)!} \left(\frac{\sigma^2}{2}\right)^i m_{k-2i}^*, \\
&\vdots \\
m_{2N-1}^* &= m_{2N-1} - \sum_{i=1}^{N-1} \frac{(2N-1)!}{i!(2N-2i-1)!} \left(\frac{\sigma^2}{2}\right)^i m_{2N-2i-1}^*,
\end{aligned} \tag{3.93}$$

and

$$m_{2N}^* = m_{2N} - \sum_{i=1}^{N} \frac{(2N)!}{i!(2N-2i)!} \left(\frac{\sigma^2}{2}\right)^i m_{2N-2i}^*. \tag{3.94}$$

In order to solve this system of $2N + 1$ equations to find the N weights, N abscissas, and σ^2, the following iterative method can be used.

1. Starting with an initial guess $\sigma^2 = 0$, use the Wheeler algorithm and the $2N$ moments $\{m_0^*, m_1^*, \ldots, m_{2N-1}^*\}$ computed from Eq. (3.93) to find N weights w_α and N abscissas ξ_α.

2. Using the weights and abscissas (for the current value of σ^2), compute the moment m_{2N}^* from

$$m_{2N}^* = \sum_{\alpha=1}^{N} w_\alpha \xi_\alpha^{2N}.$$

3. Use a nonlinear equation solver (e.g. the bounded-secant method) to find the smallest (real) $\sigma^2 \geq 0$ such that the objective function

$$f(\sigma^2) = m_{2N} - \sum_{i=0}^{N} \frac{(2N)!}{(N-i)!(2i)!} \left(\frac{\sigma^2}{2}\right)^{N-i} m_{2i}^*$$

is null. Note that, each time σ^2 is changed, the $2N$ moments $\{m_0^*, m_1^*, \dots, m_{2N-1}^*\}$ must be recomputed from Eq. (3.93), the Wheeler algorithm must be re-applied to find the weights and abscissas (i.e. step 1), and the moment m_{2N}^* must be reevaluated (i.e. step 2).

When applying this algorithm, the moment set $\{m_0, m_1, \dots, m_{2N}\}$ must be known (and realizable). From the definition of the objective function and the properties of Gaussian quadrature,[14] we have $f(0) \geq 0$. Thus, as a first step in the bounded-search algorithm, an upper bound σ^+ can be determined such that $f(\sigma^+) < 0$ under the condition that the moment set $\{m_0^*, m_1^*, \dots, m_{2N-1}^*\}$ found from Eq. (3.93) using $\sigma^2 = \sigma^+$ is realizable. If no such σ^+ exists, then σ^2 can be chosen such that it minimizes $|f(\sigma^2)|$ and the moment set $\{m_0^*, m_1^*, \dots, m_{2N-1}^*\}$ is realizable.

For the case $N = 2$ (i.e. a bi-Gaussian NDF (Chalons *et al.*, 2010)), it is possible to find a third-order polynomial from Eq. (3.94) with a unique real root that determines σ^2. The moment set in Eqs. (3.93) and (3.94) with $N = 2$ are as follows:

$$\begin{aligned} m_0^* &= m_0, \\ m_1^* &= m_1, \\ m_2^* &= m_2 - m_0\sigma^2, \\ m_3^* &= m_3 - 3m_1\sigma^2, \\ m_4^* &= m_4 - 6m_2\sigma^2 + 3m_0\sigma^4. \end{aligned} \tag{3.95}$$

Assuming that the five moments $\{m_0, \dots, m_4\}$ are realizable, this system of five equations can be solved analytically for the five unknowns $\{w_1, w_2, \xi_1, \xi_2, \sigma^2\}$. For two-node quadrature, the weights and abscissas depend on $\{m_0^*, \dots, m_3^*\}$ (Desjardins *et al.*, 2008):

$$\begin{gathered} w_1 = m_0^*\left(\frac{1}{2} + \gamma\right), \quad w_2 = m_0^*\left(\frac{1}{2} - \gamma\right), \\ \xi_1 = \langle\xi^*\rangle - \left(\frac{w_2}{w_1}\langle\xi^{*2}\rangle\right)^{1/2}, \quad \xi_2 = \langle\xi^*\rangle + \left(\frac{w_1}{w_2}\langle\xi^{*2}\rangle\right)^{1/2}, \end{gathered} \tag{3.96}$$

where $\langle\xi^*\rangle$, $\langle\xi^{*2}\rangle$, and γ are defined by

$$\langle\xi^*\rangle = \frac{m_1^*}{m_0^*}, \quad \langle\xi^{*2}\rangle = \frac{m_0^* m_2^* - \left(m_1^*\right)^2}{\left(m_0^*\right)^2}, \quad \gamma = \frac{q^*/2}{\left(q^{*2} + 4\langle\xi^{*2}\rangle^3\right)^{1/2}} \tag{3.97}$$

with

$$q^* = \frac{1}{m_0^*}\left(m_3^* - 3m_0^*\langle\xi^*\rangle\langle\xi^{*2}\rangle - m_0^*\langle\xi^*\rangle^3\right). \tag{3.98}$$

Note that the weights and abscissas are implicit functions of σ^2 through their dependence on m_2^* and m_3^*.

The weights and abscissas in Eq. (3.96) can then be used to compute m_4^*:

$$m_4^* = w_1\xi_1^4 + w_2\xi_2^4, \tag{3.99}$$

[14] A Gaussian quadrature computed from the first $2N$ moments results in the smallest realizable value of m_{2N}^* (Dette & Studden, 1997).

which leads to the following expression that must be solved for σ^2:

$$0 = w_1\xi_1^4 + w_2\xi_2^4 - m_4 + 6m_2\sigma^2 - 3m_0\sigma^4. \tag{3.100}$$

After some tedious algebra (Chalons *et al.*, 2010), Eq. (3.100) can be rewritten as a third-order polynomial:

$$P(s) = 2s^3 + \left(\eta^\dagger - 3\right) s - (q^\dagger)^2 = 0, \tag{3.101}$$

where $\eta^\dagger = \eta/\langle\xi'^2\rangle^2$, $q^\dagger = q/\langle\xi'^2\rangle^{3/2}$,

$$\begin{aligned}\eta &= \frac{1}{m_0}\left(m_4 - 4m_3\langle\xi\rangle + 6\langle\xi\rangle^2 m_2 - 3m_0\langle\xi\rangle^4\right),\\ q &= \frac{1}{m_0}\left(m_3 - 3m_0\langle\xi\rangle\langle\xi'^2\rangle - m_0\langle\xi\rangle^3\right),\end{aligned} \tag{3.102}$$

with

$$\langle\xi\rangle = \frac{m_1}{m_0}, \quad \langle\xi'^2\rangle = \frac{m_0 m_2 - (m_1)^2}{(m_0)^2}. \tag{3.103}$$

Under the conditions that $\eta^\dagger > 1+(q^\dagger)^2$ and, if $q^\dagger = 0$, $\eta^\dagger \leq 3$, the polynomial in Eq. (3.101) has a unique real root[15] in the interval $0 \leq s \leq 1$:

$$s = c_3 - \frac{c_1}{c_3}, \tag{3.104}$$

where

$$c_1 = \frac{1}{6}\left(\eta^\dagger - 3\right), \quad c_2 = \frac{1}{4}\left(q^\dagger\right)^2 \quad c_3 = \left[\left(c_1^3 + c_2^2\right)^{1/2} + c_2\right]^{1/3}.$$

The desired value of σ^2 is related to this root by

$$\sigma^2 = \langle\xi'^2\rangle\left(1 - c_3 + \frac{c_1}{c_3}\right). \tag{3.105}$$

Examples of dependences of σ^2 on $\eta^\dagger$ for selected values of $q^\dagger$ are shown in Figure 3.8. Note that the coefficients of the polynomial depend only on the known moments m_k, and not on the weights and abscissas.[16] Thus, σ^2 can be computed first, followed by the moments m_k^*, using Eq. (3.95), from which the weights and abscissas are found using the Wheeler algorithm. In the limit where $s = 0$ (i.e. $\eta^\dagger = 3$ and $q^\dagger = 0$), all of the moments m_k are Gaussian and only one quadrature node is required (i.e. $\xi_1 = \xi_2 = \langle\xi\rangle$). In the limit $s = 1$ (i.e. $\eta^\dagger = 1 + (q^\dagger)^2$), the moment set corresponds exactly to two weighted Dirac delta functions.

[15] The condition $\eta^\dagger \geq 1 + (q^\dagger)^2$ will be satisfied by any realizable moment set with the equality holding for the case of two weighted Dirac delta functions. When $q^\dagger = 0$, the NDF is symmetric about its mean value and $\eta^\dagger = 3$ is the largest possible value for $\eta^\dagger$ that can be achieved using a Gaussian distribution for δ_σ. If the true value of $\eta^\dagger$ is larger than this value, the objective function $f(\sigma^2)$ will never be zero (i.e. the polynomial $P(s)$ does not hold), but the minimum of $f(\sigma^2)$ occurs at $\sigma^2 = \langle\xi'^2\rangle$. It is possible to find analytical expressions for the three roots of $P(s)$ in Eq. (3.101). In the Gaussian limit, there is a triple root at $s = 0$, otherwise two of the roots are complex numbers and the desired root is real.

[16] To the best of the authors' knowledge, it is currently unknown whether an analogous polynomial can be found for $N > 2$, thereby allowing a direct solution to find σ^2. Nonetheless, the iterative algorithm described above for determining σ^2 has been found to be robust.

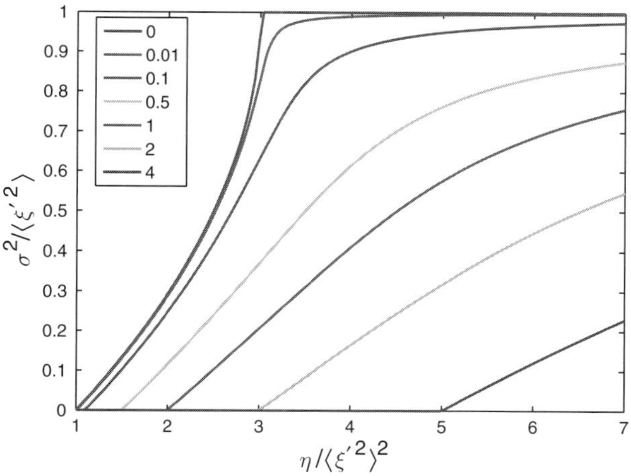

Figure 3.8. The parameter $\sigma^2/\langle\xi'^2\rangle$ appearing in the bi-Gaussian distribution versus $\eta^\dagger$ for selected value of $(q^\dagger)^2$.

The gamma distribution

The moment-inversion algorithm described above for a Gaussian distribution can be applied for other choices of δ_σ (e.g. for NDF defined on a semi-infinite or a finite interval). The first step is to compute the integer moments of $\delta_\sigma(x, y)$ with respect to x analytically. For example, with the definition in Eq. (3.84) for $x \in [0, \infty)$ (Yuan *et al.*, 2012), these integer moments with $y = \xi_\alpha$ are

$$m_k^{(\alpha)} = \frac{\Gamma(k_\alpha + k)}{\Gamma(k_\alpha)} \left(\frac{\xi_\alpha}{k_\alpha}\right)^k, \tag{3.106}$$

and thus

$$m_k = \sum_{\alpha=1}^{N} w_\alpha \frac{\Gamma(\xi_\alpha/\sigma + k)}{\Gamma(\xi_\alpha/\sigma)} \sigma^k = \sum_{\alpha=1}^{N} w_\alpha G_k(\xi_\alpha, \sigma), \tag{3.107}$$

where

$$G_k(\xi_\alpha, \sigma) = \begin{cases} 1 & \text{if } k = 0, \\ \prod_{i=0}^{k-1} (\xi_\alpha + i\sigma) & k \geq 1. \end{cases} \tag{3.108}$$

Noting that G_k can be written as $G_k(\xi_\alpha, \sigma) = \xi_\alpha^k + H_{k-1}(\xi_\alpha, \sigma)$, where $H_{k-1}(x, y)$ is a homogeneous polynomial of order $k - 1$ in x and y, we can rewrite the integer moments as

$$m_k = m_k^* + \sum_{\alpha=1}^{N} w_\alpha H_{k-1}(\xi_\alpha, \sigma), \tag{3.109}$$

where the second term on the left-hand side can be expressed in terms of σ and the moments $m_{k-1}^*, \ldots, m_0^*$. In other words, Eq. (3.109) has a form similar to Eq. (3.91) wherein the higher-order moments are known in terms of the lower-order moments and σ. For example, up to $k = 4$ we have

$$
\begin{aligned}
m_0 &= m_0^*, \\
m_1 &= m_1^*, \\
m_2 &= m_2^* + \sigma m_1^*, \\
m_3 &= m_3^* + 3\sigma m_2^* + 2\sigma^2 m_1^*, \\
m_4 &= m_4^* + 6\sigma m_3^* + 11\sigma^2 m_2^* + 6\sigma^3 m_1^*.
\end{aligned} \tag{3.110}
$$

As discussed above for a Gaussian distribution, this system of equations can be solved iteratively to determine σ and $\{w_\alpha, \xi_\alpha\}$ for $\alpha = 1, 2$ given a realizable moment set m_k, $k = 0, \ldots, 4$. The extension to larger N uses the same iterative procedure.

In summary, we should note that the form of Eq. (3.109) (and Eq. (3.110)) where the right-hand side is closed in terms of the known moments is not guaranteed for all possible choices for δ_σ. Indeed, the choice of Eq. (3.84) using $t = \xi/\sigma$ was made specifically so that the resulting moment system would be closed. The reader should keep this important technical detail in mind when defining other choices for δ_σ (i.e. using the weight functions for other types of orthogonal polynomials). An example of a δ_σ with this property on finite intervals is discussed next.

The beta distribution

Univariate EQMOM can be applied to an NDF defined on a finite interval (Yuan *et al.*, 2012). For example, if the NDF is nonzero only on the interval $\xi \in [0, 1]$, then we can define the kernel density functions using a beta distribution:

$$
\delta_\sigma(x, y) = \frac{x^{y/\sigma - 1}(1 - x)^{(1-y)/\sigma - 1}}{B(y/\sigma, (1 - y)/\sigma)} \tag{3.111}
$$

where $\sigma > 0$ and B is the beta function. Example NDF found using Eq. (3.111) and $N = 2$ for four values of σ are shown in Figure 3.9. Note that σ controls the width of the peaks such that small values of σ yield peaks closer to delta functions. The right-hand side of Eq. (3.111) corresponds to the (scaled and shifted) weight function for the Jacobi

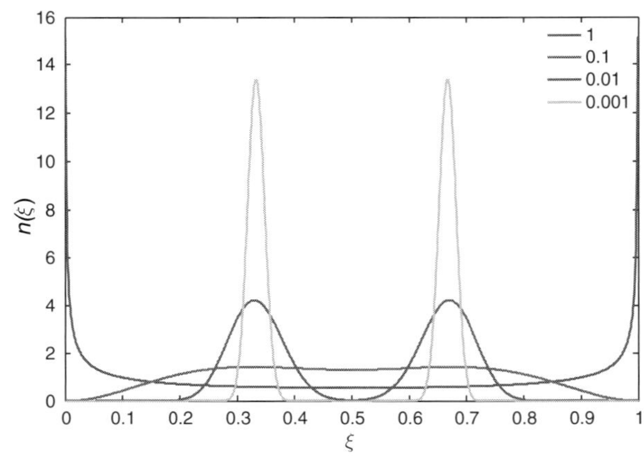

Figure 3.9. Examples of $n(\xi)$ found with Eq. (3.111) for four values of σ. The other parameters are $w_1 = w_2 = 0.5$, $\xi_1 = 1/3$, and $\xi_2 = 2/3$.

polynomials,[17] and integrals like that in Eq. (3.87) can be evaluated using the corresponding recursion coefficients (Gautschi, 2004). For the beta distribution in Eq. (3.111), the integer moments can be found from a recursion formula:

$$m_k^{(\alpha)} = \left[\frac{\xi_\alpha + (k-1)\sigma}{1 + (k-1)\sigma}\right] m_{k-1}^{(\alpha)} \text{ for } k > 0 \tag{3.112}$$

and $m_0^{(\alpha)} = 1$. We can thus express the integer moments of the NDF as

$$m_k = \sum_{\alpha=1}^{N} w_\alpha G_k(\xi_\alpha, \sigma), \tag{3.113}$$

where

$$G_k(\xi_\alpha, \sigma) = \begin{cases} 1 & \text{if } k = 0, \\ \prod_{i=0}^{k-1} (\xi_\alpha + i\sigma)/(1 + i\sigma) & k \geq 1, \end{cases} \tag{3.114}$$

which is very similar to Eq. (3.108). Because ξ_α appears only in the numerator in Eq. (3.114), the product will yield polynomials in ξ_α of order k. Thus, we can express the integer moments of the NDF as $m_k = \gamma_k m_k^* + \gamma_{k-1} m_{k-1}^* + \cdots + \gamma_1 m_1^*$, where the coefficients γ_k depend only on the parameter σ. For example, up to $k = 4$ we have

$$\begin{aligned} m_0 &= m_0^*, \\ m_1 &= m_1^*, \\ m_2 &= \frac{1}{1+\sigma}(m_2^* + \sigma m_1^*), \\ m_3 &= \frac{1}{(1+2\sigma)(1+\sigma)}\left(m_3^* + 3\sigma m_2^* + 2\sigma^2 m_1^*\right), \\ m_4 &= \frac{1}{(1+3\sigma)(1+2\sigma)(1+\sigma)}\left(m_4^* + 6\sigma m_3^* + 11\sigma^2 m_2^* + 6\sigma^3 m_1^*\right), \end{aligned} \tag{3.115}$$

which are very similar to Eq. (3.110). As discussed above for a Gaussian distribution, this system of equations can be solved iteratively to determine σ and $\{w_\alpha, \xi_\alpha\}$ for $\alpha = 1, 2$ given a realizable moment set m_k, $k = 0, \ldots, 4$.

3.3.3 Evaluation of integrals with the EQMOM

While EQMOM allows us to capture an additional moment, the use of kernel density functions can lead to a closure problem when evaluating integrals such as Eq. (3.9):

$$\int_{\Omega_\xi} n(\xi) g(\xi) \mathrm{d}\xi \approx \sum_{\alpha=1}^{N} w_\alpha \int_{\Omega_\xi} g(\xi) \delta_\sigma(\xi, \xi_\alpha) \mathrm{d}\xi. \tag{3.116}$$

As noted earlier, if δ_σ is defined using the weight function of a known family of orthogonal polynomials, Eq. (3.116) can be closed as described in Eq. (3.86) using the recursion

[17] The Jacobi polynomials $P_n^{(\alpha,\beta)}(t)$ defined for $t \in [-1, 1]$ have weight functions $w(t) = (1-t)^\alpha(1+t)^\beta$ (Gautschi, 2004). By shifting and scaling, they can be used for any finite interval $\xi \in [a, b]$: $t = (a + b - 2\xi)/(a - b)$. Thus, if t_α are the abscissas in the interval $[-1, 1]$, then $\xi_\alpha = [a + b - (a - b)t_\alpha]/2$ are the abscissas in $[a, b]$.

coefficients. Otherwise, assuming that the integer moments of $\delta_\sigma(\xi, \xi_\alpha)$ can be calculated analytically:[18]

$$m_k^{(\alpha)} = \int_{\Omega_\xi} \xi^k \delta_\sigma(\xi, \xi_\alpha) \mathrm{d}\xi, \tag{3.117}$$

the closure problem can be handled using Gaussian quadrature as will be described next. We should note, however, that using the recursion coefficients allows much more accurate integral evaluations since the moment-inversion problem is often ill-conditioned (Gautschi, 2004). Consider the set of $2N'$ (realizable) moments $\{m_0^{(\alpha)}, m_1^{(\alpha)}, \ldots, m_{2N'-1}^{(\alpha)}\}$ for $\alpha \in (1, 2, \ldots, N)$ defined by Eq. (3.117). Using the adaptive Wheeler algorithm, for each value of α this set of moments leads to N' weights and N' abscissas:[19] $\{w_1^{(\alpha)}, w_2^{(\alpha)}, \ldots, w_{N'}^{(\alpha)}\}$ and $\{\xi_1^{(\alpha)}, \xi_2^{(\alpha)}, \ldots, \xi_{N'}^{(\alpha)}\}$. The integral in Eq. (3.116) can then be written using a dual-quadrature representation of the NDF:

$$\int_{\Omega_\xi} n(\xi) g(\xi) \mathrm{d}\xi \approx \sum_{\alpha=1}^{N} \sum_{\beta=1}^{N'} w_\alpha w_\beta^{(\alpha)} g\left(\xi_\beta^{(\alpha)}\right). \tag{3.118}$$

Note that N' can be chosen independently of N so that the order of the quadrature can be increased while keeping the number of moments fixed (i.e. equal to $2N + 1$). This fact can be used advantageously for limiting cases by choosing δ_σ to have a form close to the true limiting NDF (e.g. if the limiting NDF is a weighted Gaussian, then Eq. (3.88) is a good choice for $\delta_\sigma(x, y)$). In these cases, N can be small (e.g. $N = 1$) while still retaining the accuracy of the Gaussian quadrature approximation by adjusting N' in Eq. (3.118).

As an example, consider the case in which δ_σ is the Gaussian distribution in Eq. (3.88). In this case, σ functions as a scaling parameter, and thus the integral in Eq. (3.118) can be written as

$$\int n(\xi) g(\xi) \mathrm{d}\xi \approx \sum_{\alpha=1}^{N} \sum_{\beta=1}^{N'} w_\alpha w_\beta^* g\left(\sigma \varphi_\beta + \xi_\alpha\right), \tag{3.119}$$

where the N' weights w_β^* and N' abscissas φ_β are found from the first $2N'$ standard normal moments:[20]

$$\mu_k = \begin{cases} k!/(2^{k/2}(k/2)!) & \text{if } k \text{ is even,} \\ 0 & \text{if } k \text{ is odd.} \end{cases} \tag{3.120}$$

(See Exercise 3.2 for an example of the weights and abscissas for $N' = 4$.) Note that, because the moments μ_k are fixed, the weights w_β^* and abscissas φ_β need be computed only once for a given value of N'. The additional computational cost of employing the EQMOM versus the QMOM is thus mainly associated with the computation of σ. In the limiting case $\sigma = 0$, Eq. (3.119) simplifies to

$$\int n(\xi) g(\xi) \mathrm{d}\xi \approx \sum_{\alpha=1}^{N} \sum_{\beta=1}^{N'} w_\alpha w_\beta^* g(\xi_\alpha) = \sum_{\alpha=1}^{N} w_\alpha g(\xi_\alpha), \tag{3.121}$$

---

[18] This step is the principal reason why $\delta_\sigma(x, y)$ should be chosen to have easily computable moments.

[19] For some cases, the weights do not depend on α and thus can be computed for $\xi_\alpha = 0$. Likewise, if σ enters the definition of δ_σ as a scaling parameter (i.e. $\sigma\delta_\sigma(\xi, \xi_\alpha) = \delta_{\sigma=1}((\xi - \xi_\alpha)/\sigma, 0)$), then the weights and abscissas can be found from the integer moments of $\delta_{\sigma=1}(x, 0)$: if φ_β are the resulting abscissas, then $\xi_\beta^{(\alpha)} = \sigma\varphi_\beta + \xi_\alpha$. Note that the distribution in Eq. (3.84) does not have this form, which explains why the abscissas must be computed separately for each value of α.

[20] A even better alternative is to use the recursion coefficients for Hermite polynomials to form the Jacobi matrix directly (Gautschi, 2004).

which is the QMOM approximation. In the limiting case $\sigma^2 = \langle \xi'^2 \rangle$, we have $\xi_\alpha = \langle \xi \rangle$, and thus

$$\int n(\xi) g(\xi) \mathrm{d}\xi \approx m_0 \sum_{\beta=1}^{N'} w_\beta^* g\left(\sigma\varphi_\beta + \langle \xi \rangle\right), \tag{3.122}$$

which is the N'-node Gaussian quadrature approximation which is based on a Gaussian NDF (Gautschi, 2004). To conclude, we can note that Eq. (3.119) is equivalent to approximating the univariate NDF as a dual-quadrature representation:

$$n(t, \mathbf{x}, \xi) \approx \sum_{\alpha=1}^{N} \sum_{\beta=1}^{N'} w_\alpha(t, \mathbf{x}) w_\beta^* \delta\left(\xi - \xi_\alpha(t, \mathbf{x}) - \sigma(t, \mathbf{x})\varphi_\beta\right), \tag{3.123}$$

wherein we show explicitly the quantities that depend on time t and location $\mathbf{x}$. The extension of Eq. (3.123) to other kernel density functions besides Gaussian ones (e.g. Eq. (3.87)) is straightforward. From the dual-quadrature representation of the NDF, the role of σ in determining the EQMOM approximation of unclosed terms in the moment-transport equation is clearly linked to the underlying weight function through the time/space-independent[21] weights w_β^* and nodes φ_β.

3.3.4 Multivariate EQMOM

In principle, the EQMOM introduced in Section 3.3.2 can be generalized to include multiple internal coordinates. However, depending on the assumed form of the kernel density functions, it may be necessary to use a multivariate nonlinear-equation solver to find the parameters (i.e. similar to the brute-force QMOM discussed in Section 3.2.1). An interesting alternative is to extend the CQMOM algorithm described in Section 3.2.3. Here we consider examples using both methods.

Brute-force QMOM

In the literature (Chalons *et al.*, 2010), only a bivariate EQMOM with four abscissas represented by weighted Gaussian distributions with a diagonal covariance matrix[22] has been considered. However, it is likely that brute-force QMOM algorithms can be developed for other distribution functions. Using the multi-Gaussian representation as an example, the approximate NDF can be written as

$$n(\boldsymbol{\xi}) = \sum_{\alpha=1}^{N} w_\alpha \delta_\Sigma(\boldsymbol{\xi}, \boldsymbol{\xi}_\alpha), \tag{3.124}$$

where, for M internal coordinates, the kernel density function is defined by

$$\delta_\Sigma(\mathbf{x}, \mathbf{y}) = \frac{1}{(2\pi|\boldsymbol{\Sigma}|)^{M/2}} \exp\left(-\frac{1}{2}(\mathbf{x} - \mathbf{y})^{\mathrm{T}} \boldsymbol{\Sigma}^{-1} (\mathbf{x} - \mathbf{y})\right) \tag{3.125}$$

and the $M \times M$ symmetric covariance matrix $\boldsymbol{\Sigma}$ contains the scaling components σ_{ij}. The total number of unknown scaling components is $M(M+1)/2$, and these must be determined

[21] In the dual-quadrature representation corresponding to Eq. (3.87), the weights and nodes are neither time- nor space-independent since k_α depends on the transported moments.

[22] For a diagonal covariance matrix, $\boldsymbol{\Sigma} = \mathrm{diag}(\sigma_{11}, \sigma_{22})$.

by fixing an equal number of multivariate moments. Recall that the optimal moment set used in the CQMOM contains $(M+1)n^M$ moments, where n is the number of nodes in the univariate quadrature[23] and thus $N = n^M$. Since the additional moments needed in order to determine $\boldsymbol{\Sigma}$ must be independent of the optimal moment set, the multivariate EQMOM will use a total of $(M+1)(n^M + M/2)$ multivariate moments. Hereafter, we will refer to this moment set as the *extended* optimal moment set.

In order to be consistent with the univariate EQMOM discussed in Section 3.3.2, the diagonal components of $\boldsymbol{\Sigma}$ are determined from the pure moments of order $2n$ (e.g. in 2D phase space $m(2n, 0)$ and $m(0, 2n)$ are used to find σ_{11} and σ_{22}, respectively). The off-diagonal components of $\boldsymbol{\Sigma}$ can then be determined by using the cross moments of total order $2n$, but of order n in the i and j components. For example, in 2D phase space, the moment $m(n, n)$ is used to find σ_{12}, and in 3D phase space the moment $m(n, 0, n)$ is used to find σ_{13}. The reader can verify that none of these moments is contained in the optimal moment set and, hence, they are available for use in the extended optimal moment set. The multivariate moments $m(\mathbf{k})$ computed from Eq. (3.124) can be written as

$$m(\mathbf{k}) = m^*(\mathbf{k}) + \sum_{\substack{i_1=0 \\ i_1+\cdots+i_M>0}}^{k_1} \cdots \sum_{i_M=0}^{k_M} \binom{k_1}{i_1} \cdots \binom{k_M}{i_M} m^*(\mathbf{k}-\mathbf{i}) m_{\Sigma}(\mathbf{i}), \tag{3.126}$$

where the integer moments with respect to $\boldsymbol{\xi}$ are defined by

$$m_{\Sigma}(\mathbf{i}) = \int \boldsymbol{\xi}^{\mathbf{i}} \delta_{\Sigma}(\boldsymbol{\xi}, \mathbf{0}) \mathrm{d}\boldsymbol{\xi}, \tag{3.127}$$

and the QMOM moments are defined by

$$m^*(\mathbf{k}) = \sum_{\alpha=1}^{N} w_\alpha \boldsymbol{\xi}_\alpha^{\mathbf{k}}. \tag{3.128}$$

In these expressions, we have used the index notation $\mathbf{i} = (i_1, \ldots, i_M)$ such that $\boldsymbol{\xi}^{\mathbf{i}} = \xi_1^{i_1} \ldots \xi_M^{i_M}$. Note that, in the limit where $|\boldsymbol{\Sigma}| \to 0$, only the zeroth-order moment $m_{\Sigma}(\mathbf{0}) = 1$ is nonzero, and Eq. (3.126) reverts to the QMOM moments. The central Gaussian moments $m_{\Sigma}(\mathbf{i})$ are known functions of the covariance matrix $\boldsymbol{\Sigma}$. For a given moment order $\gamma = k_1 + \cdots + k_M$, Eq. (3.126) has a lower triangular form that can be inverted using forward substitution:

$$\begin{aligned} m^*(\mathbf{0}) &= m(\mathbf{0}), \\ m^*(\mathbf{e}_i) &= m(\mathbf{e}_i), \\ &\vdots \\ m^*(\mathbf{k}) &= m(\mathbf{k}) - \sum_{\substack{i_1=0 \\ i_1+\cdots+i_M>0}}^{k_1} \cdots \sum_{i_M=0}^{k_M} \binom{k_1}{i_1} \cdots \binom{k_M}{i_M} m^*(\mathbf{k}-\mathbf{i}) m_{\Sigma}(\mathbf{i}), \end{aligned} \tag{3.129}$$

where $\mathbf{e}_i$ is the ith unit vector. Thus, given a known extended optimal moment set $m(\mathbf{k})$ and a known covariance matrix $\boldsymbol{\Sigma}$, Eq. (3.129) can be used to find the optimal moment set

[23] Here, to simplify the notation, in comparison with that reported in Section 3.2, we assume that $N_1 = N_2 = \cdots = N_M = n$.

$m^*(\mathbf{k})$ used for multivariate quadrature (e.g. the CQMOM). In other words, once $m^*(\mathbf{k})$ is known, we can compute a set of N weights w_α and N abscissas $\boldsymbol{\xi}_\alpha$. It therefore remains to explain how $\boldsymbol{\Sigma}$ is determined from the extended optimal moments.

The algorithm for computing $\boldsymbol{\Sigma}$ is analogous to the one used in the univariate case. Using a root-finding (or minimization) scheme, we must vary the components of $\boldsymbol{\Sigma}$ until the objective function is minimized. The components of the objective function found from the extended optimal moments are

$$\begin{aligned}
f_1(\boldsymbol{\Sigma}) &= m(2n, 0, \dots, 0) - \sum_{i=0}^{2n} \binom{2n}{i} m^*(2n - i, 0, \dots, 0) m_{\boldsymbol{\Sigma}}(i, 0, \dots, 0), \\
&\vdots \\
f_M(\boldsymbol{\Sigma}) &= m(0, \dots, 0, 2n) - \sum_{i=0}^{2n} \binom{2n}{i} m^*(0, \dots, 0, 2n - i) m_{\boldsymbol{\Sigma}}(0, \dots, 0, i), \\
f_{1,2}(\boldsymbol{\Sigma}) &= m(n, n, 0, \dots, 0) - \sum_{i,j=0}^{n} \binom{n}{i}\binom{n}{j} m^*(n - i, n - j, 0, \dots, 0) m_{\boldsymbol{\Sigma}}(i, j, 0, \dots, 0), \\
&\vdots \\
f_{M-1,M}(\boldsymbol{\Sigma}) &= m(0, \dots, 0, n, n) - \sum_{i,j=0}^{n} \binom{n}{i}\binom{n}{j} m^*(0, \dots, 0, n - i, n - j) m_{\boldsymbol{\Sigma}}(0, \dots, 0, i, j).
\end{aligned} \tag{3.130}$$

Using $f_1(\boldsymbol{\Sigma})$ and $f_{1,2}(\boldsymbol{\Sigma})$ as examples, the moments m^* on the right-hand sides of Eqs. (3.130) with order $\gamma < 2n$ are found from Eqs. (3.129) for a given value of $\boldsymbol{\Sigma}$. From these moments, the weights and abscissas are computed and then used to compute the moments of order $\gamma = 2n$:

$$m^*(2n, 0, \dots, 0) = \sum_{\alpha=1}^{N} w_\alpha \left(\xi_{\alpha 1}\right)^{2n}, \quad m^*(n, n, 0, \dots, 0) = \sum_{\alpha=1}^{N} w_\alpha \left(\xi_{\alpha 1}\xi_{\alpha 2}\right)^{n}. \tag{3.131}$$

Following this procedure, all of the moments needed to compute the objective function from Eqs. (3.130) are now available. By iterating on $\boldsymbol{\Sigma}$, the sum of the squared objective function components can be minimized, under the constraint that the moments m^* remain realizable. Note that, although only a few components of $\boldsymbol{\Sigma}$ appear in moments such as $m_{\boldsymbol{\Sigma}}(i, 0, \dots, 0)$ and $m_{\boldsymbol{\Sigma}}(i, j, 0, \dots, 0)$, the weights and abscissas involve all components. Thus, in general, it is not possible to uncouple the solution algorithm for the separate components.

The application of multivariate EQMOM to evaluate integrals is very similar to that of univariate EQMOM. For the multi-Gaussian EQMOM, we have

$$\int_{\mathbb{R}^M} n(\boldsymbol{\xi}) g(\boldsymbol{\xi}) \mathrm{d}\boldsymbol{\xi} \approx \sum_{\alpha=1}^{N} \sum_{\beta=1}^{N'} w_\alpha w_\beta^* g\left(\boldsymbol{\Sigma}^{1/2} \boldsymbol{\varphi}_\beta + \boldsymbol{\xi}_\alpha\right), \tag{3.132}$$

where $N' = (n')^M$ and n' is the number of quadrature points used to represent each univariate Gaussian distribution. The square-root matrix $\boldsymbol{\Sigma}^{1/2}$, defined such that $\boldsymbol{\Sigma} = \left(\boldsymbol{\Sigma}^{1/2}\right)^2$, is a convenient (but not unique) choice for transforming the tensor-product abscissas $\boldsymbol{\varphi}_\beta$. The latter (along with the weights w^*_β) are found using an M-dimensional Gaussian distribution with unit covariance matrix (or the recursion coefficients for the Hermite polynomials). (See Figure 3.5 for related examples with $n = 2, 3$.) As shown in Exercise 3.8, for a unit covariance matrix all multivariate quadrature methods with Gaussian moments will yield the same values for w^*_β and $\boldsymbol{\varphi}_\beta$ (which are related to the zeros of a Hermite polynomial (Gautschi, 2004)). As in the univariate case, w^*_β and $\boldsymbol{\varphi}_\beta$ are constant and can be pre-computed for given values of M and n'. The dual-quadrature representation of the multivariate NDF is thus

$$n(t, \mathbf{x}, \boldsymbol{\xi}) = \sum_{\alpha=1}^{N} \sum_{\beta=1}^{N'} w_\alpha(t, \mathbf{x}) w^*_\beta \delta\left(\boldsymbol{\xi} - \boldsymbol{\xi}_\alpha(t, \mathbf{x}) - \boldsymbol{\Sigma}^{1/2}(t, \mathbf{x})\boldsymbol{\varphi}_\beta\right), \tag{3.133}$$

and we can again clearly see the role that $\boldsymbol{\Sigma}$ plays in the quadrature approximation. In conclusion, although multivariate EQMOM is well defined in terms of the moments, the crucial step is the computation of $\boldsymbol{\Sigma}$ for a given (realizable) extended optimal moment set. To the best of the authors' knowledge, the brute-force QMOM described in this section has been applied to solve moment transport equations only for the bivariate case ($M = n = n' = 2$) using the CQMOM with $\boldsymbol{\Sigma} = \text{diag}(\sigma_{11}, \sigma_{22})$ (Chalons *et al.*, 2010).

Extended CQMOM

In order to avoid the need for an iterative solver, we can extend the CQMOM algorithm described in Section 3.2.3 to include the additional moments needed to compute the parameters $\sigma_{\alpha\beta}$. However, in order to apply the algorithm, we must assume that the conditional PDF has the same form for all values of the conditioning variable. As an example, consider the bivariate case with (ξ_1, ξ_2). Conditioning on ξ_1, the bivariate NDF can be written as $f(\xi_1, \xi_2) = f(\xi_1) f(\xi_2|\xi_1)$. With extended CQMOM (ECQMOM) for ξ_2 conditioned on ξ_1, we can assume[24] that the NDF is be approximated by

$$f(\xi_1, \xi_2) = \sum_{\alpha=1}^{N} \sum_{\beta=1}^{N} w_{\alpha\beta} \delta_{\sigma_1}(\xi_1, \xi_{1,\alpha}) \delta_{\sigma_{2,\alpha}}(\xi_2, \xi_{2,\alpha\beta}). \tag{3.134}$$

Note that, while the spread parameter σ_1 is the same for all terms in the summation, the conditional parameter $\sigma_{2,\alpha}$ can depend on α. Also, the functional form used for δ_{σ_1} need not be the same as that used for $\delta_{\sigma_{2,\alpha}}$ (e.g. ξ_1 could use beta EQMOM, while ξ_2 uses Gaussian EQMOM). Although the form in Eq. (3.134) is not as general as that in Eq. (3.124), we shall see that it leads to a direct method for moment inversion that is very similar to the one used in the CQMOM. The bivariate moments found from Eq. (3.134) have the form

$$\begin{aligned} m_{i,j} &= \sum_{\alpha=1}^{N} \sum_{\beta=1}^{N} w_{\alpha\beta} \left(\int \xi_1^i \delta_{\sigma_1}(\xi_1, \xi_{1,\alpha}) \mathrm{d}\xi_1 \right) \left(\int \xi_2^j \delta_{\sigma_{2,\alpha}}(\xi_2, \xi_{2,\alpha\beta}) \mathrm{d}\xi_2 \right) \\ &= \sum_{\alpha=1}^{N} \sum_{\beta=1}^{N} w_{\alpha\beta} m_{1,i}^{(\alpha)} m_{2,j}^{(\alpha\beta)}, \end{aligned} \tag{3.135}$$

where $m_{1,i}^{(\alpha)}$ and $m_{2,j}^{(\alpha\beta)}$ are defined by the integrals on the right-hand side. Thus, for $j = 0$, the univariate moments with respect to ξ_1 reduce to

[24] More general forms are possible for $\delta_{\sigma_{2,\alpha}}(\xi_2, \xi_{2,\alpha\beta})$. For example, it could depend on ξ_1, since $\delta_{\sigma_1}(\xi_1, \xi_{1,\alpha})$ will be known when solving for the parameters in $\delta_{\sigma_{2,\alpha}}(\xi_2, \xi_{2,\alpha\beta})$. Nevertheless, the numerical inversion algorithm can quickly become intractable if too complicated a dependence on ξ_1 is attempted.

$$m_{i,0} = \sum_{\alpha=1}^{N} w_\alpha m_{1,i}^{(\alpha)}, \tag{3.136}$$

where $w_\alpha = \sum_{\beta=1}^{N} w_{\alpha\beta}$. Note that Eq. (3.136) has the same form as in univariate EQMOM and thus the same algorithm can be used to find the parameters w_α, $\xi_{1,\alpha}$, and σ_1. Moreover, by construction, $m_{1,i}^{(\alpha)}$ for each value of α is a known function of these parameters. Following the same logic as that which is used in the CQMOM, we next consider moments with fixed j and $i \geq 0$:

$$\begin{aligned} m_{0,j} &= \sum_{\alpha=1}^{N} w_\alpha \mu_\alpha^j, \\ m_{1,j} &= \sum_{\alpha=1}^{N} w_\alpha \xi_{1,\alpha} \mu_\alpha^j, \\ m_{2,j} &= \sum_{\alpha=1}^{N} w_\alpha m_{1,2}^{(\alpha)} \mu_\alpha^j, \\ &\vdots \\ m_{N-1,j} &= \sum_{\alpha=1}^{N} w_\alpha m_{1,N-1}^{(\alpha)} \mu_\alpha^j, \end{aligned} \tag{3.137}$$

where we have introduced the notation

$$\mu_\alpha^j \equiv \sum_{\beta=1}^{N} w_{\alpha\beta}^* m_{2,j}^{(\alpha\beta)} \tag{3.138}$$

and $w_{\alpha\beta}^* = w_{\alpha\beta}/w_\alpha$. In Eq. (3.137), the weights w_α and abscissas $\xi_{1,\alpha}$ (as well as the moments $m_{1,i}^{(\alpha)}$) are known. Likewise, the bivariate moments on the left-hand side of Eq. (3.137) are also known (i.e. they are found by solving the moment-transport equations). The new variables μ_α^j correspond to the conditional moments of ξ_2 given ξ_1. Moreover, the form of Eq. (3.138) for fixed α is exactly the same as the moment system used in the EQMOM. Thus, assuming that μ_α^j can be computed from Eq. (3.137) for $j = 0, 1, \ldots, 2N$, Eq. (3.138) can be inverted to find $w_{\alpha\beta}^*$, $\xi_{2,\alpha\beta}$, and $\sigma_{2,\alpha}$. Analogously to the CQMOM, the conditional moments are found be solving a linear system:

$$\begin{bmatrix} 1 & \cdots & 1 \\ \xi_{1,1} & \cdots & \xi_{1,N} \\ m_{1,2}^{(1)} & \cdots & m_{1,2}^{(N)} \\ \vdots & & \vdots \\ m_{1,N-1}^{(1)} & \cdots & m_{1,N-1}^{(N)} \end{bmatrix} \begin{bmatrix} w_1 & & \\ & \ddots & \\ & & w_N \end{bmatrix} \begin{bmatrix} \mu_1^1 & \cdots & \mu_1^{2N} \\ \mu_2^1 & \cdots & \mu_2^{2N} \\ \mu_3^1 & \cdots & \mu_3^{2N} \\ \vdots & & \vdots \\ \mu_N^1 & \cdots & \mu_N^{2N} \end{bmatrix} = \begin{bmatrix} m_{0,1} & \cdots & m_{0,2N} \\ m_{1,1} & \cdots & m_{1,2N} \\ m_{2,1} & \cdots & m_{2,2N} \\ \vdots & & \vdots \\ m_{N-1,1} & \cdots & m_{N-1,2N} \end{bmatrix} \tag{3.139}$$

with $\mu_\alpha^0 = 1$. The first matrix on the left-hand side of Eq. (3.139) is a generalized Vandermonde matrix with the integer moments $m_{1,i}^{(\alpha)}$ replacing the abscissas. In the limit where $\sigma_1 \to 0$, these moments become $m_{1,i}^{(\alpha)} = \xi_{1,\alpha}^i$, and thus the coefficient matrices are the same as in the CQMOM in this limit. In general, as long as the moment vectors $\boldsymbol{\phi}_\alpha = [m_{1,0}^{(\alpha)}, m_{1,1}^{(\alpha)}, \ldots, m_{1,N-1}^{(\alpha)}]^{\mathrm{T}}$ are linearly independent, the system in Eq. (3.139) can be solved to find the conditional moments.

As an example, consider a four-node bi-Gaussian distribution given by

$$f(\xi_1, \xi_2) = \sum_{\alpha=1}^{2} \sum_{\beta=1}^{2} w_{\alpha\beta} g(\xi_1; \xi_{1,\alpha}, \sigma_1) g(\xi_2; \xi_{2,\alpha\beta}, \sigma_{2,\alpha}), \tag{3.140}$$

where the Gaussian function is

$$g(x; \mu, \sigma) = \frac{1}{\sigma \sqrt{2\pi}} \exp\left(-\frac{(x-\mu)^2}{2\sigma^2} \right). \tag{3.141}$$

In Eq. (3.140), there are 13 parameters, ($w_{\alpha\beta}$, $\xi_{1,\alpha}$, $\xi_{2,\alpha\beta}$; $\alpha, \beta = 1, 2$) and (σ_1, $\sigma_{2,\alpha}$; $\alpha = 1, 2$). These parameters must be found by solving a nonlinear system corresponding to the following 13 bivariate moments up to fifth order:

$$\begin{bmatrix} m_{0,0} & m_{0,1} & m_{0,2} & m_{0,3} & m_{0,4} \\ m_{1,0} & m_{1,1} & m_{1,2} & m_{1,3} & m_{1,4} \\ m_{2,0} & & & & \\ m_{3,0} & & & & \\ m_{4,0} & & & & \end{bmatrix}.$$

As in the 1D EQMOM, $m_{4,0}$ is used to find σ_1, while $m_{0,4}$ and $m_{1,4}$ are needed in order to find $\sigma_{2,\alpha}$ for $\alpha = 1, 2$. For $j = 0$, the five moments found from Eq. (3.136) with $i = 0, \ldots, 4$ are

$$\begin{aligned}
m_{0,0} &= w_1 + w_2, \\
m_{1,0} &= w_1 \xi_{1,1} + w_2 \xi_{1,2}, \\
m_{2,0} &= w_1(\sigma_1^2 + \xi_{1,1}^2) + w_2(\sigma_1^2 + \xi_{1,2}^2), \\
m_{3,0} &= w_1 \xi_{1,1}(3\sigma_1^2 + \xi_{1,1}^2) + w_2 \xi_{1,2}(3\sigma_1^2 + \xi_{1,2}^2), \\
m_{4,0} &= w_1 \xi_{1,1}^2(6\sigma_1^2 + \xi_{1,1}^2) + w_2 \xi_{1,2}^2(6\sigma_1^2 + \xi_{1,2}^2) + 3\sigma_1^4(w_1 + w_2).
\end{aligned}$$

This system has exactly the same form as that in the 1D bi-Gaussian quadrature in Section 3.3.2 and, hence, can be inverted directly (i.e. without iterations) to find the five parameters (w_1, w_2, $\xi_{1,1}$, $\xi_{1,2}$, and σ_1^2). The linear system for the conditional moments found from Eq. (3.139) is

$$\begin{bmatrix} 1 & 1 \\ \xi_{1,1} & \xi_{1,2} \end{bmatrix} \begin{bmatrix} w_1 & \\ & w_2 \end{bmatrix} \begin{bmatrix} \mu_1^1 & \mu_1^2 & \mu_1^3 & \mu_1^4 \\ \mu_2^1 & \mu_2^2 & \mu_2^3 & \mu_2^4 \end{bmatrix} = \begin{bmatrix} m_{0,1} & m_{0,2} & m_{0,3} & m_{0,4} \\ m_{1,1} & m_{1,2} & m_{1,3} & m_{1,4} \end{bmatrix}$$

with $\mu_1^0 = \mu_2^0 = 1$. As long as $\xi_{1,1} \neq \xi_{1,2}$ and $w_1, w_2 > 0$, this system can be inverted to find μ_α^j. From the definition of the conditional moments, we then have

$$\begin{aligned}
1 &= w_{\alpha 1}^* + w_{\alpha 2}^*, \\
\mu_\alpha^1 &= w_{\alpha 1}^* \xi_{2,\alpha 1} + w_{\alpha 2}^* \xi_{2,\alpha 2}, \\
\mu_\alpha^2 &= w_{\alpha 1}^*(\sigma_{2,\alpha}^2 + \xi_{2,\alpha 1}^2) + w_{\alpha 2}^*(\sigma_{2,\alpha}^2 + \xi_{2,\alpha 2}^2), \\
\mu_\alpha^3 &= w_{\alpha 1}^* \xi_{2,\alpha 1}(3\sigma_{2,\alpha}^2 + \xi_{2,\alpha 1}^2) + w_{\alpha 2}^* \xi_{2,\alpha 2}(3\sigma_{2,\alpha}^2 + \xi_{2,\alpha 2}^2), \\
\mu_\alpha^4 &= w_{\alpha 1}^* \xi_{2,\alpha 1}^2(6\sigma_{2,\alpha}^2 + \xi_{2,\alpha 1}^2) + w_{\alpha 2}^* \xi_{2,\alpha 2}^2(6\sigma_{2,\alpha}^2 + \xi_{2,\alpha 2}^2) + 3\sigma_{2,\alpha}^4.
\end{aligned}$$

For each α, this system again has exactly the same form as that in 1D bi-Gaussian quadrature and, hence, can be inverted to find the five parameters ($w_{\alpha 1}$, $w_{\alpha 2}$, $\xi_{2,\alpha 1}$, $\xi_{2,\alpha 2}$, and $\sigma_{2,\alpha}^2$). Once this has been done, the 13 parameters appearing in Eq. (3.140) are known.

Except in degenerate cases, the 2D moment-inversion algorithm described above is able to recover 13 bivariate moments. In practice, there exist several possible degenerate cases, for example, when the true distribution f can be exactly represented by the sum of fewer than four Gaussian distributions. Another possible degeneracy occurs when the variance of ξ_1 is null, but the variance of ξ_2 is finite (i.e. the support of f is a line perpendicular to the ξ_1 axis). By using the adaptive Wheeler algorithm to handle all possible degenerate cases, a robust 2D moment-inversion algorithm can be constructed using the ECQMOM to find the unknown parameters, provided that the input bivariate moment set is realizable. Because the ECQMOM requires only multiple 1D EQMOM solves, it is much faster to compute than the brute-force QMOM algorithm.

3.4 The direct quadrature method of moments (DQMOM)

The direct quadrature method of moments (DQMOM) is technically not a moment-inversion algorithm. Instead, with the DQMOM a moment-inversion algorithm is used to find initial values for the weights $w_\alpha(0, \mathbf{x})$ and abscissas $\boldsymbol{\xi}_\alpha(0, \mathbf{x})$, and then transport equations are solved for $w_\alpha(t, \mathbf{x})$ and $\boldsymbol{\xi}_\alpha(t, \mathbf{x})$. Obviously, the DQMOM is not compatible with the EQMOM because it relies on the dual-quadrature representation of the approximate NDF. The formal derivation of these transport equations is described in Marchisio & Fox (2005) and in Fox (2003), starting from the NDF transport equation. Another alternative is to derive the DQMOM transport equations starting from the transport equations for integer moments by making the nonlinear change of variables

$$m_{\mathbf{k}}(t, \mathbf{x}) = \sum_{\alpha=1}^{N} w_\alpha(t, \mathbf{x}) \xi_{1,\alpha}^{k_1}(t, \mathbf{x}) \dots \xi_{M,\alpha}^{k_M}(t, \mathbf{x}).$$

The original motivation for developing the DQMOM was the lack of a robust multivariate moment-inversion algorithm for $M > 1$ (i.e. which is now overcome by the development of the CQMOM). In principle, the DQMOM requires only a multivariate moment-inversion algorithm to find the initial conditions, after which the time evolution of the weights and abscissas is found by solving their transport equations. In fact, for $M = 1$ the univariate QMOM algorithm that solves transport equations for the moments is equivalent to the DQMOM (up to numerical error) for parabolic systems (i.e. systems with diffusion terms). For purely hyperbolic systems, the solutions to the DQMOM equations can be different because of characteristic crossings.[25] (See Appendix C for details.) Thus, for hyperbolic systems, solving the moment-transport equations is the only physically valid option when the number of abscissas N is fixed.

In order to understand the DQMOM, we consider here only the univariate equations associated with the NDF transport equation:

$$\frac{\partial f}{\partial t} + \frac{\partial u f}{\partial x} + \frac{\partial S f}{\partial \xi} = \Gamma \frac{\partial^2 f}{\partial x^2}, \tag{3.142}$$

where $S(\xi)$ is a generic source term and $\Gamma > 0$ is a (constant) spatial diffusion coefficient. The velocity $u(t, x)$ is assumed to be known, as is the initial NDF $f(0, x, \xi)$. The

[25] When characteristics cross, the number N of nonzero weights must increase locally in order to have a physically valid solution using the DQMOM. However, since the number of crossings could tend to infinity, the numerical complexity can quickly become intractable.

moment-transport equation corresponding to Eq. (3.142) is

$$\frac{\partial m_k}{\partial t} + \frac{\partial u m_k}{\partial x} = k \int \xi^{k-1} S f \, \mathrm{d}\xi + \Gamma \frac{\partial^2 m_k}{\partial x^2}, \tag{3.143}$$

wherein the source-term integral must be closed by quadrature:

$$\int \xi^{k-1} S f \, \mathrm{d}\xi \approx \sum_{\alpha=1}^{N} w_\alpha \xi_\alpha^{k-1} S(\xi_\alpha).$$

In this expression, the weights and abscissas are found from the moments at every time step and at every spatial location using a moment-inversion algorithm. Following the method described in Marchisio & Fox (2005), the DQMOM equations corresponding to Eq. (3.143) are

$$\begin{aligned} \frac{\partial w_\alpha}{\partial t} + \frac{\partial u w_\alpha}{\partial x} &= \Gamma \frac{\partial^2 w_\alpha}{\partial x^2} + a_\alpha, \\ \frac{\partial w_\alpha \xi_\alpha}{\partial t} + \frac{\partial u w_\alpha \xi_\alpha}{\partial x} &= \Gamma \frac{\partial^2 w_\alpha \xi_\alpha}{\partial x^2} + w_\alpha S(\xi_\alpha) + b_\alpha, \end{aligned} \tag{3.144}$$

where a_α and b_α are source terms found by forcing Eq. (3.144) to agree with the moment-transport equation for $k = 0, 1, \ldots, 2N - 1$. For clarity, we will consider the case with $N = 2$. The linear system for the moment constraints then yields

$$\begin{bmatrix} 1 & 1 & 0 & 0 \\ 0 & 0 & 1 & 1 \\ -\xi_1^2 & -\xi_2^2 & 2\xi_1 & 2\xi_2 \\ -2\xi_1^3 & -2\xi_2^3 & 3\xi_1^2 & 3\xi_2^2 \end{bmatrix} \begin{bmatrix} a_1 \\ a_2 \\ b_1 \\ b_2 \end{bmatrix} = \begin{bmatrix} 0 \\ 0 \\ 2w_1\Gamma(\partial\xi_1/\partial x)^2 + 2w_2\Gamma(\partial\xi_2/\partial x)^2 \\ 6w_1\xi_1\Gamma(\partial\xi_1/\partial x)^2 + 6w_2\xi_2\Gamma(\partial\xi_2/\partial x)^2 \end{bmatrix}. \tag{3.145}$$

The coefficient matrix[26] on the left-hand side of Eq. (3.145) will be non-singular if $\xi_1 \neq \xi_2$, which corresponds to a nonzero variance. The vector on the right-hand side arises due to the diffusion term and involves the squared gradients of the abscissas. In practice, these gradients can become quite large, making the source terms very stiff. Obviously, special care must be taken if the initial NDF has zero variance (see Marchisio & Fox (2005) for an example), or if the abscissas become nearly equal, causing the coefficient matrix to be ill-conditioned. Nonetheless, as long as $\Gamma > 0$ and the numerical algorithm fully resolves the spatial gradients, the solution to Eq. (3.144) will remain quite accurate compared with the solution to Eq. (3.143).

On the other hand, when Γ is very small (or zero) the accuracy of the DQMOM solutions can be seriously degraded by numerical errors. An important point to recognize when designing numerical methods for Eq. (3.144) is that the weights and abscissas are not conserved quantities, while the moments are conserved in the absence of source terms. Thus, conservative finite-volume schemes are very well adapted to solving Eq. (3.143), but do not guarantee conservation for the DQMOM equations due to the nonzero source terms a_α and b_α. The obvious disadvantage of solving the moment-transport equations relative to the DQMOM is that a moment-inversion algorithm must be applied at every time step and in every grid cell to find the weights and abscissas. However, because high-order flux reconstruction with moment-transport equations will inevitably lead to unrealizable moments

[26] As has already been mentioned, this coefficient matrix, also indicated as **A**, corresponds to the Jacobian of the nonlinear system reported in Eq. (3.43) and employed in brute-force QMOM for calculating the quadrature approximation.

even when the moment-transport equations are closed, a moment-inversion algorithm cannot be avoided if high-order fluxes are used (see Appendix B for details). This fact is especially true for multivariate cases, for which it is even more difficult to ensure moment realizability than it is in univariate cases. Finally, for hyperbolic and nearly hyperbolic systems, the EQMOM moment-inversion algorithm provides a higher-fidelity representation of the spatial fluxes. Because the DQMOM is not compatible with the EQMOM, moment systems with very small spatial diffusion should be treated by solving the moment-transport equations using the EQMOM.

4

The generalized population-balance equation

The goal of this chapter is to provide a relatively rigorous derivation of the generalized population-balance equation (GPBE) starting from the microscale description of a disperse multiphase system. We begin by defining the number-density function (NDF) for a system of discrete particles using a probability-density-function (PDF) approach. Once the NDF has been defined, we proceed to the derivation of the GPBE by introducing the concept of conditional expected values. The latter contain the mesoscale representation of the microscale physics and thus contain the mesoscale models needed to close the GPBE. Next, we provide a detailed explanation of how the transport equations are found from the GPBE for selected multivariate moments of interest in later chapters. The chapter concludes with a short description of moment closures in the context of moment-transport equations.

4.1 Particle-based definition of the NDF

In the particle-based definition of the NDF, we begin at the microscale and write a dynamic equation for the rate of change of the disperse-phase particle properties at the mesoscale. The simplest system, which we consider first, is a collection of interacting particles in a vacuum wherein the particles interact through collisions and short-range forces. Such a system is referred to as a *granular* system. We then consider a disperse two-phase system, wherein the particles are dispersed in a fluid.

4.1.1 Definition of the NDF for granular systems

Consider first a granular system of N_p particles in the absence of a surrounding fluid phase. The dynamics of the nth particle can be described by the position of its center of mass $\mathbf{X}^{(n)}$,[1]

[1] Because particles have finite size, the volume occupied by a particle will be nonzero. Nevertheless, insofar as the particle dynamics are concerned, we can label a particle by its center of mass, which is a point with no volume.

its velocity $\mathbf{U}^{(n)}$,[2] and its M internal coordinates $\boldsymbol{\xi}^{(n)}$.[3] For simplicity, we will assume for now that the particles have identical masses. Without loss of generality, the dynamics of the particles can be described by a system of differential equations:

$$\frac{\mathrm{d}\mathbf{X}^{(n)}}{\mathrm{d}t} = \mathbf{U}^{(n)}, \tag{4.1}$$

$$\frac{\mathrm{d}\mathbf{U}^{(n)}}{\mathrm{d}t} = \mathbf{A}^{(n)} + \mathbf{C}_U^{(n)}, \tag{4.2}$$

and

$$\frac{\mathrm{d}\boldsymbol{\xi}^{(n)}}{\mathrm{d}t} = \mathbf{G}^{(n)} + \mathbf{C}_\xi^{(n)}. \tag{4.3}$$

The operators on the right-hand sides represent, respectively, continuous acceleration $\mathbf{A}^{(n)}$ due to body forces, changes due to discontinuous collisions (or particle–particle interactions) $\mathbf{C}_U^{(n)}$ and $\mathbf{C}_\xi^{(n)}$, and changes due to "growth" $\mathbf{G}^{(n)}$ (i.e. continuous changes in the internal coordinates). In general, these operators depend on the complete set of variables for all particles, and are intrinsic functions of the volume of the system containing the particles. In other words, the system volume must be large enough to contain all particles with nonzero volumes. Equivalently, we can say that the operators are intrinsic functions of the particle-number density. In order to simplify the notation, we will denote the set of all particle variables as follows:

$$\left\{\mathbf{X}^{(n)}\right\} \equiv \left\{\mathbf{X}^{(1)}, \ldots, \mathbf{X}^{(N_\mathrm{p})}\right\}, \tag{4.4}$$

$$\left\{\mathbf{U}^{(n)}\right\} \equiv \left\{\mathbf{U}^{(1)}, \ldots, \mathbf{U}^{(N_\mathrm{p})}\right\}, \tag{4.5}$$

and

$$\left\{\boldsymbol{\xi}^{(n)}\right\} \equiv \left\{\boldsymbol{\xi}^{(1)}, \ldots, \boldsymbol{\xi}^{(N_\mathrm{p})}\right\}. \tag{4.6}$$

Note that this system has a total of $N_\mathrm{d} = N_\mathrm{p}(3 + 3 + M)$ degrees of freedom.

For each set of initial conditions, Eqs. (4.1)–(4.3) can be solved to find $\{\mathbf{X}^{(n)}\}$, $\{\mathbf{U}^{(n)}\}$, and $\{\boldsymbol{\xi}^{(n)}\}$. The initial conditions are randomly selected from known distribution functions, and we can assume that there is an infinite number of possible combinations. Each combination is called a *realization* of the granular flow, and the set of all possible realizations forms an *ensemble*. Note that, because the particles have finite size, they cannot be located at the same point: thus $\mathbf{X}^{(n)} \neq \mathbf{X}^{(m)}$ for $n \neq m$. Also, the collision operator $\mathbf{C}_U^{(n)}$ will generate chaotic trajectories and thus the particle positions will become uncorrelated after a relatively small number of collisions. In contrast, for particles suspended in a fluid the collisions are suppressed and correlations can be long-lived and of long range. We will make these concepts more precise when we introduce fluid–particle systems later. While the exact nature of the particle correlations is not a factor in the definition of the multi-particle joint PDF introduced below, it is important to keep in mind that they will have

[2] We could use the particle momentum instead of the velocity. However, since the particle mass can be one of the internal coordinates, there is no loss of generality by using the velocity.

[3] The notion of an internal coordinate is very general and can include particle mass, radius, geometry, angular momentum, etc. As we describe later, modeling how the internal coordinates change due to the microscale physics is usually nontrivial.

a strong influence on the particle dynamics. Given that we have an infinite ensemble of realizations, we can define a *multi-particle joint PDF* at time t by[4]

$$f_{N_\mathrm{p}}\left(t, \{\mathbf{x}^{(n)}\}, \{\mathbf{V}^{(n)}\}, \{\boldsymbol{\eta}^{(n)}\}\right) \mathrm{d}\{\mathbf{x}^{(n)}\}\mathrm{d}\{\mathbf{V}^{(n)}\}\,\mathrm{d}\{\boldsymbol{\eta}^{(n)}\} \equiv P\Big[\cap_{n=1}^{N_\mathrm{p}} \Big\{\left(\mathbf{x}^{(n)} < \mathbf{X}^{(n)}(t) \le \mathbf{x}^{(n)} + \mathrm{d}\mathbf{x}^{(n)}\right) \cap \left(\mathbf{V}^{(n)} < \mathbf{U}^{(n)}(t) \le \mathbf{V}^{(n)} + \mathrm{d}\mathbf{V}^{(n)}\right) \cap \left(\boldsymbol{\eta}^{(n)} < \boldsymbol{\xi}^{(n)}(t) \le \boldsymbol{\eta}^{(n)} + \mathrm{d}\boldsymbol{\eta}^{(n)}\right)\Big\}\Big], \tag{4.7}$$

where the probability on the right-hand side is computed using all realizations in the ensemble, and the N_d state-space variables are $\{\mathbf{x}^{(n)}\}$, $\{\mathbf{V}^{(n)}\}$, and $\{\boldsymbol{\eta}^{(n)}\}$. Although it has too many degrees of freedom to be useful for modeling practical systems, in principle this joint PDF completely describes the dynamics of the granular system. The multi-particle joint PDF can be reduced to a *single-particle joint PDF* by integrating out all of the state variables for the other particles. Formally for the nth particle this yields

$$f_1^{(n)}\left(t, \mathbf{x}^{(n)}, \mathbf{V}^{(n)}, \boldsymbol{\eta}^{(n)}\right) = \int_{m\neq n} f_{N_\mathrm{p}}\left(t, \{\mathbf{x}^{(m)}\}, \{\mathbf{V}^{(m)}\}, \{\boldsymbol{\eta}^{(m)}\}\right) \mathrm{d}\{\mathbf{x}^{(m)}\}\mathrm{d}\{\mathbf{V}^{(m)}\}\mathrm{d}\{\boldsymbol{\eta}^{(m)}\}, \tag{4.8}$$

where the integral is over all $m \neq n$. In the special case in which all particles are *identically distributed*, the single-particle joint PDF will be the same for all n and we can denote it simply by f_1.[5] Note that, unlike f_{N_p}, $f_1^{(n)}$ does not offer a complete description of the particle dynamics, since two-particle (or more) correlations cannot in general be found from the single-particle PDF. Moreover, the number of phase-space variables in $f_1^{(n)}$ is $3 + 3 + M$, which is $1/N_\mathrm{p}$ times smaller than N_d. The general NDF introduced in Chapter 2 is defined in terms of an expected value with respect to f_{N_p}:[6]

$$\begin{aligned} n(t, \mathbf{x}, \mathbf{v}, \boldsymbol{\xi}) &\equiv \sum_{n=1}^{N_\mathrm{p}} \left\langle \delta\left(\mathbf{X}^{(n)}(t) - \mathbf{x}\right) \delta\left(\mathbf{U}^{(n)}(t) - \mathbf{v}\right) \delta\left(\boldsymbol{\xi}^{(n)}(t) - \boldsymbol{\xi}\right)\right\rangle \\ &= \sum_{n=1}^{N_\mathrm{p}} \int \delta\left(\mathbf{x}^{(n)} - \mathbf{x}\right) \delta\left(\mathbf{V}^{(n)} - \mathbf{v}\right) \delta\left(\boldsymbol{\eta}^{(n)} - \boldsymbol{\xi}\right) f_{N_\mathrm{p}}\, \mathrm{d}\{\mathbf{x}^{(m)}\}\mathrm{d}\{\mathbf{V}^{(m)}\}\mathrm{d}\{\boldsymbol{\eta}^{(m)}\} \\ &= \sum_{n=1}^{N_\mathrm{p}} \int \delta\left(\mathbf{x}^{(n)} - \mathbf{x}\right) \delta\left(\mathbf{V}^{(n)} - \mathbf{v}\right) \delta\left(\boldsymbol{\eta}^{(n)} - \boldsymbol{\xi}\right) f_1^{(n)}\, \mathrm{d}\mathbf{x}^{(n)}\, \mathrm{d}\mathbf{V}^{(n)}\, \mathrm{d}\boldsymbol{\eta}^{(n)} \\ &= \sum_{n=1}^{N_\mathrm{p}} f_1^{(n)}(t, \mathbf{x}, \mathbf{v}, \boldsymbol{\xi}). \end{aligned} \tag{4.9}$$

Thus, if all particles are identically distributed, we find

$$n(t, \mathbf{x}, \mathbf{v}, \boldsymbol{\xi}) = N_\mathrm{p} f_1(t, \mathbf{x}, \mathbf{v}, \boldsymbol{\xi}), \tag{4.10}$$

---

[4]The probability statement $P[A \cap B \cap C]$ is interpreted as the probability that events A, B, and C occur together. The probability is computed as the fraction of all realizations in the ensemble for which the statement $A \cap B \cap C$ is true.

[5]One can argue that, because the particle-numbering system is arbitrary, the particles must be identically distributed. See Subramaniam (2000) for a discussion of examples for which this does not hold. In any case, when deriving models for the NDF in practical systems, it will be convenient to assume that $f_1^{(n)} = f_1$.

[6]Recall that the random variables in this equation are the complete set of particle positions, velocities, and internal coordinates. The state-space variables $\mathbf{x}$, $\mathbf{v}$, and $\boldsymbol{\xi}$ are fixed.

where the state-space variables $\mathbf{x}$, $\mathbf{v}$, and $\boldsymbol{\xi}$ are fixed. The first line of Eq. (4.9) is the definition of the NDF, and the second line is the definition of the expected value in terms of the multi-particle joint PDF. In the third line, we integrate out all particle information except for the nth particle, leaving only the single-particle joint PDF. In the final line, we apply the definition of the multivariate delta function. The other NDF defined in Chapter 2 (i.e. n_{ξ} and n_{U}) can be found from Eq. (4.9) by integrating out the appropriate internal coordinates. It is now evident from Eq. (4.9) that the general NDF is defined in terms of the following probability statement:[7]

$$n(t, \mathbf{x}, \mathbf{v}, \boldsymbol{\xi})\mathrm{d}\mathbf{x}\,\mathrm{d}\mathbf{v}\,\mathrm{d}\boldsymbol{\xi} \equiv P\left[\cup_{n=1}^{N_{\mathrm{p}}}\left\{\left(\mathbf{x} < \mathbf{X}^{(n)}(t) \le \mathbf{x} + \mathrm{d}\mathbf{x}\right) \cap \left(\mathbf{v} < \mathbf{U}^{(n)}(t) \le \mathbf{v} + \mathrm{d}\mathbf{v}\right)\right.\right.$$
$$\left.\left.\cap\left(\boldsymbol{\xi} < \boldsymbol{\xi}^{(n)}(t) \le \boldsymbol{\xi} + \mathrm{d}\boldsymbol{\xi}\right)\right\}\right]. \tag{4.11}$$

The probability on the right-hand side is again defined with respect to the ensemble of all realizations of the granular flow. The dynamical behavior of the NDF will be determined by the right-hand sides of Eqs. (4.1)–(4.3), and we will return to this aspect later. First, however, we will look at methods to *estimate* the NDF that are based on a *single* realization of the granular flow.

4.1.2 NDF estimation methods

The formal definition of the NDF given in Eq. (4.11) is mathematically consistent, but difficult to implement in practice. It is therefore useful to define methods for estimating the NDF from a single realization of the granular flow. Note that mathematically a statistical estimate is a *random variable*, and thus should not be confused with the NDF, which is deterministic. In order to distinguish the estimated NDF from n, we will denote the estimate by $\hat{n}$. Thus, for example, if the estimate is unbiased then $\langle \hat{n} \rangle = n$, where the expected value is taken with respect to the multi-particle joint PDF $f_{N_{\mathrm{p}}}$ defined in Eq. (4.7).

Volume averages

Because the particles are discrete, the probability that a particle is located at a given point $\mathbf{x}$ is null. Thus, in order to have a finite sample of particles to estimate the NDF, we need to introduce a kernel density function $h_W(\mathbf{x})$ centered at $\mathbf{x}$ with bandwidth W. For example, a constant kernel density function defined by

$$h_W(\mathbf{x}) = \begin{cases} 3/(4\pi W^3) & \text{if } |\mathbf{x}| \le W \\ 0 & \text{otherwise} \end{cases} \tag{4.12}$$

can be employed for this purpose. Note that this kernel is isotropic, nonnegative, and defined such that the volume integral is normalized:

$$\int h_W(\mathbf{x})\mathrm{d}\mathbf{x} = 1. \tag{4.13}$$

An estimator for the NDF can now be defined as

$$\hat{n}(t, \mathbf{x}, \mathbf{v}, \boldsymbol{\xi}) = \sum_{n=1}^{N_p} \delta\left(\mathbf{v} - \mathbf{U}^{(n)}(t)\right) \delta\left(\boldsymbol{\xi} - \boldsymbol{\xi}^{(n)}(t)\right) h_W\left(\mathbf{x} - \mathbf{X}^{(n)}(t)\right). \tag{4.14}$$

[7] The probability statement $P[A \cup B \cup C]$ is interpreted as the probability that event A, or event B, or event C occurs. The probability is computed as the fraction of all realizations in the ensemble for which the statement $A \cup B \cup C$ is true.

Note that $\hat{n}$ is essentially a *volume average* over all particles in a neighborhood of radius W around $\mathbf{x}$. The statistical properties of the NDF estimator can be found in terms of the multi-particle joint PDF. For example, the mean of the NDF estimator is

$$\begin{aligned}\langle \hat{n} \rangle &= \sum_{n=1}^{N_\text{p}} \left\langle \delta\left(\mathbf{v} - \mathbf{U}^{(n)}(t)\right) \delta\left(\boldsymbol{\xi} - \boldsymbol{\xi}^{(n)}(t)\right) h_W\left(\mathbf{x} - \mathbf{X}^{(n)}(t)\right) \right\rangle \\ &= \sum_{n=1}^{N_\text{p}} \int h_W\left(\mathbf{x} - \mathbf{x}^{(n)}\right) f_1^{(n)}\left(t, \mathbf{x}^{(n)}, \mathbf{v}, \boldsymbol{\xi}\right) \mathrm{d}\mathbf{x}^{(n)}.\end{aligned} \tag{4.15}$$

Thus the estimator will be unbiased if and only if

$$f_1^{(n)}(t, \mathbf{x}) = \int h_W\left(\mathbf{x} - \mathbf{x}^{(n)}\right) f_1^{(n)}\left(t, \mathbf{x}^{(n)}\right) \mathrm{d}\mathbf{x}^{(n)}. \tag{4.16}$$

This would occur, for example, when all particles are uniformly distributed. In this limit, the normalized standard deviation of the estimator will scale like

$$\sigma = \frac{\sigma_{\hat{n}}}{\langle \hat{n} \rangle} \sim \left(\frac{V}{V_W N_\text{p}}\right)^{1/2}, \tag{4.17}$$

where V is the volume of the domain and V_W is the sample volume. Since we must have $\sigma \ll 1$ for accurate estimates, the total number of particles in the system, N_p, must be large compared with V/V_W.[8] This constraint puts a lower limit on the sample volume that can be used to estimate the NDF. Finally, note that, if the particles are identically distributed, Eq. (4.15) becomes

$$\langle \hat{n} \rangle = N_\text{p} \int h_W(\mathbf{x} - \mathbf{y}) f_1(t, \mathbf{y}, \mathbf{v}, \boldsymbol{\xi}) \mathrm{d}\mathbf{y}, \tag{4.18}$$

which relates the volume average of the single-particle joint PDF to the expected value of the estimator for the NDF. This result is the volume-average analog to Eq. (4.10). In practice, Eq. (4.18) is of no practical use since f_1 is unknown. However, it does show that, at least in a statistical sense, volume averaging is consistent with the formal definition of the NDF.

Time averages

For systems for which volume averages are insufficient for estimating the NDF (due to the value for σ found from Eq. (4.17) being too large), we can attempt to improve the statistics by defining a time average. Theoretically, time averages should be applied only to systems that are *statistically stationary*. For the NDF, this restricts time averaging to systems for which $f_1^{(n)}$ appearing in Eq. (4.10) does not depend on t. However, in practice, time averaging can also be used for systems in which $f_1^{(n)}$ varies slowly compared with the time-averaging interval Δt. The time average of the estimator in Eq. (4.14) is defined by

$$\bar{\hat{n}}(t, \mathbf{x}, \mathbf{v}, \boldsymbol{\xi}) \equiv \sum_{n=1}^{N_\text{p}} \frac{1}{\Delta t} \int_t^{t+\Delta t} \delta\left(\mathbf{v} - \mathbf{U}^{(n)}(s)\right) \delta\left(\boldsymbol{\xi} - \boldsymbol{\xi}^{(n)}(s)\right) h_W\left(\mathbf{x} - \mathbf{X}^{(n)}(s)\right) \mathrm{d}s. \tag{4.19}$$

[8] The reader will recognize N_p/V to be the *average* number density of particles in the system. In particulate systems of practical interest, the average number density ranges from extremely large for systems of nanoparticles suspended in a fluid to very small for large particles (e.g. ore particles in mineral processing). For the former, volume averaging is useful for estimating the NDF because V_W can be chosen very small relative to the smallest scales in the fluid flow.

Note that, just like $\hat{n}$, the time average $\overline{\hat{n}}$ is a *random variable*. Thus, it is not identical to the NDF. Instead its expected value is

$$\langle \overline{\hat{n}} \rangle = \sum_{n=1}^{N_\mathrm{p}} \int h_W \left(\mathbf{x} - \mathbf{x}^{(n)} \right) \frac{1}{\Delta t} \int_t^{t+\Delta t} f_1^{(n)} \left(s, \mathbf{x}^{(n)}, \mathbf{v}, \boldsymbol{\xi} \right) \mathrm{d}s \, \mathrm{d}\mathbf{x}^{(n)}. \tag{4.20}$$

If the system is statistically stationary, then $\langle \overline{\hat{n}} \rangle = \langle \hat{n} \rangle$. In addition, if the estimator is unbiased, then $\langle \overline{\hat{n}} \rangle = n$. It is in this *statistical sense* that the time average in Eq. (4.19) represents the NDF, and hence one should not confuse $\overline{\hat{n}}$ with n by assuming that they are equivalent. When applying the time average to estimate n, the length of the time interval Δt must be chosen large enough to reduce the variance of the estimator to an acceptable level. The variance of $\overline{\hat{n}}$ can be found from the multi-particle joint PDF. In general, its normalized standard deviation will scale like

$$\frac{\sigma_{\overline{\hat{n}}}}{\langle \overline{\hat{n}} \rangle} \sim \left(\frac{V_W N_\mathrm{p}}{V} + \frac{\Delta t}{\tau_W} \right)^{-1/2}, \tag{4.21}$$

where τ_W is the autocorrelation time for the random variable $h_W(\mathbf{x} - \mathbf{X}^{(n)}(t))$. Physically, τ_W is a measure of how much time must elapse before the particles in the volume of radius W around $\mathbf{x}$ are replaced by new particles. Thus, the ratio $N_{\Delta t} = \Delta t / \tau_W$ is the effective sample size used in the time-averaging part of the estimator and $N_W = V_W N_\mathrm{p} / V$ is the effective sample size used in the volume average. The overall effective sample size for the estimator in Eq. (4.19) is thus

$$N_\mathrm{eff} = N_{\Delta t} + N_W = \frac{\Delta t}{\tau_W} + \frac{V_W N_\mathrm{p}}{V}, \tag{4.22}$$

and the parameters W and Δt must be chosen such that $N_\mathrm{eff} \gg 1$ in order to control the estimation error. Two limiting cases wherein the estimator defined in Eq. (4.19) works well are as follows.

(i) When $N_W \gg 1$ due to a large average particle number density (e.g. fine-particle and nanoparticle systems), we can take $\Delta t = 0$ and use only volume averaging.

(ii) When $\tau_W \ll 1$ due (for example) to frequent particle–particle collisions, we can set $W = 0$ and use only time averaging.

On the other hand, for systems with low particle-number density and low collision frequencies the estimator $\overline{\hat{n}}$ will yield a poor representation of the NDF. Nonetheless, this *does not imply* that the NDF cannot be defined for such systems. Indeed, it is still precisely defined by Eq. (4.9). Instead it simply states that it will be extremely difficult to estimate the NDF using a *single realization* of the granular flow. The practical consequence of this statement is that it will be difficult to validate closure models for the terms in the GPBE (using either DNS of the microscale system or experimental measurements) for systems for which the standard deviation of the estimator is large.

4.1.3 Definition of the NDF for fluid–particle systems

The definition of the NDF derived above was carried out in the limit where the particles are surrounded by a vacuum. In the more general case where they are surrounded by a fluid, each particle will be accelerated by fluid stresses acting on its surface. The exact description of the resulting surface forces is very complicated and would involve the complete microscale solution to the coupled fluid–particle system.[9] However, in order to define the NDF, we do not need to know the exact form of the surface forces. It suffices to know that they exist, and that in general they will depend not only on the surrounding fluid, but also on the complete set of random variables describing the particle system: particle position $\{\mathbf{X}^{(n)}\}$, particle velocity $\{\mathbf{U}_\mathrm{p}^{(n)}\}$, and particle internal coordinate $\{\boldsymbol{\xi}_\mathrm{p}^{(n)}\}$. Formally, the presence of the fluid will generate a new continuous acceleration term in Eq. (4.2),[10]

$$\frac{\mathrm{d}\mathbf{U}_\mathrm{p}^{(n)}}{\mathrm{d}t} = \mathbf{A}_\mathrm{fp}^{(n)} + \mathbf{A}_\mathrm{p}^{(n)} + \mathbf{C}_{\mathrm{p}U}^{(n)}, \tag{4.23}$$

and the particle internal coordinates

$$\frac{\mathrm{d}\boldsymbol{\xi}_\mathrm{p}^{(n)}}{\mathrm{d}t} = \mathbf{G}_\mathrm{p}^{(n)} + \mathbf{C}_{\mathrm{p}\xi}^{(n)} \tag{4.24}$$

can include mass transfer between phases. In most descriptions of fluid–particle flows, the mesoscale fluid–particle acceleration $\mathbf{A}_\mathrm{fp}^{(n)}$ includes a drag term of the form

$$\mathbf{A}_\mathrm{fp}^{(n)} \propto C_\mathrm{D}^{(n)} \left|\mathbf{U}_\mathrm{f}^{(n)} - \mathbf{U}_\mathrm{p}^{(n)}(t)\right| \left(\mathbf{U}_\mathrm{f}^{(n)} - \mathbf{U}_\mathrm{p}^{(n)}(t)\right), \tag{4.25}$$

where $C_\mathrm{D}^{(n)}$ is a drag coefficient and $\mathbf{U}_\mathrm{f}^{(n)}$ is a representative fluid velocity in the neighborhood of $\mathbf{X}^{(n)}(t)$.[11] The correct choice of $\mathbf{U}_\mathrm{f}^{(n)}$ will be system-dependent. However, a suitable choice might be the surface average of the fluid velocity at the edge of the momentum boundary layer surrounding the particle.[12] It should be evident to the reader that the major challenge in developing a mesoscale model for $\mathbf{A}_\mathrm{fp}^{(n)}$ is to adequately account for its dependence on the other particles without having to solve for the full multi-particle joint PDF.

For the present discussion, we do not need to know the exact form of the mesoscale model for the fluid velocity and characteristic properties (e.g. composition and enthalpy). Instead, we will simply assume that $\mathbf{U}_\mathrm{f}^{(n)}$ and $\boldsymbol{\xi}_\mathrm{f}^{(n)}$ obey[13]

$$\frac{\mathrm{d}\mathbf{U}_\mathrm{f}^{(n)}}{\mathrm{d}t} = \mathbf{A}_\mathrm{pf}^{(n)} + \mathbf{A}_\mathrm{f}^{(n)} \tag{4.26}$$

and

$$\frac{\mathrm{d}\boldsymbol{\xi}_\mathrm{f}^{(n)}}{\mathrm{d}t} = \mathbf{G}_\mathrm{f}^{(n)} + \mathbf{C}_{\mathrm{f}\xi}^{(n)}, \tag{4.27}$$

---

[9] For example, the fluid could obey the Navier–Stokes equation in the volume of the domain not occupied by particles.

[10] The fluid could also change the internal coordinates so that additional source terms for the internal coordinates are needed. For the present discussion, we will ignore this possibility.

[11] In drag models for gas–solid flows, $\mathbf{U}_\mathrm{f}^{(n)}$ is called the "fluid velocity seen by the nth solid particle." In all cases, it is important to recognize that the statistical properties of $\mathbf{U}_\mathrm{f}^{(n)}$ will in general not be the same as for a Lagrangian fluid particle. Only in the limit of zero Stokes number where all solid particles exactly follow the fluid will the statistical properties be identical. This can occur, for example, for nanoparticles submerged in a viscous fluid.

[12] It should be obvious that we cannot use the fluid velocity at the particle location $\mathbf{x} = \mathbf{X}^{(n)}(t)$ since, by definition, there is no fluid at that point. Using the momentum boundary layer to define $\mathbf{U}_\mathrm{f}^{(n)}$ has the advantage that the location will depend on the particle Reynolds number.

[13] Unlike in Peirano & Minier (2002), the time derivatives here are in a Lagrangian reference frame *following the nth particle.*

where the mesoscale particle–fluid acceleration $\mathbf{A}_{\mathrm{pf}}^{(n)}$ depends on $\mathbf{A}_{\mathrm{fp}}^{(n)}$ through the law of conservation of momentum, while $\mathbf{A}_{\mathrm{f}}^{(n)}$ depends on the pressure, stresses, and body forces in the fluid phase. Likewise, $\mathbf{G}_{\mathrm{f}}^{(n)}$ and $\mathbf{C}_{\mathrm{f}\xi}^{(n)}$ depend on $\mathbf{G}_{\mathrm{p}}^{(n)}$ and $\mathbf{C}_{\mathrm{p}\xi}^{(n)}$, respectively, through the law of conservation of mass and chemical species.[14] The set of random variables needed to describe a fluid–particle system with N_{p} particles will then be augmented to include the fluid velocity and the additional characteristic properties of the fluid (or fluid internal coordinates): $\{\mathbf{X}^{(n)}\}$, $\{\mathbf{U}_{\mathrm{p}}^{(n)}\}$, $\{\boldsymbol{\xi}_{\mathrm{p}}^{(n)}\}$, $\{\mathbf{U}_{\mathrm{f}}^{(n)}\}$, and $\{\boldsymbol{\xi}_{\mathrm{f}}^{(n)}\}$. However, the definition of the ensemble remains basically unchanged. A single realization in the ensemble is still defined by one possible set of initial conditions, but now extended to include $\{\mathbf{U}_{\mathrm{f}}^{(n)}(0)\}$ and $\{\boldsymbol{\xi}_{\mathrm{f}}^{(n)}(0)\}$. In practice, the initial microscale fluid velocity field $\mathbf{U}(0, \mathbf{x})$ can be specified in a manner that is consistent with zero-slip boundary conditions on the particle surfaces. The surface terms $\{\mathbf{A}_{\mathrm{fp}}^{(n)}\}$ can then be computed, and used with Eq. (4.25) to define $\{\mathbf{U}_{\mathrm{f}}^{(n)}(0)\}$. Given the infinite ensemble of realizations, we can now define a *multi-particle–fluid joint PDF* by

$$\begin{aligned} f_{N_{\mathrm{p}}}&\left(t, \{\mathbf{x}^{(n)}\}, \{\mathbf{V}_{\mathrm{p}}^{(n)}\}, \{\boldsymbol{\eta}_{\mathrm{p}}^{(n)}\}, \{\mathbf{V}_{\mathrm{f}}^{(n)}\}, \{\boldsymbol{\eta}_{\mathrm{f}}^{(n)}\}\right) \mathrm{d}\{\mathbf{x}^{(n)}\}\mathrm{d}\{\mathbf{V}_{\mathrm{p}}^{(n)}\}\mathrm{d}\{\boldsymbol{\eta}_{\mathrm{p}}^{(n)}\}\mathrm{d}\{\mathbf{V}_{\mathrm{f}}^{(n)}\}\mathrm{d}\{\boldsymbol{\eta}_{\mathrm{f}}^{(n)}\} \\ &= P\Big[\cap_{n=1}^{N_{\mathrm{p}}} \Big\{\left(\mathbf{x}^{(n)} < \mathbf{X}^{(n)}(t) \le \mathbf{x}^{(n)} + \mathrm{d}\mathbf{x}^{(n)}\right) \cap \left(\mathbf{V}_{\mathrm{p}}^{(n)} < \mathbf{U}_{\mathrm{p}}^{(n)}(t) \le \mathbf{V}_{\mathrm{p}}^{(n)} + \mathrm{d}\mathbf{V}_{\mathrm{p}}^{(n)}\right) \\ &\qquad \cap \left(\boldsymbol{\eta}_{\mathrm{p}}^{(n)} < \boldsymbol{\xi}_{\mathrm{p}}^{(n)}(t) \le \boldsymbol{\eta}_{\mathrm{p}}^{(n)} + \mathrm{d}\boldsymbol{\eta}_{\mathrm{p}}^{(n)}\right) \cap \left(\mathbf{V}_{\mathrm{f}}^{(n)} < \mathbf{U}_{\mathrm{f}}^{(n)}(t) \le \mathbf{V}_{\mathrm{f}}^{(n)} + \mathrm{d}\mathbf{V}_{\mathrm{f}}^{(n)}\right) \\ &\qquad \cap \left(\boldsymbol{\eta}_{\mathrm{f}}^{(n)} < \boldsymbol{\xi}_{\mathrm{f}}^{(n)}(t) \le \boldsymbol{\eta}_{\mathrm{f}}^{(n)} + \mathrm{d}\boldsymbol{\eta}_{\mathrm{f}}^{(n)}\right)\Big\}\Big], \end{aligned} \tag{4.28}$$

where the state-space variables for the fluid's velocity are $\{\mathbf{V}_{\mathrm{f}}^{(n)}\}$ and those for the fluid's characteristic properties are $\{\boldsymbol{\eta}_{\mathrm{f}}^{(n)}\}$. Then, following the same steps as were used to obtain Eq. (4.9), we can define the fluid–particle NDF by

$$n(t, \mathbf{x}, \mathbf{v}_{\mathrm{p}}, \boldsymbol{\xi}_{\mathrm{p}}, \mathbf{v}_{\mathrm{f}}, \boldsymbol{\xi}_{\mathrm{f}}) \equiv \sum_{n=1}^{N_{\mathrm{p}}} f_1^{(n)}(t, \mathbf{x}, \mathbf{v}_{\mathrm{p}}, \boldsymbol{\xi}_{\mathrm{p}}, \mathbf{v}_{\mathrm{f}}, \boldsymbol{\xi}_{\mathrm{f}}), \tag{4.29}$$

where $\mathbf{v}_{\mathrm{f}}$ is the phase-space variable for the fluid velocity and $\boldsymbol{\xi}_{\mathrm{f}}$ for fluid composition. In the limit of identically distributed particles, Eq. (4.29) becomes $n(t, \mathbf{x}, \mathbf{v}_{\mathrm{p}}, \boldsymbol{\xi}_{\mathrm{p}}, \mathbf{v}_{\mathrm{f}}, \boldsymbol{\xi}_{\mathrm{f}}) = N_{\mathrm{p}} f_1(t, \mathbf{x}, \mathbf{v}_{\mathrm{p}}, \boldsymbol{\xi}_{\mathrm{p}}, \mathbf{v}_{\mathrm{f}}, \boldsymbol{\xi}_{\mathrm{f}})$. Thus, we now have a mathematically consistent definition of the fluid–particle NDF in terms of a well-defined single-particle joint PDF.[15] For practical applications, the major remaining challenge is to find an adequate mesoscale model for the dynamics of $f_1^{(n)}$. We look at this question in Chapter 5.

In order to understand the range of physics contained in the fluid–particle NDF, it is instructive to consider two limiting cases.

(i) In the limit where the particles are completely uncoupled from the fluid, the fluid–particle NDF is separable: $n(t, \mathbf{x}, \mathbf{v}_{\mathrm{p}}, \boldsymbol{\xi}_{\mathrm{p}}, \mathbf{v}_{\mathrm{f}}, \boldsymbol{\xi}_{\mathrm{f}}) = n(t, \mathbf{x}, \mathbf{v}_{\mathrm{p}}, \boldsymbol{\xi}_{\mathrm{p}}) f_{\mathrm{f}}(t, \mathbf{x}, \mathbf{v}_{\mathrm{f}}, \boldsymbol{\xi}_{\mathrm{f}})$, where $f_{\mathrm{f}}(t, \mathbf{x}, \mathbf{v}_{\mathrm{f}}, \boldsymbol{\xi}_{\mathrm{f}})$ is the one-point PDF of the fluid defined by $f_{\mathrm{f}}(t, \mathbf{x}, \mathbf{v}_{\mathrm{f}}, \boldsymbol{\xi}_{\mathrm{f}}) = \langle \delta(\mathbf{U}_{\mathrm{f}}(t, \mathbf{x}) - \mathbf{v}_{\mathrm{f}}) \delta(\boldsymbol{\phi}_{\mathrm{f}}(t, \mathbf{x}) - \boldsymbol{\xi}_{\mathrm{f}}) \rangle$, in which $\mathbf{U}_{\mathrm{f}}(t, \mathbf{x})$ and $\boldsymbol{\phi}_{\mathrm{f}}(t, \mathbf{x})$ are the fluid velocity and characteristic-property fields. The ensemble used to compute the expected value

[14] The model for $\mathbf{G}_{\mathrm{f}}^{(n)}$ will have contributions due to the difference between the particle and fluid velocities, due to mass transfer between phases, and due to chemical reactions in the fluid phase. We will focus here on the first two contributions. The term $\mathbf{C}_{\mathrm{p}\xi}^{(n)}$ represents primarily discontinuous changes in the particle number due to particle nucleation, where the corresponding change in mass in the fluid phase is $\mathbf{C}_{\mathrm{f}\xi}^{(n)}$.

[15] The extension of this result to cases in which N_{p} changes with time due to, for example, aggregation and breakage is straightforward. One must simply condition on the value of N_{p}, and include a dynamical description of the processes that change the number of particles in the system.

on the right-hand side is the collection of all possible realizations of the fluid flow[16] (in the absence of interaction with the particles). This limit would occur when the drag coefficient in Eq. (4.25) is null, and the presence of the particles can be ignored when solving for the fluid velocity (i.e. one-way coupling).

(ii) In the limit where the particles all have the same velocity as the fluid and the fluid flow is not affected by the presence of the particles,[17] the fluid–particle NDF becomes

$$\begin{aligned} n(t, \mathbf{x}, \mathbf{v}_\mathrm{p}, \boldsymbol{\xi}_\mathrm{p}, \mathbf{v}_\mathrm{f}, \boldsymbol{\xi}_\mathrm{f}) &= \sum_{n=1}^{N_\mathrm{p}} f_1^{(n)}(t, \mathbf{x}, \mathbf{v}_\mathrm{p}, \boldsymbol{\xi}_\mathrm{p}) f_\mathrm{f}(t, \mathbf{x}, \mathbf{v}_\mathrm{f}, \boldsymbol{\xi}_\mathrm{f}) \\ &= \sum_{n=1}^{N_\mathrm{p}} f_1^{(n)}(t, \mathbf{x}, \boldsymbol{\xi}_\mathrm{p} | \mathbf{v}_\mathrm{p}) \delta(\mathbf{v}_\mathrm{p} - \mathbf{v}_\mathrm{f}) f_\mathrm{f}(t, \mathbf{x}, \mathbf{v}_\mathrm{f}, \boldsymbol{\xi}_\mathrm{f}), \end{aligned} \tag{4.30}$$

where $f_1^{(n)}(t, \mathbf{x}, \boldsymbol{\xi}_\mathrm{p} | \mathbf{v}_\mathrm{f})$ is the one-particle conditional PDF given that $\mathbf{U}_\mathrm{p}^{(n)} = \mathbf{v}_\mathrm{f}$. Thus, the general NDF can be expressed as

$$n(t, \mathbf{x}, \mathbf{v}_\mathrm{f}, \boldsymbol{\xi}_\mathrm{p}) = \sum_{n=1}^{N_\mathrm{p}} f_1^{(n)}(t, \mathbf{x}, \boldsymbol{\xi}_\mathrm{p} | \mathbf{v}_\mathrm{f}) f_\mathrm{f}(t, \mathbf{x}, \mathbf{v}_\mathrm{f}, \boldsymbol{\xi}_\mathrm{f}) \tag{4.31}$$

or, for identically distributed particles, as $n(t, \mathbf{x}, \mathbf{v}_\mathrm{f}, \boldsymbol{\xi}_\mathrm{p}) = N_\mathrm{p} f_1(t, \mathbf{x}, \boldsymbol{\xi}_\mathrm{p} | \mathbf{v}_\mathrm{f}) f_\mathrm{f}(t, \mathbf{x}, \mathbf{v}_\mathrm{f}, \boldsymbol{\xi}_\mathrm{f})$. In words, the term $N_\mathrm{p} f_1(t, \mathbf{x}, \boldsymbol{\xi}_\mathrm{p} | \mathbf{v}_\mathrm{f})$ is the NDF for the internal coordinates at time t and point $\mathbf{x}$ *given* that all particles have the same velocity as the fluid. This limit would occur in a very dilute system when the drag coefficient in Eq. (4.25) is infinite and there is no mass transfer.

The form of the fluid–particle NDF between these two limits will depend on the exact form of the surface-force models and the interstitial fluid dynamics. Thus, in theory, the only way to compute the NDF exactly is to simulate the complete microscale fluid–particle system using direct numerical simulations (DNS). Nevertheless, useful approximations for the NDF can be obtained by providing mesoscale closures to the multi-particle statistics that arise in the transport equation for the NDF (i.e. in the GPBE). In summary, we have demonstrated that by employing a well-defined multi-particle joint PDF it is straightforward to define the NDF in its various forms. Likewise, we have also shown how volume and time averages can be used, under certain conditions, to estimate the NDF from a single realization of the flow. Now we turn to the much more difficult task of deriving the GPBE that describes the evolution of the NDF starting from the differential equations that define the particle positions, velocities, internal coordinates, and surface forces. This GPBE will have unclosed terms, and thus we review the various mesoscale models in Chapter 5.

4.2 From the multi-particle–fluid joint PDF to the GPBE

In this section we give a streamlined derivation of the GPBE for the fluid–particle NDF starting from the transport equation for a multi-particle–fluid PDF. This will require three steps.

[16] In laminar flow, f_f reduces to a delta function. In turbulent flow, f_f can be modeled using PDF methods.
[17] Thus, by definition, there is no mass transfer between phases.

1. Starting from the mesoscale models for the particle properties (Eqs. (4.1), (4.3), (4.23), and (4.26)), we first derive a generalized transport equation for f_{N_p}.

2. Given the transport equation for f_{N_p}, we will find the transport equation for $f_1^{(n)}$ by integrating out all degrees of freedom except those associated with the nth particle. Owing to the loss of information, this step will generate several unclosed terms that must be modeled.

3. Finally, using the definition of the fluid–particle NDF given in Eq. (4.29), we will derive the GPBE.

Readers familiar with the kinetic theory of gas dynamics should find many similarities with the derivation of the Boltzmann kinetic equation.

4.2.1 The transport equation for the multi-particle joint PDF

The general form of the transport equation for the multi-particle joint PDF (which is a generalization of the Liouville equation) is

$$\frac{\partial f_{N_\mathrm{p}}}{\partial t} + \sum_n \frac{\partial}{\partial \mathbf{x}^{(n)}} \cdot \left(\mathbf{V}_\mathrm{p}^{(n)} f_{N_\mathrm{p}}\right) + \sum_n \frac{\partial}{\partial \mathbf{V}_\mathrm{p}^{(n)}} \cdot \left[\left(\mathbf{A}_\mathrm{fp}^{(n)} + \mathbf{A}_\mathrm{p}^{(n)}\right) f_{N_\mathrm{p}}\right] + \sum_n \frac{\partial}{\partial \boldsymbol{\eta}_\mathrm{p}^{(n)}} \cdot \left(\mathbf{G}_\mathrm{p}^{(n)} f_{N_\mathrm{p}}\right)$$
$$+ \sum_n \frac{\partial}{\partial \mathbf{V}_\mathrm{f}^{(n)}} \cdot \left[\left(\mathbf{A}_\mathrm{pf}^{(n)} + \mathbf{A}_\mathrm{f}^{(n)}\right) f_{N_\mathrm{p}}\right] + \sum_n \frac{\partial}{\partial \boldsymbol{\eta}_\mathrm{f}^{(n)}} \cdot \left(\mathbf{G}_\mathrm{f}^{(n)} f_{N_\mathrm{f}}\right) = C_{N_\mathrm{p}}. \quad (4.32)$$

Note that left-hand side of this expression is, in fact, a continuity equation for f_{N_p}, which states that the multi-particle joint PDF is constant along trajectories in phase space. The term on the right-hand side of Eq. (4.32) has a contribution due to the N_p-particle collision operator, which generates discontinuous changes in particle velocities $\mathbf{U}_\mathrm{p}^{(n)}$ and internal coordinates $\boldsymbol{\xi}_\mathrm{p}^{(n)}$, and to particle nucleation or evaporation. The first term on the left-hand side is accumulation of f_{N_p}. The remaining terms on the left-hand side represent continuous changes in particle-position space $\mathbf{x}^{(n)}$, particle-velocity space $\mathbf{V}_\mathrm{p}^{(n)}$, particle-internal-coordinate space $\boldsymbol{\eta}_\mathrm{p}^{(n)}$, fluid-velocity space $\mathbf{V}_\mathrm{f}^{(n)}$, and fluid-internal-coordinate space $\boldsymbol{\eta}_\mathrm{f}^{(n)}$. In order to simplify the notation, we have written the derivatives in a compact form that can be expanded. For example,

$$\sum_n \frac{\partial}{\partial \mathbf{x}^{(n)}} \cdot \left(\mathbf{V}_\mathrm{p}^{(n)} f_{N_\mathrm{p}}\right) \equiv \sum_{n=1}^{N_\mathrm{p}} \sum_{i=1}^{3} \frac{\partial}{\partial x_i^{(n)}} \left(V_{\mathrm{p}i}^{(n)} f_{N_\mathrm{p}}\right). \quad (4.33)$$

The total number of independent variables appearing in Eq. (4.32) is thus quite large, and in fact too large for practical applications. However, as mentioned earlier, by coupling Eq. (4.32) with the Navier–Stokes equation to find the forces on the particles due to the fluid, the N_p-particle system is completely determined. Although not written out explicitly, the reader should keep in mind that the mesoscale models for the phase-space fluxes and the collision term depend on the complete set of independent variables. For example, the surface terms depend on all of the state variables: $\mathbf{A}_\mathrm{fp}^{(n)}(\{\mathbf{x}^{(n)}\}, \{\mathbf{V}_\mathrm{p}^{(n)}\}, \{\boldsymbol{\eta}_\mathrm{p}^{(n)}\}, \{\mathbf{V}_\mathrm{f}^{(n)}\}, \{\boldsymbol{\eta}_\mathrm{f}^{(n)}\})$. The only known way to determine these functions is to perform direct numerical simulations of the microscale fluid–particle system using all possible sets of initial conditions. Obviously, such an approach is intractable. We are thus led to reduce the number of independent variables and to introduce mesoscale models that attempt to capture the "average" effect of multi-particle interactions.

4.2.2 The transport equation for the single-particle joint PDF

The simplest useful representation of the fluid–particle system is the single-particle joint PDF $f_1^{(n)}$ defined by integrating out all independent variables except those associated with the nth particle:

$$f_1^{(n)} \equiv \int_{m \neq n} f_{N_\mathrm{p}} \, \mathrm{d}\{\mathbf{x}^{(m)}\}\mathrm{d}\{\mathbf{V}_\mathrm{p}^{(m)}\}\mathrm{d}\{\boldsymbol{\eta}_\mathrm{p}^{(m)}\}\mathrm{d}\{\mathbf{V}_\mathrm{f}^{(m)}\}\mathrm{d}\{\boldsymbol{\eta}_\mathrm{f}^{(m)}\}. \tag{4.34}$$

Starting from Eq. (4.32) and integrating out all variables except those associated with the nth particle yields

$$\begin{aligned}\frac{\partial f_1^{(n)}}{\partial t} &+ \frac{\partial}{\partial \mathbf{x}^{(n)}} \cdot \left(\mathbf{V}_\mathrm{p}^{(n)} f_1^{(n)}\right) + \frac{\partial}{\partial \mathbf{V}_\mathrm{p}^{(n)}} \cdot \left[\left(\langle \mathbf{A}_\mathrm{fp}^{(n)}\rangle_1 + \langle \mathbf{A}_\mathrm{p}^{(n)}\rangle_1\right) f_1^{(n)}\right] + \frac{\partial}{\partial \boldsymbol{\eta}_\mathrm{p}^{(n)}} \cdot \left(\langle \mathbf{G}_\mathrm{p}^{(n)}\rangle_1 f_1^{(n)}\right) \\ &+ \frac{\partial}{\partial \mathbf{V}_\mathrm{f}^{(n)}} \cdot \left[\left(\langle \mathbf{A}_\mathrm{pf}^{(n)}\rangle_1 + \langle \mathbf{A}_\mathrm{f}^{(n)}\rangle_1\right) f_1^{(n)}\right] + \frac{\partial}{\partial \boldsymbol{\eta}_\mathrm{f}^{(n)}} \cdot \left(\langle \mathbf{G}_\mathrm{f}^{(n)}\rangle_1 f_1^{(n)}\right) = C_1^{(n)},\end{aligned} \tag{4.35}$$

where we have introduced the single-particle collision operator defined by

$$C_1^{(n)} \equiv \int_{m \neq n} C_{N_\mathrm{p}} \, \mathrm{d}\{\mathbf{x}^{(m)}\}\mathrm{d}\{\mathbf{V}_\mathrm{p}^{(m)}\}\mathrm{d}\{\boldsymbol{\eta}_\mathrm{p}^{(m)}\}\mathrm{d}\{\mathbf{V}_\mathrm{f}^{(m)}\}\mathrm{d}\{\boldsymbol{\eta}_\mathrm{f}^{(m)}\} \tag{4.36}$$

and the single-particle conditional expected values defined, for example, by

$$\langle \mathbf{A}_\mathrm{fp}^{(n)}\rangle_1 \equiv \frac{1}{f_1^{(n)}} \int_{m \neq n} \mathbf{A}_\mathrm{fp}^{(n)} f_{N_\mathrm{p}} \, \mathrm{d}\{\mathbf{x}^{(m)}\}\mathrm{d}\{\mathbf{V}_\mathrm{p}^{(m)}\}\mathrm{d}\{\boldsymbol{\eta}_\mathrm{p}^{(m)}\}\mathrm{d}\{\mathbf{V}_\mathrm{f}^{(m)}\}\mathrm{d}\{\boldsymbol{\eta}_\mathrm{f}^{(m)}\}. \tag{4.37}$$

Compared with $\mathbf{A}_\mathrm{fp}^{(n)}$, the left-hand side of this expression depends on a much smaller set of independent variables: $\langle \mathbf{A}_\mathrm{fp}^{(n)}\rangle_1(\mathbf{x}^{(n)}, \mathbf{V}_\mathrm{p}^{(n)}, \boldsymbol{\eta}_\mathrm{p}^{(n)}, \mathbf{V}_\mathrm{f}^{(n)}, \boldsymbol{\eta}_\mathrm{f}^{(n)})$. Thus, in principle, it should be simpler to deal with than its multi-particle counterpart. Nevertheless, due to the loss of direct information about particle–particle interactions, the single-particle conditional expected values must be represented by mesoscale closures. The development of closures that yield accurate predictions of $f_1^{(n)}$ by solving Eq. (4.35) is a nontrivial task, and the forms of the mesoscale closures will be highly dependent on the physical properties of the system. The reader should also clearly distinguish between the single-particle conditional expected values such as Eq. (4.37) and the right-hand sides of Eqs. (4.1)–(4.3). The latter can potentially involve all of the particle positions, velocities, and internal coordinates, while the former depend only on the mesoscale variables for the nth particle. In the literature, discrete-element methods (DEM) often claim to be solving the microscale problem defined by Eqs. (4.1)–(4.3), but this can only be true if the particles move in a vacuum. Once a fluid is added, modification of one particle's position/velocity can affect all other particles in the system. Thus, a DEM simulation that models the fluid forces on a particle as a drag term that depends only on that particle's velocity is not the same as the microscale DNS of Eqs. (4.1)–(4.3) and the fluid velocity field.

4.2.3 The transport equation for the NDF

The NDF is defined by Eq. (4.29), and evolves according to a GPBE. Starting from Eq. (4.35), we first rewrite the independent variables (see the right-hand side of Eq. (4.29)) as

$$\frac{\partial f_1^{(n)}}{\partial t} + \frac{\partial}{\partial \mathbf{x}} \cdot \left(\mathbf{v}_\mathrm{p} f_1^{(n)} \right) + \frac{\partial}{\partial \mathbf{v}_\mathrm{p}} \cdot \left[\left(\langle \mathbf{A}_\mathrm{fp}^{(n)} \rangle_1 + \langle \mathbf{A}_\mathrm{p}^{(n)} \rangle_1 \right) f_1^{(n)} \right] + \frac{\partial}{\partial \boldsymbol{\xi}_\mathrm{p}} \cdot \left(\langle \mathbf{G}_\mathrm{p}^{(n)} \rangle_1 f_1^{(n)} \right)$$
$$+ \frac{\partial}{\partial \mathbf{v}_\mathrm{f}} \cdot \left[\left(\langle \mathbf{A}_\mathrm{pf}^{(n)} \rangle_1 + \langle \mathbf{A}_\mathrm{f}^{(n)} \rangle_1 \right) f_1^{(n)} \right] + \frac{\partial}{\partial \boldsymbol{\xi}_\mathrm{f}} \cdot \left(\langle \mathbf{G}_\mathrm{f}^{(n)} \rangle_1 f_1^{(n)} \right) = C_1^{(n)}, \quad (4.38)$$

where all terms in Eq. (4.38) are now evaluated at the phase-space location ($\mathbf{x}$, $\mathbf{v}_\mathrm{p}$, $\boldsymbol{\xi}_\mathrm{p}$, $\mathbf{v}_\mathrm{f}$, $\boldsymbol{\xi}_\mathrm{f}$). Summing over all particles then yields the GPBE:

$$\frac{\partial n}{\partial t} + \frac{\partial}{\partial \mathbf{x}} \cdot \left(\mathbf{v}_\mathrm{p} n \right) + \frac{\partial}{\partial \mathbf{v}_\mathrm{p}} \cdot \left[\left(\langle \mathbf{A}_\mathrm{fp} \rangle_1 + \langle \mathbf{A}_\mathrm{p} \rangle_1 \right) n \right] + \frac{\partial}{\partial \boldsymbol{\xi}_\mathrm{p}} \cdot \left(\langle \mathbf{G}_\mathrm{p} \rangle_1 n \right)$$
$$+ \frac{\partial}{\partial \mathbf{v}_\mathrm{f}} \cdot \left[\left(\langle \mathbf{A}_\mathrm{pf} \rangle_1 + \langle \mathbf{A}_\mathrm{f} \rangle_1 \right) n \right] + \frac{\partial}{\partial \boldsymbol{\xi}_\mathrm{f}} \cdot \left(\langle \mathbf{G}_\mathrm{f} \rangle_1 n \right) = \mathcal{S}_1. \quad (4.39)$$

In order to account for variable particle numbers, we generalize the collision term $\mathcal{S}_1$ to include changes in N_p due to nucleation, aggregation, and breakage. These processes will also require models in order to close Eq. (4.39). This equation can be compared with Eq. (2.16) on page 37, and it can be observed that they have the same general form. However, it is now clear that the GPBE cannot be solved until mesoscale closures are provided for the conditional phase-space velocities ($\langle \mathbf{A}_\mathrm{fp} \rangle_1$, $\langle \mathbf{A}_\mathrm{p} \rangle_1$, $\langle \mathbf{G}_\mathrm{p} \rangle_1$, $\langle \mathbf{A}_\mathrm{pf} \rangle_1$, $\langle \mathbf{A}_\mathrm{f} \rangle_1$, and $\langle \mathbf{G}_\mathrm{f} \rangle_1$) and the discontinuous "source" term $\mathcal{S}_1$. Note that we have dropped the superscript $^{(n)}$ on the conditional phase-space velocities in Eq. (4.39). Formally, this implies that the definition of (for example) $\langle \mathbf{A}_\mathrm{fp} \rangle_1$ is

$$\langle \mathbf{A}_\mathrm{fp} \rangle_1 (\mathbf{x}, \mathbf{v}_\mathrm{p}, \boldsymbol{\xi}_\mathrm{p}, \mathbf{v}_\mathrm{f}, \boldsymbol{\xi}_\mathrm{f}) \equiv \frac{\sum_n \langle \mathbf{A}_\mathrm{fp}^{(n)} \rangle_1 (\mathbf{x}, \mathbf{v}_\mathrm{p}, \boldsymbol{\xi}_\mathrm{p}, \mathbf{v}_\mathrm{f}, \boldsymbol{\xi}_\mathrm{f}) f_1^{(n)} (\mathbf{x}, \mathbf{v}_\mathrm{p}, \boldsymbol{\xi}_\mathrm{p}, \mathbf{v}_\mathrm{f}, \boldsymbol{\xi}_\mathrm{f})}{\sum_n f_1^{(n)} (\mathbf{x}, \mathbf{v}_\mathrm{p}, \boldsymbol{\xi}_\mathrm{p}, \mathbf{v}_\mathrm{f}, \boldsymbol{\xi}_\mathrm{f})}.$$

Alternatively, if one argues that all particles are statistically identical, then $\langle \mathbf{A}_\mathrm{fp}^{(n)} \rangle_1$ and $f_1^{(n)}$ should not depend on n. Although this assumption is not necessary for deriving Eq. (4.39), it is usually invoked to simplify the physical modeling. In other words, if all particles are statistically identical, the mesoscale models can be validated by considering single-particle statistics from DNS for randomly selected particles.

4.2.4 The closure problem

As noted in Chapter 1, in this book we will consider computational models for cases in which Eq. (4.39) appears in closed form. In particular, this implies that all of the phase-space flux and source terms ($\langle \mathbf{A}_\mathrm{fp} \rangle_1$, $\langle \mathbf{A}_\mathrm{p} \rangle_1$, $\langle \mathbf{G}_\mathrm{p} \rangle_1$, $\langle \mathbf{A}_\mathrm{pf} \rangle_1$, $\langle \mathbf{A}_\mathrm{f} \rangle_1$, $\langle \mathbf{G}_\mathrm{f} \rangle_1$, and $\mathcal{S}_1$) in the GPBE will depend only on the independent variables (t, $\mathbf{x}$, $\mathbf{v}_\mathrm{p}$, $\boldsymbol{\xi}_\mathrm{p}$, $\mathbf{v}_\mathrm{f}$, and $\boldsymbol{\xi}_\mathrm{f}$) and on the NDF $n(t, \mathbf{x}, \mathbf{v}_\mathrm{p}, \boldsymbol{\xi}_\mathrm{p}, \mathbf{v}_\mathrm{f}, \boldsymbol{\xi}_\mathrm{f})$. We should note, however, that finding accurate mesoscale closures for these terms is a highly non-trivial task. In general, such closures will be nonlinear and can involve convolution integrals (e.g. to describe aggregation and collision processes). Thus, the accurate numerical solution of Eq. (4.39) will also be nontrivial, and will depend on the particular forms of the mesoscale closures. Another formidable complication when solving Eq. (4.39) is the fact that the NDF often depends on a large number of independent variables. In order to reduce the dimensionality of the problem, it is sometimes possible to work with selected moments of the NDF by solving their transport equations. In general, the moment-transport equations will not be closed, and further modeling approximations are necessary. In the next section, we derive the most commonly used moment-transport equations starting from Eq. (4.39).

4.3 Moment-transport equations

In this section we will derive a few example transport equations for selected moments of the NDF starting from Eq. (4.39). Because the conditional phase-space velocities appearing in Eq. (4.39) have yet to be replaced by mesoscale models, we will not enter into the details on how these terms can be approximated at the macroscale level using moment closures. However, readers familiar with the kinetic theory of gases (Chapman & Cowling, 1961) will recognize that developing macroscale closures for the moments is a major challenge. The goal here is more modest. We are primarily interested in understanding the process of finding the moment-transport equations starting from the GPBE from a mathematical perspective. However, as discussed in detail in later chapters, our overarching goal is to illustrate that a *closed* GPBE is the proper starting point for developing models for polydisperse multiphase flows. In this context, it is important that the reader understand how the moment-transport equations arise from the GPBE. As we proceed with deriving the moment-transport equations, we will also discuss the issue of consistency between the macroscale model for the fluid phase and the mesoscale models for the fluid seen by the particles (i.e., $\langle \mathbf{A}_{\mathrm{pf}} \rangle_1$, $\langle \mathbf{A}_{\mathrm{f}} \rangle_1$, and $\langle \mathbf{G}_{\mathrm{f}} \rangle_1$). This discussion will result in a set of *consistency constraints* that take the form of integrals of the fluid-phase mesoscale models over the phase-space variables. As examples, we propose some *minimal* consistent models that satisfy the consistency constraints. However, because these models are designed only to satisfy the consistency constraints (e.g. conservation of mass, species, momentum, energy, etc.), they do not rule out more complex, physics-based models. As a general rule, because the moments of interest are a linear (integral) transformation of the GPBE, the resulting transport equations are in conservative form (i.e. with accumulation and spatial-flux terms whose spatial volume integral is null for a closed system). As will be discussed in later chapters, transport equations in conservative form are ideally suited for finite-volume methods (Leveque, 2002), especially for moment systems that can generate discontinuous solutions or shocks.

4.3.1 A few words about phase-space integration

Computing moments from Eq. (4.39) basically involves integration over some of the independent variables in the NDF (i.e. $\mathbf{v}_{\mathrm{p}}$, $\boldsymbol{\xi}_{\mathrm{p}}$, $\mathbf{v}_{\mathrm{f}}$, and $\boldsymbol{\xi}_{\mathrm{f}}$) while keeping the others fixed (i.e. t and $\mathbf{x}$). In order to facilitate this task, it is useful to review some basic rules of integration.

(i) **Treatment of time and space derivatives.** Let $g(\mathbf{v}_{\mathrm{p}}, \boldsymbol{\xi}_{\mathrm{p}}, \mathbf{v}_{\mathrm{f}}, \boldsymbol{\xi}_{\mathrm{f}})$ be an arbitrary function of its variables. Then

$$\begin{aligned}
&\int \left(\frac{\partial n}{\partial t} + \frac{\partial}{\partial \mathbf{x}} \cdot \left(\mathbf{v}_{\mathrm{p}} n \right) \right) g \, \mathrm{d}\mathbf{v}_{\mathrm{p}} \, \mathrm{d}\boldsymbol{\xi}_{\mathrm{p}} \, \mathrm{d}\mathbf{v}_{\mathrm{f}} \, \mathrm{d}\boldsymbol{\xi}_{\mathrm{f}} \\
&\quad = \int \left(\frac{\partial g n}{\partial t} + \frac{\partial}{\partial \mathbf{x}} \cdot \left(\mathbf{v}_{\mathrm{p}} g n \right) \right) \mathrm{d}\mathbf{v}_{\mathrm{p}} \, \mathrm{d}\boldsymbol{\xi}_{\mathrm{p}} \, \mathrm{d}\mathbf{v}_{\mathrm{f}} \, \mathrm{d}\boldsymbol{\xi}_{\mathrm{f}} \\
&\quad = \frac{\partial}{\partial t} \left(\int g n \, \mathrm{d}\mathbf{v}_{\mathrm{p}} \, \mathrm{d}\boldsymbol{\xi}_{\mathrm{p}} \, \mathrm{d}\mathbf{v}_{\mathrm{f}} \, \mathrm{d}\boldsymbol{\xi}_{\mathrm{f}} \right) + \frac{\partial}{\partial \mathbf{x}} \cdot \left(\int \mathbf{v}_{\mathrm{p}} g n \, \mathrm{d}\mathbf{v}_{\mathrm{p}} \, \mathrm{d}\boldsymbol{\xi}_{\mathrm{p}} \, \mathrm{d}\mathbf{v}_{\mathrm{f}} \, \mathrm{d}\boldsymbol{\xi}_{\mathrm{f}} \right).
\end{aligned} \tag{4.40}$$

(ii) **Treatment of velocity derivatives.** With velocity derivatives, it is necessary to use integration by parts. For example,

$$\begin{aligned}\int g \frac{\partial}{\partial \mathbf{v}_\mathrm{p}} \cdot \left[\left(\langle \mathbf{A}_\mathrm{fp}\rangle_1 + \langle \mathbf{A}_\mathrm{p}\rangle_1\right) n\right] \mathrm{d}\mathbf{v}_\mathrm{p}\, \mathrm{d}\boldsymbol{\xi}_\mathrm{p}\, \mathrm{d}\mathbf{v}_\mathrm{f}\, \mathrm{d}\boldsymbol{\xi}_\mathrm{f} \\ = \int \frac{\partial}{\partial \mathbf{v}_\mathrm{p}} \cdot \left[\left(\langle \mathbf{A}_\mathrm{fp}\rangle_1 + \langle \mathbf{A}_\mathrm{p}\rangle_1\right) g n\right] \mathrm{d}\mathbf{v}_\mathrm{p}\, \mathrm{d}\boldsymbol{\xi}_\mathrm{p}\, \mathrm{d}\mathbf{v}_\mathrm{f}\, \mathrm{d}\boldsymbol{\xi}_\mathrm{f} \\ - \int \left(\langle \mathbf{A}_\mathrm{fp}\rangle_1 + \langle \mathbf{A}_\mathrm{p}\rangle_1\right) \cdot \frac{\partial g}{\partial \mathbf{v}_\mathrm{p}} n \,\mathrm{d}\mathbf{v}_\mathrm{p}\, \mathrm{d}\boldsymbol{\xi}_\mathrm{p}\, \mathrm{d}\mathbf{v}_\mathrm{f}\, \mathrm{d}\boldsymbol{\xi}_\mathrm{f}. \end{aligned} \tag{4.41}$$

The first term on the right-hand side of this equation is a total derivative, so it can be integrated formally to find

$$\int \frac{\partial}{\partial \mathbf{v}_\mathrm{p}} \cdot \left[\left(\langle \mathbf{A}_\mathrm{fp}\rangle_1 + \langle \mathbf{A}_\mathrm{p}\rangle_1\right) g n\right] \mathrm{d}\mathbf{v}_\mathrm{p}\, \mathrm{d}\boldsymbol{\xi}_\mathrm{p}\, \mathrm{d}\mathbf{v}_\mathrm{f}\, \mathrm{d}\boldsymbol{\xi}_\mathrm{f} = \left(\langle \mathbf{A}_\mathrm{fp}\rangle_1 + \langle \mathbf{A}_\mathrm{p}\rangle_1\right) g n\Big|_{\mathbf{v}_\mathrm{p}=\pm\infty} = 0, \tag{4.42}$$

where the second term corresponds to the flux normal to the surface bounding the velocity phase space. Because the velocity phase space extends to infinity, this flux must be null. Hence, the contributions arising due to the flux during integration by parts of the velocity variables (i.e. $\mathbf{v}_\mathrm{p}$ and $\mathbf{v}_\mathrm{f}$) are set to zero.

(iii) **Treatment of internal-coordinate derivatives.** With internal-coordinate derivatives, it is again necessary to use integration by parts:[18]

$$\begin{aligned}\int g \frac{\partial}{\partial \boldsymbol{\xi}_\mathrm{p}} \cdot \left(\langle \mathbf{G}_\mathrm{p}\rangle_1 n\right) \mathrm{d}\mathbf{v}_\mathrm{p}\, \mathrm{d}\boldsymbol{\xi}_\mathrm{p}\, \mathrm{d}\mathbf{v}_\mathrm{f}\, \mathrm{d}\boldsymbol{\xi}_\mathrm{f} \\ = \int \frac{\partial}{\partial \boldsymbol{\xi}_\mathrm{p}} \cdot \left(\langle \mathbf{G}_\mathrm{p}\rangle_1 g n\right) \mathrm{d}\mathbf{v}_\mathrm{p}\, \mathrm{d}\boldsymbol{\xi}_\mathrm{p}\, \mathrm{d}\mathbf{v}_\mathrm{f}\, \mathrm{d}\boldsymbol{\xi}_\mathrm{f} - \int \langle \mathbf{G}_\mathrm{p}\rangle_1 \cdot \frac{\partial g}{\partial \boldsymbol{\xi}_\mathrm{p}} n \,\mathrm{d}\mathbf{v}_\mathrm{p}\, \mathrm{d}\boldsymbol{\xi}_\mathrm{p}\, \mathrm{d}\mathbf{v}_\mathrm{f}\, \mathrm{d}\boldsymbol{\xi}_\mathrm{f}. \end{aligned} \tag{4.43}$$

The first term on the right-hand side of this equation is a total derivative, so it can be integrated formally. However, for internal coordinates the phase space does not usually extend to infinity. In order to see clearly what can happen, consider the case with a single internal coordinate that is bounded by zero and infinity:

$$\int \frac{\partial}{\partial \xi_\mathrm{p}} \cdot \left(\langle G_\mathrm{p}\rangle_1 g n\right) \mathrm{d}\mathbf{v}_\mathrm{p}\, \mathrm{d}\mathbf{v}_\mathrm{f}\, \mathrm{d}\boldsymbol{\xi}_\mathrm{f}\, \mathrm{d}\xi_\mathrm{p} = \langle G_\mathrm{p}\rangle_1 g n\Big|_{\xi_\mathrm{p}=\infty} - \langle G_\mathrm{p}\rangle_1 g n\Big|_{\xi_\mathrm{p}=0}. \tag{4.44}$$

The two terms on the right-hand side are the flux at infinity, which we can safely set to zero, and the flux at the origin. Depending on the forms of $\langle G_\mathrm{p}\rangle_1$ and g, the flux at the origin need not be zero. For example, if g is nonzero when $\xi_\mathrm{p} = 0$, then the product $\langle G_\mathrm{p}\rangle_1 n$ evaluated at $\xi_\mathrm{p} = 0$ would have to be zero in order for the flux term to cancel out. Unfortunately, there are important applications in which the flux term is nonzero, so one must pay attention to how the flux term is handled in the derivation of the moment-transport equations. For example, if ξ_p represents the surface of evaporating droplets and $\langle G_\mathrm{p}\rangle_1$ is constant (i.e. the evaporation rate is proportional to the surface area), then n will be nonzero at $\xi_\mathrm{p} = 0$. Physically, the nonzero flux is due to the disappearance of droplets due to evaporation, and thus it cannot be neglected.

[18] An analogous expression is found with $\boldsymbol{\xi}_\mathrm{f}$.

In summary, computing the moment-transport equations starting from Eq. (4.39) involves integration over phase space using the rules described above for particular choices of g. In the following, we will assume that the flux term at the boundary of phase space can be neglected. However, the reader should keep in mind that this assumption must be verified for particular cases.

4.3.2 Disperse-phase number transport

The total number concentration $N(t, \mathbf{x})$ corresponds to the zeroth-order moment of the NDF (i.e. $g = 1$), and is defined by

$$N \equiv \int n \, \mathrm{d}\mathbf{v}_\mathrm{p} \, \mathrm{d}\boldsymbol{\xi}_\mathrm{p} \, \mathrm{d}\mathbf{v}_\mathrm{f} \, \mathrm{d}\boldsymbol{\xi}_\mathrm{f}. \tag{4.45}$$

Its transport equation can be found from Eq. (4.39):

$$\frac{\partial N}{\partial t} + \frac{\partial}{\partial \mathbf{x}} \cdot (N \mathbf{U}_\mathrm{N}) = \mathcal{S}_\mathrm{N}, \tag{4.46}$$

where we have explicitly assumed that the flux at the boundary of phase space is null. The number-average disperse-phase velocity is defined by

$$\mathbf{U}_\mathrm{N} \equiv \frac{1}{N} \int \mathbf{v}_\mathrm{p} n \, \mathrm{d}\mathbf{v}_\mathrm{p} \, \mathrm{d}\boldsymbol{\xi}_\mathrm{p} \, \mathrm{d}\mathbf{v}_\mathrm{f} \, \mathrm{d}\boldsymbol{\xi}_\mathrm{f}, \tag{4.47}$$

and the particle-number source term by

$$\mathcal{S}_\mathrm{N} \equiv \int \mathcal{S}_1 \, \mathrm{d}\mathbf{v}_\mathrm{p} \, \mathrm{d}\boldsymbol{\xi}_\mathrm{p} \, \mathrm{d}\mathbf{v}_\mathrm{f} \, \mathrm{d}\boldsymbol{\xi}_\mathrm{f}. \tag{4.48}$$

Note that the sign of the source term will depend on whether particles are created or destroyed in the system. Note also that the spatial transport term in Eq. (4.46) will generally not be closed unless, for example, all particles have identical velocities. The transport equation in Eq. (4.46) is mainly used for systems with particle aggregation and breakage (i.e. when $N(t, \mathbf{x})$ is not constant). In such cases, it will typically be coupled to a system of moment-transport equations involving higher-order moments.

4.3.3 Disperse-phase volume transport

In order to derive a transport equation of the disperse-phase volume fraction α_p, we will let the first internal coordinate $\xi_{\mathrm{p}1}$ be equal to the particle volume (V_p). The disperse-phase volume fraction is then defined by

$$\alpha_\mathrm{p} \equiv \int \xi_{\mathrm{p}1} n \, \mathrm{d}\mathbf{v}_\mathrm{p} \, \mathrm{d}\boldsymbol{\xi}_\mathrm{p} \, \mathrm{d}\mathbf{v}_\mathrm{f} \, \mathrm{d}\boldsymbol{\xi}_\mathrm{f}. \tag{4.49}$$

Starting from Eq. (4.39) with $g = \xi_{\mathrm{p}1}$, the transport equation for α_p is

$$\frac{\partial \alpha_\mathrm{p}}{\partial t} + \frac{\partial}{\partial \mathbf{x}} \cdot \left(\alpha_\mathrm{p} \mathbf{U}_\mathrm{V} \right) = [\![G_\mathrm{p}]\!]_\mathrm{V} + \mathcal{S}_\mathrm{V}, \tag{4.50}$$

where the volume-average disperse-phase velocity is defined by

$$\mathbf{U}_\mathrm{V} \equiv \frac{1}{\alpha_\mathrm{p}} \int \xi_{\mathrm{p}1} \mathbf{v}_\mathrm{p} n \, \mathrm{d}\mathbf{v}_\mathrm{p} \, \mathrm{d}\boldsymbol{\xi}_\mathrm{p} \, \mathrm{d}\mathbf{v}_\mathrm{f} \, \mathrm{d}\boldsymbol{\xi}_\mathrm{f}. \tag{4.51}$$

For the case in which all particles have the same volume, $\mathbf{U}_{\mathrm{V}} = \mathbf{U}_{\mathrm{N}}$. Note that, like $\mathbf{U}_{\mathrm{N}}$, the volume-average disperse-phase velocity will not usually be in closed form. The particle-volume source terms are defined by

$$[\![G_{\mathrm{p}}]\!]_{\mathrm{V}} \equiv \int \langle G_{\mathrm{p}1} \rangle_1 n \, \mathrm{d}\mathbf{v}_{\mathrm{p}} \, \mathrm{d}\boldsymbol{\xi}_{\mathrm{p}} \, \mathrm{d}\mathbf{v}_{\mathrm{f}} \, \mathrm{d}\boldsymbol{\xi}_{\mathrm{f}} \tag{4.52}$$

and

$$\mathcal{S}_{\mathrm{V}} \equiv \int \xi_{\mathrm{p}1} \mathcal{S}_1 \, \mathrm{d}\mathbf{v}_{\mathrm{p}} \, \mathrm{d}\boldsymbol{\xi}_{\mathrm{p}} \, \mathrm{d}\mathbf{v}_{\mathrm{f}} \, \mathrm{d}\boldsymbol{\xi}_{\mathrm{f}}. \tag{4.53}$$

The continuous contribution $[\![G_{\mathrm{p}}]\!]_{\mathrm{V}}$ could be due to mass transfer from the fluid to the solid phase. On the other hand, the discontinuous term $\mathcal{S}_{\mathrm{V}}$ might appear due to nucleation of particles with nonzero volume from the fluid phase. For systems in which the particle volume is conserved, the right-hand side of Eq. (4.50) will be null. Finally, we should note that, by definition, the disperse-phase volume fraction cannot be greater than unity ($0 \leq \alpha_{\mathrm{p}} \leq 1$).[19] This constraint implies that the divergence of $\mathbf{U}_{\mathrm{V}}$ must be nonnegative when $\alpha_{\mathrm{p}} = 1$. In two-fluid models this condition is often enforced by adding a pressure term to the transport equation for $\mathbf{U}_{\mathrm{V}}$. The disperse-phase volume-fraction equation in Eq. (4.49) is used when the material density of the disperse phase is the same for all particles. In other words, the density ρ_{p} is not an internal coordinate, but the particle volume changes, for example, due to coalescence and breakage. For such cases, the volume-average velocity is the same as the mass-average velocity. Otherwise, a more natural choice for the mean disperse-phase velocity is found using the particle mass.

4.3.4 Fluid-phase volume transport

In order to derive a transport equation of the fluid-phase volume fraction α_{f}, we will let the internal coordinate $\xi_{\mathrm{f}1}$ be equal to the fluid volume *seen* by a particle.[20] The fluid-phase volume fraction is then defined by

$$\alpha_{\mathrm{f}} \equiv \int \xi_{\mathrm{f}1} n \, \mathrm{d}\mathbf{v}_{\mathrm{p}} \, \mathrm{d}\boldsymbol{\xi}_{\mathrm{p}} \, \mathrm{d}\mathbf{v}_{\mathrm{f}} \, \mathrm{d}\boldsymbol{\xi}_{\mathrm{f}} \tag{4.54}$$

and the volume-average fluid velocity by

$$\mathbf{U}_{\mathrm{fV}} \equiv \frac{1}{\alpha_{\mathrm{f}}} \int \xi_{\mathrm{f}1} \mathbf{v}_{\mathrm{f}} n \, \mathrm{d}\mathbf{v}_{\mathrm{p}} \, \mathrm{d}\boldsymbol{\xi}_{\mathrm{p}} \, \mathrm{d}\mathbf{v}_{\mathrm{f}} \, \mathrm{d}\boldsymbol{\xi}_{\mathrm{f}}. \tag{4.55}$$

Starting from Eq. (4.39) with $g = \xi_{\mathrm{f}1}$, the transport equation for α_{f} is

$$\frac{\partial \alpha_{\mathrm{f}}}{\partial t} + \frac{\partial}{\partial \mathbf{x}} \cdot \left(\alpha_{\mathrm{f}} \mathbf{U}_{\mathrm{pV}} \right) = [\![G_{\mathrm{f}}]\!]_{\mathrm{V}} + \mathcal{S}_{\mathrm{fV}}, \tag{4.56}$$

[19] In fact, because the shapes of the particles usually do not allow the system to be completely occupied by the disperse phase, the upper limit on α_{p} will often be less than unity.

[20] We implicitly assume that the fluid volume seen by each particle is distinct and the union of all such volumes is equal to the total fluid volume. Since n represents the number of particles per unit volume, $\xi_{\mathrm{f}1} n$ is the fluid volume per unit volume. In principle, if all particles see the same fluid volume, $\xi_{\mathrm{f}1}$ could be constant. However, we shall see that in general $\xi_{\mathrm{f}1}$ cannot be constant if the system is spatially inhomogeneous. Intuitively, this fact is consistent with the image of the decrease (increase) in fluid volume surrounding a particle when the disperse-phase volume fraction α_{p} increases (decreases). The simplest case is when $\xi_{\mathrm{f}1} = g(t, \mathbf{x}, \xi_{\mathrm{p}1})$, where g is a function that satisfies Eq. (4.54). In this case, the NDF can be written as $n(t, \mathbf{x}, \mathbf{v}_{\mathrm{p}}, \xi_{\mathrm{p}1}, \mathbf{v}_{\mathrm{f}}, \xi_{\mathrm{f}1}) = n(t, \mathbf{x}, \mathbf{v}_{\mathrm{p}}, \xi_{\mathrm{p}1}, \mathbf{v}_{\mathrm{f}}) \delta(\xi_{\mathrm{f}1} - g(t, \mathbf{x}, \xi_{\mathrm{p}1}))$. If we choose the linear function $g(t, \mathbf{x}, \xi_{\mathrm{p}1}) = \alpha_{\mathrm{f}} \xi_{\mathrm{p}1} / \alpha_{\mathrm{p}}$, then each particle has associated with it a volume of fluid that is proportional to its volume $\xi_{\mathrm{p}1}$.

where

$$\mathbf{U}_{\mathrm{pV}} \equiv \frac{1}{\alpha_{\mathrm{f}}} \int \xi_{\mathrm{f1}} \mathbf{v}_{\mathrm{p}} n \, \mathrm{d}\mathbf{v}_{\mathrm{p}} \, \mathrm{d}\boldsymbol{\xi}_{\mathrm{p}} \, \mathrm{d}\mathbf{v}_{\mathrm{f}} \, \mathrm{d}\boldsymbol{\xi}_{\mathrm{f}}. \tag{4.57}$$

The fluid-phase volume-fraction source terms are defined by

$$[\![G_{\mathrm{f}}]\!]_{\mathrm{V}} \equiv \int \langle G_{\mathrm{f1}} \rangle_1 n \, \mathrm{d}\mathbf{v}_{\mathrm{p}} \, \mathrm{d}\boldsymbol{\xi}_{\mathrm{p}} \, \mathrm{d}\mathbf{v}_{\mathrm{f}} \, \mathrm{d}\boldsymbol{\xi}_{\mathrm{f}} \tag{4.58}$$

and

$$\mathcal{S}_{\mathrm{fV}} \equiv \int \xi_{\mathrm{f1}} \mathcal{S}_1 \, \mathrm{d}\mathbf{v}_{\mathrm{p}} \, \mathrm{d}\boldsymbol{\xi}_{\mathrm{p}} \, \mathrm{d}\mathbf{v}_{\mathrm{f}} \, \mathrm{d}\boldsymbol{\xi}_{\mathrm{f}}. \tag{4.59}$$

By definition, $\alpha_{\mathrm{p}} + \alpha_{\mathrm{f}} = 1$ and, in the absence of mass transfer between phases, the right-hand side of Eq. (4.50) is null. In this limit, Eq. (4.56) should reduce to

$$\frac{\partial \alpha_{\mathrm{f}}}{\partial t} + \frac{\partial}{\partial \mathbf{x}} \cdot (\alpha_{\mathrm{f}} \mathbf{U}_{\mathrm{fV}}) = 0, \tag{4.60}$$

which simply states that the fluid-phase volume fraction is a conserved quantity (as is the disperse-phase volume fraction).

A consistent model for the change in fluid volume seen by the particles

In order for Eqs. (4.56) and (4.60) to be equivalent in the absence of mass transfer, we must have

$$[\![G_{\mathrm{f}}]\!]_{\mathrm{V}} = \frac{\partial}{\partial \mathbf{x}} \cdot \left[\alpha_{\mathrm{f}} \left(\mathbf{U}_{\mathrm{pV}} - \mathbf{U}_{\mathrm{fV}} \right) \right]. \tag{4.61}$$

Thus, it will be necessary for $\langle G_{\mathrm{f1}} \rangle_1$ to be nonzero in Eq. (4.58); or, in other words, the volume of fluid seen by a particle must change along a particle trajectory. Physically, this change is associated with the presence of other particles (i.e. it cannot be modeled using single-particle physics). A simple consistent model for $\langle G_{\mathrm{f1}} \rangle_1$ can be written as

$$\langle G_{\mathrm{f1}} \rangle_1 = \xi_{\mathrm{f1}} \left(\mathbf{v}_{\mathrm{p}} - \mathbf{v}_{\mathrm{f}} \right) \cdot \frac{\partial n}{\partial \mathbf{x}} \frac{1}{n} + \frac{1}{\tau_{\mathrm{pf}}} \left(\frac{\alpha_{\mathrm{f}}}{\alpha_{\mathrm{p}}} \xi_{\mathrm{p1}} - \xi_{\mathrm{f1}} \right), \tag{4.62}$$

where τ_{pf} is the time scale for relaxation of ξ_{f1} back to its conditional mean value $\alpha_{\mathrm{f}} \xi_{\mathrm{p1}} / \alpha_{\mathrm{p}}$. The first term on the right-hand side of Eq. (4.62) states that spatial gradients of the NDF in the direction of relative velocity between the particle and the fluid result in a change in the fluid volume seen by a particle. Using a gradient of the NDF to account for finite-volume particles is not unusual. (For point particles, $\alpha_{\mathrm{f}} = 1$ so that $[\![G_{\mathrm{f}}]\!]_{\mathrm{V}} = 0$.) For example, a spatial gradient of the NDF appears in the Boltzmann–Enskog equation for hard-sphere collisions in Chapter 6 (see, for example, Eq. (6.71) on page 235). Finally, note that, with Eq. (4.62), it is not necessary to know $\mathbf{U}_{\mathrm{pV}}$ even when mass transfer is included.

4.3.5 Disperse-phase mass transport

If the disperse phase has particles with different volumes and different masses (or material densities), at least two internal coordinates are necessary to describe a particle state. In order to derive a transport equation of the disperse-phase mass density ϱ_{p}, we will let the internal coordinate ξ_{p2} be equal to the particle mass (M_p).[21] The disperse-phase mass density is then defined by

$$\varrho_p \equiv \int \xi_{p2} n \, d\mathbf{v}_p \, d\boldsymbol{\xi}_p \, d\mathbf{v}_f \, d\boldsymbol{\xi}_f. \tag{4.63}$$

Starting from Eq. (4.39) with $g = \xi_{p2}$, the transport equation for ϱ_p is

$$\frac{\partial \varrho_p}{\partial t} + \frac{\partial}{\partial \mathbf{x}} \cdot \left(\varrho_p \mathbf{U}_M\right) = [\![G_p]\!] + \mathcal{S}_M, \tag{4.64}$$

where the mass-average disperse-phase velocity is defined by

$$\mathbf{U}_M \equiv \frac{1}{\varrho_p} \int \xi_{p2} \mathbf{v}_p n \, d\mathbf{v}_p \, d\boldsymbol{\xi}_p \, d\mathbf{v}_f \, d\boldsymbol{\xi}_f. \tag{4.65}$$

For the case in which all particles have the same mass, $\mathbf{U}_M = \mathbf{U}_N$; for the case in which all particles have the same material density, $\mathbf{U}_M = \mathbf{U}_V$. Note that, like $\mathbf{U}_N$ and $\mathbf{U}_V$, the mass-average disperse-phase velocity will not usually be in closed form. In most applications, the mean particle velocity $\mathbf{U}_p$ will be set to the mass-average disperse-phase velocity. Thus, unless noted otherwise, we will set $\mathbf{U}_p = \mathbf{U}_M$ throughout the rest of this book. The particle-mass source terms are defined by

$$[\![G_p]\!] \equiv \int \langle G_{p2} \rangle_1 n \, d\mathbf{v}_p \, d\boldsymbol{\xi}_p \, d\mathbf{v}_f \, d\boldsymbol{\xi}_f \tag{4.66}$$

and

$$\mathcal{S}_M \equiv \int \xi_{p2} \mathcal{S}_1 \, d\mathbf{v}_p \, d\boldsymbol{\xi}_p \, d\mathbf{v}_f \, d\boldsymbol{\xi}_f. \tag{4.67}$$

Since mass is conserved during transport, the continuous contribution $[\![G_p]\!]$ is due to mass transfer from the fluid to the solid phase. Likewise, the discontinuous term $\mathcal{S}_M$ might appear due to nucleation of particles with nonzero mass from the fluid phase. For systems with no mass transfer between the disperse and continuous phases, the right-hand side of Eq. (4.64) will be null.

The fluid-phase continuity equation

Since the total mass of the fluid–particle system is conserved, we can define the fluid-phase mass density ϱ_f by

$$\frac{\partial \varrho_f}{\partial t} + \frac{\partial}{\partial \mathbf{x}} \cdot (\varrho_f \mathbf{U}_f) = -[\![G_p]\!] - \mathcal{S}_M, \tag{4.68}$$

where the mass-average fluid velocity $\mathbf{U}_f$ is defined by a conservation equation for the total momentum as discussed below. Note that ϱ_p and ϱ_f are not equal to the actual material densities ρ_p and ρ_f, respectively. Rather, they are equal to the mass of solid (fluid) per unit volume of the fluid–particle system. In order to distinguish between the two densities, when the material densities are constant we can relate them to the volume fractions α_p and α_f: $\rho_p \alpha_p = \varrho_p$ and $\rho_f \alpha_f = \varrho_f$. These relations *do not* hold in general. If the material densities vary from particle to particle, then ρ_f depends on the internal coordinates, but α_p and ϱ_p do not.

[21] The particle mass can be written as $M_p = \rho_p V_p$, where ρ_p is the material density of the particle and V_p is its volume. Note that, in addition to mass, either material density or volume could be included in the internal-coordinate vector. (In general, we will use mass and volume as the internal coordinates.) Thus, for example, fixing the particle masses to be equal does not imply that all particles have the same volume.

The mixture continuity equation

In general, a mixture mass density can be defined by $\varrho_{\text{mix}} \equiv \varrho_{\text{p}} + \varrho_{\text{f}}$, and satisfies the mixture continuity equation:

$$\frac{\partial \varrho_{\text{mix}}}{\partial t} + \frac{\partial}{\partial \mathbf{x}} \cdot (\varrho_{\text{mix}} \mathbf{U}_{\text{mix}}) = 0, \tag{4.69}$$

where the mass-average mixture velocity is defined by

$$\mathbf{U}_{\text{mix}} \equiv \frac{1}{\varrho_{\text{mix}}} \left(\varrho_{\text{p}} \mathbf{U}_{\text{p}} + \varrho_{\text{f}} \mathbf{U}_{\text{f}} \right). \tag{4.70}$$

The transport equations for ϱ_{p} and ϱ_{f} are used in two-fluid models for multiphase flows. A k-fluid model can be developed by treating all particles with the same internal coordinates as a "fluid." Thus, for example, if all particles are identical except that some have mass M_1 and the others have mass M_2 (which implies that they have different solid densities), then we can treat the particle phase as two "fluids." Mathematically, this follows directly from the form of the NDF for this case.

Constant volume and mass

Because it will be useful when developing numerical methods in Chapter 5, we will consider next the case in which $\xi_{\text{p}1} = V_{\text{p}}$ and $\xi_{\text{p}2} = M_{\text{p}}$ are constant. This case corresponds to all particles having the same volume and mass, so that the only phase-space variables are the velocities $\mathbf{v}_{\text{p}}$ and $\mathbf{v}_{\text{f}}$. Starting from Eq. (4.63), we find that for this case $\varrho_{\text{p}} = M_{\text{p}} N = \rho_{\text{p}} \alpha_{\text{p}}$. Likewise, the average particle velocities are all equal, so we can set them equal to the mean particle velocity: $\mathbf{U}_{\text{p}} \equiv \mathbf{U}_M = \mathbf{U}_V = \mathbf{U}_N$. Under these conditions, Eq. (4.64) becomes

$$\frac{\partial \alpha_{\text{p}}}{\partial t} + \frac{\partial}{\partial \mathbf{x}} \cdot \left(\alpha_{\text{p}} \mathbf{U}_{\text{p}} \right) = 0. \tag{4.71}$$

Note that, in order to be consistent with the assumption that all particles have the same volume and mass, the right-hand side of this expression is null. If the material density of the fluid is also constant (i.e. the fluid is incompressible), then

$$\frac{\partial \alpha_{\text{f}}}{\partial t} + \frac{\partial}{\partial \mathbf{x}} \cdot (\alpha_{\text{f}} \mathbf{U}_{\text{f}}) = 0. \tag{4.72}$$

Adding together Eqs. (4.71) and (4.72) yields a realizability constraint for the velocity fields, namely $\nabla_{\mathbf{x}} \cdot \mathbf{U}_{\text{vol}} = 0$, where $\mathbf{U}_{\text{vol}} \equiv \alpha_{\text{p}} \mathbf{U}_{\text{p}} + \alpha_{\text{f}} \mathbf{U}_{\text{f}}$. As mentioned earlier, this constraint must be incorporated into the conditional source terms in the disperse-phase momentum transport equation. Note that, in general, $\mathbf{U}_{\text{vol}} \neq \mathbf{U}_{\text{mix}}$ unless the fluid and the particles have the same material density.

Multicomponent disperse phase

In many applications, the particles will be composed of multiple chemical species. In such cases, it is necessary to introduce a vector of internal coordinates $\boldsymbol{\xi}_{\text{p}}$ whose components are the mass of each chemical species. Obviously, the sum of these internal coordinates is equal to the particle mass. By definition, if $\xi_{\text{p}\alpha}$ is the mass of component α, then integration over phase space leads to a component disperse-phase mass density:

$$\varrho_{\text{p}\alpha} \equiv \int \xi_{\text{p}\alpha} n \, \text{d}\mathbf{v}_{\text{p}} \, \text{d}\boldsymbol{\xi}_{\text{p}} \, \text{d}\mathbf{v}_{\text{f}} \, \text{d}\boldsymbol{\xi}_{\text{f}}. \tag{4.73}$$

Likewise, since all particle components move with the particle, the transport equation for $\varrho_{p\alpha}$ has the form

$$\frac{\partial \varrho_{p\alpha}}{\partial t} + \frac{\partial}{\partial \mathbf{x}} \cdot \left(\varrho_{p\alpha} \mathbf{U}_p\right) = [\![G_p]\!]_\alpha + \mathcal{S}_\alpha, \tag{4.74}$$

where the mass-transfer and nucleation terms on the right-hand side can be different for each component. Summing Eq. (4.74) over all components α leads to Eq. (4.64).

4.3.6 Fluid-phase mass transport

In the literature on turbulent two-phase flow (Minier & Peirano, 2001; Peirano & Minier, 2002; Simonin *et al.*, 1993), the fluid phase is usually treated using a separate distribution function whose integral over phase space leads to the fluid-phase mass density. Here, we use a different approach starting from $n(t, \mathbf{x}, \mathbf{v}_p, \boldsymbol{\xi}_p, \mathbf{v}_f, \boldsymbol{\xi}_f)$. In this approach, we let the internal coordinate ξ_{f2} be equal to the fluid mass seen by a particle.[22] The fluid-phase mass density is then given by

$$\varrho_f = \int \xi_{f2} n \, d\mathbf{v}_p \, d\boldsymbol{\xi}_p \, d\mathbf{v}_f \, d\boldsymbol{\xi}_f. \tag{4.75}$$

We also define a corresponding fluid-mass-average disperse-phase velocity by

$$\mathbf{U}_{pM} \equiv \frac{1}{\varrho_f} \int \xi_{f2} \mathbf{v}_p n \, d\mathbf{v}_p \, d\boldsymbol{\xi}_p \, d\mathbf{v}_f \, d\boldsymbol{\xi}_f. \tag{4.76}$$

Using this definition and starting from Eq. (4.39) with $g = \xi_{f2}$, the transport equation for ϱ_f is

$$\frac{\partial \varrho_f}{\partial t} + \frac{\partial}{\partial \mathbf{x}} \cdot \left(\varrho_f \mathbf{U}_{pM}\right) = [\![G_f]\!] - \mathcal{S}_M, \tag{4.77}$$

where

$$[\![G_f]\!] \equiv \int \langle G_{f2} \rangle_1 n \, d\mathbf{v}_p \, d\boldsymbol{\xi}_p \, d\mathbf{v}_f \, d\boldsymbol{\xi}_f \tag{4.78}$$

and $\mathcal{S}_M$ represents the loss of fluid mass due to particle nucleation (i.e. non-constant N). By forcing Eq. (4.77) to be equal to Eq. (4.68), we can deduce a consistent model for $\langle G_{f2} \rangle_1$ in Eq. (4.78).

A consistent model for the change in fluid mass seen by the particles

Formally, in order for Eqs. (4.77) and (4.68) to be consistent, we must have

$$[\![G_f]\!] + [\![G_p]\!] = \frac{\partial}{\partial \mathbf{x}} \cdot \left[\varrho_f \left(\mathbf{U}_{pM} - \mathbf{U}_f\right)\right], \tag{4.79}$$

which is very similar to Eq. (4.61). A simple consistent model for $\langle G_{f2} \rangle_1$ is[23]

$$\langle G_{f2} \rangle_1 = -\langle G_{p2} \rangle_1 + \xi_{f2} \left(\mathbf{v}_p - \mathbf{v}_f\right) \cdot \frac{\partial n}{\partial \mathbf{x}} \frac{1}{n} + \frac{1}{\tau_{pf}} \left(\frac{\varrho_f}{\varrho_p} \xi_{p2} - \xi_{f2}\right), \tag{4.80}$$

[22] As with the fluid volume seen by a particle, we implicitly assume that the fluid mass seen by each particle is distinct. In this context, Eqs. (4.75) and (4.81) are consistency constraints on the definition of ξ_{f2}. The simplest case is when $\xi_{f2} = g(t, \mathbf{x}, \xi_{p2})$, where g is a function that satisfies Eq. (4.75). In this case, the NDF can be written as $n(t, \mathbf{x}, \mathbf{v}_p, \xi_{p2}, \mathbf{v}_f, \xi_{f2}) = n(t, \mathbf{x}, \mathbf{v}_p, \xi_{p2}, \mathbf{v}_f) \delta(\xi_{f2} - g(t, \mathbf{x}, \xi_{p2}))$. If we choose the linear function $g(t, \mathbf{x}, \xi_{p2}) = \varrho_f \xi_{p2} / \varrho_p$, then each particle has associated with it a mass of fluid that is proportional to its mass ξ_{p2}.

[23] An even simpler model is

$$\langle G_{f1} \rangle_1 = -\langle G_{p2} \rangle_1 + \frac{1}{N} \frac{\partial}{\partial \mathbf{x}} \cdot \left[\varrho_f \left(\mathbf{U}_p - \mathbf{U}_f\right)\right] + \frac{1}{\tau_{pf}} \left(\frac{\varrho_f}{\varrho_p} \xi_{p2} - \xi_{f2}\right),$$

where the first term on the right-hand side represents mass transfer between the phases and the last term is a model for the relaxation of ξ_{f2} back to its conditional mean value $\varrho_f \xi_{p2}/\varrho_p$. The mass-average fluid velocity, introduced in Eq. (4.68), is given in terms of the NDF by

$$\mathbf{U}_f = \frac{1}{\varrho_f} \int \xi_{f2} \mathbf{v}_f n \, d\mathbf{v}_p \, d\boldsymbol{\xi}_p \, d\mathbf{v}_f \, d\boldsymbol{\xi}_f. \tag{4.81}$$

As in Eq. (4.62), it is noteworthy that the right-hand side of Eq. (4.80) is nonzero for systems without mass transfer. In fact, whenever the particles and fluid have different velocities (i.e. nonzero particle Stokes number), it will be necessary to include the model given in Eq. (4.80) in the GPBE. Note that the fluid seen by the particles is defined only when particles are present. In practice, $\mathbf{U}_f$ is found by solving a fluid-phase momentum-transport equation that is well defined even when particles are absent, in which case Eq. (4.81) is interpreted as a consistency requirement for the GPBE (Eq. (4.39)). Nonetheless, knowing $\langle G_{f2} \rangle_1$ from Eq. (4.80) is useful for deriving the higher-order moments involving $\mathbf{v}_f$, such as the *fluctuating* fluid velocity seen by the particles.

Multicomponent fluid phase

If the particles are composed of multiple chemical species, then usually the fluid phase will be also. In such cases, it is necessary to introduce a vector of internal coordinates $\boldsymbol{\xi}_f$ whose components are the mass of each chemical species seen by a particle. Obviously, the sum of these internal coordinates is equal to the fluid mass seen by a particle. By definition, if $\xi_{f\alpha}$ is the mass of component α, then integration over phase space leads to a component fluid-phase mass density:

$$\varrho_{f\alpha} \equiv \int \xi_{f\alpha} n \, d\mathbf{v}_p \, d\boldsymbol{\xi}_p \, d\mathbf{v}_f \, d\boldsymbol{\xi}_f. \tag{4.82}$$

Assuming all fluid components move with the same fluid velocity,[24] the transport equation for $\varrho_{f\alpha}$ has the form

$$\frac{\partial \varrho_{f\alpha}}{\partial t} + \frac{\partial}{\partial \mathbf{x}} \cdot (\varrho_{f\alpha} \mathbf{U}_f) = [\![G_f]\!]_{M\alpha} - \mathcal{S}_{M\alpha}, \tag{4.83}$$

where the mass-transfer and nucleation terms on the right-hand side can be different for each component. Summing Eq. (4.83) over all components α leads to Eq. (4.77). Note that Eq. (4.83) may contain an additional term on the right-hand side corresponding to homogeneous chemical reactions between species in the fluid phase. For each component, the model for $\langle G_{f\alpha} \rangle_1$ can have the form of Eq. (4.80) with $\varrho_{f\alpha}$ in place of ϱ_f. In summary,

---

which distributes the "correction term" evenly among all particles present regardless of their masses or velocities. We rule out this type of model because it does not account for spatial gradients in the NDF (only in the statistics). In the limit where there are no fluid velocity fluctuations, the model in Eq. (4.80) becomes

$$\langle G_{f2} \rangle_1 = -\langle G_{p2} \rangle_1 + \frac{1}{n} \frac{\partial}{\partial \mathbf{x}} \cdot \left[\xi_{f2} \left(\mathbf{v}_p - \mathbf{U}_f \right) n \right] + \frac{1}{\tau_{pf}} \left(\frac{\varrho_f}{\varrho_p} \xi_{p2} - \xi_{f2} \right)$$

where n no longer includes $\mathbf{v}_f$ and $\mathbf{U}_f(t, \mathbf{x})$ is found by solving Eq. (4.92) on page 124.

[24] It would also be possible to assume that each component moves with the same *average* velocity plus a *diffusive* component. For example, Eq. (4.83) could be replaced by

$$\frac{\partial \varrho_{f\alpha}}{\partial t} + \frac{\partial}{\partial \mathbf{x}} \cdot (\varrho_{f\alpha} \mathbf{U}_f) = \frac{\partial}{\partial \mathbf{x}} \cdot \left(\varrho_f \Gamma_f \frac{\partial Y_{f\alpha}}{\partial \mathbf{x}} \right) + [\![G_f]\!]_{\alpha} - \mathcal{S}_{\alpha},$$

where $Y_{f\alpha}$ is the species mass fraction in the fluid phase and Γ_f is an *effective* diffusion coefficient. Because of the presence of the particles, the effective diffusion coefficient may be significantly larger than the molecular diffusion coefficient in the pure fluid.

due to the importance of mass conservation in models for physical systems, Eqs. (4.64) and (4.68) should always be included in the set of moment-transport equations used to model a polydisperse multiphase flow. Generally speaking, several additional moment-transport equations will be required in order to describe adequately the physics of the system. Most importantly, descriptions of the particle-phase and fluid-phase momenta are needed, as will be described next.

4.3.7 Disperse-phase momentum transport

The disperse-phase momentum density is defined by

$$\varrho_{\mathrm{p}} \mathbf{U}_{\mathrm{p}} \equiv \int \xi_{\mathrm{p}2} \mathbf{v}_{\mathrm{p}} n \, \mathrm{d}\mathbf{v}_{\mathrm{p}} \, \mathrm{d}\boldsymbol{\xi}_{\mathrm{p}} \, \mathrm{d}\mathbf{v}_{\mathrm{f}} \, \mathrm{d}\boldsymbol{\xi}_{\mathrm{f}}, \tag{4.84}$$

where the internal coordinate $\xi_{\mathrm{p}2}$ again corresponds to the particle mass. Starting from Eq. (4.39) with $g = \xi_{\mathrm{p}2} \mathbf{v}_{\mathrm{p}}$, the transport equation for the disperse-phase momentum density is

$$\frac{\partial \varrho_{\mathrm{p}} \mathbf{U}_{\mathrm{p}}}{\partial t} + \frac{\partial}{\partial \mathbf{x}} \cdot \left(\varrho_{\mathrm{p}} \mathbf{F}_{\mathrm{p}} \right) = \varrho_{\mathrm{p}} [\![\mathbf{A}_{\mathrm{fp}}]\!]_{\mathrm{p}} + \varrho_{\mathrm{p}} [\![\mathbf{A}_{\mathrm{p}}]\!]_{\mathrm{p}} + [\![\mathbf{G}_{\mathrm{p}}]\!]_{\mathrm{p}} + \varrho_{\mathrm{p}} [\![\mathbf{S}]\!]_{\mathrm{p}}, \tag{4.85}$$

where the (unclosed) momentum-convection term (a second-order tensor) is defined by

$$\varrho_{\mathrm{p}} \mathbf{F}_{\mathrm{p}} \equiv \int \xi_{\mathrm{p}2} (\mathbf{v}_{\mathrm{p}} \otimes \mathbf{v}_{\mathrm{p}}) n \, \mathrm{d}\mathbf{v}_{\mathrm{p}} \, \mathrm{d}\boldsymbol{\xi}_{\mathrm{p}} \, \mathrm{d}\mathbf{v}_{\mathrm{f}} \, \mathrm{d}\boldsymbol{\xi}_{\mathrm{f}}. \tag{4.86}$$

Note that, in general, the transport and source terms in Eq. (4.85) will not be in closed form. From the definition in Eq. (4.84), we can identify three limiting cases.

1. If the particle mass (M_{p}) is constant, then $\varrho_{\mathrm{p}} \mathbf{U}_{\mathrm{p}} = M_{\mathrm{p}} N \mathbf{U}_{\mathrm{N}}$.
2. If the particle material density (ρ_{p}) is constant, then $\varrho_{\mathrm{p}} \mathbf{U}_{\mathrm{p}} = \rho_{\mathrm{p}} \alpha_{\mathrm{p}} \mathbf{U}_{\mathrm{V}}$.
3. If M_{p} and ρ_{p} are constant, then $\varrho_{\mathrm{p}} \mathbf{U}_{\mathrm{p}} = \rho_{\mathrm{p}} \alpha_{\mathrm{p}} \mathbf{U}_{\mathrm{N}}$.

These cases would suggest that the best averaged velocity to use depends on which internal coordinates are held constant. However, because conservation of momentum is of fundamental importance, it is always best to use the mass-average velocity in Eq. (4.85). For this reason, Eq. (4.85) should always be included in the set of moment-transport equations used to model the disperse phase. The disperse-phase momentum source terms appearing on the right-hand side of Eq. (4.85) are defined as follows. The first term,

$$\varrho_{\mathrm{p}} [\![\mathbf{A}_{\mathrm{fp}}]\!]_{\mathrm{p}} \equiv \int \xi_{\mathrm{p}2} \langle \mathbf{A}_{\mathrm{fp}2} \rangle_1 n \, \mathrm{d}\mathbf{v}_{\mathrm{p}} \, \mathrm{d}\boldsymbol{\xi}_{\mathrm{p}} \, \mathrm{d}\mathbf{v}_{\mathrm{f}} \, \mathrm{d}\boldsymbol{\xi}_{\mathrm{f}}, \tag{4.87}$$

describes surface forces on the particles due to the surrounding fluid (e.g. the drag, lift, and pressure forces). The second term,

$$\varrho_{\mathrm{p}} [\![\mathbf{A}_{\mathrm{p}}]\!]_{\mathrm{p}} \equiv \int \xi_{\mathrm{p}2} \langle \mathbf{A}_{\mathrm{p}2} \rangle_1 n \, \mathrm{d}\mathbf{v}_{\mathrm{p}} \, \mathrm{d}\boldsymbol{\xi}_{\mathrm{p}} \, \mathrm{d}\mathbf{v}_{\mathrm{f}} \, \mathrm{d}\boldsymbol{\xi}_{\mathrm{f}}, \tag{4.88}$$

is due to body forces (such as gravity) acting on individual particles. The third term,

$$[\![\mathbf{G}_{\mathrm{p}}]\!]_{\mathrm{p}} \equiv \int \mathbf{v}_{\mathrm{p}} \langle G_{\mathrm{p}2} \rangle_1 n \, \mathrm{d}\mathbf{v}_{\mathrm{p}} \, \mathrm{d}\boldsymbol{\xi}_{\mathrm{p}} \, \mathrm{d}\mathbf{v}_{\mathrm{f}} \, \mathrm{d}\boldsymbol{\xi}_{\mathrm{f}}, \tag{4.89}$$

describes momentum added to the particle phase due to mass transfer from the liquid phase. The fourth term,

$$\varrho_{\mathrm{p}}[\![\mathbf{S}]\!]_{\mathrm{p}} \equiv \int \xi_{\mathrm{p}2}\mathbf{v}_{\mathrm{p}}\mathcal{S}_1 \, \mathrm{d}\mathbf{v}_{\mathrm{p}} \, \mathrm{d}\boldsymbol{\xi}_{\mathrm{p}} \, \mathrm{d}\mathbf{v}_{\mathrm{f}} \, \mathrm{d}\boldsymbol{\xi}_{\mathrm{f}}, \tag{4.90}$$

describes discontinuous changes in particle momentum due to collisions (see Chapter 6) and particle nucleation. In general, none of these terms will appear in closed form (even if the GPBE is closed), and macroscale models must be provided in order to close Eq. (4.85).

Constant volume and mass

Consider again a system wherein all particles have the same volume and mass, and the disperse-phase momentum density is $\varrho_{\mathrm{p}} = \rho_{\mathrm{p}}\alpha_{\mathrm{p}}$. The transport equation for the disperse-phase momentum density for this case is[25]

$$\frac{\partial \rho_{\mathrm{p}}\alpha_{\mathrm{p}}\mathbf{U}_{\mathrm{p}}}{\partial t} + \frac{\partial}{\partial \mathbf{x}} \cdot \left(\rho_{\mathrm{p}}\alpha_{\mathrm{p}}\mathbf{U}_{\mathrm{p}} \otimes \mathbf{U}_{\mathrm{p}}\right) + \frac{\partial}{\partial \mathbf{x}} \cdot \left(\rho_{\mathrm{p}}\alpha_{\mathrm{p}}[\![\mathbf{u}_{\mathrm{p}}\mathbf{u}_{\mathrm{p}}]\!]_{\mathrm{N}}\right) = \rho_{\mathrm{p}}\alpha_{\mathrm{p}}[\![\mathbf{A}_{\mathrm{fp}}]\!]_{\mathrm{N}} + \rho_{\mathrm{p}}\alpha_{\mathrm{p}}[\![\mathbf{A}_{\mathrm{p}}]\!]_{\mathrm{N}}, \tag{4.91}$$

where the disperse-phase stress tensor is defined by $[\![\mathbf{u}_{\mathrm{p}}\mathbf{u}_{\mathrm{p}}]\!]_{\mathrm{N}} = \mathbf{F}_{\mathrm{N}} - \mathbf{U}_{\mathrm{p}} \otimes \mathbf{U}_{\mathrm{p}}$. The disperse-phase stress tensor is a second-order moment, and thus is not closed. In Section 4.4 we review some classical closures for $[\![\mathbf{u}_{\mathrm{p}}\mathbf{u}_{\mathrm{p}}]\!]_{\mathrm{N}}$ that are valid for particular limiting behavior of the NDF. However, in general, the closure of the disperse-phase stress tensor is nontrivial, but extremely important in determining the nature of solutions to Eq. (4.91). In fact, it is the failure to provide an accurate closure for this term that results in the two-fluid model being ill-posed (i.e. the conservative form of the system is not hyperbolic). As we shall see in Chapter 8, accurate closures for the disperse-phase stress tensor always result in (weakly) hyperbolic moment equations. Finally, note that the term for collisions can generate a separate spatial flux term (known as the *collisional flux*), which would be added to the disperse-phase stress tensor. More details on this term can be found in Chapter 6.

Multicomponent disperse phase

For multicomponent particles, all components in a particle have the same velocity and thus $\mathbf{U}_{\mathrm{p}}$ can be found from Eq. (4.85). The only difference from the single-component case is that the terms for mass transfer and nucleation may be different for each component (e.g. just one component may change phases), in which case the right-hand side of Eq. (4.85) would include sums over the contributions for each component.

4.3.8 Fluid-phase momentum transport

For the fluid-phase momentum density, we can postulate a momentum balance of the form

$$\frac{\partial \varrho_{\mathrm{f}}\mathbf{U}_{\mathrm{f}}}{\partial t} + \frac{\partial}{\partial \mathbf{x}} \cdot \left(\varrho_{\mathrm{f}}[\![\mathbf{U}_{\mathrm{f}} \otimes \mathbf{U}_{\mathrm{f}}]\!]_{\mathrm{f}}\right) = \frac{\partial}{\partial \mathbf{x}} \cdot \mathbf{S}_{\mathrm{f}} - \varrho_{\mathrm{p}}[\![\mathbf{A}_{\mathrm{fp}}]\!]_{\mathrm{p}} - [\![\mathbf{G}_{\mathrm{p}}]\!]_{\mathrm{p}} - \varrho_{\mathrm{p}}[\![\mathbf{S}]\!]_{\mathrm{p}} + \varrho_{\mathrm{f}}\mathbf{A}_{\mathrm{f}}, \tag{4.92}$$

where $\mathbf{S}_{\mathrm{f}}$ denotes the viscous and pressure stress tensor in the fluid phase, and $\mathbf{A}_{\mathrm{f}}$ represents the body forces on the fluid phase. The convection term $[\![\mathbf{U}_{\mathrm{f}} \otimes \mathbf{U}_{\mathrm{f}}]\!]_{\mathrm{f}}$ would require a separate model to describe possible fluid-velocity fluctuations due, for example, to microscale turbulence. In the limit where there are no fluctuations in the fluid velocity,

[25] For this case, the mass average is replaced by the number average denoted by subscript $_{\mathrm{N}}$.

$[\![\mathbf{U}_\mathrm{f} \otimes \mathbf{U}_\mathrm{f}]\!]_\mathrm{f} = \mathbf{U}_\mathrm{f} \otimes \mathbf{U}_\mathrm{f}$. However, in general, the fluid velocity seen by a particle may fluctuate due, for example, to the momentum coupling with the particle phase. As in Eq. (4.85), the terms $[\![\mathbf{G}_\mathrm{p}]\!]_\mathrm{p}$ and $[\![\mathbf{S}]\!]_\mathrm{p}$ result from mass transfer between phases. Letting $\xi_{\mathrm{f}2}$ again be the fluid mass seen by a particle, the convection term can be written as

$$[\![\mathbf{U}_\mathrm{f} \otimes \mathbf{U}_\mathrm{f}]\!]_\mathrm{f} = \frac{1}{\varrho_\mathrm{f}} \int \xi_{\mathrm{f}2} \mathbf{v}_\mathrm{f} \otimes \mathbf{v}_\mathrm{f} n \, \mathrm{d}\mathbf{v}_\mathrm{p} \, \mathrm{d}\boldsymbol{\xi}_\mathrm{p} \, \mathrm{d}\mathbf{v}_\mathrm{f} \, \mathrm{d}\boldsymbol{\xi}_\mathrm{f}, \tag{4.93}$$

which can be viewed as a consistency constraint on the definition of $\xi_{\mathrm{f}2}$. Note that Eq. (4.92) should be consistent with the moment-transport equation found directly from the GPBE in Eq. (4.39), which provides information about the particle phase and the fluid phase *seen* by the particle phase. In other words, Eq. (4.92) should be consistent with the result found from deriving the moment-transport equation for $g = \xi_{\mathrm{f}2} \mathbf{v}_\mathrm{f}$ from Eq. (4.39). In fact, if we neglect mass transfer and particle nucleation, multiplying Eq. (4.39) by $\xi_{\mathrm{f}2} \mathbf{v}_\mathrm{f}$, followed by integration over phase space, yields

$$\frac{\partial \varrho_\mathrm{f} \mathbf{U}_\mathrm{f}}{\partial t} + \frac{\partial}{\partial \mathbf{x}} \cdot \left(\varrho_\mathrm{f} [\![\mathbf{U}_\mathrm{p} \otimes \mathbf{U}_\mathrm{f}]\!]_\mathrm{f} \right) = \varrho_\mathrm{f} [\![\mathbf{A}_\mathrm{pf}]\!]_\mathrm{f} + \varrho_\mathrm{f} [\![\mathbf{A}_\mathrm{f}]\!]_\mathrm{f} + [\![\mathbf{G}_\mathrm{f}]\!]_\mathrm{f}, \tag{4.94}$$

where $\mathbf{U}_\mathrm{f}$ is given by Eq. (4.81). The remaining terms in Eq. (4.94) are defined by

$$[\![\mathbf{U}_\mathrm{p} \otimes \mathbf{U}_\mathrm{f}]\!]_\mathrm{f} \equiv \frac{1}{\varrho_\mathrm{f}} \int \xi_{\mathrm{f}2} (\mathbf{v}_\mathrm{p} \otimes \mathbf{v}_\mathrm{f}) n \, \mathrm{d}\mathbf{v}_\mathrm{p} \, \mathrm{d}\boldsymbol{\xi}_\mathrm{p} \, \mathrm{d}\mathbf{v}_\mathrm{f} \, \mathrm{d}\boldsymbol{\xi}_\mathrm{f}, \tag{4.95}$$

$$[\![\mathbf{A}_\mathrm{pf}]\!]_\mathrm{f} \equiv \frac{1}{\varrho_\mathrm{f}} \int \xi_{\mathrm{f}2} \langle \mathbf{A}_{\mathrm{pf}2} \rangle_1 n \, \mathrm{d}\mathbf{v}_\mathrm{p} \, \mathrm{d}\boldsymbol{\xi}_\mathrm{p} \, \mathrm{d}\mathbf{v}_\mathrm{f} \, \mathrm{d}\boldsymbol{\xi}_\mathrm{f}, \tag{4.96}$$

$$[\![\mathbf{A}_\mathrm{f}]\!]_\mathrm{f} \equiv \frac{1}{\varrho_\mathrm{f}} \int \xi_{\mathrm{f}2} \langle \mathbf{A}_{\mathrm{f}2} \rangle_1 n \, \mathrm{d}\mathbf{v}_\mathrm{p} \, \mathrm{d}\boldsymbol{\xi}_\mathrm{p} \, \mathrm{d}\mathbf{v}_\mathrm{f} \, \mathrm{d}\boldsymbol{\xi}_\mathrm{f}, \tag{4.97}$$

and

$$[\![\mathbf{G}_\mathrm{f}]\!]_\mathrm{f} \equiv \int \mathbf{v}_\mathrm{f} \langle G_{\mathrm{f}2} \rangle_1 n \, \mathrm{d}\mathbf{v}_\mathrm{p} \, \mathrm{d}\boldsymbol{\xi}_\mathrm{p} \, \mathrm{d}\mathbf{v}_\mathrm{f} \, \mathrm{d}\boldsymbol{\xi}_\mathrm{f}. \tag{4.98}$$

The term $[\![\mathbf{A}_\mathrm{pf}]\!]_\mathrm{f}$ is the mass-average acceleration of the fluid seen by the particles due in part to momentum transfer between phases, and $[\![\mathbf{A}_\mathrm{f}]\!]_\mathrm{f}$ is due to forces in the fluid phase. The term $[\![\mathbf{G}_\mathrm{f}]\!]_\mathrm{f}$ can be evaluated using Eq. (4.80):

$$[\![\mathbf{G}_\mathrm{f}]\!]_\mathrm{f} = -[\![\mathbf{G}_\mathrm{p}]\!]_\mathrm{f} + \frac{\partial}{\partial \mathbf{x}} \cdot \left[\varrho_\mathrm{f} \left([\![\mathbf{U}_\mathrm{p} \otimes \mathbf{U}_\mathrm{f}]\!]_\mathrm{f} - [\![\mathbf{U}_\mathrm{f} \otimes \mathbf{U}_\mathrm{f}]\!]_\mathrm{f} \right) \right] + \frac{\varrho_\mathrm{f}}{\tau_\mathrm{pf}} \left([\![\mathbf{U}_\mathrm{f}]\!]_\mathrm{p} - \mathbf{U}_\mathrm{f} \right), \tag{4.99}$$

where the first term on the right-hand side results from the exchange of momentum between phases due to mass transfer, the second term replaces the convection term on the left-hand side of Eq. (4.94) with the corresponding term in Eq. (4.92), and the last term contains the particle-mass-average fluid velocity:

$$[\![\mathbf{U}_\mathrm{f}]\!]_\mathrm{p} \equiv \frac{1}{\varrho_\mathrm{p}} \int \xi_{\mathrm{p}2} \mathbf{v}_\mathrm{f} n \, \mathrm{d}\mathbf{v}_\mathrm{p} \, \mathrm{d}\boldsymbol{\xi}_\mathrm{p} \, \mathrm{d}\mathbf{v}_\mathrm{f} \, \mathrm{d}\boldsymbol{\xi}_\mathrm{f}. \tag{4.100}$$

Remarkably, no additional models are required to obtain consistent convection of the fluid-phase momentum density. For consistency, we can require that $[\![\mathbf{U}_\mathrm{f}]\!]_\mathrm{p} = \mathbf{U}_\mathrm{f}$ so that the final term on the right-hand side of Eq. (4.99) is null;[26] however, due to conservation of momentum the final term will be cancelled out due to a contribution from $\langle \mathbf{A}_{\mathrm{pf}2} \rangle_1$, as will be shown next.

[26] In the limit $\tau_\mathrm{pf} \to 0$, this consistency condition will be exact, since in this limit $\xi_{\mathrm{f}2} \to \varrho_\mathrm{f} \xi_{\mathrm{p}2} / \varrho_\mathrm{p}$.

A consistent model for the fluid acceleration seen by the particles

By comparing Eq. (4.94) with Eq. (4.92), after inserting Eq. (4.99), we see that

$$[\![\mathbf{G}_\mathrm{p}]\!]_\mathrm{f} = [\![\mathbf{G}_\mathrm{p}]\!]_\mathrm{p}, \tag{4.101}$$

$$[\![\mathbf{A}_\mathrm{f}]\!]_\mathrm{f} = \frac{1}{\varrho_\mathrm{f}}\frac{\partial}{\partial \mathbf{x}} \cdot \mathbf{S}_\mathrm{f} + \mathbf{A}_\mathrm{f}, \tag{4.102}$$

and

$$[\![\mathbf{A}_\mathrm{pf}]\!]_\mathrm{f} = -\frac{\varrho_\mathrm{p}}{\varrho_\mathrm{f}}[\![\mathbf{A}_\mathrm{fp}]\!]_\mathrm{p}. \tag{4.103}$$

The constraint in Eq. (4.101) states that momentum must be conserved during mass transfer between phases, and it can be combined with the constraint in Eq. (4.103). *Minimal* consistent models for the fluid acceleration terms in Eq. (4.39) can then be written as

$$\langle \mathbf{A}_\mathrm{f2} \rangle_1 = \frac{1}{\varrho_\mathrm{f}}\frac{\partial}{\partial \mathbf{x}} \cdot \mathbf{S}_\mathrm{f} + \mathbf{A}_\mathrm{f} + \frac{1}{\tau_\mathrm{f}}(\mathbf{U}_\mathrm{f} - \mathbf{v}_\mathrm{f}) \tag{4.104}$$

and[27]

$$\langle \mathbf{A}_\mathrm{pf2} \rangle_1 = \frac{1}{\xi_\mathrm{f2}}\left(\mathbf{v}_\mathrm{f} - \mathbf{v}_\mathrm{p}\right)\langle G_\mathrm{p2}\rangle_1 - \frac{\xi_\mathrm{p2}}{\xi_\mathrm{f2}}\langle \mathbf{A}_\mathrm{fp2}\rangle_1 + \frac{1}{\tau_\mathrm{pf}}\left(1 - \frac{\varrho_\mathrm{f}\xi_\mathrm{p2}}{\varrho_\mathrm{p}\xi_\mathrm{f2}}\right)\mathbf{v}_\mathrm{f}. \tag{4.105}$$

The first two terms on the right-hand side of Eq. (4.105) represent momentum transfer between phases, and it is noteworthy that no additional terms involving the velocity differences or the spatial gradient of the NDF are required for consistency.[28] The third term results from the conservation of momentum during the relaxation process with time scale τ_pf, and its contribution to the mean fluid-phase momentum balance exactly cancels out the final term in Eq. (4.99). Note, however, that the first term on the right-hand side of Eq. (4.105) generates the term $[\![\mathbf{G}_\mathrm{p}]\!]_\mathrm{p}$ in Eq. (4.92), and the second term corresponds to momentum transfer in the absence of mass transfer. The last term in Eq. (4.104) is a simple fluid-velocity-fluctuation-dissipation model wherein τ_f is a characteristic relaxation time for the fluid velocity to return to its mass-average value.[29] More sophisticated models that account for *macroscale* fluid turbulence can be found in the literature (Minier & Peirano, 2001; Simonin *et al.*, 1993). However, the reader should keep in mind that the mesoscale models discussed in this book do not require (or need) models for the macroscale turbulence. At most, the mesoscale models will include the effects of microscale turbulence

[27] Particle nucleation generates a contribution similar to the mass-transfer term involving $\langle G_\mathrm{p2}\rangle_1$ in Eq. (4.105). It is also important to recognize that the model for $\langle \mathbf{A}_\mathrm{fp2}\rangle_1$ will have a mass-transfer contribution of the form

$$\langle \mathbf{A}_\mathrm{fp2}\rangle_1\Big|_\mathrm{mass} = \frac{1}{\xi_\mathrm{p2}}\left(\mathbf{v}_\mathrm{f} - \mathbf{v}_\mathrm{p}\right)\max\left(\langle G_\mathrm{p2}\rangle_1, 0\right).$$

Thus, $\langle \mathbf{A}_\mathrm{pf2}\rangle_1$ will have a contribution due to mass transfer only when $\langle G_\mathrm{p2}\rangle_1 < 0$, i.e. for mass transfer from the particles to the fluid.

[28] For turbulent flows (Minier & Peirano, 2001; Simonin *et al.*, 1993), a correction term that depends on the velocity difference $\mathbf{U}_\mathrm{f} - \mathbf{U}_\mathrm{p}$ is added to $\langle \mathbf{A}_\mathrm{pf2}\rangle_1$. However, these models do not include the velocity-difference term in the model for $\langle G_\mathrm{f2}\rangle_1$. From Eq. (4.80), we see that, even when n is very small (i.e. $\alpha_\mathrm{p} \ll 1$), its spatial gradients can be large enough that $\langle G_\mathrm{f2}\rangle_1$ cannot be neglected.

[29] In the mesoscale model, setting $\tau_\mathrm{f} = 0$ forces the fluid velocity seen by the particles to be equal to the mass-average fluid velocity. This would be appropriate, for example, for one-way coupling wherein the particles do not disturb the fluid. In general, fluctuations in the fluid generated by the presence of other particles or microscale turbulence could be modeled by adding a phase-space diffusion term for $\mathbf{v}_\mathrm{f}$, similar to those used for macroscale turbulence (Minier & Peirano, 2001). The time scale τ_f would then correspond to the dissipation time scale of the microscale turbulence.

(such as particle wakes) on the mesoscale variables. In the *monokinetic fluid limit*, we take $\tau_f = 0$ so that $\mathbf{v}_f = \mathbf{U}_f$, which is appropriate when the particle phase is very dilute and momentum coupling is not important.

An effective viscosity closure for the fluid phase

A simple closure for $[\![\mathbf{U}_f \otimes \mathbf{U}_f]\!]_f$ in Eq. (4.92) that includes the effects of the microscale fluid-velocity fluctuations is to replace it with $\mathbf{U}_f \otimes \mathbf{U}_f$, and then to model $\mathbf{S}_f$ as a Newtonian fluid for the fluid velocity $\mathbf{U}_f$ with an effective viscosity equal to the sum of the molecular kinematic viscosity of the fluid (ν_f) and a microscale effective kinematic viscosity (ν_{eff}). This leads to $\mathbf{S}_f = -p_f\mathbf{I} + \boldsymbol{\tau}_f$, where p_f is the pressure in the fluid phase (e.g. that found from an equation of state) and $\boldsymbol{\tau}_f = 2\varrho_f(\nu_f + \nu_{\text{eff}})[\mathbf{D}_f - \frac{1}{3}\,\text{trace}(\mathbf{D}_f)\mathbf{I}]$ with $\mathbf{D}_f = \frac{1}{2}[\nabla_{\mathbf{x}}\mathbf{U}_f + (\nabla_{\mathbf{x}}\mathbf{U}_f)^{\text{T}}]$. For equal-volume particles, the effective viscosity could be described by a mixing-length model: $\nu_{\text{eff}} = C_\nu \alpha_p d_p |\mathbf{U}_p - \mathbf{U}_f|$, where d_p is the particle diameter and C_ν is a model constant of order unity. With this closure, the mass-average fluid velocity can be found by solving Eq. (4.92) coupled to the moment-transport equations for the disperse phase. Elsewhere in the moment-transport equations, the fluctuating fluid velocity $\mathbf{v}_f$ is replaced by $\mathbf{U}_f$, which is similar to the monokinetic fluid limit.

The mixture momentum equation

The mixture continuity equation given in Eq. (4.69) can be combined with a mixture momentum equation found by adding together Eqs. (4.85) and (4.92):

$$\frac{\partial \varrho_{\text{mix}}\mathbf{U}_{\text{mix}}}{\partial t} + \frac{\partial}{\partial \mathbf{x}} \cdot \left(\varrho_p \mathbf{F}_p + \varrho_f [\![\mathbf{U}_f \otimes \mathbf{U}_f]\!]_f\right) = \frac{\partial}{\partial \mathbf{x}} \cdot \mathbf{S}_f + \varrho_p [\![\mathbf{A}_p]\!]_p + \varrho_f \mathbf{A}_f. \tag{4.106}$$

While this expression does not contain terms for momentum transfer between phases (and, hence, would seem to be much simpler), it is of little practical value because the spatial flux term on the left-hand side cannot be written in terms of $\mathbf{U}_{\text{mix}}$. Only in the limiting case where $\mathbf{U}_{\text{mix}}$ is very close to $\mathbf{U}_f$ (i.e. very small Stokes numbers) would it be possible to model accurately a disperse multiphase flow using a closed form of Eq. (4.106). For finite Stokes numbers, it is best to solve the separate momentum equations for the disperse and fluid phases.

4.3.9 Higher-order moment transport

The moment-transport equations that we have derived up to this point are of first order in the velocity variables. In order to describe fluctuations about the first-order moments, it is necessary to derive transport equations for second- and sometimes higher-order moments. Just as before, this is accomplished by starting from Eq. (4.39) with a particular choice for g. In order to illustrate how this is done, we will consider the function $g = \xi_{p2} v_{pi}^2$ which results in the particle-mass-average moment

$$\varrho_p [\![U_{pi}^2]\!]_p \equiv \int \xi_{p2} v_{pi}^2 n \, \mathrm{d}\mathbf{v}_p \, \mathrm{d}\boldsymbol{\xi}_p \, \mathrm{d}\mathbf{v}_f \, \mathrm{d}\boldsymbol{\xi}_f, \tag{4.107}$$

where ξ_{p2} is again equal to the particle mass. Note that this moment is one component of the total disperse-phase energy:

$$E_p = \frac{1}{2} \sum_{i=1}^{3} [\![U_{pi}^2]\!]_p, \tag{4.108}$$

which is used to define the granular temperature, a measure of the magnitude of the velocity fluctuations in the disperse phase. In general, the second-order disperse-phase velocity moment will not be equal to the square of the first-order disperse-phase velocity moment ($[\![U_{\mathrm{p}i}^2]\!]_\mathrm{p} \neq U_{\mathrm{p}i}^2$), and the same is true for the second-order fluid-velocity moment ($[\![U_{\mathrm{f}i}^2]\!]_\mathrm{f} \neq U_{\mathrm{f}i}^2$). For example, mesoscale velocity fluctuations might be introduced through a stochastic component in the mesoscale model for the fluid-particle drag as described in Section 5.1, or by particle trajectory crossing (PTC) for inertial particles. In such cases, it may be necessary to solve transport equations for the higher-order moments.

Second-order moment of disperse-phase velocity

Starting from Eq. (4.39), the moment-transport equation corresponding to Eq. (4.107) is (no summation over i is implied)

$$\frac{\partial \varrho_\mathrm{p}[\![U_{\mathrm{p}i}^2]\!]_\mathrm{p}}{\partial t} + \frac{\partial}{\partial \mathbf{x}} \cdot \left(\varrho_\mathrm{p}[\![U_{\mathrm{p}i}^2 \mathbf{U}_\mathrm{p}]\!]_\mathrm{p}\right) = 2\varrho_\mathrm{p}[\![U_{\mathrm{p}i}A_{\mathrm{fp}i}]\!]_\mathrm{p} + 2\varrho_\mathrm{p}[\![U_{\mathrm{p}i}A_{\mathrm{p}i}]\!]_\mathrm{p} + [\![U_{\mathrm{p}i}^2 G_\mathrm{p}]\!] + \varrho_\mathrm{p}[\![U_{\mathrm{p}i}^2 S]\!]_\mathrm{p}, \tag{4.109}$$

where the mass-transfer term is defined by

$$[\![U_{\mathrm{p}i}^2 G_\mathrm{p}]\!] = \int v_{\mathrm{p}i}^2 \langle G_{\mathrm{p}2}\rangle_1 n \, \mathrm{d}\mathbf{v}_\mathrm{p}\, \mathrm{d}\boldsymbol{\xi}_\mathrm{p}\, \mathrm{d}\mathbf{v}_\mathrm{f}\, \mathrm{d}\boldsymbol{\xi}_\mathrm{f} \tag{4.110}$$

and the convection term by

$$\varrho_\mathrm{p}[\![U_{\mathrm{p}i}^2 \mathbf{U}_\mathrm{p}]\!]_\mathrm{p} \equiv \int \xi_{\mathrm{p}2} v_{\mathrm{p}i}^2 \mathbf{v}_\mathrm{p} n \, \mathrm{d}\mathbf{v}_\mathrm{p}\, \mathrm{d}\boldsymbol{\xi}_\mathrm{p}\, \mathrm{d}\mathbf{v}_\mathrm{f}\, \mathrm{d}\boldsymbol{\xi}_\mathrm{f}. \tag{4.111}$$

The other terms involving $[\![\cdot]\!]_\mathrm{p}$ are defined similarly. Note that, as is usually the case in moment methods, the convection term is unclosed and involves even higher-order moments. The remaining terms on the right-hand side of Eq. (4.109) are defined in a manner very similar to their counterparts in Eq. (4.85). Thus we will not discuss them in detail except to say that they will usually not appear in closed form. Only in the special case discussed in Section 4.4 where the velocity NDF is nearly Gaussian can we provide an accurate closure for Eq. (4.111) in terms of the velocity moments of up to second order.

The second-order moment of the fluid-phase velocity

Consider next the fluid-mass-average second-order moment of the fluid-phase velocity defined by

$$[\![U_{\mathrm{f}i}^2]\!]_\mathrm{f} \equiv \frac{1}{\varrho_\mathrm{f}} \int \xi_{\mathrm{f}2} v_{\mathrm{f}i}^2 n \, \mathrm{d}\mathbf{v}_\mathrm{p}\, \mathrm{d}\boldsymbol{\xi}_\mathrm{p}\, \mathrm{d}\mathbf{v}_\mathrm{f}\, \mathrm{d}\boldsymbol{\xi}_\mathrm{f}, \tag{4.112}$$

where $\xi_{\mathrm{f}2}$ is the fluid mass seen by a particle. Starting from Eq. (4.39) with $g = \xi_{\mathrm{f}2} v_{\mathrm{f}i}^2$, the moment-transport equation corresponding to Eq. (4.112) has the form (no summation over i is implied)

$$\frac{\partial \varrho_\mathrm{f}[\![U_{\mathrm{f}i}^2]\!]_\mathrm{f}}{\partial t} + \frac{\partial}{\partial \mathbf{x}} \cdot \left(\varrho_\mathrm{f}[\![U_{\mathrm{f}i}^2 \mathbf{U}_\mathrm{p}]\!]_\mathrm{f}\right) = 2\varrho_\mathrm{f}[\![U_{\mathrm{f}i}A_{\mathrm{pf}i}]\!]_\mathrm{f} + 2\varrho_\mathrm{f}[\![U_{\mathrm{f}i}A_{\mathrm{f}i}]\!]_\mathrm{f} + [\![U_{\mathrm{f}i}^2 G_\mathrm{f}]\!] + \varrho_\mathrm{f}[\![U_{\mathrm{f}i}^2 S]\!]_\mathrm{f}, \tag{4.113}$$

where the mass-transfer term is defined by

$$[\![U_{\mathrm{f}i}^2 G_\mathrm{f}]\!] = \int v_{\mathrm{f}i}^2 \langle G_{\mathrm{f}2}\rangle_1 n \, \mathrm{d}\mathbf{v}_\mathrm{p}\, \mathrm{d}\boldsymbol{\xi}_\mathrm{p}\, \mathrm{d}\mathbf{v}_\mathrm{f}\, \mathrm{d}\boldsymbol{\xi}_\mathrm{f} \tag{4.114}$$

and the convection term by

$$[\![U_{fi}^2 \mathbf{U}_p]\!]_f \equiv \frac{1}{\varrho_f} \int \xi_{f2} v_{fi}^2 \mathbf{v}_p n \, d\mathbf{v}_p \, d\boldsymbol{\xi}_p \, d\mathbf{v}_f \, d\boldsymbol{\xi}_f. \tag{4.115}$$

The other terms involving $[\![\cdot]\!]_f$ are defined similarly. As expected, the convection term depends on the particle velocity because it is written for the fluid seen by the particle. Using the consistent models for $\langle G_{f2} \rangle_1$ (Eq. (4.80)), $\langle \mathbf{A}_{f2} \rangle_1$ (Eq. (4.104)), and $\langle \mathbf{A}_{pf2} \rangle_1$ (Eq. (4.105)), the terms on the right-hand side of Eq. (4.113) can be rewritten in a more familiar form (with summation over j, but not over i):

$$\begin{aligned} \frac{\partial \varrho_f [\![U_{fi}^2]\!]_f}{\partial t} + \frac{\partial}{\partial \mathbf{x}} \cdot \left(\varrho_f [\![U_{fi}^2 \mathbf{U}_f]\!]_f \right) = {} & 2U_{fi} \frac{\partial S_{fji}}{\partial x_j} + 2\varrho_f U_{fi} A_{fi} - 2\varrho_p [\![U_{fi} A_{fpi}]\!]_f \\ & + [\![U_{fi}^2 G_p]\!] - 2[\![U_{fi} U_{pi} G_p]\!] + \varrho_f [\![U_{fi}^2 S]\!]_f \\ & - \frac{2\varrho_f}{\tau_f} \left([\![U_{fi}^2]\!]_f - U_{fi}^2 \right) + \frac{\varrho_f}{\tau_{pf}} \left([\![U_{fi}^2]\!]_f - [\![U_{fi}^2]\!]_p \right), \end{aligned} \tag{4.116}$$

where, remarkably, the convection term on the left-hand side is now written in terms of the fluid velocity. In Eq. (4.116) the terms on the right-hand side have the following origins.

(i) The first two result from the stresses and body forces in the fluid phase.

(ii) The third term is due to momentum transfer between phases (including the mass-transfer contribution to $\langle \mathbf{A}_{fp2} \rangle_1$ when $\langle G_{p2} \rangle_1 > 0$).

(iii) The fourth, fifth, and sixth terms result from mass transfer between phases.

(iv) The seventh term, involving τ_f, is the fluid-velocity-variance-dissipation term. Note that, as $\tau_f \to 0$ (i.e. the monokinetic fluid limit), we have $[\![U_{fi}^2]\!]_f \to U_{fi}^2$.

(v) The final term will be null in the limit $\tau_{pf} \to 0$. Thus, for consistency, we can require that ξ_{f2} be defined such that $[\![\mathbf{U}_f \otimes \mathbf{U}_f]\!]_f = [\![\mathbf{U}_f \otimes \mathbf{U}_f]\!]_p$.

In Eq. (4.116) there appear to be no *explicit* homogeneous fluid-phase-velocity-variance source terms. Nonetheless, the terms for mass and momentum transfer are all potential sources of fluid-phase velocity variance. For example, a fluid-drag term can be a source or a sink of fluid-phase velocity variance, depending on the magnitude of the mixed moments $[\![U_{pi} U_{fi}]\!]_p$ and $[\![U_{pi} U_{fi}]\!]_f$, which we will describe next.

The second-order moments of disperse-phase and fluid-phase velocities

The mesoscale interphase momentum-exchange terms are typically written in terms of the velocity difference $(\mathbf{v}_p - \mathbf{v}_f)$. Thus, in the simplest cases, in which these terms are linear in the velocity difference, the second-order moment-transport equations in Eqs. (4.109) and (4.116) will contain source terms involving the following mixed moments:

$$\varrho_p [\![U_{pi} (\mathbf{U}_f - \mathbf{U}_p)]\!]_p = \int \xi_{p2} v_{pi} (\mathbf{v}_f - \mathbf{v}_p) n \, d\mathbf{v}_p \, d\boldsymbol{\xi}_p \, d\mathbf{v}_f \, d\boldsymbol{\xi}_f,$$

$$\varrho_f [\![U_{fi} (\mathbf{U}_f - \mathbf{U}_p)]\!]_f = \int \xi_{f2} v_{fi} (\mathbf{v}_f - \mathbf{v}_p) n \, d\mathbf{v}_p \, d\boldsymbol{\xi}_p \, d\mathbf{v}_f \, d\boldsymbol{\xi}_f,$$

which, when expanded out in components, lead to two new second-order mixed moments:

$$\varrho_{\rm p}[\![U_{{\rm p}i}U_{{\rm f}j}]\!]_{\rm p} = \int \xi_{\rm p2} v_{{\rm p}i} v_{{\rm f}j} n \, {\rm d}\mathbf{v}_{\rm p} \, {\rm d}\boldsymbol{\xi}_{\rm p} \, {\rm d}\mathbf{v}_{\rm f} \, {\rm d}\boldsymbol{\xi}_{\rm f},$$

$$\varrho_{\rm f}[\![U_{{\rm f}i}U_{{\rm p}j}]\!]_{\rm f} = \int \xi_{\rm f2} v_{{\rm f}i} v_{{\rm p}j} n \, {\rm d}\mathbf{v}_{\rm p} \, {\rm d}\boldsymbol{\xi}_{\rm p} \, {\rm d}\mathbf{v}_{\rm f} \, {\rm d}\boldsymbol{\xi}_{\rm f}.$$

Note that these mass-average moments are defined using different masses (i.e. the particle mass, $\xi_{\rm p2}$, versus the fluid mass seen by a particle, $\xi_{\rm f2}$). However, it is possible to introduce a consistency constraint[30] on the definition of $\xi_{\rm f2}$ such that $[\![\mathbf{U}_{\rm f} \otimes \mathbf{U}_{\rm p}]\!]_{\rm f} = [\![\mathbf{U}_{\rm f} \otimes \mathbf{U}_{\rm p}]\!]_{\rm p}$ (which is exact in the limit $\tau_{\rm pf} \to 0$), in which case only one new moment-transport equation is needed. This transport equation can be found starting from Eq. (4.39) with $g = \xi_{\rm p2} v_{{\rm p}i} v_{{\rm f}j}$, and will have a form similar to Eq. (4.109). Finally, it is worth noting that, in the monokinetic fluid limit (i.e. $\tau_{\rm f} \to 0$), the velocity mixed moments should simplify to $[\![\mathbf{U}_{\rm p} \otimes \mathbf{U}_{\rm f}]\!]_{\rm p} \to \mathbf{U}_{\rm p} \otimes \mathbf{U}_{\rm f}$, in which case only the higher-order moments of the particle velocity are of interest. To conclude this section, we should remind the reader that, as a general rule, the moment-transport equations are never closed either due to the spatial fluxes or to nonlinearities in the mesoscale models. Here, we have derived transport equations for a few low-order moments that have specific physical meaning (e.g. density, velocity, etc.). However, since none of the transport equations are closed, it will be necessary to provide moment closures. In the next section, we briefly review a few moment closures that are valid for particular cases as determined by the source terms in the GPBE.

4.4 Moment closures for the GPBE

In this chapter, we have introduced the basic steps needed to derive the mesoscale model starting from a microscale description of the polydisperse multiphase flow. The first step is to identify a set of microscale variables, denoted here by $(\mathbf{X}^{(n)}, \mathbf{U}_{\rm p}^{(n)}, \boldsymbol{\xi}_{\rm p}^{(n)}, \mathbf{U}_{\rm f}^{(n)}, \boldsymbol{\xi}_{\rm f}^{(n)})$, that suffice to characterize the mesoscale properties of the flow. The time evolution of these variables (and thus their source terms) can, in principle, be investigated using a direct numerical simulation of the microscale model. The next (and undoubtedly the most difficult) step is to postulate mesoscale models for the conditional expectations of the microscale source terms given a fixed set of mesoscale variables $(\mathbf{X}^{(n)} = \mathbf{x}, \mathbf{U}_{\rm p}^{(n)} = \mathbf{v}_{\rm p}, \boldsymbol{\xi}_{\rm p}^{(n)} = \boldsymbol{\xi}_{\rm p}, \mathbf{U}_{\rm f}^{(n)} = \mathbf{v}_{\rm f}, \boldsymbol{\xi}_{\rm f}^{(n)} = \boldsymbol{\xi}_{\rm f})$. In other words, the complete multi-particle description of the fully coupled system must be reduced to a single-particle description by postulating (or, in rare cases, directly deriving) mesoscale models that appear in the GPBE. In general, mesoscale models contain both deterministic and probabilistic components (e.g. mean drag and Brownian motion representing microscale velocity fluctuations). Chapter 5 is devoted to a detailed description of mesoscale models for multiphase systems. However, the reader should appreciate the fundamental importance of the mesoscale models in determining the predictive capability of the final macroscale description of polydisperse multiphase flow. In short, if important physics is missing at the mesoscale level, it will be impossible to obtain a satisfactory description at the macroscale. Indeed, as discussed in Chapter 1, a major advantage of using a kinetics-based modeling approach is the separation of the physical

[30]Recall that Eqs. (4.75), (4.81), and (4.93) also represent consistency constraints on the definition of $\xi_{\rm f2}$. In general, the system of moment-transport equations takes on the simplest form in the limit $\tau_{\rm pf} \to 0$, in which case weighting with respect to particle mass or fluid mass is equivalent since the two will be linearly dependent in this limit (see the last term in Eq. (4.80)).

modeling step (i.e. microscale → mesoscale) from the purely mathematical modeling step (i.e. mesoscale → macroscale). For the latter, we are of course referring to the moment closures needed to close the moment-transport equations. The mathematical step of deriving moment closures for a given GPBE is by no means obvious or even guaranteed to work. If we consider the GPBE given by

$$\underbrace{\frac{\partial n}{\partial t} + \frac{\partial}{\partial \mathbf{x}} \cdot \left(\mathbf{v}_{\mathrm{p}} n\right)}_{\text{accumulation \& free transport}} = \underbrace{-\frac{\partial}{\partial \mathbf{v}_{\mathrm{p}}} \cdot \left[\left(\langle \mathbf{A}_{\mathrm{fp}} \rangle_1 + \langle \mathbf{A}_{\mathrm{p}} \rangle_1\right) n\right] - \frac{\partial}{\partial \boldsymbol{\xi}_{\mathrm{p}}} \cdot \left(\langle \mathbf{G}_{\mathrm{p}} \rangle_1 n\right)}_{\text{particle phase-space transport}}$$
$$\underbrace{- \frac{\partial}{\partial \mathbf{v}_{\mathrm{f}}} \cdot \left[\left(\langle \mathbf{A}_{\mathrm{pf}} \rangle_1 + \langle \mathbf{A}_{\mathrm{f}} \rangle_1\right) n\right] - \frac{\partial}{\partial \boldsymbol{\xi}_{\mathrm{f}}} \cdot \left(\langle \mathbf{G}_{\mathrm{f}} \rangle_1 n\right) + \mathcal{S}_1}_{\text{fluid phase-space transport \& source terms}}$$

we can distinguish terms for accumulation and free transport, on the one hand, and phase-space transport and source terms, on the other. Generally speaking, the free-transport term tends to complicate the mathematical form of the NDF because it transports information throughout the spatial domain at a velocity that appears in the NDF. Indeed, if the phase-space transport and source terms are absent, there is virtually no hope of finding a moment closure that will work for any arbitrary initial NDF. (Fortunately, for *particular* initial NDF the situation is not so hopeless!) In fact, the most successful moment closures have been developed for systems in which the phase-space transport and/or source terms dominate the free-transport term. In the following, we briefly review some of the more relevant examples for multiphase flows.

Monokinetic fluid phase

The first example is the case in which $\mathbf{v}_{\mathrm{f}} = \mathbf{U}_{\mathrm{f}}$ and $\boldsymbol{\xi}_{\mathrm{f}} = \boldsymbol{\phi}_{\mathrm{f}}$ so that there are no microscale fluctuations in the fluid phase. The NDF in this case is $n(\mathbf{v}_{\mathrm{p}}, \boldsymbol{\xi}_{\mathrm{p}}, \mathbf{v}_{\mathrm{f}}, \boldsymbol{\xi}_{\mathrm{f}}) = n(\mathbf{v}_{\mathrm{p}}, \boldsymbol{\xi}_{\mathrm{p}})\delta(\mathbf{v}_{\mathrm{f}} - \mathbf{U}_{\mathrm{f}})\delta(\boldsymbol{\xi}_{\mathrm{f}} - \boldsymbol{\phi}_{\mathrm{f}})$, where $\mathbf{U}_{\mathrm{f}}$ is found by solving Eq. (4.92) and $\boldsymbol{\phi}_{\mathrm{f}}$ is the mean fluid-composition vector[31] for the fluid phase. The GPBE reduces to

$$\frac{\partial n}{\partial t} + \frac{\partial}{\partial \mathbf{x}} \cdot \left(\mathbf{v}_{\mathrm{p}} n\right) = -\frac{\partial}{\partial \mathbf{v}_{\mathrm{p}}} \cdot \left[\left(\langle \mathbf{A}_{\mathrm{fp}} \rangle_1 + \langle \mathbf{A}_{\mathrm{p}} \rangle_1\right) n\right] - \frac{\partial}{\partial \boldsymbol{\xi}_{\mathrm{p}}} \cdot \left(\langle \mathbf{G}_{\mathrm{p}} \rangle_1 n\right) + \mathcal{S}_1, \tag{4.117}$$

wherein all of the mesoscale models are evaluated at $\mathbf{v}_{\mathrm{f}} = \mathbf{U}_{\mathrm{f}}$ and $\boldsymbol{\xi}_{\mathrm{f}} = \boldsymbol{\phi}_{\mathrm{f}}$. Here the term *monokinetic* applies to mesoscale variables that have a unique value at a given time and location. The monokinetic fluid limit of the GBPE is found by taking $\tau_{\mathrm{f}} \to 0$ in the fluid-acceleration model in Eq. (4.104).

Monokinetic disperse phase

The second example is the case in which the disperse-phase velocity is equal to the conditional expected disperse-phase velocity given the other mesoscale variables: $\mathbf{v}_{\mathrm{p}} = \langle \mathbf{U}_{\mathrm{p}} | \boldsymbol{\xi}_{\mathrm{p}}, \mathbf{v}_{\mathrm{f}}, \boldsymbol{\xi}_{\mathrm{f}} \rangle$. In this case, fluctuations in the disperse phase are "slaved" to the fluid-phase fluctuations and can depend on the particle internal coordinates (e.g. particle size). The NDF in this case is $n(\mathbf{v}_{\mathrm{p}}, \boldsymbol{\xi}_{\mathrm{p}}, \mathbf{v}_{\mathrm{f}}, \boldsymbol{\xi}_{\mathrm{f}}) = n(\boldsymbol{\xi}_{\mathrm{p}}, \mathbf{v}_{\mathrm{f}}, \boldsymbol{\xi}_{\mathrm{f}})\delta(\mathbf{v}_{\mathrm{p}} - \langle \mathbf{U}_{\mathrm{p}} | \boldsymbol{\xi}_{\mathrm{p}}, \mathbf{v}_{\mathrm{f}}, \boldsymbol{\xi}_{\mathrm{f}} \rangle)$ and the GPBE reduces to

[31] The only exception is for the fluid mass seen by the particle, which was denoted earlier in this chapter as ξ_{f2}. In the monokinetic fluid limit, we will let $\xi_{\mathrm{f2}} = \varrho_{\mathrm{f}} \xi_{\mathrm{p2}} / \varrho_{\mathrm{p}}$, where ξ_{p2} is the particle mass.

$$\frac{\partial n}{\partial t} + \frac{\partial}{\partial \mathbf{x}} \cdot \left(\langle \mathbf{U}_\mathrm{p} | \xi_\mathrm{p}, \mathbf{v}_\mathrm{f}, \xi_\mathrm{f} \rangle n \right) = -\frac{\partial}{\partial \xi_\mathrm{p}} \cdot \left(\langle \mathbf{G}_\mathrm{p} \rangle_1 n \right) - \frac{\partial}{\partial \mathbf{v}_\mathrm{f}} \cdot \left[\left(\langle \mathbf{A}_\mathrm{pf} \rangle_1 + \langle \mathbf{A}_\mathrm{f} \rangle_1 \right) n \right] - \frac{\partial}{\partial \xi_\mathrm{f}} \cdot \left(\langle \mathbf{G}_\mathrm{f} \rangle_1 n \right) + \mathcal{S}_1$$

wherein all of the mesoscale models are evaluated at $\mathbf{v}_\mathrm{p} = \langle \mathbf{U}_\mathrm{p} | \xi_\mathrm{p}, \mathbf{v}_\mathrm{f}, \xi_\mathrm{f} \rangle$. A monokinetic disperse phase is found for particles with very small Stokes number. Usually, for small Stokes numbers the fluid phase is also assumed to be monokinetic, leading to $\mathbf{v}_\mathrm{p} = \langle \mathbf{U}_\mathrm{p} | \xi_\mathrm{p}, \xi_\mathrm{f} \rangle$ and additional simplification of the GPBE. Note that, for a monokinetic disperse phase, the conditional velocity must be found using a consistent model equation derived from the GPBE. For both of the monokinetic cases described above the GPBE is simplified, but still must be solved to find the NDF. Hence, additional closures are usually required in order to proceed further. For this purpose, one of the closure methods which are based on the moments described next can sometimes be applied.

Hydrodynamic limit

As discussed in Chapter 1, the hydrodynamic limit refers to cases in which the collision term $\mathcal{S}_1$ is strong enough to determine the form of the velocity-distribution function. In its original formulation (Chapman & Cowling, 1961), the hydrodynamic limit uses the fact that during elastic collisions the disperse-phase mass ϱ_p, momentum $\varrho_\mathrm{p}\mathbf{U}_\mathrm{p}$, and total energy $\varrho_\mathrm{p} E_\mathrm{p}$ are conserved. This, in turn, implies that in the corresponding moment-transport equations the terms resulting from $\mathcal{S}_1$ are null. In the limit of infinitely fast collisions (i.e. zero Knudsen number), the five equations for the moments (ϱ_p, $\varrho_\mathrm{p}\mathbf{U}_\mathrm{p}$, $\varrho_\mathrm{p} E_\mathrm{p}$) are the Euler equations for an inviscid fluid, and the velocity-distribution function is a Maxwellian distribution. For very small, but finite, Knudsen numbers, the system of moment-transport equations leads to the Navier–Stokes equation of gas dynamics. More details on the Knudsen-number expansion procedure used to develop the hydrodynamic closures can be found in Struchtrup (2005). An important limitation of this procedure is that the resulting transport equations are useful only for small Knudsen numbers, and in many polydisperse multiphase flows the disperse-phase Knudsen number is not small. For this reason, methods that do not depend on a truncated Knudsen-number (or similar) expansion are needed for closing the moment-transport equations.

Parameterized NDF

A popular method for closing a system of moment-transport equations is to assume a functional form for the NDF in terms of the mesoscale variables. Preferably, the parameters of the functional form can be written in closed form in terms of a few lower-order moments. It is then possible to solve only the transport equations for the lower-order moments which are needed in order to determine the parameters in the presumed NDF. The functional form of the NDF is then known, and can be used to evaluate the integrals appearing in the moment-transport equations. As an example, consider a case in which the velocity NDF is assumed to be Gaussian:[32]

$$n(t, \mathbf{x}, \mathbf{v}_\mathrm{p}) = \frac{\varrho_\mathrm{p}}{(2\pi |\boldsymbol{\Sigma}_\mathrm{p}|)^{3/2}} \exp\left[-\frac{1}{2} \left(\mathbf{v}_\mathrm{p} - \mathbf{U}_\mathrm{p} \right) \cdot \boldsymbol{\Sigma}_\mathrm{p} \cdot \left(\mathbf{v}_\mathrm{p} - \mathbf{U}_\mathrm{p} \right) \right]$$

[32]Recall that for a Maxwellian distribution $\boldsymbol{\Sigma}_\mathrm{p}$ is isotropic. Thus, the Gaussian NDF allows five additional degrees of freedom as compared with the Maxwellian distribution.

and $\mathbf{\Sigma}_p$ is the particle-velocity covariance matrix (i.e. a symmetric second-order tensor). The first ten moments of this presumed NDF up to second order are ϱ_p, $\varrho_p \mathbf{U}_M$, and $\varrho_p(\mathbf{\Sigma}_p + \mathbf{U}_p \otimes \mathbf{U}_p)$; thus the NDF is completely determined by solving the ten transport equations for these moments. An obvious advantage of using a presumed NDF is that the NDF is guaranteed to be realizable (nonnegative) for all values of the phase-space variables. On the other hand, an obvious disadvantage is that, once the form of the NDF has been fixed, there is no room for describing more complex functional shapes. For example, the Gaussian NDF is unimodal (i.e. n has one maximum at $\mathbf{v} = \mathbf{U}_p$ for all choices of $\mathbf{\Sigma}_p$). Thus, it cannot describe bimodal physics such as particle trajectory crossing (PTC) or specular reflections at walls, for which it will simply represent the bimodal NDF by a unimodal Gaussian. In general, since we do not know *a priori* whether the solution to a GPBE is close to a particular known NDF, it is usually difficult to prescribe a parameterized NDF for a given problem. The only exception to this rule is when one of the phase-space source terms in the GPBE dominates the other terms so that the NDF remains close to a known form. For example, this could occur if the microscale fluid-velocity fluctuations force the particle-velocity fluctuations to obey a Gaussian form in the limit of small Stokes numbers. Such cases are rare and, thus, in order to provide more flexibility in the shape of the NDF, the idea of using a truncated infinite series would seem attractive.

Functional expansion for NDF

In order to increase the number of degrees of freedom in a systematic manner, a functional expansion can be used to represent the NDF (Grad, 1949b). Using the velocity distribution as an example, the formal expansion is

$$n(t, \mathbf{x}, \mathbf{v}_p) = \sum_{\alpha=0}^{\infty} C_\alpha(t, \mathbf{x}) p_\alpha(\mathbf{v}) w(t, \mathbf{x}, \mathbf{v}_p),$$

where C_α are the expansion coefficients, p_α is a known family of orthogonal polynomials of order α, and w is the corresponding weight function for the family. For example, if w is the Maxwellian distribution (whose parameters depend on the five hydrodynamic moments of the NDF), the tensor-product Hermite polynomials are used for p_α. In theory, almost any NDF can be expanded in this way with a unique set of expansion coefficients. In practice, only a finite expansion is used to represent the NDF:

$$n^*(t, \mathbf{x}, \mathbf{v}_p) = \sum_{\alpha=0}^{N} C_\alpha(t, \mathbf{x}) p_\alpha(\mathbf{v}_p) w(t, \mathbf{x}, \mathbf{v}_p)$$

and the finite set of expansion coefficients is determined from a finite set of lower-order moments. Using the definition of the moments, the reconstructed NDF n^* can be used to evaluate the moments:

$$m_{\mathbf{k}} = \sum_{\alpha=0}^{N} \left(\int v_{p1}^{k_1} v_{p2}^{k_2} v_{p3}^{k_3} p_\alpha(\mathbf{v}_p) w(\mathbf{v}_p) \, d\mathbf{v}_p \right) C_\alpha.$$

The integral terms in parentheses are known for the family of orthogonal polynomials. With finite N and known moments, this system of linear equations has the form $\mathbf{M} = \mathbf{AC}$ and can be solved to find the expansion coefficients $C_\alpha(t, \mathbf{x})$. Thus, the presumed NDF $n^*(t, \mathbf{x}, \mathbf{v}_p)$ is a unique function of a finite set of moments, and the latter are found by solving the moment-transport equations using n^* to close the unclosed terms. The fact that

$C_\alpha(t, \mathbf{x})$ is found from a (small) linear system makes the method computationally very tractable, especially since the coefficient matrix **A** can be computed and stored in advance (using a change of variables, **A** can be written in terms of the hydrodynamic parameters). While mathematically very attractive, the finite functional expansion suffers from a major shortcoming: for finite N the NDF n^* can be unrealizable (negative) for some values of $\mathbf{v}_\mathrm{p}$. In other words, given a *realizable* moment set $m_\mathbf{k}$, the linear functional expansion will yield an *unrealizable* NDF n^*. This shortcoming is present whenever the coefficients are determined from the moments by solving a linear system. As noted in Chapter 1, when an unrealizable NDF is used to close the moment-transport equation, there is a high probability that the solution to the moment-transport equations will become unrealizable. When this happens, the numerical simulation will fail in an unrecoverable manner. In practice, the likelihood of generating an unrealizable NDF is high unless the true solution to the GBPE is close to the weight function $w(t, \mathbf{x}, \mathbf{v}_\mathrm{p})$. For this reason, we do not consider using a functional expansion of the NDF to be a viable alternative for closing the moment-transport equations for most systems.

Discrete NDF

A increasingly popular method for closing the moment-transport equations is to assume a discrete form for the phase-space variables. Taking the velocity NDF as an example, the velocity phase space can be discretized on a uniform, symmetric lattice centered at $\mathbf{v}_\mathrm{p} = \mathbf{0}$. For illustration purposes, let us assume that $N = 16$ lattice points are used and denote the corresponding velocities as $\mathbf{U}_\alpha$. The formal definition of the discrete NDF is

$$n^*(t, \mathbf{x}, \mathbf{v}_\mathrm{p}) = \sum_{\alpha=1}^{16} \rho_\alpha(t, \mathbf{x}) \delta\left(\mathbf{v}_\mathrm{p} - \mathbf{U}_\alpha\right).$$

It is important to note that the 16 velocities $\mathbf{U}_\alpha$ are fixed, and hence do not vary with space and time.[33] Thus, the moments of the discrete NDF will change only due to changes in the weights ρ_α. Given a set of 16 moments (or linear combinations of moments), we can evaluate them using n^*, which yields a linear system of the form $\mathbf{M} = \mathbf{A}\boldsymbol{\rho}$, where **A** depends on the velocities and **M** are the known moments. This linear system can be solved to find ρ_α, and therefore the discrete NDF is uniquely determined by its moments. In practice, the 16 moment transport equations can be rewritten as a system of equations for the 16 weights. Nevertheless, conceptually speaking, the discrete NDF is just a reconstructed NDF with a particular form that is used to close the moment-transport equations. As with the functional expansion method described above, there is no guarantee that the weights in the discrete NDF will be nonnegative and, hence, the reconstructed NDF can be unrealizable. Furthermore, the possibility of getting negative weights increases when the number of abscissas is increased (as one might do in order to find a better reconstruction). In the literature, the discrete NDF is used with the lattice Boltzmann method (Aidun & Clausen, 2010) to simulate fluid flow in the hydrodynamic limit (i.e. for small Knudsen numbers), for which the weights remain positive (because the true NDF is nearly Maxwellian). Unfortunately, the possibility of negative weights makes the discrete NDF with fixed abscissas unsuitable as a general closure method for moment-transport equations.

[33] Although we do not need to use it here, in practice n^* is defined only on lattice nodes in physical space denoted by $\mathbf{x}_\beta$. In one time step, the free-transport term in the GPBE moves information between the discrete spatial lattice nodes given the discrete velocities $\mathbf{U}_\alpha$. Thus, if n is initially a point distribution on the spatial lattice, it will remain so for all time.

Quadrature-based moment methods

The reconstruction of the NDF using quadrature-based moment methods (QBMM) is described in detail in Chapter 3. In comparison with the moment closures introduced above, QBMM have the following similarities and advantages.

(i) The reconstructed NDF is always *realizable*.

(ii) The moment-inversion algorithm is *direct*.

(iii) The reconstructed NDF reproduces *exactly* a particular set of moments.

(iv) The accuracy of the reconstructed NDF can be increased by using more moments.

(v) The reconstructed NDF can be discrete (QMOM) or continuous (EQMOM).

(vi) The continuous reconstructed NDF has a discrete quadrature representation with arbitrary accuracy.

It is the authors' opinion that QBMM constitute the best methods for closing the moment-transport equations resulting from a GPBE that is not close to a known limiting regime. Nonetheless, it should be obvious to the reader that, as the number of internal coordinates grows, so does the difficulty of finding an accurate reconstruction of the NDF. Therefore, it is always advantageous to first attempt to reduce the number of mesoscale degrees of freedom in the NDF before attempting to solve the macroscale moment problem. For example, if the microscale physics consists of several distinct interacting particle populations, it is better to write multiple low-dimensional GBPE (i.e. one for each population) rather than a single high-dimensional GBPE to describe the system. In the following chapters, we provide a detailed discussion of how to use QBMM to solve moment-transport equations, along with many examples. In order to understand these examples, the reader should first master the methodology of deriving the moment-transport equations from the GBPE presented in Section 4.3. In addition, a good understanding of the moment-inversion algorithms presented in Chapter 3 will be required in order to follow the discussion of the solution methods presented in Chapters 7 and 8.

5

Mesoscale models for physical and chemical processes

This chapter illustrates and discusses some of the most widely used mesoscale models for describing particulate processes. The reader should keep in mind that the chapter is not a comprehensive discussion of all the possible mesoscale models, but is simply a collection of some example models, used in this context to highlight the major issues typically encountered in the simulation of multiphase systems. Although some of the models illustrated in the next sections have general validity, they typically assume slightly different forms when applied to the simulation of solid–liquid systems (e.g. crystallization and precipitation processes), solid–gas systems (e.g. fluidized suspensions, aerosol reactors), liquid–liquid systems (e.g. emulsions), gas–liquid systems (e.g. bubble columns and gas–liquid stirred tanks), and liquid–gas systems (e.g. evaporating and non-evaporating sprays). In what follows we will often refer to the elements of the disperse phase as "particles" to indicate both solid particles (such as crystals, solid amorphous particles, and solid aggregates) and fluid particles (such as droplets and bubbles). The remainder of the chapter is organized as follows. After providing an overview of the philosophy behind the development of mesoscale models in Section 5.1, specific examples of phase- and real-space advection processes (and the consequent diffusion processes) are discussed in Sections 5.2–5.5. Subsequently, phase-space point (discontinuous) processes are presented in Sections 5.6–5.8. For each of these processes, the corresponding functional form of the mesoscale model appearing in the final GPBE is derived and discussed. The discussion includes also the most common approaches for the quantification of the rate of each of these processes, together with the ranges of applicability for different flow regimes.

5.1 An overview of mesoscale modeling

In this section, we provide a brief overview of the mesoscale modeling approach and its significance in closing the GPBE. We begin by describing the physical significance of the various terms appearing in the GPBE derived in Chapter 4, followed by a discussion of how the mesoscale model is derived from the microscale physics.

5.1.1 Mesoscale models in the GPBE

As has already been discussed in Chapter 4, the state of the disperse phase is completely identified by the internal-coordinate vector $\boldsymbol{\xi}_\mathrm{p}$ and velocity $\mathbf{v}_\mathrm{p}$, whereas the particle position is defined by the external-coordinate vector $\mathbf{x}$. Particles are, moreover, considered as indistinguishable elements of the disperse phase. Typically the continuous rate of change in the directions of the external and internal coordinates is assumed to be a function not only of $\boldsymbol{\xi}_\mathrm{p}$ and $\mathbf{v}_\mathrm{p}$, but also of the state of the suspending fluid *seen by the particle* (i.e. the primary continuous phase). The state of the suspending fluid seen by the particle is identified by its velocity $\mathbf{v}_\mathrm{f}$ and by a vector grouping together all other relevant fluid-phase variables seen by the particles $\boldsymbol{\xi}_\mathrm{f}$. The assumption that particles are indistinguishable and that the rate of change of their state is a function only of $\mathbf{v}_\mathrm{p}$, $\boldsymbol{\xi}_\mathrm{p}$, $\mathbf{v}_\mathrm{f}$, and $\boldsymbol{\xi}_\mathrm{f}$ implies that particles with the same state and same velocity, surrounded by a fluid in the same state and velocity, will inevitably evolve in the same direction with the same rate. However, we shall show below that diffusion processes representing very rapid forces or fluxes at the microscale that are not resolved at the mesoscale can be modeled as random changes in the particle state. Thus, when diffusion processes are present, two particles with identical initial states need not evolve in exactly the same manner.

Mesoscale description of a polydisperse multiphase flow

As derived from the microscale model in Chapter 4, the GPBE for a polydisperse multiphase flow has the following form:[1]

$$
\begin{aligned}
\frac{\partial n}{\partial t} + \frac{\partial}{\partial \mathbf{x}} \cdot \left[\mathbf{v}_\mathrm{p} n + \left(\mathbf{v}_\mathrm{p} - \mathbf{v}_\mathrm{f} \right) \frac{\partial (\xi_\mathrm{f1} n)}{\partial \xi_\mathrm{f1}} \right] = &-\frac{\partial}{\partial \mathbf{v}_\mathrm{p}} \cdot \left[\left(\langle \mathbf{A}_\mathrm{fp} \rangle_1 + \langle \mathbf{A}_\mathrm{p} \rangle_1 \right) n \right] - \frac{\partial}{\partial \boldsymbol{\xi}_\mathrm{p}} \cdot \left(\langle \mathbf{G}_\mathrm{p} \rangle_1 n \right) \\
&- \frac{\partial}{\partial \mathbf{v}_\mathrm{f}} \cdot \left[\left(\langle \mathbf{A}_\mathrm{pf} \rangle_1 + \langle \mathbf{A}_\mathrm{f} \rangle_1 \right) n \right] - \frac{\partial}{\partial \boldsymbol{\xi}_\mathrm{f}} \cdot \left(\langle \mathbf{G}_\mathrm{f} \rangle_1 n \right) + \mathcal{S}_1 \\
&- \frac{\partial}{\partial \mathbf{v}_\mathrm{f}} \cdot \left[\frac{1}{\tau_\mathrm{pf}} \left(1 - \frac{\varrho_\mathrm{f} \xi_\mathrm{p1}}{\varrho_\mathrm{p} \xi_\mathrm{f1}} \right) \mathbf{v}_\mathrm{f} n \right] \\
&- \frac{\partial}{\partial \xi_\mathrm{f1}} \left[\frac{1}{\tau_\mathrm{pf}} \left(\frac{\varrho_\mathrm{f}}{\varrho_\mathrm{p}} \xi_\mathrm{p1} - \xi_\mathrm{f1} \right) n \right],
\end{aligned}
\tag{5.1}
$$

where ξ_p1 is the particle mass and ξ_f1 is the fluid mass seen by a particle.[2] The remaining components in the phase-space vector $\boldsymbol{\xi}_\mathrm{f} = (\xi_\mathrm{f1}, \ldots, \xi_{\mathrm{f}M_\mathrm{f}})$ represent the fluid properties seen by the particle (e.g. temperature, chemical concentrations, etc.), which will be needed in order to describe mass and energy transfer between the phases and finite-size particles (see Section 4.3.4). The GPBE accounts for the effect of the microscale physical and chemical processes on the evolution of the number-density function (NDF) $n(t, \mathbf{x}, \mathbf{v}_\mathrm{p}, \boldsymbol{\xi}_\mathrm{p}, \mathbf{v}_\mathrm{f}, \boldsymbol{\xi}_\mathrm{f})$. The terms on the left-hand side of Eq. (5.1) represent accumulation, free transport in real space due to the particle velocity $\mathbf{v}_\mathrm{p} = (v_\mathrm{p1}, v_\mathrm{p2}, v_\mathrm{p3})$, and the flux model needed to account for spatial gradients in the fluid properties seen by a particle. The first term on the right-hand side represents continuous phase-space advection processes caused by fluid–particle

[1] Here we explicitly include the consistency terms discussed in Section 4.3.6 by specifying ξ_f1 to be the fluid mass seen by a particle. Thus, $\langle \mathbf{G}_\mathrm{f} \rangle_1$ and $\langle \mathbf{A}_\mathrm{pf} \rangle_1$ contain only contributions arising from mass/energy/momentum transfer between phases.

[2] The simplest manner to specify the fluid mass seen by a particle is $\xi_\mathrm{f1} = \varrho_\mathrm{f} \xi_\mathrm{p1} / \varrho_\mathrm{p}$, where ξ_p1 is the particle mass. If all particles have the same mass, then ξ_p1 is constant. Nevertheless, for deriving the moment-transport equations, it is easiest to treat ξ_f1 and ξ_p1 as independent variables during the derivations and then to substitute $\xi_\mathrm{f1} = \varrho_\mathrm{f} \xi_\mathrm{p1} / \varrho_\mathrm{p}$ in the final result, which is equivalent to taking the limit $\tau_\mathrm{pf} \to 0$.

momentum exchange and body forces, resulting in the phase-space velocities $\langle \mathbf{A}_{\text{fp}} \rangle_1$ and $\langle \mathbf{A}_{\text{p}} \rangle_1$, respectively. The second term on the right-hand side represents instead the effect on the NDF of continuous changes for the internal-coordinate vector $\boldsymbol{\xi}_{\text{p}} = (\xi_{\text{p}1}, \dots, \xi_{\text{p}M_{\text{p}}})$ due to the phase-space velocity $\langle \mathbf{G}_{\text{p}} \rangle_1$ (e.g. mass exchange between phases). The third term quantifies the changes in the NDF related to continuous changes in the fluid velocity seen by the particle $\mathbf{v}_{\text{f}}$ due to the phase-space velocities $\langle \mathbf{A}_{\text{pf}} \rangle_1$ and $\langle \mathbf{A}_{\text{f}} \rangle_1$. The fourth term represents the effect of continuous changes in the fluid properties seen by the particle $\boldsymbol{\xi}_{\text{f}}$ due to the phase-space velocity $\langle \mathbf{G}_{\text{f}} \rangle_1$. The fifth term represents the changes in the NDF induced by discontinuous or point processes such as particle nucleation and particle–particle collisions. Finally, the last two terms in Eq. (5.1) enforce the mesoscale conservation of mass and momentum for the fluid seen by the particles for arbitrary values of the relaxation time τ_{pf}. As discussed in Chapter 4, we are mainly interested in the behavior of the GPBE in the limit $\tau_{\text{pf}} \rightarrow 0$.

Mesoscale advection models

In general, the phase-space advection terms are modeled as the sum of contributions due to pure advection and to phase-space diffusion (Gardiner, 2004):[3]

$$\begin{aligned}
\langle \mathbf{A}_{\text{fp}} \rangle_1 n &= \mathbf{A}_{\text{fp}} n - \frac{\partial}{\partial \mathbf{v}_{\text{p}}} \cdot \left(\mathbf{B}_{\text{pv}} n\right) - \frac{\partial}{\partial \boldsymbol{\xi}_{\text{p}}} \cdot \left(\mathbf{B}_{\text{p}\xi} n\right) - \frac{\partial}{\partial \mathbf{v}_{\text{f}}} \cdot \left(\mathbf{B}_{\text{pv}_{\text{f}}} n\right) - \frac{\partial}{\partial \boldsymbol{\xi}_{\text{f}}} \cdot \left(\mathbf{B}_{\text{p}\xi_{\text{f}}} n\right), \\
\langle \mathbf{G}_{\text{p}} \rangle_1 n &= \mathbf{G}_{\text{p}} n - \frac{\partial}{\partial \mathbf{v}_{\text{p}}} \cdot \left(\mathbf{C}_{\text{pv}} n\right) - \frac{\partial}{\partial \boldsymbol{\xi}_{\text{p}}} \cdot \left(\mathbf{C}_{\text{p}\xi} n\right) - \frac{\partial}{\partial \mathbf{v}_{\text{f}}} \cdot \left(\mathbf{C}_{\text{pv}_{\text{f}}} n\right) - \frac{\partial}{\partial \boldsymbol{\xi}_{\text{f}}} \cdot \left(\mathbf{C}_{\text{p}\xi_{\text{f}}} n\right), \\
\langle \mathbf{A}_{\text{pf}} \rangle_1 n &= \mathbf{A}_{\text{pf}} n - \frac{\partial}{\partial \mathbf{v}_{\text{p}}} \cdot \left(\mathbf{B}_{\text{fv}} n\right) - \frac{\partial}{\partial \boldsymbol{\xi}_{\text{p}}} \cdot \left(\mathbf{B}_{\text{f}\xi} n\right) - \frac{\partial}{\partial \mathbf{v}_{\text{f}}} \cdot \left(\mathbf{B}_{\text{fv}_{\text{f}}} n\right) - \frac{\partial}{\partial \boldsymbol{\xi}_{\text{f}}} \cdot \left(\mathbf{B}_{\text{f}\xi_{\text{f}}} n\right), \\
\langle \mathbf{G}_{\text{f}} \rangle_1 n &= \mathbf{G}_{\text{f}} n - \frac{\partial}{\partial \mathbf{v}_{\text{p}}} \cdot \left(\mathbf{C}_{\text{fv}} n\right) - \frac{\partial}{\partial \boldsymbol{\xi}_{\text{p}}} \cdot \left(\mathbf{C}_{\text{f}\xi} n\right) - \frac{\partial}{\partial \mathbf{v}_{\text{f}}} \cdot \left(\mathbf{C}_{\text{fv}_{\text{f}}} n\right) - \frac{\partial}{\partial \boldsymbol{\xi}_{\text{f}}} \cdot \left(\mathbf{C}_{\text{f}\xi_{\text{f}}} n\right).
\end{aligned} \tag{5.2}$$

The pure advection velocities ($\mathbf{A}_{\text{fp}}$, $\mathbf{G}_{\text{p}}$, $\mathbf{A}_{\text{pf}}$, $\mathbf{G}_{\text{f}}$) are vectors whose lengths depend on the lengths of the phase-space vectors $\mathbf{v}_{\text{p}}$, $\boldsymbol{\xi}_{\text{p}}$, $\mathbf{v}_{\text{f}}$, and $\boldsymbol{\xi}_{\text{f}}$, respectively. In general, these velocities can depend on all of the independent variables in the NDF, namely (t, $\mathbf{x}$, $\mathbf{v}_{\text{p}}$, $\boldsymbol{\xi}_{\text{p}}$, $\mathbf{v}_{\text{f}}$, $\boldsymbol{\xi}_{\text{f}}$). One of the principal tasks when developing a mesoscale model is to determine the correct functional dependences of the pure advection velocities that are consistent with the microscale model. Likewise, the phase-space diffusion coefficients appearing in the partial derivatives on the right-hand side of Eq. (5.2) are matrix functions (i.e. second-order tensors) of the independent variables. The size of each matrix is determined by its context, and the matrices are, in general, not square.[4] We shall see in Section 5.5 that the diffusion terms arise as models for "random" fluctuating microscale processes that occur on short time scales relative to the pure advection velocities. As a general rule, the diffusion matrices will not be defined independently of each other, hence the necessity of deriving them from a consistent microscale model. Furthermore, because the diffusion terms do not contribute to the transport equations for zeroth- and first-order moments of the NDF, they are often neglected in the direct derivation of macroscale models for polydisperse multiphase systems that do not consider the mesoscale description. However, this can lead to inconsistent macroscale models when, for example, the pure advection velocities are negligible relative

[3]The model for $\langle \mathbf{A}_{\text{p}} \rangle_1$ could include a random component due, for example, to fluctuations in the external force fields. However, we shall assume in this chapter that $\langle \mathbf{A}_{\text{p}} \rangle_1 = \mathbf{A}_{\text{p}}$.

[4]For example, $\mathbf{B}_{\text{pv}}$ is square, with size determined by the length of $\mathbf{v}_{\text{p}}$, while $\mathbf{B}_{\text{p}\xi}$ has a number of rows equal to the length of $\boldsymbol{\xi}_{\text{p}}$ and a number of columns equal to the length of $\mathbf{v}_{\text{p}}$.

to the diffusion terms. Mathematically, the phase-space diffusion terms tend to "smooth out" discontinuities in the NDF (i.e. the GPBE has parabolic directions in phase space). Thus, the diffusion terms should be included whenever the underlying microscale physics generates them. As we shall see in Section 5.5, many of the diffusion matrices in Eq. (5.2) will be zero because of the independence of certain state-space variables. For example, if momentum exchange between phases does not depend on the internal coordinates $\boldsymbol{\xi}_\mathrm{p}$, then $\mathbf{B}_{\mathrm{p}\xi}$ will be null.

Pure advection processes

With pure advection processes, we refer to continuous phenomena that cause continuous changes in the external and internal coordinates. Continuous changes of the particle's position in real space are quantified by the real-space advection (or free-transport) term:

$$\frac{\partial}{\partial \mathbf{x}} \cdot \left(\mathbf{v}_\mathrm{p} n\right) = \frac{\partial}{\partial x_1}\left(v_{\mathrm{p}1} n\right) + \frac{\partial}{\partial x_2}\left(v_{\mathrm{p}2} n\right) + \frac{\partial}{\partial x_3}\left(v_{\mathrm{p}3} n\right), \tag{5.3}$$

where the particle velocity vector, $\mathbf{v}_\mathrm{p} = (v_{\mathrm{p}1}, v_{\mathrm{p}2}, v_{\mathrm{p}3})$, can be calculated using different strategies, as will be outlined in the following sections. Continuous changes in particle velocity also appear in the pure advection terms:

$$\frac{\partial}{\partial \mathbf{v}_\mathrm{p}} \cdot \left[\left(\mathbf{A}_\mathrm{fp} + \mathbf{A}_\mathrm{p}\right) n\right] = \frac{\partial}{\partial v_{\mathrm{p}1}}\left[\left(A_{\mathrm{fp}1} + A_{\mathrm{p}1}\right) n\right] + \frac{\partial}{\partial v_{\mathrm{p}2}}\left[\left(A_{\mathrm{fp}2} + A_{\mathrm{p}2}\right) n\right] + \frac{\partial}{\partial v_{\mathrm{p}3}}\left[\left(A_{\mathrm{fp}3} + A_{\mathrm{p}3}\right) n\right], \tag{5.4}$$

where $\mathbf{A}_\mathrm{fp} = (A_{\mathrm{fp}1}, A_{\mathrm{fp}2}, A_{\mathrm{fp}3})$ is the pure advection velocity due to fluid–particle momentum exchange and $\mathbf{A}_\mathrm{p} = (A_{\mathrm{p}1}, A_{\mathrm{p}2}, A_{\mathrm{p}3})$ is the pure advection velocity due to body forces on particles. Continuous changes in the internal coordinates (e.g. particle volume, surface area, length, etc.) also appear in the phase-space advection term:

$$\frac{\partial}{\partial \boldsymbol{\xi}_\mathrm{p}} \cdot \left(\mathbf{G}_\mathrm{p} n\right) = \frac{\partial}{\partial \xi_{\mathrm{p}1}}\left(G_{\mathrm{p}1} n\right) + \frac{\partial}{\partial \xi_{\mathrm{p}2}}\left(G_{\mathrm{p}2} n\right) + \cdots + \frac{\partial}{\partial \xi_{\mathrm{p}M_\mathrm{p}}}\left(G_{\mathrm{p}M} n\right), \tag{5.5}$$

where $\mathbf{G}_\mathrm{p} = (G_{\mathrm{p}1}, G_{\mathrm{p}2}, \ldots, G_{\mathrm{p}M})$ is a vector representing the continuous rate of change of the internal-coordinate vector. Similar terms in the GPBE result from the pure advection velocities for the fluid seen by the particle: $\mathbf{A}_\mathrm{pf}$ and $\mathbf{G}_\mathrm{f}$. We will return to the subject of modeling pure advection processes due to fluid–particle interactions in Sections 5.2–5.4. However, as shown in Eq. (4.80), when the disperse-phase volume fraction is non-negligible, the vector $\boldsymbol{\xi}_\mathrm{f}$ must have at least one component corresponding to the fluid mass seen by a particle.

Phase-space diffusion processes

The phase-space diffusion terms in Eq. (5.2) generate a very large number of terms in the GPBE (many of which are zero). For example, considering only the fluid–particle interaction term in the limiting case in which particle-velocity fluctuations are due to microscale fluid turbulence[5] (i.e. $\mathbf{B}_{\mathrm{p}\xi} = \mathbf{0}$, $\mathbf{B}_{\mathrm{p}\xi_\mathrm{f}} = \mathbf{0}$) yields the diffusion terms in velocity phase space

[5]Microscale fluid turbulence is, by definition, present only when the continuous fluid phase is present. The coefficients $\mathbf{B}_\mathrm{pv}$ describe the interaction of the particle phase with the continuous phase. In contrast, $\mathbf{B}_{\mathrm{pv}_\mathrm{f}}$ models rapid fluctuations in the fluid velocity seen by the particle that are not included in the mesoscale drag term $\mathbf{A}_\mathrm{p}$. In the mesoscale particle momentum balance, the term that generates $\mathbf{B}_{\mathrm{pv}_\mathrm{f}}$ will depend on the fluid-phase mass density ϱ_f and, hence, will be null when the fluid material density (ρ_f) is null. In any case, $\mathbf{B}_{\mathrm{pv}_\mathrm{f}}$ models momentum transfer to/from the particle phase in fluid–particle systems for which the total momentum is conserved (see discussion leading to Eq. (5.17)).

due to gradients in the directions of fluid-velocity phase space,

$$\begin{aligned}\frac{\partial^2}{\partial \mathbf{v}_\mathrm{p}\,\partial \mathbf{v}_\mathrm{f}} : \left(\mathbf{B}_{\mathrm{p}\mathbf{v}_\mathrm{f}} n\right) = {} & \frac{\partial}{\partial v_{\mathrm{p}1}}\left(\frac{\partial B_{\mathrm{p}\mathbf{v}_\mathrm{f}11} n}{\partial v_{\mathrm{f}1}} + \frac{\partial B_{\mathrm{p}\mathbf{v}_\mathrm{f}12} n}{\partial v_{\mathrm{f}2}} + \frac{\partial B_{\mathrm{p}\mathbf{v}_\mathrm{f}13} n}{\partial v_{\mathrm{f}3}}\right) \\ & + \frac{\partial}{\partial v_{\mathrm{p}2}}\left(\frac{\partial B_{\mathrm{p}\mathbf{v}_\mathrm{f}21} n}{\partial v_{\mathrm{f}1}} + \frac{\partial B_{\mathrm{p}\mathbf{v}_\mathrm{f}22} n}{\partial v_{\mathrm{f}2}} + \frac{\partial B_{\mathrm{p}\mathbf{v}_\mathrm{f}23} n}{\partial v_{\mathrm{f}3}}\right) \\ & + \frac{\partial}{\partial v_{\mathrm{p}3}}\left(\frac{\partial B_{\mathrm{p}\mathbf{v}_\mathrm{f}31} n}{\partial v_{\mathrm{f}1}} + \frac{\partial B_{\mathrm{p}\mathbf{v}_\mathrm{f}32} n}{\partial v_{\mathrm{f}2}} + \frac{\partial B_{\mathrm{p}\mathbf{v}_\mathrm{f}33} n}{\partial v_{\mathrm{f}3}}\right),\end{aligned} \tag{5.6}$$

and ($\mathbf{B}_{\mathrm{f}\xi} = \mathbf{0}$, $\mathbf{B}_{\mathrm{f}\xi_\mathrm{f}} = \mathbf{0}$) the diffusion terms in fluid-velocity phase space due to gradients in fluid-velocity phase space,

$$\begin{aligned}\frac{\partial^2}{\partial \mathbf{v}_\mathrm{f}\,\partial \mathbf{v}_\mathrm{f}} : \left(\mathbf{B}_{\mathrm{f}\mathbf{v}_\mathrm{f}} n\right) = {} & \frac{\partial}{\partial v_{\mathrm{f}1}}\left(\frac{\partial B_{\mathrm{f}\mathbf{v}_\mathrm{f}11} n}{\partial v_{\mathrm{f}1}} + \frac{\partial B_{\mathrm{f}\mathbf{v}_\mathrm{f}12} n}{\partial v_{\mathrm{f}2}} + \frac{\partial B_{\mathrm{f}\mathbf{v}_\mathrm{f}13} n}{\partial v_{\mathrm{f}3}}\right) \\ & + \frac{\partial}{\partial v_{\mathrm{f}2}}\left(\frac{\partial B_{\mathrm{f}\mathbf{v}_\mathrm{f}21} n}{\partial v_{\mathrm{f}1}} + \frac{\partial B_{\mathrm{f}\mathbf{v}_\mathrm{f}22} n}{\partial v_{\mathrm{f}2}} + \frac{\partial B_{\mathrm{f}\mathbf{v}_\mathrm{f}23} n}{\partial v_{\mathrm{f}3}}\right) \\ & + \frac{\partial}{\partial v_{\mathrm{f}3}}\left(\frac{\partial B_{\mathrm{f}\mathbf{v}_\mathrm{f}31} n}{\partial v_{\mathrm{f}1}} + \frac{\partial B_{\mathrm{f}\mathbf{v}_\mathrm{f}32} n}{\partial v_{\mathrm{f}2}} + \frac{\partial B_{\mathrm{f}\mathbf{v}_\mathrm{f}33} n}{\partial v_{\mathrm{f}3}}\right).\end{aligned} \tag{5.7}$$

The diffusion matrices $\mathbf{B}_{\mathrm{p}\mathbf{v}_\mathrm{f}}$ and $\mathbf{B}_{\mathrm{f}\mathbf{v}_\mathrm{f}}$ are symmetric and, most importantly, conservation of momentum at the microscale will require that they be *dependent* and, hence, at most only six diffusion coefficients need be determined from the microscale model (see the discussion leading to Eq. (5.17)). The simplest case occurs when the diffusion matrices are *isotropic*:

$$\begin{aligned}\frac{\partial^2}{\partial \mathbf{v}_\mathrm{p}\,\partial \mathbf{v}_\mathrm{f}} : \left(\mathbf{B}_{\mathrm{p}\mathbf{v}_\mathrm{f}} n\right) &= \frac{\partial^2 B_{\mathrm{p}\mathbf{v}_\mathrm{f}} n}{\partial v_{\mathrm{p}1}\,\partial v_{\mathrm{f}1}} + \frac{\partial^2 B_{\mathrm{p}\mathbf{v}_\mathrm{f}} n}{\partial v_{\mathrm{p}2}\,\partial v_{\mathrm{f}2}} + \frac{\partial^2 B_{\mathrm{p}\mathbf{v}_\mathrm{f}} n}{\partial v_{\mathrm{p}3}\,\partial v_{\mathrm{f}3}}, \\ \frac{\partial^2}{\partial \mathbf{v}_\mathrm{f}\,\partial \mathbf{v}_\mathrm{f}} : \left(\mathbf{B}_{\mathrm{f}\mathbf{v}_\mathrm{f}} n\right) &= \frac{\partial^2 B_{\mathrm{f}\mathbf{v}_\mathrm{f}} n}{\partial v_{\mathrm{f}1}^2} + \frac{\partial^2 B_{\mathrm{f}\mathbf{v}_\mathrm{f}} n}{\partial v_{\mathrm{f}2}^2} + \frac{\partial^2 B_{\mathrm{f}\mathbf{v}_\mathrm{f}} n}{\partial v_{\mathrm{f}3}^2},\end{aligned} \tag{5.8}$$

so that only one diffusion coefficient is required to close Eqs. (5.6) and (5.7). These different types of mesoscale diffusion models will be clearly formulated when we return to the subject of modeling diffusion processes due to fluid–particle interactions in Section 5.5.

Point processes

The point-process term $\mathcal{S}_1$ in Eq. (5.1) can represent a large range of different physical processes. However, the terminology *point process* is meant to distinguish it from the other terms in the GPBE that can be written in terms of derivatives. In general, we can further classify point processes into three categories, depending on how the NDF appears in the mathematical expression.

(i) **Zeroth order.** Terms in this category *do not depend* on the NDF. A common example is nucleation of new particles due to exchange of mass from the continuous phase.

(ii) **First order.** Terms in this category *depend linearly* on the NDF. A common example is breakage of particles due to surface stresses generated by the continuous phase.

(iii) **Second order.** Terms in this category *depend quadratically* on the NDF. A common example is aggregation of two particles with different sizes.

In general, zeroth-order point processes are modeled by a particle source term in a subset of phase space corresponding to the properties of the particles added to the system. On the other hand, first-order point processes typically result in integral terms over subsets of phase space involving the NDF. For example, with particle breakage there is a linear integral source term over all larger sizes in phase space and a linear sink term. Finally, second-order point processes involve integral source and sink terms involving the two NDF evaluated at different values of the phase-space variables. From a mathematical point of view, dealing with second-order point processes in the GPBE is more complicated than the other types because it leads to a nonlinear dependence on the NDF. We will return to the subject of modeling point processes due to fluid–particle and particle–particle interactions in Sections 5.6–5.8, and those due to particle–particle collisions in Chapter 6.

5.1.2 Formulation of mesoscale models

As discussed in Chapter 4, the conditional phase-space velocities ($\langle \mathbf{A}_{\text{fp}} \rangle_1$, $\langle \mathbf{A}_{\text{p}} \rangle_1$, $\langle \mathbf{G}_{\text{p}} \rangle_1$, $\langle \mathbf{A}_{\text{pf}} \rangle_1$, $\langle \mathbf{A}_{\text{f}} \rangle_1$, and $\langle \mathbf{G}_{\text{f}} \rangle_1$) and the discontinuous "source" term $\mathcal{S}_1$ need mesoscale closures by making use of appropriate "averaged" models for the microscale chemical and physical processes. The simplest procedure to derive these closure models is to assume that all particles are statistically identical. Under this assumption the phase-space velocities can be derived from single-particle statistics (i.e. the rate of change of the particle characteristics for a single particle) interacting with mean-field quantities, which model the effects of all other particles. The effect of multi-particle statistics on the phase-space velocities is generally introduced in the form of corrections, which are based, for example, on the disperse-phase volume fraction, that will reduce to isolated-particle statistics in the dilute limit. A useful starting point is to consider the changes in a particle's properties in a Lagrangian frame of reference moving with the particle velocity.

The Lagrangian single-particle model

Following a single particle, the rates of change of the mesoscale variables can be written in a Lagrangian form:[6]

$$
\begin{aligned}
\dot{\mathbf{X}}^* &= \frac{\mathrm{d}\mathbf{X}_{\text{p}}^*}{\mathrm{d}t} = \mathbf{U}_{\text{p}}^*, \\
\dot{\mathbf{U}}_{\text{p}}^* &= \frac{\mathrm{d}\mathbf{U}_{\text{p}}^*}{\mathrm{d}t} = \mathbf{A}_{\text{fp}}^* + \mathbf{A}_{\text{p}}^* + \mathbf{C}_{\text{pU}}^*, \\
\dot{\boldsymbol{\xi}}_{\text{p}}^* &= \frac{\mathrm{d}\boldsymbol{\xi}_{\text{p}}^*}{\mathrm{d}t} = \mathbf{G}_{\text{p}}^* + \mathbf{C}_{\text{p}\xi}^*, \\
\dot{\mathbf{U}}_{\text{f}}^* &= \frac{\mathrm{d}\mathbf{U}_{\text{f}}^*}{\mathrm{d}t} = \mathbf{A}_{\text{pf}}^* + \mathbf{A}_{\text{f}}^* + \mathbf{C}_{\text{fU}}^*, \\
\dot{\boldsymbol{\xi}}_{\text{f}}^* &= \frac{\mathrm{d}\boldsymbol{\xi}_{\text{f}}^*}{\mathrm{d}t} = \mathbf{G}_{\text{f}}^* + \mathbf{C}_{\text{f}\xi}^*,
\end{aligned}
\tag{5.9}
$$

where the left-hand side in each case is the time derivative following the particle, the first terms on the right-hand sides correspond to continuous changes in the particle variables,

[6]This formulation is particularly convenient when Euler–Lagrange simulations are used to approximate the disperse multiphase flow in terms of a *finite* sample of particles. As discussed in Sections 5.2 and 5.3, although some of the mesoscale variables are *intensive* (i.e. independent of the particle mass), it is usually best to start with a conserved *extensive* variable (e.g. particle mass or particle momentum) when deriving the single-particle models. For example, in Chapter 4 we found that $\boldsymbol{\xi}_{\text{f}}$ must have at least one component, corresponding to the fluid mass seen by a particle, in order to describe cases in which the disperse-phase volume fraction is not constant.

and the final terms on the right-hand sides correspond to discontinuous changes (such as collisions). The goal of the mesoscale modeler is to provide functional forms for the terms on the right-hand sides of Eq. (5.9) that depend on

(i) the mesoscale variables describing the particle, namely $\mathbf{X}_p^*(t)$, $\mathbf{U}_p^*(t)$, $\boldsymbol{\xi}_p^*(t)$, $\mathbf{U}_f^*(t)$, and $\boldsymbol{\xi}_f^*(t)$; and

(ii) the (time-dependent) moments of the NDF evaluated at the particle location: $\mathbf{x} = \mathbf{X}_p^*(t)$.

Note that the dependence of the mesoscale model on the moments of the NDF is used to introduce multi-particle effects through the "mean-field" variables such as the disperse-phase volume fraction. For this reason, the right-hand sides of Eq. (5.9) are different from the exact microscale models that were introduced in Chapter 4 (i.e. Eqs. (4.1)–(4.3) on page 103). Formally, given these dependences, the mesoscale models on the right-hand sides of Eq. (5.9) can be expressed explicitly (for example) as $\mathbf{A}_{fp}^*(t, \mathbf{X}^*, \mathbf{U}_p^*, \boldsymbol{\xi}_p^*, \mathbf{U}_f^*, \boldsymbol{\xi}_f^*)$, where the functional arguments indicate the variables needed to determine the single-particle properties. As described in detail in Chapter 4, these functions are usually related to the conditional phase-space velocities as follows: $\langle \mathbf{A}_{fp} \rangle_1 = \mathbf{A}_{fp}^*(t, \mathbf{x}, \mathbf{v}_p, \boldsymbol{\xi}_p, \mathbf{v}_f, \boldsymbol{\xi}_f)$, i.e. by evaluating the functions at the phase-space variables appearing in the GPBE. However, if $\mathbf{A}_{fp}^*$ includes a stochastic component to model rapidly changing processes on the microscale, an additional step is needed in order to derive the conditional phase-space velocities (Gardiner, 2004). Finally, the discontinuous terms in Eq. (5.9) are typically modeled using stochastic point processes.

Stochastic acceleration models

As mentioned above, the mesoscale model can contain stochastic processes, which lead to the diffusion terms appearing in Eq. (5.2). Using $\mathbf{A}_{fp}^*$ as an example, we can write

$$\mathbf{A}_{fp}^* = \mathbf{A}_{fp} + \mathbf{b}_{pv_p} \mathbf{W}_{pv_p}(t) + \mathbf{b}_{p\xi_p} \mathbf{W}_{p\xi_p}(t) + \mathbf{b}_{pv_f} \mathbf{W}_{pv_f}(t) + \mathbf{b}_{p\xi_f} \mathbf{W}_{p\xi_f}(t), \tag{5.10}$$

where we introduced the delta-correlated Wiener process $W(t)$ to model the fast time scales (Gardiner, 2004). For example, the random vector $\mathbf{W}_{pv_p}(t) = [W_{pv_p1}(t), W_{pv_p2}(t), W_{pv_p3}(t)]^T$ has independent components $W_{pv_pi}(t)$ (with zero mean value $\langle W_{pv_pi}(t) \rangle = 0$) whose expected values at different times are given by $\langle W_{pv_pi}(t) W_{pv_pi}(s) \rangle = \delta(t - s)$. The lengths of the random vectors on the right-hand side of Eq. (5.10) are the same as the vector in the subscript (i.e. $\mathbf{W}_{p\xi_p}(t)$ has the same length as $\boldsymbol{\xi}_p$). For completeness, we have included all possible random vectors needed to find the diffusion terms in Eq. (5.2); however, the reader should realize that the coefficients (e.g. $\mathbf{b}_{pv_p}$) may be zero if the underlying microscale model does not introduce a stochastic component into the mesoscale model. Also, all of the coefficients appearing on the right-hand side of Eq. (5.10) can depend on the same variables as $\mathbf{A}_{fp}^*$ as described above.[7] Using the Wiener processes (Gardiner, 2004), the complete set of single-particle models that generate the mesoscale models in Eq. (5.2) is[8]

[7] In practice, it is often sufficient to assume that the diffusion coefficients depend only on t and $\mathbf{x}$. The underlying stochastic processes are then approximated as Gaussian-distributed.

[8] We again assume that $\mathbf{A}_p^* = \mathbf{A}_p$, so that the body forces have no fluctuating components.

$$\begin{aligned}
\mathbf{A}^*_{\mathrm{fp}} &= \mathbf{A}_{\mathrm{fp}} + \mathbf{b}_{\mathrm{pv_p}}\mathbf{W}_{\mathrm{pv_p}} + \mathbf{b}_{\mathrm{p}\xi_\mathrm{p}}\mathbf{W}_{\mathrm{p}\xi_\mathrm{p}} + \mathbf{b}_{\mathrm{pv_f}}\mathbf{W}_{\mathrm{pv_f}} + \mathbf{b}_{\mathrm{p}\xi_\mathrm{f}}\mathbf{W}_{\mathrm{p}\xi_\mathrm{f}}, \\
\mathbf{G}^*_{\mathrm{p}} &= \mathbf{G}_{\mathrm{p}} + \mathbf{c}_{\mathrm{pv_p}}\mathbf{W}_{\mathrm{pv_p}} + \mathbf{c}_{\mathrm{p}\xi_\mathrm{p}}\mathbf{W}_{\mathrm{p}\xi_\mathrm{p}} + \mathbf{c}_{\mathrm{pv_f}}\mathbf{W}_{\mathrm{pv_f}} + \mathbf{c}_{\mathrm{p}\xi_\mathrm{f}}\mathbf{W}_{\mathrm{p}\xi_\mathrm{f}}, \\
\mathbf{A}^*_{\mathrm{pf}} &= \mathbf{A}_{\mathrm{pf}} + \mathbf{b}_{\mathrm{fv_p}}\mathbf{W}_{\mathrm{pv_p}} + \mathbf{b}_{\mathrm{f}\xi_\mathrm{p}}\mathbf{W}_{\mathrm{p}\xi_\mathrm{p}} + \mathbf{b}_{\mathrm{fv_f}}\mathbf{W}_{\mathrm{pv_f}} + \mathbf{b}_{\mathrm{f}\xi_\mathrm{f}}\mathbf{W}_{\mathrm{p}\xi_\mathrm{f}}, \\
\mathbf{G}^*_{\mathrm{f}} &= \mathbf{G}_{\mathrm{f}} + \mathbf{c}_{\mathrm{fv_p}}\mathbf{W}_{\mathrm{pv_p}} + \mathbf{c}_{\mathrm{f}\xi_\mathrm{p}}\mathbf{W}_{\mathrm{p}\xi_\mathrm{p}} + \mathbf{c}_{\mathrm{fv_f}}\mathbf{W}_{\mathrm{pv_f}} + \mathbf{c}_{\mathrm{f}\xi_\mathrm{f}}\mathbf{W}_{\mathrm{p}\xi_\mathrm{f}}.
\end{aligned} \tag{5.11}$$

In order to make the connection between the single-particle model and the mesoscale diffusion coefficients, we can rewrite Eq. (5.11) using vectors and matrices: $\mathbf{Z}^* = \mathbf{Z} + \mathbf{BW}$, where $\mathbf{Z}^* = [\mathbf{A}^*_{\mathrm{fp}}, \mathbf{G}^*_{\mathrm{p}}, \mathbf{A}^*_{\mathrm{pf}}, \mathbf{G}^*_{\mathrm{f}}]^{\mathrm{T}}$ and $\mathbf{Z} = [\mathbf{A}_{\mathrm{fp}}, \mathbf{G}_{\mathrm{p}}, \mathbf{A}_{\mathrm{pf}}, \mathbf{G}_{\mathrm{f}}]^{\mathrm{T}}$ are column vectors of length $\mathcal{N}$, $\mathbf{W} = [\mathbf{W}_{\mathrm{pv_p}}, \mathbf{W}_{\mathrm{p}\xi_\mathrm{p}}, \mathbf{W}_{\mathrm{pv_f}}, \mathbf{W}_{\mathrm{p}\xi_\mathrm{f}}]^{\mathrm{T}}$ is a column vector of length $\mathcal{N}$, and $\mathbf{B}$ is a matrix of size $\mathcal{N} \times \mathcal{N}$ defined by

$$\mathbf{B} = \begin{bmatrix} \mathbf{b}_{\mathrm{pv_p}} & \mathbf{b}_{\mathrm{p}\xi_\mathrm{p}} & \mathbf{b}_{\mathrm{pv_f}} & \mathbf{b}_{\mathrm{p}\xi_\mathrm{f}} \\ \mathbf{c}_{\mathrm{pv_p}} & \mathbf{c}_{\mathrm{p}\xi_\mathrm{p}} & \mathbf{c}_{\mathrm{pv_f}} & \mathbf{c}_{\mathrm{p}\xi_\mathrm{f}} \\ \mathbf{b}_{\mathrm{fv_p}} & \mathbf{b}_{\mathrm{f}\xi_\mathrm{p}} & \mathbf{b}_{\mathrm{fv_f}} & \mathbf{b}_{\mathrm{f}\xi_\mathrm{f}} \\ \mathbf{c}_{\mathrm{fv_p}} & \mathbf{c}_{\mathrm{f}\xi_\mathrm{p}} & \mathbf{c}_{\mathrm{fv_f}} & \mathbf{c}_{\mathrm{f}\xi_\mathrm{f}} \end{bmatrix}.$$

The reader should note that the microscale model is used to determine the nonzero terms in $\mathbf{B}$, and thus for the following discussion $\mathbf{B}$ can be assumed to be known. Using matrix notation and the properties of the Wiener process (Gardiner, 2004), a symmetric $\mathcal{N} \times \mathcal{N}$ diffusion matrix $\mathbf{D}$ can be defined by

$$\mathbf{D} = \frac{1}{2}\mathbf{B}\mathbf{B}^{\mathrm{T}}. \tag{5.12}$$

If we denote the mesoscale advection terms on the left-hand side of Eq. (5.2) by

$$\langle \mathbf{Z} \rangle_1 = [\langle \mathbf{A}_{\mathrm{fp}} \rangle_1, \langle \mathbf{G}_{\mathrm{p}} \rangle_1, \langle \mathbf{A}_{\mathrm{pf}} \rangle_1, \langle \mathbf{G}_{\mathrm{f}} \rangle_1]^{\mathrm{T}},$$

we then have $\langle \mathbf{Z} \rangle_1 n = \mathbf{Z} n - \nabla_\eta \cdot (\mathbf{D} n)$, where the phase-space gradient vector of length $\mathcal{N}$ is defined by

$$\nabla_\eta = \left[\frac{\partial}{\partial \mathbf{v}_{\mathrm{p}}}, \frac{\partial}{\partial \xi_{\mathrm{p}}}, \frac{\partial}{\partial \mathbf{v}_{\mathrm{f}}}, \frac{\partial}{\partial \xi_{\mathrm{f}}} \right]^{\mathrm{T}}.$$

On comparing the definition of $\langle \mathbf{Z} \rangle_1$ with Eq. (5.2), the reader can observe that $\mathbf{D}$ is directly related to the phase-space diffusion coefficients in the GPBE. Furthermore, these diffusion coefficients are determined from the single-particle mesoscale model by Eq. (5.12). In other words, the form of the advection terms in the GPBE can be exactly determined from a mesoscale (Lagrangian) model for a single particle in the form of Eq. (5.9). For this reason, the preferred approach for deriving the closed form for the GPBE is to start with a Lagrangian particle model. This choice offers the additional advantage of connecting directly to a Euler–Lagrange formulation of the GPBE that can be used to simulate the disperse multiphase flow. Exceptions to this rule are the mesoscale models for the fluid properties seen by a particle. In Chapter 4 we found that to ensure consistency with the macroscale models for the fluid phase $\langle \mathbf{G}_{\mathrm{f}} \rangle_1$ must include spatial-gradient terms as seen in Eqs. (4.62) and (4.80). The need for such nonzero terms would not be obvious on observing only the Lagrangian behavior of a single disperse-phase particle.

Microscale momentum conservation

As a concrete example of the diffusion coefficients, let us consider a case in which the only mesoscale variables are $\mathbf{U}^*_{\mathrm{p}}$ and $\mathbf{U}^*_{\mathrm{f}}$. Moreover, let us assume that the Lagrangian single-particle model for these variables is

$$\begin{aligned}\frac{\mathrm{d}\mathbf{U}_\mathrm{p}^*}{\mathrm{d}t} &= \mathbf{A}_\mathrm{fp} + \mathbf{b}_{\mathrm{p}\mathbf{v}_\mathrm{p}}\mathbf{W}_{\mathrm{p}\mathbf{v}_\mathrm{p}}(t) + \mathbf{b}_{\mathrm{p}\mathbf{v}_\mathrm{f}}\mathbf{W}_{\mathrm{p}\mathbf{v}_\mathrm{f}}(t),\\ \frac{\mathrm{d}\mathbf{U}_\mathrm{f}^*}{\mathrm{d}t} &= \mathbf{A}_\mathrm{pf} + \mathbf{b}_{\mathrm{f}\mathbf{v}_\mathrm{p}}\mathbf{W}_{\mathrm{p}\mathbf{v}_\mathrm{p}}(t) + \mathbf{b}_{\mathrm{f}\mathbf{v}_\mathrm{f}}\mathbf{W}_{\mathrm{p}\mathbf{v}_\mathrm{f}}(t).\end{aligned} \tag{5.13}$$

If we then denote the (constant) particle and fluid masses[9] by m_p and m_f, respectively, the conservation of momentum at the microscale can be expressed as

$$m_\mathrm{p}\frac{\mathrm{d}\mathbf{U}_\mathrm{p}^*}{\mathrm{d}t} + m_\mathrm{f}\frac{\mathrm{d}\mathbf{U}_\mathrm{f}^*}{\mathrm{d}t} = \mathbf{0}. \tag{5.14}$$

Thus, in terms of the model coefficients in Eq. (5.13), we must have

$$\begin{aligned} m_\mathrm{p}\mathbf{A}_\mathrm{fp} + m_\mathrm{f}\mathbf{A}_\mathrm{pf} &= \mathbf{0},\\ m_\mathrm{p}\mathbf{b}_{\mathrm{p}\mathbf{v}_\mathrm{p}} + m_\mathrm{f}\mathbf{b}_{\mathrm{f}\mathbf{v}_\mathrm{p}} &= \mathbf{0},\\ m_\mathrm{p}\mathbf{b}_{\mathrm{p}\mathbf{v}_\mathrm{f}} + m_\mathrm{f}\mathbf{b}_{\mathrm{f}\mathbf{v}_\mathrm{f}} &= \mathbf{0}.\end{aligned} \tag{5.15}$$

The first constraint is the usual deterministic condition for the inter-phase momentum transfer. The remaining two constraints affect the diffusion matrix appearing in the GPBE. From Eq. (5.13) and Eq. (5.12), we see that the diffusion matrix appearing in the GPBE for this example is (after applying the constraints in Eq. (5.15)) defined by[10]

$$\mathbf{D} = \frac{1}{2}\begin{bmatrix}\mathbf{b}_{\mathrm{p}\mathbf{v}_\mathrm{p}} & \mathbf{b}_{\mathrm{p}\mathbf{v}_\mathrm{f}}\\ \mathbf{b}_{\mathrm{f}\mathbf{v}_\mathrm{p}} & \mathbf{b}_{\mathrm{f}\mathbf{v}_\mathrm{f}}\end{bmatrix}\begin{bmatrix}\mathbf{b}_{\mathrm{p}\mathbf{v}_\mathrm{p}}^\mathrm{T} & \mathbf{b}_{\mathrm{f}\mathbf{v}_\mathrm{p}}^\mathrm{T}\\ \mathbf{b}_{\mathrm{p}\mathbf{v}_\mathrm{f}}^\mathrm{T} & \mathbf{b}_{\mathrm{f}\mathbf{v}_\mathrm{f}}^\mathrm{T}\end{bmatrix} = \begin{bmatrix}1 & -\varrho_\mathrm{p}/\varrho_\mathrm{f}\\ -\varrho_\mathrm{p}/\varrho_\mathrm{f} & (\varrho_\mathrm{p}/\varrho_\mathrm{f})^2\end{bmatrix}\otimes\mathbf{D}_\mathrm{pf}, \tag{5.16}$$

where the symmetric 3×3 diffusion matrix is defined by $\mathbf{D}_\mathrm{pf} = \frac{1}{2}(\mathbf{b}_{\mathrm{p}\mathbf{v}_\mathrm{p}}\mathbf{b}_{\mathrm{p}\mathbf{v}_\mathrm{p}}^\mathrm{T} + \mathbf{b}_{\mathrm{p}\mathbf{v}_\mathrm{f}}\mathbf{b}_{\mathrm{p}\mathbf{v}_\mathrm{f}}^\mathrm{T})$. In order to proceed further, it would be necessary to introduce the mesoscale models for $\mathbf{b}_{\mathrm{p}\mathbf{v}_\mathrm{p}}$ and $\mathbf{b}_{\mathrm{p}\mathbf{v}_\mathrm{f}}$. The simplest possible models would be isotropic, leading, for example, to $\mathbf{b}_{\mathrm{p}\mathbf{v}_\mathrm{p}} = \mathbf{0}$ and $\mathbf{b}_{\mathrm{p}\mathbf{v}_\mathrm{f}} \propto (\varrho_\mathrm{f}/\varrho_\mathrm{p})\mathbf{I}$, where $\mathbf{I}$ is a 3×3 identity matrix. A simple consistent model for the diffusion matrix in the GPBE that conserves microscale momentum would thus be[11]

$$\mathbf{D} = \begin{bmatrix}\mathbf{B}_{\mathrm{p}\mathbf{v}_\mathrm{p}} & \mathbf{B}_{\mathrm{p}\mathbf{v}_\mathrm{f}}\\ \mathbf{B}_{\mathrm{f}\mathbf{v}_\mathrm{p}} & \mathbf{B}_{\mathrm{f}\mathbf{v}_\mathrm{f}}\end{bmatrix} = D_\mathrm{pf}\begin{bmatrix}(\varrho_\mathrm{f}/\varrho_\mathrm{p})^2 & -\varrho_\mathrm{f}/\varrho_\mathrm{p}\\ -\varrho_\mathrm{f}/\varrho_\mathrm{p} & 1\end{bmatrix}\otimes\mathbf{I}, \tag{5.17}$$

where D_pf is a scalar coefficient. Note that, although $\mathbf{D}$ is a 6×6 matrix, its maximum rank is three. Thus, in order to conserve momentum at the microscale, the diffusion in phase space is present in only three of the six dimensions. An analogous result will be found in

[9] In the GPBE, the masses will be replaced by the phase densities ϱ_p and ϱ_f, which must be constant if the only mesoscale variables are $\mathbf{U}_\mathrm{p}^*$ and $\mathbf{U}_\mathrm{f}^*$.

[10] The matrix product $\mathbf{A}\otimes\mathbf{B}$ is defined by

$$\mathbf{A}\otimes\mathbf{B} = \begin{bmatrix}a_{11} & \dots & a_{1N}\\ \vdots & \ddots & \vdots\\ a_{N1} & \dots & a_{NN}\end{bmatrix}\otimes\mathbf{B} = \begin{bmatrix}a_{11}\mathbf{B} & \dots & a_{1N}\mathbf{B}\\ \vdots & \ddots & \vdots\\ a_{N1}\mathbf{B} & \dots & a_{NN}\mathbf{B}\end{bmatrix}.$$

[11] Here, as an example, we have assumed that the velocity fluctuations are due to microscale turbulence in the continuous phase, and thus D_pf will be proportional to the velocity variance of the fluid seen by the particle $[\![\mathbf{u}_\mathrm{f}\cdot\mathbf{u}_\mathrm{f}]\!]_\mathrm{f}$.

the more general context of Eq. (5.2) due to the conservation of mass, momentum, energy, etc., at the microscale. We shall discuss this point in more detail in Section 5.5.

Deterministic acceleration models

Because the construction of a full-blown Lagrangian model that includes both deterministic and stochastic components is quite challenging, often the first step is to concentrate on modeling the deterministic part. This will be our approach when developing fluid–particle interaction models in Sections 5.2–5.4. Here, as an example, we consider deterministic models for the body-force acceleration term $\mathbf{A}_{\mathrm{p}}$. These forces result from the existence of a force field, and typical examples are electrical and gravitational fields. As is well known, body forces act also when the continuous phase is absent (i.e. particles moving in vacuum). In the case of gravity, the final force per unit mass acting on a single particle can be written as $\mathbf{A}_{\mathrm{p}} = \mathbf{g}$. In the case of an electrical field, the force can be computed as the product of the intensity of the electrical field $\mathbf{E}$ and the charge q_{p} associated with the particle divided by the mass of the particle: $\mathbf{A}_{\mathrm{p}} = q_{\mathrm{p}}\mathbf{E}/M_{\mathrm{p}}$. Note that (in contrast to gravity, which always has the same sign) the electrical-field force depends on the sign of the charge associated with the particle. We should also note that body forces also appear in the mesoscale model for the fluid velocity seen by the particle $\mathbf{A}_{\mathrm{f}}$. In general, because such forces act on each phase independently of whether the other phase is present, it is usually straightforward to find an appropriate mesoscale model. Also, multiple body forces can be present in the same system, in which case $\mathbf{A}_{\mathrm{p}}$ will be the sum of all such forces. In order to avoid cumbersome notation, in this chapter we will discuss each force individually (with the implicit assumption that the total force is found by summation) unless otherwise necessary.[12] Throughout most of this chapter, we will neglect fluctuations in the fluid-phase properties seen by a particle, which amounts to assuming that the phase-space variables for the fluid are "slaved" to their average values, i.e. the monokinetic fluid limit. (See Section 4.4 for more details.) In practice, this implies that $\mathbf{v}_{\mathrm{f}} = \mathbf{U}_{\mathrm{f}}$ and $\boldsymbol{\xi}_{\mathrm{f}} = \boldsymbol{\phi}_{\mathrm{f}}$, where the mean velocity and mean properties of the fluid phase are found from separate conservation equations of the form of Eq. (4.92). Nevertheless, when the particle material density is small compared with the fluid material density (e.g. bubbly flows), neglecting the fluid-phase fluctuations is not physically justified even in laminar flow (Cartellier *et al.*, 2009). For such cases, a stochastic component should be added to the phase-space velocity models.

5.1.3 Relation to macroscale models

In Section 4.3, example macroscale transport equations are derived for selected moments of the NDF. Having introduced the precise forms of the mesoscale advection models in Eq. (5.2), it is of interest to derive explicitly some example moment source terms resulting from these models. In order to do so, we will use the rules presented in Section 4.3.1 for phase-space integration. For simplicity, we consider only the advection term involving $\langle \mathbf{A}_{\mathrm{fp}} \rangle_1$ and assume that the only phase-space variables of interest are $\mathbf{v}$ and $\mathbf{v}_{\mathrm{f}}$, and that the model in Eq. (5.2) reduces to

$$\langle \mathbf{A}_{\mathrm{fp}} \rangle_1 n = \mathbf{A}_{\mathrm{fp}} n - \frac{\partial n}{\partial \mathbf{v}_{\mathrm{p}}} \cdot \left(\mathbf{B}_{\mathrm{pv_p}} n \right) - \frac{\partial n}{\partial \mathbf{v}_{\mathrm{f}}} \cdot \left(\mathbf{B}_{\mathrm{pv_f}} n \right).$$

[12] The virtual-mass force is the exception to this rule because it involves the time derivative $\dot{U}^*$. For this case, $\mathbf{A}_{\mathrm{fp}}$ must be defined in a Lagrangian sense by taking into account the virtual mass in Eq. (5.9).

Letting $g(\mathbf{v}, \mathbf{v}_\mathrm{f})$ be our arbitrary function, the integral in Eq. (4.41) can be rewritten as

$$\begin{aligned}\int g \frac{\partial}{\partial \mathbf{v}_\mathrm{p}} \cdot \left(\langle \mathbf{A}_\mathrm{fp} \rangle_1 n \right) \mathrm{d}\mathbf{v}_\mathrm{p} \, \mathrm{d}\mathbf{v}_\mathrm{f} &= - \int \left(\langle \mathbf{A}_\mathrm{fp} \rangle_1 \cdot \frac{\partial g}{\partial \mathbf{v}_\mathrm{p}} \right) n \, \mathrm{d}\mathbf{v}_\mathrm{p} \, \mathrm{d}\mathbf{v}_\mathrm{f} \\ &= - \int \left[\mathbf{A}_\mathrm{fp} n - \frac{\partial}{\partial \mathbf{v}_\mathrm{p}} \cdot \left(\mathbf{B}_{\mathrm{p}\mathbf{v}_\mathrm{p}} n \right) - \frac{\partial}{\partial \mathbf{v}_\mathrm{f}} \cdot \left(\mathbf{B}_{\mathrm{p}\mathbf{v}_\mathrm{f}} n \right) \right] \cdot \frac{\partial g}{\partial \mathbf{v}_\mathrm{p}} \, \mathrm{d}\mathbf{v}_\mathrm{p} \, \mathrm{d}\mathbf{v}_\mathrm{f}.\end{aligned} \tag{5.18}$$

By applying integration by parts, the diffusion terms can be further reduced to find

$$\begin{aligned}\int g \frac{\partial}{\partial \mathbf{v}_\mathrm{p}} \cdot \left(\langle \mathbf{A}_\mathrm{fp} \rangle_1 n \right) \mathrm{d}\mathbf{v}_\mathrm{p} \, \mathrm{d}\mathbf{v}_\mathrm{f} &= - \int \left(\mathbf{A}_\mathrm{fp} \cdot \frac{\partial g}{\partial \mathbf{v}_\mathrm{p}} \right) n \, \mathrm{d}\mathbf{v}_\mathrm{p} \, \mathrm{d}\mathbf{v}_\mathrm{f} \\ &\quad - \int \left(\mathbf{B}_{\mathrm{p}\mathbf{v}_\mathrm{p}} : \frac{\partial^2 g}{\partial \mathbf{v}_\mathrm{p} \, \partial \mathbf{v}_\mathrm{p}} + \mathbf{B}_{\mathrm{p}\mathbf{v}_\mathrm{f}} : \frac{\partial^2 g}{\partial \mathbf{v}_\mathrm{f} \, \partial \mathbf{v}_\mathrm{p}} \right) n \, \mathrm{d}\mathbf{v}_\mathrm{p} \, \mathrm{d}\mathbf{v}_\mathrm{f}.\end{aligned} \tag{5.19}$$

On observing the right-hand side of this equation, it is clear that the mesoscale model for $\langle \mathbf{A}_\mathrm{fp} \rangle_1$ will generate source terms in the macroscale moment-transport equations that depend on the order of the moments, which depend on the powers of the phase-space variables appearing in g.

The zeroth-order moment

For the zeroth-order moment, the arbitrary function is set to $g = 1$. From Eq. (5.19), this choice yields

$$\int \frac{\partial}{\partial \mathbf{v}_\mathrm{p}} \cdot \left(\langle \mathbf{A}_\mathrm{fp} \rangle_1 n \right) \mathrm{d}\mathbf{v}_\mathrm{p} \, \mathrm{d}\mathbf{v}_\mathrm{f} = 0$$

as expected, since particle acceleration does not change the particle number.

First-order moments

For the first-order velocity moments, the arbitrary function is set to $g = v_{\mathrm{p}i}$. From Eq. (5.19), this choice yields

$$\int v_{\mathrm{p}i} \frac{\partial}{\partial \mathbf{v}_\mathrm{p}} \cdot \left(\langle \mathbf{A}_\mathrm{fp} \rangle_1 n \right) \mathrm{d}\mathbf{v}_\mathrm{p} \, \mathrm{d}\mathbf{v}_\mathrm{f} = - \int A_{\mathrm{fp}i} n \, \mathrm{d}\mathbf{v}_\mathrm{p} \, \mathrm{d}\mathbf{v}_\mathrm{f} = - [\![A_{\mathrm{fp}i}]\!]_\mathrm{N} N,$$

where N is the total number concentration defined by Eq. (4.45). This expression is the (negative) macroscale particle acceleration in the x_i direction due to fluid–particle interactions. Note that, if we set $g = v_{\mathrm{f}i}$ (i.e. the phase-space fluid velocity seen by the particle), then this integral is null; however, the corresponding integral for $\mathbf{A}_\mathrm{pf}$ will be nonzero. As stated earlier, here we can observe that the mesoscale models for the diffusion terms in Eq. (5.2) do not affect the first-order moments, but the mesoscale model for $\mathbf{A}_\mathrm{fp}$ determines the macroscale momentum exchange between the disperse and continuous phases.

Second-order moments

There are many second-order moments involving the components of $\mathbf{v}_\mathrm{p}$ and $\mathbf{v}_\mathrm{f}$. Nevertheless, the three types that lead to nonzero acceleration terms are $g_1 = v_{\mathrm{p}i}^2$, $g_2 = v_{\mathrm{p}i} v_{\mathrm{p}j}$, and $g_3 = v_{\mathrm{p}i} v_{\mathrm{f}j}$. Inserting these functions into Eq. (5.19) leads to

$$\int g_1 \frac{\partial}{\partial \mathbf{v}_\mathrm{p}} \cdot \left(\langle \mathbf{A}_\mathrm{fp} \rangle_1 n \right) \mathrm{d}\mathbf{v}_\mathrm{p} \, \mathrm{d}\mathbf{v}_\mathrm{f} = -2 \left([\![U_{\mathrm{p}i} A_{\mathrm{fp}i}]\!]_\mathrm{N} + B_{\mathrm{pv_p}ii} \right) N,$$

$$\int g_2 \frac{\partial}{\partial \mathbf{v}_\mathrm{p}} \cdot \left(\langle \mathbf{A}_\mathrm{fp} \rangle_1 n \right) \mathrm{d}\mathbf{v}_\mathrm{p} \, \mathrm{d}\mathbf{v}_\mathrm{f} = - \left([\![U_{\mathrm{p}j} A_{\mathrm{fp}i}]\!]_\mathrm{N} + [\![U_{\mathrm{p}i} A_{\mathrm{fp}j}]\!]_\mathrm{N} + B_{\mathrm{pv_p}ij} \right) N,$$

and

$$\int g_3 \frac{\partial}{\partial \mathbf{v}_\mathrm{p}} \cdot \left(\langle \mathbf{A}_\mathrm{fp} \rangle_1 n \right) \mathrm{d}\mathbf{v}_\mathrm{p} \, \mathrm{d}\mathbf{v}_\mathrm{f} = - \left([\![U_{\mathrm{f}j} A_{\mathrm{fp}i}]\!]_\mathrm{N} + B_{\mathrm{pv_f}ij} \right) N.$$

The reader can observe that the phase-space diffusion coefficients appear as source (sink[13]) terms in the transport equations for the second-order moments. In many mesoscale models for the acceleration terms, the diffusion coefficients are simply neglected. This would be appropriate, for example, when the disperse-phase material density is much larger than that of the continuous phase ($\rho_\mathrm{f} \ll \rho_\mathrm{p}$) such as for gas–particle flows. However, when this is not the case (e.g. bubbly flows), microscale turbulent fluctuations in the continuous phase can lead to significant velocity fluctuations in the disperse phase. In such cases, the spatial dispersion of the particle phase will be significantly larger when the diffusion terms are included in the mesoscale model. We will discuss this point in more detail when we introduce mesoscale models for diffusive momentum transport in Section 5.5. As with the second-order moments, the transport equations for the third- and higher-order moments will contain contributions from the pure advection and the diffusion terms. By following the derivation starting from the mesoscale models introduced in Section 5.1.2, these contributions will be known explicitly (i.e. no additional *physical* modeling is needed to close the transport equations).

5.2 Phase-space advection: mass and heat transfer

When a particle belonging to a disperse multiphase system exchanges mass and heat with the surrounding fluid phase, some changes in the internal coordinates identifying the state of the particle are induced. In the case of mass transfer, the driving force is usually related to a difference in chemical potential between the particle and the surrounding fluid. For example, in a supersaturated solution, solute molecules will diffuse from the solution to the particles and, after being included in the particle lattice, will increase its mass, and consequently its volume, surface area, and "size." Another popular example comes from gas bubbles. When a bubble containing different gas molecules exchanges some of these molecules with the surrounding fluid phase, its mass changes according to the rate of mass transfer. In the case of heat transfer, the driving force is usually related to a difference in temperature between the particle and the surrounding fluid. In Section 5.1.2, we introduced the general formulation of the mesoscale model in Eq. (5.9). Here, we are interested in mesoscale models for $\mathbf{G}_\mathrm{p}^*$ describing changes in $\boldsymbol{\xi}_\mathrm{p}^*$.[14] In particular, we are interested in the deterministic component $\mathbf{G}_\mathrm{p}$ in Eq. (5.11). In the following, we will denote the internal-coordinate vector as $\boldsymbol{\xi}_\mathrm{p}^*$, but will often denote particular components of $\boldsymbol{\xi}_\mathrm{p}^*$ by a specific symbol for clarity. For example, if $\xi_{\mathrm{p}1}^*$ is the particle mass, we use the symbol M_p to represent it. However, the reader should keep in mind that in all cases we are modeling the changes in the internal coordinates following a particle as described in Section 5.1.2.

[13] The symmetric square matrix $\mathbf{B}_{\mathrm{pv_p}}$ is nonnegative, leading to a source term. The non-symmetric square matrix $\mathbf{B}_{\mathrm{pv_f}}$ corresponds to cross correlations between the particle and fluid velocities and thus need not have a particular sign.

[14] In Section 5.3 we will discuss mesoscale models for the change in momentum due to mass transfer, where we shall see that $\mathbf{A}_\mathrm{fp}$ contains a contribution proportional to $\mathbf{G}_\mathrm{p}$.

Particle-mass balance

In general, continuity statements are written in terms of particle mass ($\xi^*_{p1} = M_p$), from which all the other expressions are derived. The rate of change of particle mass due to the exchange of molecules with the fluid phase is usually written as

$$\frac{dM_p}{dt} = M_w A_p J, \tag{5.20}$$

where $J(\mathbf{U}^*_p, \boldsymbol{\xi}^*_p, \mathbf{U}^*_f, \boldsymbol{\xi}^*_f)$ is the molar flux quantifying the number of molecules approaching (or leaving) the particle per unit time and per unit of particle surface area (which can be a function of all of the mesoscale variables), M_w is the molecular weight of the molecule and A_p is the particle surface area. To further elaborate on Eq. (5.20) it is necessary to relate the particle mass M_p to the particle surface area A_p. The first step is usually to write the particle mass as the product of particle material density ρ_p and particle volume V_p. When the material density of the elements of the disperse phase is constant, all the changes in particle mass are reflected in the particle volume, and the use of the two variables is perfectly equivalent. When the material density of the elements of the disperse phase is not constant, then some changes in particle volume are possible even at constant particle mass. A very common example is that of bubbles moving in vessels with large pressure gradients (for example due to hydrostatics), where, without modifying the mass of the single bubble, pressure differences result in significant changes in the bubble volume. In the following section, we look at various choices for the internal coordinates for describing the particle "size." Also, note that in the single-particle model both A_p and M_p could be included in the definition of $\boldsymbol{\xi}^*_p$ if they differ from particle to particle. In a general context, one can introduce multiple internal coordinates to account for changes in the volume and density of the particle (i.e. $\xi^*_{p1} = V_p$ and $\xi^*_{p2} = M_p$), which would be appropriate when the volume and density change independently due, for example, to chemical transformations. Likewise, if the chemical composition of the particle is important, it may be necessary to introduce internal coordinates for the particle species mass fractions $\mathbf{Y}_p$ and the fluid-phase species mass fractions $\mathbf{Y}_f$, which may be coupled through mass-exchange relations like Eq. (5.20) or chemical reactions. Conceptually, the important point is that the modeler is free to define the internal coordinates $\boldsymbol{\xi}^*_p$ and $\boldsymbol{\xi}^*_f$ to include whichever variables are required to describe the chemistry and physics at the single-particle level. In the remainder of this chapter, we will attempt to keep the notation tractable by not introducing new components for $\boldsymbol{\xi}^*_p$ at every step. Nevertheless, the reader should keep in mind that any particle characteristic (e.g. diameter, surface area, chemical composition, etc.) is potentially polydisperse and thus can be included in the definition of $\boldsymbol{\xi}^*_p$.

Particle-enthalpy balance

Likewise, in the case of heat exchange between phases, continuous changes for the internal coordinates are induced. If, for example, one of the internal coordinates is the temperature of the particle, T_p, the rate of change of particle enthalpy can be calculated as

$$\frac{d}{dt}(M_p c^p_p T_p) = A_p q, \tag{5.21}$$

where c^p_p is the specific heat of the particle, and $q(\mathbf{U}^*_p, \boldsymbol{\xi}^*_p, \mathbf{U}^*_f, \boldsymbol{\xi}^*_f)$ is the mesoscale model for the heat flux from the continuous phase to the particle. Obviously, in the case of particles with constant mass and specific heat,[15] the following mesoscale model for particle temperature is obtained:

$$\frac{\mathrm{d}T_\mathrm{p}}{\mathrm{d}t} = \frac{A_\mathrm{p}}{M_\mathrm{p} c_\mathrm{p}^\mathrm{p}} q. \tag{5.22}$$

However, in general, a particle can be described by other mesoscale variables in addition to T_p. Likewise, the heat flux can be due to convection, radiation, and internal heating, and thus can depend on multiple mesoscale variables. We briefly consider the problem of convective transport in Section 5.2.3, and refer the reader to the book of Crowe *et al.* (1998) for a detailed discussion of other models. Also, as noted above, in the single-particle model both A_p and M_p could be included in the definition of ξ_p^*. In a more general context, the single-particle enthalpy balance in Eq. (5.21) can include source/sink terms on the right-hand side due to mass transfer, chemical reactions, radiation, etc., which would couple the single-particle energy balance to the mass balance. Thus, as with the single-particle mass balance, additional internal coordinates will usually be necessary for multicomponent and chemically reacting systems. Finally, it is important to recall that there are two types of terms in the single-particle models for $\mathbf{G}_\mathrm{p}^*$ and $\mathbf{G}_\mathrm{f}^*$ appearing in Eq. (5.9). The first type represents mass/energy/species transfer between phases, and the second type represents changes of the internal coordinates inside one phase (e.g. homogeneous chemical reactions). For the first type, the conservation of microscale mass/species/energy will couple the single-particle models for $\mathbf{G}_\mathrm{p}^*$ and $\mathbf{G}_\mathrm{f}^*$ as described for momentum in Section 5.1.2. In the following sections, we will not consider explicitly the single-particle models for the fluid phase seen by the particle. Nevertheless, the reader should keep in mind that these models must be consistent with the single-particle models for the particle properties as shown in Chapter 4.

5.2.1 Mesoscale variables for particle size

As a matter of practice, it is useful to limit the number of internal coordinates required to describe a particle's "size." Thus, in this section, we review the available descriptions of the particle size that use either one or two internal coordinates.

Equidimensional particles

When the elements of the disperse phase can be classified as *equidimensional*, namely they have nearly the same size or spread in multiple directions, and have constant material density, typically a single internal coordinate is used to identify the "size" of the elements. This could be particle mass (or volume), particle surface area or particle length. In fact, in the case of equidimensional particles these quantities are all related to each other. For example, in the trivial cases of spherical or cubic particles, particle volume and particle surface area can be easily written as $V_\mathrm{p} = k_\mathrm{V} d_\mathrm{p}^3$ and $A_\mathrm{p} = k_\mathrm{A} d_\mathrm{p}^2$, or, in other words, as functions of a characteristic length, d_p (i.e. the diameter for the sphere and the edge for the cube), a volume shape factor, k_V, and a surface-area shape factor, k_A. For equidimensional objects the choice of the characteristic length is straightforward and the ratio between k_A and k_V is always equal to six. The approach can, however, be extended also to non-equidimensional objects. In this context, the extension turns out to be very useful only if

[15] If, for example, the particle mass is changing along with its temperature, the heat flux must include both a component corresponding to the difference in temperature between the particle and the fluid and a component for the change in the total enthaply due to mass transfer.

the rate of change in each dimension of the particle is such that when the particle shrinks or grows its shape and morphology is not changed. Although in this case the particle is not strictly an equidimensional object, its size can still be quantified by a single internal coordinate. In fact, for a generic object (with a specific shape) it is always possible to freely define one characteristic length, d_p^*, among the many possible,[16] and to express the real particle volume and particle surface area as functions of this characteristic length: $V_p = k_V^*(d_p^*)^3$ and $A_p = k_A^*(d_p^*)^2$. In general the ratio between the real shape factors k_A^* and k_V^* will not be six, but, after introducing an equivalent characteristic length $d_p = \phi_c d_p^*$, with the shape correction factor $\phi_c = 6k_V^*/k_A^*$, the equivalent or averaged shape factors (whose ratio is consequently forced to be equal to six) are obtained:

$$k_V = \frac{\left(k_A^*\right)^3}{216\left(k_V^*\right)^2} \quad \text{and} \quad k_A = \frac{\left(k_A^*\right)^3}{36\left(k_V^*\right)^2}.$$

Under these hypotheses and if particle size is the (first) internal coordinate (i.e. $\xi_{p1}^* = d_p$) and ρ_p is assumed to be constant, Eq. (5.20) becomes

$$\dot{d}_p = \frac{M_w}{\rho_p}\frac{k_A}{3k_V}J. \tag{5.23}$$

The rate of change of particle size is also indicated as G_p and can be positive, in the case of growing particles, or negative, in the case of shrinking particles. This is probably the most popular way to indicate the rate of phase-space advection due to mass exchange, perhaps because it is quite easy to measure the change in particle size at different instants, for example by simple imaging techniques. If the internal coordinate is instead particle volume (i.e. $\xi_{p1}^* = V_p$), the definition becomes

$$\dot{V}_p = \frac{M_w}{\rho_p}\frac{k_A}{(k_V)^{2/3}}V_p^{2/3}J,$$

whereas, if the internal coordinate is the particle surface area (i.e. $\xi_{p1}^* = A_p$),

$$\dot{A}_p = \frac{M_w}{\rho_p}\frac{2(k_A)^{3/2}}{3k_V}A_p^{1/2}J.$$

Since M_w, ρ_p, k_A, and k_V are typically constants, the only relevant mesoscale dependences are those with respect to ξ_{p1}^* and with respect to the molar flux $J(\mathbf{U}_p^*, \xi_p^*, \mathbf{U}_f^*, \xi_f^*)$ from the continuous phase to the particle. The approach described above for defining the particle size is the simplest to simulate advection in phase space due to mass transfer, and readers interested in the details are referred to the specialized literature (Gavi *et al.*, 2007, 2008; Marchisio *et al.*, 2001a, 2001b, 2001c, 2002). This approach has been successfully applied to the simulation of fluid–particle systems (i.e. fluidized suspensions, crystallization processes) in which solid particles can reasonably be assumed to be equidimensional and to the simulation of fluid–fluid systems (i.e. sprays, gas bubbles, droplets) in which particles are actually spherical or slightly elliptical.

[16] The characteristic length is usually chosen in order to represent an easily measurable quantity. For example, if sieving is used to measure particle size then this characteristic length could be the second largest. In fact, when sieving or screening particles the second largest dimension (rather than the largest dimension) determines whether the particle passes through the sieve or not. When instead static light scattering is used to determine particle size, an intermediate dimension between the largest and the second largest could be chosen.

Non-equidimensional particles

In the case of non-equidimensional particles, typically more than one internal coordinate is used to describe the shape, morphology, and size of the particles. To illustrate the issues related to phase-space advection due to mass transfer, let us analyze a simple example. We consider needle-like particles described as rectangular parallelepipeds with length ξ^*_{p1} and equal width and depth ξ^*_{p2}. Clearly the particle volume can easily be calculated from these two quantities (i.e. $V_p = \xi^*_{p1}(\xi^*_{p2})^2$), as can the particle surface area (i.e. $A_p = 2(\xi^*_{p2})^2 + 4\xi^*_{p1}\xi^*_{p2}$). The problem can be described in terms of these two internal coordinates, after having introduced the rates of change for the two characteristic sizes, which are $G_{p1} = \dot{\xi}^*_{p1}$ and $G_{p2} = \dot{\xi}^*_{p2}$, respectively. This can easily be done by quantifying the molar flux of molecules approaching the different faces of the particles. If we denote by $J_{12}(\mathbf{U}^*_p, \boldsymbol{\xi}^*_p, \mathbf{U}^*_f, \boldsymbol{\xi}^*_f)$ the molar flux on the surface delimited by ξ^*_{p1} and ξ^*_{p2} and by $J_{22}(\mathbf{U}^*_p, \boldsymbol{\xi}^*_p, \mathbf{U}^*_f, \boldsymbol{\xi}^*_f)$ the molar flux on the surface delimited by ξ^*_{p2} and ξ^*_{p2}, the following rates are obtained: $G_{p1} = 2M_w J_{22}/\rho_p$ and $G_{p2} = 4M_w J_{12}/\rho_p$. An alternative description could be, for example, based on particle volume and surface area, and in this case the rates of change of these two new internal coordinates would become $\dot{V}_p = (\xi^*_{p2})^2 G_{p1} + 2\xi^*_{p1}\xi^*_{p2}G_{p2}$ and $\dot{A}_p = (2\xi^*_{p2} + 4\xi^*_{p1})G_{p2} + 4\xi^*_{p2}G_{p1}$. Readers interested in more details are referred to the specialized literature (Heineken *et al.*, 2011; Immanuel & Doyle, 2005; Ma *et al.*, 2007; Pinto *et al.*, 2007; Qamar *et al.*, 2011b; Qamar & Seidel-Morgenstern, 2009; Wang *et al.*, 2008).

Fractal particles

In many practical cases (for example during aggregation) the particles constituting the disperse phase cannot be considered to be either equidimensional or non-equidimensional filled objects. Instead, because of some voids contained inside them, they are more conveniently described as fractal objects. In this case, a particle is considered as composed of primary particles, and typically its structure is identified by the size (or volume) of the primary particle, $\xi^*_{p1} = d_p$, and the number of primary particles per particle, $\xi^*_{p2} = N_p$. In this context, it is useful to distinguish between primary particles, namely those aggregating together forming the fractal object, and secondary particles, corresponding to the fractal object itself. Particles characterized by specific values of d_p and N_p can still have very different apparent sizes according to the structure of the particle. This structure is quantified by the fractal dimension, and one possible definition is

$$R_g = \frac{d_p}{2}\left(\frac{N_p}{k_g}\right)^{1/D_f},$$

where R_g is the radius of gyration of the particle, typically defined as the root-mean-square distance of the center of the primary particle from the center of mass of the secondary particle, and k_g is a scaling factor of the order of magnitude of unity. It is clear that, for a given d_p and N_p, smaller fractal dimensions correspond to larger radii of gyration and vice versa. The fractal dimension typically ranges from one to three, representing very different structures ranging from loose linear aggregates to compact completely filled objects. When phase-space advection due to mass exchange involves fractal objects, different scenarios are possible. Typically mass transfer causes a modification both in the size of the primary particles and in the morphology and structure of the overall secondary particle. The rate of change of the size of the primary particles can be calculated from Eq. (5.23) by assuming that primary particles behave as isolated particles. A corrective "crowding" factor, k_c (ranging from zero to unity), can be introduced to account for the presence of the other primary particles,[17] as shown by Marchisio (2009). One possible assumption is that during mass transfer the entire secondary particle grows, keeping constant the number of primary particles. In this first case mass transfer induces a change in the size of the primary particles but does not affect the number of primary particles per secondary particle, resulting in the following expressions for the continuous rate of change of the internal coordinates:

$$\dot{\xi}^*_{\text{p1}} = \dot{d}_{\text{p}} = \frac{M_{\text{w}}}{\rho_{\text{p}}} \frac{k_{\text{A}} k_{\text{c}}}{3 k_{\text{V}}} J \quad \text{and} \quad \dot{\xi}^*_{\text{p2}} = \dot{N}_{\text{p}} = 0,$$

where k_c is the above-mentioned crowding factor, and k_V and k_A are the volume and surface area shape factors of the primary particles. Another possible situation is that, due to mass exchange, primary particles join together, losing their identity. This seems to be more likely in the case of positive molar flux, $J > 0$, or, in other words, in the case of growing particles. In this second case (often referred to as obliteration), mass transfer induces a change both in primary-particle size and in the number of primary particles per aggregate. If during primary-particle growth the radius of gyration of the secondary particle and its fractal dimension are assumed to remain nearly constant, the following expressions for the continuous rates of change of the two internal coordinates are obtained:

$$\dot{\xi}^*_{\text{p1}} = \dot{d}_{\text{p}} = \frac{M_{\text{w}}}{\rho_{\text{p}}} \frac{k_{\text{A}} k_{\text{c}}}{3 k_{\text{V}}} \frac{1}{3 - D_{\text{f}}} J \quad \text{and} \quad \dot{\xi}^*_{\text{p2}} = \dot{N}_{\text{p}} = -\frac{M_{\text{w}}}{\rho_{\text{p}}} \frac{k_{\text{A}} k_{\text{c}}}{3 k_{\text{V}}} \frac{D_{\text{f}}}{3 - D_{\text{f}}} \frac{N_{\text{p}}}{d_{\text{p}}} J.$$

Another possible scenario is that as a consequence of mass transfer only the number of primary particles changes, whereas their size remains more or less constant. This hypothesis seems to be realistic in the case of negative molar flux, $J < 0$, or, in other words, in the case of shrinking particles. In fact, in this case it is more likely that the external particles will be consumed before the internal ones. The resulting expressions for the continuous rate of change of the two internal coordinates therefore read as

$$\dot{\xi}^*_{\text{p1}} = \dot{d}_{\text{p}} = 0 \quad \text{and} \quad \dot{\xi}^*_{\text{p2}} = \dot{N}_{\text{p}} = \frac{M_{\text{w}}}{\rho_{\text{p}}} \frac{N_{\text{p}}}{d_{\text{p}}} \frac{k_{\text{A}} k_{\text{c}}}{k_{\text{V}}} J.$$

Different particulate processes will behave very differently, and, in each case, a microscale physical scenario must be hypothesized in order to formulate the mesoscale model. For more details on these and other examples, readers are referred to the work of Lee *et al.* (1962), Marchisio & Barresi (2009), and Zucca *et al.* (2006).

5.2.2 Size change for crystalline and amorphous particles

The process of growth of crystalline particles involves several steps, such as diffusion of solute molecules from the bulk of the solution to the crystal surface, adsorption on the crystal surface, diffusion over the surface, attachment to a step, diffusion along a step, and integration into a crystal kink site. The rates of these different processes depend on the type of material and on the operating conditions. The final overall molar flux of solute molecules, which has previously been indicated with $J(\mathbf{U}^*_{\text{p}}, \xi^*_{\text{p}}, \mathbf{U}^*_{\text{f}}, \xi^*_{\text{f}})$, will of course be

[17] In many particulate systems it has been shown that the crowding factor is related to the reduction of available surface area. From simple geometric considerations, it is possible to show that this factor scales as $k_c \propto N_p^{-1/3}$, whereas from fitting with experiments different scaling exponents have been found.

influenced by the rate of each single process, typically with one or two processes controlling the entire phenomenon. For equidimensional particles, the size-growth rate G_p is related to J by $G_p = M_w J/\rho_p$. Generally, the final molar flux turns out to be a function of the fluid-phase temperature seen by the particle T_f, the supersaturation,[18] S, and some particle properties such as size, d_p, or surface area, A_p. For clarity, we will write the mesoscale models in terms of these variables, but the reader should keep in mind that in the single-particle model these variables correspond to the internal coordinates for the fluid phase ξ_f^* and the particles ξ_p^*, respectively.

The kinetic regime

Different theories have been developed to explain the process of crystal growth and to derive the relative rates. One popular theory is that of two-dimensional growth. According to this theory, when molecules are absorbed onto the crystal surface they form a new two-dimensional circular nucleus that, as in classical three-dimensional nucleation, when it has dimensions larger than a critical size, grows further, covering the entire crystal surface. Depending on the relative rates of formation of new nuclei and growth of the existing ones, different scenarios are possible. If the rate with which the nuclei spread to form a complete layer is much larger than the rate of formation of a new nucleus, we are in the *mononuclear growth limit*, resulting in the following rate:

$$G_p = C_1 h A_p [\ln(S)]^{1/2} \exp\left(-\frac{C_2}{T_f^2 \ln(S)}\right), \tag{5.24}$$

where h is the step height, A_p is the crystal surface area, and C_1 and C_2 have exact expressions derived from the theory but, being functions of parameters that are rarely known and difficult to measure, are often used as empirical constants obtained by fitting experimental data. More details can be found in the works of Ohara & Reid (1973) and Myerson (2002). In the opposite situation, when the two-dimensional nucleation is much faster than the spreading and covering of the face, we are in the *polynuclear growth limit*, resulting in the following rate:

$$G_p = \left(\frac{C_3}{T_f^2 [\ln(S)]^{3/2}}\right) \exp\left(-\frac{C_2}{T_f^2 \ln(S)}\right), \tag{5.25}$$

where again C_3 can be derived from the theory but is often obtained by fitting data from experiments. Alternative growth models, which are based on a birth-and-spread mechanism that assumes a finite rate for nucleus spread as well as that nuclei can form at any location, including incomplete layers, and that there is no inter-growth between the nuclei, have also been formulated, resulting in the following expression:

$$G_p = C_4 (S-1)^{2/3} [\ln(S)]^{1/6} \exp\left(-\frac{C_5}{T_f^2 \ln(S)}\right), \tag{5.26}$$

where again C_4 and C_5 are typically obtained by fitting data from experiments. Readers interested in the details are referred to the literature (Garside *et al.*, 1975; Janssen Van Rosmalen *et al.*, 1975). The models given above are very useful in formulating and

[18] The supersaturation, S, is typically defined as the ratio of the local instantaneous solute concentration to the equilibrium concentration or solubility in the fluid phase. In the case of precipitation supersaturation is instead defined as the ratio of the product of the concentration of ions constituting the precipitate and the solubility product in the fluid phase.

testing different mechanisms, but all of them fail in predicting quantitatively (and for a reasonably large range of operating conditions) the correct behavior observed experimentally. An interesting, and more reliable approach, is the so-called Burton–Cabrera–Frank model (Nyvlt *et al.*, 1985), which assumes a continuous growth process on the crystal face along a screw dislocation, resulting in the following expression:

$$G_\mathrm{p} = C_6 T_\mathrm{f}(S - 1)\ln(S)\tanh\left(\frac{C_7}{T_\mathrm{f}\ln(S)}\right), \tag{5.27}$$

where again C_6 and C_7 are derived either from the theory or from experiments. Equation (5.27) describes well the change of dependence, from nearly parabolic to linear, of the growth rate versus supersaturation and is therefore very much used in practical applications.

The empirical growth model

An alternative (very popular) approach is based on the very simple idea of calculating the growth rate of particles with the following empirical expression: $G_\mathrm{p} = k_\mathrm{g}(S - 1)^g$, where g is derived from experiments, as well as k_g, which is in turn very often described with an Arrhenius-type law:

$$k_\mathrm{g} = k_\mathrm{g}^\mathrm{o} \exp\left(-\frac{E_\mathrm{g}}{RT_\mathrm{f}}\right). \tag{5.28}$$

These empirical laws have limited validity and should not be used outside the range of supersaturation values from which they were derived.

The mass-transfer-limited regime

The growth models presented previously are valid if the rate of integration of the solute molecule in the crystal lattice and the building up of the face are the rate-controlling steps. This is, for example, almost always true for crystalline particles suspended in a vapor. However, when crystals are suspended in a liquid, or when the growing particle is amorphous, and therefore there is no barrier (or a very low barrier) for the incorporation of the solute molecules, then the process of diffusion of solute molecules from the bulk of the solution to the particle surface might become the controlling step. As is very well known, around the particle there is formed a concentration (as well as a momentum and a thermal) boundary layer that can be described with simple models. These models assume a linear concentration profile in the boundary layer, and introduce a mass-transfer coefficient, k_d. The resulting growth rate (in the case of diffusion-controlled growth for an equidimensional object) becomes

$$G_\mathrm{p} = \frac{k_\mathrm{A}}{3k_\mathrm{V}}\frac{M_\mathrm{w}}{\rho_\mathrm{p}} k_\mathrm{d} c_\mathrm{eq}(S - 1), \tag{5.29}$$

where c_eq is the equilibrium solute concentration (or solubility) in the fluid phase and the mass-transfer coefficient k_d is calculated from correlations expressed in terms of the Sherwood number:

$$\mathrm{Sh} = \frac{k_\mathrm{d} d_\mathrm{p}}{D}, \tag{5.30}$$

where D is the molecular diffusion coefficient of solute molecules in the bulk solution. Under creeping (or Stokes) flow conditions, and in the case of a rigid sphere, the following correlation for the Sherwood number is readily derived from the theory:

$$\mathrm{Sh} = 1 + (1 + \mathrm{Pe})^{1/3} \tag{5.31}$$

where the Péclet number is defined by

$$\mathrm{Pe} = \mathrm{Re_p Sc}, \tag{5.32}$$

in which the Schmidt number for the fluid phase is defined by

$$\mathrm{Sc} = \frac{\nu_\mathrm{f}}{D},$$

and the particle Reynolds number by

$$\mathrm{Re_p} = \frac{d_\mathrm{p}|\mathbf{U}_\mathrm{p}^* - \mathbf{U}_\mathrm{f}^*|}{\nu_\mathrm{f}}. \tag{5.33}$$

It is important to remind the reader that $\mathbf{U}_\mathrm{f}^*$ is the velocity of the fluid phase seen by the particle, $|\mathbf{U}_\mathrm{p}^* - \mathbf{U}_\mathrm{f}^*|$ is the slip velocity, d_p is the particle diameter, and ν_f is the kinematic viscosity of the fluid phase. Note that Eq. (5.33) depends on the particle velocity $\mathbf{U}_\mathrm{p}^*$ and is valid in the zero-Stokes-number limit where $\mathbf{U}_\mathrm{p}^* = \mathbf{U}_\mathrm{f}^*$ so that particles follow the fluid. The correlation in Eq. (5.31) is valid only for $\mathrm{Re_p} < 1$ and $\mathrm{Sc} > 200$. For larger particle Reynolds numbers the following correlations can be used: $\mathrm{Sh} = 2 + 0.724\mathrm{Re_p^{0.48}Sc^{1/3}}$, which is valid for $100 < \mathrm{Re_p} \leq 2000$, and $\mathrm{Sh} = 2 + 0.425\mathrm{Re_p^{0.55}Sc^{1/3}}$, which is valid for $2000 < \mathrm{Re_p} \leq 10^5$. Among the other correlations available, it is important to cite the one proposed by Ranz & Marshall (1952) for macroparticles: $\mathrm{Sh} = 2.0 + 0.6\mathrm{Re_p^{1/2}Sc^{1/3}}$. These expressions assume that the fluid velocity $\mathbf{U}_\mathrm{f}^*$ is known. For micron-sized (or smaller) particles moving in turbulent fluids for which only the *ensemble-mean* fluid velocity $\langle \mathbf{U}_\mathrm{f} \rangle$ is known, it is instead better to employ the mesoscale model derived by Armenante & Kirwan (1989): $\mathrm{Sh} = 2.0 + 0.52\left(\mathrm{Re_p^*}\right)^{0.52}\mathrm{Sc^{1/3}}$, where $\mathrm{Re_p^*} = \varepsilon_\mathrm{f}^{1/3} d_\mathrm{p}^{4/3}/\nu_\mathrm{f}$ is the modified particle Reynolds number and ε_f is the fluid turbulence-dissipation rate. However, we should remind the reader that in the mesoscale model only microscale turbulence generated by the presence of the particles in the fluid is included, in which case the macroscale turbulence correlation is not applicable. The fact that Eq. (5.29) for diffusion-controlled growth is size-dependent (through the Sherwood number) has important implications, as will be highlighted in the next chapters (especially when discussing the method of moments in Section 7.3).

The dissolution model

The inverse process of dissolution, namely solute molecules leaving the lattice of the crystalline particles or the amorphous particles because the particle is in contact with an under-saturated solution (i.e. $S < 1$), is typically controlled by diffusion. Therefore, the mesoscale model reported in Eq. (5.29) can be used to calculate the rate of the size change due to dissolution.

5.2.3 Non-isothermal systems

Similarly to mass transfer, the convective heat flux for non-isothermal particles (in the absence of mass transfer) is modeled as

$$q = k_\mathrm{h}(T_\mathrm{f} - T_\mathrm{p}), \tag{5.34}$$

where k_h is the heat-transfer coefficient, T_f is the temperature of the fluid seen by the particle, and T_p is the particle temperature. The coefficient k_h is generally calculated with a correlation based on the Nusselt number:

$$\mathrm{Nu} = \frac{k_\mathrm{h} d_\mathrm{p}}{\alpha_\mathrm{p}^\mathrm{c}},$$

where $\alpha_\mathrm{p}^\mathrm{c}$ is the thermal conductivity of the disperse phase. In writing Eq. (5.34), we have assumed that no thermal gradients are present inside the particle. The Nusselt number is calculated with correlations, such as the one proposed by Ranz & Marshall (1952): $\mathrm{Nu} = 2 + 0.6\mathrm{Re}_\mathrm{p}^{1/2}\mathrm{Pr}^{1/3}$, where the Prandtl number is defined by

$$\mathrm{Pr} = \frac{\nu_\mathrm{f}}{\alpha_\mathrm{f}^\mathrm{c}}$$

and $\alpha_\mathrm{f}^\mathrm{c}$ is the thermal conductivity of the fluid phase. The correlation reported above provides good agreement with experiments for $\mathrm{Re}_\mathrm{p} < 5 \times 10^4$. Corrections are available for non-uniform field effects (Michaelides & Feng, 1994), blowing effects (Renksizbulut & Yuen, 1983), and turbulence effects (Yearling & Gould, 1993). In the presence of mass transfer, the model in Eq. (5.34) must be modified to include the gain/loss of particle enthalpy due to the addition/loss of mass.

5.2.4 Mass transfer to gas bubbles

Another interesting example is that of gas bubbles dispersed in a continuous liquid phase with which mass is exchanged. Also for this case the rate of change of the internal coordinates due to mass transfer is written starting from a simple mass balance for a single bubble. Following the standard notation for gas–liquid systems, the single-particle mass balance becomes

$$\frac{\mathrm{d}M_\mathrm{p}}{\mathrm{d}t} = M_\mathrm{w} k_\mathrm{d} A_\mathrm{p}\, \Delta c, \tag{5.35}$$

where k_d is the mass-transfer coefficient, A_p is the surface area of the bubble, and Δc is the mass-transfer driving force (i.e. the difference in concentration of the active component between the bubble and the equilibrium concentration in the liquid phase seen by the particle). Also, in this case, the underlying assumption is that the bubble is homogeneous and no significant concentration gradients are detectable inside the bubble. The overall mass-transfer rate is therefore quantified by estimation of both k_d and A_p. In fact, one important complication in the case of gas bubbles is that their shape is not fixed but, depending on the nature of the primary and disperse phases and depending on their size, they can assume spherical, ellipsoidal, dimpled ellipsoidal-cap, skirted, and spherical-cap shapes (Tomiyama *et al.*, 2002a). The regimes corresponding to the different shapes are identified by the values assumed by the particle Reynolds number in Eq. (5.33), namely the Eötvös (or Bond) number,

$$\mathrm{Eo} = \frac{g\, \Delta\rho\, d_\mathrm{p}}{\sigma}, \tag{5.36}$$

and the Morton number,

$$\mathrm{Mo} = \frac{g \mu_\mathrm{f}^4\, \Delta\rho}{\rho_\mathrm{f}^2 \sigma^3}, \tag{5.37}$$

where μ_f and ρ_f are the viscosity and the material density of the liquid phase, g is the acceleration due to gravity, σ is the interface tension between the liquid and gas phases, and

$\Delta\rho$ is the difference in material density between the liquid and gas phases. This situation is typically handled by describing the bubbles with a single internal coordinate (i.e. the equivalent-sphere diameter) and by introducing an aspect ratio, defined as the ratio between the minor and the major axes of the bubble. This aspect ratio E can be calculated by using the empirical equation proposed by Moore (1965) as a function of the Morton number: $E = 1/(1 + 0.043\text{Re}_\text{p}^{5/3}\text{Mo}^{1/3})$. An alternative to this is the use of the correlation proposed by Wellek *et al.* (1966) for liquid–liquid droplets: $E = 1/(1 + 0.1613\text{Eo}^{0.757})$, which is valid for $\text{Eo} < 40$ and $\text{Mo} \leq 10^{-6}$, whereas for $\text{Eo} > 40$ and $\text{Re}_\text{p} > 1.2$ fluid particles are typically of spherical shape. Once the characteristic E value is known, the ratio of the real area of the bubble A_p and the area A_eq of a sphere with an equivalent volume can be calculated as follows:

$$\frac{A_\text{p}}{A_\text{eq}} = \frac{1}{2E^{2/3}}\left[1 + \frac{E^2}{2\sqrt{1-E^2}}\ln\left(\frac{1+\sqrt{1-E^2}}{1-\sqrt{1-E^2}}\right)\right]. \tag{5.38}$$

Insofar as the mass-transfer coefficient for clean bubbles is concerned (see, for example, the review by Clift *et al.* (1978)), in the case of spherical bubbles moving under creeping (or Stokes) flow conditions, the following correlation has been proposed: $\text{Sh} = 1 + (1 + 0.564\text{Pe}^{2/3})^{3/4}$. For spherical particles with $\text{Re}_\text{p} \geq 70$ the Sherwood number can be expressed by the following relationship (Lochiel & Calderbank, 1964):

$$\text{Sh} = \frac{2}{\sqrt{\pi}}\left(1 - \frac{2.96}{\text{Re}_\text{p}^{1/2}}\right)^{1/2}\text{Pe}^{1/2}, \tag{5.39}$$

where, as has already been mentioned, D is the molecular diffusivity of the molecules transferring from one phase to the other. If the Reynolds number is sufficiently large, the term in parentheses is small compared with unity and Eq. (5.39) reduces to the classical result proposed by Higbie (1935): $\text{Sh} = 2\text{Pe}^{1/2}/\sqrt{\pi}$. At higher Reynolds numbers the bubble assumes approximately the shape of an oblate spheroid. In this case, the Sherwood number (which is based on the real bubble surface) becomes

$$\text{Sh} = \frac{2}{\sqrt{\pi}}\text{Pe}^{1/2}\frac{\dfrac{8(1-E^2)^{3/2}E^{1/3}}{3\arcsin(\sqrt{1-E^2} - E\sqrt{1-E^2})}}{1 + \dfrac{E^2}{2\sqrt{1-E^2}}\ln\left(\dfrac{1+\sqrt{1-E^2}}{1-\sqrt{1-E^2}}\right)}. \tag{5.40}$$

Closer observation of Eq. (5.40) shows that the effect of the eccentricity on the mass-transfer coefficient is minor (a few percent at very low E). Nevertheless, the net mass-transfer rate may increase substantially above the value for a sphere of the same volume because of the increase in the transfer area due to the fattening of the bubble, as already described by Eq. (5.38). In the case of bubbles in a turbulent liquid phase, an alternative description is that offered by Danckwerts (1951), who suggested a refinement of the penetration model by assuming that k_d is related to the average surface-renewal rate resulting from exposure of the bubble interface to turbulent eddies with a variable contact time, resulting in $k_\text{d} = \sqrt{Ds}$, where s is the rate of replacement of surface elements. As suggested by Lamont & Scott (1970), the latter quantity can be assumed to be related to the microscale turbulent motions in the liquid phase, ranging from the smallest viscous scales to the inertial ones. By using Kolmogorov's theory of isotropic turbulence the final turbulent mass-transfer coefficient is $k_\text{d} = KD^{0.5}(\varepsilon_\text{f}/\nu_\text{f})^{0.25}$, where ε_f is the turbulence-dissipation rate in the liquid phase and K is a model constant of the order of unity that is often taken to be equal to 1.13. However, we must again remind the reader that the mesoscale model does not describe the macroscale turbulence effects and, thus, correlations involving ε_f are not sufficient for describing the mesoscale physics.

5.2.5 Heat/mass transfer to liquid droplets

It is interesting to discuss simultaneous heat and mass transfer occurring when liquid droplets dispersed in a gas grow or shrink due to condensation or evaporation, respectively. In this case, the rate of change of particle size (surface area or volume) is strongly coupled with the exchange of energy due to the latent heat of evaporation and condensation. To illustrate the main issues, let us consider the case of a droplet composed of a single-component liquid that evaporates creating mass exchange with the surrounding gas phase. (The corresponding condensation case can be treated analogously by inverting signs on the mass and heat fluxes.) Following the discussion in Chapter 5 of Marchisio & Fox (2007), we let Y_1 denote the molar fraction in the gas phase of the chemical component that evaporates from the droplet, Y_2 denote the molar fraction of the other (stagnant) component, and T_f denote the gas-phase temperature seen by the droplet. The evaporating droplet is surrounded by material and thermal boundary layers. We denote the molar fractions of the evaporating component in the bulk of the gas phase (at large distances from the droplet surface) and right on the droplet surface by Y_{f1} and Y_{s1}, respectively. Analogously, we define T_f and T_s as the temperature in the gas phase at large distances from the droplet surface and right on the droplet surface, respectively. Of these four quantities, Y_{f1} and T_f are generally assumed to be known (and are usually grouped together in the internal-coordinate vector $\boldsymbol{\xi}_f^*$ for the gas-phase properties seen by a droplet).

By imposing thermodynamic equilibrium (with the Clausius–Clapeyron law whose validity holds far away from the critical point) at the droplet surface, the conservation of mass and enthalpy (considering that changes in the droplet temperature must be related to the latent heat of evaporation) yields the following expressions:

$$\begin{aligned}\frac{c_f^p(T_f - T_s)}{L_v} &= \frac{Y_{s1} - Y_{f1}}{1 - Y_{s1}},\\ p_{s1} &= p_{\text{ref}} \exp\left[-\frac{L_v}{R}\left(\frac{1}{T_p} - \frac{1}{T_{\text{ref}}}\right)\right],\\ Y_{s1} &= \left[1 + \frac{M_{w2}}{M_{w1}}\left(\frac{p_f}{p_{s1}} - 1\right)\right]^{-1},\end{aligned} \tag{5.41}$$

where c_f^p is the specific heat of the gas phase, L_v is the latent heat of evaporation of the evaporating component, p_f is the pressure of the gas phase in the bulk, p_{s1} is the partial pressure of the evaporating component in the gas phase, and R is the ideal-gas constant. The parameters p_{ref} and T_{ref} are the reference pressure and temperature for liquid boiling. The constants M_{w1} and M_{w2} are the molecular weights of the two chemical species. Once all the physical constants have been defined and once T_p, p_f, T_f, and Y_{f1} have been specified (through the solution of transport equations for the gas phase seen by the droplet), the three equations can be solved in the three unknowns Y_{s1}, T_s, and p_{s1}. Knowledge of these three quantities suffices for the calculation of the mass and enthalpy transfer from the droplets to the gas phase. The rate of change of droplet mass (under the assumption of spherical droplets) is given by

$$\frac{dM_p}{dt} = -\pi \rho_f D \text{Sh} d_p \ln(1 + B_m), \tag{5.42}$$

where ρ_f is the material density of the gas phase, D is the diffusion coefficient in the gas phase of the molecules of the evaporating species, and the factor B_m is defined as

$$B_m = \frac{Y_{s1} - Y_{f1}}{1 - Y_{s1}}. \tag{5.43}$$

The Sherwood number for this particular system can be calculated by using empirical correlations, such as the one proposed by Faeth and Fendell (Kuo, 1986):

$$\mathrm{Sh} = 2 + \frac{0.55\mathrm{Re_p Sc}}{\left(1.232 + \mathrm{Re_p Sc}^{4/3}\right)^{1/2}}. \tag{5.44}$$

To eliminate the explicit droplet size dependence on the right-hand side of Eq. (5.42), very often the continuous rate of change due to mass exchange is written in terms of the droplet surface area (always under the assumption of spherical shape):

$$\frac{\mathrm{d}A_p}{\mathrm{d}t} = -4\pi \frac{\rho_f D}{\rho_p} \mathrm{Sh} \ln(1 + B_m). \tag{5.45}$$

The implicit assumption in the determination of Eqs. (5.42) and (5.45) is that the mass- and heat-transfer resistances within the droplet are null. This is a common assumption that is based on the idea of quasi-stationary conditions within the droplet that often goes under the name of the *rapid-mixing model*. In the special case in which B_m is approximately constant, the right-hand side of Eq. (5.45) is constant, and thus the rate of change of the surface area is independent of the droplet diameter (i.e. the d^2 evaporation law).

Alternatively, the *thin-skin model*, which assumes an infinite diffusion resistance with only a thin layer on the droplet surface being heated and evaporated, can be used. This second assumption is more appropriate for cases in which the droplet reaches its boiling temperature immediately after the start of the evaporation process. All the other approaches (such as the *diffusion-limited model*), which are recommended in the case of multicomponent droplets, require some form of spatial discretization inside the droplet, resulting in much higher computational costs. To close the set of equations, it is necessary to describe the evolution of the droplet temperature T_p as discussed in Section 5.2.3. For example, upon including the effect of evaporation on the heat flux for the rapid-mixing model, the particle-enthalpy balance yields

$$\frac{\mathrm{d}T_p}{\mathrm{d}t} = 6\mathrm{LeNu}\frac{c_f^p \rho_f D}{c_p^p \rho_p A_p} \frac{\ln(1 + B_m)}{B_m} \left(T_f - T_p - \frac{B_m L_v}{c_f^p} \right), \tag{5.46}$$

where the final term on the right-hand side depends on Eq. (5.45) and c_p^p is the specific heat of the particle. The Lewis number is defined by

$$\mathrm{Le} = \frac{\alpha_f^c}{D} = \frac{\mathrm{Sc}}{\mathrm{Pr}},$$

and the Nusselt number for this particular system can be calculated by using the empirical correlation proposed by Faeth and Fendell (Kuo, 1986):

$$\mathrm{Nu} = 2 + \frac{0.55\mathrm{Re_p Pr}}{\left(1.232 + \mathrm{Re_p Pr}^{4/3}\right)^{1/2}}. \tag{5.47}$$

Given T_p, the expression for p_{s1} is closed, thereby fixing the mass-transfer rate. The discussion above is applicable to single-component droplets. In many applications, the liquid/gas phase will contain multiple chemical species, for which additional internal coordinates will be necessary in order to describe the physics of evaporation (Sazhin, 2006). In the context of a single-particle model for a multicomponent droplet, the simplest mesoscale model must include the particle mass M_p, the component mass fractions $\mathbf{Y}_p$ and $\mathbf{Y}_f$, and the temperatures T_p and T_f.

In conclusion, we remind the reader that in the single-particle models for mass and energy transfer any one of the coefficients and dimensionless numbers could potentially be different for different particles. Thus, in general, all of the coefficients and dimensionless numbers are functions of the mesoscale variables ($\mathbf{U}_p^*$, $\boldsymbol{\xi}_p^*$, $\mathbf{U}_f^*$, and $\boldsymbol{\xi}_f^*$). Nevertheless, in practice, many of the correlations described in this section have been developed in the context of macroscale models for which only the *average* values of the state-space variables are used. Thus, unless the correlation has been fitted using monodisperse data (which is usually not the case for experimental data), it is likely that some adjustments in the numerical constants appearing in the correlations will be necessary when they are applied to polydisperse systems. On the other hand, correlations developed using direct numerical simulations of the microscale model often apply specifically to monodisperse particles (e.g. equal-diameter spheres). Ideally, separate correlations could be developed for each value of the internal coordinates (i.e. particle diameter) from a simulation of interacting polydisperse particles. In any case, the reader should be acutely aware of the range of applicability of the available correlations before applying them in the context of single-particle models. Finally, the reader should recall that the mesoscale model for the fluid-phase advection $\mathbf{G}_f^*$ must be consistent with the particle-phase model for $\mathbf{G}_p^*$ in order for the overall system to conserve mass and energy at the mesoscale.

5.2.6 Momentum change due to mass transfer

When the velocity of the particle phase is different than that of the fluid phase, the transfer of mass between phases will also result in the transfer of momentum. For example, if we let ξ_{p1}^* be the mass of a particle and ξ_{f1}^* be the fluid mass seen by the particle, then conservation of mass at the mesoscale leads to

$$\frac{\mathrm{d}\xi_{p1}^*}{\mathrm{d}t} = G_{p1}^* \quad \text{and} \quad \frac{\mathrm{d}\xi_{f1}^*}{\mathrm{d}t} = -G_{p1}^*, \tag{5.48}$$

where G_{p1}^* is the mass-transfer rate *from* the fluid to the particle (i.e. a negative sign indicates mass transfer from the particle to the fluid).[19] The corresponding change in momentum for each phase can be written as

$$\begin{aligned} \frac{\mathrm{d}\xi_{p1}^*\mathbf{U}_p^*}{\mathrm{d}t} &= \mathbf{U}_f^* \max(G_{p1}^*, 0) + \mathbf{U}_p^* \min(G_{p1}^*, 0), \\ \frac{\mathrm{d}\xi_{f1}^*\mathbf{U}_f^*}{\mathrm{d}t} &= -\mathbf{U}_f^* \max(G_{p1}^*, 0) - \mathbf{U}_p^* \min(G_{p1}^*, 0), \end{aligned} \tag{5.49}$$

where the sign of G_{p1}^* determines the momentum contribution due to the transferred mass and, by construction, momentum is conserved at the mesoscale. By manipulating

[19] If particle nucleation/dissolution were also present, its contribution to mass/momentum transfer would simply be added to G_{p1}^* in all of the formulas appearing in this section.

Eqs. (5.48) and (5.49), the following expressions are found for the accelerations due to mass transfer:

$$\begin{aligned}\frac{\mathrm{d}\mathbf{U}_\mathrm{p}^*}{\mathrm{d}t} &= \mathbf{A}_\mathrm{fp}^*\Big|_\mathrm{mass} = \frac{1}{\xi_\mathrm{p1}^*}\left(\mathbf{U}_\mathrm{f}^* - \mathbf{U}_\mathrm{p}^*\right)\max(G_\mathrm{p1}^*, 0),\\ \frac{\mathrm{d}\mathbf{U}_\mathrm{f}^*}{\mathrm{d}t} &= \mathbf{A}_\mathrm{pf}^*\Big|_\mathrm{mass} = \frac{1}{\xi_\mathrm{f1}^*}\left(\mathbf{U}_\mathrm{f}^* - \mathbf{U}_\mathrm{p}^*\right)\min(G_\mathrm{p1}^*, 0).\end{aligned} \tag{5.50}$$

Thus, as mentioned above, the mesoscale models $\mathbf{A}_\mathrm{fp}^*$ and $\mathbf{A}_\mathrm{pf}^*$ for momentum transfer will contain a contribution due to mass transfer whenever $\mathbf{U}_\mathrm{f}^* \neq \mathbf{U}_\mathrm{p}^*$.

5.3 Phase-space advection: momentum transfer

The change of momentum for a particle in the disperse phase is typically due to body forces and fluid–particle interaction forces. Among body forces, gravity is probably the most important. However, because body forces act on each phase individually, they do not result in momentum transfer between phases. In contrast, fluid–particle forces result in momentum transfer between the continuous phase and the disperse phase. The most important of these are the buoyancy and drag forces, which, for reasons that will become clearer below, must be defined in a consistent manner. However, as detailed in the work of Maxey & Riley (1983), additional forces affect the motion of a particle in the disperse phase, such as the added-mass or virtual-mass force (Auton *et al.*, 1988), the Saffman lift force (Saffman, 1965), the Basset history term, and the Brownian and thermophoretic forces.[20] All these forces will be discussed in the following sections, and the equations for their quantification will be presented and discussed.

The mesoscale models for momentum transfer between phases differ quite substantially depending on the multiphase system under investigation, and different semi-empirical relationships have been developed for different systems. Since the nature of the disperse phase is particularly important, the available mesoscale models are generally divided into those valid for fluid–fluid and those valid for fluid–solid systems. The main difference is that in fluid–fluid systems the elements of the disperse phase are deformable particles (i.e. bubbles or droplets), whereas in fluid–solid systems the disperse phase is constituted by particles of constant shape. Typical fluid–fluid systems for which the mesoscale models reported below apply are gas–liquid, liquid–liquid, and liquid–gas systems. The mesoscale models reported for fluid–solid systems are valid both for gas–solid and for liquid–solid systems. As a general rule, the mesoscale model for $\mathbf{A}_\mathrm{fp}$ should be derived starting from a single-particle momentum balance:

$$\frac{\mathrm{d}M_\mathrm{p}\mathbf{U}_\mathrm{p}^*}{\mathrm{d}t} = \sum \{\text{Forces acting on particle}\} + \{\text{Momentum change due to mass transfer}\},$$

[20] It is worth mentioning that Brownian and thermophoretic forces are often neglected as actual forces, with just their effect on particle motion being quantified. For example, since the Brownian force is a random force that induces a random motion of the particle, it is perceived as a diffusion process. For this reason, this phenomenon is often accounted for as a physical-space diffusion process and details will be given in the next sections. Analogously, since the thermophoretic force results in a next flux of particles in real space driven by temperature gradients, it is often represented as a real-space advection process. Also for this term additional details are reported in the next sections.

where M_p is the particle mass. However, in the following we shall assume that the particle mass is constant and that there is no net momentum exchange due to mass transfer.[21] Even in this case, where we can write $M_p \dot{\mathbf{U}}_p^* = \sum \{\text{Forces acting on particle}\}$, care must be taken when defining the single-particle model for $\mathbf{A}_{fp}$. This is because the forces on the right-hand side (e.g. the virtual-mass and history forces) can involve the time derivative $\dot{\mathbf{U}}_p^*$, and thus the momentum balance may be an implicit function of $\dot{\mathbf{U}}_p^*$. However, it is often possible to express the forces on the right-hand side as at most a linear function of $\dot{\mathbf{U}}_p^*$, in which case it is possible to write the single-particle momentum balance in an explicit form as $\dot{\mathbf{U}}_p^* = $ Function of $\mathbf{U}_p^*(t)$ and other mesoscale variables, which is the form implied by Eq. (5.9). In the following, we discuss the individual forces appearing in the single-particle momentum balance separately. Because most of these forces are explicit functions of $\mathbf{U}_p^*(t)$, they lead directly (in the absence of other forces) to a model for $\mathbf{A}_{fp}$. Nonetheless, the reader should keep in mind that when the virtual-mass force (which depends on $\dot{\mathbf{U}}_p^*$) is included, it will be necessary to modify $\mathbf{A}_{fp}$ to include the added-mass factor.

5.3.1 Buoyancy and drag forces

Among the many relevant fluid–particle interaction forces that must be accounted for in disperse multiphase systems, the most important is produced by the stresses of the continuous phase on the particle surface, resulting in the drag and buoyancy forces. In the case of an isolated particle of spherical shape flowing under laminar conditions characterized by very small slip velocities,[22] the problem has an analytical solution, resulting in the following expression for the buoyancy and drag forces acting on a sphere of diameter d_p:

$$\mathbf{F}_{fp} = -\frac{\pi d_p^3}{6} \left.\frac{\partial p_f}{\partial \mathbf{x}}\right|_{\mathbf{X}^*} + \pi \mu_f d_p (\mathbf{U}_f^* - \mathbf{U}_p^*) + 2\pi \mu_f d_p (\mathbf{U}_f^* - \mathbf{U}_p^*), \tag{5.51}$$

where the first and second terms are caused by the normal component of the stress tensor acting on the particle surface whereas the last one is caused by the tangential component. The first term is the buoyancy force; it is equal to the weight of the fluid displaced by the particle when the fluid and particle have the same velocity (i.e. null slip velocity, or $\mathbf{U}_f^* = \mathbf{U}_p^*$). In the buoyancy force, the fluid pressure gradient is evaluated at the particle location $\mathbf{X}^*(t)$. The second and third terms are instead called the form and friction drag, respectively. These two terms are often grouped together into the kinetic drag force, resulting in the following expression:

$$\mathbf{F}_{fp} = -\frac{\pi d_p^3}{6} \left.\frac{\partial p_f}{\partial \mathbf{x}}\right|_{\mathbf{X}^*} + \frac{\pi d_p^2}{4} \frac{24 \nu_f}{|\mathbf{U}_f^* - \mathbf{U}_p^*| d_p} \frac{1}{2} \rho_f |\mathbf{U}_f^* - \mathbf{U}_p^*| \left(\mathbf{U}_f^* - \mathbf{U}_p^*\right), \tag{5.52}$$

which is valid, as has been mentioned, only for an isolated sphere characterized by a very small slip velocity. The dimensionless term $24\nu_f/(|\mathbf{U}_f^* - \mathbf{U}_p^*| d_p)$ in Eq. (5.52) represents the friction factor and is often referred to as the drag coefficient and denoted by C_D. Note that, if the particle volume V_p and cross-section surface area exposed to the flow A_D are different than those for a sphere, we can generalize the drag force as

[21] A constant particle mass does not imply zero net momentum change due to mass transfer. For example, the particle could be losing and gaining mass at the same rates so that $dM_p/dt = 0$, but the momentum of the lost and gained masses need not be the same due to differences in their velocities.

[22] This condition of very slow flow under which inertial terms are unimportant is also known as creeping flow or Stokes flow.

$$\mathbf{F}_{\text{fp}} = -V_{\text{p}} \left.\frac{\partial p_{\text{f}}}{\partial \mathbf{x}}\right|_{\mathbf{X}^*} + C_{\text{D}} A_{\text{D}} \frac{1}{2} \rho_{\text{f}} |\mathbf{U}_{\text{f}}^* - \mathbf{U}_{\text{p}}^*| \left(\mathbf{U}_{\text{f}}^* - \mathbf{U}_{\text{p}}^*\right). \tag{5.53}$$

However, it would then be necessary to relate V_{p} and A_{D} to the internal coordinates used to describe the particle size (see Section 5.2.1). By dividing Eq. (5.53) by the particle mass ($\rho_{\text{p}} V_{\text{p}}$), the particle acceleration due to buoyancy and drag is readily calculated, and by assuming that all the particles are statistically identical the following expression for the pure advection velocity is obtained for an isolated sphere ($\alpha_{\text{f}} = 1$):

$$\mathbf{A}_{\text{fp}} = -\frac{1}{\rho_{\text{p}}} \left.\frac{\partial p_{\text{f}}}{\partial \mathbf{x}}\right|_{\mathbf{X}^*} + \frac{C_{\text{D}} A_{\text{D}}}{2 V_{\text{p}}} \frac{\rho_{\text{f}}}{\rho_{\text{p}}} |\mathbf{U}_{\text{f}}^* - \mathbf{U}_{\text{p}}^*| (\mathbf{U}_{\text{f}}^* - \mathbf{U}_{\text{p}}^*), \tag{5.54}$$

where $A_{\text{D}}/V_{\text{p}} = 3/(2d_{\text{p}})$ for spheres. It is clear that in the case of creeping (or Stokes) flow, when Eq. (5.52) is valid, the following definition for the drag coefficient holds:

$$C_{\text{D}} = \frac{24 \nu_{\text{f}}}{|\mathbf{U}_{\text{f}}^* - \mathbf{U}_{\text{p}}^*| d_{\text{p}}} = \frac{24}{\text{Re}_{\text{p}}}, \tag{5.55}$$

and it is important to note that C_{D} depends on the mesoscale velocities $\mathbf{U}_{\text{p}}^*$ and $\mathbf{U}_{\text{f}}^*$, as well as on the internal coordinate d_{p} for polydisperse particles. To avoid cumbersome notation, we do not denote the explicit dependence of $\mathbf{A}_{\text{fp}}$ on all of the internal coordinates. Nonetheless, the reader should keep in mind that any of the particle/fluid parameters appearing in Eq. (5.54) could potentially be different for different particles. When applying the single-particle model in Eq. (5.54) to a polydisperse system, the main challenge is that of finding an accurate expression for the drag coefficient that takes into account the polydispersity of the particle properties (i.e. shape, size, density, etc.).

The expression in Eq. (5.55) is of course valid only for a sphere characterized by very small slip velocity, and it is generally assumed to be valid for $\text{Re}_{\text{p}} \leq 0.1$. For spherical particles at higher particle Reynolds numbers, the following corrections can be used:

$$C_{\text{D}} = \begin{cases} [(24/\text{Re}_{\text{p}})^{1/2} + 0.5407]^2 & \text{for } 0.1 < \text{Re}_{\text{p}} < 6000, \\ 0.445 & \text{for } 6000 < \text{Re}_{\text{p}} < 10^5. \end{cases} \tag{5.56}$$

These corrections take into account the fact that the flow of the primary continuous phase around the particle is not slow enough to satisfy the hypothesis used to derive the Stokes drag law and that at a certain point the flow begins to separate and vortices are formed behind the sphere. The presence of the vortices reduces the wake and increases the form drag. In the flat portion of the C_{D} versus Re_{p} curve, the drag force is almost entirely due to the form drag. In this region, known as the region of applicability of Newton's drag law, the drag coefficient can be approximated by a constant value of 0.445. On further increasing the particle Reynolds number, the boundary layer becomes turbulent and the separation point moves rearward, sharply reducing the drag coefficient. This transition occurs at the critical Reynolds number, which isusually assumed to be equal to 3.5×10^5. The flow characterized by higher Re_{p} is generally labeled as supercritical flow, and the drag coefficient drops sharply from 0.445 to 0.07. Another very simple correlation similar to that reported in Eq. (5.56) is the drag model of Schiller & Naumann (1935):

$$C_{\text{D}} = \begin{cases} (24/\text{Re}_{\text{p}})\left(1 + 0.15\text{Re}_{\text{p}}^{0.687}\right) & \text{for } \text{Re}_{\text{p}} \leq 1000, \\ 0.445 & \text{for } \text{Re}_{\text{p}} > 1000, \end{cases} \tag{5.57}$$

where the particle Reynolds number is defined as usual. An alternative, more detailed, description of the variation of the drag coefficient over the particle Reynolds number is the one proposed by Morsi & Alexander (1972):

$$C_{\mathrm{D}} = a_1 + \frac{a_2}{\mathrm{Re}_{\mathrm{p}}} + \frac{a_3}{\mathrm{Re}_{\mathrm{p}}^2}, \tag{5.58}$$

where

$$(a_1, a_2, a_3) = \begin{cases} (0, 24, 0) & \text{for } \mathrm{Re}_{\mathrm{p}} \leq 0.1, \\ (3.69, 22.73, 0.0903) & \text{for } 0.1 < \mathrm{Re}_{\mathrm{p}} \leq 1, \\ (1.22, 29.17, -3.89) & \text{for } 1 < \mathrm{Re}_{\mathrm{p}} \leq 10, \\ (0.617, 46.50, -116.67) & \text{for } 10 < \mathrm{Re}_{\mathrm{p}} \leq 100, \\ (0.364, 98.33, -2778) & \text{for } 100 < \mathrm{Re}_{\mathrm{p}} \leq 1000, \\ (0.357, 148.62, -47\,500) & \text{for } 1000 < \mathrm{Re}_{\mathrm{p}} \leq 5000, \\ (0.46, -490.55, 578\,700) & \text{for } 5000 < \mathrm{Re}_{\mathrm{p}} \leq 10\,000, \\ (0.5191, -1662.5, 5\,416\,700) & \text{for } \mathrm{Re}_{\mathrm{p}} > 10\,000. \end{cases} \tag{5.59}$$

These equations are valid for isolated spherical particles when the surrounding continuous phase can be treated as a continuum (as opposed to a rarefied gas). In fact, it is important to quantify the ratio between the mean free path (i.e. average time interval between two subsequent collisions) of the molecules constituting the primary phase and the particle size:

$$\mathrm{Kn} = \frac{\lambda_{\mathrm{f}}}{d_{\mathrm{p}}}. \tag{5.60}$$

When this quantity is much smaller than unity, the continuum approach for describing the interactions between the primary phase and the elements of the secondary phase is appropriate. When this ratio is bigger than unity, the interactions between the primary phase and the elements of the secondary phase must be described in terms of individual molecules impacting onto and rebounding from the particle surface; this is the so-called free-molecular regime. For intermediate values the interactions between the primary and secondary phases are in the slip regime, where the continuum approach for the primary phase can still be used, albeit with some corrections to account for the velocity jump between the fluid adjacent to the surface of the particle. This correction results in the following expression for the drag coefficient:

$$\frac{C_{\mathrm{D}}^*}{C_{\mathrm{D}}} = \frac{1}{1 + \mathrm{Kn}\,[2.49 + 0.84\exp(-1.74/\mathrm{Kn})]}, \tag{5.61}$$

which is commonly referred to as the Cunningham correction factor (Cunningham, 1910; Knudsen & Weber, 1911; Millikan, 1910). We refer to the more recent work of Crowe (2006) for a review of the newest corrections.[23] Additional corrections have been derived

[23] It is important to remind readers that non-continuum effects are related to the compressibility of the continuous phase. In fact, for an ideal gas, kinetic theory predicts that

$$\mathrm{Kn} = \sqrt{\frac{\pi k}{2}}\frac{\mathrm{Ma}}{\mathrm{Re}_{\mathrm{p}}}, \tag{5.62}$$

where Ma is the Mach number for the continuous phase (i.e. the ratio between the slip velocity to the speed of sound) and k is the ratio between the specific heats of the continuous phase. The corrected drag coefficient C_{D}^* is therefore a function of two of these three dimensionless numbers, e.g. Kn and Re_{p}. For practical purposes any isothermal flow with $\mathrm{Ma} < 0.2$ is considered to be incompressible and if $\mathrm{Kn} < 10^{-3}$ the continuous phase is considered to be a continuum.

to account for the shape of the particle, the presence of other particles ($\alpha_f < 1$), and wall effects, and differ substantially depending on the type of multiphase system (i.e. fluid–solid, fluid–fluid, etc.). Thus, next we discuss correlations for the drag coefficient for some selected example systems.

The drag coefficient for fluid–fluid systems

The calculation of the drag coefficient, C_D, in fluid–fluid systems is complicated by a number of factors. The first one is that in fluid–fluid systems it is very common that due to the accumulation of fluid elements in one region coalescence occurs, resulting in very high (close to unity) volume fraction values for the disperse phase.[24] Another issue is related to the fact that, being deformable, depending on their size, fluid particles can assume different shapes, resulting in different types of interactions. In what follows the discussion will focus on gas–liquid systems, but with some caution the conclusions can be applied also to liquid–liquid and liquid–gas systems. In the case of gas bubbles dispersed in a continuous liquid phase, a spherical shape is assumed when the bubble size is very small or the surface tension between the gas and liquid is very high. In this case, bubble motion is more or less rectilinear and the drag force is controlled by viscous forces and therefore depends on the bubble Reynolds number. On the other hand, bubbles take non-spherical shapes when the bubble size is large or the surface tension is low. In this case bubble motion is no longer rectilinear and the drag force is controlled by surface-tension forces; it no longer depends on Re_p, but rather on the Eötvös number.

Another important element is related to the presence of impurities that accumulate on the gas–liquid interface. In fact, experimental evidence (collected on the basis of the resulting terminal velocity) shows that some of these impurities do not significantly change macroscopic properties, such as the density or viscosity of primary or disperse phases or the interface energy between the two phases, but their presence results in important changes in the drag coefficient. The latter factor is particularly important for spherical gas bubbles and is quite difficult to take into account. With respect to impurities, gas–liquid systems can be roughly divided into three categories: (i) pure systems, (ii) slightly contaminated systems, and (iii) fully contaminated systems. Although the distinction might seem ambiguous, it is very much used in practice. For example, a pure air–water system could correspond to water carefully distilled two or more times (i.e. bi-distilled water), a contaminated air–water system could correspond to bubbles moving in tap water, and a partially contaminated air–water system could correspond to intermediate levels of contamination

[24] It is important to remind the reader that the multiphase systems treated in this book are disperse multiphase systems, meaning that together with the primary continuous phase there is a secondary phase constituted by particles, droplets, or bubbles. The situation depicted here (with volume-fraction values close to unity for the disperse phase) is not included in this category; in fact, if all the particles of the disperse phase coalesce together the disperse phase is no longer dispersed. This situation is generally reflected by the values of the moments and generally labeled as gelling or phase inversion. During gelling the moment of order zero of the NDF goes to zero, whereas the other moments diverge to non-physical values. One practical way to overcome this limitation is to freeze the evolution of the moments to their pre-gelling values in those regions where the disperse-phase volume fraction becomes very high; moreover, in order to bridge these regions to the others where the disperse phase evolves as expected, corrections to the drag coefficient can be conveniently adopted. As has already been mentioned, these situations are quite common in fluid–fluid systems and can be encountered for example in gas–liquid stirred tanks in low-pressure regions (behind the stirrer blades or the baffles) or whenever phase inversion occurs (for example at the interface between gas and liquid). Of course, in the case of non-coalescing systems, such as in the case of solid particles that do not aggregate together, this situation is avoided by introducing a force that keeps particles apart. This is generally done by introducing a maximum packing limit for the disperse-phase volume fraction and a force proportional to the difference between the actual disperse-phase volume fraction and this limit.

(Tomiyama *et al.*, 1998). The fully contaminated bubble behaves as a solid particle; in fact, the accumulation of impurities at the gas–liquid interface makes the interface rigid. In contrast, in pure systems the absence of impurities allows the formation of a recirculation pattern inside the bubble, which decreases the viscous drag. For pure systems and for an isolated bubble, Hadamard (1911) proposed the following relationship:

$$C_\mathrm{D} = \frac{16}{\mathrm{Re_p}}, \tag{5.63}$$

which is valid only for very small $\mathrm{Re_p}$. If Eq. (5.63) is compared with Eq. (5.55), which was developed for solid particles, it is clear that it results in smaller values of the drag force. The expression contained in Eq. (5.63) is valid for small $\mathrm{Re_p}$, whereas for higher $\mathrm{Re_p}$ the correction developed by linearizing the convective terms in the Navier–Stokes equation must be applied, resulting in the following expression:

$$C_\mathrm{D} = \frac{48}{\mathrm{Re_p}}, \tag{5.64}$$

and, using an approach very similar to that of Eq. (5.56), the following expression with a more general validity can be used:

$$C_\mathrm{D} = \min\left[\frac{16}{\mathrm{Re_p}}\left(1 + 0.15\mathrm{Re_p^{0.687}}\right), \frac{48}{\mathrm{Re_p}}\right]. \tag{5.65}$$

A similar expression has been developed for spherical isolated bubbles in contaminated systems:[25]

$$C_\mathrm{D} = \min\left[\frac{24}{\mathrm{Re_p}}\left(1 + 0.15\mathrm{Re_p^{0.687}}\right), \frac{72}{\mathrm{Re_p}}\right]. \tag{5.66}$$

In the case of larger non-spherical bubbles the drag coefficient no longer depends on the level of contamination and on $\mathrm{Re_p}$, but only on the Eötvös number. One commonly adopted dependence is

$$C_\mathrm{D} = \frac{8}{3}\left(\frac{\mathrm{Eo}}{\mathrm{Eo} + 4}\right). \tag{5.67}$$

When these different drag coefficients are grouped together, they are often referred to as Tomiyama's drag law (Tomiyama, 2004). In summary, for clean systems the following expression is obtained:

$$C_\mathrm{D} = \max\left\{\min\left[\frac{16}{\mathrm{Re_p}}(1 + 0.15\mathrm{Re_p^{0.687}}), \frac{48}{\mathrm{Re_p}}\right], \frac{8}{3}\left(\frac{\mathrm{Eo}}{\mathrm{Eo} + 4}\right)\right\}. \tag{5.68}$$

The correlation used for slightly contaminated systems is

$$C_\mathrm{D} = \max\left\{\min\left[\frac{24}{\mathrm{Re_p}}(1 + 0.15\mathrm{Re_p^{0.687}}), \frac{72}{\mathrm{Re_p}}\right], \frac{8}{3}\left(\frac{\mathrm{Eo}}{\mathrm{Eo} + 4}\right)\right\}, \tag{5.69}$$

whereas for fully contaminated systems the following correlation is employed:

$$C_\mathrm{D} = \max\left[\frac{24}{\mathrm{Re_p}}(1 + 0.15\mathrm{Re_p^{0.687}}), \frac{8}{3}\left(\frac{\mathrm{Eo}}{\mathrm{Eo} + 4}\right)\right]. \tag{5.70}$$

[25] As just mentioned, impurities and surfactants tend to accumulate at the interface, making it rigid and therefore damping out internal motion. For this reason the drag coefficient of a contaminated system for fluid particles is the same as that for solid particles.

These relationships are valid for isolated bubbles moving under laminar flow conditions. In the case of turbulent flow, the effect of turbulent eddies impinging on the bubble surface is to increase the drag forces. This is typically accounted for by introducing an effective fluid viscosity (rather than the molecular viscosity of the continuous phase, μ_{f}) defined as $\mu_{\mathrm{f,eff}} = \mu_{\mathrm{f}} + C_{\mathrm{l}}\rho_{\mathrm{f}}\varepsilon_{\mathrm{f}}^{1/3}d_{\mathrm{p}}^{4/3}$, where ε_{f} is the turbulence-dissipation rate in the fluid phase and C_{l} is a constant that is usually taken equal to 0.02. This effective viscosity, which is used for the calculation of the bubble/particle Reynolds number (Bakker & van den Akker, 1994), accounts for the turbulent reduction of slip due to the increased momentum transport around the bubble, which is in turn related to the ratio of bubble size and turbulence length scale. However, the reader is reminded that the mesoscale model does not include macroscale turbulence and, hence, using an effective viscosity that is based on the macroscale turbulence is not appropriate.

All the correlations mentioned previously work quite well in dilute systems when each bubble can be considered as isolated. However, when the bubbles start perceiving the presence of one another[26] then none of the correlations presented will result in acceptable predictions. Global performances are generally evaluated in terms of the drag law's ability to predict the overall concentration of the disperse phase (e.g. global gas hold-up) in realistic vessels or to predict transition from one regime to another[27] as reported by Montante *et al.* (2007) and Petitti *et al.* (2009).

A purely empirical approach to define a drag coefficient capable of reproducing experimentally measured behavior is to resort to the concept of terminal velocity. Let us imagine a single particle on which only gravity, buoyancy, and drag forces act. The particle moves (driven solely by gravity and buoyancy) with the continuous phase at rest, and after a certain time a steady state is reached. The constant velocity with which the particle moves when the steady state is reached is called the terminal velocity and will be denoted here by U_∞. As is well known, on setting $\mathbf{A}_{\mathrm{fp}} = 0$ in Eq. (5.54) and considering the continuous phase to be at rest (i.e. $\mathbf{U}_{\mathrm{f}}^* = 0$), the velocity difference between the disperse and continuous phases is the terminal velocity, namely $U_\infty = |\mathbf{U}_{\mathrm{p}}^* - \mathbf{U}_{\mathrm{f}}^*|$. This results in the following approximation for the drag coefficient:

$$C_{\mathrm{D}} = \frac{4d_{\mathrm{p}}(\rho_{\mathrm{f}} - \rho_{\mathrm{p}})g}{3\rho_{\mathrm{f}}U_\infty^2}. \tag{5.71}$$

Equation (5.71) allows us to calculate the drag coefficient at steady state for a specific terminal velocity. This formulation is particularly useful because it suggests a strategy for calculating C_{D} that can easily be applied to very different fluid–fluid systems. In fact, given the particular system under investigation, we can identify the terminal velocity and

[26] This situation is particularly important in fluid–fluid systems where the elements of the disperse phase start interacting before entering into physical contact. When the total particle number density (and the disperse-phase volume fraction) is large enough to allow this to happen, the regime is labeled as three-way coupling. According to this definition, when the disperse phase is so dilute that it does not affect the primary continuous phase, the system is under one-way coupling conditions: the continuous phase affects the disperse phase but not vice versa. In contrast, when the secondary-phase concentration increases and some effects on the continuous phase are detected, the system is said to be under two-way coupling conditions. When the elements of the disperse phase reach a concentration high enough to allow interactions through the primary continuous phase (e.g. particles/bubbles crossing each others' wakes) then the system is defined as being under three-way coupling conditions. Eventually, when the concentration of the disperse phase is such that interactions among elements of the disperse phase are relevant, the system is said to be under four-way coupling conditions.

[27] Typical examples refer to gas–liquid stirred tanks, where, according to the gas distribution in the vessel, three regimes are detected, namely complete dispersion, loading, and flooding. Other important regime transitions in gas–liquid stirred tanks refer to the number and shape of the gas cavities behind the stirrer's blades, namely vortex-clinging cavities, large cavities, and ragged cavities.

then use this information to compute the drag coefficient, ensuring a tighter control on the predicted bubble hold-up in the multiphase system. For example, it is quite easy to quantify the changes in terminal velocity due to the presence of other bubbles or due to any other effects (e.g. the presence of fine or ultra-fine particles). Experimental evidence shows that, for example, if an isolated bubble moves with a specific terminal velocity, the presence of other bubbles decreases the terminal velocity, highlighting a sort of hindrance or crowding effect that is related to a larger drag coefficient. An important shortcoming of using an empirical drag coefficient is that it is valid only for a homogeneous system wherein the disperse-phase volume fraction α_p is spatially uniform. If the true drag coefficient is strongly dependent on the local disperse-phase volume fraction, and the system is spatially non-uniform, Eq. (5.71) will not capture the correct dependence on the disperse-phase volume fraction. Ideally, C_D could be measured at different disperse-phase volume fractions and a correlation could be developed to correctly capture the dependence of C_D on α_p. Although the empirical approach is trivial to implement and not very sophisticated, it can be quite effective in practice because the terminal velocity of the bubbles can be determined in many different ways. For very popular systems, such as the classical clean air–water system, the air–bubble terminal velocity is given by the correlation proposed by Mendelson (1967) and its extensions:

$$U_\infty = \begin{cases} (1/18)(d_p^2/\mu_f)(\rho_f - \rho_p)g & \text{for } d_p < 0.8 \text{ mm}, \\ (1/18)[(0.8\times 10^{-3})^2/\mu_f](\rho_f - \rho_p)g & \text{for } 0.8 \le d_p < 1.3 \text{ mm}, \\ [2.14\sigma/(\rho_f d_p) + 0.505 g d_p]^{1/2} & \text{for } 1.3 \le d_p < 18.8 \text{ mm}, \\ 0.71\left[g d_p(\rho_f - \rho_p)/\rho_f\right]^{1/2} & \text{for } d_p > 18.8 \text{ mm}. \end{cases} \tag{5.72}$$

This is of course valid only for isolated bubbles moving in clean water, and corrections (i.e. reductions) of the terminal velocity are applied to account for the presence of the other bubbles or partial contamination. Alternatively the terminal velocity can be determined experimentally or by resorting to detailed simulations. Simulations for the determination of the terminal velocity of single bubbles or groups (swarms) of bubbles can be based on the volume-of-fluid (VOF) method or similar methods (Bertola *et al.*, 2004; Petitti *et al.*, 2007).

The drag coefficient for fluid–solid systems

As has already been reported, the case of solid particles can be treated with a simpler approach than that for fluid–fluid systems. One of the most popular correlations for the calculation of the drag coefficient for solid particles suspended in a fluid is that proposed by Wen & Yu (1966):

$$C_D = \frac{24}{\mathrm{Re}_p}\left[1 + 0.15(\alpha_f \mathrm{Re})^{0.687}\right]\alpha_f^{-\beta}, \tag{5.73}$$

where $\beta = 3.65$ is a correction factor to account for the presence of other particles. In this case, when the particulate system is very dilute (i.e. $\alpha_f \approx 1$), Eq. (5.73) reduces to Eq. (5.57). A similar expression was proposed by Di Felice (1994), where the exponent is now a function of the particle Reynolds number:

$$\beta = 3.65 - 0.65 \exp\left[-\frac{(1.5 - \log \mathrm{Re}_p)^2}{2}\right]. \tag{5.74}$$

When the concentration of the particles increases beyond a certain critical value Eq. (5.73) might become inaccurate. In these situations (which are common in dense almost-packed fluid–solid systems) the drag force is calculated by using the Ergun equation, which estimates the pressure drops experienced by a fluid in flowing through a packed bed of spheres:

$$C_D = \frac{1}{\alpha_f}\left[\frac{200}{\mathrm{Re_p}}\frac{1-\alpha_f}{\alpha_f} + \frac{7}{3}\right]. \tag{5.75}$$

In practice, a combination of Eqs. (5.73) and (5.75) is employed to cover all values of α_f. Another widely used drag model is the one proposed by Syamlal & O'Brien (1989):

$$C_D = \frac{1}{U_\infty^2}\left(0.63 + \frac{4.8}{\sqrt{\mathrm{Re_p}/U_\infty}}\right)^2,$$

where the particle terminal velocity is calculated from the correlation proposed by Garside & Al-Dibouni (1977), $U_\infty = 0.5\{A - 0.06\mathrm{Re_p} + [(0.06\mathrm{Re_p})^2 + 0.12\mathrm{Re_p}(2B - A) + A^2]^{1/2}\}$, in which $A = \alpha_f^{4.14}$ and

$$B = \begin{cases} 0.8\alpha_f^{1.28} & \text{for } \alpha_f \le 0.85, \\ \alpha_f^{2.65} & \text{for } \alpha_f > 0.85. \end{cases}$$

Although both expressions are commonly used, they fail to predict some important macroscopic properties of solid–fluid suspensions, such as the expansion and sedimentation profiles. To overcome this limitation Mazzei & Lettieri (2007) developed a relationship for the drag coefficient that is based on the empirical correlation by Richardson & Zaki (1954) describing the expansion profiles of homogeneous fluid–solid suspensions. Its main feature resides in the fact that the expression is consistent with the Richardson and Zaki correlation over the whole range of fluid-dynamic regimes and for any value of the suspension void fraction. It has the following formulation:

$$C_D = \left(0.63 + \frac{4.8}{\sqrt{\mathrm{Re_p}\alpha_f}}\right)^2 \alpha_f^{-\psi-1}, \tag{5.76}$$

where

$$\psi = -\frac{\ln\phi}{\ln\alpha_f}, \tag{5.77}$$

and

$$\phi = \frac{(0.63 + 4.8/\sqrt{\mathrm{Re_p}\alpha_f})^2}{(0.63 + 4.8/\sqrt{\mathrm{Re_p}\alpha_f^{1-n}})^2}\alpha_f^{2(1-n)}, \tag{5.78}$$

in which

$$n = \frac{4.8 + 0.42(\mathrm{Re_p}\alpha_f)^{3/4}}{1 + 0.175(\mathrm{Re_p}\alpha_f)^{3/4}}. \tag{5.79}$$

It is important to note that the fluid–solid drag coefficients discussed above are valid only for monodisperse particles (i.e. particles with equal diameters and material densities). Using direct numerical simulations (DNS) of the microscale equations for fluid–particle flows, several authors (Beetstra *et al.*, 2007; Buhrer-Skinner *et al.*, 2009; Holloway *et al.*, 2010; Tenneti *et al.*, 2010, 2012; Yin & Sundaresan, 2009) have proposed improved drag coefficients to account for polydisperse particles.

Drag-coefficient corrections due to turbulence

As has already been mentioned, corrections are generally implemented to account for the effect of turbulence. It is beyond the scope of this book to give a detailed account of the different treatments and corrections developed in the last few decades (Balachandar & Eaton, 2010). Among the different approaches, it is important to mention fully resolved simulations,[28] Lagrangian point-particle simulations,[29] and the different Eulerian approaches (Prosperetti & Tryggvason, 2009). However, the reader should recall that the mesoscale models discussed in this chapter do not account for the macroscale turbulence. Thus, the correlations discussed below which have been proposed for use with classical turbulence models should be used with caution in the context of mesoscale modeling. In general, two main factors are used to correct the drag coefficient, namely the turbulence intensity and the length-scale ratio between the turbulence length scales and the particle size. When particles are smaller than the Kolmogorov length scale (i.e. $d_\mathrm{p} \ll \eta_\mathrm{k}$) the standard coefficients reported above are generally employed, since the ambient flow appears to the particles as nearly laminar. For larger particles of size comparable to the fluid-phase flow scales (i.e. $d_\mathrm{p} \gtrapprox \eta_\mathrm{k}$) the ambient flow seen by the particle can be more complex. In fact, since in a turbulent flow there is a cascade of length scales, multiple effects are typically present. Intuitively, turbulent structures smaller than the particle size will affect the boundary layer around the particle, whereas those much larger than the particle size will result in particle accumulation, also called inertial clustering. Depending on the approach employed, only the first effect, or both effects, must be accounted for in corrections to the drag coefficient. The presence of turbulence generally decreases the critical particle Reynolds number (Crowe *et al.*, 1998). The critical Reynolds number,[30] $\mathrm{Re}_\mathrm{p}^\mathrm{c}$, is conventionally considered as the number at which the drag coefficient reaches the value of 0.3. Further increases in Re_p cause first a decrease in the drag coefficient, until a minimum in reached, then the drag coefficient starts to increase again. The value of Re_p at which C_D again reaches 0.3 is called the meta-critical Reynolds number, and is often written as $\mathrm{Re}_\mathrm{p}^\mathrm{M}$. The critical particle Reynolds number under the effect of turbulent flow can be calculated according to the following expression (Crowe *et al.*, 1998):

$$\log \mathrm{Re}_\mathrm{p}^\mathrm{c} = \begin{cases} 5.477 - 15.8 I_\mathrm{r} & \text{for } I_\mathrm{r} \leq 0.15, \\ 3.371 - 1.75 I_\mathrm{r} & \text{for } I_\mathrm{r} > 0.15, \end{cases} \tag{5.80}$$

where I_r is the relative turbulence intensity, which is defined as the ratio of the turbulence kinetic energy and the slip-velocity magnitude. A similar expression is valid for the meta-critical particle Reynolds number:

$$\log \mathrm{Re}_\mathrm{p}^\mathrm{M} = \begin{cases} 6.876 - 23.2 I_\mathrm{r} & \text{for } I_\mathrm{r} \leq 0.15, \\ 3.663 - 1.8 I_\mathrm{r} & \text{for } I_\mathrm{r} > 0.15. \end{cases} \tag{5.81}$$

[28] With fully resolved simulations, we generally refer to a detailed description of the fluid–solid interface in which all the scales of ambient turbulence, and the flow scales introduced by the particles, are completely resolved by employing DNS for the fluid phase and Lagrangian tracking for the particles.

[29] With this approach the fluid phase is generally treated with DNS and the disperse phase is described with point particles, whose evolution is generally treated with Lagrangian tracking. In this case, since the fluid–solid interface is not resolved, some form of correction is needed.

[30] When dealing with turbulent flows all the relevant dimensionless numbers are evaluated with the available quantities. For example, in DNS, the fluid and particle instantaneous velocities will be employed, whereas in large-eddy simulation (LES) or in Reynolds-average Navier–Stokes-equations (RANS) simulations the filtered or Reynolds-average values will be used.

The resulting drag coefficient can be calculated as $C_D = 162 I_r^{1/3} \mathrm{Re}_p^{-1}$ for $\mathrm{Re}_p < 50$ and $0.05 < I_r < 0.5$, and as $C_D = 0.133(1 + 150/\mathrm{Re}_p)^{1.565} + 4I_r$ for $50 < \mathrm{Re}_p < 700$ and $0.05 < I_r < 0.5$. For particle Reynolds numbers with intermediate values between Re_p^c and Re_p^M the following expression can be used: $C_D = 0.3(\mathrm{Re}_p/\mathrm{Re}_p^c)^{-3}$ for $0.9\mathrm{Re}_p^c < \mathrm{Re}_p < \mathrm{Re}_p^m$, where Re_p^m can be calculated from the expression $(\mathrm{Re}_p^m)^{(3.45+20I_r)} = (\mathrm{Re}_p^c)^3(\mathrm{Re}_p^M)^{0.45+20I_r}$. For this case, the following expressions are used: for $\mathrm{Re}_p^m < \mathrm{Re}_p < \mathrm{Re}_p^M$, $C_D = 0.3(\mathrm{Re}_p/\mathrm{Re}_p^c)^{0.45+20I_r}$; and for $\mathrm{Re}_p^M < \mathrm{Re}_p < 3 \times 10^4$ and $I_r > 0.07$, $C_D = 399\mathrm{Re}_p^{-6.10} - 4.47 \times 10^5 I_r^{-0.97}\mathrm{Re}_p^{-1.80}$. There is considerable discrepancy among the different data concerning the dependence of the drag coefficient on turbulence, but the correlations developed by Torobin & Gauvin (1960) and Clamen & Gauvin (1969) seem to be the most appropriate. Among the different approaches developed, it is worth mentioning the idea that due to inertial clustering (which is exactly captured by the mesoscale model) a fictitious particle size must be considered in the macroscale model (Xu & Kato, 1999).

To conclude our discussion on drag coefficients, it is important to remind the reader that almost all of the existing correlations have been developed in the context of macroscale models for the average momentum transfer between phases. Thus, when used in the context of a single-particle model, we cannot expect them to fully capture all of the dependences on the mesoscale variables. Nevertheless, recent work using DNS of the microscale model (Beetstra *et al.*, 2007; Buhrer-Skinner *et al.*, 2009; Holloway *et al.*, 2010; Yin & Sundaresan, 2009) holds great promise for improving the available drag correlations. The work of Tenneti *et al.* (2010) on incorporating a stochastic component into the drag model is particularly noteworthy.

5.3.2 Virtual-mass and lift forces

The virtual-mass (or added-mass) force corresponds to the inertia added to the particle because of the volume of surrounding fluid that accelerates or decelerates with the particle as it moves through the fluid. For simplicity, the virtual-mass force is often modeled using the mass of fluid moving with the particle; however, in reality a large fraction of the fluid is accelerated, although at different degrees. Historically, the virtual-mass force was first reported after the investigation of the motion of a pendulum in a fluid (Stokes, 1851). The resulting acceleration of the particle and therefore the rate of change of particle velocity (assuming all particles are statistically equivalent) is quantified by the following force balance (Auton, 1987):

$$\rho_p V_p \frac{d\mathbf{U}_p^*}{dt} = \rho_f C_{vm} V_p \left(\frac{D\mathbf{U}_f^*}{Dt} - \frac{d\mathbf{U}_p^*}{dt} \right), \tag{5.82}$$

where $C_{vm} V_p$ is the volume of fluid accelerated with a particle of volume V_p and $D\mathbf{U}_f^*/Dt$ is the rate of change of the fluid velocity seen by the particle *along the trajectory of a fluid particle*, which we can approximate by

$$\frac{D\mathbf{U}_f^*}{Dt} = \frac{d\mathbf{U}_f^*}{dt} + \left(\mathbf{U}_f^* - \mathbf{U}_p^*\right) \cdot \frac{\partial \mathbf{U}_f}{\partial \mathbf{x}}. \tag{5.83}$$

Invoking conservation of momentum at the mesoscale then leads to the following expression for the fluid velocity seen by the particle:

$$\rho_f V_f \frac{d\mathbf{U}_f^*}{dt} = -\rho_f C_{vm} V_p \left(\frac{D\mathbf{U}_f^*}{Dt} - \frac{d\mathbf{U}_p^*}{dt} \right), \tag{5.84}$$

where V_f is the fluid volume seen by the particle (which is not the same as $C_{vm}V_p$). As discussed in Section 4.3.4 (where we denote V_f by ξ_{f1}), the fluid-phase volume fraction is related to V_f by Eq. (4.54) on page 117, as is the mean fluid velocity $\mathbf{U}_f$ through Eq. (4.81) on page 122. In the monokinetic-fluid limit, we can approximate $\mathrm{D}\mathbf{U}_f^*/\mathrm{D}t$ by the rate of change of the mean fluid velocity $\mathbf{U}_f(t, \mathbf{x})$ along the trajectory of a *fluid* particle:

$$\frac{\mathrm{D}\mathbf{U}_f}{\mathrm{D}t} = \left(\frac{\partial \mathbf{U}_f}{\partial t} + \mathbf{U}_f \cdot \frac{\partial \mathbf{U}_f}{\partial \mathbf{x}} \right)_{\mathbf{x}=\mathbf{X}^*(t)}. \tag{5.85}$$

Collecting the terms in Eq. (5.82) involving the time derivative of $\mathbf{U}_p^*$ leads to the mesoscale model for the virtual-mass force (in the absence of other forces[31]) in the monokinetic-fluid limit:

$$\mathbf{A}_{fp} = \left(\frac{C_{vm}\rho_f}{\rho_p + C_{vm}\rho_f} \right) \frac{\mathrm{D}\mathbf{U}_f}{\mathrm{D}t}, \tag{5.86}$$

where C_{vm} is the virtual-mass coefficient, which is generally shape-dependent (i.e. it depends on the internal coordinates describing the shape of the particle), and often includes a dependence on the fluid-phase volume fraction α_f (Auton, 1987). The most common value is 0.5, as proposed by Drew *et al.* (1979) for rigid spherical particles, whereas 0.25 was employed for bubbles in water (Cook & Harlow, 1986). The coefficient was also found to vary with the local concentration of elements of the disperse phase (Biesheuvel & Spoelstra, 1989; Homsy *et al.*, 1980). For example, Sankaranarayanan *et al.* (2002) developed a model for the virtual-mass coefficient and observed that it increases nearly linearly with increasing volume fraction of the disperse phase. As is obvious from Eq. (5.86), the virtual-mass force is particularly important when the density of the continuous phase is higher than that of the disperse phase (i.e. $\rho_f \gg \rho_p$), and gas–liquid systems are the classical example. Among gas–liquid systems, the virtual-mass force was found to be very important in correctly predicting regime transitions and fluid-dynamic behavior especially in bubble columns (see, for example, the work of Monahan & Fox (2007) and Monahan *et al.* (2005)). In the case of gas–liquid stirred tanks drag seems to be the dominant force.

Particles moving in a fluid with mean shear experience a lift force perpendicular to the direction of fluid flow. The shear lift originates from inertia effects in the viscous flow around the particle and depends on the mean vorticity of the fluid phase[32] evaluated at the particle location $\mathbf{x} = \mathbf{X}^*(t)$. For a spherical particle, the particle acceleration due to the lift force (also known as the Saffman lift force) is equal to (Auton, 1987; Drew & Lahey, 1993; Drew & Passman, 1999; Saffman, 1965)

$$\mathbf{A}_{fp} = C_L \frac{\rho_f}{\rho_p} \left(\mathbf{U}_f^* - \mathbf{U}_p^* \right) \times (\nabla \times \mathbf{U}_f), \tag{5.87}$$

where C_L is the lift coefficient that can be derived from the theory (Maxey & Riley, 1983). Conflicting conclusions concerning the importance and the significance of the lift force can be found in the literature. For example, Delnoij *et al.* (1997) reported that the lift

[31] This is a very important point because the added-mass term will modify the model for $\mathbf{A}_{fp}$ when other forces are included. In fact, for the general formulation, one should start with the single-particle momentum balance in Eq. (5.82) and add the other forces on the right-hand side. The final mesoscale model for $\mathbf{A}_{fp}$ will have all of the terms on the right-hand side multiplied by the added-mass factor $C_{vm}\rho_f/(\rho_p + C_{vm}\rho_f)$. In other words, due to the added mass, $\mathbf{A}_{fp}$ cannot be found by simply adding together the models for the individual forces. See Section 5.3.4 for more details.

[32] The mean vorticity of the fluid phase is the curl of the mean fluid velocity field, $\omega_c = \nabla \times \mathbf{U}_f$, and is a measure of the mean rotation of the continuous phase.

force seems to be necessary for the accurate representation of flow behavior in gas–liquid systems and the work of Monahan *et al.* (2005) demonstrated the importance of the lift force to the stability of homogeneous flow predictions when using the Euler–Euler two-fluid model. Auton (1987) showed that C_L should be taken equal to 0.5 to ensure good agreement with experiments. The same value has been reported elsewhere (Deen *et al.*, 2001; Lain *et al.*, 2002), whereas Drew & Passman (1999) suggested using $C_L = 0.25$. Tomiyama *et al.* (2002b) determined an empirical correlation for the net transverse lift coefficient that is valid for gas–liquid systems. For $d_p < 4.4$ mm, C_L was found to be a function of Re_p, whereas for $d_p > 4.4$ mm, C_L was found to be a function of the Eötvös number. Also, the sign of C_L changed from positive to negative when $d_p = 5.8$ mm. Other similar correlations that are based on the capillary number have been proposed (Sankaranarayanan & Sundaresan, 2002), showing that for the majority of the applications C_L should be positive.

Summarizing the forces introduced above, tests carried out in different multiphase systems have shown that the order of importance of the different forces involved typically ranks buoyancy and drag in the first positions[33] and then lift and virtual-mass forces for fluid–solid systems and virtual-mass and lift forces for fluid–fluid systems (see, for example, the studies on non-drag forces by Diaz *et al.* (2008) and Barton (1995)), whereas the most common values for the corresponding constants are $C_L = 0.25$ and $C_{vm} = 0.5$ both for fluid–fluid and for fluid–solid systems. Naturally, since it is straightforward to implement all the forces in a computational code (Vikas *et al.*, 2011b), it is best to include them all for the sake of generality.

5.3.3 Boussinesq–Basset, Brownian, and thermophoretic forces

As has already been mentioned, the acceleration of a body moving in a fluid generates a force referred to as the virtual-mass force. The virtual-mass force is due to the instantaneous acceleration imparted to the fluid surrounding the particle. However, particle acceleration generates additional effects related to the transient development of the boundary layer around the particle and its wake. These additional effects, which were first identified in an accelerating Stokes flow by Boussinesq (1885) and Basset (1888), are generally quantified into a single force that is sometimes referred to as the Boussinesq–Basset or history force (Crowe *et al.*, 1998). This term describes the force due to the lagging boundary-layer development with acceleration of particles moving through a fluid. In other words, it accounts for unsteady viscous drag effects due to the temporal development of the particle boundary layer and wake, and therefore it addresses the temporal delay in boundary-layer development as the relative fluid–particle velocity changes with time. The Boussinesq–Basset force (in terms of the corresponding acceleration) can be written as

$$\frac{\mathrm{d}\mathbf{U}_p^*}{\mathrm{d}t} = \frac{9}{d_p}\frac{\rho_f}{\rho_p}\sqrt{\frac{\nu_f}{\pi}}\int_{-\infty}^{t}\frac{1}{\sqrt{t-\tau}}\left(\frac{\mathrm{D}\mathbf{U}_f^*}{\mathrm{D}\tau}-\frac{\mathrm{d}\mathbf{U}_p^*}{\mathrm{d}\tau}\right)\mathrm{d}\tau. \tag{5.88}$$

This expression is often rewritten in order to highlight the presence of a history-force kernel:

$$\frac{\mathrm{d}\mathbf{U}_p^*}{\mathrm{d}t} = \frac{18\nu_f}{d_p}\frac{\rho_f}{\rho_p}\int_{-\infty}^{t}K_B(t-\tau)\left(\frac{\mathrm{D}\mathbf{U}_f^*}{\mathrm{D}\tau}-\frac{\mathrm{d}\mathbf{U}_p^*}{\mathrm{d}\tau}\right)\mathrm{d}\tau, \tag{5.89}$$

[33] For gas–particle systems where $\rho_f \ll \rho_p$, only the drag force is important, since all other forces are proportional to the ratio ρ_f/ρ_p.

where the history-force kernel is $K_B(s) = d_p/(4\pi \nu_f s)^{1/2}$ and represents a sort of filter that weights the history of the particle boundary layer and wake. The Boussinesq–Basset force is difficult to implement in the single-particle model because it is not explicit in $\dot{\mathbf{U}}_p^*(t)$. Thus, although some efficient strategies to calculate it have been developed (Dorgan & Loth, 2007; van Hinsberg *et al.*, 2011), it is commonly neglected for practical reasons in multiphase flow codes. Nonetheless, it should be pointed that the Boussinesq–Basset force can be very large when a particle is accelerated at an extremely high rate.

The Brownian force is the well-known force that becomes important in the case of very small particles suspended in a continuous phase. The Brownian force can be defined as the instantaneous momentum exchange due to collisions between the molecules of the continuous phase with a suspended particle. When the particle is small enough to perceive the molecular nature (and motion) of the continuous phase (i.e. when the particle Knudsen number is large enough), it exhibits a random motion, which was observed as early as 2000 years ago by the Roman Lucretius. The Brownian force is typically described as a stochastic process (Gardiner, 2004), and it can be modeled as a Wiener process:[34]

$$\mathbf{A}_{fp}^* = \frac{1}{M_p} \left(12 d_p \mu_f k_B T_f \right)^{1/2} \mathbf{W}(t), \tag{5.90}$$

where M_p is the particle mass, k_B is the Boltzmann constant, and T_f is the absolute temperature of the continuous phase. The Wiener process $\mathbf{W}(t)$ was introduced in Section 5.1.2. On comparing Eq. (5.90) with Eq. (5.10), we can observe that Eq. (5.90) represents an isotropic diffusion coefficient in velocity phase space. In practice, because Brownian motion is significant only for very small particles, it is usually included as a real-space diffusion term for the particle concentration in the limit of zero particle Stokes number. For this reason, Eq. (5.90) is rarely used in practical computations.

Thermophoresis (or the Soret effect) is a phenomenon observed when particles are carried in a fluid phase subjected to a large temperature gradient. The physical explanation is very simple. As is well known and as has already been discussed, particles exchange momentum with the surrounding fluid thanks to collisions between the molecules of the primary phase and the particles. The intensity and the frequency of these collisions are of course related to the temperature of the fluid phase. Particles in a region of low fluid-phase temperature next to a region of high fluid-phase temperature will exchange more momentum with the molecules in the hot region than with the molecules in the cold region. This imbalance in the momentum exchange is translated into a force pushing particles from high-temperature to low-temperature regions. This force is particularly important for aerosols and, in general, small particles moving in gases with large temperature gradients. A typical example is a particle formed and/or transported in a flame. The quantification of the thermophoresis depends on the particle-size-to-mean-free-path (of the molecules of the fluid phase) ratio. This ratio is the inverse of the fluid-phase Knudsen number introduced in Eq. (5.60). When the ratio is smaller than unity (Knudsen number greater than unity), then the single-particle model for the thermophoretic acceleration is generally expressed as[35]

$$\mathbf{A}_{fp} = -\frac{6 p_f \lambda_f}{\pi T_f \rho_p d_p} \frac{\partial T_f}{\partial \mathbf{x}}, \tag{5.91}$$

[34] Technically speaking, the Brownian force is not a pure-advection term and could be included in the discussion of diffusion processes in Section 5.5.

[35] Technically speaking, thermophoresis is not a true pure-advection term and could be included in the discussion on collision processes in Chapter 6.

where p_f and T_f are the fluid-phase pressure and temperature evaluated at the particle location, and λ_f is the mean free path of the fluid-phase molecules. This equation is valid for spherical particles, but can be extended to different shapes. When the particle-size-to-mean-free-path ratio is bigger than unity (or, conversely, the Knudsen number is smaller than unity), the mechanism is more complicated because a temperature gradient is established within the particle. For such cases, the following expression is typically employed:

$$\mathbf{A}_{fp} = -\frac{36\nu_f^2 C_s}{T_f d_p^2}\frac{\rho_f}{\rho_p}\frac{1}{1+6C_m \mathrm{Kn}}\frac{k_f/k_p + 2C_t\mathrm{Kn}}{1+2k_f/k_p+4C_t\mathrm{Kn}}\frac{\partial T_f}{\partial \mathbf{x}}, \tag{5.92}$$

where k_f and k_p are the thermal conductivities of the fluid and particle, respectively, $C_s = 1.17$ is the thermal slip coefficient, $C_t = 2.18$ is the thermal exchange coefficient, and $C_m = 1.14$ is the momentum exchange coefficient. More details on this topic can be found in Talbot (1981). As with the Brownian force, the thermophoretic force is significant for very small particles and is often included as a real-space diffusion term in the particle-concentration transport equation.

In summary, the Boussinesq–Basset, Brownian, and thermophoretic forces are rarely used in disperse multiphase flow simulations for different reasons. The Boussinesq–Basset force is neglected because it is needed only for rapidly accelerating particles and because its form makes its simulation difficult to implement. The Brownian and thermophoretic forces are important for very small particles, which usually implies that the particle Stokes number is near zero. For such particles, it is not necessary to solve transport equations for the disperse-phase momentum density. Instead, the Brownian and thermophoretic forces generate real-space diffusion terms in the particle-concentration transport equation (which is coupled to the fluid-phase momentum equation).

5.3.4 Final expressions for the mesoscale acceleration models

In order to complete our discussion on momentum transfer, we must consider the final forms of the mesoscale acceleration models in the presence of all the fluid–particle forces. When the virtual-mass force is included, the mesoscale acceleration models must be derived starting from the force balance on a single particle:[36]

$$\rho_p V_p \frac{\mathrm{d}\mathbf{U}_p^*}{\mathrm{d}t} = \rho_f V_p C_{vm}\left(\frac{\mathrm{d}\mathbf{U}_f^*}{\mathrm{d}t} - \frac{\mathrm{d}\mathbf{U}_p^*}{\mathrm{d}t}\right) + \sum_p \left\{\text{all forces } except\ C_{vm}\frac{\mathrm{d}}{\mathrm{d}t}\right\},$$

where we have implicitly assumed that the particle density and velocity do not change with time. A similar expression is found for the fluid seen by a particle:

$$\rho_f V_f \frac{\mathrm{d}\mathbf{U}_f^*}{\mathrm{d}t} = -\rho_f V_p C_{vm}\left(\frac{\mathrm{d}\mathbf{U}_f^*}{\mathrm{d}t} - \frac{\mathrm{d}\mathbf{U}_p^*}{\mathrm{d}t}\right) + \sum_f \left\{\text{all forces } except\ C_{vm}\frac{\mathrm{d}}{\mathrm{d}t}\right\}.$$

On the right-hand sides of these force balances, we include the body and fluid–particle forces discussed in the previous sections. By moving the time derivatives from the virtual-mass force to the right-hand side, we can combine the balance of forces in matrix form:

$$\left(\begin{bmatrix} \left(\rho_p + \rho_f C_{vm}\right) V_p & -\rho_f V_p C_{vm} \\ -\rho_f V_p C_{vm} & \rho_f\left(V_f + V_p C_{vm}\right)\end{bmatrix} \otimes \mathbf{I}\right)\frac{\mathrm{d}}{\mathrm{d}t}\begin{bmatrix}\mathbf{U}_p^* \\ \mathbf{U}_f^*\end{bmatrix} = \begin{bmatrix}\mathbf{F}_{fp} + \mathbf{F}_p \\ \mathbf{F}_{pf} + \mathbf{F}_f\end{bmatrix},$$

[36] The remaining term in the virtual-mass force is $\rho_f V_p C_{vm}\left(\mathbf{U}_f^* - \mathbf{U}_p^*\right)\cdot\nabla\mathbf{U}_f$.

where $\mathbf{I}$ is a 3×3 identity matrix, and the forces have been denoted by inter-phase forces (for which $\mathbf{F}_{\mathrm{pf}} + \mathbf{F}_{\mathrm{fp}} = 0$) and intra-phase forces ($\mathbf{F}_{\mathrm{p}}$ and $\mathbf{F}_{\mathrm{f}}$). The matrix on the left-hand side can be inverted to find

$$\kappa \frac{\mathrm{d}}{\mathrm{d}t} \begin{bmatrix} \mathbf{U}_{\mathrm{p}}^* \\ \mathbf{U}_{\mathrm{f}}^* \end{bmatrix} = \left(\begin{bmatrix} \rho_{\mathrm{f}} \left(V_{\mathrm{f}} + V_{\mathrm{p}} C_{\mathrm{vm}} \right) & \rho_{\mathrm{f}} V_{\mathrm{p}} C_{\mathrm{vm}} \\ \rho_{\mathrm{f}} V_{\mathrm{p}} C_{\mathrm{vm}} & \left(\rho_{\mathrm{p}} + \rho_{\mathrm{f}} C_{\mathrm{vm}} \right) V_{\mathrm{p}} \end{bmatrix} \otimes \mathbf{I} \right) \begin{bmatrix} \mathbf{F}_{\mathrm{fp}} + \mathbf{F}_{\mathrm{p}} \\ \mathbf{F}_{\mathrm{pf}} + \mathbf{F}_{\mathrm{f}} \end{bmatrix},$$

where $\kappa = \rho_{\mathrm{f}} V_{\mathrm{p}} [\rho_{\mathrm{p}} V_{\mathrm{f}} + (\rho_{\mathrm{p}} V_{\mathrm{p}} + \rho_{\mathrm{f}} V_{\mathrm{f}}) C_{\mathrm{vm}}]$. Thus, the final expressions for the mesoscale acceleration models are

$$\frac{\mathrm{d}\mathbf{U}_{\mathrm{p}}^*}{\mathrm{d}t} = \frac{1}{\rho_{\mathrm{p}} V_{\mathrm{p}} \gamma} \mathbf{F}_{\mathrm{fp}} + \frac{1}{\gamma} \left[\left(1 + \frac{V_{\mathrm{p}}}{V_{\mathrm{f}}} C_{\mathrm{vm}} \right) \frac{1}{\rho_{\mathrm{p}} V_{\mathrm{p}}} \mathbf{F}_{\mathrm{p}} + \left(\frac{\rho_{\mathrm{f}}}{\rho_{\mathrm{p}}} C_{\mathrm{vm}} \right) \frac{1}{\rho_{\mathrm{f}} V_{\mathrm{f}}} \mathbf{F}_{\mathrm{f}} \right] \tag{5.93}$$

and

$$\frac{\mathrm{d}\mathbf{U}_{\mathrm{f}}^*}{\mathrm{d}t} = \frac{1}{\rho_{\mathrm{f}} V_{\mathrm{f}} \gamma} \mathbf{F}_{\mathrm{pf}} + \frac{1}{\gamma} \left[\left(1 + \frac{\rho_{\mathrm{f}}}{\rho_{\mathrm{p}}} C_{\mathrm{vm}} \right) \frac{1}{\rho_{\mathrm{f}} V_{\mathrm{f}}} \mathbf{F}_{\mathrm{f}} + \left(\frac{V_{\mathrm{p}}}{V_{\mathrm{f}}} C_{\mathrm{vm}} \right) \frac{1}{\rho_{\mathrm{p}} V_{\mathrm{p}}} \mathbf{F}_{\mathrm{p}} \right], \tag{5.94}$$

where the effective volume coefficient is given by

$$\gamma = 1 + \left(\frac{V_{\mathrm{p}}}{V_{\mathrm{f}}} + \frac{\rho_{\mathrm{f}}}{\rho_{\mathrm{p}}} \right) C_{\mathrm{vm}} \approx 1 + \left(\frac{\alpha_{\mathrm{p}}}{\alpha_{\mathrm{f}}} + \frac{\rho_{\mathrm{f}}}{\rho_{\mathrm{p}}} \right) C_{\mathrm{vm}}. \tag{5.95}$$

From Eqs. (5.93) and (5.94), it is very interesting to note that while the fluid–particle forces are the same as in the absence of the virtual-mass force, the body forces for the particle and the fluid are now coupled whenever $C_{\mathrm{vm}} \neq 0$. However, body forces that are linearly proportional to mass such as gravity (i.e. $\mathbf{F}_{\mathrm{p}} = \rho_{\mathrm{p}} V_{\mathrm{p}} \mathbf{g}$ and $\mathbf{F}_{\mathrm{f}} = \rho_{\mathrm{f}} V_{\mathrm{f}} \mathbf{g}$) will remain uncoupled. In the very dilute limit, we have $V_{\mathrm{p}} \ll V_{\mathrm{f}}$ and thus Eq. (5.93) reduces to the classical particle model in which the virtual-mass force affects only the particle mass. The same is true for the fluid seen by the particle in the limit where $\rho_{\mathrm{f}} \ll \rho_{\mathrm{p}}$.

The principal fluid–particle forces discussed in this chapter lead to the following expression for the fluid–particle force:[37]

$$\begin{aligned} \mathbf{F}_{\mathrm{fp}} = & -V_{\mathrm{p}} \gamma \boldsymbol{\nabla} p_{\mathrm{f}} + \frac{1}{2} \rho_{\mathrm{f}} C_{\mathrm{D}} A_{\mathrm{d}} |\mathbf{U}_{\mathrm{f}}^* - \mathbf{U}_{\mathrm{p}}^*| \left(\mathbf{U}_{\mathrm{f}}^* - \mathbf{U}_{\mathrm{p}}^* \right) + \rho_{\mathrm{f}} V_{\mathrm{p}} C_{\mathrm{vm}} \left(\mathbf{U}_{\mathrm{f}}^* - \mathbf{U}_{\mathrm{p}}^* \right) \cdot \boldsymbol{\nabla} \mathbf{U}_{\mathrm{f}} \\ & + \rho_{\mathrm{f}} V_{\mathrm{p}} C_{\mathrm{L}} \left(\mathbf{U}_{\mathrm{f}}^* - \mathbf{U}_{\mathrm{p}}^* \right) \times (\boldsymbol{\nabla} \times \mathbf{U}_{\mathrm{f}}), \end{aligned} \tag{5.96}$$

where the terms on the right-hand side are, respectively, the buoyancy, drag, virtual-mass, and lift forces. On comparing Eqs. (5.93) and (5.94) with Eq. (5.9), we can observe that with Eq. (5.96) and gravity, the mesoscale acceleration models are given by[38]

$$\begin{aligned} \mathbf{A}_{\mathrm{p}} &= \mathbf{g}, \\ \mathbf{A}_{\mathrm{f}} &= \mathbf{g} - \frac{1}{\varrho_{\mathrm{f}}} (\boldsymbol{\nabla} p_{\mathrm{f}} - \boldsymbol{\nabla} \cdot \boldsymbol{\tau}_{\mathrm{f}}) + \frac{1}{\tau_{\mathrm{f}}} \left(\mathbf{U}_{\mathrm{f}} - \mathbf{U}_{\mathrm{f}}^* \right), \end{aligned} \tag{5.97}$$

$$\begin{aligned} \mathbf{A}_{\mathrm{fp}} = & -\frac{1}{\rho_{\mathrm{p}}} \boldsymbol{\nabla} p_{\mathrm{f}} + \frac{\rho_{\mathrm{f}} C_{\mathrm{D}} A_{\mathrm{d}}}{2 \rho_{\mathrm{p}} V_{\mathrm{p}} \gamma} |\mathbf{U}_{\mathrm{f}}^* - \mathbf{U}_{\mathrm{p}}^*| \left(\mathbf{U}_{\mathrm{f}}^* - \mathbf{U}_{\mathrm{p}}^* \right) + \frac{\rho_{\mathrm{f}} C_{\mathrm{vm}}}{\rho_{\mathrm{p}} \gamma} \left(\mathbf{U}_{\mathrm{f}}^* - \mathbf{U}_{\mathrm{p}}^* \right) \cdot \boldsymbol{\nabla} \mathbf{U}_{\mathrm{f}} \\ & + \frac{\rho_{\mathrm{f}} C_{\mathrm{L}}}{\rho_{\mathrm{p}} \gamma} \left(\mathbf{U}_{\mathrm{f}}^* - \mathbf{U}_{\mathrm{p}}^* \right) \times (\boldsymbol{\nabla} \times \mathbf{U}_{\mathrm{f}}), \end{aligned} \tag{5.98}$$

[37] We have written the buoyancy force in terms of a virtual volume $V_{\mathrm{p}}\gamma$ in order to account for the fluid volume displaced with the particle.

[38] The mesoscale fluid pressure must be independent of C_{vm} in order for the hydrostatic pressure to depend only on the mixture density.

and

$$\mathbf{A}_{\mathrm{pf}} = \frac{V_{\mathrm{p}}}{\rho_{\mathrm{f}} V_{\mathrm{f}}} \nabla p_{\mathrm{f}} - \frac{C_{\mathrm{D}} A_{\mathrm{d}}}{2 V_{\mathrm{f}} \gamma} |\mathbf{U}_{\mathrm{f}}^{*} - \mathbf{U}_{\mathrm{p}}^{*}| \left(\mathbf{U}_{\mathrm{f}}^{*} - \mathbf{U}_{\mathrm{p}}^{*} \right) - \frac{V_{\mathrm{p}} C_{\mathrm{vm}}}{V_{\mathrm{f}} \gamma} \left(\mathbf{U}_{\mathrm{f}}^{*} - \mathbf{U}_{\mathrm{p}}^{*} \right) \cdot \nabla \mathbf{U}_{\mathrm{f}} - \frac{V_{\mathrm{p}} C_{\mathrm{L}}}{V_{\mathrm{f}} \gamma} \left(\mathbf{U}_{\mathrm{f}}^{*} - \mathbf{U}_{\mathrm{p}}^{*} \right) \times (\nabla \times \mathbf{U}_{\mathrm{f}}). \tag{5.99}$$

It is very interesting to note that the virtual-mass force mainly acts through the effective volume coefficient γ, and tends to reduce the momentum-exchange terms. The final step is to use the definitions of the mass-average moments introduced in Chapter 4 to derive the force terms in the macroscale momentum balance. For the disperse-phase momentum balance (see Eq. (4.85) on page 123), this procedure leads to

$$\varrho_{\mathrm{p}} [\![\mathbf{A}_{\mathrm{fp}}]\!]_{\mathrm{p}} + \varrho_{\mathrm{p}} [\![\mathbf{A}_{\mathrm{p}}]\!]_{\mathrm{p}} = -\alpha_{\mathrm{p}} \nabla p_{\mathrm{f}} + \varrho_{\mathrm{p}} \mathbf{g} + \frac{\varrho_{\mathrm{p}}}{2\gamma} [\![\rho_{\mathrm{f}} C_{\mathrm{D}} A_{\mathrm{d}} |\mathbf{U}_{\mathrm{f}}^{*} - \mathbf{U}_{\mathrm{p}}^{*}| (\mathbf{U}_{\mathrm{f}}^{*} - \mathbf{U}_{\mathrm{p}}^{*})]\!]_{\mathrm{p}} + \frac{\varrho_{\mathrm{p}} C_{\mathrm{vm}} \rho_{\mathrm{f}}}{\gamma \rho_{\mathrm{p}}} (\mathbf{U}_{\mathrm{f}} - \mathbf{U}_{\mathrm{p}}) \cdot \nabla \mathbf{U}_{\mathrm{f}} + \frac{\varrho_{\mathrm{p}} C_{\mathrm{L}} \rho_{\mathrm{f}}}{\gamma \rho_{\mathrm{p}}} (\mathbf{U}_{\mathrm{f}} - \mathbf{U}_{\mathrm{p}}) \times (\nabla \times \mathbf{U}_{\mathrm{f}}). \tag{5.100}$$

For the fluid-phase momentum balance (see Eq. (4.94) on page 125), the macroscale acceleration terms are

$$\varrho_{\mathrm{f}} [\![\mathbf{A}_{\mathrm{pf}}]\!]_{\mathrm{f}} + \varrho_{\mathrm{f}} [\![\mathbf{A}_{\mathrm{f}}]\!]_{\mathrm{f}} = -\alpha_{\mathrm{f}} \nabla p_{\mathrm{f}} + \nabla \cdot \boldsymbol{\tau}_{\mathrm{f}} + \varrho_{\mathrm{f}} \mathbf{g} - \frac{\varrho_{\mathrm{p}}}{2\gamma} [\![\rho_{\mathrm{f}} C_{\mathrm{D}} A_{\mathrm{d}} |\mathbf{U}_{\mathrm{f}}^{*} - \mathbf{U}_{\mathrm{p}}^{*}| (\mathbf{U}_{\mathrm{f}}^{*} - \mathbf{U}_{\mathrm{p}}^{*})]\!]_{\mathrm{p}} - \frac{\varrho_{\mathrm{p}} C_{\mathrm{vm}} \rho_{\mathrm{f}}}{\gamma \rho_{\mathrm{p}}} (\mathbf{U}_{\mathrm{f}} - \mathbf{U}_{\mathrm{p}}) \cdot \nabla \mathbf{U}_{\mathrm{f}} - \frac{\varrho_{\mathrm{p}} C_{\mathrm{L}} \rho_{\mathrm{f}}}{\gamma \rho_{\mathrm{p}}} (\mathbf{U}_{\mathrm{f}} - \mathbf{U}_{\mathrm{p}}) \times (\nabla \times \mathbf{U}_{\mathrm{f}}). \tag{5.101}$$

As can be seen from Eq. (5.100), the virtual-mass force reduces the drag and lift forces by a factor of $1/\gamma$. The buoyancy force is not modified because we have chosen to define it in terms of the effective volume $V_{\mathrm{p}}\gamma$. We remind the reader that the mesoscale acceleration model for the fluid seen by the particle $\mathbf{A}_{\mathrm{pf}}^{*}$ must be consistent with the mesoscale model for the particle phase $\mathbf{A}_{\mathrm{fp}}^{*}$ in order to ensure that the overall system conserves momentum at the mesoscale. (See Section 4.3.8 for more details.) As discussed near Eq. (5.14) on page 144, this is accomplished in the single-particle model by constraining the model for $\mathbf{A}_{\mathrm{pf}}$ given the model for $\mathbf{A}_{\mathrm{fp}}$ (which is derived from the force terms introduced in this section). Thus, as in Eqs. (5.98) and (5.99), it is not necessary to derive separate models for the momentum-transfer terms appearing in $\mathbf{A}_{\mathrm{pf}}$.

5.4 Real-space advection

Real-space advection represents the movement of the particles in real space. As already reported in Eq. (5.3), it is represented by the second term on the left-hand side of the GPBE (see Eq. (5.1)):

$$\frac{\partial}{\partial \mathbf{x}} \cdot \left(\mathbf{v}_{\mathrm{p}} n \right) = \frac{\partial}{\partial x_1} \left(v_{\mathrm{p}1} n \right) + \frac{\partial}{\partial x_2} \left(v_{\mathrm{p}2} n \right) + \frac{\partial}{\partial x_3} \left(v_{\mathrm{p}3} n \right).$$

Although the particle velocity is often included as one of the variables of the internal-coordinate vector, it clearly plays an important and different role, which is related to the concept of passive and active internal variables. The particle velocity is an active variable since it generates spatial transport, whereas other internal coordinates, such as for example the particle size, are passive variables. Passive variables are simply carried along with a specific velocity, and therefore the real-space advection term can be solved for fixed values of the passive variables. This is not the case for active variables, as described in detail in Section 8.1.3. In what follows different strategies to evaluate the real-space advection term will be discussed. It is important to remind the reader that all these strategies require knowledge of the fluid velocity seen by the particle. In the present context (i.e. the monokinetic-fluid limit), this information is typically found by solving a fluid-phase momentum-balance equation, such as Eq. (4.92):

$$\frac{\partial \varrho_{\mathrm{f}} \mathbf{U}_{\mathrm{f}}}{\partial t} + \frac{\partial}{\partial \mathbf{x}} \cdot (\varrho_{\mathrm{f}} \mathbf{U}_{\mathrm{f}} \otimes \mathbf{U}_{\mathrm{f}}) = \frac{\partial}{\partial \mathbf{x}} \cdot \mathbf{S}_{\mathrm{f}} - \varrho_{\mathrm{p}} [\![\mathbf{A}_{\mathrm{fp}}]\!]_{\mathrm{p}} - [\![\mathbf{G}_{\mathrm{p}}]\!]_{\mathrm{p}} - \varrho_{\mathrm{p}} [\![\mathbf{S}]\!]_{\mathrm{p}} + \varrho_{\mathrm{f}} \mathbf{A}_{\mathrm{f}},$$

where $\mathbf{S}_{\mathrm{f}}$ denotes the viscous and pressure-stress tensor in the fluid phase and $\mathbf{A}_{\mathrm{f}}$ represents the body forces in the fluid phase. The second term on the right-hand side represents the effect of the presence of the solid particles on the continuous phase. It can be neglected in the case of one-way coupling, but must be included in the case of two- or three-way coupling.

If the NDF is a function of the particle velocity then the solution of the GPBE provides the modeler with the essential information for calculating the real-space advection term. This approach is used whenever the particle Stokes number is not small, and will result in the development of a particle-velocity distribution. More details on this topic can be found in Chapter 8. An alternative approach consists of integrating the NDF with respect to the particle velocity. Let us consider, for example, a generic NDF $n(t, \mathbf{x}, \mathbf{v}_{\mathrm{p}}, \boldsymbol{\xi}_{\mathrm{p}})$, which is a function of the time t, space $\mathbf{x}$, particle velocity $\mathbf{v}_{\mathrm{p}}$, and internal coordinates $\boldsymbol{\xi}_{\mathrm{p}}$. By integrating out the particle velocity the following NDF is obtained:

$$n^*(t, \mathbf{x}, \boldsymbol{\xi}_{\mathrm{p}}) \equiv \int n(t, \mathbf{x}, \mathbf{v}_{\mathrm{p}}, \boldsymbol{\xi}_{\mathrm{p}}) \mathrm{d}\mathbf{v}_{\mathrm{p}}. \tag{5.102}$$

Likewise, if the particle velocity is integrated out from Eq. (5.3), the following expression for the real-space advection term is obtained:

$$\frac{\partial}{\partial \mathbf{x}} \cdot \left(\langle \mathbf{U}_{\mathrm{p}} | \boldsymbol{\xi}_{\mathrm{p}} \rangle n^* \right) = \frac{\partial}{\partial x_1} \left(\langle U_{\mathrm{p}1} | \boldsymbol{\xi}_{\mathrm{p}} \rangle n^* \right) + \frac{\partial}{\partial x_2} \left(\langle U_{\mathrm{p}2} | \boldsymbol{\xi}_{\mathrm{p}} \rangle n^* \right) + \frac{\partial}{\partial x_3} \left(\langle U_{\mathrm{p}3} | \boldsymbol{\xi}_{\mathrm{p}} \rangle n^* \right). \tag{5.103}$$

By definition, the conditional velocity is

$$\langle \mathbf{U}_{\mathrm{p}} | \boldsymbol{\xi}_{\mathrm{p}} \rangle = \frac{\int \mathbf{v}_{\mathrm{p}} n \, \mathrm{d}\mathbf{v}_{\mathrm{p}}}{\int n \, \mathrm{d}\mathbf{v}_{\mathrm{p}}} = \frac{1}{n^*} \left(\int v_{\mathrm{p}1} n \, \mathrm{d}\mathbf{v}_{\mathrm{p}}, \int v_{\mathrm{p}2} n \, \mathrm{d}\mathbf{v}_{\mathrm{p}}, \int v_{\mathrm{p}3} n \, \mathrm{d}\mathbf{v}_{\mathrm{p}} \right), \tag{5.104}$$

which represents the expected mean velocity of particles with internal coordinates equal to $\boldsymbol{\xi}_{\mathrm{p}}$, or, equivalently, the average particle velocity conditioned on the internal coordinates. The calculation of the real-space advection term then requires knowledge of the conditioned particle velocity $\langle \mathbf{U}_{\mathrm{p}} | \boldsymbol{\xi}_{\mathrm{p}} \rangle$. This can be determined (in an Eulerian framework) with different approaches, namely the pseudo-homogeneous or dusty-gas model, the equilibrium or algebraic Eulerian model, and the Eulerian two-fluid model (Balachandar, 2009; Balachandar & Eaton, 2010; Ferry & Balachandar, 2001). The three approaches are adequate in different flow regimes that depend on the type of phase coupling between the primary and disperse phases and are determined by the state of the primary and disperse phases. The flow regime is often characterized by the particle Stokes number defined in Eq. (1.20) on page 10 as the ratio between the characteristic particle relaxation time (τ_p) and the fluid characteristic time (τ_f), and by the relative importance of particle–particle collisions. It is important to point out here that, for many solution methods used to calculate the moments of the GPBE, knowledge of the conditional particle velocity $\langle \mathbf{U}_p | \boldsymbol{\xi}_p \rangle$ for a small number of values of $\boldsymbol{\xi}_p$ is often sufficient for computing the spatial fluxes. In fact, as will become clearer in Chapter 7, the population of particles can often be represented by N particle classes (or groups, or nodes) with disperse-phase volume fractions α_{pk} and values of the internal coordinates $\boldsymbol{\xi}_{pk}$ with $k = 1, \ldots, N$, where N is relatively small.

5.4.1 The pseudo-homogeneous or dusty-gas model

With the dusty-gas model, the particle velocity is assumed to be equal to that of the fluid:[39] $\mathbf{U}_p^* = \mathbf{U}_f$. This situation is realistic when particles relax very quickly to the velocity of the surrounding fluid. As is well known, this is determined by evaluating the particle relaxation time, which is generally defined by considering the Lagrangian momentum-balance equation for a single particle experiencing viscous drag and gravity forces:

$$\frac{d\mathbf{U}_p^*}{dt} = \frac{1}{\tau_p(\boldsymbol{\xi}_p^*)} \left(\mathbf{U}_f - \mathbf{U}_p^* \right) + \mathbf{g} \tag{5.105}$$

with $\tau_p(\boldsymbol{\xi}_p^*) = \rho_p d_p^2 / (18\mu_f)$. The momentum balance in Eq. (5.105) does not consider instantaneous changes due to particle–particle collisions. The definition of the particle relaxation time results in the following definition for the particle Stokes number:

$$\mathrm{St}_p = \frac{\rho_p d_p^2}{18 \rho_f \nu_f \tau_f}. \tag{5.106}$$

For laminar flow, the characteristic time of the fluid phase τ_f can be defined as the ratio between a characteristic velocity U_f and a characteristic dimension L. For example, in the case of channel flows confined within two parallel plates, L can be taken equal to the distance between the plates, whereas U_f can be the friction velocity. Another common choice is to base this calculation on the viscous scale, by dividing the kinematic viscosity of the fluid phase by the friction velocity squared. For turbulent flow, τ_f is usually assumed to be the Kolmogorov time scale in the fluid phase. The dusty-gas model can be applied only when the particle relaxation time tends to zero (i.e. $\mathrm{St}_p \ll 1$). Under these conditions, Eq. (5.105) yields $\langle \mathbf{U}_p | \boldsymbol{\xi}_p \rangle = \mathbf{U}_f + O(\tau_p(\boldsymbol{\xi}_p))$. As has already been mentioned, this model is clearly appropriate only when the particles immediately adapt to the local fluid flow. This typically happens when particles are very small and/or the continuous phase is highly viscous and/or the disperse-to-primary-phase density ratio is very small. The dusty-gas model assumes that there is only one particle velocity field, which is identical to that of the fluid. With this approach, preferential accumulation and segregation effects are clearly not predicted since particles are transported as scalars in the continuous phase. If the system is very dilute (one-way coupling), the properties of the continuous phase (i.e. density and viscosity) are assumed to be equal to those of the fluid. If the solid-particle concentration starts to have an influence on the fluid phase (two-way coupling), a modified density and viscosity for the continuous phase are generally introduced in Eq. (4.92).

[39] The dusty-gas model neglects Brownian motion of the particles, and thus is not intended for describing diffusive transport of sub-micron particles.

5.4.2 The equilibrium or algebraic Eulerian model

For very small but finite particle Stokes numbers ($0 < \mathrm{St_p} \ll 1$), particles move with a velocity that is different from that of the fluid, but all particles at the same location have a unique velocity[40] that depends on the values of the internal coordinates. In fact, under these flow conditions, the difference between two initially different particle velocities decreases and, when the transient dies out, all particles with the same internal coordinates will relax to a unique velocity. One very intuitive example is that of monodisperse, non-interacting particles falling in a stagnant fluid. Regardless of their initial velocities, after some time all the particles will relax to the particle terminal settling velocity, which is different from the fluid velocity. The final particle terminal velocity represents, in this case, the equilibrium velocity, which can depend on the internal coordinates, i.e. on particle size and shape. For very small particle Stokes numbers, one can therefore expect a unique particle velocity field when particles relax undisturbed and very quickly to their equilibrium velocity. The characteristic time scale on which this occurs is proportional to the time scale of particle velocity relaxation (see Eq. (5.105)). This unique particle velocity is therefore realized for particulate systems characterized by finite particle relaxation times that are much smaller than the characteristic time scale of the fluid, and for particulate systems in which particle–particle collisions are negligible. In fact, for these circumstances any initially different particle velocities will converge (or decay) exponentially fast to the equilibrium velocity.[41] In addition to the condition on the particle Stokes number, a second important condition is that particle–particle collisions are not very frequent and do not influence the particle velocity. Since the particle Stokes number is related to the particle size and density and the collision frequency is related to the particle concentration, very often the different possible scenarios are parameterized in terms of these quantities, namely small versus large particles and dilute versus dense systems.

The calculation of the particle velocity with the equilibrium model is based on a simple expansion valid for small particle relaxation times. In fact, we know that when $\tau_\mathrm{p} \to 0$ the particle velocity $\mathbf{U}_\mathrm{p}^*$ tends to the fluid velocity $\mathbf{U}_\mathrm{f}$. If the particle velocity on the left-hand side of Eq. (5.105) is replaced with the fluid velocity, the derivative is replaced with a total (or substantial) derivative (due to the fact that $\mathbf{U}_\mathrm{f}$ is the Eulerian fluid velocity), and the resulting equation is solved with respect to the particle velocity appearing on the right-hand side, the following equation is obtained (Maxey, 1987):

$$\langle \mathbf{U}_\mathrm{p} | \boldsymbol{\xi}_\mathrm{p} \rangle = \mathbf{U}_\mathrm{f} + \left(\frac{\mathrm{D}\mathbf{U}_\mathrm{f}}{\mathrm{D}t} - \mathbf{g} \right) \tau_\mathrm{p}(\boldsymbol{\xi}_\mathrm{p}) + O\left(\tau_\mathrm{p}^2\right), \tag{5.107}$$

where $\mathrm{D}/\mathrm{D}t$ is the total fluid derivative. This algebraic equation presents the main advantage of directly relating the particle velocity to the fluid velocity and its local and temporal derivatives. Thus, in any computational procedure, once the Eulerian fluid velocity has been evaluated, the corresponding particle velocity can be obtained relatively easily without resorting to the solution of additional differential equations. Moreover, it is equally easy to obtain the conditional velocities of particles with different relaxation times or, in other

[40]This statement is true only when Brownian motion is neglected, which is an implicit assumption in the derivation of the equilibrium model. Since Brownian motion is important for sub-micron particles suspended in a fluid, the equilibrium model is not intended for describing, for example, nanoparticle transport.

[41]The convergence of particle velocities should not be confused with the convergence of particle paths. In fact, even with an equilibrium velocity, two particles initially placed at adjacent points with different velocities will quickly converge to a single velocity but at different spatial positions! This is, for example, very important in turbulent flows, where particle paths always diverge. However, as is well known, also very simple flow fields can generate chaotic particle paths.

words, with different sizes and densities, generally appearing as internal coordinates $\boldsymbol{\xi}_\text{p}$. The expansion method with which Eq. (5.107) was derived can also include higher-order terms, resulting in the following expression (Druzhinin, 1987):

$$\langle \mathbf{U}_\text{p} | \boldsymbol{\xi}_\text{p} \rangle = \mathbf{U}_\text{f} - \mathbf{a}_\text{f} \tau_\text{p}(\boldsymbol{\xi}_\text{p}) + \left(\frac{\text{D}\mathbf{a}_\text{f}}{\text{D}t} + \mathbf{a}_\text{f} \cdot \frac{\partial \mathbf{U}_\text{f}}{\partial \mathbf{x}} \right) \tau_\text{p}^2(\boldsymbol{\xi}_\text{p}) + O\left(\tau_\text{p}^3\right), \tag{5.108}$$

where $\mathbf{a}_\text{f} = \text{D}\mathbf{U}_\text{f}/\text{D}t - \mathbf{g}$. Equation (5.108) was derived only for particles moving under the effects of drag and gravity forces. If other forces (for example the ones reported in Section 5.3) are included, Eq. (5.108) assumes a similar form. For example when gravity, buoyancy, drag, virtual-mass, lift, and Basset forces are accounted for, the following expression can be used:

$$\begin{aligned} \langle \mathbf{U}_\text{p} | \boldsymbol{\xi}_\text{p} \rangle = \mathbf{U}_\text{f} &+ \left(1 - \frac{\rho_\text{f}}{\rho_\text{p}}\right) \\ &\times \left[-\mathbf{a}_\text{f} \tau_\text{p}(\boldsymbol{\xi}_\text{p}) + \mathcal{L}[\mathbf{a}_\text{f}] \tau_\text{p}^{3/2}(\boldsymbol{\xi}_\text{p}) + \left(\frac{\text{D}\mathbf{a}_\text{f}}{\text{D}t} + \mathbf{a}_\text{f} \cdot \frac{\partial \mathbf{U}_\text{f}}{\partial \mathbf{x}} - \frac{\rho_\text{f}}{\rho_\text{p}} \mathcal{L}^2[\mathbf{a}_\text{f}] \right) \tau_\text{p}^2(\boldsymbol{\xi}_\text{p}) + O\left(\tau_\text{p}^{5/2}\right) \right], \end{aligned} \tag{5.109}$$

where the fluid acceleration is now defined by

$$\mathbf{a}_\text{f} = \left(\frac{\rho_\text{p} - C_\text{vm} \rho_\text{f}}{\rho_\text{p} - \rho_\text{f}} \right) \frac{\text{D}\mathbf{U}_\text{f}}{\text{D}t} - \mathbf{g}, \tag{5.110}$$

$\mathcal{L}$ is a linear operator defined by

$$\mathcal{L}[\cdot] = -C_\text{L} \frac{\rho_\text{f}}{\rho_\text{p}} [\cdot] \times (\nabla \times \mathbf{U}_\text{f}) + \frac{9 \rho_\text{f}}{d_\text{p} \rho_\text{p}} \sqrt{\frac{\nu_\text{f}}{\pi}} \int_{-\infty}^{t} \frac{\text{d}[\cdot]/\text{d}\tau}{\sqrt{t - \tau}} \, \text{d}\tau, \tag{5.111}$$

C_vm is the virtual-mass coefficient, and C_L is the lift coefficient. In summary, the equilibrium or algebraic Eulerian model can be used for very small St_p, and cases with one-, two-, and three-way coupling can be easily treated. In the case of one-way coupling (dilute systems), a momentum-balance equation for the continuous phase is solved (see Eq. (4.92)), and then Eq. (5.109) (or any of its simplified forms) is used to calculate the conditional particle velocity of the different particle classes: $\mathbf{U}_{\text{p}k} \equiv \langle \mathbf{U}_\text{p} | \boldsymbol{\xi}_\text{p} = \boldsymbol{\xi}_{\text{p}k} \rangle$. For denser particulate systems (two- or three-way coupling), a momentum-balance equation for the mixture (whose density and velocity are defined, respectively, in Eqs. (4.69) and (4.70)) is solved and then, as for the one-way coupling case, Eq. (5.109) (or any of its simplified forms) is used to calculate the conditional particle velocity. It is important to remind the reader that, although the equilibrium model is capable of accounting for two- and three-way coupling, even moderately dense particulate systems are generally not well described due to the importance in such systems of particle–particle collisions.

5.4.3 The Eulerian two-fluid model

When the particle Stokes number is not small, the truncated expansion for the particle velocity $\langle \mathbf{U}_\text{p} | \boldsymbol{\xi}_\text{p} \rangle$ is no longer valid. Under these conditions, the mean particle velocity must be calculated from the disperse-phase momentum equation described in Section 4.3.7. Let us for the time being consider a very dilute population of identical particles. The mean velocity of these particles can be found by solving Eq. (4.91). For small particle

Stokes numbers, particle trajectory crossings (PTC) are not important, in which case it is possible to neglect velocity fluctuations (i.e. $[\![\mathbf{u}_\mathrm{p}\mathbf{u}_\mathrm{p}]\!] \ll 1$). In this limit, the disperse-phase momentum balance reduces to

$$\frac{\partial \varrho_\mathrm{p}\mathbf{U}_\mathrm{p}}{\partial t} + \nabla \cdot \left(\varrho_\mathrm{p}\mathbf{U}_\mathrm{p} \otimes \mathbf{U}_\mathrm{p}\right) = \varrho_\mathrm{p}\left([\![\mathbf{A}_\mathrm{fp}]\!]_\mathrm{p} + [\![\mathbf{A}_\mathrm{p}]\!]_\mathrm{p}\right).$$

If the buoyancy, gravity, drag, virtual-mass, and lift forces are accounted for, the disperse-phase momentum balance becomes (see Section 5.3.4)

$$\begin{aligned}\frac{\partial \varrho_\mathrm{p}\mathbf{U}_\mathrm{p}}{\partial t} + \nabla \cdot (\varrho_\mathrm{p}\mathbf{U}_\mathrm{p} \otimes \mathbf{U}_\mathrm{p}) &= -\alpha_\mathrm{p}\,\nabla p_\mathrm{f} + \varrho_\mathrm{p}\mathbf{g} + \frac{\varrho_\mathrm{p}\rho_\mathrm{f}C_\mathrm{D}A_\mathrm{d}}{2\gamma}|\mathbf{U}_\mathrm{f} - \mathbf{U}_\mathrm{p}|(\mathbf{U}_\mathrm{f} - \mathbf{U}_\mathrm{p}) \\ &\quad + \frac{\varrho_\mathrm{p}C_\mathrm{vm}\rho_\mathrm{f}}{\gamma\rho_\mathrm{p}}(\mathbf{U}_\mathrm{f} - \mathbf{U}_\mathrm{p}) \cdot \nabla\mathbf{U}_\mathrm{f} + \frac{\varrho_\mathrm{p}C_\mathrm{L}\rho_\mathrm{f}}{\gamma\rho_\mathrm{p}}(\mathbf{U}_\mathrm{f} - \mathbf{U}_\mathrm{p}) \times (\nabla \times \mathbf{U}_\mathrm{f}).\end{aligned} \tag{5.112}$$

Likewise, in the monokinetic-fluid limit the fluid-phase momentum balance is given by

$$\begin{aligned}\frac{\partial \varrho_\mathrm{f}\mathbf{U}_\mathrm{f}}{\partial t} + \nabla \cdot (\varrho_\mathrm{f}\mathbf{U}_\mathrm{f} \otimes \mathbf{U}_\mathrm{f}) &= \nabla \cdot \boldsymbol{\tau}_\mathrm{f} - \alpha_\mathrm{f}\,\nabla p_\mathrm{f} + \varrho_\mathrm{f}\mathbf{g} + \frac{\varrho_\mathrm{p}\rho_\mathrm{f}C_\mathrm{D}A_\mathrm{d}}{2\gamma}|\mathbf{U}_\mathrm{f} - \mathbf{U}_\mathrm{p}|(\mathbf{U}_\mathrm{p} - \mathbf{U}_\mathrm{f}) \\ &\quad + \frac{\varrho_\mathrm{p}C_\mathrm{vm}\rho_\mathrm{f}}{\gamma\rho_\mathrm{p}}(\mathbf{U}_\mathrm{p} - \mathbf{U}_\mathrm{f}) \cdot \nabla\mathbf{U}_\mathrm{f} + \frac{\varrho_\mathrm{p}C_\mathrm{L}\rho_\mathrm{f}}{\gamma\rho_\mathrm{p}}(\mathbf{U}_\mathrm{p} - \mathbf{U}_\mathrm{f}) \times (\nabla \times \mathbf{U}_\mathrm{f}).\end{aligned} \tag{5.113}$$

In summary, the Eulerian two-fluid model is represented by Eqs. (5.112) and (5.113) in addition to a constitutive model for the fluid stress tensor $\boldsymbol{\tau}_\mathrm{f}$. As already mentioned, Eq. (5.112) was derived under the assumption that the particle-velocity distribution is very narrow (i.e. small particle Stokes number), and the particles must have the same internal coordinates. If these simplifications do not hold, for example under dense conditions when particle–particle collisions become important, then particle-velocity fluctuations must be taken into account, as discussed at the end of Chapter 4.

5.4.4 Guidelines for real-space advection

There are three parameters that can be used to identify the best approach to calculate the real-space advection term: particle concentration, particle Stokes number, and polydispersity. When the particle loading, defined by $\varrho_\mathrm{p}/\varrho_\mathrm{f}$, is much smaller than unity the system can be considered dilute with one-way momentum coupling, whereas when $\varrho_\mathrm{p}/\varrho_\mathrm{f} \geq 0.1$ two-way momentum coupling is required. We consider the small particle Stokes number limit to be identified by $\mathrm{St}_\mathrm{p} \ll 1$, whereas moderate and large particle Stokes numbers are identified by $\mathrm{St}_\mathrm{p} \approx 1$ and $\mathrm{St}_\mathrm{p} \gg 1$, respectively. It is important to keep in mind that in general the particle Stokes number can vary from physical point to physical point in the system, depending on the definition of the characteristic time scale for the continuous phase. Polydispersity (insofar as particle advection is concerned) can instead be identified by a polydispersity index (PDI) defined as the ratio between the particle Stokes numbers calculated for the largest/heaviest particles and the smallest/lightest particles.[42] Given the expected polydispersity of the multiphase systems and the type of processes that the multiphase system undergoes, the corresponding particle velocity can be calculated using one of the following models.

[42] There are several ways to calculate the PDI. If there are no density inhomogeneities and size is the only variable, then the polydispersity is completely defined by the particle-size distribution. If this is known, then the particle Stokes numbers for the smaller and largest particles can be calculated in terms of $d_{10\%}$ and $d_{90\%}$. These

1. **Diffusion equation:** sub-micron particles subject to Brownian motion.
2. **The pseudo-homogeneous or dusty-gas model:** very small particle Stokes number and limited polydispersity (momentum-balance equation only for the continuous phase if the system is dilute or for the mixture of continuous and disperse phases if the system is dense).
3. **The equilibrium or algebraic Eulerian model with a single conditional velocity that is based on the mean particle size:** small particle Stokes number and limited polydispersity (momentum-balance equation only for the continuous phase if the system is dilute or for the mixture of continuous and disperse phases if the system is dense).
4. **The equilibrium or algebraic Eulerian model with conditional velocity:** small particle Stokes number and non-negligible polydispersity (momentum-balance equation only for the continuous phase if the system is dilute or for the mixture of continuous and disperse phases if the system is dense).
5. **The Eulerian two-fluid model with particle-phase velocity that is based on the mean particle size:** small particle Stokes number and limited polydispersity (both in dilute and in dense systems).

For all other cases, it will be necessary to solve the moment-transport equations derived from the GPBE as described in Chapter 4. In Chapter 8 the numerical algorithms used to find approximation solutions to the GPBE using quadrature-based moment methods are presented in detail.

5.5 Diffusion processes

As discussed in Section 5.1, diffusion processes appear in the GPBE as a result of mesoscale approximations of unresolved microscale physics. These phenomena generate advection in both physical and phase space that is characterized by a random chaotic nature at time and length scales smaller then the pure-advection terms discussed in the proceeding sections. Examples include Brownian forces, random lattice defects in crystals, and unsteady fluid flow around individual particles. In terms of the single-particle models at the mesoscale, diffusion processes appear as stochastic processes (see Section 5.1.2 for details). The general mathematical formulation in Eq. (5.11) introduces a large number of model coefficients that potentially must be specified by employing physical reasoning. However, as was noted in Section 5.1.2, the most basic models use only a small subset of the possible diffusion processes and, hence, we shall consider only the most important examples here. The single-particle mesoscale models in Eq. (5.11) on page 143 lead directly to the diffusion coefficients in Eq. (5.2) through the relation given in Eq. (5.12). As a general rule, it is preferable to start with the mesoscale models in the form of Eq. (5.11), for which we have already introduced models for the pure-advection terms (i.e. $\mathbf{G}_\mathrm{p}$, $\mathbf{A}_\mathrm{fp}$,

two numbers can be identified from the cumulative particle-size distribution with the following criterion: $d_{10\%}$ is the particle size such that 10% of the particles are smaller than $d_{10\%}$, and analogously for $d_{90\%}$. If the population of particles is distributed over particle size and density, we can similarly define $d_{10\%}$ and $\rho_{\mathrm{p},10\%}$ and $d_{90\%}$ and $\rho_{\mathrm{p},90\%}$ as the sizes and densities that isolate in the cumulative distribution 10% and 90% of the particles. The particle Stokes numbers for the smaller and lighter particles and for the larger and heavier particles can therefore be calculated using these numbers.

etc.) in Sections 5.2 and 5.3. The one exception to this rule is the diffusion term for the fluid mass seen by a particle appearing in Eq. (4.80) on page 121, which results from the presence of multiple particles in the neighborhood of the single particle being modeled. Thus, for this term, we can attempt to work backwards to determine the coefficient in Eq. (5.11) needed to generate the diffusion coefficient in Eq. (4.80). Likewise, using the concepts of conservation of mass and momentum at the mesoscale introduced in Section 5.1.2, we can specify the forms for the coefficients in Eq. (5.11) for the fluid seen by a particle (i.e. $\mathbf{A}^*_{\mathrm{pf}}$ and $\mathbf{G}^*_{\mathrm{f}}$) by using the stochastic models developed for $\mathbf{A}^*_{\mathrm{fp}}$ and $\mathbf{G}^*_{\mathrm{p}}$.

5.5.1 Phase-space diffusion

In this section, we consider two examples that are modeled using phase-space diffusion. The first example deals with the particle-size distribution of growing particles, and the second deals with particle-velocity fluctuations.

Growth-rate dispersion

One example of phase-space diffusion is growth-rate dispersion in crystallization processes. Many experiments have shown that, when crystals grow under the effect of supersaturation, the resulting NDF moves to higher particle sizes as expected, but at the same time spreads in particle-size space. This observation suggests that the rate of change of particle size G^*_{p} must contain both a pure-advection component G_{p} (see Section 5.2.2) and a self-diffusion component $c_{\mathrm{p}\xi_{\mathrm{p}}}$, where ξ_{p} is the particle mass. This mechanism was first proposed by Larson *et al.* (1985) and Zumstein & Rousseau (1987), although for a certain period of time growth-rate dispersion was confused with size-dependent growth, since the latter has a similar effect on the evolution of the NDF (see for example the work of Girolami & Rousseau (1985) and of Mydlarz & Briedis (1992)). In most published works, the growth-rate dispersion coefficient is calculated by fitting with data from experiments. There is, in fact, a significant lack of information on how this quantity is influenced by the operating conditions (most notably supersaturation, slip velocity, solid concentration, and temperature). It is believed that growth-rate dispersion is caused by the random formation of crystal-lattice defects, which makes each crystal grow with a slightly different growth rate.

In terms of the fluid mass seen by the particle ξ_{f}, conservation of mass at the mesoscale leads to the following mesoscale models in the limit of zero particle Stokes number (i.e. $\mathbf{U}^*_{\mathrm{p}} = \mathbf{U}^*_{\mathrm{f}} = \mathbf{U}_{\mathrm{f}}$):

$$\begin{aligned} G^*_{\mathrm{p}} &= G_{\mathrm{p}} + c_{\mathrm{p}\xi_{\mathrm{p}}} W_{\mathrm{p}\xi_{\mathrm{p}}}(t) \\ G^*_{\mathrm{f}} &= -G_{\mathrm{p}} - c_{\mathrm{p}\xi_{\mathrm{p}}} W_{\mathrm{p}\xi_{\mathrm{p}}}(t) \end{aligned} \quad \Longrightarrow \quad \mathbf{B} = \begin{bmatrix} c_{\mathrm{p}\xi_{\mathrm{p}}} & 0 \\ -c_{\mathrm{p}\xi_{\mathrm{p}}} & 0 \end{bmatrix}, \tag{5.114}$$

where $\mathbf{B}$ is the matrix appearing in the definition of the diffusion coefficients in Eq. (5.12). The resulting GPBE is then given by[43]

$$\frac{\partial n}{\partial t} = -\frac{\partial G_{\mathrm{p}} n}{\partial \xi_{\mathrm{p}}} + \frac{\partial G_{\mathrm{p}} n}{\partial \xi_{\mathrm{f}}} + \frac{1}{2}\frac{\partial^2}{\partial \xi_{\mathrm{p}}^2}\left(c^2_{\mathrm{p}\xi_{\mathrm{p}}} n\right) - \frac{\partial^2}{\partial \xi_{\mathrm{p}}\,\partial \xi_{\mathrm{f}}}\left(c^2_{\mathrm{p}\xi_{\mathrm{p}}} n\right) + \frac{1}{2}\frac{\partial^2}{\partial \xi_{\mathrm{f}}^2}\left(c^2_{\mathrm{p}\xi_{\mathrm{p}}} n\right),$$

where $n(t, \xi_{\mathrm{p}}, \xi_{\mathrm{f}})$ is the NDF. Because $\mathbf{B}$ is of rank one, the diffusion coefficients in this GPBE allow dispersion in only one direction of phase space. It is instructive to calculate the lower-order moments of the GPBE in Eq. (5.114) using the methods described

[43] For clarity, in this section we will omit the spatial advection terms when writing the GPBE.

in Section 4.3. For clarity, we will assume that G_p and $c_{p\xi_p}$ are constants. The zeroth-order moment, corresponding to the particle-number density N_p, is constant. The first-order moments with respect to ξ_p and ξ_f yield

$$\frac{d\varrho_p}{dt} = G_p N_p \quad \text{and} \quad \frac{d\varrho_f}{dt} = -G_p N_p,$$

which reflect the conservation of mass in the system. The second-order moments with respect to ξ_p^2 and ξ_f^2 yield

$$\frac{d[\![\xi_p^2]\!]}{dt} = 2G_p\varrho_p + c_{p\xi_p}^2 N_p \quad \text{and} \quad \frac{d[\![\xi_f^2]\!]}{dt} = -2G_p\varrho_f + c_{p\xi_p}^2 N_p,$$

where $[\![(\cdot)]\!] = \int (\cdot) n \, d\xi_p \, d\xi_f$ (e.g. $[\![1]\!] = N_p$, $[\![\xi_p]\!] = \varrho_p$, $[\![\xi_f]\!] = \varrho_f$), and the second-order mixed moment yields

$$\frac{d[\![\xi_p\xi_f]\!]}{dt} = G_p(\varrho_f - \varrho_p) - c_{p\xi_p}^2 N_p.$$

By defining the mass variances and covariance as

$$[\![\xi_p^2]\!]^* = \frac{1}{N_p}[\![\xi_p^2]\!] - \frac{\varrho_p^2}{N_p^2}, \qquad [\![\xi_p\xi_f]\!]^* = \frac{1}{N_p}[\![\xi_p\xi_f]\!] - \frac{\varrho_p\varrho_f}{N_p^2}, \qquad [\![\xi_f^2]\!]^* = \frac{1}{N_p}[\![\xi_f^2]\!] - \frac{\varrho_f^2}{N_p^2}$$

it is straightforward to show that

$$\frac{d[\![\xi_p^2]\!]^*}{dt} = c_{p\xi_p}^2, \qquad \frac{d[\![\xi_p\xi_f]\!]^*}{dt} = -c_{p\xi_p}^2, \qquad \frac{d[\![\xi_f^2]\!]^*}{dt} = c_{p\xi_p}^2,$$

and thus that the correlation coefficient between the particle and fluid masses goes to -1 for large t (or is equal to -1 for all t if the initial variances are zero). This result follows directly as a consequence of mass conservation at the mesoscale (i.e. Eq. (5.114)), and would be true for all choices of G_p and $c_{p\xi_p}$.

Fluid- and disperse-phase velocity fluctuations

Fluctuations in the disperse-phase velocity can arise from a number of different physical processes, but here we focus on fluctuations arising due to interactions with the fluid phase, for example, due to unsteady boundary layers and wakes around a particle (Tenneti *et al.*, 2012). Such fluctuations will be present when the particle Reynolds number defined in Table 1.1 is larger than unity. The simplest single-particle mesoscale model can be written as $\mathbf{A}_{fp}^* = \mathbf{A}_{fp} + \mathbf{b}_{pv_f}\mathbf{W}_{pv_f}(t)$, where the pure-advection term $\mathbf{A}_{fp}$ is as described in Section 5.3. Letting ξ_p and ξ_f be the mass of the particle and of the fluid seen by the particle, respectively, the conservation of momentum at the mesoscale leads to $\mathbf{A}_{pf}^* = -\xi_p\mathbf{A}_{fp}/\xi_f - \xi_p\mathbf{b}_{pv_f}\mathbf{W}_{pv_f}(t)/\xi_f$ and hence $\mathbf{A}_{pf} = -\xi_p\mathbf{A}_{fp}/\xi_f$ and $\mathbf{b}_{fv_f} = -\xi_p\mathbf{b}_{pv_f}/\xi_f$ in Eq. (5.11). Here, for clarity, we will assume that G_p^* and G_f^* have no fluctuating components due to stochastic processes. Moreover, for the time being, we will ignore the "diffusion" contribution in Eq. (4.80) (see Section 5.5.3 below), and thus the diffusion matrix found from Eq. (5.12) will depend only on $\mathbf{b}_{pv_f}$ and $\mathbf{b}_{fv_f}$:

$$\mathbf{B} = \begin{bmatrix} \mathbf{b}_{pv_f} & \mathbf{0} \\ \mathbf{b}_{fv_f} & \mathbf{0} \end{bmatrix} \implies \mathbf{D} = \frac{1}{2}\begin{bmatrix} \mathbf{b}_{pv_f}\mathbf{b}_{pv_f}^T & \mathbf{b}_{pv_f}\mathbf{b}_{fv_f}^T \\ \mathbf{b}_{fv_f}\mathbf{b}_{pv_f}^T & \mathbf{b}_{fv_f}\mathbf{b}_{fv_f}^T \end{bmatrix} = \frac{1}{2}\begin{bmatrix} 1 & -\xi_p/\xi_f \\ -\xi_p/\xi_f & \xi_p^2/\xi_f^2 \end{bmatrix} \otimes \mathbf{B}_{fp},$$

where the 3 × 3 diffusion matrix is defined by $\mathbf{B}_{\text{fp}} = \mathbf{b}_{\text{pv}_\text{f}} \mathbf{b}_{\text{pv}_\text{f}}^{\text{T}}$. On denoting the (homogeneous) NDF by $n(t, \mathbf{v}_\text{p}, \mathbf{v}_\text{f})$, the resulting GPBE is

$$
\begin{aligned}
\frac{\partial n}{\partial t} = & -\frac{\partial}{\partial \mathbf{v}_\text{p}} \cdot (\mathbf{A}_{\text{fp}} n) + \frac{\xi_\text{p}}{\xi_\text{f}} \frac{\partial}{\partial \mathbf{v}_\text{f}} \cdot (\mathbf{A}_{\text{fp}} n) \\
& + \frac{1}{2} \frac{\partial^2}{\partial \mathbf{v}_\text{p} \partial \mathbf{v}_\text{p}} : \left(\mathbf{B}_{\text{fp}} n \right) - \frac{\xi_\text{p}}{\xi_\text{f}} \frac{\partial^2}{\partial \mathbf{v}_\text{p} \partial \mathbf{v}_\text{f}} : \left(\mathbf{B}_{\text{fp}} n \right) + \frac{1}{2} \frac{\xi_\text{p}^2}{\xi_\text{f}^2} \frac{\partial^2}{\partial \mathbf{v}_\text{f} \partial \mathbf{v}_\text{f}} : \left(\mathbf{B}_{\text{fp}} n \right).
\end{aligned}
$$

The positive-definite matrix $\mathbf{B}_{\text{fp}}$ will be determined by the structure of the fluid flow around the particle, and will more than likely be significantly anisotropic (Tenneti *et al.*, 2012). Note that we have not included the fluid-velocity-fluctuation-dissipation model in Eq. (4.104) when writing the GPBE. Here, for clarity, we will focus exclusively on the interdependence of the fluid and particle velocities, for which it suffices to consider a 1D velocity phase space for v_p and v_f. The GPBE for this case is given by

$$
\frac{\partial n}{\partial t} = -\frac{\partial A_{\text{fp}} n}{\partial v_\text{p}} + \frac{\xi_\text{p}}{\xi_\text{f}} \frac{\partial A_{\text{fp}} n}{\partial v_\text{f}} + \frac{B_{\text{fp}}}{2} \frac{\partial^2 n}{\partial v_\text{p}^2} - \frac{\xi_\text{p} B_{\text{fp}}}{\xi_\text{f}} \frac{\partial^2 n}{\partial v_\text{p} \partial v_\text{f}} + \frac{B_{\text{fp}}}{2} \frac{\xi_\text{p}^2}{\xi_\text{f}^2} \frac{\partial^2 n}{\partial v_\text{f}^2}, \tag{5.115}
$$

where we will assume that $A_{\text{fp}} = (1/\tau_\text{p})(v_\text{f} - v_\text{p})$ and that B_{fp} is uniform in phase space. The first-order moments of Eq. (5.115) with respect to ξ_p and ξ_f, respectively, confirm that the phase densities ϱ_p and ϱ_f are constant. The second-order moments with respect to $\xi_\text{p} v_\text{p}$ and $\xi_\text{f} v_\text{f}$ yield, respectively,

$$
\frac{\text{d}\varrho_\text{p} U_\text{p}}{\text{d}t} = \frac{\varrho_\text{p}}{\tau_\text{p}} (U_\text{f} - U_\text{p}) \quad \text{and} \quad \frac{\text{d}\varrho_\text{f} U_\text{f}}{\text{d}t} = \frac{\varrho_\text{p}}{\tau_\text{p}} (U_\text{p} - U_\text{f}),
$$

and express the conservation of momentum at the macroscale. The third-order moments of Eq. (5.115) with respect to $\xi_\text{p} v_\text{p}^2$, $\xi_\text{f} v_\text{f}^2$, and $\xi_\text{p} v_\text{p} v_\text{f}$ yield (see Section 4.3.9 for details), respectively,

$$
\begin{aligned}
\frac{\text{d}[\![U_\text{p}^2]\!]}{\text{d}t} &= \frac{2}{\tau_\text{p}} \left([\![U_\text{p} U_\text{f}]\!] - [\![U_\text{p}^2]\!] \right) + B_{\text{fp}}, \\
\frac{\text{d}[\![U_\text{f}^2]\!]}{\text{d}t} &= \frac{2\varrho_\text{p}}{\tau_\text{p} \varrho_\text{f}} \left([\![U_\text{p} U_\text{f}]\!] - [\![U_\text{f}^2]\!] \right) + \frac{\varrho_\text{p}^2}{\varrho_\text{f}^2} B_{\text{fp}},
\end{aligned}
$$

and

$$
\frac{\text{d}[\![U_\text{p} U_\text{f}]\!]}{\text{d}t} = \frac{1}{\tau_\text{p}} \left[[\![U_\text{f}^2]\!] + \frac{\varrho_\text{p}}{\varrho_\text{f}} [\![U_\text{p}^2]\!] - \left(1 + \frac{\varrho_\text{p}}{\varrho_\text{f}} \right) [\![U_\text{p} U_\text{f}]\!] \right] - \frac{\varrho_\text{p}}{\varrho_\text{f}} B_{\text{fp}},
$$

where we have divided through by the constant phase densities. It is straightforward to show that the equations for the variances and covariance (e.g. $[\![U_\text{p}^2]\!]^* = [\![U_\text{p}^2]\!] - U_\text{p}^2$) have exactly the same forms, and hence B_{fp} can be seen to function as a variance source term. However, it can also be shown that the fluctuations leave the following quantity unchanged:[44]

$$
2[\![U_\text{p} U_\text{f}]\!]^* + \frac{\varrho_\text{p}}{\varrho_\text{f}} [\![U_\text{p}^2]\!]^* + \frac{\varrho_\text{f}}{\varrho_\text{p}} [\![U_\text{f}^2]\!]^*.
$$

[44] When the fluid-velocity-fluctuation-dissipation model in Eq. (4.104) is included, an additional variance reduction mechanism will be present.

Thus, if the initial variances are null, then $[\![U_p U_f]\!]^*$ is completely determined by $[\![U_p^2]\!]^*$ and $[\![U_f^2]\!]^*$. At steady state, the velocity fluctuations are related to B_{fp} by

$$[\![U_p^2]\!]^* = \frac{\varrho_f \tau_p B_{fp}}{2(\varrho_p + \varrho_f)}, \qquad [\![U_p U_f]\!]^* = -\frac{\varrho_p \tau_p B_{fp}}{2(\varrho_p + \varrho_f)}, \qquad [\![U_f^2]\!]^* = \frac{\varrho_p^2 \tau_p B_{fp}}{2\varrho_f(\varrho_p + \varrho_f)};$$

which implies that $[\![U_p U_f]\!]^* = -([\![U_p^2]\!]^* [\![U_f^2]\!]^*)^{1/2}$ (i.e. the fluid–particle velocity correlation coefficient is -1, implying that the fluctuations are anti-correlated due to the conservation of mesoscale momentum[45]). The remaining mesoscale modeling task is to determine how the dimensionless mesoscale parameter $C(\mathrm{Re_p}, \mathrm{St_p}) = \tau_p B_{fp}/|\mathbf{U}_f - \mathbf{U}_p|^2$ (and, more generally, $\mathbf{B}_{fp}$) depends on the particle Reynolds and Stokes numbers using, for example, DNS of the microscale equations (Tenneti *et al.*, 2012). In summary, the two examples discussed above illustrate how unresolved microscale phenomena can generate phase-space diffusion terms at the mesoscale. In general, the starting point for developing mesoscale models for a single particle will be DNS-derived correlations using dimensionless parameters such as the particle Reynolds and Stokes numbers. As shown in the two examples above, the mesoscale models for the fluid seen by the particle can often be derived by invoking the conservation of mass, chemical species, momentum, etc. at the mesoscale.

5.5.2 Physical-space diffusion

The most common example of physical-space diffusion is Brownian motion, which has already been discussed in Section 5.3.3. Let us consider a population of particles suspended in a fluid. As is well known, the fluid molecules interact with each other, resulting in a velocity distribution that, under equilibrium conditions, is the classical Maxwellian distribution. The fluid molecules also interact (and exchange momentum) with the suspended particles. When particle inertia is large (as measured by a finite particle Stokes number), the fluid–particle forces (i.e. $\mathbf{A}^*_{fp}$ and $\mathbf{A}^*_{pf}$) are the relevant mesoscale models for these interactions. However, when particle inertia is very small, particles are sensitive also to small, nearly instantaneous, momentum exchanges with the surrounding fluid molecules. The resulting forces yield random particle velocity fluctuations (Gardiner, 2004) that can be modeled either as a stochastic process in particle-velocity phase space when $1 \le \mathrm{St_p}$, or as a random walk in physical space when $\mathrm{St_p} = 0$. For the latter, Brownian motion is described with a positive-definite isotropic diffusion tensor: $\mathbf{\Gamma} = \Gamma \mathbf{I} \implies \nabla \cdot (\Gamma \nabla n)$, where the diffusion coefficient is calculated using the Stokes–Einstein equation:

$$\Gamma = \frac{k_B T_f}{3\pi \mu_f d_p}, \tag{5.116}$$

where k_B is the Boltzmann constant, T_f the fluid temperature, μ_f the fluid viscosity, and d_p the particle diffusion diameter. It is important to note that, in the limit $\mathrm{St_p} = 0$, the GPBE reduces to a PBE for the NDF $n(t, \mathbf{x}, \boldsymbol{\xi}_p)$ with advection velocity $\mathbf{U}_f$ and the diffusivity coefficient given in Eq. (5.116). The fluid velocity $\mathbf{U}_f$ and fluid properties $\boldsymbol{\phi}_f$ (including T_f) are found by solving separate transport equations. Indeed, as a general rule, when the GPBE is used to find the NDF $n(t, \mathbf{x}, \mathbf{v}_p, \boldsymbol{\xi}_p, \mathbf{v}_f, \boldsymbol{\xi}_f)$, the Brownian-force terms will appear in the models for $\mathbf{A}^*_{fp}$ and $\mathbf{A}^*_{pf}$ as velocity phase-space diffusion terms. (See Chapter 1 of Gardiner (2004) for details on the Langevin model for Brownian motion.)

[45]This result is derived in the limit where velocity fluctuations in the fluid phase are caused by the presence of particles. In a system where fluid-phase velocity fluctuations are generated by multiple sources, the fluid–particle velocity correlation coefficient would move towards zero and depend on the relative magnitude of the source terms.

5.5.3 Mixed phase- and physical-space diffusion

Some models (Reeks, 2005a, 2005b) for unclosed terms in the GPBE equation use mixed second-order derivatives involving **x** and some of the phase-space variables. As an example, let us consider the following GPBE for $n(t, x, v_\mathrm{p}, v_\mathrm{f}, \xi_\mathrm{f})$:

$$\frac{\partial n}{\partial t} + \frac{\partial}{\partial x}\left[v_\mathrm{p} n + (v_\mathrm{p} - v_\mathrm{f})\frac{\partial(\xi_\mathrm{f} n)}{\partial \xi_\mathrm{f}}\right] = -\frac{\partial}{\partial v_\mathrm{p}}\left[\frac{1}{\tau_\mathrm{p}}(v_\mathrm{f} - v_\mathrm{p})n\right] + \frac{\xi_\mathrm{p}}{\xi_\mathrm{f}}\frac{\partial}{\partial v_\mathrm{f}}\left[\frac{1}{\tau_\mathrm{p}}(v_\mathrm{f} - v_\mathrm{p})n\right]$$
$$+ \frac{B_\mathrm{fp}}{2}\frac{\partial^2 n}{\partial v_\mathrm{p}^2} - \frac{\xi_\mathrm{p} B_\mathrm{fp}}{\xi_\mathrm{f}}\frac{\partial^2 n}{\partial v_\mathrm{p}\partial v_\mathrm{f}} + \frac{B_\mathrm{fp}}{2}\frac{\xi_\mathrm{p}^2}{\xi_\mathrm{f}^2}\frac{\partial^2 n}{\partial v_\mathrm{f}^2}, \tag{5.117}$$

which is a spatially inhomogeneous version of Eq. (5.115). Here, ξ_f is the fluid mass seen by a particle, and its rate of change is given by Eq. (4.62). The latter introduces the mixed derivative between x and ξ_f on the left-hand side of Eq. (5.117) (see Eq. (5.1)). Now suppose that we would like to simplify the GPBE in Eq. (5.117) by integrating out the fluid variables v_f and ξ_f (and thereby eliminating the mixed derivative). On defining a new NDF as

$$n^*(t, x, v_\mathrm{p}) = \int n \, \mathrm{d}v_\mathrm{f} \, \mathrm{d}\xi_\mathrm{f},$$

the formal integration of Eq. (5.117) leads to

$$\frac{\partial n^*}{\partial t} + \frac{\partial v_\mathrm{p} n^*}{\partial x} = -\frac{\partial}{\partial v_\mathrm{p}}\left[\frac{1}{\tau_\mathrm{p}}(\langle U_\mathrm{f} | v_\mathrm{p}\rangle - v_\mathrm{p})n^*\right] + \frac{B_\mathrm{fp}}{2}\frac{\partial^2 n^*}{\partial v_\mathrm{p}^2}, \tag{5.118}$$

where the unknown conditional fluid velocity is defined by

$$\langle U_\mathrm{f} | v_\mathrm{p}\rangle n^* = \int v_\mathrm{f} n \, \mathrm{d}v_\mathrm{f} \, \mathrm{d}\xi_\mathrm{f}.$$

Note that, although we have eliminated the mixed derivative in Eq. (5.118), the new GPBE is no longer closed. In order to close Eq. (5.118), we follow the example of Reeks (2005b) and let[46]

$$\langle U_\mathrm{f} | v_\mathrm{p}\rangle n^* = U_\mathrm{f} n^* - \frac{\partial \mu n^*}{\partial x},$$

where μ is a nonnegative coefficient that depends on the correlation between the particle position and the fluid velocity seen by the particle. This model leads to a closed GPBE:

$$\frac{\partial n^*}{\partial t} + \frac{\partial}{\partial x}\left(v_\mathrm{p} n^* - \frac{1}{\tau_\mathrm{p}}\frac{\partial \mu n^*}{\partial v_\mathrm{p}}\right) = -\frac{\partial}{\partial v_\mathrm{p}}\left[\frac{1}{\tau_\mathrm{p}}(U_\mathrm{f} - v_\mathrm{p})n^*\right] + \frac{B_\mathrm{fp}}{2}\frac{\partial^2 n^*}{\partial v_\mathrm{p}^2}; \tag{5.119}$$

however, the closure has introduced a new mixed derivative between x and v_p. From a mathematical perspective, we can ask whether the mixed derivative terms can be modeled by a mesoscale stochastic process such as those given in Eq. (5.11). By isolating the terms involving the variables in the mixed derivatives both in Eq. (5.117) (i.e. x and ξ_f) and in Eq. (5.119) (i.e. x and v_p), it is possible to show that both of these equations are hyperbolic (not parabolic as required for a diffusion model). Thus, unfortunately, it does not appear that the mesoscale models needed to describe ξ_f (i.e. for non-dilute systems wherein the fluid-phase volume fraction is non-uniform) and $\mathbf{v}_\mathrm{p}$ (i.e. for systems with unresolved fluid

[46] In addition to the spatial gradient, the standard closure includes a term with a derivative with respect to v_p. A similar term is already included as the last term in Eq. (5.118). Indeed, since B_fp models forcing due to fluid-velocity fluctuations, it can be argued that μ should be proportional to B_fp.

velocity fluctuations that are not included in the NDF and the GPBE) can be easily formulated in terms of Lagrangian single-particle models.[47] Fortunately, however, such effects can be accounted for in Eulerian moment methods in a straightforward manner starting from the GPBE in Eq. (5.1).

5.6 Zeroth-order point processes

Point processes affect the state of the particles through discontinuous events. The principal difference between continuous and discontinuous events is that for the former during an infinitesimal time interval the internal coordinates are subject to infinitesimal changes, whereas for the latter the changes are finite. A classical example is the difference between the effect of the drag force and a collision event on particle velocity. The drag force acts continuously, causing a smooth change in particle velocity, whereas a collision instantaneously changes the particle velocity from the pre-collision value to the post-collision value.[48] Note that in the GPBE given in Eq. (5.1) the point processes are all lumped together into $\mathcal{S}_1$, but here we will separate them out, introducing new notation for each type.

5.6.1 Formation of the disperse phase

In many multiphase systems the disperse phase is not present at the beginning, but is produced during the evolution of the multiphase system. However, very different scenarios are possible. In the case of systems constituted by a single chemical component, the disperse phase can be spontaneously formed because local thermodynamic conditions allow a state transition. One such example is boiling water, where, for example, water flows on a hot surface and, due to local temperature values, some water-vapor bubbles can be formed. In this case the driving force is the distance (often expressed as the ratio) between the local water temperature and the equilibrium temperature for the liquid–vapor transition. Another important example is crystallization from a melt, where solid crystalline particles are formed within a liquid phase. In such systems, the formation of new disperse-phase elements generally occurs through a nucleation process. Important practical examples are crystallization and precipitation processes, where, for example, due to the supersaturation of the continuous phase with respect to one (or more) components, solid particles (crystals) are nucleated. Another interesting example is that of particle formation in flames. In this case some chemical reactions occurring in the gas phase result in the build up of supersaturation and the subsequent formation of a solid product (e.g. titanium dioxide particles, soot, etc.). Very often in the specialized literature this latter process is referred to as inception. Many other examples exist for fluid–fluid systems, notably demixing of immiscible liquids and vapor formation in boiling liquid systems.

[47] In Euler–Lagrange simulations, the Lagrangian particles are simulated according to a closed GPBE for $n^*(t, \mathbf{x}, \mathbf{v}_\mathrm{p})$ that takes $\langle \mathbf{U}_\mathrm{f} | \mathbf{v}_\mathrm{p} \rangle = \mathbf{U}_\mathrm{f}$ (i.e. they ignore the contribution due to μ). Because μ is proportional to $1/\mathrm{St}_\mathrm{p}$ (Reeks, 2005a, 2005b) such codes are not suitable for approximating disperse-phase flows with small or even moderate particle Stokes numbers.

[48] This is formally true only for interactions between particles involving hard-sphere potentials. In fact, under these circumstances, the interacting particles perceive each others' presence only at the point of contact. When the interactions are governed by a smooth potential, theoretically collisions should not be treated as point processes. Very often, however, even in these cases collisions are described as discontinuous processes occurring at the shortest distance between interacting particles, as in the case of the Maxwell molecules, as explained in the original work of Maxwell (1867). For more details readers are referred to Chapter 6.

In the case of systems constituted by two or more chemical components, local conditions can allow the spontaneous formation of a new (or multiple) thermodynamic phase(s) constituted by part of the chemical components. In this case the driving force is identified by the ratio of the concentration of the chemical component involved in the formation of the new phase to the corresponding equilibrium concentration. This ratio is often referred to as supersaturation. With multiple components the formation of the new phase can occur through either a nucleation or a spinodal decomposition process. In fact, depending on the local thermodynamic conditions, different scenarios are possible. The formation of a new thermodynamic phase is generally related to small molecular fluctuations that can bring together a number of molecules of the secondary phase. Although the new phase is thermodynamically stable, there may be an energy barrier to overcome, which is often represented by the sign of the second derivative of the free energy with respect to a compositional parameter. When this energy barrier is present, particles are formed through nucleation, meaning that only particles larger than a critical size (called nuclei) are locally stable and can further grow, forming the new disperse-phase elements. Therefore a nucleation event occurs only when random fluctuations bring together the critical number of molecules (to form a nucleus). In this case particle formation is localized at specific sites and its rate is controlled by the energy barrier and the thermal energy of the molecules of the newly formed phase. When there is no energy barrier to overcome, because the new phase is locally and globally stable, even two molecules of the disperse phase are more stable together than separated. In this case the disperse-phase elements are formed through spinodal decomposition, whose rate is controlled by diffusion. Spinodal decomposition is generally less localized than nucleation and often results in complete phase separation rather than in a stable disperse system.

Below, we will discuss two important examples of zeroth-order point processes *as seen from the perspective of the particle-phase NDF*. However, some zeroth-order point processes such as the formation of the disperse phase from the fluid phase are accompanied by a change of state in the fluid phase (i.e. the total mass and momentum of the two phases are conserved). Thus, seen in the perspective of the particle and fluid phase, the overall process requires a decrease in the mass of the fluid phase equal to the mass of the formed particle, which is represented by the term $\mathcal{S}_{\mathrm{M}}$ in Eq. (4.68) on page 119. In other words, the term $\mathcal{S}_{\mathrm{M}}$ in Eq. (4.77) representing the rate of addition of mass to the particle phase from the fluid phase must follow from the source term for the zeroth-order point process for formation of the disperse phase. Using the properties of delta functions, we can formally write the source term in Eq. (5.1) for a zeroth-order point process as

$$\mathcal{S}_1 = J(\boldsymbol{\phi}_{\mathrm{f}})\delta(\mathbf{v}_{\mathrm{p}} - \mathbf{U}_{\mathrm{f}})\delta(\boldsymbol{\xi}_{\mathrm{p}} - \boldsymbol{\xi}_{\mathrm{pc}})\delta(\mathbf{v}_{\mathrm{f}} - \mathbf{U}_{\mathrm{f}})\delta(\boldsymbol{\xi}_{\mathrm{f}} - \boldsymbol{\xi}_{\mathrm{fc}}), \tag{5.120}$$

where $J(\boldsymbol{\phi}_{\mathrm{f}}) \geq 0$ is the rate of formation of particles with properties $\boldsymbol{\xi}_{\mathrm{pc}}$, which depends on the mean composition of the fluid phase $\boldsymbol{\phi}_{\mathrm{f}}$. Here we assume that both the velocity of the new particles and the velocity of the fluid seen by these particles are equal to the mean fluid velocity $\mathbf{U}_{\mathrm{f}}$. The properties of the new particle $\boldsymbol{\xi}_{\mathrm{pc}}$ and of the fluid seen by these particles $\boldsymbol{\xi}_{\mathrm{fc}}$ must be specified, and the latter must be consistent with the mass balance in Eq. (4.77). For example, if ξ_{p1} and ξ_{f1} are masses, then $\xi_{\mathrm{pc1}} = m_{\mathrm{c}}$ is the mass of a new particle and $\xi_{\mathrm{fc1}} = -m_{\mathrm{c}}$ for consistency with the overall mass balance.[49] Indeed, using the notation from Chapter 4, the corresponding changes in number,

[49] In many cases, m_{c} is extremely small and thus has a negligible affect on the mass and momentum balances. However, for consistency, it is best to include it when computing the moment-transport equations from the GPBE.

$$\mathcal{S}_{\mathrm{N}} = \int \mathcal{S}_1 \, \mathrm{d}\mathbf{v}_{\mathrm{p}} \, \mathrm{d}\boldsymbol{\xi}_{\mathrm{p}} \, \mathrm{d}\mathbf{v}_{\mathrm{f}} \, \mathrm{d}\boldsymbol{\xi}_{\mathrm{f}} = J,$$

mass,

$$\mathcal{S}_{\mathrm{M}} = \int \xi_{\mathrm{p}1} \mathcal{S}_1 \, \mathrm{d}\mathbf{v}_{\mathrm{p}} \, \mathrm{d}\boldsymbol{\xi}_{\mathrm{p}} \, \mathrm{d}\mathbf{v}_{\mathrm{f}} \, \mathrm{d}\boldsymbol{\xi}_{\mathrm{f}} = m_{\mathrm{c}} J,$$

$$\int \xi_{\mathrm{f}1} \mathcal{S}_1 \, \mathrm{d}\mathbf{v}_{\mathrm{p}} \, \mathrm{d}\boldsymbol{\xi}_{\mathrm{p}} \, \mathrm{d}\mathbf{v}_{\mathrm{f}} \, \mathrm{d}\boldsymbol{\xi}_{\mathrm{f}} = -m_{\mathrm{c}} J,$$

and momentum,

$$\varrho_{\mathrm{p}} [\![\mathbf{S}]\!]_{\mathrm{p}} = \int \xi_{\mathrm{p}1} \mathbf{v}_{\mathrm{p}} \mathcal{S}_1 \, \mathrm{d}\mathbf{v}_{\mathrm{p}} \, \mathrm{d}\boldsymbol{\xi}_{\mathrm{p}} \, \mathrm{d}\mathbf{v}_{\mathrm{f}} \, \mathrm{d}\boldsymbol{\xi}_{\mathrm{f}} = m_{\mathrm{c}} \mathbf{U}_{\mathrm{f}} J,$$

$$\int \xi_{\mathrm{f}1} \mathbf{v}_{\mathrm{f}} \mathcal{S}_1 \, \mathrm{d}\mathbf{v}_{\mathrm{p}} \, \mathrm{d}\boldsymbol{\xi}_{\mathrm{p}} \, \mathrm{d}\mathbf{v}_{\mathrm{f}} \, \mathrm{d}\boldsymbol{\xi}_{\mathrm{f}} = -m_{\mathrm{c}} \mathbf{U}_{\mathrm{f}} J,$$

verify the overall conservation of mass and momentum between the phases. In the next two sections, we give examples of how the rate of formation J is found for two different systems.

5.6.2 Nucleation of crystals from solution

Insofar as the solution of the GPBE is concerned, it is sufficient to define the functional form of the rate of formation of new elements of the disperse phase $J(\boldsymbol{\phi}_{\mathrm{f}})$. For example, the formation of particles in solution is typically classified as homogeneous or heterogeneous. For the former, particles are formed in the bulk solution without the support of any surface, whereas for the latter an external surface (provided, for example, by particles that have already formed or by foreign particles) is involved. According to the classical theory (Myerson, 2002), the nucleation rate prescribes a burst of monodisperse nuclei with size d_{c}. The resulting nucleation rate is

$$J = \frac{2D}{d^5} \exp\left(-\frac{16\pi\sigma^3 \tilde{v}^2}{3k_{\mathrm{B}}^3 T_{\mathrm{f}}^3 [\ln S]^2} \right), \tag{5.121}$$

where D is the molecular diffusion coefficient of the solute (the explicit expression can be found in Bird *et al.*, 1960), d its molecular diameter, k_{B} the Boltzmann constant, T_{f} the absolute fluid temperature, σ the interfacial tension between the particles and the solution (in the case of homogeneous nucleation) or the foreign surface (in the case of heterogeneous nucleation), S the supersaturation in the fluid, and $\tilde{v}$ the solute molecular volume. The size of the nuclei is estimated by application of nucleation theory:

$$d_{\mathrm{c}} = \frac{4\sigma\tilde{v}}{k_{\mathrm{B}} T_{\mathrm{f}}} \frac{1}{\ln S} \quad \Longrightarrow \quad m_{\mathrm{c}} = \rho_{\mathrm{p}} k_{\mathrm{V}} d_{\mathrm{c}}^3. \tag{5.122}$$

More details on the theory and on the relevant applications of Eq. (5.121) can be found in the handbook edited by Myerson (2002).

5.6.3 Nucleation of vapor bubbles in a boiling liquid

It is generally assumed that the formation of vapor (within a bulk liquid) starts when the liquid temperature becomes greater than the equilibrium temperature (i.e. the saturation temperature, T_{sat}). Under this hypothesis, the rate of evaporation (for $T_{\text{f}} > T_{\text{sat}}$) can be written as

$$J = \frac{6\dot{m}_{\text{cp}}}{\pi\rho_{\text{p}}d_{\text{c}}^3}, \tag{5.123}$$

where $\dot{m}_{\text{cp}}$ (expressed in kg m^{-3} s^{-1}) represents the mass of liquid evaporating per unit volume of liquid–vapor mixture per unit time and d_{c} is the size of a nucleated bubble (i.e. $m_{\text{c}} = \rho_{\text{p}}k_{\text{V}}d_{\text{c}}^3$). Different regimes are possible, corresponding to the evaporation rate. For nucleate boiling the following equation can be used:

$$\dot{m}_{\text{cp}} = \varrho_{\text{f}}\beta\frac{|T_{\text{f}} - T_{\text{sat}}|}{T_{\text{sat}}}, \tag{5.124}$$

where β is the evaporation frequency and ϱ_{f} is the phase density of the liquid phase. The evaporation frequency β (s^{-1}) can be derived from theoretical considerations and strong simplifying assumptions:

$$\beta = \frac{6L_{\text{v}}}{d_{\text{c}}}\sqrt{\frac{M_{\text{w}}}{2\pi RT_{\text{sat}}}}\left(\frac{\rho_{\text{f}}}{\rho_{\text{f}} - \rho_{\text{p}}}\right), \tag{5.125}$$

where M_{w} is the molecular weight of water molecules and L_{v} is the latent heat of evaporation. This equation is derived under the assumption of a flat interface and the formation of disperse bubbles with constant diameter. Nevertheless, the same equation is used under very different operating conditions by multiplying the rate by a correction factor ranging from zero to infinity. More details on bubble nucleation are reported in the work of Lee (1979) and Tanasawa (1991).

5.7 First-order point processes

First-order point processes group together all the discontinuous processes involving a single particle. Because only one particle takes part in the discontinuous event, the rate can be assumed to be proportional to the NDF. A common example is particle breakage: when a particle (solid, droplet or bubble) undergoes some sort of stress it can be broken into fragments. In this case positive and negative source terms must be included, since one particle will disappear, while other particles (fragments) will simultaneously enter the phase space. Another example is particle deposition on walls and boundaries, wherein only the negative source terms are present (Bensaid *et al.*, 2009, 2010). Here, we will write the source term in the GPBE as

$$\mathcal{S}_1 = h^+ - h^-,$$

where h^+ is the rate of production and h^- is the rate of loss of particles due to first-order point processes. First-order processes are generally quantified by $b(\mathbf{v}_{\text{p}}, \boldsymbol{\xi}_{\text{p}}, \mathbf{v}_{\text{f}}, \boldsymbol{\xi}_{\text{f}})\text{d}t$, which is the probability that a particle with velocity $\mathbf{v}_{\text{p}}$ and internal coordinates $\boldsymbol{\xi}_{\text{p}}$ in a fluid with velocity $\mathbf{v}_{\text{f}}$ and internal coordinates $\boldsymbol{\xi}_{\text{f}}$ undergoes the process under investigation in the infinitesimal time interval dt. The quantity b is the frequency of the process and has units of inverse time, and its dependence on the particle and fluid velocities and on the fluid properties is used, for example, in shear-induced breakage. According to this definition the number of particles per unit volume and unit time disappearing because of the first-order process can be readily written as (suppressing the t and $\mathbf{x}$ dependences for clarity)

$$h^-(\mathbf{v}_{\text{p}}, \boldsymbol{\xi}_{\text{p}}, \mathbf{v}_{\text{f}}, \boldsymbol{\xi}_{\text{f}}) = b(\mathbf{v}_{\text{p}}, \boldsymbol{\xi}_{\text{p}}, \mathbf{v}_{\text{f}}, \boldsymbol{\xi}_{\text{f}})n(\mathbf{v}_{\text{p}}, \boldsymbol{\xi}_{\text{p}}, \mathbf{v}_{\text{f}}, \boldsymbol{\xi}_{\text{f}}). \tag{5.126}$$

In order to quantify the rate of formation of particles due to first-order processes an additional function must be defined. This is usually introduced as a conditional probability density function (PDF) that states the probability of formation of a daughter particle with $\mathbf{v}'_\mathrm{p}$ and $\boldsymbol{\xi}'_\mathrm{p}$ from a mother particle with $\mathbf{v}_\mathrm{p}$ and $\boldsymbol{\xi}_\mathrm{p}$. This PDF has to satisfy some constraints, such as the normalization condition:[50]

$$\int P(\mathbf{v}'_\mathrm{p}, \boldsymbol{\xi}'_\mathrm{p} | \mathbf{v}_\mathrm{p}, \boldsymbol{\xi}_\mathrm{p}) \mathrm{d}\mathbf{v}'_\mathrm{p} \, \mathrm{d}\boldsymbol{\xi}'_\mathrm{p} = 1, \tag{5.127}$$

which states that the probability of producing the daughter particles is unity. (Note that, if the daughter particles always have the same velocity as the mother particle, then the dependence on velocity is not required either for P or in the following formulas.)

In many cases, instead of the PDF in Eq. (5.127), an equivalent formulation in terms of a corresponding conditional NDF $N(\mathbf{v}'_\mathrm{p}, \boldsymbol{\xi}'_\mathrm{p} | \mathbf{v}_\mathrm{p}, \boldsymbol{\xi}_\mathrm{p})$ is used. The corresponding mathematical object is such that the integral over phase space now yields

$$\int N(\mathbf{v}'_\mathrm{p}, \boldsymbol{\xi}'_\mathrm{p} | \mathbf{v}_\mathrm{p}, \boldsymbol{\xi}_\mathrm{p}) \mathrm{d}\mathbf{v}'_\mathrm{p} \, \mathrm{d}\boldsymbol{\xi}'_\mathrm{p} = \nu(\mathbf{v}_\mathrm{p}, \boldsymbol{\xi}_\mathrm{p}), \tag{5.128}$$

where ν is the number of new particles formed by the first-order process. Clearly the relationship between the two functions is a simple renormalization. The function N is often referred to as the daughter distribution function. The rate of formation of new particles due to first-order discontinuous processes can now be expressed as

$$h^+(\mathbf{v}_\mathrm{p}, \boldsymbol{\xi}_\mathrm{p}, \mathbf{v}_\mathrm{f}, \boldsymbol{\xi}_\mathrm{f}) = \int N(\mathbf{v}_\mathrm{p}, \boldsymbol{\xi}_\mathrm{p} | \mathbf{v}'_\mathrm{p}, \boldsymbol{\xi}'_\mathrm{p}) b(\mathbf{v}'_\mathrm{p}, \boldsymbol{\xi}'_\mathrm{p}, \mathbf{v}_\mathrm{f}, \boldsymbol{\xi}_\mathrm{f}) n(\mathbf{v}'_\mathrm{p}, \boldsymbol{\xi}'_\mathrm{p}, \mathbf{v}_\mathrm{f}, \boldsymbol{\xi}_\mathrm{f}) \mathrm{d}\mathbf{v}'_\mathrm{p} \, \mathrm{d}\boldsymbol{\xi}'_\mathrm{p} \tag{5.129}$$

or, in terms of P, as

$$h^+ = \int \nu(\mathbf{v}'_\mathrm{p}, \boldsymbol{\xi}'_\mathrm{p}) P(\mathbf{v}_\mathrm{p}, \boldsymbol{\xi}_\mathrm{p} | \mathbf{v}'_\mathrm{p}, \boldsymbol{\xi}'_\mathrm{p}) b(\mathbf{v}'_\mathrm{p}, \boldsymbol{\xi}'_\mathrm{p}, \mathbf{v}_\mathrm{f}, \boldsymbol{\xi}_\mathrm{f}) n(\mathbf{v}'_\mathrm{p}, \boldsymbol{\xi}'_\mathrm{p}, \mathbf{v}_\mathrm{f}, \boldsymbol{\xi}_\mathrm{f}) \mathrm{d}\mathbf{v}'_\mathrm{p} \, \mathrm{d}\boldsymbol{\xi}'_\mathrm{p}. \tag{5.130}$$

Note that these expressions are written under the assumption that the fluid velocity and composition do not change during the first-order process. Thus, henceforth we will suppress $\mathbf{v}_\mathrm{f}$ and $\boldsymbol{\xi}_\mathrm{f}$ in the notation when they are not needed. In the following sections, we review some of the most common functional forms for b and N.

5.7.1 Particle filtration and deposition

When particles flow near solid walls, then, due to collisions with the walls, deposition can occur. Particle deposition is a surface process, since it occurs on the surface of the wall and depends, among other things, on the extent of the surface. However, in many practical applications, such as particle filtration, particles are forced through a complex structure of pores and channels (as, for example, in ceramic or fiber filters) that has to be described as a porous medium. For such cases the deposition process is modeled as a volumetric first-order process. Let us consider a system of packed objects with a particle-laden fluid flowing through the voids left between the objects. This is a common situation in catalytic processes, ground-water flows, and filtration processes. Let us now describe the group of objects as a porous medium characterized by its porosity, ϵ, and by an average object size,

[50] If the daughter particles depend on the fluid properties, a dependence on $\mathbf{v}_\mathrm{f}$ and $\boldsymbol{\xi}_\mathrm{f}$ can be included in the definitions of P and N.

D_G. The porosity is readily defined as the amount of space unoccupied by the objects and free to be filled with fluid, whereas the average object size is often interpreted as the equivalent spherical diameter $D_G = 6/s$, which is determined in turn from the specific surface area s of the porous medium. If the system is described as a continuum with superficial particle velocity[51] $\mathbf{v}_p$, the coefficient for the rate of particle deposition, or in other words the coefficient for the first-order process, is (Elimelech *et al.*, 1998)

$$b = \frac{3}{2}\frac{|\mathbf{v}_p|}{D_G}\frac{1-\epsilon}{\epsilon}. \tag{5.131}$$

This expression is derived by considering one isolated spherical object of size D_G, with particles flowing at very low velocity $\mathbf{v}_p$ around it, in the absence of repulsive forces between the particles and the object (often called the collector).[52] In the case of particle deposition, the daughter NDF N is identically null, since when particles hit the collector they disappear from the multiphase system without generating new particles. Particles are instead generated if, after attaching to the collector, they detach (or are released) back into the multiphase system. Since in real systems the conditions under which Eq. (5.131) is derived are not respected, a transport efficiency η is introduced in order to account for non-ideal situations related to the flow regime of particles around the collector and other factors. The overall transport efficiency is expressed as $\eta = \eta_D + \eta_I - \eta_D\eta_I$, where η_D is the efficiency related to Brownian motion and η_I is related to interception. Since particles are subject to a random Brownian force, they can be brought into close proximity of the collector and adhere to it. The efficiency of this process can be quantified as

$$\eta_D = 3.5g(\epsilon)\mathrm{Pe}^{-2/3} = 3.5g(\epsilon)\left(\frac{d_p|\mathbf{v}_p|}{\Gamma}\right)^{-2/3}, \tag{5.132}$$

where here a different definition of the Péclet number is used, Γ is the Stokes–Einstein diffusion coefficient defined in Eq. (5.116), and the Kuwabara function is

$$g(\epsilon) = \left(\frac{\epsilon}{2-\epsilon-\frac{9}{5}(1-\epsilon)^{1/3}-\frac{1}{5}(1-\epsilon)^2}\right). \tag{5.133}$$

Equation (5.132) states that when the particle velocity increases (or when the particle size increases) the theoretical conditions under which the deposition rate coefficient was derived are no longer valid and the actual deposition rate is smaller than the theoretical one (i.e. $\eta_D \ll 1$). Moreover, because particles have a finite size d_p, the deposition rate coefficient (derived assuming point particles) must be corrected. This correction is introduced through the interception efficiency:

$$\eta_I = 1.5\frac{M_G^2 g(\epsilon)^3}{(1+M_G)^s}, \tag{5.134}$$

where $M_G = d_p/D_G$ and $s = (3-2\epsilon)/\epsilon$. More accurate expressions for these efficiencies (including efficiencies accounting for repulsive particle–collector forces) have been

[51] The velocity that the particles would have if they occupied the entire space without the objects forming the porous medium.

[52] This assumption implies that when particles hit the surface of the object they attach to it immediately. If repulsive forces arise, then an efficiency, accounting for the fact that not all collisions between particles and the collector are successful, must be introduced.

formulated (Ma *et al.*, 2009; Weroński & Elimelech, 2008). Concluding the discussion on particle deposition h^-, the overall rate appearing in Eq. (5.126) is

$$b = -\left(\frac{3}{2}\frac{|\mathbf{v}_\mathrm{p}|}{D_\mathrm{G}}\frac{1-\epsilon}{\epsilon}\eta(\mathbf{v}_\mathrm{p}, d_\mathrm{p})\right), \tag{5.135}$$

and h^+ is null when particles remain stuck to solid surfaces.

5.7.2 Particle breakage

Among first-order processes, particle breakage is one of the most common. Particle breakage occurs when some form of stress interferes with the forces that keep the particle intact. The dynamics of fragmentation is very different for solid particles (i.e. crystals, aggregates, etc.) than for fluid particles (i.e. bubbles or droplets). In general, two types of stresses can cause breakage: (i) stresses induced by collisions and (ii) stresses caused by velocity gradients in the fluid. In the case of collisions, the relevant forces are generally induced by particle–particle collisions or by particle–wall collisions. The latter can in turn be divided into impact with stationary and moving walls (Ramachandran *et al.*, 2009). Crystals impacting on impeller blades in crystallizers (also known as secondary nucleation) or particles impacting on pipe walls in bends are very common examples. The amount of energy involved in these cases is generally enough to break chemical bonds (i.e. ionic, covalent, or hydrogen bonds). For such cases, since it is very difficult to derive from first principles the breakage kernel, purely empirical or semi-empirical expressions are used (Myerson, 2002; Randolph & Larson, 1971). In the case of stresses caused by velocity gradients in the fluid, the amount of energy involved is much smaller and therefore only weaker bonds are broken, namely those induced by Van der Waals or other attractive forces of electrostatic origin, which typically keep together aggregates of separate primary particles. For this case, it is relatively easy to derive theoretical expressions for the breakage kernels.

In general, the breakage kernel is written as a nonlinear function of the ratio between fragmenting and cohesive forces. Different functional forms (e.g. power-law and exponential) have been adopted in the literature, resulting in different expressions. By highlighting the functional dependence on particle size and the fragmentation driving force, different expressions are obtained. In general, the fragmenting driving force can be identified with the impact, collisional, or relative velocity U_r for particles fragmented by collision with a surface; or with the fluid shear rate G_f in the case of particles fragmented by the stresses induced by the fluid. For example, the power-law breakage kernel is

$$b = C_1 G_\mathrm{f}^{C_2} d_\mathrm{p}^{C_3}, \tag{5.136}$$

whereas the exponential kernel is

$$b = C_4 G_\mathrm{f} \exp\left(-\frac{C_5 G_\mathrm{f}^{C_6}}{G_\mathrm{f}^2 d_\mathrm{p}}\right). \tag{5.137}$$

In these expressions, C_1–C_6 are constants whose values are generally determined through fitting with data from experiments (Marchisio *et al.*, 2006). For the case of particles breaking up because of impact on stationary walls or moving surfaces, the shear rate G_f is replaced with the impact velocity U_r and the corresponding constants have different units and different meanings. The power-law kernel has been largely used purely on empirical grounds, but is also supported by the mechanistic model of Ghadiri & Zhang (2002), which predicts that the extent of breakage is proportional to particle size d_p and to U_r^2, and by the model of Moreno-Atanasio & Ghadiri (2006) that predicts a damage ratio, defined as the ratio of broken contacts to the initial number of bonds, that is proportional to $d_p^{5/3}$ and to U_r^2. Additional examples and more details on the derivations can be found in the works of Ramachandran *et al.* (2009) and Rajniak *et al.* (2008). The functional form of the exponential kernel has a theoretical basis that will be discussed below for the particular case of particle fragmentation in turbulent flows.

Breakage kernels in turbulent flow

In dilute particle-laden turbulent flow, the fluid shear rate acting on the particles is approximated as $G_f = \sqrt{\varepsilon_f/\nu_f}$, where ε_f is the turbulence-dissipation rate in the continuous phase, which must be found from a turbulence model. The latter introduces a great deal of uncertainty because for many disperse multiphase flows (e.g. gas–liquid flows) accurate turbulence models do not exist. Nevertheless, for particles undergoing breakage in a turbulent fluid the literature is particularly rich, but almost all of the available models are highly empirical. One of the most popular engineering models is (Coulaloglou & Tavlarides, 1977)

$$b = C_1 d_p^{-2/3} \varepsilon_f^{1/3} \exp\left(-\frac{C_2 \sigma}{\rho_p \varepsilon_f^{2/3} d_p^{5/3}}\right), \tag{5.138}$$

where C_1 and C_2 are two constants that are determined by fitting to experimental data. This kernel was originally derived for fluid droplets under the hypothesis of local isotropy and assuming that the droplet size is within the inertial sub-range. A breakage event occurs if the kinetic energy transmitted by an eddy to the particle is greater than the droplet surface energy. The breakup frequency is derived by considering the fraction of eddies with sufficient kinetic energy to cause breakage present in the system. This can be done by estimating the fraction of particles breaking and by estimating the characteristic time required for breakage (t_b) through the second-order structure function of the fluid velocity $D_{LL}(r)$. This function represents the covariance of the difference in velocity differences between two points at a given distance r and, according to turbulence theory, is proportional to $(\varepsilon_f r)^{2/3}$. Assuming that the motion of the daughter particle is similar to the relative motion of two lumps of fluid in a turbulent flow and that the distance r is the particle size, the following approximation is found: $(d_p/t_b)^{2/3} \approx D_{LL}(d_p) \propto (\varepsilon_f L)^{2/3}$. This can be used to solve for t_b, resulting in Eq. (5.138). Another popular engineering kernel, derived for liquid droplets dispersed in another liquid, was proposed by Narsimhan *et al.* (1979):

$$b = C_3 \operatorname{erfc}\left(3.5\left(\frac{d_p}{d_p^*}\right)^{-5/6}\right). \tag{5.139}$$

It was derived by considering the arrival of turbulent eddies on the surface of a droplet with maximum stable diameter d_p^* as a Poisson process. The original expression was modified by Alopaeus *et al.* (2002) by adding the dependence on $\varepsilon_f^{1/3}$ in the eddy collisions frequency and the viscous stresses in the energy balance, considered in the probability of a collision turning into breakup. Laakkonen *et al.* (2006) used this kernel to describe bubble breakup, considering the fact that the mechanisms of breakup of drops and bubbles are closely related. They further modified the expression by replacing the disperse-phase viscosity with the continuous-phase viscosity, considering that one component of the forces opposing bubble breakup is given by the viscous stresses of the surrounding liquid. The final expression for the breakage kernel is

$$b = C_4 \varepsilon_\mathrm{f}^{1/3} \,\mathrm{erfc}\left(\sqrt{C_5 \frac{\sigma}{\rho_\mathrm{f} \varepsilon_\mathrm{f}^{2/3} d_\mathrm{p}^{5/3}} + C_6 \frac{\mu_\mathrm{f}}{\sqrt{\rho_\mathrm{f}\rho_\mathrm{p}} \varepsilon_\mathrm{f}^{1/3} d_\mathrm{p}^{4/3}}} \right), \tag{5.140}$$

where also in this case C_4, C_5, and C_6 are constants that can be estimated through the theory, but are then tuned by fitting with data from experiments. It appears that the constant C_4 has the dimension of length to the minus two thirds and, according to the theory, should be linked to an integral length scale. This means that C_4 is not a true kinetic constant, but contains some scaling parameter; hence, great caution should be exercised when employing Eq. (5.140). There is no reliable way to calculate C_4 *a priori*; therefore its value is determined by fitting experimental data. For C_5 and C_6 the values adopted by Laakkonen *et al.* (2006) are generally used, namely 0.04 and 0.01, respectively. Under the same phenomenological simplifications of the above-mentioned models, and considering the different contributions of the population of eddies in the inertial sub-range, Luo & Svendsen (1996) derived the following engineering correlation for gas bubbles dispersed in a turbulent liquid:

$$bN = 0.92 \left(\frac{\varepsilon_\mathrm{f}}{d_\mathrm{p}'^2} \right)^{1/3} \int_{\zeta_\mathrm{min}}^{1} \frac{(1+\zeta)^2}{\zeta^{11/3}} \exp\left(-\frac{12 c_\mathrm{f} \sigma}{\beta \rho_\mathrm{p} \varepsilon_\mathrm{f}^{2/3} d_\mathrm{p}'^{5/3} \zeta^{11/3}} \right) \mathrm{d}\zeta, \tag{5.141}$$

where $\zeta = \eta_\mathrm{k}/d_\mathrm{p}'$, $\beta = 3/2$, $c_\mathrm{f} = f^{2/3} + (1-f)^{2/3} - 1$ is the increase coefficient for the surface area, $f = (d_\mathrm{p}/d_\mathrm{p}')^3$ is the volume fraction of one of the two fragments, d_p is the size of the daughter particle, and d_p' is the size of the parent particle. Equation (5.141) is not a pure breakage kernel, but is instead the product of the breakage kernel and a daughter distribution function valid only for binary breakage (i.e. formation of two fragments). It is worth mentioning that for this kernel the only parameter that has to be fixed arbitrarily is ζ_min. This quantity represents the minimum length scale of the turbulent eddies (normalized with respect to the parent particle size) which contribute to the breakup of the particle. This quantity is often taken equal to zero (Sanyal *et al.*, 2005). Other kernels proposed in the literature are those of Martinez-Bazán *et al.* (1999) and of Sathyagal & Ramkrishna (1996). For a review and comparison of the different kernels (especially for fluid–fluid systems) readers are referred to the work of Patruno *et al.* (2009).

As with breakage of liquid droplets, breakage of solid aggregates depends strongly on the ratio between the particle size and the smallest turbulent eddy. When the size of a particle is greater than the size of the smallest eddies, breakage is likely to occur by means of instantaneous normal stresses due to pressure fluctuations acting on the surface of the particle. For particle sizes smaller than the turbulence microscale, breakage is likely to be caused by shear stresses originating from the turbulent dynamic velocity differences acting on the opposite sides of the particle. Under these conditions the breakage kernel can be written as a function of the aggregate strength (τ_s), which can be estimated as $\tau_\mathrm{s} = 9k_\mathrm{co}\phi_\mathrm{p}F/(8\pi d_\mathrm{o})$, where d_o is the diameter of a primary particle, k_co is the coordination number, and F is the inter-particle force between two primary particles (Ayazi Shamlou *et al.*, 1994). The resulting breakage kernel is

$$b = \frac{1}{\sqrt{15}} \left(\frac{\varepsilon_\mathrm{f}}{\nu_\mathrm{f}} \right)^{1/2} \exp\left(-\frac{\tau_\mathrm{s}}{\mu_\mathrm{f}(\varepsilon_\mathrm{f}/\nu_\mathrm{f})^{1/2}} \right). \tag{5.142}$$

The inter-particle force F can be computed as $F = A_{\rm H}d_{\rm o}/(12H_{\rm o}^2)$, where $A_{\rm H}$ is the Hamaker parameter for the liquid–particle system and $H_{\rm o}$ is the distance between two primary particles. The coordination number is based on experimental observation and can be calculated as $k_{\rm c} \approx 15\phi_{\rm p}^{1.2}$, where $\phi_{\rm p}$ is the volume fraction of solid within the aggregates. In the case of compact (or solid) particles $\phi_{\rm p}$ is close to unity, whereas in the case of fractal aggregates $\phi_{\rm p}$ can be determined once the fractal dimension $D_{\rm f}$ of the aggregates is known: $\phi_{\rm p} = (0.414D_{\rm f} - 0.211)(d_{\rm p}/d_{\rm o})^{D_{\rm f}-3}$, where $d_{\rm p}$ is the size of the particle and $d_{\rm o}$ is the size of the primary particle (Vanni, 2000b).

Starting from the same basic considerations, but using a highly simplified phenomenological approach, Kramer & Clark (1999) developed a hydrodynamic breakage model for solid aggregates. They classified failure modes by defining the manifestation, induction, and location of the failure, and stated that aggregates break when the maximum eigenvalue of the stress tensor is greater than the aggregate strength ($\tau_{\rm p}$). Using the assumption that the number of aggregate bonds with a strength at or below the failure strength is not a linear function of the strain rate and that the size dependence should also be considered in a breakup model, the resulting form of the breakage kernel is $b = C_7\nu_{\rm f}^{\alpha}\varepsilon_{\rm f}^{\beta}d_{\rm p}^{\gamma}$, where C_7, α, β, and γ are model constants. By invoking semi-theoretical considerations and fitting to experimental data, several values of the exponent γ have been proposed, namely $\gamma = 0, 1/3, 2/3$, and 1 (Luo & Svendsen, 1996). Pandya & Spielman (1982) found that $\gamma = 1$, whereas Peng & Williams (1994), by fitting the model to experimental data, found that γ can assume values ranging between 1 and 3, and that $\beta = 1/2$. Serra & Casamitjana (1998) fit experimental data and found that the relationship between the breakage kernel and the turbulence-dissipation rate depends on the solid fraction. When the solid fraction is low, $\beta = 0.9$, whereas for higher values of the solid fraction β is much higher. This can be explained by the fact that particle–particle collisions are more effective if the solid fraction is higher. Their results confirm the findings of Spicer & Pratsinis (1996), who used data from Oles (1991) and found $\beta = 0.8$ for low values of the solid concentration. In their investigation of nucleation, growth, agglomeration, and disruption kinetics for calcium oxalate, Zauner & Jones (2000) found a linear dependence between the breakage rate and the turbulence-dissipation rate $\varepsilon_{\rm f}$ (i.e. $\beta = 1$), and the breakage kernel was size-independent ($\gamma = 0$). Wojcik & Jones (1998), using a size-dependent breakage kernel, found a linear relationship between the kernel and the turbulence-dissipation rate ($\beta = 1$).

Daughter distribution function

For the daughter distribution function, very different scenarios are possible depending on the nature of the disperse phase (i.e. solid versus fluid particles) as explained by Zaccone *et al.* (2007) and Patruno *et al.* (2009). In general, the simplest functional form is a summation of delta functions:

$$N(\boldsymbol{\xi}_{\rm p}|\boldsymbol{\xi}'_{\rm p}) = \sum_{i=1}^{\nu} \delta\left[\boldsymbol{\xi}_{\rm p} - \boldsymbol{\xi}_i(\boldsymbol{\xi}'_{\rm p})\right], \tag{5.143}$$

where ν is the total number of daughters and $\boldsymbol{\xi}_i$ is a function stating the relationship between the internal coordinates of the ith daughter and the internal coordinates of the parent $\boldsymbol{\xi}'_{\rm p}$. Equation (5.143) states that when a particle breaks ν fragments are formed and that the internal coordinates characterizing each fragment are equal to $\boldsymbol{\xi}_i$. It is important to note here that when formulating the shape of the daughter distribution function some additional constraints have to be respected. For example, if the internal coordinate considered is the particle mass (i.e. M_p), then during breakage the summation of the masses of all the daughter fragments must be equal to that of the parent particle. In the case of binary breakage ($\nu = 2$), the situation when the two fragments are identical is labeled as symmetric breakage, as opposed to asymmetric breakage when the masses are unequal. If f is the mass fraction of one of the two daughter particles, when f is very different from 0.5 the fragmentation process is referred to as erosion. All the above-mentioned functions refer to the formation of fragments with known characteristics, meaning that, on adopting this description, any time a breakage event occurs, the same fragments are formed. It is, however, more plausible that, due to the diversity of the many elements composing the disperse phase, sometimes one type of fragmentation occurs (e.g. symmetric) whereas other times another mechanism is followed (i.e. erosion). To represent this situation a continuous function, rather than a summation of delta functions, can be adopted. The simplest continuous function is the uniform distribution which states that all the possible fragments are equally probable. More details on the functional form of this distribution can be discussed only if the internal coordinates are specified. For example, in the case of a single internal coordinate $m = \xi_p$ that assumes only positive values and exhibits additive properties during fragmentation (i.e. the sum of the internal coordinate of the fragments equals that of the parent particle), as in the case of particle mass, the following expression is obtained:

$$N(m|m') = \begin{cases} 0 & m < 0, \\ \nu/m' & 0 \le m \le m', \\ 0 & m > m', \end{cases} \tag{5.144}$$

where of course the probability of forming fragments with internal coordinate m larger than that of the parent particle m' is null. By imposing that the mass of the formed fragments sum up to the mass of the parent particle,

$$\int_0^{m'} mN(m|m')\mathrm{d}m = \frac{\nu m'}{2}, \tag{5.145}$$

it is easy to show that such a function is consistent only for binary breakage (i.e. $\nu = 2$).

Another very popular daughter distribution function is the parabolic distribution defined using three coefficients. By imposing the positiveness of the distribution over the range of interest (i.e. $0 < m < m'$) and by imposing the conservation of the total mass of the fragments, the following function is obtained:

$$N(m|m') = \begin{cases} 0 & m < 0, \\ C + (1 - C/2)\left[24\,(m/m')^2 - 24\,(m/m') + 6\right] & 0 \le m \le m', \\ 0 & m > m', \end{cases} \tag{5.146}$$

where the coefficient C represents the third degree of freedom ($0 \le C \le 3$). Equation (5.146) always results in a total number of formed fragments equal to two, and on varying the coefficient C different shapes are obtained. For $0 \le C < 2$ the formation of differently sized fragments (i.e. erosion) is likely, whereas for $2 < C \le 3$ symmetric breakage is more probable. $C = 2$ recovers the uniform distribution in Eq. (5.144). The daughter distribution functions reported above can be readily employed when the internal coordinate is the particle mass. In this case the daughter distribution function has units of inverse mass (i.e. kg^{-1}). The same distribution can be written in terms of the particle size d_p once the relationship between the particle mass and the particle size has been derived: $N(d_{\rm p}|d'_{\rm p}) = N(m|m')\partial m/\partial d_{\rm p}$, where $\partial m/\partial d_{\rm p} = 3\rho_{\rm p}k_{\rm V}d_{\rm p}^2$ in the case of equidimensional constant-density particles. In this case the daughter distribution function has units of inverse particle size. With this distribution the constraints representing the number of fragments formed and the conservation of mass are now formulated in terms of the moments of order zero and three with respect to the daughter-particle size of the distribution. This version of the daughter distribution function is often used in practical problems. For example, Coulaloglou & Tavlarides (1977), starting from the earlier work of Valentas *et al.* (1966), formulated the following function:

$$N(d_{\rm p}|d'_{\rm p}) = \frac{135 d_{\rm p}^2}{{d'_{\rm p}}^3\sqrt{2\pi}}\exp\left(-\frac{25}{2}\left[3\left(\frac{d_{\rm p}}{d'_{\rm p}}\right)^3\right]^2\right), \tag{5.147}$$

resulting in the formation of three fragments, which according to Konno *et al.* (1983) is the most likely event for liquid–liquid dispersions in stirred tanks. This distribution has as most probable event that of fragments corresponding to a size of 69% of that of the parent particle, or in other words, fragments with a volume (or mass) of one third that of the parent particle.

Another functional form that is very much used is that of a beta distribution (Hsia & Tavlarides, 1983; Konno *et al.*, 1983; Lee *et al.*, 1987), mainly due to its flexibility in representing very different mechanisms (i.e. symmetric breakage, uniform distribution, erosion) by changing only a few parameters. The beta distribution has been generalized by Diemer & Olson (2002b), who adjusted the previous model of Hill & Ng (1996) by formulating it in terms of two free parameters. On changing these parameters the shape of the distribution changes, but the conservation of mass (or volume) is ensured. The final expression is[53]

$$N(d_{\rm p}|d'_{\rm p}) = \frac{\Gamma(q\nu)}{\Gamma(q)\Gamma(q(\nu-1))}\left[1-\left(\frac{d_{\rm p}}{d'_{\rm p}}\right)^3\right]^{q(\nu-1)-1}\left(\frac{d_{\rm p}}{d'_{\rm p}}\right)^{3q-1}\frac{3\nu}{d'_{\rm p}}, \tag{5.148}$$

where ν is the number of fragments formed and q is a shape parameter. When q is greater than unity the most likely event is the formation of ν equally sized fragments, whereas for q smaller than unity the formation of one large fragment with $\nu - 1$ satellite fragments is favored. The distribution in Eq. (5.148) is very similar to that proposed by Laakkonen *et al.*

[53] The β distribution for a positive and normalized variable z is

$$\beta(z) = \frac{\Gamma(\alpha+\beta)}{\Gamma(\alpha)\Gamma(\beta)}z^{\alpha-1}(1-z)^{\beta-1}.$$

If z is the daughter-to-parent mass ratio the corresponding daughter distribution function is

$$N(m|m') = \frac{\Gamma(\alpha+\beta)}{\Gamma(\alpha)\Gamma(\beta)}\left(\frac{m}{m'}\right)^{\alpha-1}\left(1-\frac{m}{m'}\right)^{\beta-1}\frac{\nu}{m'}.$$

If we now take $\alpha = q$ and $\beta = q(\nu-1)$ we obtain

$$N(m|m') = \frac{\Gamma(q\nu)}{\Gamma(q)\Gamma(q(\nu-1))}\left(\frac{m}{m'}\right)^{q-1}\left(1-\frac{m}{m'}\right)^{q(\nu-1)-1}\frac{\nu}{m'}.$$

In the case of compact particles we have that $m = \rho_{\rm p}k_{\rm V}d_{\rm p}^3$, and

$$N(d_{\rm p}|d'_{\rm p}) = \frac{\Gamma(q\nu)}{\Gamma(q)\Gamma(q(\nu-1))}\left[\left(\frac{d_{\rm p}}{d'_{\rm p}}\right)^3\right]^{q-1}\left[1-\left(\frac{d_{\rm p}}{d'_{\rm p}}\right)^3\right]^{q(\nu-1)-1}3\nu\frac{{d'_{\rm p}}^2}{d_{\rm p}^3},$$

which can be easily simplified into Eq. (5.148).

(2007). Among turbulent-breakage models derived from physical reasonings and incorporating physical parameters, one of the most popular was formulated by Luo & Svendsen (1996). The daughter distribution function is included in the breakage kernel and it can be extracted and isolated by proceeding to standard normalization. The model contains as unique free parameter the lower limit of integration on the eddy scale, namely the smallest length scale of eddies, normalized with respect to particle size, that are able to induce effective particle breakup. This quantity, denoted $\zeta_{\min}$ in Eq. (5.141), exerts a strong influence on the final prediction and is often taken equal to zero. This distribution is valid only for binary breakage, and shows at low turbulence-dissipation rates a form indicating the formation of asymmetric fragments, with a minimum corresponding to symmetric breakage $f = 0.5$ and a probability approaching infinity for $f = 0$ and $f = 1$. As anticipated, the shape of the distribution is a function of local turbulence conditions through the turbulence-dissipation rate, and the curves flatten out with increasing turbulence intensity. The probability of symmetric breakage increases very rapidly, but the dependence on this parameter seems to be highly non-monotonic and some authors (Lasheras *et al.*, 2002) contest that this behavior does not accurately describe physical reality. A recent model that originates from the approach introduced by Luo & Svendsen (1996), but overcomes some of its limitation, is the one developed by Wang *et al.* (2003). The final shape of the daughter distribution function is upper-bounded and predicts zero probability when the daughter-to-parent size ratio equals zero or unity and minimum probability for daughter particles characterized by identical size. However, also for this approach, the implementation in simulation codes is cumbersome due to the mathematical complexity of its formulation, which requires the solution of a three-dimensional integral. Much simpler is the distribution proposed by Martinez-Bazán *et al.* (1999), which was derived on the basis of purely mechanical principles. The distribution is derived by establishing a proportionality relationship between the probability of a pair of particles being formed and the difference between the turbulence kinetic energy and surface energy that would act on the two daughter particles. With some modifications, the original idea assuming binary breakage is then extended to ternary breakage as well. A shortcoming present in many models described above, with the exceptions of Coulaloglou & Tavlarides (1977), Diemer & Olson (2002b), and Luo & Svendsen (1996), is the fact that the conservation of mass is not always respected. Models that do not predict mass conservation, once implemented in simulation codes for closed systems, lead to an evolution of the predicted volume fraction that is not meaningful from a physical point of view.

Most daughter distribution functions can be easily extended to bivariate problems. Let us consider two examples. In the first example particles with two components A and B are described. The particulate system is defined in terms of the size of these particles $d_{\rm p}$ and the composition of the particles ϕ, expressed for example as the mass fraction of component A in the particle. When a particle breaks we can assume for example that the amount of component A is partitioned among the daughters proportionally to the mass of the fragments. Under these hypotheses, and the additional assumption of binary breakage following the beta distribution, the resulting bivariate distribution is

$$N(d_{\rm p}, \phi | d'_{\rm p}, \phi') = 180 \frac{d_{\rm p}^2}{{d'_{\rm p}}^3} \left(\frac{d_{\rm p}^3}{{d'_{\rm p}}^3} \right)^2 \left(1 - \frac{d_{\rm p}^3}{{d'_{\rm p}}^3} \right)^2 \delta \left(\phi - \frac{d_{\rm p}^3}{{d'_{\rm p}}^3} \phi' \right), \tag{5.149}$$

where $d'_{\rm p}$ and ϕ' represent the properties of the parent particle, while $d_{\rm p}$ and ϕ represent the properties of the daughter particles (Buffo *et al.*, 2011, 2012). A second example is that of fragmentation of fractal aggregates. In this case aggregates are composed of primary particles with infinitesimally small contact points. As an aggregate breaks into two equal fragments both the mass and the surface area of the aggregate are halved. If the particles are described in terms of particle volume V_p and surface area A_p, the daughter distribution function is (Zucca *et al.*, 2007)

$$P(V_p, A_p|V'_p, A'_p) = 2\delta\left(V_p - \frac{V'_p}{2}\right)\delta\left(A_p - \frac{A'_p}{2}\right). \tag{5.150}$$

Other forms that do not assume that the daughter particles are equal are also possible. In conclusion, it is very difficult to define *a priori* what daughter distribution function can best describe breakage. This is particularly true for solid particles, in which case, depending on the nature of the particle (e.g. compact growing crystal versus loose fractal aggregate), very different mechanisms control the overall process. In these cases it is always advisable to carry out an extensive validation through comparison with experimental data. In the case of fluid particles the situation is slightly simpler. In fact, for this case it seems that the beta distribution is the most adequate. Also in this case, however, the identification of the two parameters v and q has to be carried out by comparison with data from experiments. Experiments involving fluids of different characteristics (see, for example, Andersson & Andersson (2006) and Zaccone *et al.* (2007)) have shown that during breakup under turbulent conditions the fluid droplet is deformed, its surface being extended, until a sort of active state is formed. Shortly before the fluid particle breaks the internal fluid-redistribution mechanism defines the size and numbers of daughters formed. In the case of bubbles, the resistance offered by the fluid constituting the disperse phase is very small and therefore two equally sized bubbles are formed (i.e. $v = 2$ and $q > 1$). In the case of liquid droplets, the much higher viscosity of the disperse phase hinders this process, resulting in more droplets ($v = 3$ or 4) with a U-shaped distribution ($q < 1$). All this is valid for clean systems; as soon as impurities and other substances that are capable of accumulating at the fluid–fluid interface and altering the bubble or droplet surface properties occur completely different scenarios are possible. Unfortunately, there exists at present no reliable model to describe such systems.

5.8 Second-order point processes

Second-order point processes are discontinuous events involving two particles. As has already been mentioned, by definition these are processes that result in an instantaneous change in the particle internal coordinates. Collision and aggregation between particles are typically treated as second-order point processes, meaning that in a collision event only two particles are involved[54] and that the change in particle momentum is assumed to be instantaneous and to occur right at the time of collision. Aggregation, agglomeration, and coalescence are other common examples. If after colliding the two particles stick together forming a new single particle, not only the particle velocity but also other properties (such as the particle mass) will instantaneously change. We will thus begin by deriving the source term for the GPBE due to a generic second-order point process. For simplicity, we will consider a granular system (i.e. without the fluid phase) and denote the particle velocity and

[54]This assumption has a wide range of applicability, since in most practical cases collisions are actual interactions between two particles only. This is certainly true for dilute systems, in which the probability of finding three particles in close proximity is very low, and can be considered true also for denser systems, since any interaction involving more than two particles can be reasonably described as many subsequent collisions between two particles.

properties by a single phase-space vector $\boldsymbol{\eta}$. As before, we will write the source term in the GPBE as

$$S_1 = h^+ - h^-,$$

where h^+ is the rate of production and h^- is the rate of loss of particles due to second-order point processes.[55]

5.8.1 Derivation of the source term

Let us imagine that one particle has its center of mass at position $\mathbf{x}'$ and is characterized by phase-space vector $\boldsymbol{\eta}'$. Moreover, let us imagine that another particle has its center of mass at the physical point $\tilde{\mathbf{x}}$ and has the phase-space vector $\tilde{\boldsymbol{\eta}}$. We define the frequency[56] of this second-order point process, $a(\mathbf{x}', \boldsymbol{\eta}'; \tilde{\mathbf{x}}, \tilde{\boldsymbol{\eta}})$, so that the quantity $a(\mathbf{x}', \boldsymbol{\eta}'; \tilde{\mathbf{x}}, \tilde{\boldsymbol{\eta}})\mathrm{d}t$ represents the fraction of particles undergoing the point process in a time interval $\mathrm{d}t$.[57] This frequency is symmetric with respect to permutations in the ordering of the particles, resulting in $a(\mathbf{x}', \boldsymbol{\eta}'; \tilde{\mathbf{x}}, \tilde{\boldsymbol{\eta}}) = a(\tilde{\mathbf{x}}, \tilde{\boldsymbol{\eta}}; \mathbf{x}', \boldsymbol{\eta}')$. We now define a pair number-density function such that $n^{(2)}(t, \mathbf{x}', \boldsymbol{\eta}', \tilde{\mathbf{x}}, \tilde{\boldsymbol{\eta}})\mathrm{d}\mathbf{x}'\, \mathrm{d}\boldsymbol{\eta}'\, \mathrm{d}\tilde{\mathbf{x}}\, \mathrm{d}\tilde{\boldsymbol{\eta}}$ represents the expected number of particle pairs with $(\mathbf{x}', \boldsymbol{\eta}')$ and $(\tilde{\mathbf{x}}, \tilde{\boldsymbol{\eta}})$. As discussed in Chapter 4, since knowledge of this function would require the solution of an equation involving $n^{(3)}$, the following closure is often used: $n^{(2)}(t, \mathbf{x}', \boldsymbol{\eta}', \tilde{\mathbf{x}}, \tilde{\boldsymbol{\eta}}) \approx n(t, \mathbf{x}', \boldsymbol{\eta}')n(t, \tilde{\mathbf{x}}, \tilde{\boldsymbol{\eta}})$. Given these definitions, we can calculate the number of events per unit time involving particle pairs $(\mathbf{x}', \boldsymbol{\eta}')$ and $(\tilde{\mathbf{x}}, \tilde{\boldsymbol{\eta}})$ as $a(\mathbf{x}', \boldsymbol{\eta}'; \tilde{\mathbf{x}}, \tilde{\boldsymbol{\eta}})n(t, \mathbf{x}', \boldsymbol{\eta}')n(t, \tilde{\mathbf{x}}, \tilde{\boldsymbol{\eta}})\mathrm{d}\mathbf{x}'\, \mathrm{d}\boldsymbol{\eta}'\, \mathrm{d}\tilde{\mathbf{x}}\, \mathrm{d}\tilde{\boldsymbol{\eta}}$. If we now want to calculate the number of events occurring per unit time and unit volume involving particles located near $(\tilde{\mathbf{x}}, \tilde{\boldsymbol{\eta}})$, we must account for all the possible interactions between *test* and *field* particles. In the nomenclature of binary particle encounters, particles located at $(\tilde{\mathbf{x}}, \tilde{\boldsymbol{\eta}})$ are called test particles, whereas particles located at $(\mathbf{x}', \boldsymbol{\eta}')$ are called field particles. The final result for the collision frequency is

$$N_\mathrm{e}(t, \tilde{\mathbf{x}}, \tilde{\boldsymbol{\eta}}) = \int a(\mathbf{x}', \boldsymbol{\eta}'; \tilde{\mathbf{x}}, \tilde{\boldsymbol{\eta}})n(t, \mathbf{x}', \boldsymbol{\eta}')n(t, \tilde{\mathbf{x}}, \tilde{\boldsymbol{\eta}})\mathrm{d}\mathbf{x}'\, \mathrm{d}\boldsymbol{\eta}', \tag{5.151}$$

where the integrals are over all locations $\mathbf{x}'$ and phase-space variables $\boldsymbol{\eta}'$ of the field particle. In most practical cases, the distance in physical space between test and field particles when the encounter occurs (i.e. $|\mathbf{x}' - \tilde{\mathbf{x}}|$) is of the order of magnitude of the particle diameter[58] and within this distance the NDF does not change significantly. Therefore, the following closure is often invoked: $n(t, \mathbf{x}', \boldsymbol{\eta}')n(t, \tilde{\mathbf{x}}, \tilde{\boldsymbol{\eta}}) \approx n(t, \tilde{\mathbf{x}}, \boldsymbol{\eta}')n(t, \tilde{\mathbf{x}}, \tilde{\boldsymbol{\eta}})$. If this simplified expression is substituted into Eq. (5.151) and the order of the integrals is inverted, we obtain

$$N_\mathrm{e}(t, \tilde{\mathbf{x}}, \tilde{\boldsymbol{\eta}}) = \int \left(\int a(\mathbf{x}', \boldsymbol{\eta}'; \tilde{\mathbf{x}}, \tilde{\boldsymbol{\eta}})\mathrm{d}\mathbf{x}' \right) n(t, \tilde{\mathbf{x}}, \boldsymbol{\eta}')n(t, \tilde{\mathbf{x}}, \tilde{\boldsymbol{\eta}})\mathrm{d}\boldsymbol{\eta}'. \tag{5.152}$$

The integral of the frequency appearing between parentheses in Eq. (5.152) is called the kernel:

$$\beta(\tilde{\mathbf{x}}; \boldsymbol{\eta}', \tilde{\boldsymbol{\eta}}) = \int a(\mathbf{x}', \boldsymbol{\eta}'; \tilde{\mathbf{x}}, \tilde{\boldsymbol{\eta}})\mathrm{d}\mathbf{x}'. \tag{5.153}$$

---

[55] For second-order point processes such as hard-sphere collisions, the total number of particles remains unchanged. However, the number of particles with a specific phase-space vector will always increase or decrease.

[56] As a standard frequency this quantity has units of inverse time.

[57] This quantity can be regarded as the probability that the second-order point process occurs within the infinitesimal time step $\mathrm{d}t$.

[58] In the case of particles interacting with hard-sphere potentials this distance is exactly equal to the sum of the radii of the interacting particles. When particles interact with other potentials this distance is larger but remains of this order of magnitude.

The kernel for a specific second-order point process has dimensions of spatial volume per unit time and it roughly represents the volumetric flux of particles undergoing the second-order point process located at the physical point $\tilde{\mathbf{x}}$ and with internal coordinates near $\boldsymbol{\eta}'$ and $\tilde{\boldsymbol{\eta}}$. This expression is generally used to derive kernels for particulate systems undergoing different physical processes (i.e. collisions of hard-spheres, Brownian motion, differential settling, etc.). Because particle encounters are dominated by local transport phenomena at distances larger than the particle size and by interaction potentials at shorter distances, the kernels are naturally split into two terms. The first one accounts for particle transport in the region of interaction, whereas the second one, often referred to as the efficiency, accounts for the short-range interactions. In what follows a few examples will be treated, and readers interested in more details are referred to Ramkrishna (2000).

Let us now derive the source term appearing in the GPBE due to second-order point processes (corresponding to the last term on the right-hand side of Eq. (5.1) on page 137). The GPBE operates on the NDF $n(t, \mathbf{x}, \mathbf{v}_\text{p}, \boldsymbol{\xi}_\text{p}, \mathbf{v}_\text{f}, \boldsymbol{\xi}_\text{f})$; therefore, the source term must express the net rate at which particles change their number density. Similarly to what was previously illustrated for granular systems, the derivation is done by considering the interaction of test particles with internal coordinates at about $\tilde{\boldsymbol{\eta}}$ with field particles located at about $\boldsymbol{\eta}'$, whose encounters result in particles characterized by $\boldsymbol{\eta}$. However, now the definition of $\boldsymbol{\eta}$ includes all of the phase-space variables ($\mathbf{v}_\text{p}$, $\boldsymbol{\xi}_\text{p}$, $\mathbf{v}_\text{f}$, and $\boldsymbol{\xi}_\text{f}$). This source term is constituted by a positive term and a negative term. The positive term is (Ramkrishna, 2000)

$$h^+(t, \mathbf{x}, \boldsymbol{\eta}) = \frac{1}{\delta} \int \beta(\boldsymbol{\eta}', \tilde{\boldsymbol{\eta}}) n(t, \mathbf{x}, \boldsymbol{\eta}') n(t, \mathbf{x}, \tilde{\boldsymbol{\eta}}) J(\tilde{\boldsymbol{\eta}}, \boldsymbol{\eta}) \mathrm{d}\boldsymbol{\eta}', \tag{5.154}$$

when written in terms of the kernel, where δ is the symmetry factor to avoid multiple counting and $J(\tilde{\boldsymbol{\eta}}, \boldsymbol{\eta})$ is the Jacobian of the variable transformation relating the phase-space variables before collision to the values after collision. The symmetry factor is two for identical particles and one for different particles, in the case of second-order point processes resulting in a single particle (e.g. aggregation, coalescence, etc.), whereas it is always equal to one in the case of second-order point processes that leave the number of particles unchanged. The Jacobian of the variable transformation is the determinant of the matrix containing the partial derivatives of the internal-coordinate vector for the test particle $\tilde{\boldsymbol{\eta}}$ with respect to the final value of the internal coordinate $\boldsymbol{\eta}$, while the internal coordinates for the field particle $\boldsymbol{\eta}'$ are held constant:

$$J(\tilde{\boldsymbol{\eta}}, \boldsymbol{\eta}) = \left| \frac{\partial \tilde{\boldsymbol{\eta}}}{\partial \boldsymbol{\eta}} \right| = \begin{vmatrix} \partial \tilde{\eta}_1 / \partial \eta_1 & \dots & \partial \tilde{\eta}_1 / \partial \eta_M \\ \vdots & \ddots & \vdots \\ \partial \tilde{\eta}_M / \partial \eta_1 & \dots & \partial \tilde{\eta}_M / \partial \eta_M \end{vmatrix}. \tag{5.155}$$

The relationship among $\tilde{\boldsymbol{\eta}}$, $\boldsymbol{\eta}'$, and $\boldsymbol{\eta}$ can be derived by relating continuity statements written in terms of pre- and post-event values for the internal coordinates and depends on the second-order point process under investigation. The positive source term can also be written in terms of the frequency (rather than the kernel):

$$h^+(t, \mathbf{x}, \boldsymbol{\eta}) = \frac{1}{\delta} \int a(\mathbf{x}', \boldsymbol{\eta}'; \mathbf{x}, \tilde{\boldsymbol{\eta}}) n(t, \mathbf{x}, \boldsymbol{\eta}') n(t, \mathbf{x}, \tilde{\boldsymbol{\eta}}) J(\tilde{\boldsymbol{\eta}}; \boldsymbol{\eta}) \mathrm{d}\mathbf{x}' \, \mathrm{d}\boldsymbol{\eta}'. \tag{5.156}$$

The negative source term is more readily found:

$$h^-(t, \mathbf{x}, \boldsymbol{\eta}) = \int \beta(\boldsymbol{\eta}', \boldsymbol{\eta}) n(t, \mathbf{x}, \boldsymbol{\eta}') n(t, \mathbf{x}, \boldsymbol{\eta}) \mathrm{d}\boldsymbol{\eta}', \tag{5.157}$$

when written in terms of the kernel, or

$$h^{-}(t, \mathbf{x}, \boldsymbol{\xi}) = \int a(\mathbf{x}', \boldsymbol{\eta}'; \mathbf{x}, \boldsymbol{\eta}) n(t, \mathbf{x}, \boldsymbol{\eta}') n(t, \mathbf{x}, \boldsymbol{\eta}) \mathrm{d}\mathbf{x}' \, \mathrm{d}\boldsymbol{\eta}', \tag{5.158}$$

when written in terms of the frequency. In what follows a few examples will be analyzed and discussed. The source terms for hard-sphere collisions are presented in Chapter 6.

5.8.2 Source terms for aggregation and coalescence

Let us now consider the case of two particles colliding and sticking together, forming a single particle. In general, mass and momentum will be conserved and the phase-space variables after collision will hence be related by algebraic constraints to the values for the two particles before collision. Depending on the number of conserved quantities, the Jacobian $J(\tilde{\boldsymbol{\eta}}; \boldsymbol{\eta})$ can be complicated to compute. For simplicity, we will start with an example for which the particles are very small, such that their velocities are the same as that of the fluid. We will then consider a more complicated case in which the particle velocities before collision are different.

Aggregation of fine particles

Consider a population of particles characterized by an NDF that is a function of time, physical space, and particle mass, i.e. $n(t, \mathbf{x}, \xi_\mathrm{p})$. We define as ξ'_p and $\tilde{\xi}_\mathrm{p}$ as the masses of the particles that, after sticking together, will result in a particle of mass ξ_p. In this case, is a single conservation statement, namely $\xi_\mathrm{p} = \xi'_\mathrm{p} + \tilde{\xi}_\mathrm{p}$, resulting in a Jacobian equal to unity. The final source term (written here in terms of the kernel) is

$$S_1(\xi_\mathrm{p}) = \frac{1}{2} \int_0^{\xi_\mathrm{p}} \beta(\xi', \tilde{\xi}_\mathrm{p}) n(\xi'_\mathrm{p}) n(\tilde{\xi}_\mathrm{p}) \mathrm{d}\xi'_\mathrm{p} - \int_0^{\infty} \beta(\xi'_\mathrm{p}, \xi_\mathrm{p}) n(\xi'_\mathrm{p}) n(\xi_\mathrm{p}) \mathrm{d}\xi'_\mathrm{p}, \tag{5.159}$$

where the limits of integration are explicitly indicated. The source term in Eq. (5.159) is often written only in terms of the final mass ξ_p and the field particle mass ξ'_p:

$$S_1(\xi_\mathrm{p}) = \frac{1}{2} \int_0^{\xi_\mathrm{p}} \beta(\xi'_\mathrm{p}, \xi_\mathrm{p} - \xi'_\mathrm{p}) n(\xi'_\mathrm{p}) n(\xi_\mathrm{p} - \xi') \mathrm{d}\xi' - \int_0^{\infty} \beta(\xi'_\mathrm{p}, \xi_\mathrm{p}) n(\xi'_\mathrm{p}) n(\xi_\mathrm{p}) \mathrm{d}\xi'_\mathrm{p}. \tag{5.160}$$

If the particle density is constant, this description is equivalent to one based on the particle volume. On the other hand, consider a system with two internal coordinates (such as, for example, particle mass ξ_p1 and particle surface area ξ_p2) that are additive during an aggregation event: $\boldsymbol{\xi}_\mathrm{p} = \boldsymbol{\xi}'_\mathrm{p} + \tilde{\boldsymbol{\xi}}_\mathrm{p}$, where $\boldsymbol{\xi}_\mathrm{p} = (\xi_\mathrm{p1}, \xi_\mathrm{p2})$, $\boldsymbol{\xi}'_\mathrm{p} = (\xi'_\mathrm{p1}, \xi'_\mathrm{p2})$, and $\tilde{\boldsymbol{\xi}}_\mathrm{p} = (\tilde{\xi}_\mathrm{p1}, \tilde{\xi}_\mathrm{p2})$. This linear transformation yields a unit Jacobian, and the corresponding source term is

$$S_1(\boldsymbol{\xi}_\mathrm{p}) = \frac{1}{2} \int_0^{\boldsymbol{\xi}_\mathrm{p}} \beta(\boldsymbol{\xi}'_\mathrm{p}, \tilde{\boldsymbol{\xi}}_\mathrm{p}) n(\boldsymbol{\xi}'_\mathrm{p}) n(t, \tilde{\boldsymbol{\xi}}_\mathrm{p}) \mathrm{d}\boldsymbol{\xi}'_\mathrm{p} - \int_0^{\infty} \beta(\boldsymbol{\xi}'_\mathrm{p}, \boldsymbol{\xi}_\mathrm{p}) n(\boldsymbol{\xi}'_\mathrm{p}) n(\boldsymbol{\xi}_\mathrm{p}) \mathrm{d}\boldsymbol{\xi}'_\mathrm{p}, \tag{5.161}$$

which is the multivariate version of Eq. (5.160). If the relevant internal coordinate is the particle size, $\xi_\mathrm{p} = d_\mathrm{p}$, then some assumption on the shape of the particles must be made. If it is assumed that after aggregation the shape of the resulting particle is similar to that of the original ones,[59] and if the particle mass (which is the conserved quantity during

[59] This is reasonable during, for example, coalescence of bubble or droplets, since, after the aggregation of two spherical fluid particles, another spherical fluid particle will be formed. This assumption is also considered valid during agglomeration (i.e. aggregation of crystals in supersaturated solutions) since, due to molecular growth, the voids left between the contact points of the aggregating crystals are filled by new crystallized material.

aggregation) is related to the particle size by the standard power law of Eulerian objects, the following continuity statement can be formulated: $\xi_p^3 = (\xi_p')^3 + \tilde{\xi}_p^3$, or, equivalently, $\tilde{\xi}_p = [\xi_p^3 - (\xi_p')^3]^{1/3}$. This nonlinear transformation results in a Jacobian equal to $\xi_p^2[\xi_p^3 - (\xi_p')^3]^{-2/3}$. The source term for aggregation is then

$$S_1(\xi_p) = \frac{\xi_p^2}{2}\int_0^{\xi_p}\left[\xi_p^3 - (\xi_p')^3\right]^{-2/3}\beta(\xi_p', [\xi_p^3 - (\xi_p')^3]^{1/3})n(\xi_p')n([\xi_p^3 - (\xi_p')^3]^{1/3})\mathrm{d}\xi_p' \\ - \int_0^{\infty}\beta(\xi_p', \xi_p)n(\xi_p')n(\xi_p)\mathrm{d}\xi_p', \tag{5.162}$$

which is entirely equivalent to Eq. (5.160). The examples given above explicitly assume that the particles have the same velocity as the fluid (i.e. the particle Stokes number is null). In the next example, we consider the more general case of particles with finite Stokes number.

Coalescence of droplets

As a final example, let us consider a system of coalescing droplets characterized by particle mass ξ_p and particle velocity $\mathbf{v}_p$. In this case, the continuity statements are written in terms of particle mass and momentum, respectively, as $\xi_p = \xi_p' + \tilde{\xi}_p$ and $\xi_p\mathbf{v}_p = \xi_p'\mathbf{v}_p' + \tilde{\xi}_p\tilde{\mathbf{v}}_p$ or, equivalently, as $\tilde{\xi}_p = \xi_p - \xi_p'$ and $\tilde{\mathbf{v}}_p = (\xi_p\mathbf{v}_p - \xi_p'\mathbf{v}_p')/(\xi_p - \xi_p')$. The Jacobian for this nonlinear transformation is

$$J = \frac{\partial(\tilde{\xi}_p, \tilde{\mathbf{v}}_p)}{\partial(\xi_p, \mathbf{v}_p)} = \left(\frac{\xi_p}{\xi_p - \xi_p'}\right)^3 \tag{5.163}$$

and the resulting source term is

$$S_1(\xi_p, \mathbf{v}_p) = \frac{1}{2}\int\left(\int_0^{\xi_p}\beta(\xi_p', \mathbf{v}_p', \tilde{\xi}_p, \tilde{\mathbf{v}}_p)n(\xi_p', \mathbf{v}_p')n(\tilde{\xi}_p, \tilde{\mathbf{v}}_p)\frac{\xi_p^3}{(\xi_p - \xi_p')^3}\,\mathrm{d}\xi_p'\right)\mathrm{d}\mathbf{v}_p' \\ - \int\beta(\xi_p', \mathbf{v}_p', \xi_p, \mathbf{v}_p)n(\xi_p', \mathbf{v}_p')n(\xi_p, \mathbf{v}_p)\mathrm{d}\xi_p'\,\mathrm{d}\mathbf{v}_p'. \tag{5.164}$$

As we will discuss next, the coalescence kernel used in Eq. (5.164) normally depends on the velocity difference: $\beta(\xi_p', \mathbf{v}_p', \xi_p, \mathbf{v}_p) = \beta(\xi_p', \xi_p, |\mathbf{v}_p' - \mathbf{v}_p|)$. Since the different kernels are primarily functions of the particle size d_p and some additional variables, only these functional dependences will be highlighted and discussed.

5.8.3 Aggregation kernels for fine particles

The frequency of aggregation of fine particles (i.e. $\mathrm{St}_p = 0$) strongly depends on the transport phenomenon that brings the particles into close proximity. Once the frequency has been defined the calculation of the kernel is carried out by employing its definition in Eq. (5.153). We limit the discussion to the final results and readers interested in the details of the procedure can consult Ramkrishna (2000). Recall, however, that for fine particles only the internal coordinates $\boldsymbol{\xi}_p$ appear in the NDF. Brownian motion is one important phenomenon responsible for bringing fine particles into close contact and for inducing collision (and possibly aggregation) events. As already explained, particles moving due to Brownian motion exhibit the diffusion coefficient in Eq. (5.116). The resulting aggregation kernel[60] (of particles of size d_p and d'_p) can easily be derived by calculating the net volumetric flux of particles of size d'_p crossing a sphere centered on the particle of size d_p with radius equal to $(d_p + d'_p)/2$, resulting in the following well-known equation (Elimelech *et al.*, 1998):

$$\beta(d_p, d'_p) = \frac{2k_B T_f}{3\mu_f}\left(\frac{1}{d_p} + \frac{1}{d'_p}\right)\left(d_p + d'_p\right) = \frac{2k_B T_f}{3\mu_f}\frac{\left(d_p + d'_p\right)^2}{d_p d'_p}, \tag{5.165}$$

where μ_f is the fluid viscosity. If the aggregating particles are of about the same size, then a size-independent kernel results: $\beta(d_p, d'_p) = 8k_B T_f/(3\mu_f)$. These kernels were derived for the continuum regime, in other words, assuming that the continuous phase is perceived by the particles as a continuum. As has already been discussed, this hypothesis is valid only if the particle Knudsen number (relative to the continuous phase), defined here as

$$\mathrm{Kn}^* = \frac{2\lambda_f}{d_p}, \tag{5.166}$$

is smaller than unity, where λ_f is the mean free path for the molecules of the continuous phase. When the Knudsen number is much smaller than unity, we are in the continuum (or Stokes–Einstein) regime.

For higher values of Kn^*, we move into the free-molecular (or Epstein) regime where the following aggregation kernel is used:

$$\beta(d_p, d'_p) = \left(\frac{3}{4\pi}\right)^{1/6}\left(\frac{6k_B T_f}{\rho_p}\right)^{1/2}\left(d_p + d'_p\right)^2\left(\frac{1}{d_p^3} + \frac{1}{d_p'^3}\right)^{1/2}. \tag{5.167}$$

In order to work with a single expression that covers both regimes as well as the transition between them, very often the Fuchs (1964) equation is used:

$$\beta = 4\pi(\Gamma + \Gamma')\left(d_p + d'_p\right)\left[\frac{d_p + d'_p}{d_p + d'_p + \sqrt{g^2 + g'^2}} + \frac{4(\Gamma + \Gamma')}{(d_p + d'_p)\sqrt{c^2 + c'^2}}\right]^{-1}, \tag{5.168}$$

where the corrected diffusion coefficient[61] is

$$\Gamma = \frac{k_B T_f}{6\pi\mu_f d_p}\left[\frac{5 + 4\mathrm{Kn}^* + 6\mathrm{Kn}^{*2} + 18\mathrm{Kn}^{*3}}{5 - \mathrm{Kn}^* + (8 + \pi)\mathrm{Kn}^{*2}}\right] \tag{5.169}$$

with $g = [(2d_p + l)^3 - (4d_p^2 + l^2)^{3/2}]/(6d_p l) - 2d_p$, $l = 8\Gamma/(\pi c)$, and $c = \sqrt{8k_B T_f/(\pi M_p)}$, where M_p is the particle mass.

Another mechanism that brings fine particles into close proximity and can induce aggregation is the presence of spatial gradients in fluid velocity. Let us consider two groups of particles moving in parallel motion, following two lines placed at a distance smaller then their size. If they move with the same velocity they will never collide, whereas a velocity difference will result in a collision. From this simple argument and calculations very similar to those used for the perikinetic kernel, the following aggregation kernel[62] is obtained (von Smoluchowski, 1917):

$$\beta(d_p, d'_p) = \frac{4}{3}G_f\left(d_p + d'_p\right)^3, \tag{5.170}$$

[60] This aggregation kernel and, in general, the phenomenon of particles aggregating because of Brownian motion, is labeled as perikinetic.

[61] It is useful to compare the standard Stokes–Einstein equation in Eq. (5.116) with the corrected expression of Eq. (5.169).

[62] This is often referred to as the orthokinetic aggregation kernel.

where G_f is the fluid shear rate. For very simple flows, the shear rate is assumed to be equal to the main local fluid-velocity gradient. In general, for a laminar flow, the fluid shear rate is calculated as the second invariant of the rate-of-strain tensor.[63] In turbulent systems, the fluid shear rate is generally calculated through the values of the turbulence-dissipation rate and the kinematic viscosity of the continuous phase, resulting in $G_\mathrm{f} = \sqrt{\varepsilon_\mathrm{f}/\nu_\mathrm{f}}$, as explained in Saffman & Turner (1956). This is valid only if the particles are smaller than the Kolmogorov length scale $\eta_\mathrm{k} = (\nu_\mathrm{f}^3/\epsilon_\mathrm{f})^{1/4}$, resulting in the following expression:

$$\beta(d_\mathrm{p}, d'_\mathrm{p}) = \left(\frac{8\pi}{15}\right)^{1/2} \left(\frac{\nu_\mathrm{f}}{\varepsilon_\mathrm{f}}\right)^{1/2} \left(d_\mathrm{p} + d'_\mathrm{p}\right)^3 \approx 1.29 \left(\frac{\nu_\mathrm{f}}{\varepsilon_\mathrm{f}}\right)^{1/2} \left(d_\mathrm{p} + d'_\mathrm{p}\right)^3, \tag{5.171}$$

since in this case particles collide while entrapped in the same eddy. When the particles are instead larger than the Kolmogorov length scale, the following expression is often used:

$$\beta(d_\mathrm{p}, d'_\mathrm{p}) = \left(\frac{8\pi}{3}\right)^{1/2} \varepsilon_\mathrm{f}^{1/3} \left(d_\mathrm{p} + d'_\mathrm{p}\right)^{7/3}. \tag{5.172}$$

These kernels are valid for inertialess particles (i.e. $\mathrm{St}_\mathrm{p} = 0$) and can be extended to finite Stokes numbers only by employing *ad hoc* corrections. For example, Ammar & Reeks (2009) derived for the kernel proposed by Saffman & Turner (1956) a correction that is based on the local Stokes number. The relative importance of perikinetic aggregation versus orthokinetic aggregation is quantified by a Péclet number:

$$\mathrm{Pe} = \frac{3\pi\mu_\mathrm{f} d_\mathrm{p}^3 G_\mathrm{f}}{k_\mathrm{B} T_\mathrm{f}}. \tag{5.173}$$

In fact, when $\mathrm{Pe} \leq 0.001$ only perikinetic aggregation need be considered, whereas when $\mathrm{Pe} \geq 10$ the perikinetic contribution is negligible.

The total aggregation kernel is usually written as the summation of the kernels described above, as well as other possible contributions arising from other collision mechanisms. The quantity obtained in this way, however, accounts only for the total particle volumetric flux due to collisions. In order to account for the fact that not all collisions are successful at producing a new aggregate, an efficiency factor α is introduced. This efficiency factor is a number between zero and unity (although in some cases it can also assume values larger than unity) that quantifies the number of successful collisions leading to an aggregation event in comparison with the total number of collisions, calculated with the kernels reported above. There are many factors that impede the successful aggregation of two particles. In fact, even if two particles are brought into close proximity, only the balance between repulsive and attractive forces will determine whether particles will stick together. Indeed, for particles to stick together, an attractive force is needed and a potential repulsive barrier generally has to be overcome. Typical attractive (Van der Waals) forces in particulate systems are those resulting from attractive forces between permanent

[63] As is well known, the components of the rate-of-strain tensor are defined by

$$s_{ij} = \frac{1}{2}\left(\frac{\partial U_{\mathrm{f}i}}{\partial x_j} + \frac{\partial U_{\mathrm{f}j}}{\partial x_i}\right).$$

The shear rate is often calculated as the second invariant (the first invariant is the trace) of the rate-of-strain tensor:

$$G_\mathrm{f} = \frac{1}{2}\left(s_{ij}s_{ij} - s_{kk}^2\right),$$

where repeated indices imply summation (i.e. Einstein notation).

dipoles (the Keesom force), between a permanent dipole and a corresponding induced dipole (the Debye force), and between two instantaneously induced dipoles (the London dispersion force). Other attractive forces refer instead to the presence of a stable charge on the particle surface (electrostatic force) or the possibility of forming a chemical bond on the particle surface, as in the case of hydrogen bonds. Other notable cases are growing crystals suspended in supersaturated solutions and stable necks formed on the contact points of particles due to the presence of a binder. The latter factors are very important in crystallization, as well as in granulation, as discussed in Marshall *et al.* (2011).

Repulsive forces are instead caused by the presence of the continuous phase. In the case of particles suspended in a gaseous primary phase, very weak (if not completely negligible) repulsive forces arise. In the case of particles suspended in a liquid, repulsive forces are often caused by the electrical double layer of ions that surrounds the particles. In fact, when a charged particle is suspended in a liquid (aqueous) phase, it becomes immediately surrounded by alternating layers of ions and counter-ions. When the two particles get close to one another, then the two clouds of ions have to overlap and, due to the osmotic pressure, a repulsive force is generated. Another cause is the viscous drainage of the liquid film between the two colliding particles (immediately before they come into physical contact), which generates resistance forces. In this case the fact that the fluid can (in fractal porous aggregates) or cannot (in solid aggregates) penetrate the colliding particles plays a very important role. Depending on the relevant forces acting on the particles, different expressions for the aggregation efficiency can be derived. These expressions are typically derived by writing the aggregation efficiency as the inverse of the Fuchs stability ratio (i.e. $\alpha = 1/W$), which for two particles of size d_p and d'_p is

$$W = \left(\frac{d_\mathrm{p} + d'_\mathrm{p}}{2}\right) \int_0^\infty \exp\left(\frac{V(h)}{k_\mathrm{B}T_\mathrm{f}}\right) \left[\left(\frac{d_\mathrm{p} + d'_\mathrm{p}}{2}\right) + h\right]^{-2} \mathrm{d}h, \tag{5.174}$$

where h is the inter-particle distance and $V(h)$ is the inter-particle potential. The inter-particle potential is calculated as the summation of the potentials referring to attractive, $V_\mathrm{a}(h)$, and repulsive, $V_\mathrm{r}(h)$, forces, which for conservative systems expresses the force acting on the particles (since the inter-particle force is $F = -\partial V/\partial h$). Van der Waals attractive forces are generally expressed in terms of the following potential function:

$$V_\mathrm{a} = -\frac{A_\mathrm{H}}{6}\left[\frac{d_\mathrm{p}d'_\mathrm{p}/2}{h^2 + (d_\mathrm{p} + d'_\mathrm{p})h} + \frac{d_\mathrm{p}d'_\mathrm{p}/2}{h^2 + (d_\mathrm{p} + d'_\mathrm{p})h + d_\mathrm{p}d'_\mathrm{p}} + \ln\left(\frac{h^2 + 2(d_\mathrm{p} + d'_\mathrm{p})h}{h^2 + 2(d_\mathrm{p} + d'_\mathrm{p})h + 4d_\mathrm{p}d'_\mathrm{p}}\right)\right], \tag{5.175}$$

where A_H is the Hamaker parameter, and d_p and d'_p are the sizes of the two colliding particles. Typically, A_H lies in the range 10^{-21}–10^{-19} J, and can be estimated with different methods. According to the Lifshitz theory it is possible to determine the Hamaker parameter from the interfacial tension (Israelachvili, 1992): $A_\mathrm{H} \approx 24\pi\gamma D_0$, where D_0 is the cut-off distance, which can generally be assumed to be around 1–2 Å. The value of the Hamaker parameter can be calculated for systems of nonconducting particles according to the following equation (Venables & Schmuttenmaer, 2000): $A_\mathrm{H} = A_\mathrm{H}(\omega = 0) + A_\mathrm{H}(\omega > 0)f(h)$, where $A_\mathrm{H}\,(\omega = 0)$ and $A_\mathrm{H}\,(\omega > 0)$ are two constants that summarize the effect of the zero- and nonzero-frequency components of the interaction, respectively, and h is the distance between the atomic centers. The effect of retardation is included in the function $f(h)$ (Vanni & Baldi, 2002). The three terms depend on the composition of the solution, and details on these calculations can be found in Lince *et al.* (2008). The potential related to repulsive forces (of double-layer nature) is usually written in terms of the surface potential of the particle, Ψ_s. Different expressions can be derived under different simplification hypotheses (for a detailed review readers are referred to Elimelech *et al.* (1998)). One of the most popular approaches is the one developed by Gregory (Elimelech *et al.*, 1998) for particles of sizes d_p and d'_p:

$$V_r(h) = 64\pi\varepsilon_0\varepsilon_r \left(\frac{k_B T_f}{ze}\right)^2 \tan\left(\frac{ze\Psi_s}{4k_B T_f}\right)^2 \left(\frac{d_p d'_p}{d_p + d'_p}\right) e^{-\kappa h}, \tag{5.176}$$

where ε_0 and ε_r are the absolute and relative permittivity constants, κ is the Debye–Hückel parameter, z is the valency of the ions (a symmetric z–z electrolyte is assumed in this example) and e is the elementary charge. The surface potential can in turn be written as a function of the concentration in solution of some potential-determining ions (PDI) as follows:

$$\Psi_s = \frac{k_B T_f}{ze}\left[\ln(C_i) - \ln(C_i^{pzc})\right], \tag{5.177}$$

where C_i is the concentration of the PDI and C_i^{pzc} is the concentration of the PDI at the point of zero charge (when the surface of the particles has zero charge).[64] Examples of typical calculations for different chemical systems can be found in Gavi *et al.* (2007, 2010).

Equation (5.174) is generally used for particles brought into close proximity by Brownian motion and neglecting any type of short-range fluid-dynamic interaction. In fact, the only interactions considered are those related to attractive Van der Waal forces and repulsive double-layer forces. Fluid-dynamic effects are accounted for by modifying Eq. (5.174) as follows:

$$W = \left(\frac{d_p + d'_p}{2}\right)\int_0^{+\infty} \exp\left(\frac{V(h)}{k_B T_f}\right)\left[\left(\frac{d_p + d'_p}{2}\right) + h\right]^{-2} \frac{dh}{G_f^o(h)}, \tag{5.178}$$

where G_f^o is a function expressing the fluid-dynamic interactions between the particles. The function $G_f^o(h)$ can be, for example, assumed equal to unity in the case of very porous particles[65] or can be, instead, approximated by

$$G_f^o(h) = \frac{6(2h/d_p)^2 + 4(2h/d_p)}{6(2h/d_p)^2 + 13(2h/d_p) + 2} \tag{5.179}$$

in the case of two completely impenetrable particles (Stokes creeping flow) of equal size d_p. In the case of particles of different sizes d_p and d'_p, only approximate expressions are available. For example, for h that tends to zero, we obtain $G_f(h) = 2h(d_p + d'_p)/(d_p d'_p)$. Once the particle interaction potential $V(h)$ has been defined, Eq. (5.174) can be integrated. Integration is typically carried out numerically for different values of the input parameters (the sizes of the colliding particles, namely d_p and d'_p, surface potential Ψ_s, etc). The results are then collected in a look-up table that defines the dependence of the collision efficiency on these parameters: $\alpha = \alpha(d_p, d'_p, \Psi_s)$. For some simplified cases, however, integration of Eq. (5.174) is carried out analytically, resulting in a functional form for the efficiency. The number of plausible scenarios is very large, so a comprehensive and complete discussion is impossible. In general, every practical case must be analyzed separately, and the different

[64] The term $k_B T_f/(ze)$ often appears as $RT/(zF)$, where F is the Faraday constant, equal to the Avogadro number multiplied by the elementary electrical charge.

[65] In this case, in fact, the fluid resistance does not play any role and can be completely neglected, resulting in Eq. (5.174).

forces acting on the particles and mechanisms driving their collisions must be identified and quantified. In what follows only a few examples that result in analytical expressions for the aggregation efficiency will be presented. The first example concerns very small (sub-micron or Brownian) particles interacting through some well-defined attractive and repulsive potentials. The integral appearing in Eq. (5.174) can be analytically solved by using the Reerink & Overbeek (1954) approximation, resulting in the following expression for the aggregation efficiency:

$$\alpha = \frac{d_\mathrm{p} + d'_\mathrm{p}}{2\kappa} \exp\left(-\frac{\phi_\mathrm{max}}{k_\mathrm{B} T_\mathrm{f}}\right), \tag{5.180}$$

where κ is the Debye–Hückel parameter and ϕ_max is the energy barrier for particle aggregation (i.e. the maximum in the plot of potential energy versus inter-particle distance). This maximum is a function of the concentration of PDI, the total ionic strength of the solution, and other operating parameters. The second example refers to particles undergoing shear-induced aggregation under the action of attractive Van der Waals forces and repulsive viscous forces. In this case the aggregation efficiency results in (Wang *et al.*, 2005a) $\alpha = k\mathrm{Fl}^{-0.18}$, where k is a pre-factor of order of unity, generally determined by fitting with data from experiments, and Fl is the flow number,[66] which, for impermeable particles, is defined as $\mathrm{Fl} = 6\pi\mu_\mathrm{f}(d_\mathrm{p} + d'_\mathrm{p})G_\mathrm{f}/A_\mathrm{H}$, where μ_f is the fluid viscosity and G_f is the fluid shear. In the case of permeable particles (as in the case of fractal objects) similar (but more complicated) equations are employed, more details of which can be found in Wang *et al.* (2005b).

Another interesting case worth mentioning is that of large particles aggregating in a supersaturated solution. In the case of large particles, Van der Waals attractive forces are not very important, and the aggregation efficiency is generally dictated by the ability of the growing particles to form a stable connecting neck. This is possible only if the time required to grow this connecting neck, τ_n, is smaller than the time during which the two particles remain in close contact, τ_f. In the case of particles aggregating under the effect of shear, this time is related to the inverse of the shear rate, $\tau_\mathrm{f} = 1/G_\mathrm{f}$. The time for growing a connecting neck is instead calculated as $\tau_\mathrm{n} = D_\mathrm{b}/(f(\delta)\dot{d}_\mathrm{p})$, where $\dot{d}_\mathrm{p}$ is the particle-growth rate (see for example Eq. (5.23)), D_b is the diameter of the stable bridge, and $f(\delta)$ is a shape function that depends on the size ratio δ of the colliding particles. Most of these parameters can be determined by independent experiments or by fitting. The aggregation (or, more properly, agglomeration) efficiency is then calculated as $\alpha = \mathrm{e}^{-\tau_\mathrm{n}/\tau_\mathrm{c}}$. As is evident from this expression, when particles stay close to each other for the time interval τ_c, much longer than the time required to grow a stable neck τ_n, their ratio tends to zero and the efficiency tends to unity. In contrast, the efficiency tends to zero when $\tau_\mathrm{c} \ll \tau_\mathrm{n}$. More details for a particular example can be found in Marchisio *et al.* (2003a). In many cases, the derivation of an aggregation efficiency from first principles is impossible, and empirical efficiencies are instead adopted.[67] Since in these cases the efficiency depends on the composition of the particles, this will have to be included as part of the internal coordinates. One example is that of binary aggregation, namely particles of type A aggregating with particles of type B (Lee *et al.*, 2008; Matsoukas *et al.*, 2006, 2009,). A highly empirical form used to express this efficiency that provides a qualitative analogy with experimentally investigated binary systems (Rajniak

[66]The flow number can be interpreted as the ratio between shear and Van der Waals forces.

[67]This is very common, for example, in granulation. During granulation, particle aggregation is induced by adding a sticky substance called a binder.

et al., 2007) is $\alpha(c, c') = \exp[-a_{AA}cc' - a_{BB}(1 - c)(1 - c') - a_{AB}(c + c' - 2cc')]$, where c and c' are the concentrations (for example mass fractions) of component A within the two aggregating particles. The three parameters a_{AA}, a_{BB}, and a_{AB} control the strength of self-aggregation (i.e. A–A and B–B aggregations involving particles of the same type) versus cross aggregation (i.e. aggregation between particle A and particle B). Evidently with $a_{AA} = a_{BB} = a_{AB} = 0$ the composition dependence is cancelled out and the aggregation efficiency reduces to unity. If, for example, $a_{AA} = -3$, $a_{BB} = +2$, and $a_{AB} = -2$, as for the case discussed by Rajniak *et al.* (2007), self-aggregation of component A is favored over self-aggregation of component B, while the strength of cross aggregation lies in between. The values of these coefficients are usually identified by fitting experimental data.

5.8.4 Coalescence kernels for droplets and bubbles

The treatment of coalescing fluid particles is complicated by the fact that droplets and bubbles are deformable. Nonetheless, also for these cases, the available kernels are often written as the product of a collision frequency and anefficiency. The collision frequency is generally derived in analogy with the kinetic theory of gases (see Chapter 6 for details), and its calculation is based on the number of particles swept by one particle moving in a collisional cylinder with diameter equal to the sum of the radii of the two colliding particles. For two particles of sizes d_p and d'_p, the collision frequency can be approximated by

$$\beta(d_p, d'_p, \mathbf{v}_r) = \frac{\pi}{4}\left(d_p + d'_p\right)^2 |\mathbf{v}_r|\eta(|\mathbf{v}_r|), \tag{5.181}$$

where the first term represents the section of the collisional cylinder, and the second term, namely that involving $\mathbf{v}_r = \mathbf{v}_p - \mathbf{v}'_p$, quantifies the relative velocity of the two colliding particles. The function $\eta \leq 1$ is a coalescence efficiency that will be described below. For particles with finite Stokes numbers (e.g. liquid sprays), $\mathbf{v}_r$ is known from the GPBE. Otherwise, $\mathbf{v}_r$ must be approximated by some other means. Once the two particles are brought into close proximity and a collision event is induced, they may, or might not, coalesce. The possible scenarios are very different, depending on the nature of the primary phase. If the primary phase is a gas, the probability that the colliding particles successfully coalesce is mainly determined by the Weber number and an impact parameter. The latter quantity is generally assumed to be the sine of the angle formed by the vector uniting the centers of the colliding particles and the particle-velocity difference $\mathbf{v}_r$. According to this definition, the impact parameter is null when the velocity difference is parallel to the vector uniting the centers of the droplets and is, conversely, equal to unity when the two are orthogonal. The Weber number is defined as

$$\mathrm{We} = \frac{\rho_p|\mathbf{v}_r|d_p}{\sigma}, \tag{5.182}$$

where ρ_p is the material density of the disperse phase and σ is the interfacial tension between the continuous and disperse phases. The Weber number is a measure of the relative importance of the fluid inertia compared with the interfacial tension, and is very small for very high interfacial tension. When the Weber number is smaller than 10, coalescence is very likely and the efficiency η is very close to unity. In contrast, as We increases coalescence becomes possible only for impact factors close to 0.2, corresponding to impact angles of about 10–15°. Under these conditions coalescence is unlikely to happen (i.e. $\eta \approx 0$), and the droplets after colliding are said to undergo lateral or head-on bouncing and reflective or stretching separation.

If the continuous phase is a liquid, the main obstacle to coalescence is the drainage of the film of liquid in the small space in between the two particles. The efficiency is in these cases usually quantified as a function (generally a negative exponential function) of the ratio of the characteristic time for droplet contact and film drainage. For example, in the case of small bubbles coalescing due to turbulent velocity fluctuations the coalescence kernel assumes the form (Buffo *et al.*, 2012; Laakkonen *et al.*, 2006; Petitti *et al.*, 2010)

$$\beta(d_\mathrm{p}, d'_\mathrm{p}) = 0.88\epsilon_\mathrm{f}^{1/3}(d_\mathrm{p} + d'_\mathrm{p})^2(d_\mathrm{p}^{2/3} + d'^{\,2/3}_\mathrm{p})^{1/2}\eta(d_\mathrm{p}, d'_\mathrm{p}), \tag{5.183}$$

whereas the coalescence efficiency is written as follows:

$$\eta(d_\mathrm{p}, d'_\mathrm{p}) = \exp\left(-6\times 10^9 \frac{\mu_\mathrm{f}\rho_\mathrm{f}\epsilon_\mathrm{f}}{\sigma^2}\left(\frac{d_\mathrm{p}d'_\mathrm{p}}{d_\mathrm{p} + d'_\mathrm{p}}\right)^4\right). \tag{5.184}$$

Readers interested in more details for particular systems should consult the specialized literature.

6

Hard-sphere collision models

In this chapter we consider models for collisions between smooth (i.e. frictionless) spherical particles with identical (monodisperse) or different (polydisperse) densities and diameters. For simplicity, we consider only collisions during which the particle mass and diameter are conserved, and exclude other processes that might change these properties (e.g. surface condensation or aggregation (Fox *et al.*, 2008)). Likewise, assuming smooth spherical particles means that the particle angular momentum does not change during a collision, and hence only the particle velocity need be accounted for in the kinetic equation. We limit the discussion to hard-sphere collisions, which implies that the particle velocities after a collision can be written as explicit functions of the particle velocities before the collision, but also consider inelastic collisions with a constant coefficient of restitution and finite-size particle effects. More details on hard-sphere collisions can be found in the books by Cercignani (Cercignani, 1975, 1988, 2000; Cercignani *et al.*, 1994). We will also briefly discuss simpler collision models that are often used to approximate the hard-sphere collision model in the dilute limit. These include the Maxwell model (Maxwell, 1879) and two linearized collision models (i.e. BGK (Bhatnagar *et al.*, 1954) and ES-BGK (Holway, 1966)) for monodisperse particles, and an extension of the ES-BGK model to polydisperse particles. A discussion of kinetic models for collisions that are not of hard-sphere type can be found in Struchtrup (2005).

As an example, in Section 6.4 we will consider the kinetic equation governing the velocity number density function $f(t, \mathbf{x}, \mathbf{v})$:

$$\partial_t f + \mathbf{v} \cdot \partial_{\mathbf{x}} f = \mathbb{C}, \tag{6.1}$$

where t is time, $\mathbf{x}$ is the spatial location, $\mathbf{v}$ is the particle velocity, and $\mathbb{C}$ represents particle–particle collisions. In order to simplify the notation, we will include the independent variables in the formulas only when they are different from the set $(t, \mathbf{x}, \mathbf{v})$. The density function in Eq. (6.1) is normalized such that

$$\alpha_{\mathrm{p}} = \int_{\mathbb{R}^3} f \, \mathrm{d}\mathbf{v},$$

where α_{p} is the disperse-phase volume fraction. Since collisions do not change the number of particles, the collision models have the property that

$$\int_{\mathbb{R}^3} \mathbb{C} \, \mathrm{d}\mathbf{v} = 0.$$

In addition, conservation of momentum implies that

$$\int_{\mathbb{R}^3} \mathbf{v}\mathbb{C}\, d\mathbf{v} = 0.$$

Finally, the change of kinetic energy due to collisions requires that

$$\int_{\mathbb{R}^3} v^2\mathbb{C}\, d\mathbf{v} \leq 0,$$

where the equality holds for elastic collisions. By deriving the moment-transport equations for each of the collisions models, it will be possible to verify that these conditions hold for all models. More generally, the change in the moments due to collisions can be found by integrating a functional $\psi(\mathbf{v})$ multiplied by $\mathbb{C}$ over the phase space. One of the main goals of this chapter is to find the change in integer moments due to collisions for monodisperse and polydisperse particles. For integer moments, the functional can be expressed as $\psi(\mathbf{v}) = v_1^{l_1} v_2^{l_2} v_3^{l_3}$, where l_1, l_2, and l_3 are nonnegative integers. A detailed introduction to the use of velocity moments for deriving macroscopic transport equations can be found in Struchtrup (2005).

The rest of this chapter is organized as follows. First, in Section 6.1, we consider the collision term for monodisperse hard-sphere collisions both for elastic and for inelastic particles. We introduce the kinetic closures due to Boltzmann (1872) and Enksog (1921) for the pair correlation function, and then derive the exact source terms for the velocity moments of arbitrary order and then for integer moments. Second, in Section 6.2, we consider the exact source terms for polydisperse hard-sphere collisions, deriving exact expressions for arbitrary and integer-order moments. Next, in Section 6.3, we consider simplified kinetic models for monodisperse and polydisperse systems that are derived from the exact collision source terms, and discuss their properties *vis-à-vis* the hard-sphere collision models. In Section 6.4, we discuss properties of the moment-transport equations derived from Eq. (6.1) with the hard-sphere collision models. Finally, in Section 6.5 we briefly describe how quadrature-based moment methods are applied to close the collision source terms for the velocity moments.

6.1 Monodisperse hard-sphere collisions

In this section we will briefly review the collision model for binary hard-sphere collisions using the notation in Fox & Vedula (2010). The change in the number-density function due to elastic hard-sphere collisions (Boltzmann, 1872; Cercignani, 1988; Chapman & Cowling, 1961; Enksog, 1921) obeys an (unclosed) integral expression of the form[1]

$$\mathbb{C} = \frac{6}{\pi d_p} \int_{\mathbb{R}^3} \int_{\mathbb{S}^+} \left[f^{(2)}(\mathbf{x}, \mathbf{v}_1'; \mathbf{x} - d_p \mathbf{x}_{12}, \mathbf{v}_2') - f^{(2)}(\mathbf{x}, \mathbf{v}_1; \mathbf{x} + d_p \mathbf{x}_{12}, \mathbf{v}_2) \right] |\mathbf{v}_{12} \cdot \mathbf{x}_{12}| d\mathbf{x}_{12}\, d\mathbf{v}_2, \tag{6.2}$$

where d_p is the particle diameter, $f^{(2)}$ is the pair correlation function,

$$\mathbf{x}_{12} = -\frac{\mathbf{x}_1 - \mathbf{x}_2}{|\mathbf{x}_1 - \mathbf{x}_2|} \tag{6.3}$$

[1] Although it is not strictly necessary for the monodisperse case, we employ a notation with subscripts $_{12}$ to denote the two particles involved in the collision. The reason for doing so is that in the polydisperse case, where particles 1 and 2 have different properties, it will be straightforward to modify the collision integrals with very little change in the notation.

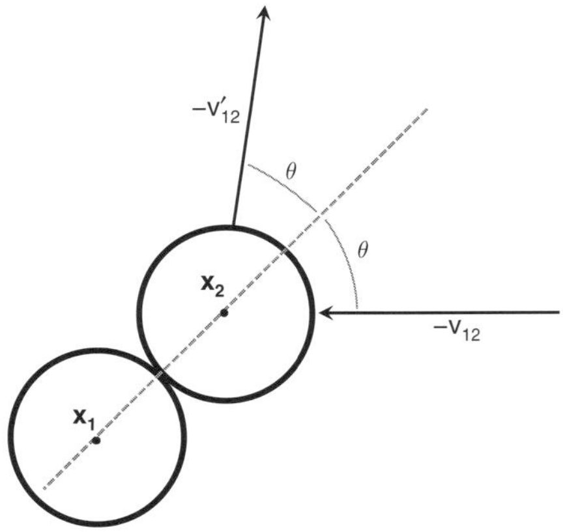

Figure 6.1. A direct collision between two hard spheres with diameters d_p.

is the unit vector in the direction between the particle centers (the so-called collision vector), and $\mathbf{v}_{12} = \mathbf{v}_1 - \mathbf{v}_2$ (with magnitude v_{12}) is the velocity difference before a direct collision. (See Figure 6.1 for a geometric interpretation of these symbols.) The pre-collision velocities for direct collisions ($\mathbf{v}_1$ and $\mathbf{v}_2$) are related to the post-collision velocities ($\mathbf{v}'_1$ and $\mathbf{v}'_2$) by (Cercignani, 1988; Chapman & Cowling, 1961)

$$\begin{aligned} \mathbf{v}_1 &= \mathbf{v}'_1 - (\mathbf{v}'_{12} \cdot \mathbf{x}_{12})\mathbf{x}_{12}, \\ \mathbf{v}_2 &= \mathbf{v}'_2 + (\mathbf{v}'_{12} \cdot \mathbf{x}_{12})\mathbf{x}_{12}, \end{aligned} \tag{6.4}$$

where $\mathbf{v}'_{12} = \mathbf{v}'_1 - \mathbf{v}'_2$ is the velocity difference after a direct collision. The surface $\mathbb{S}^+$ is the unit half sphere on which $\mathbf{v}_{12} \cdot \mathbf{x}_{12} = -\mathbf{v}'_{12} \cdot \mathbf{x}_{12} = v_{12} \cos\theta > 0$ (i.e. velocity differences that result in collisions), where θ is the collision angle (whose value is such that $\cos\theta$ is positive when collisions occur).

Because the collisions are elastic, the pre-collision velocities before *inverse* collisions are the same as the values after a direct collision (which is not the case for inelastic collisions, as will be discussed below). As a result, conservation of momentum and energy follows directly from the definitions of the pre-collision velocities since

$$\begin{aligned} \mathbf{v}_1 + \mathbf{v}_2 &= \mathbf{v}'_1 + \mathbf{v}'_2, \\ \mathbf{v}_1 \cdot \mathbf{v}_1 + \mathbf{v}_2 \cdot \mathbf{v}_2 &= \mathbf{v}'_1 \cdot \mathbf{v}'_1 + \mathbf{v}'_2 \cdot \mathbf{v}'_2. \end{aligned}$$

The normalized pair correlation function

$$\frac{f^{(2)}(\mathbf{x}, \mathbf{v}_1; \mathbf{y}, \mathbf{v}_2)}{\alpha_p(\mathbf{x})\alpha_p(\mathbf{y})}$$

is the probability of finding a particle with velocity $\mathbf{v}_1$ at location $\mathbf{x}$ *and* a particle with velocity $\mathbf{v}_2$ at location $\mathbf{y}$ at the same time t. This information about the *correlation* between particle locations and velocities is not contained in the one-particle number-density function, and thus must be modeled before we can make use of Eq. (6.2).

For inelastic collisions, the hard-sphere collision model is given by (Fox & Vedula, 2010; Jenkins & Mancini, 1989; Jenkins & Richman, 1985)

$$\mathbb{C} = \frac{6}{\pi d_p} \int_{\mathbb{R}^3} \int_{\mathbb{S}^+} \left[\chi f^{(2)}(\mathbf{x}, \mathbf{v}''_1; \mathbf{x} - d_p\mathbf{x}_{12}, \mathbf{v}''_2) - f^{(2)}(\mathbf{x}, \mathbf{v}_1; \mathbf{x} + d_p\mathbf{x}_{12}, \mathbf{v}_2) \right] |\mathbf{v}_{12} \cdot \mathbf{x}_{12}| \mathrm{d}\mathbf{x}_{12} \, \mathrm{d}\mathbf{v}_2, \tag{6.5}$$

where $\chi = 1/e^2$ is a factor relating the pre-collision velocities for *direct* collisions $(\mathbf{v}_1, \mathbf{v}_2)$ to those for *inverse* collisions $(\mathbf{v}_1'', \mathbf{v}_2'')$ and $0 \le e \le 1$ is the coefficient of restitution. These velocities are related by (Cercignani, 1988)

$$\begin{aligned} \mathbf{v}_1 &= \mathbf{v}_1'' - \omega(\mathbf{v}_{12}'' \cdot \mathbf{x}_{12})\mathbf{x}_{12}, \\ \mathbf{v}_2 &= \mathbf{v}_2'' + \omega(\mathbf{v}_{12}'' \cdot \mathbf{x}_{12})\mathbf{x}_{12}, \end{aligned} \tag{6.6}$$

where $\omega = (1 + e)/2$ and $\mathbf{v}_{12}'' = \mathbf{v}_1'' - \mathbf{v}_2''$ (with magnitude v_{12}'') is the velocity difference before an inverse collision. Note that the post-collision velocities after direct collisions are given by

$$\begin{aligned} \mathbf{v}_1' &= \mathbf{v}_1 - \omega(\mathbf{v}_{12} \cdot \mathbf{x}_{12})\mathbf{x}_{12}, \\ \mathbf{v}_2' &= \mathbf{v}_2 + \omega(\mathbf{v}_{12} \cdot \mathbf{x}_{12})\mathbf{x}_{12}. \end{aligned} \tag{6.7}$$

Thus, each direct collision results in a loss of energy, the amount of which depends on e:

$$\mathbf{v}_1 \cdot \mathbf{v}_1 + \mathbf{v}_2 \cdot \mathbf{v}_2 = \mathbf{v}_1' \cdot \mathbf{v}_1' + \mathbf{v}_2' \cdot \mathbf{v}_2' - \frac{1 - e^2}{2} |\mathbf{v}_{12} \cdot \mathbf{x}_{12}|^2.$$

As in the elastic case, it will be necessary to provide a closure for the pair correlation functions appearing in Eq. (6.5).

6.1.1 The Boltzmann collision model

Owing to the presence of the pair correlation function, the collision model in Eq. (6.2) is unclosed. Thus, in order to close the kinetic equation (Eq. 6.1), we must provide a closure for $f^{(2)}$ written in terms of f. The simplest closure is the Boltzmann *Stoßzahlansatz* (Boltzmann, 1872):

$$\begin{aligned} f^{(2)}(\mathbf{x}, \mathbf{v}_1'; \mathbf{x} - d_\mathrm{p}\mathbf{x}_{12}, \mathbf{v}_2') &\approx f(\mathbf{x}, \mathbf{v}_1')f(\mathbf{x}, \mathbf{v}_2'), \\ f^{(2)}(\mathbf{x}, \mathbf{v}_1; \mathbf{x} + d_\mathrm{p}\mathbf{x}_{12}, \mathbf{v}_2) &\approx f(\mathbf{x}, \mathbf{v}_1)f(\mathbf{x}, \mathbf{v}_2), \end{aligned} \tag{6.8}$$

which amounts to assuming that the velocities of the colliding particles are uncorrelated *before* collisions and that d_p is much smaller than the average distance between the particles. With this assumption, the (elastic) Boltzmann collision model is written as

$$\mathbb{C} = \frac{6}{\pi d_\mathrm{p}} \int_{\mathbb{R}^3} \int_{\mathbb{S}^+} [f(\mathbf{v}_1')f(\mathbf{v}_2') - f(\mathbf{v}_1)f(\mathbf{v}_2)] \, |\mathbf{v}_{12} \cdot \mathbf{x}_{12}| \mathrm{d}\mathbf{x}_{12} \, \mathrm{d}\mathbf{v}_2, \tag{6.9}$$

where we have dropped the $\mathbf{x}$ dependence since all terms in the integral are evaluated at the same spatial location.

The elastic Boltzmann collision model has a unique equilibrium solution defined by $\mathbb{C} = 0$. This *Maxwellian* distribution is given by

$$f_\mathrm{eq}(\mathbf{v}) = \frac{\alpha_\mathrm{p}}{(2\pi\Theta_\mathrm{p})^{3/2}} \exp\left(-\frac{|\mathbf{v} - \mathbf{U}_\mathrm{p}|^2}{2\Theta_\mathrm{p}^2}\right), \tag{6.10}$$

where $\mathbf{U}_\mathrm{p}$ is the mean velocity defined by

$$\mathbf{U}_\mathrm{p} = \frac{1}{\alpha_\mathrm{p}} \int \mathbf{v} f \, \mathrm{d}\mathbf{v}$$

and Θ_p is the particle energy[2] defined by

$$\Theta_p = \frac{1}{3\alpha_p} \int |\mathbf{v} - \mathbf{U}_p|^2 f \, d\mathbf{v}.$$

Note that the transport term on the left-hand side of Eq. (6.1) can be larger or smaller in magnitude than the collision term. For cases in which the collision term is much more important than the transport term, the solution to Eq. (6.1) with the Boltzmann collision model is a *local Maxwellian* wherein α_p, $\mathbf{U}_p$, and Θ_p depend on space and time but f is well approximated by Eq. (6.10). In this limit, the particles behave as an ideal gas and the mean velocity obeys the Euler equation.

In general, the elastic Boltzmann collision model provides a good description of molecular gases over a wide range of concentrations, including very-low-concentration or rarefied gases. However, for macroscopic particles or dense gases the assumption that d_p is much smaller than the average distance between the particles no longer holds. Furthermore, collisions between macroscopic particles are usually inelastic. Thus, we will consider next the change in the velocity moments due to inelastic, finite-size, hard-sphere collisions starting from Eq. (6.5).

6.1.2 The collision term for arbitrary moments

The source term for the velocity moments due to Boltzmann–Enskog collisions can be expressed in a compact form using the properties of the hard-sphere collision integral (Cercignani, 1988). For the sake of generality, we consider the change in property $\psi \equiv \psi(\mathbf{v})$ due to the collision term in Eq. (6.5). The corresponding collision source term is given by

$$\mathbb{C}(\psi) \equiv \int_{\mathbb{R}^3} \psi(\mathbf{v}_1)\mathbb{C} \, d\mathbf{v}_1. \tag{6.11}$$

Using the relation $\chi|\mathbf{v}_{12} \cdot \mathbf{x}_{12}| d\mathbf{v}_2 \, d\mathbf{v}_1 = |\mathbf{v}''_{12} \cdot \mathbf{x}_{12}| d\mathbf{v}''_2 \, d\mathbf{v}''_1$, the right-hand side of Eq. (6.11) can be rewritten as

$$\begin{aligned} \mathbb{C}(\psi) = {} & \frac{6}{\pi d_p} \int_{\mathbb{R}^6} \int_{\mathbb{S}^+} \psi(\mathbf{v}_1) f^{(2)}(\mathbf{x}, \mathbf{v}''_1; \mathbf{x} - d_p \mathbf{x}_{12}, \mathbf{v}''_2) |\mathbf{v}''_{12} \cdot \mathbf{x}_{12}| d\mathbf{x}_{12} \, d\mathbf{v}''_2 \, d\mathbf{v}''_1 \\ & - \frac{6}{\pi d_p} \int_{\mathbb{R}^6} \int_{\mathbb{S}^+} \psi(\mathbf{v}_1) f^{(2)}(\mathbf{x}, \mathbf{v}_1; \mathbf{x} + d_p \mathbf{x}_{12}, \mathbf{v}_2) |\mathbf{v}_{12} \cdot \mathbf{x}_{12}| d\mathbf{x}_{12} \, d\mathbf{v}_2 \, d\mathbf{v}_1, \end{aligned} \tag{6.12}$$

where $\psi(\mathbf{v}_1) = \psi\left(\mathbf{v}''_1 - \omega(\mathbf{v}''_{12} \cdot \mathbf{x}_{12})\mathbf{x}_{12}\right)$ in the first integral on the right-hand side.

We can now use a change of notation in the first multidimensional integral appearing on the right-hand side of Eq. (6.12) to replace the double-primed variables (which refer to values before an inverse collision) with the corresponding variables without the primes (which refer to values after an inverse collision or before a direct collision) and simultaneously replace the unprimed velocity variables (values after an inverse collision or before a direct collision) with the corresponding single-primed variables (which refer to values after a direct collision). In other words, a consistent change of notation allows us to express the source term for velocity moments due to inverse collisions in terms of velocity variables before and after direct collisions as

$$\mathbb{C}(\psi) = \frac{6}{\pi d_p} \int_{\mathbb{R}^6} \int_{\mathbb{S}^+} [\psi(\mathbf{v}'_1) - \psi(\mathbf{v}_1)] \, f^{(2)}(\mathbf{x}, \mathbf{v}_1; \mathbf{x} + d_p \mathbf{x}_{12}, \mathbf{v}_2) |\mathbf{v}_{12} \cdot \mathbf{x}_{12}| d\mathbf{x}_{12} \, d\mathbf{v}_2 \, d\mathbf{v}_1, \tag{6.13}$$

[2] For molecular systems, the definition of the temperature includes the particle mass. For granular systems, we will refer to Θ_p as the granular temperature.

where $\psi(\mathbf{v}_1') = \psi(\mathbf{v}_1 - \omega(\mathbf{v}_{12} \cdot \mathbf{x}_{12})\mathbf{x}_{12})$. Note that the pair correlation function for direct collisions is evaluated at $\mathbf{x} + d_p\mathbf{x}_{12}$ for the second particle, and thus a consistent change of variables uses only the form of $f^{(2)}$ for direct collisions.

The goal of the next series of steps is to obtain a form of Eq. (6.13) wherein the pair correlation function depends on only one spatial location (i.e. the point where the particles touch during a collision). We start by adding and subtracting the pair distribution function $f^{(2)}(\mathbf{x} - d_p\mathbf{x}_{12}, \mathbf{v}_1; \mathbf{x}, \mathbf{v}_2)$ to $f^{(2)}(\mathbf{x}, \mathbf{v}_1; \mathbf{x} + d_p\mathbf{x}_{12}, \mathbf{v}_2)$ in Eq. (6.13):

$$\begin{aligned}\mathbb{C}(\psi) = \frac{3}{\pi d_p} \int_{\mathbb{R}^6} \int_{\mathbb{S}^+} & [\psi(\mathbf{v}_1') - \psi(\mathbf{v}_1)] \\ & \times \Big[\left(f^{(2)}(\mathbf{x}, \mathbf{v}_1; \mathbf{x} + d_p\mathbf{x}_{12}, \mathbf{v}_2) + f^{(2)}(\mathbf{x} - d_p\mathbf{x}_{12}, \mathbf{v}_1; \mathbf{x}, \mathbf{v}_2)\right) \\ & \quad + \left(f^{(2)}(\mathbf{x}, \mathbf{v}_1; \mathbf{x} + d_p\mathbf{x}_{12}, \mathbf{v}_2) - f^{(2)}(\mathbf{x} - d_p\mathbf{x}_{12}, \mathbf{v}_1; \mathbf{x}, \mathbf{v}_2)\right)\Big] |\mathbf{v}_{12} \cdot \mathbf{x}_{12}| \mathrm{d}\mathbf{x}_{12}\, \mathrm{d}\mathbf{v}_2\, \mathrm{d}\mathbf{v}_1 .\end{aligned} \tag{6.14}$$

(Note the factor of 1/2 included in this procedure.)

Using Taylor-series expansions[3] about the collision contact point,

$$(\mathbf{x}_1, \mathbf{x}_2) = \left(\mathbf{x} - \frac{1}{2} d_p\mathbf{x}_{12}, \mathbf{x} + \frac{1}{2} d_p\mathbf{x}_{12}\right),$$

we can now rewrite Eq. (6.14) in the form[4]

$$\mathbb{C}(\psi) = C(\psi) - \nabla_x \cdot \mathbf{G}(\psi), \tag{6.15}$$

[3] We will make repeated use of multivariate Taylor-series expansions involving a vector $\mathbf{x} = (x_1, x_2, x_3)$ and the gradient operator $\nabla_y = (\partial_{y_1}, \partial_{y_2}, \partial_{y_3})$. In the expansions, a term of the form $(\mathbf{x} \cdot \nabla_y)^n \psi(\mathbf{y})$ will appear for $n = 0, 1, \ldots$ and is defined as follows:

$$\text{for } n = 0, \quad (\mathbf{x} \cdot \nabla_y)^0 \psi(\mathbf{y}) = 1;$$

$$\text{for } n = 1, \quad (\mathbf{x} \cdot \nabla_y)^1 \psi(\mathbf{y}) = \sum_{i=1}^{3} x_i \frac{\partial \psi(\mathbf{y})}{\partial y_i};$$

$$\text{for } n = 2, \quad (\mathbf{x} \cdot \nabla_y)^2 \psi(\mathbf{y}) = \sum_{i=1}^{3} \sum_{j=1}^{3} x_i x_j \frac{\partial^2 \psi(\mathbf{y})}{\partial y_i\, \partial y_j};$$

$$\text{for } n = 3, \quad (\mathbf{x} \cdot \nabla_y)^3 \psi(\mathbf{y}) = \sum_{i=1}^{3} \sum_{j=1}^{3} \sum_{k=1}^{3} x_i x_j x_k \frac{\partial^3 \psi(\mathbf{y})}{\partial y_i\, \partial y_j\, \partial y_k}.$$

$$\vdots$$

Note that the sums on the right-hand sides can be rewritten as multinomial expansions. We will make use of this fact at several points in the derivation.

[4] A simple example may help clarify the two Taylor-series expansions used in this step. Consider a function $h(x)$ evaluated at $x \pm a$. The Taylor-series expansion is

$$h(x \pm a) = h(x) \pm a\, \partial_x h(x) + \frac{a^2}{2} \partial_x^2 h(x) \pm \frac{a^3}{6} \partial_x^3 h(x) + \cdots .$$

Adding and subtracting the two expansions yields

$$h(x + a) + h(x - a) = 2h(x) + a^2\, \partial_x^2 h(x) + \cdots = 2\left[1 + \frac{1}{2}(a\, \partial_x)^2 + \cdots\right] h(x),$$

$$h(x + a) - h(x + a) = 2a\, \partial_x h(x) + \frac{a^3}{3} \partial_x^3 h(x) + \cdots = 2a\, \partial_x \left[1 + \frac{1}{6}(a\, \partial_x)^2 + \cdots\right] h(x),$$

which have the same forms as the multivariate expansions used to find Eq. (6.16) and Eq. (6.17), respectively.

where the collision source term is given by

$$C(\psi) = \frac{6}{\pi d_p} \int_{\mathbb{R}^6} \int_{\mathbb{S}^+} [\psi(\mathbf{v}_1') - \psi(\mathbf{v}_1)] \left[1 + \sum_{n=1}^{\infty} \frac{1}{(2n)!} \left(\frac{1}{2} d_p \mathbf{x}_{12} \cdot \nabla_x \right)^{2n} \right] \\ \times f^{(2)} \left(\mathbf{x} - \frac{1}{2} d_p \mathbf{x}_{12}, \mathbf{v}_1; \mathbf{x} + \frac{1}{2} d_p \mathbf{x}_{12}, \mathbf{v}_2 \right) |\mathbf{v}_{12} \cdot \mathbf{x}_{12}| \mathrm{d}\mathbf{x}_{12} \, \mathrm{d}\mathbf{v}_2 \, \mathrm{d}\mathbf{v}_1, \tag{6.16}$$

and the collisional-flux term is given by

$$\mathbf{G}(\psi) = -\frac{3}{\pi} \int_{\mathbb{R}^6} \int_{\mathbb{S}^+} \mathbf{x}_{12} [\psi(\mathbf{v}_1') - \psi(\mathbf{v}_1)] \left[1 + \sum_{n=1}^{\infty} \frac{1}{(2n+1)!} \left(\frac{1}{2} d_p \mathbf{x}_{12} \cdot \nabla_x \right)^{2n} \right] \\ \times f^{(2)} \left(\mathbf{x} - \frac{1}{2} d_p \mathbf{x}_{12}, \mathbf{v}_1; \mathbf{x} + \frac{1}{2} d_p \mathbf{x}_{12}, \mathbf{v}_2 \right) |\mathbf{v}_{12} \cdot \mathbf{x}_{12}| \mathrm{d}\mathbf{x}_{12} \, \mathrm{d}\mathbf{v}_2 \, \mathrm{d}\mathbf{v}_1. \tag{6.17}$$

In most applications, the summation terms for $n = 1, \ldots, \infty$ are neglected in the definitions of C and $\mathbf{G}$, and henceforth we will do the same. This approximation amounts to assuming that d_p is much smaller than the characteristic length scale of the spatial gradients of $f^{(2)}$ (i.e. $|d_p \nabla f^{(2)}| \ll \alpha_p^2$) and thus that only terms up to order d_p are important.

The next step is to provide a closure for the pair correlation function appearing in the collision source and collisional-flux terms. For moderately dense flows, the collision frequency for finite-size particles is known to be larger than that found using the Boltzmann *Stoßzahlansatz* (Carnahan & Starling, 1969; Enksog, 1921). In order to account for this effect, the pair correlation function can be modeled as the product of two single-particle velocity distribution functions and a radial distribution function:

$$f^{(2)} \left(\mathbf{x} - \frac{1}{2} d_p \mathbf{x}_{12}, \mathbf{v}_1; \mathbf{x} + \frac{1}{2} d_p \mathbf{x}_{12}, \mathbf{v}_2 \right) = g_0 \left(\mathbf{x} - \frac{1}{2} d_p \mathbf{x}_{12}, \mathbf{x} + \frac{1}{2} d_p \mathbf{x}_{12} \right) f \left(\mathbf{x} - \frac{1}{2} d_p \mathbf{x}_{12}, \mathbf{v}_1 \right) \\ \times f \left(\mathbf{x} + \frac{1}{2} d_p \mathbf{x}_{12}, \mathbf{v}_2 \right). \tag{6.18}$$

Note that this assumption simply transforms the problem of modeling the pair correlation function into the new problem of modeling g_0.

The usual model for g_0 assumes that the radial distribution function depends neither explicitly on the collision angle (i.e. on $\mathbf{x}_{12}$) nor explicitly on $\mathbf{x}$. The former amounts to assuming that the particle with velocity $\mathbf{v}_2$ has no preferential spatial direction relative to the particle with velocity $\mathbf{v}_1$.[5] The radial distribution function can then be modeled as a function of the disperse-phase volume fraction. For example, a typical model is (Carnahan & Starling, 1969)

$$g_0(\alpha_p) = \frac{2 - c}{2(1 - c)^3},$$

where $c = \alpha_p / \alpha_p^*$ and $\alpha_p^* \approx 0.63$ for close-packed spheres. Neglecting the spatial gradients of g_0 in Eq. (6.18) and using Taylor-series approximations of the single-particle density

[5]This assumption is reasonable for collisions between particles with the same properties (e.g. mass, diameter, etc.). However, for collisions between particles with different properties it is unlikely to hold when external forces affect unlike particles differently. For example, the fluid drag will be stronger for small particles than for large particles, causing each type to have a different mean velocity. It would then be more likely for collisions to occur on the upstream side (i.e. in the direction of the velocity difference) of the faster moving particles. In such cases, the model for g_0 would need to include such information in order to accurately model the collision frequency.

functions about spatial point **x** yields

$$f^{(2)}\left(\mathbf{x}-\frac{1}{2}d_\text{p}\mathbf{x}_{12},\mathbf{v}_1;\mathbf{x}+\frac{1}{2}d_\text{p}\mathbf{x}_{12},\mathbf{v}_2\right)=g_0(\alpha_\text{p})f(\mathbf{v}_1)f(\mathbf{v}_2)\times\left[1+\frac{1}{2}d_\text{p}\mathbf{x}_{12}\cdot\nabla_x\ln\left(\frac{f(\mathbf{v}_2)}{f(\mathbf{v}_1)}\right)+O(d_\text{p}^2)\right].\quad(6.19)$$

With this expression, the change in arbitrary velocity moments due to collisions (i.e. Eq. (6.15)) in now closed in terms of the single-particle density function.

The final step is to write $\psi(\mathbf{v}_1')$ as an explicit function of $\mathbf{x}_{12}$ using the definition $\mathbf{v}_1'=\mathbf{v}_1-\omega(\mathbf{v}_{12}\cdot\mathbf{x}_{12})\mathbf{x}_{12}$. Employing a Taylor-series expansion about $\mathbf{v}_1$, we find

$$\psi(\mathbf{v}_1')-\psi(\mathbf{v}_1)=\sum_{n=1}^{\infty}\frac{(-\omega)^n}{n!}(\mathbf{v}_{12}\cdot\mathbf{x}_{12})^n(\mathbf{x}_{12}\cdot\nabla_v)^n\psi(\mathbf{v}_1).\quad(6.20)$$

Keeping only terms up to first-order spatial derivatives in Eq. (6.19) (i.e. up to order d_p overall), the collision source term and collisional-flux term are

$$C(\psi)=\frac{6g_0}{\pi d_\text{p}}\int_{\mathbb{R}^6}\int_{\mathbb{S}^+}\left[1+\frac{1}{2}d_\text{p}\mathbf{x}_{12}\cdot\nabla_x\ln\left(\frac{f(\mathbf{v}_2)}{f(\mathbf{v}_1)}\right)\right]\times\left[\sum_{n=1}^{\infty}\frac{(-\omega)^n}{n!}(\mathbf{v}_{12}\cdot\mathbf{x}_{12})^n(\mathbf{x}_{12}\cdot\nabla_v)^n\psi(\mathbf{v}_1)\right]|\mathbf{v}_{12}\cdot\mathbf{x}_{12}|f(\mathbf{v}_1)f(\mathbf{v}_2)\mathrm{d}\mathbf{x}_{12}\,\mathrm{d}\mathbf{v}_2\,\mathrm{d}\mathbf{v}_1\quad(6.21)$$

and

$$\mathbf{G}(\psi)=-\frac{3g_0}{\pi}\int_{\mathbb{R}^6}\int_{\mathbb{S}^+}\mathbf{x}_{12}\left[\sum_{n=1}^{\infty}\frac{(-\omega)^n}{n!}(\mathbf{v}_{12}\cdot\mathbf{x}_{12})^n(\mathbf{x}_{12}\cdot\nabla_v)^n\psi(\mathbf{v}_1)\right]\times|\mathbf{v}_{12}\cdot\mathbf{x}_{12}|f(\mathbf{v}_1)f(\mathbf{v}_2)\mathrm{d}\mathbf{x}_{12}\,\mathrm{d}\mathbf{v}_2\,\mathrm{d}\mathbf{v}_1.\quad(6.22)$$

In these definitions, we can observe that there are two types of integrals involving $\mathbf{x}_{12}$:

$$\int_{\mathbb{S}^+}(\mathbf{v}_{12}\cdot\mathbf{x}_{12})^{n+1}(\mathbf{x}_{12}\cdot\nabla_v)^n\,\psi(\mathbf{v}_1)\mathrm{d}\mathbf{x}_{12}\text{ and }\int_{\mathbb{S}^+}\mathbf{x}_{12}(\mathbf{v}_{12}\cdot\mathbf{x}_{12})^{n+1}(\mathbf{x}_{12}\cdot\nabla_v)^n\psi(\mathbf{v}_1)\mathrm{d}\mathbf{x}_{12}\quad(6.23)$$

for $n\geq 1$, where the second integral is defined for each component of $\mathbf{x}_{12}$. In the next two sections, we will show that it is possible to compute these integrals analytically. The integrals with respect to the density functions in Eq. (6.21) and Eq. (6.22) can then be treated using, for example, quadrature (see Section 6.5), or, in the special case of the Maxwell model, they can be done analytically. Then, in Section 6.1.5, we will look at a specific example in which ψ corresponds to the integer velocity moments. For this case, the Taylor series in Eq. (6.20) is finite.

6.1.3 Collision angles and the transformation matrix

The integrals over the collision cross section $\mathbb{S}^+$ can be computed analytically by defining a linear transformation $L(\mathbf{v}_{12})$ from the laboratory frame of reference into the collision frame of reference (Fox & Vedula, 2010). The linear transformation does not depend on $\mathbf{x}_{12}$ (only on $\mathbf{v}_{12}$) and thus it can be taken outside of the integrals. We can define a transformation matrix **L** between the laboratory and collision frames. At this point, it is not necessary to know the exact expression for **L** (which will be given below). Using **L**, the unit vector $\mathbf{x}_{12}$ transforms to $\mathbf{x}_{12}^{\dagger} = \mathbf{L}\mathbf{x}_{12}$, where

$$\mathbf{x}_{12}^{\dagger} = \begin{bmatrix} \sin\theta\,\cos\phi \\ \sin\theta\,\sin\phi \\ \cos\theta \end{bmatrix} \tag{6.24}$$

and $\mathrm{d}\mathbf{x}_{12}^{\dagger} = \sin\theta\,\mathrm{d}\theta\,\mathrm{d}\phi$, where $0 \le \theta \le \pi/2$ and $0 \le \phi \le 2\pi$ are the collision angles. In the collision frame of reference, $\mathbf{v}_{12} \cdot \mathbf{x}_{12} = v_{12}\cos\theta$, where $v_{12} = |\mathbf{v}_1 - \mathbf{v}_2|$ is independent of the collision angles.

The transformation matrix from the laboratory frame ($\mathbf{x}$) into the collision frame of reference ($\mathbf{x}^{\dagger}$) depends only on the relative velocity vector $\mathbf{v}_{12} = \mathbf{v}_1 - \mathbf{v}_2$. By convention, $\mathbf{v}_{12}$ is aligned with the $x_3^{\dagger}$ axis. Thus, the orthonormal transformation matrix defined such that $\mathbf{x}^{\dagger} = \mathbf{L}\mathbf{x}$ can be written as

$$\mathbf{L} = \begin{bmatrix} \sin\phi_1 & -\cos\phi_1 & 0 \\ \cos\theta_1\,\cos\phi_1 & \cos\theta_1\,\sin\phi_1 & -\sin\theta_1 \\ \sin\theta_1\,\cos\phi_1 & \sin\theta_1\,\sin\phi_1 & \cos\theta_1 \end{bmatrix}, \tag{6.25}$$

where the spherical angles $0 \le \theta_1 \le \pi$ and $0 \le \phi_1 \le 2\pi$ are related to the three components of $\mathbf{v}_{12}$ by

$$\begin{aligned} v_{12,1} &= v_{12}\sin\theta_1\,\cos\phi_1, \\ v_{12,2} &= v_{12}\sin\theta_1\,\sin\phi_1, \\ v_{12,3} &= v_{12}\cos\theta_1. \end{aligned} \tag{6.26}$$

Note that the ordering of the first and second rows of **L** is not important, and that **L** is only defined for $v_{12} > 0$. The components of **L** can be written as functions of $\mathbf{v}_{12}$, but they could equally well be denoted by $L_{ij}(\theta_1, \phi_1)$, i.e. as functions of the spherical angles θ_1 and ϕ_1.

The following pseudo-code illustrates how the spherical angles can be computed from the vector $y = \mathbf{v}_{12}$:

```
if norm(y) == 0 return
yn = y/norm(y) ;
theta1 = acos(yn(3)) ;
sth = sin(theta1) ;
phi1 = 0 ;
if sth ~= 0
    if yn(2) > 0
        phi1 = acos(yn(1)/sth) ;
    elseif yn(2) < 0
        phi1 = pi + acos(-yn(1)/sth) ;
    elseif yn(1) < 0
        phi1 = pi ;
    end
end
```

The reader should not confuse the spherical angles (θ_1 and ϕ_1) (which parameterize $\mathbf{v}_{12}$) with the collision angles (θ and ϕ) (which parameterize $\mathbf{x}_{12}$). Indeed, the integrals over the collision angles are done with fixed values of the spherical angles (i.e. fixed values of $\mathbf{v}_{12}$).

In summary, for a given value of $\mathbf{v}_{12}$ we can compute the spherical angles (θ_1 and ϕ_1), and then compute the transformation matrix **L** from Eq. (6.25). The components of the transformation matrix appear in the definitions of the collision term for the velocity moments. In the next section, we will show that the integrals in Eq. (6.23) for the collision terms can be written as explicit functions of the components of $\mathbf{v}_{12}$ and hence it will not be necessary to compute the spherical angles in order to evaluate the integrals. Nevertheless, it is sometimes useful to express the linear transformation in terms of the spherical angles to simplify the manipulations (e.g. when computing the exact expressions for moments using a symbolic math program).

6.1.4 Integrals over collision angles

We would now like to evaluate explicitly the integrals in the collision terms appearing in Eq. (6.23). We will do this separately for each term involving the collision angles, and then reconstruct the final result at the end. The term $\mathbf{x}_{12} \cdot \nabla_v \psi(\mathbf{v})$ can be written in the collision frame of reference using $\mathbf{x}_{12} = \mathbf{L}^{\mathrm{T}} \mathbf{x}_{12}^{\dagger}$ as

$$\begin{aligned} \mathbf{x}_{12} \cdot \nabla_v \psi(\mathbf{v}) = {} & (L_{11} \sin\theta \cos\phi + L_{21} \sin\theta \sin\phi + L_{31} \cos\theta) \frac{\partial \psi(\mathbf{v})}{\partial v_1} \\ & + (L_{12} \sin\theta \cos\phi + L_{22} \sin\theta \sin\phi + L_{32} \cos\theta) \frac{\partial \psi(\mathbf{v})}{\partial v_2} \\ & + (L_{13} \sin\theta \cos\phi + L_{23} \sin\theta \sin\phi + L_{33} \cos\theta) \frac{\partial \psi(\mathbf{v})}{\partial v_3}. \end{aligned} \tag{6.27}$$

Upon applying the multinomial theorem[6] multiple times, we then find

$$\begin{aligned} (\mathbf{x}_{12} \cdot \nabla_v)^n \psi(\mathbf{v}) = {} & \sum_{i+j+k=n} \binom{n}{i,\ j,\ k} \frac{\partial^n \psi(\mathbf{v})}{\partial v_1^i \, \partial v_2^j \, \partial v_3^k} \\ & \times (L_{11} \sin\theta \cos\phi + L_{21} \sin\theta \sin\phi + L_{31} \cos\theta)^i \\ & \times (L_{12} \sin\theta \cos\phi + L_{22} \sin\theta \sin\phi + L_{32} \cos\theta)^j \\ & \times (L_{13} \sin\theta \cos\phi + L_{23} \sin\theta \sin\phi + L_{33} \cos\theta)^k, \end{aligned} \tag{6.28}$$

which can be further rewritten as

$$\begin{aligned} (\mathbf{x}_{12} \cdot \nabla_v)^n \psi(\mathbf{v}) = {} & \sum_{i+j+k=n} \sum_{i_1+i_2+i_3=i} \sum_{j_1+j_2+j_3=j} \sum_{k_1+k_2+k_3=k} \frac{\partial^n \psi(\mathbf{v})}{\partial v_1^i \, \partial v_2^j \, \partial v_3^k} \\ & \times \binom{n}{i,\ j,\ k} \binom{i}{i_1,\ i_2,\ i_3} \binom{j}{j_1,\ j_2,\ j_3} \binom{k}{k_1,\ k_2,\ k_3} \\ & \times (L_{11} \sin\theta \cos\phi)^{i_1} (L_{21} \sin\theta \sin\phi)^{i_2} (L_{31} \cos\theta)^{i_3} \\ & \times (L_{12} \sin\theta \cos\phi)^{j_1} (L_{22} \sin\theta \sin\phi)^{j_2} (L_{32} \cos\theta)^{j_3} \\ & \times (L_{13} \sin\theta \cos\phi)^{k_1} (L_{23} \sin\theta \sin\phi)^{k_2} (L_{33} \cos\theta)^{k_3}, \end{aligned} \tag{6.29}$$

---

[6] The multinomial coefficient is defined, for example, by

$$\binom{n}{i,\ j,\ k} = \frac{n!}{i!j!k!},$$

where $i + j + k = n$. The multinomial theorem states that for any positive integer m and nonnegative integer n

$$(x_1 + x_2 + \cdots + x_m)^n = \sum_{k_1+\cdots+k_m=n} \binom{n}{k_1,\ k_2, \ldots,\ k_m} x_1^{k_1} x_2^{k_2} \ldots x_m^{k_m},$$

where the summation is taken over all sequences of nonnegative integer indices k_1 through k_m such that the sum of all k_i is n.

and finally as

$$(\mathbf{x}_{12}\cdot\nabla_v)^n\psi(\mathbf{v}) = \sum_{i+j+k=n} \frac{\partial^n \psi(\mathbf{v})}{\partial v_1^i\,\partial v_2^j\,\partial v_3^k} \binom{n}{i_1,\, i_2,\, i_3,\, j_1,\, j_2,\, j_3,\, k_1,\, k_2,\, k_3} \times L_{11}^{i_1} L_{21}^{i_2} L_{31}^{i_3} L_{12}^{j_1} L_{22}^{j_2} L_{32}^{j_3} L_{13}^{k_1} L_{23}^{k_2} L_{33}^{k_3} \times (\cos\phi)^{n_1}(\sin\phi)^{n_2}(\cos\theta)^{n_3}(\sin\theta)^{n-n_3}, \quad (6.30)$$

where we have defined $i = i_1 + i_2 + i_3$, $j = j_1 + j_2 + j_3$, $k = k_1 + k_2 + k_3$, $n_1 = i_1 + j_1 + k_1$, $n_2 = i_2 + j_2 + k_2$, and $n_3 = i_3 + j_3 + k_3$. The first integral over the collision angles in Eq. (6.23) involves only the sines and cosines in the last line of Eq. (6.30), multiplied by $(\mathbf{v}_{12}\cdot\mathbf{x}_{12})^{n+1}\,\mathrm{d}\mathbf{x}_{12}^{\dagger} = (v_{12}\cos\theta)^{n+1}\sin\theta\,\mathrm{d}\theta\,\mathrm{d}\phi$.

The first integral

Considering the first integral in Eq. (6.23) and using the result in Eq. (6.30), after a change of reference into the collision frame, the two integrals over the collision angles are

$$\int_0^{2\pi} (\cos\phi)^{n_1}(\sin\phi)^{n_2}\,\mathrm{d}\phi = \frac{1}{2}[1+(-1)^{n_1}][1+(-1)^{n_2}]B\left(\frac{n_1+1}{2}, \frac{n_2+1}{2}\right) \quad (6.31)$$

and

$$\int_0^{\pi/2} (\cos\theta)^{n+n_3+1}(\sin\theta)^{n-n_3+1}\,\mathrm{d}\theta = \frac{1}{2}B\left(\frac{n-n_3+2}{2}, \frac{n+n_3+2}{2}\right), \quad (6.32)$$

where $B(x, y)$ is the beta function, which is defined in terms of the gamma function by

$$B(x, y) = \frac{\Gamma(x)\Gamma(y)}{\Gamma(x+y)}.$$

Multiplying together the two results in Eq. (6.31) and Eq. (6.32) and rearranging the gamma functions yields

$$\int_0^{2\pi} (\cos\phi)^{n_1}(\sin\phi)^{n_2}\,\mathrm{d}\phi \int_0^{\pi/2} (\sin\theta)^{n-n_3+1}(\cos\theta)^{n+n_3+1}\,\mathrm{d}\theta = \frac{1}{4(n+1)!}[1+(-1)^{n_1}][1+(-1)^{n_2}]\Gamma\left(\frac{n_1+1}{2}\right)\Gamma\left(\frac{n_2+1}{2}\right)\Gamma\left(\frac{n+n_3+2}{2}\right), \quad (6.33)$$

where we have used the identity $\Gamma(x+1) = x!$. Note that this expression is nonzero only when n_1 and n_2 are even integers. We can then use the identity

$$\Gamma\left(\frac{n+1}{2}\right) = \frac{\sqrt{\pi}n!}{2^n(n/2)!} \quad \text{for nonnegative even integer } n \quad (6.34)$$

to rewrite the integral term as

$$\int_0^{2\pi} (\cos\phi)^{n_1}(\sin\phi)^{n_2}\,\mathrm{d}\phi \int_0^{\pi/2} (\cos\theta)^{n+n_3+1}(\sin\theta)^{n-n_3+1}\,\mathrm{d}\theta = \frac{\pi}{(n+1)!}\frac{n_1!n_2![n-(n_1+n_2)/2]!}{2^{n_1+n_2}(n_1/2)!(n_2/2)!}, \quad (6.35)$$

where n_1 and n_2 are nonnegative even integers.

From Eq. (6.30) and Eq. (6.35), we can observe that the following identity will be needed in order to evaluate the first integral in Eq. (6.23):

$$\begin{aligned}\binom{n}{i_1,\, i_2,\, i_3,\, j_1,\, j_2,\, j_3,\, k_1,\, k_2,\, k_3}&\frac{\pi}{(n+1)!}\frac{n_1!n_2![n-(n_1+n_2)/2]!}{2^{n_1+n_2}(n_1/2)!(n_2/2)!}\\ &= \frac{\pi}{(n+1)2^{n_1+n_2}}\binom{(n_1+n_2)/2+n_3}{n_1/2,\, n_2/2,\, n_3}\binom{n_1}{i_1,\, j_1,\, k_1}\binom{n_2}{i_2,\, j_2,\, k_2}\binom{n_3}{i_3,\, j_3,\, k_3}. \end{aligned} \tag{6.36}$$

On collecting the various terms, we can now write the first integral as

$$\begin{aligned}\int_{\mathbb{S}^+}(\mathbf{v}_{12}\cdot\mathbf{x}_{12})^{n+1}(\mathbf{x}_{12}\cdot\nabla_v)^n\psi(\mathbf{v})\mathrm{d}\mathbf{x}_{12} = \frac{\pi v_{12}^{n+1}}{n+1}&\sum_{i+j+k=n}\frac{1}{2^{n_1+n_2}}\binom{(n_1+n_2)/2+n_3}{n_1/2,\, n_2/2,\, n_3}\frac{\partial^n\psi(\mathbf{v})}{\partial v_1^i\,\partial v_2^j\,\partial v_3^k}\\ &\times\binom{n_1}{i_1,\, j_1,\, k_1}\binom{n_2}{i_2,\, j_2,\, k_2}\binom{n_3}{i_3,\, j_3,\, k_3}\\ &\times L_{11}^{i_1}L_{21}^{i_2}L_{31}^{i_3}L_{12}^{j_1}L_{22}^{j_2}L_{32}^{j_3}L_{13}^{k_1}L_{23}^{k_2}L_{33}^{k_3}, \end{aligned} \tag{6.37}$$

which can be rewritten as

$$\begin{aligned}\int_{\mathbb{S}^+}&(\mathbf{v}_{12}\cdot\mathbf{x}_{12})^{n+1}(\mathbf{x}_{12}\cdot\nabla_v)^n\psi(\mathbf{v})\mathrm{d}\mathbf{x}_{12}\\ &= \frac{\pi v_{12}^{n+1}}{n+1}\sum_{n_1+n_2+n_3=n}\frac{1}{2^{n_1+n_2}}\binom{(n_1+n_2)/2+n_3}{n_1/2,\, n_2/2,\, n_3}\\ &\times\sum_{i_1+j_1+k_1=n_1}\binom{n_1}{i_1,\, j_1,\, k_1}\left(L_{11}\frac{\partial}{\partial v_1}\right)^{i_1}\left(L_{12}\frac{\partial}{\partial v_2}\right)^{j_1}\left(L_{13}\frac{\partial}{\partial v_3}\right)^{k_1}\\ &\times\sum_{i_2+j_2+k_2=n_2}\binom{n_2}{i_2,\, j_2,\, k_2}\left(L_{21}\frac{\partial}{\partial v_1}\right)^{i_2}\left(L_{22}\frac{\partial}{\partial v_2}\right)^{j_2}\left(L_{23}\frac{\partial}{\partial v_3}\right)^{k_2}\\ &\times\sum_{i_3+j_3+k_3=n_3}\binom{n_3}{i_3,\, j_3,\, k_3}\left(L_{31}\frac{\partial}{\partial v_1}\right)^{i_3}\left(L_{32}\frac{\partial}{\partial v_2}\right)^{j_3}\left(L_{33}\frac{\partial}{\partial v_3}\right)^{k_3}\\ &\times\psi(\mathbf{v}), \end{aligned} \tag{6.38}$$

where we have distributed the differential operator using the definitions of i, j, and k.

Using the multinomial theorem, this expression simplifies to

$$\begin{aligned}\int_{\mathbb{S}^+}(\mathbf{v}_{12}\cdot\mathbf{x}_{12})^{n+1}(\mathbf{x}_{12}\cdot\nabla_v)^n\psi(\mathbf{v})\mathrm{d}\mathbf{x}_{12} = \frac{\pi v_{12}^{n+1}}{n+1}&\sum_{n_1+n_2+n_3=n}\frac{1}{2^{n_1+n_2}}\binom{(n_1+n_2)/2+n_3}{n_1/2,\, n_2/2,\, n_3}\\ &\times\left(L_{11}\frac{\partial}{\partial v_1}+L_{12}\frac{\partial}{\partial v_2}+L_{13}\frac{\partial}{\partial v_3}\right)^{n_1}\\ &\times\left(L_{21}\frac{\partial}{\partial v_1}+L_{22}\frac{\partial}{\partial v_2}+L_{23}\frac{\partial}{\partial v_3}\right)^{n_2}\\ &\times\left(L_{31}\frac{\partial}{\partial v_1}+L_{32}\frac{\partial}{\partial v_2}+L_{33}\frac{\partial}{\partial v_3}\right)^{n_3}\psi(\mathbf{v}). \end{aligned} \tag{6.39}$$

Using the fact that n_1 and n_2 are even integers, this expression can then be rewritten as

$$\int_{\mathbb{S}^+} (\mathbf{v}_{12} \cdot \mathbf{x}_{12})^{n+1} (\mathbf{x}_{12} \cdot \nabla_v)^n \psi(\mathbf{v}) \mathrm{d}\mathbf{x}_{12} = \frac{\pi v_{12}^{n+1}}{n+1} \sum_{2m_1+2m_2+m_3=n} \frac{1}{4^{m_1+m_2}} \binom{m_1+m_2+m_3}{m_1,\, m_2,\, m_3}$$
$$\times \left(L_{11} \frac{\partial}{\partial v_1} + L_{12} \frac{\partial}{\partial v_2} + L_{13} \frac{\partial}{\partial v_3} \right)^{2m_1}$$
$$\times \left(L_{21} \frac{\partial}{\partial v_1} + L_{22} \frac{\partial}{\partial v_2} + L_{23} \frac{\partial}{\partial v_3} \right)^{2m_2}$$
$$\times \left(L_{31} \frac{\partial}{\partial v_1} + L_{32} \frac{\partial}{\partial v_2} + L_{33} \frac{\partial}{\partial v_3} \right)^{m_3} \psi(\mathbf{v}), \qquad (6.40)$$

where m_1, m_2, and m_3 are nonnegative integers and, for fixed n, the summation in Eq. (6.40) extends over all values of m_i such that $n = 2m_1 + 2m_2 + m_3$.

Next we would like to show that the right-hand side of Eq. (6.40) does not depend on the spherical angles ϕ_1 and θ_1. Consider first the operators

$$\left(L_{11} \frac{\partial}{\partial v_1} + L_{12} \frac{\partial}{\partial v_2} + L_{13} \frac{\partial}{\partial v_3} \right)^2 = L_{11}^2 \frac{\partial^2}{\partial v_1^2} + L_{12}^2 \frac{\partial^2}{\partial v_2^2} + L_{13}^2 \frac{\partial^2}{\partial v_3^2}$$
$$+ 2L_{11}L_{12} \frac{\partial^2}{\partial v_1 \, \partial v_2} + 2L_{11}L_{13} \frac{\partial^2}{\partial v_1 \, \partial v_3} + 2L_{12}L_{13} \frac{\partial^2}{\partial v_2 \, \partial v_3} \qquad (6.41)$$

and

$$\left(L_{21} \frac{\partial}{\partial v_1} + L_{22} \frac{\partial}{\partial v_2} + L_{23} \frac{\partial}{\partial v_3} \right)^2 = L_{21}^2 \frac{\partial^2}{\partial v_1^2} + L_{22}^2 \frac{\partial^2}{\partial v_2^2} + L_{23}^2 \frac{\partial^2}{\partial v_3^2}$$
$$+ 2L_{21}L_{22} \frac{\partial^2}{\partial v_1 \, \partial v_2} + 2L_{21}L_{23} \frac{\partial^2}{\partial v_1 \, \partial v_3} + 2L_{22}L_{23} \frac{\partial^2}{\partial v_2 \, \partial v_3}. \qquad (6.42)$$

By virtue of the symmetry of m_1 and m_2, Eq. (6.41) and Eq. (6.42) will appear separately in the summation, multiplied by the same factor. From the definition of the transformation matrix in Eq. (6.25), we have the following relations:

$$\begin{aligned}
v_{12}^2 \left(L_{11}^2 + L_{21}^2 \right) &= v_{12}^2 - v_{12,1}^2, \\
v_{12}^2 \left(L_{12}^2 + L_{22}^2 \right) &= v_{12}^2 - v_{12,2}^2, \\
v_{12}^2 \left(L_{13}^2 + L_{23}^2 \right) &= v_{12}^2 - v_{12,3}^2, \\
v_{12}^2 \left(L_{11}L_{12} + L_{21}L_{22} \right) &= -v_{12,1}v_{12,2}, \\
v_{12}^2 \left(L_{11}L_{13} + L_{21}L_{23} \right) &= -v_{12,1}v_{12,3}, \\
v_{12}^2 \left(L_{12}L_{13} + L_{22}L_{23} \right) &= -v_{12,2}v_{12,3},
\end{aligned} \qquad (6.43)$$

which are the only terms involving L_{1j} and L_{2j} which appear in the summation. Note that this implies that all of the terms in the summation in Eq. (6.40) involving m_1 and m_2 are quadratically dependent on the components of $\mathbf{v}_{12}$, while those involving m_3 are linearly dependent.

In order to simplify the notation, we define the right-hand side of Eq. (6.40) using a summation–differentiation operator:

$$\sum_{\mathbf{0}}^{n}(\mathbf{v}_{12})\{\psi(\mathbf{v})\} \equiv \sum_{2m_1+2m_2+m_3=n} \frac{1}{4^{m_1+m_2}}\binom{m_1+m_2+m_3}{m_1,\ m_2,\ m_3}$$
$$\times\left(\frac{v_{12}^2 v_{12,2}^2}{v_{12,1}^2+v_{12,2}^2}\frac{\partial^2}{\partial v_1^2} - 2\frac{v_{12}^2 v_{12,1}v_{12,2}}{v_{12,1}^2+v_{12,2}^2}\frac{\partial^2}{\partial v_1\,\partial v_2} + \frac{v_{12}^2 v_{12,1}^2}{v_{12,1}^2+v_{12,2}^2}\frac{\partial^2}{\partial v_2^2}\right)^{m_1}$$
$$\times\left(\frac{v_{12,1}^2 v_{12,3}^2}{v_{12,1}^2+v_{12,2}^2}\frac{\partial^2}{\partial v_1^2} + \frac{v_{12,2}^2 v_{12,3}^2}{v_{12,1}^2+v_{12,2}^2}\frac{\partial^2}{\partial v_2^2} + (v_{12,1}^2+v_{12,2}^2)\frac{\partial^2}{\partial v_3^2}\right.$$
$$\left.+2\frac{v_{12,1}v_{12,2}v_{12,3}^2}{v_{12,1}^2+v_{12,2}^2}\frac{\partial^2}{\partial v_1\,\partial v_2} - 2v_{12,1}v_{12,3}\frac{\partial^2}{\partial v_1\,\partial v_3} - 2v_{12,2}v_{12,3}\frac{\partial^2}{\partial v_2\,\partial v_3}\right)^{m_2}$$
$$\times\left(v_{12,1}\frac{\partial}{\partial v_1} + v_{12,2}\frac{\partial}{\partial v_2} + v_{13,3}\frac{\partial}{\partial v_3}\right)^{m_3}\psi(\mathbf{v}), \tag{6.44}$$

where m_1, m_2, and m_3 are nonnegative integers. The final form of the first integral in the collision source term is

$$\int_{\mathbb{S}^+}(\mathbf{v}_{12}\cdot\mathbf{x}_{12})^{n+1}(\mathbf{x}_{12}\cdot\nabla_v)^n\psi(\mathbf{v})\mathrm{d}\mathbf{x}_{12} = \frac{v_{12}\pi}{n+1}\sum_{\mathbf{0}}^{n}(\mathbf{v}_{12})\{\psi(\mathbf{v})\}. \tag{6.45}$$

Note that the "extra" factor of v_{12} was not included in the definition of the operator in Eq. (6.44) on purpose. We shall see that without this factor, the integer moment collision integral would be closed, and thus, if we replace the factor by a constant (i.e. the Maxwell model), the moment equations are closed at every order.

The second integral

The second integral in Eq. (6.23) can be evaluated for each component of $\mathbf{x}_{12} = (x_{12,1}, x_{12,2}, x_{12,3})$. For example, for the mth component, we find

$$\int_{\mathbb{S}^+} x_{12,m}(\mathbf{v}_{12}\cdot\mathbf{x}_{12})^{n+1}(\mathbf{x}_{12}\cdot\nabla_v)^n\psi(\mathbf{v})\mathrm{d}\mathbf{x}_{12}$$
$$= v_{12}^{n+1}\sum_{i+j+k=n}\frac{\partial^n\psi(\mathbf{v})}{\partial v_1^i\,\partial v_2^j\,\partial v_3^k}\binom{n}{i_1,\ j_1,\ k_1,\ i_2,\ j_2,\ k_2,\ i_3,\ j_3,\ k_3}$$
$$\times L_{11}^{i_1}L_{21}^{i_2}L_{31}^{i_3}L_{12}^{j_1}L_{22}^{j_2}L_{32}^{j_3}L_{13}^{k_1}L_{23}^{k_2}L_{33}^{k_3}$$
$$\times\left\{\int_0^{\pi/2}\int_0^{2\pi}(\sin\theta)^{n-n_3+1}(\cos\theta)^{n+n_3+1}(\sin\phi)^{n_2}(\cos\phi)^{n_1}\right.$$
$$\left.\times(L_{1m}\sin\theta\cos\phi + L_{2m}\sin\theta\sin\phi + L_{3m}\cos\theta)\mathrm{d}\phi\,\mathrm{d}\theta\right\}. \tag{6.46}$$

The terms on the right-hand side of this expression involve three new collision-angle integrals:

$$\begin{aligned}
\int_0^{2\pi} & (\cos\phi)^{n_1+1}(\sin\phi)^{n_2}\,\mathrm{d}\phi \int_0^{\pi/2} (\sin\theta)^{n-n_3+2}(\cos\theta)^{n+n_3+1}\,\mathrm{d}\theta \\
&= \frac{1}{4}[1-(-1)^{n_1}][1+(-1)^{n_2}]B\left(\frac{n_1+2}{2}, \frac{n_2+1}{2}\right) B\left(\frac{n-n_3+3}{2}, \frac{n+n_3+2}{2}\right) \\
&= \frac{\pi}{2^{n+2}(2n+3)!!}[1-(-1)^{n_1}][1+(-1)^{n_2}] \\
&\quad \times \frac{(n_1+1)!n_2!(n+n_3+1)!}{[(n_1+1)/2]!(n_2/2)!\,[(n+n_3+1)/2]!},
\end{aligned} \tag{6.47}$$

$$\begin{aligned}
\int_0^{2\pi} & (\cos\phi)^{n_1}(\sin\phi)^{n_2+1}\,\mathrm{d}\phi \int_0^{\pi/2} (\sin\theta)^{n-n_3+2}(\cos\theta)^{n+n_3+1}\,\mathrm{d}\theta \\
&= \frac{1}{4}[1+(-1)^{n_1}][1-(-1)^{n_2}]B\left(\frac{n_1+1}{2}, \frac{n_2+2}{2}\right) B\left(\frac{n-n_3+3}{2}, \frac{n+n_3+2}{2}\right) \\
&= \frac{\pi}{2^{n+2}(2n+3)!!}[1+(-1)^{n_1}][1-(-1)^{n_2}] \\
&\quad \times \frac{n_1!(n_2+1)!(n+n_3+1)!}{(n_1/2)!\,[(n_2+1)/2]!\,[(n+n_3+1)/2]!},
\end{aligned} \tag{6.48}$$

and

$$\begin{aligned}
\int_0^{2\pi} & (\cos\phi)^{n_1}(\sin\phi)^{n_2}\,\mathrm{d}\phi \int_0^{\pi/2} (\sin\theta)^{n-n_3+1}(\cos\theta)^{n+n_3+2}\,\mathrm{d}\theta \\
&= \frac{1}{4}[1+(-1)^{n_1}][1+(-1)^{n_2}]B\left(\frac{n_1+1}{2}, \frac{n_2+1}{2}\right) B\left(\frac{n-n_3+2}{2}, \frac{n+n_3+3}{2}\right) \\
&= \frac{\pi}{2^{n+2}(2n+3)!!}[1+(-1)^{n_1}][1+(-1)^{n_2}] \\
&\quad \times \frac{n_1!n_2!(n+n_3+2)!}{(n_1/2)!(n_2/2)!\,[(n+n_3+2)/2]!},
\end{aligned} \tag{6.49}$$

where the double factorial is defined for odd positive integers by $n!! = n(n-2)(n-4)\ldots(3)(1)$ and we used the identity $(2k)! = (2k-1)!!\,k!\,2^k$. Note that Eq. (6.47) is nonzero only when n_1 is odd and n_2 is even, whereas Eq. (6.48) is nonzero only when n_1 is even and n_2 is odd. The result in Eq. (6.49) is nonzero only when both n_1 and n_2 are even.

As was done above for the first integral, it is possible to rewrite the summations in Eq. (6.46) in terms of three summation–differentiation operators:

$$\begin{aligned}
\int_{\mathbb{S}^+} & x_{12,m}(\mathbf{v}_{12}\cdot\mathbf{x}_{12})^{n+1}(\mathbf{x}_{12}\cdot\nabla_v)^n\psi(\mathbf{v})\mathrm{d}\mathbf{x}_{12} \\
&= \frac{\pi n!}{2^n(2n+3)!!}\left(v_{12}L_{1m}\sum_{\mathbf{1}}^{n}(\mathbf{v}_{12})\{\psi(\mathbf{v})\} + v_{12}L_{2m}\sum_{\mathbf{2}}^{n}(\mathbf{v}_{12})\{\psi(\mathbf{v})\} + v_{12,m}\sum_{\mathbf{3}}^{n}(\mathbf{v}_{12})\{\psi(\mathbf{v})\}\right).
\end{aligned} \tag{6.50}$$

Using manipulations similar to those described above for the first integral, it is possible to show that the first operator is defined by

$$\begin{aligned}\sum_{\mathbf{1}}^{n}(\mathbf{v}_{12})\{\psi(\mathbf{v})\} \equiv & \sum_{2m_1+2m_2+m_3+1=n} \frac{2m_1(m_1+m_2+m_3+2)}{(m_1+1)}\binom{2m_1+2m_2+2m_3+2}{m_1,\, m_2,\, m_3,\, m_1+m_2+m_3+2} \\ & \times\left(v_{12}L_{11}\frac{\partial}{\partial v_1}+v_{12}L_{12}\frac{\partial}{\partial v_2}\right)^{2m_1+1} \\ & \times\left(\frac{v_{12,1}^2 v_{12,3}^2}{v_{12,1}^2+v_{12,2}^2}\frac{\partial^2}{\partial v_1^2}+\frac{v_{12,2}^2 v_{12,3}^2}{v_{12,1}^2+v_{12,2}^2}\frac{\partial^2}{\partial v_2^2}+(v_{12,1}^2+v_{12,2}^2)\frac{\partial^2}{\partial v_3^2}\right. \\ & \left. +2\frac{v_{12,1}v_{12,2}v_{12,3}^2}{v_{12,1}^2+v_{12,2}^2}\frac{\partial^2}{\partial v_1\,\partial v_2}-2v_{12,1}v_{12,3}\frac{\partial^2}{\partial v_1\,\partial v_3}-2v_{12,2}v_{12,3}\frac{\partial^2}{\partial v_2\,\partial v_3}\right)^{m_2} \\ & \times\left(v_{12,1}\frac{\partial}{\partial v_1}+v_{12,2}\frac{\partial}{\partial v_2}+v_{13,3}\frac{\partial}{\partial v_3}\right)^{m_3}\psi(\mathbf{v}),\end{aligned} \tag{6.51}$$

where m_1, m_2, and m_3 are nonnegative integers. Likewise, the second operator is defined by

$$\begin{aligned}\sum_{\mathbf{2}}^{n}(\mathbf{v}_{12})\{\psi(\mathbf{v})\} \equiv & \sum_{2m_1+2m_2+m_3+1=n} \frac{2m_2(m_1+m_2+m_3+2)}{(m_2+1)}\binom{2m_1+2m_2+2m_3+2}{m_1,\, m_2,\, m_3,\, m_1+m_2+m_3+2} \\ & \times\left(\frac{v_{12}^2 v_{12,2}^2}{v_{12,1}^2+v_{12,2}^2}\frac{\partial^2}{\partial v_1^2}-2\frac{v_{12}^2 v_{12,1}v_{12,2}}{v_{12,1}^2+v_{12,2}^2}\frac{\partial^2}{\partial v_1\,\partial v_2}+\frac{v_{12}^2 v_{12,1}^2}{v_{12,1}^2+v_{12,2}^2}\frac{\partial^2}{\partial v_2^2}\right)^{m_1} \\ & \times\left(v_{12}L_{21}\frac{\partial}{\partial v_1}+v_{12}L_{22}\frac{\partial}{\partial v_2}+v_{12}L_{23}\frac{\partial}{\partial v_3}\right)^{2m_2+1} \\ & \times\left(v_{12,1}\frac{\partial}{\partial v_1}+v_{12,2}\frac{\partial}{\partial v_2}+v_{13,3}\frac{\partial}{\partial v_3}\right)^{m_3}\psi(\mathbf{v}),\end{aligned} \tag{6.52}$$

where m_1, m_2, and m_3 are nonnegative integers. Finally, the third operator is defined by

$$\begin{aligned}\sum_{\mathbf{3}}^{n}(\mathbf{v}_{12})\{\psi(\mathbf{v})\} \equiv & \sum_{2m_1+2m_2+m_3=n} (m_1+m_2+m_3+2)\binom{2m_1+2m_2+2m_3+2}{m_1,\, m_2,\, m_3,\, m_1+m_2+m_3+2} \\ & \times\left(\frac{v_{12}^2 v_{12,2}^2}{v_{12,1}^2+v_{12,2}^2}\frac{\partial^2}{\partial v_1^2}-2\frac{v_{12}^2 v_{12,1}v_{12,2}}{v_{12,1}^2+v_{12,2}^2}\frac{\partial^2}{\partial v_1\,\partial v_2}+\frac{v_{12}^2 v_{12,1}^2}{v_{12,1}^2+v_{12,2}^2}\frac{\partial^2}{\partial v_2^2}\right)^{m_1} \\ & \times\left(\frac{v_{12,1}^2 v_{12,3}^2}{v_{12,1}^2+v_{12,2}^2}\frac{\partial^2}{\partial v_1^2}+\frac{v_{12,2}^2 v_{12,3}^2}{v_{12,1}^2+v_{12,2}^2}\frac{\partial^2}{\partial v_2^2}+(v_{12,1}^2+v_{12,2}^2)\frac{\partial^2}{\partial v_3^2}\right. \\ & \left. +2\frac{v_{12,1}v_{12,2}v_{12,3}^2}{v_{12,1}^2+v_{12,2}^2}\frac{\partial^2}{\partial v_1\,\partial v_2}-2v_{12,1}v_{12,3}\frac{\partial^2}{\partial v_1\,\partial v_3}-2v_{12,2}v_{12,3}\frac{\partial^2}{\partial v_2\,\partial v_3}\right)^{m_2} \\ & \times\left(v_{12,1}\frac{\partial}{\partial v_1}+v_{12,2}\frac{\partial}{\partial v_2}+v_{13,3}\frac{\partial}{\partial v_3}\right)^{m_3}\psi(\mathbf{v}),\end{aligned} \tag{6.53}$$

where m_1, m_2, and m_3 are nonnegative integers.

Final forms

Using the analytical expressions for the integrals over the collision angles, we can now rewrite the collision source term in Eq. (6.21) as the sum of two contributions,

$$C(\psi) = C^{(0)}(\psi) + C^{(1)}(\psi), \tag{6.54}$$

where

$$C^{(0)}(\psi) = \frac{6g_0}{d_\mathrm{p}} \sum_{n=1}^{\infty} \frac{(-\omega)^n}{(n+1)!} \int_{\mathbb{R}^6} \sum_{\mathbf{0}}^{n} (\mathbf{v}_{12})\{\psi(\mathbf{v}_1)\} v_{12} f(\mathbf{v}_1) f(\mathbf{v}_2) \mathrm{d}\mathbf{v}_2 \, \mathrm{d}\mathbf{v}_1 \tag{6.55}$$

and

$$\begin{aligned} C^{(1)}(\psi) = 3g_0 \sum_{n=1}^{\infty} \frac{(-\omega)^n}{2^n(2n+3)!!} \\ \times \sum_{m=1}^{3} \int_{\mathbb{R}^6} \Bigg(v_{12} L_{1m} \sum_{\mathbf{1}}^{n} (\mathbf{v}_{12})\{\psi(\mathbf{v}_1)\} + v_{12} L_{2m} \sum_{\mathbf{2}}^{n} (\mathbf{v}_{12})\{\psi(\mathbf{v}_1)\} \\ + v_{12,m} \sum_{\mathbf{3}}^{n} (\mathbf{v}_{12})\{\psi(\mathbf{v}_1)\} \Bigg) \\ \times \left(f(\mathbf{v}_1) \frac{\partial f(\mathbf{v}_2)}{\partial x_m} - f(\mathbf{v}_2) \frac{\partial f(\mathbf{v}_1)}{\partial x_m} \right) \mathrm{d}\mathbf{v}_2 \, \mathrm{d}\mathbf{v}_1. \end{aligned} \tag{6.56}$$

Likewise, the components of the collisional-flux term in Eq. (6.22) can be written as

$$\begin{aligned} G_m(\psi) = -3g_0 \sum_{n=1}^{\infty} \frac{(-\omega)^n}{2^n(2n+3)!!} \\ \times \int_{\mathbb{R}^6} \Bigg(v_{12} L_{1m} \sum_{\mathbf{1}}^{n} (\mathbf{v}_{12})\{\psi(\mathbf{v}_1)\} + v_{12} L_{2m} \sum_{\mathbf{2}}^{n} (\mathbf{v}_{12})\{\psi(\mathbf{v}_1)\} + v_{12,m} \sum_{\mathbf{3}}^{n} (\mathbf{v}_{12})\{\psi(\mathbf{v}_1)\} \Bigg) \\ \times f(\mathbf{v}_1) f(\mathbf{v}_2) \mathrm{d}\mathbf{v}_2 \, \mathrm{d}\mathbf{v}_1. \end{aligned} \tag{6.57}$$

At this point, the remaining integrals in the collision source terms are with respect to the velocity density functions. We will now consider the special case in which ψ represents the integer moments with respect to the velocity components.

6.1.5 The collision term for integer moments

For quadrature-based moment methods, the collision terms for integer moments are needed. For this special case, the method developed in Fox & Vedula (2010) leads to closed expressions for the integrals in terms of finite sums. For integer moments, we have $\psi(\mathbf{v}) = v_1^{l_1} v_2^{l_2} v_3^{l_3}$ and the terms in the sums over n in Eqs. (6.55)–(6.57) are zero for $n > \max(l_1, l_2, l_3)$. For this particular case, we will define $I^{(m)}_{l_1 l_2 l_3}$ for $m = 0$ by

$$I^{(0)}_{l_1 l_2 l_3}(\omega, \mathbf{v}_1, \mathbf{v}_{12}) \equiv \sum_{n=1}^{\max(l_1,l_2,l_3)} \frac{(-\omega)^n}{(n+1)!} \sum_{\mathbf{0}}^{n} (\mathbf{v}_{12}) \left\{ v_{1,1}^{l_1} v_{1,2}^{l_2} v_{1,3}^{l_3} \right\} \tag{6.58}$$

and, for $m = 1, 2, 3$, we define

$$I^{(m)}_{l_1 l_2 l_3} \equiv - \sum_{n=1}^{\max(l_1,l_2,l_3)} \frac{(-\omega)^n}{2^n(2n+3)!!} \left(v_{12} L_{1m} \sum_{\mathbf{1}}^{n} (\mathbf{v}_{12}) \left\{ v_{1,1}^{l_1} v_{1,2}^{l_2} v_{1,3}^{l_3} \right\} + v_{12} L_{2m} \sum_{\mathbf{2}}^{n} (\mathbf{v}_{12}) \left\{ v_{1,1}^{l_1} v_{1,2}^{l_2} v_{1,3}^{l_3} \right\} \right.$$
$$\left. + v_{12,m} \sum_{\mathbf{3}}^{n} (\mathbf{v}_{12}) \left\{ v_{1,1}^{l_1} v_{1,2}^{l_2} v_{1,3}^{l_3} \right\} \right). \tag{6.59}$$

Using the definitions of the operators in Section 6.1.4, we can rewrite these terms as polynomials in the components of $\mathbf{v}_1$:

$$I^{(m)}_{l_1 l_2 l_3} = \sum_{i_1=0}^{l_1} \sum_{i_2=0}^{l_2} \sum_{i_3=0}^{l_3} (-\omega)^{i_1+i_2+i_3} S^{(m)}_{i_1 i_2 i_3} \binom{l_1}{i_1} \binom{l_2}{i_2} \binom{l_3}{i_3} v_{1,1}^{l_1-i_1} v_{1,2}^{l_2-i_2} v_{1,3}^{l_3-i_3}, \tag{6.60}$$

where the coefficients $S^{(m)}_{i_1 i_2 i_3}$ are defined in terms of the components of $\mathbf{v}_{12}$ (i.e. the components of $\mathbf{L}$) for $m = 0$ by

$$S^{(0)}_{i_1 i_2 i_3} \equiv \sum^{i_1 i_2 i_3} (v_{12} L_{11})^{i_1-j_1} (v_{12} L_{12})^{i_2-j_2} (v_{12} L_{21})^{j_1-k_1} (v_{12} L_{22})^{j_2-k_2} (v_{12} L_{23})^{i_3-k_3}$$
$$\times (v_{12} L_{31})^{k_1} (v_{12} L_{32})^{k_2} (v_{12} L_{33})^{k_3} K^{(0)}_{ijk}, \tag{6.61}$$

and, for $m = 1, 2, 3$, by

$$S^{(m)}_{i_1 i_2 i_3} \equiv - \sum^{i_1 i_2 i_3} (v_{12} L_{11})^{i_1-j_1} (v_{12} L_{12})^{i_2-j_2} (v_{12} L_{21})^{j_1-k_1} (v_{12} L_{22})^{j_2-k_2} (v_{12} L_{23})^{i_3-k_3}$$
$$\times (v_{12} L_{31})^{k_1} (v_{12} L_{32})^{k_2} (v_{12} L_{33})^{k_3} \left(v_{12} L_{1m} K^{(1)}_{ijk} + v_{12} L_{2m} K^{(2)}_{ijk} + v_{12,m} K^{(3)}_{ijk} \right). \tag{6.62}$$

In these expressions, $i = i_1 + i_2 + i_3$, $j = j_1 + j_2 + i_3$, and $k = k_1 + k_2 + k_3$, and the summation operator is defined by[7]

$$\sum^{i_1 i_2 i_3} \equiv \sum_{j_1=0}^{i_1} \sum_{j_2=0}^{i_2} \sum_{k_1=0}^{j_1} \sum_{k_2=0}^{j_2} \sum_{k_3=0}^{i_3} \binom{i_1}{j_1} \binom{i_2}{j_2} \binom{j_1}{k_1} \binom{j_2}{k_2} \binom{i_3}{k_3}. \tag{6.63}$$

For the special case in which $i_1 = i_2 = i_3 = 0$, we define $S^{(m)}_{000} = 0$.

The reader will recognize that the coefficients $K^{(m)}_{ijk}$ are related to the integrals over the collision angles discussed in Section 6.1.4. The exact definitions are

$$K^{(0)}_{ijk} \equiv \frac{1}{4} [1 + (-1)^{i-j}][1 + (-1)^{j-k}] \frac{(i-j)!(j-k)![(i+k)/2]!}{2^{i-k}(i+1)![(i-j)/2]![(j-k)/2]!}, \tag{6.64}$$

[7] We use the fact that $L_{13} = 0$ so that $j_3 = i_3$.

$$K^{(1)}_{ijk} \equiv [1-(-1)^{i-j}][1+(-1)^{j-k}] \times \frac{(i-j+1)!(j-k)!(i+k-1)!}{2^i(2i+3)!![(i-j+1)/2]![(j-k)/2]![(i+k-1)/2]!}, \tag{6.65}$$

$$K^{(2)}_{ijk} \equiv [1+(-1)^{i-j}][1-(-1)^{j-k}] \times \frac{(i-j)!(j-k+1)!(i+k-1)!}{2^i(2i+3)!![(i-j)/2]![(j-k+1)/2]![(i+k-1)/2]!}, \tag{6.66}$$

$$K^{(3)}_{ijk} \equiv \frac{1}{4}[1+(-1)^{i-j}][1+(-1)^{j-k}] \times \frac{(i-j)!(j-k)!(i+k+2)!}{2^k(2i+3)!![(i-j)/2]![(j-k)/2]![(i+k+2)/2]!}. \tag{6.67}$$

These coefficients, and the summations in Eq. (6.60), are most easily computed using a symbolic math program.[8] Some examples of $I^{(m)}_{l_1l_2l_3}(\omega, \mathbf{v}, \mathbf{g})$ for selected moments up to fifth order[9] are given in Tables 6.1–6.9. For clarity, in these tables we have denoted the velocity difference vector by $\mathbf{g} = (g_1, g_2, g_3)$ and $g^2 = g_1^2 + g_2^2 + g_3^2$.

The final expressions for the collision source terms for integer moments of order $\gamma = l_1 + l_2 + l_3$ can now be written in the form of Eq. (6.54) as

$$C^{(0)}_{l_1l_2l_3} = \frac{6g_0}{d_\mathrm{p}} \int_{\mathbb{R}^6} I^{(0)}_{l_1l_2l_3}(\omega, \mathbf{v}_1, \mathbf{v}_1 - \mathbf{v}_2)|\mathbf{v}_1 - \mathbf{v}_2| f(\mathbf{v}_1) f(\mathbf{v}_2) \mathrm{d}\mathbf{v}_2 \, \mathrm{d}\mathbf{v}_1, \tag{6.68}$$

$$C^{(1)}_{l_1l_2l_3} = 3g_0 \sum_{m=1}^{3} \int_{\mathbb{R}^6} I^{(m)}_{l_1l_2l_3}(\omega, \mathbf{v}_1, \mathbf{v}_1 - \mathbf{v}_2) \left(f(\mathbf{v}_2) \frac{\partial f(\mathbf{v}_1)}{\partial x_m} - f(\mathbf{v}_1) \frac{\partial f(\mathbf{v}_2)}{\partial x_m} \right) \mathrm{d}\mathbf{v}_2 \, \mathrm{d}\mathbf{v}_1 \tag{6.69}$$

and the collisional-flux terms as

$$G_{m,l_1l_2l_3} = 3g_0 \int_{\mathbb{R}^6} I^{(m)}_{l_1l_2l_3}(\omega, \mathbf{v}_1, \mathbf{v}_1 - \mathbf{v}_2) f(\mathbf{v}_1) f(\mathbf{v}_2) \mathrm{d}\mathbf{v}_2 \, \mathrm{d}\mathbf{v}_1. \tag{6.70}$$

The reader can verify that the source terms for the zeroth- and first-order moments are exactly zero (i.e. the conservation of mass and mean momentum). It is interesting to

Table 6.1. $I^{(0)}_{l_1l_2l_3}(\omega, \mathbf{v}, \mathbf{g})$ terms for moments of up to second order

$I^{(0)}_{000} = 0$	$I^{(0)}_{100} = -(\omega/2)g_1$	$I^{(0)}_{010} = -(\omega/2)g_2$	$I^{(0)}_{001} = -(\omega/2)g_3$
$I^{(0)}_{200} = (\omega^2/12)g^2 + (\omega^2/4)g_1^2 - \omega g_1 v_1$		$I^{(0)}_{110} = (\omega^2/4)g_1g_2 - (\omega/2)g_2v_1 - (\omega/2)g_1v_2$	
$I^{(0)}_{020} = (\omega^2/12)g^2 + (\omega^2/4)g_2^2 - \omega g_2 v_2$		$I^{(0)}_{101} = (\omega^2/4)g_1g_3 - (\omega/2)g_3v_1 - (\omega/2)g_1v_3$	
$I^{(0)}_{002} = (\omega^2/12)g^2 + (\omega^2/4)g_3^2 - \omega g_3 v_3$		$I^{(0)}_{011} = (\omega^2/4)g_2g_3 - (\omega/2)g_3v_2 - (\omega/2)g_2v_3$	

[8] In order to efficiently simplify the resulting expressions, it may be helpful to define the components of $\mathbf{L}$ in terms of the spherical angles as given in Eq. (6.25). If this is done correctly, the final expression for $S^{(m)}_{i_1i_2i_3}$ should not depend on the spherical angles, but rather will be a homogeneous polynomial in powers of the components of $\mathbf{v}_{12}$.

[9] These moments (and their permutations) are needed for eight-node quadrature.

Table 6.2. $I^{(0)}_{l_1 l_2 l_3}(\omega, \mathbf{v}, \mathbf{g})$ terms for third-order moments

$$I^{(0)}_{300} = -(\omega^3/8)\left(g^2 + g_1^2\right) g_1 + (\omega^2/4)\left(g^2 + 3g_1^2\right) v_1 - (3\omega/2) g_1 v_1^2$$

$$I^{(0)}_{210} = -(\omega^3/24)\left(g^2 + 3g_1^2\right) g_2 + (\omega^2/2) g_1 g_2 v_1 + (\omega^2/12)\left(g^2 + 3g_1^2\right) v_2 - (\omega/2) g_2 v_1^2 - \omega g_1 v_1 v_2$$

$$I^{(0)}_{201} = -(\omega^3/24)\left(g^2 g_3 + 3g_1^2\right) g_3 + (\omega^2/2) g_1 g_3 v_1 + (\omega^2/12)\left(g^2 + 3g_1^2\right) v_3 - (\omega/2) g_3 v_1^2 - \omega g_1 v_1 v_3$$

$$I^{(0)}_{120} = -(\omega^3/24)\left(g^2 + 3g_2^2\right) g_1 + (\omega^2/12)\left(g^2 + 3g_2^2\right) v_1 + (\omega^2/2) g_1 g_2 v_2 - \omega g_2 v_1 v_2 - (\omega/2) g_1 v_2^2$$

$$I^{(0)}_{111} = -(\omega^3/8) g_1 g_2 g_3 + (\omega^2/4) g_2 g_3 v_1 + (\omega^2/4) g_1 g_3 v_2 + (\omega^2/4) g_1 g_2 v_3 - (\omega/2) g_3 v_1 v_2 - (\omega/2) g_2 v_1 v_3 - (\omega/2) g_1 v_2 v_3$$

$$I^{(0)}_{102} = -(\omega^3/24)\left(g^2 + 3g_3^2\right) g_1 + (\omega^2/12)\left(g^2 + 3g_3^2\right) v_1 + (\omega^2/2) g_1 g_3 v_3 - \omega g_3 v_1 v_3 - (\omega/2) g_1 v_3^2$$

$$I^{(0)}_{030} = -(\omega^3/8)\left(g^2 + g_2^2\right) g_2 + (\omega^2/4)\left(g^2 + 3g_2^2\right) v_2 - (3\omega/2) g_2 v_2^2$$

$$I^{(0)}_{021} = -(\omega^3/24)\left(g^2 + 3g_2^2\right) g_3 + (\omega^2/2) g_2 g_3 v_2 + (\omega^2/12)\left(g^2 + 3g_2^2\right) v_3 - (\omega/2) g_3 v_2^2 - \omega g_2 v_2 v_3$$

$$I^{(0)}_{012} = -(\omega^3/24)\left(g^2 + 3g_3^2\right) g_2 + (\omega^2/12)\left(g^2 + 3g_3^2\right) v_2 + (\omega^2/2) g_2 g_3 v_3 - \omega g_3 v_2 v_3 - (\omega/2) g_2 v_3^2$$

$$I^{(0)}_{003} = -(\omega^3/8)\left(g^2 + g_3^2\right) g_3 + (\omega^2/4)\left(g^2 + 3g_3^2\right) v_3 - (3\omega/2) g_3 v_3^2$$

note that, because $I^{(m)}_{l_1 l_2 l_3}(\omega, \mathbf{v}_1, \mathbf{v}_1 - \mathbf{v}_2)$ has the form of a homogeneous polynomial in the components of $\mathbf{v}_1$ and $\mathbf{v}_2$ of order $\gamma + 1$ for $m = 1, 2, 3$, the integral in Eq. (6.70) can be expressed in terms of the integer moments up to order $\gamma + 1$. In other words, while $\mathbf{G}_{l_1 l_2 l_3}$ is not closed in terms of the moments of order γ, its value depends only on the integer moments of the next higher order. In contrast, from Eq. (6.68) we can observe that $C^{(0)}_{l_1 l_2 l_3}$ is not closed due to the appearance of $|\mathbf{v}_1 - \mathbf{v}_2|$, which is a nonlinear function of $\mathbf{v}_1$ and $\mathbf{v}_2$. However, if this nonlinear term could be approximated by a function that does not depend on $\mathbf{v}_1$ or $\mathbf{v}_2$, then the remaining term $I^{(0)}_{l_1 l_2 l_3}(\omega, \mathbf{v}_1, \mathbf{v}_1 - \mathbf{v}_2)$ would be a homogeneous polynomial of order γ and, hence, the collision source term would be closed. This approximation is known as the Maxwell model (Maxwell, 1879), and we will discuss it in more detail later.

Finally, we can note that Eq. (6.69) will be closed in terms of the moments of order $\gamma + 1$ and their spatial gradients. This result is interesting because closure of the moments of order $\gamma + 1$ given the moments of order γ using quadrature-based moment methods is usually quite accurate. Using a symmetric change of variables, it is straightforward to show

Table 6.3. $I^{(0)}_{l_1l_2l_3}(\omega, \mathbf{v}, \mathbf{g})$ terms for selected fourth- and fifth-order moments. Others can be found by permutation of the indices.

$$I^{(0)}_{400} = (\omega^4/80)\left(g^4 + 10g^2g_1^2 + 5g_1^4\right) - (\omega^3/2)\left(g^2 + g_1^2\right)g_1v_1 + (\omega^2/2)\left(g^2 + 3g_1^2\right)v_1^2 - 2\omega g_1v_1^3$$

$$I^{(0)}_{310} = (\omega^4/16)\left(g^2 + g_1^2\right)g_1g_2 - (\omega^3/8)\left(g^2 + 3g_1^2\right)g_2v_1 - (\omega^3/8)\left(g^2 + g_1^2\right)g_1v_2 + (3\omega^2/4)g_1g_2v_1^2 + (\omega^2/4)\left(g^2 + 3g_1^2\right)v_1v_2 - (3\omega/2)g_1v_1^2v_2 - (\omega/2)g_2v_1^3$$

$$I^{(0)}_{220} = (\omega^4/240)\left(g^4 + 5g^2g_1^2 + 5g^2g_2^2 + 15g_1^2g_2^2\right) - (\omega^3/12)\left(g^2 + 3g_2^2\right)g_1v_1 - (\omega^3/12)\left(g^2 + 3g_1^2\right)g_2v_2 + (\omega^2/12)\left(g^2 + 3g_2^2\right)v_1^2 + \omega^2g_1g_2v_1v_2 + (\omega^2/12)\left(g^2 + 3g_1^2\right)v_2^2 - \omega g_2v_1^2v_2 - \omega g_1v_1v_2^2$$

$$I^{(0)}_{211} = (\omega^4/48)\left(g^2 + 3g_1^2\right)g_2g_3 - (\omega^3/4)g_1g_2g_3v_1 - (\omega^3/24)\left(g^2 + 3g_1^2\right)g_3v_2 - (\omega^3/24)\left(g^2 + 3g_1^2\right)g_2v_3 + (\omega^2/4)g_2g_3v_1^2 + (\omega^2/2)g_1g_3v_1v_2 + (\omega^2/2)g_1g_2v_1v_3 + (\omega^2/12)\left(g^2 + 3g_1^2\right)v_2v_3 - (\omega/2)g_3v_1^2v_2 - (\omega/2)g_2v_1^2v_3 - \omega g_1v_1v_2v_3$$

$$I^{(0)}_{410} = -(\omega^5/60)\left(g^4 + 5g_1^4\right)g_2 + (\omega^4/4)\left(g^2 + g_1^2\right)g_1g_2v_1 + (\omega^4/80)\left(g^4 + 10g^2g_1^2 + 5g_1^4\right)v_2 - (\omega^3/4)\left(g^2 + 3g_1^2\right)g_2v_1^2 - (\omega^3/2)\left(g^2 + g_1^2\right)g_1v_1v_2 + \omega^2g_1g_2v_1^3 + (\omega^2/2)\left(g^2 + 3g_1^2\right)v_1^2v_2 - (\omega/2)g_2v_1^4 - 2\omega g_1v_1^3v_2$$

$$I^{(0)}_{311} = -(\omega^5/32)\left(g^2 + g_1^2\right)g_1g_2g_3 + (\omega^4/16)\left(g^2 + 3g_1^2\right)g_2g_3v_1 + (\omega^4/16)\left(g^2 + g_1^2\right)g_1g_3v_2 + (\omega^4/16)\left(g^2 + g_1^2\right)g_1g_2v_3 - (3\omega^3/8)g_1g_2g_3v_1^2 - (\omega^3/2)\left(g^2 + 3g_2^2 + 3g_3^2\right)g_3v_1v_2 - (\omega^3/2)\left(g^2 + 3g_2^2 + 3g_3^2\right)g_2v_1v_3 - (\omega^3/4)\left(g^2 + g_2^2 + g_3^2\right)g_1v_2v_3 + (\omega^2/4)g_2g_3v_1^3 + (3\omega^2/4)g_1g_3v_1^2v_2 + (3\omega^2/4)g_1g_2v_1^2v_3 + (\omega^2/4)\left(g^2 + 3g_1^2\right)v_1v_2v_3 - (\omega/2)g_3v_1^3v_2 - (\omega/2)g_2v_1^3v_3 - (3\omega/2)g_1v_1^2v_2v_3$$

Table 6.4. $I^{(m)}_{l_1l_2l_3}(\omega, \mathbf{v}, \mathbf{g})$ terms for first-order moments with $m = 1, 2, 3$

$I^{(1)}_{100} = (2\omega/15)\left(g^2 + 2g_1^2\right)$	$I^{(2)}_{100} = (4\omega/15)g_1g_2$	$I^{(3)}_{100} = (4\omega/15)g_1g_3$
$I^{(1)}_{010} = (4\omega/15)g_1g_2$	$I^{(2)}_{010} = (2\omega/15)\left(g^2 + 2g_2^2\right)$	$I^{(3)}_{010} = (4\omega/15)g_2g_3$
$I^{(1)}_{001} = (4\omega/15)g_1g_3$	$I^{(2)}_{001} = (4\omega/15)g_2g_3$	$I^{(3)}_{001} = (2\omega/15)\left(g^2 + 2g_3^2\right)$

Table 6.5. $I^{(1)}_{l_1 l_2 l_3}(\omega, \mathbf{v}, \mathbf{g})$ terms for second-order moments. The corresponding terms for $m = 2$ and 3 can be found by permutation of the indices.

$$I^{(1)}_{200} = -(2\omega^2/35)\left(3g^2 + 2g_1^2\right) g_1 + (4\omega/15)\left(g^2 + 2g_1^2\right) v_1$$

$$I^{(1)}_{110} = -(2\omega^2/35)\left(g^2 + 2g_1^2\right) g_2 + (4\omega/15) g_1 g_2 v_1 + (2\omega/15)\left(g^2 + 2g_1^2\right) v_2$$

$$I^{(1)}_{101} = -(2\omega^2/35)\left(g^2 + 2g_1^2\right) g_3 + (4\omega/15) g_1 g_3 v_1 + (2\omega/15)\left(g^2 + 2g_1^2\right) v_3$$

$$I^{(1)}_{020} = -(2\omega^2/35)\left(g^2 + 2g_2^2\right) g_1 + (8\omega/15) g_1 g_2 v_2$$

$$I^{(1)}_{011} = -(4\omega^2/35) g_1 g_2 g_3 + (4\omega/15) g_1 g_3 v_2 + (4\omega/15) g_1 g_2 v_3$$

$$I^{(1)}_{002} = -(2\omega^2/35)\left(g^2 + 2g_3^2\right) g_1 + (8\omega/15) g_1 g_3 v_3$$

Table 6.6. $I^{(1)}_{l_1 l_2 l_3}(\omega, \mathbf{v}, \mathbf{g})$ terms for selected third-order moments. Others can be found by permutation of the indices.

$$I^{(1)}_{300} = (2\omega^3/315)\left(3g^4 + 24g^2 g_1^2 + 8g_1^4\right) - (6\omega^2/35)\left(3g^2 + 2g_1^2\right) g_1 v_1 + (2\omega/5)\left(g^2 + 2g_1^2\right) v_1^2$$

$$\begin{aligned} I^{(1)}_{210} = {} & (8\omega^3/315)\left(3g^2 + 2g_1^2\right) g_1 g_2 - (4\omega^2/35)\left(g^2 + 2g_1^2\right) g_2 v_1 - (2\omega^2/35)\left(3g^2 + 2g_1^2\right) g_1 v_2 \\ & + (4\omega/15)\left(g^2 + 2g_1^2\right) v_1 v_2 + (4\omega/15) g_1 g_2 v_1^2 \end{aligned}$$

$$\begin{aligned} I^{(1)}_{120} = {} & (2\omega^3/315)\left(g^4 + 4g^2 g_1^2 + 4g^2 g_2^2 + 8g_1^2 g_2^2\right) - (2\omega^2/35)\left(g^2 + 2g_2^2\right) g_1 v_1 \\ & - (2\omega^2/35)\left(g^2 + 2g_1^2\right) g_2 v_2 + (2\omega/15)\left(g^2 + 2g_1^2\right) v_2^2 + (8\omega/15) g_1 g_2 v_1 v_2 \end{aligned}$$

$$\begin{aligned} I^{(1)}_{111} = {} & (8\omega^3/315)\left(g^2 + 2g_1^2\right) g_2 g_3 - (4\omega^2/35) g_1 g_2 g_3 v_1 - (2\omega^2/35)\left(g^2 + 2g_1^2\right) g_3 v_2 \\ & - (2\omega^2/35)\left(g^2 + 2g_1^2\right) g_2 v_3 + (4\omega/15) g_1 g_3 v_1 v_2 + (4\omega/15) g_1 g_2 v_1 v_3 \\ & + (2\omega/15)\left(g^2 + 2g_1^2\right) v_2 v_3 \end{aligned}$$

$$I^{(1)}_{030} = (8\omega^3/315)\left(3g^2 + 2g_2^2\right) g_1 g_2 - (6\omega^2/35)\left(g^2 + 2g_2^2\right) g_1 v_2 + (4\omega/5) g_1 g_2 v_2^2$$

$$\begin{aligned} I^{(1)}_{021} = {} & (8\omega^3/315)\left(g^2 + 2g_2^2\right) g_1 g_3 - (2\omega^2/35)\left(g^2 + 2g_2^2\right) g_1 v_3 - (8\omega^2/35) g_1 g_2 g_3 v_2 \\ & + (4\omega/15) g_1 g_3 v_2^2 + (8\omega/15) g_1 g_2 v_2 v_3 \end{aligned}$$

that Eq. (6.69) can be rewritten as

$$C^{(1)}_{l_1 l_2 l_3} = 3g_0 \sum_{m=1}^{3} \int_{\mathbb{R}^6} H_{m, l_1 l_2 l_3}(\omega, \mathbf{v}_1, \mathbf{v}_2) f(\mathbf{v}_2) \frac{\partial f(\mathbf{v}_1)}{\partial x_m} \, d\mathbf{v}_2 \, d\mathbf{v}_1, \tag{6.71}$$

where

$$H_{m, l_1 l_2 l_3}(\omega, \mathbf{v}_1, \mathbf{v}_2) = I^{(m)}_{l_1 l_2 l_3}(\omega, \mathbf{v}_1, \mathbf{v}_1 - \mathbf{v}_2) - I^{(m)}_{l_1 l_2 l_3}(\omega, \mathbf{v}_2, \mathbf{v}_2 - \mathbf{v}_1). \tag{6.72}$$

Table 6.7. $I^{(1)}_{l_1 l_2 l_3}(\omega, \mathbf{v}, \mathbf{g})$ terms for selected fourth-order moments. Others can be found by permutation of the indices.

$$I^{(1)}_{400} = -(2\omega^4/693)\left(15g^4 + 40g^2g_1^2 + 8g_1^4\right)g_1 + (8\omega^3/315)\left(3g^4 + 24g^2g_1^2 + 8g_1^4\right)v_1 - (12\omega^2/35)\left(3g^2 + 2g_1^2\right)g_1v_1^2 + (8\omega/15)\left(g^2 + 2g_1^2\right)v_1^3$$

$$I^{(1)}_{310} = -(2\omega^4/693)\left(3g^4 + 24g^2g_1^2 + 8g_1^4\right)g_2 + (2\omega^3/315)\left(3g^4 + 24g^2g_1^2 + 8g_1^4\right)v_2 + (8\omega^3/105)\left(g^2 + 2g_1^2\right)g_1g_2v_1 - (6\omega^2/35)\left(g^2 + 2g_1^2\right)g_2v_1^2 - (6\omega^2/35)\left(3g^2 + 2g_1^2\right)g_1v_1v_2 + (4\omega/15)g_1g_2v_1^3 + (6\omega/15)\left(g^2 + 2g_1^2\right)v_1^2v_2$$

$$I^{(1)}_{220} = -(2\omega^4/693)\left(3g^4 + 4g^2g_1^2 + 12g^2g_2^2 + 8g_1^2g_2^2\right)g_1 + (4\omega^3/315)\left(g^4 + 4g^2g_1^2 + 4g^2g_2^2 + 8g_1^2g_2^2\right)v_1 + (16\omega^3/315)\left(3g^2 + 2g_1^2\right)g_1g_2v_2 - (2\omega^2/35)\left(g^2 + 2g_2^2\right)g_1v_1^2 - (8\omega^2/35)\left(g^2 + 2g_1^2\right)g_2v_1v_2 - (2\omega^2/35)\left(3g^2 + 2g_1^2\right)g_1v_2^2 + (4\omega/15)\left(g^2 + 2g_1^2\right)v_1v_2^2 + (8\omega/15)g_1g_2v_1^2v_2$$

$$I^{(1)}_{211} = -(8\omega^4/693)\left(3g^2 + 2g_1^2\right)g_1g_2g_3 + (16\omega^3/315)\left(g^2 + 2g_1^2\right)g_2g_3v_1 + (8\omega^3/315)\left(3g^2 + 2g_1^2\right)g_1g_3v_2 + (8\omega^3/315)\left(3g^2 + 2g_1^2\right)g_1g_2v_3 - (4\omega^2/35)g_1g_2g_3v_1^2 - (4\omega^2/35)\left(g^2 + 2g_1^2\right)g_3v_1v_2 - (4\omega^2/35)\left(g^2 + 2g_1^2\right)g_2v_1v_3 - (2\omega^2/35)\left(3g^2 + 2g_1^2\right)g_1v_2v_3 + (4\omega/15)g_1g_2v_1^2v_3 + (4\omega/15)g_1g_3v_1^2v_2 + (4\omega/15)\left(g^2 + 2g_1^2\right)v_1v_2v_3$$

$$I^{(1)}_{130} = -(2\omega^4/693)\left(3g^4 + 12g^2g_1^2 + 4g^2g_2^2 + 8g_1^2g_2^2\right)g_2 + (2\omega^3/105)\left(g^4 + 4g^2g_1^2 + 4g^2g_2^2 + 8g_1^2g_2^2\right)v_2 + (8\omega^3/315)\left(3g^2 + 2g_2^2\right)g_1g_2v_1 - (6\omega^2/35)\left(g^2 + 2g_2^2\right)g_1v_1v_2 - (6\omega^2/35)\left(g^2 + 2g_1^2\right)g_2v_2^2 + (4\omega/5)g_1g_2v_1v_2^2 + (2\omega/15)\left(g^2 + 2g_1^2\right)v_2^3$$

Thus, since the terms for the first-order moments in Table 6.4 are the same for $\mathbf{g}$ and $-\mathbf{g}$, it follows that $C^{(1)}_{100} = C^{(1)}_{010} = C^{(1)}_{001} = 0$ and the first nonzero contribution from Eq. (6.71) is $C^{(1)}_{200}$. In other words, the first-order collision source terms $C^{(1)}_{100}$, $C^{(1)}_{010}$, and $C^{(1)}_{001}$ do not directly affect the transport of mass and mean momentum. However, the latter is affected by $\mathbf{G}_{100}$, $\mathbf{G}_{010}$, and $\mathbf{G}_{001}$, which can be interpreted as the "non-ideal-gas" contributions to the pressure tensor. We will return to this point in Section 6.4.

The analytical expressions for $I^{(m)}_{l_1 l_2 l_3}$ will be used directly with quadrature-based moment methods to evaluate the collision source and collisional-flux terms for each integer moment. The numerical implementation of these terms in the context of quadrature is discussed in Section 6.5.

6.2 Polydisperse hard-sphere collisions

We will now generalize the formulas developed in the previous section to the case of binary collisions between two particles with different diameters d_α and d_β and different masses m_α

Table 6.8. $I^{(1)}_{l_1l_2l_3}(\omega, \mathbf{v}, \mathbf{g})$ terms for selected fourth-order moments. Others can be found by permutation of the indices.

$$
\begin{aligned}
I^{(1)}_{121} = &-(2\omega^4/693)\left(g^4 + 4g^2g_1^2 + 4g^2g_2^2 + 8g_1^2g_2^2\right)g_3 \\
&+ (2\omega^3/315)\left(g^4 + 4g^2g_1^2 + 4g^2g_2^2 + 8g_1^2g_2^2\right)v_3 + (8\omega^3/315)\left(g^2 + 2g_2^2\right)g_1g_3v_1 \\
&+ (16\omega^3/315)\left(g^2 + 2g_1^2\right)g_2g_3v_2 - (8\omega^2/35)g_1g_2g_3v_1v_2 - (2\omega^2/35)\left(g^2 + 2g_2^2\right)g_1v_1v_3 \\
&- (2\omega^2/35)\left(g^2 + 2g_1^2\right)g_3v_2^2 - (4\omega^2/35)\left(g^2 + 2g_1^2\right)g_2v_2v_3 \\
&+ (4\omega/15)g_1g_3v_1v_2^2 + (8\omega/15)g_1g_2v_1v_2v_3 + (2\omega/15)\left(g^2 + 2g_1^2\right)v_2^2v_3
\end{aligned}
$$

$$
\begin{aligned}
I^{(1)}_{040} = &-(2\omega^4/693)\left(3g^4 + 24g^2g_2^2 + 8g_2^4\right)g_1 + (32\omega^3/315)\left(3g^2 + 2g_2^2\right)g_1g_2v_2 \\
&- (12\omega^2/35)\left(g^2 + 2g_2^2\right)g_1v_2^2 + (16\omega/15)g_1g_2v_2^3
\end{aligned}
$$

$$
\begin{aligned}
I^{(1)}_{031} = &-(8\omega^4/693)\left(3g^2 + 2g_2^2\right)g_1g_2g_3 \\
&+ (8\omega^3/105)\left(g^2 + 2g_2^2\right)g_1g_3v_2 + (8\omega^3/315)\left(3g^2 + 2g_2^2\right)g_1g_2v_3 \\
&- (12\omega^2/35)g_1g_2g_3v_2^2 - (6\omega^2/35)\left(g^2 + 2g_2^2\right)g_1v_2v_3 + (4\omega/5)g_1g_2v_2^2v_3 \\
&+ (4\omega/15)g_1g_3v_2^3
\end{aligned}
$$

$$
\begin{aligned}
I^{(1)}_{022} = &-(2\omega^4/693)\left(g^4 + 4g^2g_2^2 + 4g^2g_3^2 + 8g_2^2g_3^2\right)g_1 + (16\omega^3/315)\left(g^2 + 2g_3^2\right)g_1g_2v_2 \\
&+ (16\omega^3/315)\left(g^2 + 2g_2^2\right)g_1g_3v_3 - (2\omega^2/35)\left(g^2 + 2g_3^2\right)g_1v_2^2 \\
&- (2\omega^2/35)\left(g^2 + 2g_2^2\right)g_1v_3^2 - (16\omega^2/35)g_1g_2g_3v_2v_3 + (8\omega/15)g_1g_3v_2^2v_3 \\
&+ (8\omega/15)g_1g_2v_2v_3^2
\end{aligned}
$$

and m_β as illustrated in Figure 6.2. In the moment-transport equations, the collision terms will be denoted by $\mathbb{C}_{\alpha\alpha}(\psi)$ and $\mathbb{C}_{\beta\beta}(\psi)$ for collisions between like particles, and $\mathbb{C}_{\alpha\beta}(\psi)$ and $\mathbb{C}_{\beta\alpha}(\psi)$ for collisions between unlike particles. For example, for a binary system the collision term for type-1 particles is $\mathbb{C}_1 = \mathbb{C}_{11} + \mathbb{C}_{12}$, and that for type-2 particles is $\mathbb{C}_2 = \mathbb{C}_{21} + \mathbb{C}_{22}$. Following the method introduced in Fox & Vedula (2010), we will thus derive a general formula for $\mathbb{C}_{\alpha\beta}(\psi)$ that works for any combination of α and β.

6.2.1 Collision terms for arbitrary moments

For an arbitrary moment ψ the collision term is defined by (Fox & Vedula, 2010)

$$
\begin{aligned}
\mathbb{C}_{\alpha\beta}(\psi) = &\frac{6\chi_{\alpha\beta}^2}{\pi d_\beta}\int_{\mathbb{R}^6}\int_{\mathbb{S}^+}\left[\psi(\mathbf{v}_1') - \psi(\mathbf{v}_1)\right] \\
&\times f^{(2)}(\mathbf{x}, m_\alpha, \mathbf{v}_1; \mathbf{x} + d_{\alpha\beta}\mathbf{x}_{12}, m_\beta, \mathbf{v}_2)|\mathbf{v}_{12}\cdot\mathbf{x}_{12}|\mathrm{d}\mathbf{x}_{12}\,\mathrm{d}\mathbf{v}_2\,\mathrm{d}\mathbf{v}_1,
\end{aligned}
\tag{6.73}
$$

where $d_{\alpha\beta} = (d_\alpha + d_\beta)/2$, $\chi_{\alpha\beta} = d_{\alpha\beta}/d_\beta$, and $\mathbf{v}_1' = \mathbf{v}_1 - \omega_{\alpha\beta}|\mathbf{v}_{12}\cdot\mathbf{x}_{12}|\mathbf{x}_{12}$. The mass-average restitution parameter is defined by

$$
\omega_{\alpha\beta} = \frac{m_\beta(1 + e_{\alpha\beta})}{m_\alpha + m_\beta},
\tag{6.74}
$$

where $e_{\alpha\beta} = e_{\beta\alpha}$ is the restitution coefficient for collisions between particles of types α and β. Note that the subscripts 1 and 2 correspond, as before, to the collision partners, and

Table 6.9. $I^{(1)}_{l_1 l_2 l_3}(\omega, \mathbf{v}, \mathbf{g})$ terms for selected fifth-order moments. Others can be found by permutation of the indices.

$$
\begin{aligned}
I^{(1)}_{410} = {} & (4\omega^5/3003)\left(15g^4 + 40g^2g_1^2 - 7g_1^4 - 15g_2^4\right)g_1g_2 \\
& - (8\omega^4/693)\left(3g^4 + 24g^2g_1^2 + 8g_1^4\right)g_2v_1 - (2\omega^4/693)\left(15g^4 + 40g^2g_1^2 + 8g_1^4\right)g_1v_2 \\
& + (16\omega^3/105)\left(3g^2 + 2g_1^2\right)g_1g_2v_1^2 + (8\omega^3/105)\left(g^4 + 8g^2g_1^2 - 9g_1^4\right)v_1v_2 \\
& - (8\omega^2/35)\left(g^2 + 2g_1^2\right)g_2v_1^3 - (12\omega^2/35)\left(3g^2 + 2g_1^2\right)g_1v_1^2v_2 + (4\omega/15)g_1g_2v_1^4 \\
& + (8\omega/15)\left(g^2 + 2g_1^2\right)v_1^3v_2
\end{aligned}
$$

$$
\begin{aligned}
I^{(1)}_{140} = {} & (2\omega^5/3003)\left(g^6 + 6g^4g_1^2 + 12g^4g_2^2 + 48g^2g_1^2g_2^2 + 8g^2g_2^4 + 16g_1^2g_2^4\right) \\
& - (8\omega^4/693)\left(3g^4 + 12g^2g_1^2 + 4g^2g_2^2 + 8g_1^2g_2^2\right)g_2v_2 \\
& + (32\omega^3/315)\left(3g^2 + 2g_2^2\right)g_1g_2v_1v_2 + (4\omega^3/105)\left(g^4 + 4g^2g_1^2 + 4g^2g_2^2 + 8g_1^2g_2^2\right)v_2^2 \\
& - (8\omega^2/35)\left(g^2 + 2g_1^2\right)g_2v_2^3 + (16\omega/15)g_1g_2v_1v_2^3 + (2\omega/15)\left(g^2 + 2g_1^2\right)v_2^4
\end{aligned}
$$

$$
\begin{aligned}
I^{(1)}_{041} = {} & (4\omega^5/3003)\left(3g^4 + 24g^2g_2^2 + 8g_2^4\right)g_1g_3 \\
& - (32\omega^4/693)\left(3g^2 + 2g_2^2\right)g_1g_2g_3v_2 - (2\omega^4/693)\left(3g^4 + 24g^2g_2^2 + 8g_2^4\right)g_1v_3 \\
& + (16\omega^3/105)\left(g^2 + 2g_2^2\right)g_1g_3v_2^2 + (32\omega^3/315)\left(3g^2 + 2g_2^2\right)g_1g_2v_2v_3 \\
& - (16\omega^2/35)g_1g_2g_3v_2^3 - (12\omega^2/35)\left(g^2 + 2g_2^2\right)g_1v_2^2v_3 \\
& + (4\omega/15)g_1g_3v_2^4 + (16\omega/15)g_1g_2v_2^3v_3
\end{aligned}
$$

$$
\begin{aligned}
I^{(1)}_{311} = {} & (4\omega^5/3003)\left(3g^4 + 24g^2g_1^2 + 8g_1^4\right)g_2g_3 - (8\omega^4/231)\left(3g^2 + 2g_1^2\right)g_1g_2g_3v_1 \\
& - (2\omega^4/693)\left(3g^4 + 24g^2g_1^2 + 8g_1^4\right)g_3v_2 - (2\omega^4/693)\left(3g^4 + 24g^2g_1^2 + 8g_1^4\right)g_2v_3 \\
& + (8\omega^3/105)\left(g^2 + 2g_1^2\right)g_2g_3v_1^2 + (8\omega^3/105)\left(3g^2 + 2g_1^2\right)g_1g_3v_1v_2 \\
& + (8\omega^3/105)\left(3g^2 + 2g_1^2\right)g_1g_2v_1v_3 + (2\omega^3/315)\left(3g^4 + 24g^2g_1^2 + 8g_1^4\right)v_2v_3 \\
& - (4\omega^2/35)g_1g_2g_3v_1^3 - (6\omega^2/35)\left(g^2 + 2g_1^2\right)g_3v_1^2v_2 - (6\omega^2/35)\left(g^2 + 2g_1^2\right)g_2v_1^2v_3 \\
& - (6\omega^2/35)\left(3g^2 + 2g_1^2\right)g_1v_1v_2v_3 + (4\omega/15)g_1g_3v_1^3v_2 + (4\omega/15)g_1g_2v_1^3v_3 \\
& + (6\omega/15)\left(g^2 + 2g_1^2\right)v_1^2v_2v_3
\end{aligned}
$$

$$
\begin{aligned}
I^{(1)}_{131} = {} & (4\omega^5/3003)\left(3g^4 + 12g^2g_1^2 + 4g^2g_2^2 + 8g_1^2g_2^2\right)g_2g_3 - (8\omega^4/693)\left(3g^2 + 2g_2^2\right)g_1g_2g_3v_1 \\
& - (2\omega^4/231)\left(g^4 + 4g^2g_1^2 + 4g^2g_2^2 + 8g_1^2g_2^2\right)g_3v_2 \\
& - (2\omega^4/693)\left(3g^4 + 12g^2g_1^2 + 4g^2g_2^2 + 8g_1^2g_2^2\right)g_2v_3 + (8\omega^3/105)\left(g^2 + 2g_2^2\right)g_1g_3v_1v_2 \\
& + (8\omega^3/315)\left(3g^2 + 2g_2^2\right)g_1g_2v_1v_3 + (8\omega^3/105)\left(g^2 + 2g_1^2\right)g_2g_3v_2^2 \\
& + (2\omega^3/105)\left(g^4 + 4g^2g_1^2 + 4g^2g_2^2 + 8g_1^2g_2^2\right)v_2v_3 - (12\omega^2/35)g_1g_2g_3v_1v_2^2 \\
& - (6\omega^2/35)\left(g^2 + 2g_2^2\right)g_1v_1v_2v_3 - (2\omega^2/35)\left(g^2 + 2g_1^2\right)g_3v_2^3 \\
& - (6\omega^2/35)\left(g^2 + 2g_1^2\right)g_2v_2^2v_3 + (4\omega/15)g_1g_3v_1v_2^3 + (4\omega/5)g_1g_2v_1v_2^2v_3 \\
& + (2\omega/15)\left(g^2 + 2g_1^2\right)v_2^3v_3
\end{aligned}
$$

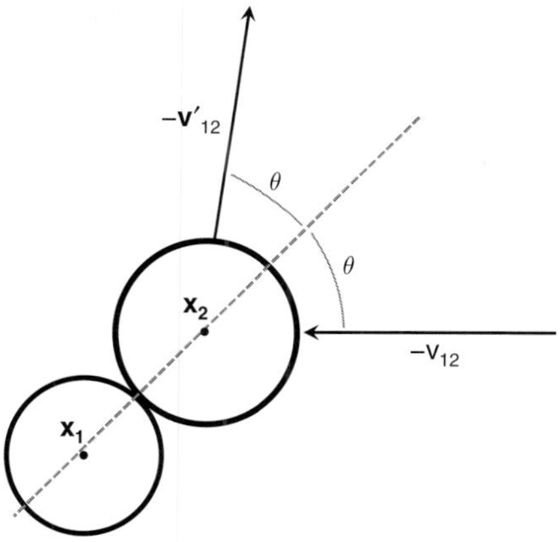

Figure 6.2. A direct collision between two hard spheres with different diameters d_1 and d_2.

the subscripts α and β distinguish between the particle properties. The pair distribution function $f^{(2)}(\mathbf{x}, m_\alpha, \mathbf{v}_1; \mathbf{x} + d_{\alpha\beta}\mathbf{x}_{12}, m_\beta, \mathbf{v}_2)$ is written such that collision partner 1 has mass m_α and collision partner 2 has mass m_β.

As was done in Section 6.1.2, we can now use a Taylor-series expansions about the collision contact point

$$(\mathbf{x}_1, \mathbf{x}_2) = \left(\mathbf{x} - \frac{1}{2}d_\alpha\mathbf{x}_{12}, \mathbf{x} + \frac{1}{2}d_\beta\mathbf{x}_{12}\right)$$

to re-express Eq. (6.73) in terms of a collision source term and a collisional-flux term:

$$\mathbb{C}_{\alpha\beta}(\psi) = C_{\alpha\beta}(\psi) - \nabla_x \cdot \mathbf{G}_{\alpha\beta}(\psi), \tag{6.75}$$

where the collision source term is given by

$$\begin{aligned} C_{\alpha\beta}(\psi) = \frac{6\chi_{\alpha\beta}^2}{\pi d_\beta} \int_{\mathbb{R}^6} \int_{\mathbb{S}^+} & [\psi(\mathbf{v}_1') - \psi(\mathbf{v}_1)] \\ & \times f^{(2)}\left(\mathbf{x} - \frac{1}{2}d_\alpha\mathbf{x}_{12}, m_\alpha, \mathbf{v}_1; \mathbf{x} + \frac{1}{2}d_\beta\mathbf{x}_{12}, m_\beta, \mathbf{v}_2\right)|\mathbf{v}_{12} \cdot \mathbf{x}_{12}| \mathrm{d}\mathbf{x}_{12}\, \mathrm{d}\mathbf{v}_2\, \mathrm{d}\mathbf{v}_1 \end{aligned} \tag{6.76}$$

and the collisional-flux term is given by

$$\begin{aligned} \mathbf{G}_{\alpha\beta}(\psi) = -\frac{3\chi_{\alpha\beta}^2 d_\alpha}{\pi d_\beta} \int_{\mathbb{R}^6} \int_{\mathbb{S}^+} & \mathbf{x}_{12} [\psi(\mathbf{v}_1') - \psi(\mathbf{v}_1)] \\ & \times f^{(2)}\left(\mathbf{x} - \frac{1}{2}d_\alpha\mathbf{x}_{12}, m_\alpha, \mathbf{v}_1; \mathbf{x} + \frac{1}{2}d_\beta\mathbf{x}_{12}, m_\beta, \mathbf{v}_2\right)|\mathbf{v}_{12} \cdot \mathbf{x}_{12}| \mathrm{d}\mathbf{x}_{12}\, \mathrm{d}\mathbf{v}_2\, \mathrm{d}\mathbf{v}_1. \end{aligned} \tag{6.77}$$

This expression is further simplified by introducing the pair correlation function $g_{\alpha\beta}$ and using a truncated Taylor-series expansion around the point $\mathbf{x}$ up to first order in $d_{\alpha\beta}$ (see the details given in Section 6.1.2):

$$\begin{aligned} & f^{(2)}\left(\mathbf{x} - \frac{1}{2}d_\alpha\mathbf{x}_{12}, m_\alpha, \mathbf{v}_1; \mathbf{x} + \frac{1}{2}d_\beta\mathbf{x}_{12}, m_\beta, \mathbf{v}_2\right) \\ & \quad \approx g_{\alpha\beta}(\mathbf{x}_{12}) f_\alpha(\mathbf{v}_1) f_\beta(\mathbf{v}_2)\left[1 - \frac{1}{2}d_\alpha\mathbf{x}_{12} \cdot \nabla_x \ln f_\alpha(\mathbf{v}_1) + \frac{1}{2}d_\beta\mathbf{x}_{12} \cdot \nabla_x \ln f_\beta(\mathbf{v}_2)\right]. \end{aligned} \tag{6.78}$$

Note that we have assumed that $g_{\alpha\beta}$ may depend on $\mathbf{x}_{12}$ when $\alpha \neq \beta$, and will discuss a possible dependence below. We have also introduced the velocity distribution functions for each particle type (f_α and f_β), which need not be the same (unless $\alpha = \beta$).

In general, $g_{\alpha\beta}$ depends on the volume fractions of each particle type and on the particle diameters. However, it can also depend on other moments of the velocity distribution function. For example, if the mean particle velocities $\mathbf{U}_\alpha$ and $\mathbf{U}_\beta$ are very different, one could expect that the collision frequency would be higher on the upstream side of the slower particle type. The unit vector $\mathbf{x}_{12}$ denotes the relative positions of the particle centers at collision. If we then consider the direction relative to the mean velocity difference, $(\mathbf{U}_\alpha - \mathbf{U}_\beta) \cdot \mathbf{x}_{12}$, we can model the dependence of the pair correlation function on the mean velocity difference as[10]

$$g_{\alpha\beta}(\mathbf{x}_{12}) = g_{0,\alpha\beta} \left[1 - \frac{c_{\alpha\beta}}{E_{\alpha\beta}^{1/2}} (\mathbf{U}_\alpha - \mathbf{U}_\beta) \cdot \mathbf{x}_{12} \right], \tag{6.79}$$

where $g_{0,\alpha\beta}$ depends only on the volume fractions and particle diameters, and $c_{\alpha\beta}$ is a model constant that must be chosen such that $g_{\alpha\beta}(\mathbf{x}_{12})$ is nonnegative[11] (i.e. $|c_{\alpha\beta}| \leq 1$). The energy scaling factor $E_{\alpha\beta}$ can be defined as

$$E_{\alpha\beta} = 3\Theta_\alpha + 3\Theta_\beta + |\mathbf{U}_\alpha - \mathbf{U}_\beta|^2, \tag{6.80}$$

where Θ_α and Θ_β are the granular temperatures of particle types α and β, respectively. Note that Eq. (6.80) corresponds to the expected square velocity difference when the velocity distributions are Maxwellian. The reader should note that the form given in Eq. (6.79) can be generalized to include the spatial gradients of the moments. For example, we might have

$$g_{\alpha\beta}(\mathbf{x}_{12}) = g_{0,\alpha\beta} \left[1 + \kappa_{\alpha\beta} d_{\alpha\beta} \mathbf{x}_{12} \cdot \nabla_x \mathcal{H}(t, \mathbf{x}) \right], \tag{6.81}$$

where $\mathcal{H}$ is a functional depending on the moments (e.g. in the hydrodynamic limit $\mathcal{H}$ would depend only on disperse-phase volume fractions, mean velocities, and granular temperatures). In order to ensure positivity, the parameter $\kappa_{\alpha\beta}$ must satisfy $|\kappa_{\alpha\beta}| < 1/|d_{\alpha\beta} \nabla_x \mathcal{H}|$. The model in Eq. (6.81) could be used to describe the buoyancy force in the limit $d_\alpha \ll d_\beta$, in which case $\mathcal{H}$ would depend on the pressure p_α. In fact, with the exception of the drag term, all of the phase-momentum exchange terms described in Chapter 5 will result from the functional form of $g_{\alpha\beta}(\mathbf{x}_{12})$ in the fluid–particle limit (i.e. $d_\alpha \ll d_\beta$).[12] Thus, when modeling multiphase systems with very different particle sizes, care should be taken to develop physically accurate expressions for $g_{\alpha\beta}(\mathbf{x}_{12})$. In the discussion below, we will use Eq. (6.79) to derive the collision terms. Nevertheless, it is straightforward to modify the resulting expressions for the form in Eq. (6.81) since the dependence on $\mathbf{x}_{12}$ is the same in both cases.

[10]This model should be interpreted as the first two terms in a truncated Taylor-series expansion in $(\mathbf{U}_\alpha - \mathbf{U}_\beta) \cdot \mathbf{x}_{12}$. The complete dependence of $g_{\alpha\beta}(\mathbf{x}_{12})$ on $(\mathbf{U}_\alpha - \mathbf{U}_\beta) \cdot \mathbf{x}_{12}$ is likely to be much more complicated.

[11]The condition $|c_{\alpha\beta}| \leq 1$ follows from the definition of $E_{\alpha\beta}$ given in Eq. (6.80). Because $\mathbf{x}_{12}$ is a unit vector, the maximum/minimum of $(\mathbf{U}_\alpha - \mathbf{U}_\beta) \cdot \mathbf{x}_{12}$ is $\pm|\mathbf{U}_\alpha - \mathbf{U}_\beta|$. The minimum of $E_{\alpha\beta}^{1/2}$ is $|\mathbf{U}_\alpha - \mathbf{U}_\beta|$ and occurs when the granular temperatures are null. Thus the sign of $g_{\alpha\beta}$ is determined by $1 \pm c_{\alpha\beta}$, which yields the bounds on $c_{\alpha\beta}$.

[12]Here, d_α would be a molecular diameter and d_β would be the particle diameter. In the fluid–particle limit, d_α tends to zero while the number of molecules tends to infinity such that the volume fraction occupied by the molecules is constant. Note that, in this limit, the fluid-phase NDF f_α can be treated as a near-Maxwellian distribution in the hydrodynamic limit.

The next step is to write $\psi(\mathbf{v}_1')$ as an explicit function of $\mathbf{x}_{12}$ using the definition $\mathbf{v}_1' = \mathbf{v}_1 - \omega_{\alpha\beta}(\mathbf{v}_{12} \cdot \mathbf{x}_{12})\mathbf{x}_{12}$. On employing a Taylor-series expansion about $\mathbf{v}_1$ we find

$$\psi(\mathbf{v}_1') - \psi(\mathbf{v}_1) = \sum_{n=1}^{\infty} \frac{(-\omega_{\alpha\beta})^n}{n!}(\mathbf{v}_{12} \cdot \mathbf{x}_{12})^n(\mathbf{x}_{12} \cdot \nabla_v)^n\psi(\mathbf{v}_1). \tag{6.82}$$

We can then express the collision source term as

$$C_{\alpha\beta}(\psi) = C_{\alpha\beta}^{(0)}(\psi) + C_{\alpha\beta}^{(1)}(\psi) + C_{\alpha\beta}^{(2)}(\psi) + C_{\alpha\beta}^{(3)}(\psi) \tag{6.83}$$

and collisional-flux term as

$$\mathbf{G}_{\alpha\beta}(\psi) = \mathbf{G}_{\alpha\beta}^{(0)}(\psi) + \mathbf{G}_{\alpha\beta}^{(1)}(\psi), \tag{6.84}$$

where

$$\begin{aligned} C_{\alpha\beta}^{(0)}(\psi) = {} & \frac{6\chi_{\alpha\beta}^2 g_{0,\alpha\beta}}{\pi d_\beta} \sum_{n=1}^{\infty} \frac{(-\omega_{\alpha\beta})^n}{n!} \\ & \times \int_{\mathbb{R}^6} \left(\int_{\mathbb{S}^+} (\mathbf{v}_{12} \cdot \mathbf{x}_{12})^{n+1} \, (\mathbf{x}_{12} \cdot \nabla_v)^n \, \psi(\mathbf{v}_1) \mathrm{d}\mathbf{x}_{12} \right) \\ & \times f_\alpha(\mathbf{v}_1) f_\beta(\mathbf{v}_2) \mathrm{d}\mathbf{v}_2 \, \mathrm{d}\mathbf{v}_1, \end{aligned} \tag{6.85}$$

$$\begin{aligned} C_{\alpha\beta}^{(1)}(\psi) = {} & \frac{3\chi_{\alpha\beta}^2 g_{0,\alpha\beta}}{\pi d_\beta} \sum_{n=1}^{\infty} \frac{(-\omega_{\alpha\beta})^n}{n!} \\ & \times \int_{\mathbb{R}^6} \left(\int_{\mathbb{S}^+} \mathbf{x}_{12} (\mathbf{v}_{12} \cdot \mathbf{x}_{12})^{n+1} (\mathbf{x}_{12} \cdot \nabla_v)^n \psi(\mathbf{v}_1) \mathrm{d}\mathbf{x}_{12} \right) \\ & \cdot \left[d_\beta f_\alpha(\mathbf{v}_1) \nabla_x f_\beta(\mathbf{v}_2) - d_\alpha f_\beta(\mathbf{v}_2) \nabla_x f_\alpha(\mathbf{v}_1) \right] \mathrm{d}\mathbf{v}_2 \, \mathrm{d}\mathbf{v}_1, \end{aligned} \tag{6.86}$$

$$\begin{aligned} C_{\alpha\beta}^{(2)}(\psi) = {} & -\frac{6\chi_{\alpha\beta}^2 g_{0,\alpha\beta} c_{\alpha\beta}}{\pi d_\beta E_{\alpha\beta}^{1/2}} \sum_{n=1}^{\infty} \frac{(-\omega_{\alpha\beta})^n}{n!} (\mathbf{U}_\alpha - \mathbf{U}_\beta) \\ & \cdot \int_{\mathbb{R}^6} \left(\int_{\mathbb{S}^+} \mathbf{x}_{12} (\mathbf{v}_{12} \cdot \mathbf{x}_{12})^{n+1} (\mathbf{x}_{12} \cdot \nabla_v)^n \psi(\mathbf{v}_1) \mathrm{d}\mathbf{x}_{12} \right) f_\alpha(\mathbf{v}_1) f_\beta(\mathbf{v}_2) \mathrm{d}\mathbf{v}_2 \, \mathrm{d}\mathbf{v}_1, \end{aligned} \tag{6.87}$$

$$\begin{aligned} C_{\alpha\beta}^{(3)}(\psi) = {} & -\frac{3\chi_{\alpha\beta}^2 g_{0,\alpha\beta} c_{\alpha\beta}}{\pi d_\beta E_{\alpha\beta}^{1/2}} \sum_{n=1}^{\infty} \frac{(-\omega_{\alpha\beta})^n}{n!} (\mathbf{U}_\alpha - \mathbf{U}_\beta) \\ & \cdot \int_{\mathbb{R}^6} \left(\int_{\mathbb{S}^+} \mathbf{x}_{12} \otimes \mathbf{x}_{12} (\mathbf{v}_{12} \cdot \mathbf{x}_{12})^{n+1} (\mathbf{x}_{12} \cdot \nabla_v)^n \psi(\mathbf{v}_1) \mathrm{d}\mathbf{x}_{12} \right) \\ & \cdot \left[d_\beta f_\alpha(\mathbf{v}_1) \nabla_x f_\beta(\mathbf{v}_2) - d_\alpha f_\beta(\mathbf{v}_2) \nabla_x f_\alpha(\mathbf{v}_1) \right] \mathrm{d}\mathbf{v}_2 \, \mathrm{d}\mathbf{v}_1, \end{aligned} \tag{6.88}$$

and

$$\begin{aligned} \mathbf{G}_{\alpha\beta}^{(0)}(\psi) = {} & -\frac{3\chi_{\alpha\beta}^2 d_\alpha g_{0,\alpha\beta}}{\pi d_\beta} \sum_{n=1}^{\infty} \frac{(-\omega_{\alpha\beta})^n}{n!} \\ & \times \int_{\mathbb{R}^6} \left(\int_{\mathbb{S}^+} \mathbf{x}_{12} (\mathbf{v}_{12} \cdot \mathbf{x}_{12})^{n+1} (\mathbf{x}_{12} \cdot \nabla_v)^n \psi(\mathbf{v}_1) \mathrm{d}\mathbf{x}_{12} \right) f_\alpha(\mathbf{v}_1) f_\beta(\mathbf{v}_2) \mathrm{d}\mathbf{v}_2 \, \mathrm{d}\mathbf{v}_1, \end{aligned} \tag{6.89}$$

$$\mathbf{G}^{(1)}_{\alpha\beta}(\psi) = \frac{3\chi^2_{\alpha\beta} d_\alpha g_{0,\alpha\beta} c_{\alpha\beta}}{\pi d_\beta E^{1/2}_{\alpha\beta}} \sum_{n=1}^{\infty} \frac{(-\omega_{\alpha\beta})^n}{n!} (\mathbf{U}_\alpha - \mathbf{U}_\beta) \cdot \int_{\mathbb{R}^6} \left(\int_{\mathbb{S}^+} \mathbf{x}_{12} \otimes \mathbf{x}_{12} (\mathbf{v}_{12} \cdot \mathbf{x}_{12})^{n+1} (\mathbf{x}_{12} \cdot \nabla_v)^n \psi(\mathbf{v}_1) \mathrm{d}\mathbf{x}_{12} \right) f_\alpha(\mathbf{v}_1) f_\beta(\mathbf{v}_2) \mathrm{d}\mathbf{v}_2 \, \mathrm{d}\mathbf{v}_1. \tag{6.90}$$

From these definitions, we can observe that, in addition to the two types of integrals involving $\mathbf{x}_{12}$ in Eq. (6.23), a third type appears when $\alpha \neq \beta$:

$$\int_{\mathbb{S}^+} \mathbf{x}_{12} \otimes \mathbf{x}_{12} (\mathbf{v}_{12} \cdot \mathbf{x}_{12})^{n+1} (\mathbf{x}_{12} \cdot \nabla_v)^n \psi(\mathbf{v}_1) \mathrm{d}\mathbf{x}_{12}, \tag{6.91}$$

which involves the tensor product $\mathbf{x}_{12} \otimes \mathbf{x}_{12}$. In the following, we will consider the value of this integral for integer moments.

6.2.2 The third integral over collision angles

The third integral, given in Eq. (6.91), can be evaluated component-wise:

$$\begin{aligned} &\int_{\mathbb{S}^+} x_{12,p} x_{12,q} (\mathbf{v}_{12} \cdot \mathbf{x}_{12})^{n+1} (\mathbf{x}_{12} \cdot \nabla_v)^n \psi(\mathbf{v}) \mathrm{d}\mathbf{x}_{12} \\ &\quad = v_{12}^{n+1} \sum_{i+j+k=n} \frac{\partial^n \psi(\mathbf{v})}{\partial v_1^i \, \partial v_2^j \, \partial v_3^k} \binom{n}{i_1,\ j_1,\ k_1,\ i_2,\ j_2,\ k_2,\ i_3,\ j_3,\ k_3} \\ &\quad \times L_{11}^{i_1} L_{21}^{i_2} L_{31}^{i_3} L_{12}^{j_1} L_{22}^{j_2} L_{32}^{j_3} L_{13}^{k_1} L_{23}^{k_2} L_{33}^{k_3} \\ &\quad \times \left\{ \int_0^{\pi/2} \int_0^{2\pi} (\sin\theta)^{n-n_3+1} (\cos\theta)^{n+n_3+1} (\sin\phi)^{n_2} (\cos\phi)^{n_1} \right. \\ &\qquad \times \left(L_{1p} \sin\theta\cos\phi + L_{2p} \sin\theta\sin\phi + L_{3p}\cos\theta \right) \\ &\qquad \left. \times \left(L_{1q} \sin\theta\cos\phi + L_{2q} \sin\theta\sin\phi + L_{3q}\cos\theta \right) \mathrm{d}\phi \, \mathrm{d}\theta \right\}, \end{aligned} \tag{6.92}$$

and involves nine different sets of integrals over the collision angles. These integrals lead to six coefficients $K^{(mn)}_{ijk} = K^{(nm)}_{ijk}$, which are defined by

$$K^{(11)}_{ijk} \equiv \frac{[1 + (-1)^{i-j}][1 + (-1)^{j-k}]}{4\pi} B\left(\frac{i-j+3}{2}, \frac{j-k+1}{2}\right) B\left(\frac{i+k+2}{2}, \frac{i-k+4}{2}\right), \tag{6.93}$$

$$K^{(12)}_{ijk} \equiv \frac{[1 - (-1)^{i-j}][1 - (-1)^{j-k}]}{4\pi} B\left(\frac{i-j+2}{2}, \frac{j-k+2}{2}\right) B\left(\frac{i+k+2}{2}, \frac{i-k+4}{2}\right), \tag{6.94}$$

$$K^{(13)}_{ijk} \equiv \frac{[1 - (-1)^{i-j}][1 + (-1)^{j-k}]}{4\pi} B\left(\frac{i-j+2}{2}, \frac{j-k+1}{2}\right) B\left(\frac{i+k+3}{2}, \frac{i-k+3}{2}\right), \tag{6.95}$$

$$K^{(22)}_{ijk} \equiv \frac{[1 + (-1)^{i-j}][1 + (-1)^{j-k}]}{4\pi} B\left(\frac{i-j+1}{2}, \frac{j-k+3}{2}\right) B\left(\frac{i+k+2}{2}, \frac{i-k+4}{2}\right), \tag{6.96}$$

$$K^{(23)}_{ijk} \equiv \frac{[1+(-1)^{i-j}][1-(-1)^{j-k}]}{4\pi} B\left(\frac{i-j+1}{2}, \frac{j-k+2}{2}\right) B\left(\frac{i+k+3}{2}, \frac{i-k+3}{2}\right), \tag{6.97}$$

$$K^{(33)}_{ijk} \equiv \frac{[1+(-1)^{i-j}][1+(-1)^{j-k}]}{4\pi} B\left(\frac{i-j+1}{2}, \frac{j-k+1}{2}\right) B\left(\frac{i+k+4}{2}, \frac{i-k+2}{2}\right), \tag{6.98}$$

which we have left in terms of the beta function. Note that in the integral in Eq. (6.92), these coefficients are associated with the components of $\mathbf{L}$ such that $K^{(mn)}_{ijk} L_{mp} L_{nq}$ appears in the final result. Also note that the factor of $1/\pi$ in Eqs. (6.93)–(6.98) has been included in order to cancel out the π-dependent terms that arise in the beta functions so that the six coefficients are rational numbers. The same factor will be removed from the collision terms when we substitute back into the definitions below.

6.2.3 Collision terms for integer moments

For the integer moments of order $\gamma = l_1 + l_2 + l_3$, the collision terms in Eqs. (6.88) and (6.90) can be written in terms of

$$I^{(p,q)}_{l_1 l_2 l_3}(\omega, \mathbf{v}, \mathbf{g}) = \sum_{i_1=0}^{l_1} \sum_{i_2=0}^{l_2} \sum_{i_3=0}^{l_3} (-\omega)^{i_1+i_2+i_3} S^{(p,q)}_{i_1 i_2 i_3} \binom{l_1}{i_1}\binom{l_2}{i_2}\binom{l_3}{i_3} v_1^{l_1-i_1} v_2^{l_2-i_2} v_3^{l_3-i_3}, \tag{6.99}$$

where the coefficients $S^{(p,q)}_{i_1 i_2 i_3} = S^{(q,p)}_{i_1 i_2 i_3}$ are defined in terms of the components of $\mathbf{g}$ for $p, q = 1, 2, 3$ by

$$S^{(p,q)}_{i_1 i_2 i_3} \equiv -\sum^{i_1 i_2 i_3} (gL_{11})^{i_1-j_1} (gL_{12})^{i_2-j_2} (gL_{21})^{j_1-k_1} (gL_{22})^{j_2-k_2} (gL_{23})^{i_3-k_3} (gL_{31})^{k_1} (gL_{32})^{k_2} (gL_{33})^{k_3}$$
$$\times \sum_{m_1,m_2=1}^{3} (gL_{m_1 p}) K^{(m_1 m_2)}_{ijk} (gL_{m_2 q}). \tag{6.100}$$

For the special case in which $i_1 = i_2 = i_3 = 0$, we define $S^{(p,q)}_{000} = 0$. Note that in this definition for $S^{(p,q)}_{i_1 i_2 i_3}$ the magnitude of $\mathbf{g}$ appears to the power $i + 2$. This is done so that $I^{(p,q)}_{l_1 l_2 l_3}(\omega, \mathbf{v}, \mathbf{g})$ will be a homogeneous polynomial of order $\gamma + 2$ in terms of the components of $\mathbf{v}$ and $\mathbf{g}$. Examples of $I^{(p,q)}_{l_1 l_2 l_3}$ are given in Tables 6.10–6.12.

We can now express the collision source term and collisional-flux term for the integer moments of order γ as

$$C_{l_1 l_2 l_3,\alpha\beta} = C^{(0)}_{l_1 l_2 l_3,\alpha\beta} + C^{(1)}_{l_1 l_2 l_3,\alpha\beta} + C^{(2)}_{l_1 l_2 l_3,\alpha\beta} + C^{(3)}_{l_1 l_2 l_3,\alpha\beta} \tag{6.101}$$

and

$$\mathbf{G}_{l_1 l_2 l_3,\alpha\beta} = \mathbf{G}^{(0)}_{l_1 l_2 l_3,\alpha\beta} + \mathbf{G}^{(1)}_{l_1 l_2 l_3,\alpha\beta}. \tag{6.102}$$

In Eq. (6.101), the four contributions on the right-hand side are defined as[13]

$$C^{(0)}_{l_1 l_2 l_3,\alpha\beta} = \frac{6\chi^2_{\alpha\beta} g_{0,\alpha\beta}}{d_\beta} \int_{\mathbb{R}^6} I^{(0)}_{l_1 l_2 l_3}\left(\omega_{\alpha\beta}, \mathbf{v}_\alpha, \mathbf{v}_\alpha - \mathbf{v}_\beta\right) |\mathbf{v}_\alpha - \mathbf{v}_\beta| f_\alpha(\mathbf{v}_\alpha) f_\beta(\mathbf{v}_\beta) \mathrm{d}\mathbf{v}_\alpha \, \mathrm{d}\mathbf{v}_\beta, \tag{6.103}$$

[13] Since it is no longer necessary to keep track of the collision partners, we now switch to denoting the integration variables as $\mathbf{v}_\alpha$ and $\mathbf{v}_\beta$.

Table 6.10. Examples of $I^{(p,q)}_{l_1 l_2 l_3}(\omega, \mathbf{v}, \mathbf{g})$ terms for first- and second-order moments. All others can be found by permutation of the indices.

$$I^{(1,1)}_{100} = (\omega/12)\left(3g^2 + g_1^2\right) g_1$$

$$I^{(1,2)}_{100} = (\omega/12)\left(g^2 + g_1^2\right) g_2$$

$$I^{(2,2)}_{100} = (\omega/12)\left(g^2 + g_2^2\right) g_1$$

$$I^{(2,3)}_{100} = (\omega/12) g_1 g_2 g_3$$

$$I^{(1,1)}_{200} = -(\omega^2/32)\left(g^4 + 6g^2 g_1^2 + g_1^4\right) + (\omega/6)\left(3g^2 + g_1^2\right) g_1 v_1$$

$$I^{(1,2)}_{200} = -(\omega^2/32)\left(3g^2 + g_1^2\right) g_1 g_2 + (\omega/6)\left(g^2 + g1^2\right) g_2 v_1$$

$$I^{(2,2)}_{200} = -(\omega^2/96)\left(g^4 + 3g^2 g_1^2 + 3g^2 g_2^2 + 3g_1^2 g_2^2\right) + (\omega/6)\left(g^2 + g_2^2\right) g_1 v_1$$

$$I^{(2,3)}_{200} = -(\omega^2/32)\left(g^2 + g_1^2\right) g_2 g_3 + (\omega/6) g_1 g_2 g_3 v_1$$

$$I^{(1,1)}_{110} = -(\omega^2/32)\left(3g^2 + g_1^2\right) g_1 g_2 + (\omega/12)\left(g^2 + g_1^2\right) g_2 v_1 + (\omega/12)\left(3g^2 + g_1^2\right) g_1 v_2$$

$$I^{(1,2)}_{110} = -(\omega^2/96)\left(g^4 + 3g^2 g_1^2 + 3g^2 g_2^2 + 3g_1^2 g_2^2\right) + (\omega/12)\left(g^2 + g_2^2\right) g_1 v_1 + (\omega/12)\left(g^2 + g_1^2\right) g_2 v_2$$

$$I^{(1,3)}_{110} = -(\omega^2/32)\left(g^2 + g_1^2\right) g_2 g_3 + (\omega/12) g_1 g_2 g_3 v_1 + (\omega/12)\left(g^2 + g_1^2\right) g_3 v_2$$

$$I^{(3,3)}_{110} = -(\omega^2/32)\left(g^2 + g_3^2\right) g_1 g_2 + (\omega/12)\left(g^2 + g_3^2\right) g_2 v_1 + (\omega/12)\left(g^2 + g_3^2\right) g_1 v_2$$

$$C^{(1)}_{l_1 l_2 l_3,\alpha\beta} = \frac{3\chi^2_{\alpha\beta} d_\alpha g_{0,\alpha\beta}}{d_\beta} \times \sum_{m=1}^{3} \int_{\mathbb{R}^6} I^{(m)}_{l_1 l_2 l_3}\left(\omega_{\alpha\beta}, \mathbf{v}_\alpha, \mathbf{v}_\alpha - \mathbf{v}_\beta\right)\left(d_\alpha f_\beta(\mathbf{v}_\beta) \frac{\partial f_\alpha(\mathbf{v}_\alpha)}{\partial x_m} - d_\beta f_\alpha(\mathbf{v}_\alpha) \frac{\partial f_\beta(\mathbf{v}_\beta)}{\partial x_m}\right) \mathrm{d}\mathbf{v}_\alpha \, \mathrm{d}\mathbf{v}_\beta, \tag{6.104}$$

$$C^{(2)}_{l_1 l_2 l_3,\alpha\beta} = \frac{6\chi^2_{\alpha\beta} g_{0,\alpha\beta} c_{\alpha\beta}}{d_\beta E^{1/2}_{\alpha\beta}} \sum_{m=1}^{3} (U_{\beta,m} - U_{\alpha,m}) \int_{\mathbb{R}^6} I^{(m)}_{l_1 l_2 l_3}\left(\omega_{\alpha\beta}, \mathbf{v}_\alpha, \mathbf{v}_\alpha - \mathbf{v}_\beta\right) f_\alpha(\mathbf{v}_\alpha) f_\beta(\mathbf{v}_\beta) \mathrm{d}\mathbf{v}_\alpha \, \mathrm{d}\mathbf{v}_\beta, \tag{6.105}$$

$$C^{(3)}_{l_1 l_2 l_3,\alpha\beta} = \frac{3\chi^2_{\alpha\beta} g_{0,\alpha\beta} c_{\alpha\beta}}{d_\beta E^{1/2}_{\alpha\beta}} \sum_{m,q=1}^{3} \int_{\mathbb{R}^6} \frac{(U_{\beta,q} - U_{\alpha,q})}{|\mathbf{v}_\alpha - \mathbf{v}_\beta|} I^{(m,q)}_{l_1 l_2 l_3}\left(\omega_{\alpha\beta}, \mathbf{v}_\alpha, \mathbf{v}_\alpha - \mathbf{v}_\beta\right) \times \left(d_\alpha f_\beta(\mathbf{v}_\beta) \frac{\partial f_\alpha(\mathbf{v}_\alpha)}{\partial x_m} - d_\beta f_\alpha(\mathbf{v}_\alpha) \frac{\partial f_\beta(\mathbf{v}_\beta)}{\partial x_m}\right) \mathrm{d}\mathbf{v}_\alpha \, \mathrm{d}\mathbf{v}_\beta. \tag{6.106}$$

The mth components of the collisional-flux terms in Eq. (6.102) are given by

Table 6.11. Examples of $I_{l_1 l_2 l_3}^{(p,q)}(\omega, \mathbf{v}, \mathbf{g})$ terms for third-order moments. Others can be found by permutation of the indices.

$$I_{300}^{(1,1)} = (\omega^3/80)\left(5g^4 + 10g^2g_1^2 + g_1^4\right)g_1 - (3\omega^2/32)\left(g^4 + 6g^2g_1^2 + g_1^4\right)v_1 + (\omega/4)\left(3g^2 + g_1^2\right)g_1v_1^3$$

$$I_{300}^{(1,2)} = (\omega^3/80)\left(g^4 + 6g^2g_1^2 + g_1^4\right)g_2 - (3\omega^2/32)\left(3g^2 + g_1^2\right)g_1g_2v_1 + (\omega/4)\left(g^2 + g_1^2\right)g_2v_1^2$$

$$I_{300}^{(2,2)} = (\omega^3/80)\left(g^4 + g^2g_1^2 + 3g^2g_1^2 + g_1^2g_2^2\right)g_1 - (\omega^2/32)\left(g^4 + 3g^2g_1^2 + 3g^2g_2^2 + 3g_1^2g_2^2\right)v_1 + (\omega/4)\left(g^2 + g_2^2\right)g_1v_1^2$$

$$I_{300}^{(2,3)} = (\omega^3/80)\left(3g^2 + g_1^2\right)g_1g_2g_3 - (3\omega^2/32)\left(g^2 + g_1^2\right)g_2g_3v_1 + (\omega/4)g_1g_2g_3v_1^2$$

$$I_{210}^{(1,1)} = (\omega^3/80)\left(g^4 + 6g^2g_1^2 + g_1^4\right)g_2 - (\omega^2/16)\left(3g^2 + g_1^2\right)g_1g_2v_1 - (\omega^2/32)\left(g^4 + 6g^2g_1^2 + g_1^4\right)v_2 + (\omega/12)\left(g^2 + g_1^2\right)g_2v_1^2 + (\omega/6)\left(3g^2 + g_1^2\right)g_1v_1v_2$$

$$I_{210}^{(1,2)} = (\omega^3/80)\left(g^4 + g^2g_1^2 + 3g^2g_1^2 + g_1^2g_2^2\right)g_1 - (\omega^2/32)\left(3g^2 + g_1^2\right)g_1g_2v_2 - (\omega^2/48)\left(g^4 + 3g^2g_1^2 + 3g^2g_2^2 + 3g_1^2g_2^2\right)v_1 + (\omega/12)\left(g^2 + g_2^2\right)g_1v_1^2 + (\omega/6)\left(g^2 + g_1^2\right)g_2v_1v_2$$

$$I_{210}^{(1,3)} = (\omega^3/80)\left(3g^2 + g_1^2\right)g_1g_2g_3 - (\omega^2/16)\left(g^2 + g_1^2\right)g_2g_3v_1 - (\omega^2/32)\left(3g^2 + g_1^2\right)g_1g_3v_2 + (\omega/12)g_1g_2g_3v_1^2 + (\omega/6)\left((g^2 + g_1^2\right)g_3v_1v_2$$

$$I_{210}^{(2,2)} = (\omega^3/80)\left(g^4 + 3g^2g_1^2 + g^2g_2^2 + g_1^2g_2^2\right)g_2 - (\omega^2/16)\left(3g^2 + g_2^2\right)g_1g_2v_1 - (\omega^2/96)\left(g^4 + 3g^2g_1^2 + 3g^2g_2^2 + 3g_1^2g_2^2\right)v_2 + (\omega/12)\left(3g^2 + g_2^2\right)g_2v_1^2 + (\omega/6)\left(g^2 + g_2^2\right)g_1v_1v_2$$

$$I_{210}^{(2,3)} = (\omega^3/240)\left(g^4 + 3g^2g_1^2 + 3g^2g_2^2 + 3g_1^2g_2^2\right)g_3 - (\omega^2/16)\left(g^2 + g_2^2\right)g_1g_3v_1 - (\omega^2/32)\left(g^2 + g_1^2\right)g_2g_3v_2 + (\omega/12)\left(g^2 + g_2^2\right)g_3v_1^2 + (\omega/6)g_1g_2g_3v_1v_2$$

$$I_{210}^{(3,3)} = (\omega^3/240)\left(g^4 + 3g^2g_1^2 + 3g^2g_3^2 + 3g_1^2g_3^2\right)g_2 - (\omega^2/16)\left(g^2 + g_3^2\right)g_1g_2v_1 - (\omega^2/96)\left(g^4 + 3g^2g_1^2 + 3g^2g_3^2 + 3g_1^2g_3^2\right)v_2 + (\omega/12)\left(g^2 + g_3^2\right)g_2v_1^2 + (\omega/6)\left(g^2 + g_3^2\right)g_1v_1v_2$$

$$G_{m,l_1l_2l_3,\alpha\beta}^{(0)} = \frac{3\chi_{\alpha\beta}^2 d_\alpha g_{0,\alpha\beta}}{d_\beta} \int_{\mathbb{R}^6} I_{l_1l_2l_3}^{(m)}\left(\omega_{\alpha\beta}, \mathbf{v}_\alpha, \mathbf{v}_\alpha - \mathbf{v}_\beta\right) f_\alpha(\mathbf{v}_\alpha) f_\beta(\mathbf{v}_\beta) \mathrm{d}\mathbf{v}_\alpha \,\mathrm{d}\mathbf{v}_\beta \tag{6.107}$$

and

$$G_{m,l_1l_2l_3,\alpha\beta}^{(1)} = \frac{3\chi_{\alpha\beta}^2 d_\alpha g_{0,\alpha\beta} c_{\alpha\beta}}{d_\beta E_{\alpha\beta}^{1/2}} \sum_{q=1}^{3} \int_{\mathbb{R}^6} \frac{(U_{\beta,q} - U_{\alpha,q})}{|\mathbf{v}_\alpha - \mathbf{v}_\beta|} I_{l_1l_2l_3}^{(m,q)} \times \left(\omega_{\alpha\beta}, \mathbf{v}_\alpha, \mathbf{v}_\alpha - \mathbf{v}_\beta\right) f_\alpha(\mathbf{v}_\alpha) f_\beta(\mathbf{v}_\beta) \mathrm{d}\mathbf{v}_\alpha \,\mathrm{d}\mathbf{v}_\beta. \tag{6.108}$$

Table 6.12. Examples of $I_{l_1 l_2 l_3}^{(p,q)}(\omega, \mathbf{v}, \mathbf{g})$ terms for third-order moments. Others can be found by permutation of the indices.

$$I_{111}^{(1,1)} = (\omega^3/80)\left(3g^2 + g_1^2\right) g_1 g_2 g_3 - (\omega^2/32)\left(g^2 + g_1^2\right) g_2 g_3 v_1 - (\omega^2/32)\left(3g^2 + g_1^2\right) g_1 g_3 v_2 - (\omega^2/32)\left(3g^2 + g_1^2\right) g_1 g_2 v_3 + (\omega/12)\left(g^2 + g_1^2\right) g_3 v_1 v_2 + (\omega/12)\left(g^2 + g_1^2\right) g_2 v_1 v_3 + (\omega/12)\left(3g^2 + g_1^2\right) g_1 v_2 v_3$$

$$I_{111}^{(1,2)} = (\omega^3/240)\left(g^4 + 3g^2 g_1^2 + 3g^2 g_2^2 + 3g_1^2 g_2^2\right) g_3 - (\omega^2/32)\left(g^2 + g_2^2\right) g_1 g_3 v_1 - (\omega^2/32)\left(g^2 + g_1^2\right) g_2 g_3 v_2 - (\omega^2/96)\left(g^4 + 3g^2 g_1^2 + 3g^2 g_2^2 + 3g_1^2 g_2^2\right) v_3 + (\omega/12)\left(g^2 + g_2^2\right) g_1 v_1 v_3 + (\omega/12)\left(g^2 + g_1^2\right) g_2 v_2 v_3 + (\omega/12) g_1 g_2 g_3 v_1 v_2$$

Note that, using the result in Eq. (6.107), we can rewrite Eq. (6.105) more compactly as

$$C_{l_1 l_2 l_3,\alpha\beta}^{(2)} = \frac{2c_{\alpha\beta}}{d_\alpha E_{\alpha\beta}^{1/2}} (\mathbf{U}_\beta - \mathbf{U}_\alpha) \cdot \mathbf{G}_{\alpha\beta}^{(0)}.$$

In Section 6.4, we will apply these results to find the moment-transport equations for a polydisperse version of Eq. (6.1).

6.3 Kinetic models

The integrals presented in the preceeding sections are rather complicated due to the form of the Boltzmann–Enskog collision integral, and this leads to quite complicated expressions for the collision source terms (especially for higher-order moments). In the literature, considerable work has been directed towards finding simpler *kinetic models* to approximate the Boltzmann collision integral (i.e. for the dilute case wherein the collisional-flux term is neglected). For monodisperse particles the most widely used kinetic model is the so-called BGK model, which was proposed, independently, by Bhatnagar *et al.* (1954) and Welander (1954). The extension of BGK-like models to polydisperse particles is more complicated due to the "cross-collision" terms involving collisions between unlike particles (see McCormack (1973) and Andries *et al.* (2002a) and references cited therein). Indeed, as discussed first by Garzó *et al.* (1989), hardly any of the polydisperse kinetic models obey the "in-differentiability principle" (Andries *et al.*, 2002a), which states that the polydisperse kinetic model should reduce to the monodisperse model when all particles are the same. In this section we will present only the simplest kinetic models for inelastic collisions. The reader can find a more exhaustive description of kinetic models in the literature on kinetic theory (Cercignani, 2000; Struchtrup, 2005).

6.3.1 Monodisperse particles

For elastic collisions, several different kinetic models have been proposed in order to close the Boltzmann hard-sphere collision term (Eq. 6.9). For inelastic collisions ($e < 1$), one must correctly account for the dependence of the dissipation of granular energy on the value of e. One method for accomplishing this task is to start from the exact (unclosed) collision integral in Eq. (6.68). From the definition of $I_{ijk}^{(0)}$ given in Eq. (6.60), it can be shown that the *approximate* collision source term for moment M^{γ}_{ijk}, defined by

$$C^{*\gamma}_{ijk} = \frac{6g_0\langle g\rangle}{d_\mathrm{p}} \int_{\mathbb{R}^6} I^{(0)}_{ijk}(\omega, \mathbf{v}_1, \mathbf{v}_1 - \mathbf{v}_2) f(\mathbf{v}_1) f(\mathbf{v}_2) \mathrm{d}\mathbf{v}_1 \, \mathrm{d}\mathbf{v}_2, \tag{6.109}$$

is closed in terms of the moments up to order γ. This expression differs from Eq. (6.68) because we have replaced $|\mathbf{v}_1 - \mathbf{v}_2|$ with the mean velocity difference defined by

$$\langle g\rangle = \frac{1}{\alpha_\mathrm{p}^2} \int_{\mathbb{R}^6} |\mathbf{v}_1 - \mathbf{v}_2| f(\mathbf{v}_1) f(\mathbf{v}_2) \mathrm{d}\mathbf{v}_1 \, \mathrm{d}\mathbf{v}_2. \tag{6.110}$$

At equilibrium, where the velocity distribution is Maxwellian, it is straightforward to show that $\langle g\rangle = 4\sqrt{\Theta_\mathrm{p}/\pi}$, where Θ_p is the granular temperature. We should note that Eq. (6.109) corresponds to an *inelastic* Maxwell particle (Maxwell, 1879), and, most importantly, it still contains the exact dependence on $\omega = (1 + e)/2$. We will therefore refer to this kinetic model as the *inelastic* Maxwell collision model.

The BGK-like kinetic models for monodisperse particles have the linearized form

$$\mathbb{C} = \frac{1}{\tau_\mathrm{col}}(f^* - f), \tag{6.111}$$

where (repeated Roman indices imply summation) the "equilibrium" distribution function is modeled as

$$f^*(\mathbf{v}) = \frac{\alpha_\mathrm{p}}{[\det(2\pi\boldsymbol{\lambda})]^{1/2}} \exp\left(-\frac{1}{2}\left(v_i - U_{\mathrm{p}i}\right)\left(\boldsymbol{\lambda}^{-1}\right)_{ij}\left(v_j - U_{\mathrm{p}j}\right)\right). \tag{6.112}$$

In this multivariate Gaussian distribution, $\mathbf{U}_\mathrm{p}$ is the mean particle velocity and $\boldsymbol{\lambda}^{-1}$ is the inverse of the second-order tensor $\boldsymbol{\lambda}$, defined such that a given set of velocity moments agrees with Eq. (6.109). Owing to conservation of mass and mean momentum, the first nonzero terms from Eq. (6.109) correspond to the second-order moments. On letting $\boldsymbol{\kappa}$ denote the symmetric second-order tensor constructed from $I^{(0)}_{ijk}(\omega, \mathbf{v}_1, \mathbf{g})$ with $i + j + k = 2$ (see Table 6.1),

$$\boldsymbol{\kappa}(\mathbf{v}_1, \mathbf{g}) = \frac{\omega}{2}\left(\frac{\omega g^2}{6}\mathbf{I} + \frac{\omega}{2}\mathbf{g}\otimes\mathbf{g} - \mathbf{g}\otimes\mathbf{v}_1 - \mathbf{v}_1\otimes\mathbf{g}\right), \tag{6.113}$$

the collision term for the second-order moments found from Eq. (6.109) is given by

$$C^{*2} = \frac{6g_0\langle g\rangle}{d_\mathrm{p}} \int_{\mathbb{R}^6} \boldsymbol{\kappa}(\mathbf{v}_1, \mathbf{v}_1 - \mathbf{v}_2) f(\mathbf{v}_1) f(\mathbf{v}_2) \mathrm{d}\mathbf{v}_1 \, \mathrm{d}\mathbf{v}_2. \tag{6.114}$$

The integral on the right-hand side of this expression can be evaluated explicitly:

$$C^{*2} = \frac{6g_0\langle g\rangle\omega\alpha_\mathrm{p}^2}{d_\mathrm{p}}\left(\frac{\omega\Theta_\mathrm{p}}{2}\mathbf{I} + \frac{\omega}{2}\boldsymbol{\sigma}_\mathrm{p} - \boldsymbol{\sigma}_\mathrm{p}\right), \tag{6.115}$$

where $\boldsymbol{\sigma}_\mathrm{p}$ is the velocity-covariance tensor and $\Theta_\mathrm{p} = \mathrm{tr}(\boldsymbol{\sigma}_\mathrm{p})/3$. Using $\omega(\omega - 1) = (e^2 - 1)/4$, it is straightforward to show that for the inelastic Maxwell collision model the homogeneous granular temperature obeys

$$\frac{\mathrm{d}\Theta_\mathrm{p}}{\mathrm{d}t} = -\frac{3g_0\langle g\rangle\alpha_\mathrm{p}(1 - e^2)}{2d_\mathrm{p}}\Theta_\mathrm{p}, \tag{6.116}$$

where $\langle g\rangle \approx 4\sqrt{\Theta_\mathrm{p}/\pi}$.

The collision term for the second-order moments found from Eq. (6.111) is

$$C^2 = \frac{\alpha_p}{\tau_{col}} \left(\lambda - \sigma_p \right). \tag{6.117}$$

Equating this expression to Eq. (6.115) and solving for the unknown model parameters, we find that

$$\tau_{col} = \frac{\zeta d_p}{3 g_0 \langle g \rangle \alpha_p} \tag{6.118}$$

and

$$\lambda = \zeta \omega^2 \Theta_p \mathbf{I} + \left(\zeta \omega^2 - 2\zeta\omega + 1 \right) \sigma_p, \tag{6.119}$$

where $0 < \zeta \leq 3/2$ is a model constant. In the elastic limit ($\omega = 1$), Eq. (6.119) corresponds to the ES-BGK model with the Prandtl number given by $\mathrm{Pr} = 1/\zeta$, and to the BGK model when $\zeta = 1$. For inelastic collisions, we will refer to the *inelastic* BGK and ES-BGK models as the linearized model given in Eq. (6.111) with matrix Eq. (6.119) using the values of ζ from the elastic limit. The only remaining unclosed term is $\langle g \rangle$. At equilibrium in the elastic limit, $\mathbf{g}$ will be Gaussian with zero mean and covariance matrix $2\Theta_p \mathbf{I}$. This fact leads to the approximation $\langle g \rangle = 4\sqrt{\Theta_p/\pi}$, and thus we can define the collision time to be $\tau_{col} = \zeta \sqrt{\pi} d_p / \left(12 g_0 \alpha_p \sqrt{\Theta_p} \right)$.

The utility of the kinetic model is most evident when evaluating the collision source terms for the moments. As noted earlier, Eq. (6.109) is closed but results in complicated polynomial expressions for higher-order moments. In contrast, the kinetic model yields

$$C_{ijk}^{*\gamma} = \frac{1}{\tau_{col}} \left(M_{ijk}^{*\gamma} - M_{ijk}^{\gamma} \right), \tag{6.120}$$

where the moments $M_{ijk}^{*\gamma}$ can be computed analytically using the definition of f^* in Eq. (6.112):

$$M_{ijk}^{*\gamma} = \int_{\mathbb{R}^3} v_1^i v_2^j v_3^k f^*(\mathbf{v}) \mathrm{d}\mathbf{v}. \tag{6.121}$$

Because f^* is a multivariate Gaussian distribution its higher-order moments can easily be computed (e.g. using the moment-generating function). However, the reader should keep in mind that the kinetic model ensures only that the moments up to second order are the same as with Eq. (6.109). Third- and higher-order moments may therefore be poorly approximated when the true velocity-distribution function is far from equilibrium.

6.3.2 Polydisperse particles

As noted earlier, there have been many attempts to extend BGK-like kinetic models to multicomponent mixtures (e.g. Andries *et al.*, 2002a; Garzó *et al.*, 1989; Goldman & Sirovich, 1967; Gross & Krook, 1956; Holway, 1973; Kosuge, 2009; McCormack, 1973; Sirovich, 1962). However, only the model of Andries *et al.* (2002a) satisfies the key *mathematical* properties of the multicomponent Boltzmann collision term (i.e. nonnegative distribution function, in-differentiability principle, mixture equilibrium, H-theorem). In general, none of the kinetic models yields results comparable to the Boltzmann collision term in all cases (Andries *et al.*, 2002b; Kosuge, 2009; Struchtrup, 2005). Nevertheless, because they are much simpler to use in practice, they may provide a useful approximation for certain applications.

For inelastic collisions, the coefficient of restitution will appear explicitly in the kinetic model as seen above in the monodisperse case. We will consider a binary case with inelastic collisions for simplicity. However, the kinetic model will be written such that it can be easily extended to more than two particle types. We will express the model in terms of the moment source terms due to collisions:

$$\begin{aligned} C^{*\gamma}_{ijk,1} &= \kappa_{11}\left(\alpha_1 G_{ijk,11} - M^{\gamma}_{ijk,1}\right) + \kappa_{12}\left(\alpha_1 G_{ijk,12} - M^{\gamma}_{ijk,1}\right), \\ C^{*\gamma}_{ijk,2} &= \kappa_{21}\left(\alpha_2 G_{ijk,21} - M^{\gamma}_{ijk,2}\right) + \kappa_{22}\left(\alpha_2 G_{ijk,22} - M^{\gamma}_{ijk,2}\right), \end{aligned} \tag{6.122}$$

where $C^{*\gamma}_{ijk,1}$ ($C^{*\gamma}_{ijk,2}$) are the collision source terms for type-1 (type-2) particles corresponding to moments $M^{\gamma}_{ijk,1}$ ($M^{\gamma}_{ijk,2}$). The parameters in the model are chosen to give the correct behavior of moments up to second order for Maxwell molecules (as was done for the monodisperse case). The term $G_{ijk,\alpha\beta}$ denotes Gaussian moments with mean velocity $\mathbf{U}_{\alpha\beta}$ and covariance matrix $\boldsymbol{\sigma}_{\alpha\beta}$ (defined below).

The collision rate parameters $\kappa_{\alpha\beta}$ are nonnegative and determine the rates of relaxation towards the Gaussian distributions:

$$\begin{aligned} \kappa_{11} &= \frac{24 g_{0,11}\alpha_1 \sqrt{\Theta_1}}{\sqrt{\pi} d_1}, & \kappa_{21} &= \frac{24 g_{0,12}\alpha_1 \chi_{21}^3 \sqrt{\Theta_{12}}}{\sqrt{\pi} d_{12}}, \\ \kappa_{12} &= \frac{24 g_{0,12}\alpha_2 \chi_{12}^3 \sqrt{\Theta_{12}}}{\sqrt{\pi} d_{12}}, & \kappa_{22} &= \frac{24 g_{0,22}\alpha_2 \sqrt{\Theta_2}}{\sqrt{\pi} d_2}, \end{aligned} \tag{6.123}$$

where $\Theta_{12} = E_{12}/3$, $d_{12} = (d_1 + d_2)/2$, and $\chi_{\alpha\beta} = d_{\alpha\beta}/d_\beta$. The mean velocities in the Gaussian distributions are defined as

$$\begin{aligned} \mathbf{U}_{11} &= \mathbf{U}_1, \quad \mathbf{U}_{12} = \mathbf{U}_1 + \frac{1}{4}(1 + e_{12})\mu_{12}(\mathbf{U}_2 - \mathbf{U}_1), \\ \mathbf{U}_{21} &= \mathbf{U}_2 + \frac{1}{4}(1 + e_{12})\mu_{21}(\mathbf{U}_1 - \mathbf{U}_2), \quad \mathbf{U}_{22} = \mathbf{U}_2, \end{aligned} \tag{6.124}$$

where

$$\mu_{\alpha\beta} = \frac{2 m_\beta}{m_\alpha + m_\beta}. \tag{6.125}$$

The covariance matrices can be expressed in the form

$$\begin{aligned} \boldsymbol{\sigma}_{11} &= \boldsymbol{\sigma}_1 + \frac{1}{2}(1 + e_{11})\left[\frac{1}{4}(1 + e_{11})\mathbf{S}_1 - \boldsymbol{\sigma}_1\right], \\ \boldsymbol{\sigma}_{12} &= \boldsymbol{\sigma}_1 + \frac{1}{2}(1 + e_{12})\mu_{12}\left[\frac{1}{4}(1 + e_{12})\mu_{12}\mathbf{S}_{12} - \boldsymbol{\sigma}_1\right], \\ \boldsymbol{\sigma}_{21} &= \boldsymbol{\sigma}_2 + \frac{1}{2}(1 + e_{12})\mu_{21}\left[\frac{1}{4}(1 + e_{12})\mu_{21}\mathbf{S}_{12} - \boldsymbol{\sigma}_2\right], \\ \boldsymbol{\sigma}_{22} &= \boldsymbol{\sigma}_2 + \frac{1}{2}(1 + e_{22})\left[\frac{1}{4}(1 + e_{22})\mathbf{S}_2 - \boldsymbol{\sigma}_2\right], \end{aligned} \tag{6.126}$$

where $\boldsymbol{\sigma}_\alpha$ is the covariance matrix for type-α particles, and the other matrices appearing on the right-hand side are defined by

$$\mathbf{S}_1 = \boldsymbol{\sigma}_1 + \Theta_1 \mathbf{I}, \quad \mathbf{S}_{12} = \frac{1}{2}(\boldsymbol{\sigma}_1 + \boldsymbol{\sigma}_2 + \Theta_{12}\mathbf{I}), \quad \mathbf{S}_2 = \boldsymbol{\sigma}_2 + \Theta_2 \mathbf{I}. \tag{6.127}$$

Note that, in order for the Gaussian distributions to be well defined, all of the covariance matrices in Eq. (6.126) must be nonnegative. This leads to the condition that $4 \geq (1 + e_{12})\mu_{21} \geq 0$, or, equivalently, that $2 \geq \mu_{21} \geq 0$, which is always true.

The polydisperse kinetic model can be compared with the monodisperse case by setting $\alpha_2 = 0$. It is straightforward to show that

$$\sigma_{11} = \frac{1}{2}\omega_{11}^2 \Theta_1 \mathbf{I} + \left(\frac{1}{2}\omega_{11}^2 - \omega_{11} + 1 \right) \sigma_1, \tag{6.128}$$

which corresponds to $\zeta = 1/2$ in Eq. (6.119). The relaxation rate in the monodisperse kinetic model is then

$$\frac{1}{\tau_{\text{col}}} = \frac{24 g_0 \alpha_{\text{p}} \sqrt{\Theta_{\text{p}}}}{\sqrt{\pi} d_{\text{p}}},$$

which agrees with κ_{11}. The polydisperse kinetic model thus reduces to a monodisperse ES-BGK model (Holway, 1966, 1973) when only one particle type is present.

In summary, we remind the reader that kinetic models offer a simplified description of the hard-sphere collision term that is exact up to second-order moments. At present, kinetic models apply only to the dilute regime where the collisional-flux terms are negligible. Insofar as the authors are aware, there has been no attempt to construct kinetic models for the collisional-flux terms, which are important for moderately dense flows.

6.4 Moment-transport equations

We will now look at the integer moment-transport equations for two systems. The first corresponds to a monodisperse case with kinetic equation given by Eq. (6.1). The second is a bidisperse system with kinetic equations

$$\begin{aligned} \partial_t f_1 + \mathbf{v}_1 \cdot \partial_{\mathbf{x}} f_1 &= \mathbb{C}_{11} + \mathbb{C}_{12}, \\ \partial_t f_2 + \mathbf{v}_2 \cdot \partial_{\mathbf{x}} f_2 &= \mathbb{C}_{21} + \mathbb{C}_{22}. \end{aligned} \tag{6.129}$$

The velocity number-density functions $f_1(t, \mathbf{x}, \mathbf{v}_1)$ and $f_2(t, \mathbf{x}, \mathbf{v}_2)$ are defined such that

$$\alpha_1 = \int_{\mathbb{R}^3} f_1 \, \mathrm{d}\mathbf{v}_1 \quad \text{and} \quad \alpha_2 = \int_{\mathbb{R}^3} f_2 \, \mathrm{d}\mathbf{v}_2,$$

where α_1 and α_2 are the disperse-phase volume fractions, and

$$\alpha_1 \mathbf{U}_1 = \int_{\mathbb{R}^3} \mathbf{v}_1 f_1 \, \mathrm{d}\mathbf{v}_1 \quad \text{and} \quad \alpha_2 \mathbf{U}_2 = \int_{\mathbb{R}^3} \mathbf{v}_2 f_2 \, \mathrm{d}\mathbf{v}_2,$$

where $\mathbf{U}_1$ and $\mathbf{U}_2$ are the mean disperse-phase velocities. Note that conservation of mass during collisions implies that

$$\int_{\mathbb{R}^3} \mathbb{C}_{\alpha\beta} \, \mathrm{d}\mathbf{v} = 0 \quad \text{for } \alpha, \beta = 1, 2,$$

while conservation of momentum implies that

$$\int_{\mathbb{R}^3} \mathbf{v} \mathbb{C}_{\alpha\alpha} \, \mathrm{d}\mathbf{v} = 0 \quad \text{for } \alpha = 1, 2,$$

and

$$\rho_1 \int_{\mathbb{R}^3} \mathbf{v}_1 \mathbb{C}_{12} \, \mathrm{d}\mathbf{v}_1 + \rho_2 \int_{\mathbb{R}^3} \mathbf{v}_2 \mathbb{C}_{21} \, \mathrm{d}\mathbf{v}_2 = 0,$$

where ρ_1 and ρ_2 are the particle material densities (i.e. the particle mass divided by the particle volume).

The integer moments of the velocity-density function will be defined as

$$M^{\gamma}_{l_1 l_2 l_3,\alpha} \equiv \int_{\mathbb{R}^3} v_1^{l_1} v_2^{l_2} v_3^{l_3} f_\alpha(\mathbf{v}) \mathrm{d}\mathbf{v} \quad \text{for } \alpha = 1, 2, \tag{6.130}$$

where $\gamma = l_1 + l_2 + l_3$ is the order of the moment. In general, the moments up to second order are associated with the "hydrodynamic" variables. For example,

$$\begin{gathered} M^0_{000,1} = \alpha_1, \\ M^1_{100,1} = \alpha_1 U_{1,1}, \quad M^1_{010,1} = \alpha_1 U_{2,1}, \quad M^1_{001,1} = \alpha_1 U_{3,1}, \\ M^2_{200,1} + M^2_{020,1} + M^2_{002,1} = 3\alpha_1 \Theta_1 + \alpha_1 \mathbf{U}_1 \cdot \mathbf{U}_1, \end{gathered}$$

where Θ_α is the granular temperature of particle type α. We will therefore look most closely at the moment-transport equations up to $\gamma = 2$. For the monodisperse case, we denote the hydrodynamic variables as α_p, $\mathbf{U}_\mathrm{p}$, and Θ_p, respectively, and the moments as $M^{\gamma}_{l_1 l_2 l_3}$.

6.4.1 Monodisperse particles

For a monodisperse system the moment-transport equation derived from Eq. (6.1) is

$$\begin{aligned} \frac{\partial M^{\gamma}_{l_1 l_2 l_3}}{\partial t} + \frac{\partial}{\partial x_1}\left(M^{\gamma+1}_{l_1+1\, l_2 l_3} + G_{1,l_1 l_2 l_3}\right) + \frac{\partial}{\partial x_2}\left(M^{\gamma+1}_{l_1 l_2+1\, l_3} + G_{2,l_1 l_2 l_3}\right) + \frac{\partial}{\partial x_3}\left(M^{\gamma+1}_{l_1 l_2 l_3+1} + G_{3,l_1 l_2 l_3}\right) \\ = C^{(0)}_{l_1 l_2 l_3} + C^{(1)}_{l_1 l_2 l_3}. \end{aligned} \tag{6.131}$$

On the left-hand side of this equation, the collisional flux, defined by Eq. (6.70), appears. On the right-hand side, the collision source terms, defined by Eqs. (6.68) and (6.69), appear. The kinetic fluxes (or free transport) correspond to the moments $M^{\gamma+1}$.

The zeroth-order moment

For the zeroth-order moment, Eq. (6.131) reduces to

$$\frac{\partial M^0_{000}}{\partial t} + \frac{\partial}{\partial x_1}\left(M^1_{1\,00}\right) + \frac{\partial}{\partial x_2}\left(M^1_{01\,0}\right) + \frac{\partial}{\partial x_3}\left(M^1_{001}\right) = 0, \tag{6.132}$$

which can be rewritten in terms of the hydrodynamic variables as a continuity equation:

$$\frac{\partial \alpha_\mathrm{p}}{\partial t} + \nabla \cdot \left(\alpha_\mathrm{p} \mathbf{U}_\mathrm{p}\right) = 0. \tag{6.133}$$

Note that this equation is closed in terms of the hydrodynamic variables, and corresponds to a compressible granular gas where α_p plays the role of the density.

First-order moments

For the first-order moments, Eq. (6.131) reduces to

$$\begin{aligned}
&\frac{\partial M^1_{100}}{\partial t} + \frac{\partial}{\partial x_1}\left(M^2_{200} + G_{1,100}\right) + \frac{\partial}{\partial x_2}\left(M^2_{110} + G_{2,100}\right) + \frac{\partial}{\partial x_3}\left(M^2_{101} + G_{3,100}\right) = 0,\\
&\frac{\partial M^1_{010}}{\partial t} + \frac{\partial}{\partial x_1}\left(M^2_{110} + G_{1,010}\right) + \frac{\partial}{\partial x_2}\left(M^2_{020} + G_{2,010}\right) + \frac{\partial}{\partial x_3}\left(M^2_{011} + G_{3,010}\right) = 0,\\
&\frac{\partial M^1_{001}}{\partial t} + \frac{\partial}{\partial x_1}\left(M^2_{101} + G_{1,001}\right) + \frac{\partial}{\partial x_2}\left(M^2_{011} + G_{2,001}\right) + \frac{\partial}{\partial x_3}\left(M^2_{002} + G_{3,001}\right) = 0,
\end{aligned} \tag{6.134}$$

which can be rewritten in terms of the hydrodynamic variables as a mean momentum balance:

$$\frac{\partial \alpha_p \mathbf{U}_p}{\partial t} + \nabla \cdot \left(\alpha_p \mathbf{U}_p \otimes \mathbf{U}_p + \mathbf{P}\right) = 0. \tag{6.135}$$

The symmetric second-order pressure tensor $\mathbf{P}$ is defined by its components:

$$\begin{aligned}
P_{11} &\equiv \left(M^2_{200} - \alpha_p U_{p,1} U_{p,1}\right) + G_{1,100},\\
P_{12} &\equiv \left(M^2_{110} - \alpha_p U_{p,1} U_{p,2}\right) + G_{2,100},\\
P_{13} &\equiv \left(M^2_{101} - \alpha_p U_{p,1} U_{p,3}\right) + G_{3,100},\\
P_{22} &\equiv \left(M^2_{020} - \alpha_p U_{p,2} U_{p,2}\right) + G_{2,020},\\
P_{23} &\equiv \left(M^2_{011} - \alpha_p U_{p,2} U_{p,3}\right) + G_{2,011},\\
P_{33} &\equiv \left(M^2_{002} - \alpha_p U_{p,3} U_{p,3}\right) + G_{3,002}.
\end{aligned} \tag{6.136}$$

The terms in parentheses on the right-hand side can be expressed in terms of the velocity covariance matrix $\boldsymbol{\sigma}_p$ defined by

$$\alpha_p \boldsymbol{\sigma}_p \equiv \int_{\mathbb{R}^3} (\mathbf{v}_1 - \mathbf{U}_p) \otimes (\mathbf{v}_1 - \mathbf{U}_p) f(\mathbf{v}_1) \mathrm{d}\mathbf{v}_1. \tag{6.137}$$

Using the definition of the collisional fluxes in Eq. (6.70) and the results in Table 6.4, we can rewrite $\mathbf{P}$ as

$$\mathbf{P} \equiv \alpha_p \boldsymbol{\sigma}_p + \tfrac{2}{5}(1+e) g_0 \int_{\mathbb{R}^6} \left[\tfrac{1}{2}|\mathbf{v}_1 - \mathbf{v}_2|^2 \mathbf{I} + (\mathbf{v}_1 - \mathbf{v}_2) \otimes (\mathbf{v}_1 - \mathbf{v}_2)\right] f(\mathbf{v}_1) f(\mathbf{v}_2) \mathrm{d}\mathbf{v}_1\, \mathrm{d}\mathbf{v}_2. \tag{6.138}$$

Using the identities

$$\begin{aligned}
(\mathbf{v}_1 - \mathbf{v}_2) \otimes (\mathbf{v}_1 - \mathbf{v}_2) = {} & (\mathbf{v}_1 - \mathbf{U}_p) \otimes (\mathbf{v}_1 - \mathbf{U}_p) + (\mathbf{v}_2 - \mathbf{U}_p) \otimes (\mathbf{v}_2 - \mathbf{U}_p)\\
& - (\mathbf{v}_1 - \mathbf{U}_p) \otimes (\mathbf{v}_2 - \mathbf{U}_p) - (\mathbf{v}_2 - \mathbf{U}_p) \otimes (\mathbf{v}_1 - \mathbf{U}_p)
\end{aligned}$$

and

$$|\mathbf{v}_1 - \mathbf{v}_2|^2 = (\mathbf{v}_1 - \mathbf{U}_p) \cdot (\mathbf{v}_1 - \mathbf{U}_p) + (\mathbf{v}_2 - \mathbf{U}_p) \cdot (\mathbf{v}_2 - \mathbf{U}_p) - 2(\mathbf{v}_1 - \mathbf{U}_p) \cdot (\mathbf{v}_2 - \mathbf{U}_p),$$

and the fact that

$$\int_{\mathbb{R}^6} (\mathbf{v}_1 - \mathbf{U}_p) \otimes (\mathbf{v}_2 - \mathbf{U}_p) f(\mathbf{v}_1) f(\mathbf{v}_2) \mathrm{d}\mathbf{v}_1\, \mathrm{d}\mathbf{v}_2 = \mathbf{0},$$

we can evaluate the integral in Eq. (6.138) to find

$$\mathbf{P} \equiv \alpha_p \boldsymbol{\sigma}_p + \tfrac{2}{5}\alpha_p^2(1+e)g_0\left(3\Theta_p\mathbf{I} + 2\boldsymbol{\sigma}_p\right), \tag{6.139}$$

where the granular temperature is defined by $\Theta_p = \frac{1}{3}\left(\sigma_{p,11} + \sigma_{p,22} + \sigma_{p,33}\right)$. Note that Eq. (6.139) depends only on the zeroth- and second-order moments.

It is common to rewrite $\mathbf{P}$ in terms of the decomposition $\mathbf{P} = \boldsymbol{\tau}_p + p\mathbf{I}$, which defines the granular pressure

$$p = \alpha_p\Theta_p\left[1 + 2\alpha_p(1+e)g_0\right], \tag{6.140}$$

and a traceless granular stress tensor:

$$\boldsymbol{\tau}_p \equiv \alpha_p\left[1 + \tfrac{4}{5}\alpha_p(1+e)g_0\right]\left(\boldsymbol{\sigma}_p - \Theta_p\mathbf{I}\right). \tag{6.141}$$

From Eq. (6.140) we can note that the granular "equation of state" is a highly nonlinear function of α_p due to the presence of g_0. In dense granular flows, this "non-ideal-gas" behavior makes the mean momentum equations in Eq. (6.135) very stiff and tightly coupled to the continuity equation in Eq. (6.133). We can also note that in the dilute regime ($\alpha_p \ll 1$) the terms involving g_0 in Eqs. (6.140) and (6.141) will be negligible. Finally, it is noteworthy that these expressions for p and $\boldsymbol{\tau}_p$ are exact in the sense that they follow directly from the hard-sphere collision model without any simplifications. However, in order to evaluate $\boldsymbol{\sigma}_p$, it is necessary to consider the six second-order moments. In hydrodynamic models, only the sum of the diagonal second-order moments is used, and the off-diagonal moments appearing in $\boldsymbol{\tau}_p$ are closed using a gradient-diffusion model in the Navier–Stokes limit.[14] In the Euler limit, $\boldsymbol{\tau}_p$ is null and the contribution due to g_0 in Eq. (6.140) is neglected so that the granular pressure behaves like an ideal gas.

---

[14] In the limit of small Knudsen number, the Chapman–Enskog expansion (Chapman, 1916; Enksog, 1921) of the elastic Boltzmann equation yields a first-order term for $\boldsymbol{\sigma}_p$ of the form

$$\sigma_{p,ij} - \Theta_p\delta_{ij} \propto -\Theta_p\tau_{col}\left(\frac{\partial U_{p,j}}{\partial x_i} + \frac{\partial U_{p,i}}{\partial x_j} - \frac{2}{3}\nabla\cdot\mathbf{U}_p\delta_{ij}\right),$$

where $\tau_{col} = \sqrt{\pi}d_p/\left(12\alpha_p\sqrt{\Theta_p}\right)$ is the collision time. In this limit, the granular stress tensor has the form

$$\boldsymbol{\tau}_p = -\mu_p\left[\nabla\mathbf{U}_p + \left(\nabla\mathbf{U}_p\right)^T - \frac{2}{3}\nabla\cdot\mathbf{U}_p\mathbf{I}\right]$$

with the granular viscosity given by

$$\mu_p \propto \frac{\sqrt{\pi\Theta_p}d_p}{12}.$$

A similar expansion can be carried out for the inelastic Boltzmann–Enskog equation (Garzó *et al.*, 2007; Jenkins & Richman, 1985) to find an approximation for $\boldsymbol{\tau}_p$ that will be valid when τ_{col} is much smaller than all other characteristic times of the system.

Second-order moments

For the second-order moments, Eq. (6.131) reduces to

$$\begin{aligned}
&\frac{\partial M^2_{200}}{\partial t} + \frac{\partial}{\partial x_1}\left(M^3_{300} + G_{1,200}\right) + \frac{\partial}{\partial x_2}\left(M^3_{210} + G_{2,200}\right) + \frac{\partial}{\partial x_3}\left(M^2_{201} + G_{3,200}\right) = C^{(0)}_{200} + C^{(1)}_{200},\\
&\frac{\partial M^2_{110}}{\partial t} + \frac{\partial}{\partial x_1}\left(M^3_{210} + G_{1,110}\right) + \frac{\partial}{\partial x_2}\left(M^3_{120} + G_{2,110}\right) + \frac{\partial}{\partial x_3}\left(M^3_{111} + G_{3,110}\right) = C^{(0)}_{110} + C^{(1)}_{110},\\
&\frac{\partial M^2_{101}}{\partial t} + \frac{\partial}{\partial x_1}\left(M^3_{201} + G_{1,101}\right) + \frac{\partial}{\partial x_2}\left(M^3_{111} + G_{2,101}\right) + \frac{\partial}{\partial x_3}\left(M^3_{102} + G_{3,101}\right) = C^{(0)}_{101} + C^{(1)}_{101},\\
&\frac{\partial M^2_{020}}{\partial t} + \frac{\partial}{\partial x_1}\left(M^3_{120} + G_{1,020}\right) + \frac{\partial}{\partial x_2}\left(M^3_{030} + G_{2,020}\right) + \frac{\partial}{\partial x_3}\left(M^3_{021} + G_{3,020}\right) = C^{(0)}_{020} + C^{(1)}_{020},\\
&\frac{\partial M^2_{011}}{\partial t} + \frac{\partial}{\partial x_1}\left(M^3_{111} + G_{1,011}\right) + \frac{\partial}{\partial x_2}\left(M^3_{021} + G_{2,011}\right) + \frac{\partial}{\partial x_3}\left(M^3_{012} + G_{3,011}\right) = C^{(0)}_{011} + C^{(1)}_{011},\\
&\frac{\partial M^2_{002}}{\partial t} + \frac{\partial}{\partial x_1}\left(M^3_{102} + G_{1,002}\right) + \frac{\partial}{\partial x_2}\left(M^3_{012} + G_{2,002}\right) + \frac{\partial}{\partial x_3}\left(M^3_{003} + G_{3,002}\right) = C^{(0)}_{002} + C^{(1)}_{002}.
\end{aligned} \tag{6.142}$$

The collisional-flux and collision source terms can be evaluated using their definitions in Section 6.1.5. The terms $\mathbf{G}_{200}$, $\mathbf{G}_{110}$, etc. depend on the moments of up to third order (and are closed in terms of these moments). In hydrodynamic models, the third-order moment terms appearing in $\mathbf{G}_{200}$, $\mathbf{G}_{110}$, etc. are closed using gradient-diffusion models involving spatial gradients of second-order central moments. The terms $C^{(1)}_{200}$, $C^{(1)}_{110}$, etc. also depend on the moments of up to third order and their spatial gradients, and hence are closed at third order.

The terms $C^{(0)}_{200}$, $C^{(0)}_{110}$, etc. are not closed in terms of any finite set of moments. Nevertheless, for the diagonal second-order moments we have

$$\begin{aligned}
C^{(0)}_{200} &= \frac{3\omega g_0}{d_p}\int_{\mathbb{R}^6}\left[\frac{\omega}{6}\left(|\mathbf{v}_1 - \mathbf{v}_2|^2 + (v_{1,1} - v_{2,1})^2\right)\right.\\
&\qquad \left. - 2(v_{1,1} - v_{2,1})v_{1,1}\right]|\mathbf{v}_1 - \mathbf{v}_2| f(\mathbf{v}_1) f(\mathbf{v}_2) \mathrm{d}\mathbf{v}_1\, \mathrm{d}\mathbf{v}_2,\\
C^{(0)}_{020} &= \frac{3\omega g_0}{d_p}\int_{\mathbb{R}^6}\left[\frac{\omega}{6}\left(|\mathbf{v}_1 - \mathbf{v}_2|^2 + (v_{1,2} - v_{2,2})^2\right)\right.\\
&\qquad \left. - 2(v_{1,2} - v_{2,2})v_{1,2}\right]|\mathbf{v}_1 - \mathbf{v}_2| f(\mathbf{v}_1) f(\mathbf{v}_2) \mathrm{d}\mathbf{v}_1\, \mathrm{d}\mathbf{v}_2,\\
C^{(0)}_{002} &= \frac{3\omega g_0}{d_p}\int_{\mathbb{R}^6}\left[\frac{\omega}{6}\left(|\mathbf{v}_1 - \mathbf{v}_2|^2 + (v_{1,3} - v_{2,3})^2\right)\right.\\
&\qquad \left. - 2(v_{1,3} - v_{2,3})v_{1,3}\right]|\mathbf{v}_1 - \mathbf{v}_2| f(\mathbf{v}_1) f(\mathbf{v}_2) \mathrm{d}\mathbf{v}_1\, \mathrm{d}\mathbf{v}_2.
\end{aligned} \tag{6.143}$$

Summing together the three expressions in Eq. (6.143) and dividing by three gives the collision source term for the granular temperature:

$$C^{(0)}(\Theta_p) = \frac{\omega g_0}{d_p}\int_{\mathbb{R}^6}\left[\omega|\mathbf{v}_1 - \mathbf{v}_2|^2 - 2(\mathbf{v}_1 - \mathbf{v}_2)\cdot\mathbf{v}_1\right]|\mathbf{v}_1 - \mathbf{v}_2| f(\mathbf{v}_1) f(\mathbf{v}_2) \mathrm{d}\mathbf{v}_1\, \mathrm{d}\mathbf{v}_2. \tag{6.144}$$

By interchanging the integration variables $\mathbf{v}_1$ and $\mathbf{v}_2$, we can find an equivalent expression:

$$C^{(0)}(\Theta_p) = \frac{\omega g_0}{d_p}\int_{\mathbb{R}^6}\left[\omega|\mathbf{v}_1 - \mathbf{v}_2|^2 + 2(\mathbf{v}_1 - \mathbf{v}_2)\cdot\mathbf{v}_2\right]|\mathbf{v}_1 - \mathbf{v}_2| f(\mathbf{v}_1) f(\mathbf{v}_2) \mathrm{d}\mathbf{v}_1\, \mathrm{d}\mathbf{v}_2. \tag{6.145}$$

Summing together Eqs. (6.144) and (6.145) and dividing the result by two then yields

$$C^{(0)}(\Theta_{\mathrm{p}}) = -\frac{(1-e^2)g_0}{4d_{\mathrm{p}}} \int_{\mathbb{R}^6} |\mathbf{v}_1 - \mathbf{v}_2|^3 f(\mathbf{v}_1) f(\mathbf{v}_2) \mathrm{d}\mathbf{v}_1 \, \mathrm{d}\mathbf{v}_2, \tag{6.146}$$

where we have used the fact that $\omega(\omega-1) = -(1-e^2)/4$. Because the result in Eq. (6.146) is always less than or equal to zero, we can observe that inelastic collisions always decrease the granular temperature. When f is Maxwellian, the integral in Eq. (6.146) is proportional to $\alpha_{\mathrm{p}}^2 \Theta_{\mathrm{p}}^{3/2}$.

In summary, we should reiterate that, in a hydrodynamic model for granular flow, only the transport equations for the sum of the diagonal elements in Eq. (6.142) are used (i.e. a transport equation for the granular temperature) in combination with the continuity (Eq. 6.133) and mean momentum equations (Eq. 6.135). The unclosed terms in the resulting equations are closed using gradient-diffusion models that are valid in the limit where f is nearly Maxwellian. In quadrature-based moment methods, all of the moments in Eq. (6.142) are used, and closure is achieved by reconstructing f from the values of its moments. In principle, since the reconstruction procedure does not rely on f being close to a Maxwellian, the quadrature-based moment method will be applicable to a wider range of flow regimes, including the case in which collisions are negligible.

6.4.2 Polydisperse particles

For a bidisperse[15] system the moment-transport equations derived from Eq. (6.129) are

$$\begin{aligned} \frac{\partial M^{\gamma}_{l_1 l_2 l_3,1}}{\partial t} &+ \frac{\partial}{\partial x_1}\left(M^{\gamma+1}_{l_1+1\, l_2 l_3,1} + G_{1,l_1 l_2 l_3,11} + G_{1,l_1 l_2 l_3,12}\right) \\ &+ \frac{\partial}{\partial x_2}\left(M^{\gamma+1}_{l_1 l_2+1\, l_3,1} + G_{2,l_1 l_2 l_3,11} + G_{2,l_1 l_2 l_3,12}\right) \\ &+ \frac{\partial}{\partial x_3}\left(M^{\gamma+1}_{l_1 l_2 l_3+1,1} + G_{3,l_1 l_2 l_3,11} + G_{3,l_1 l_2 l_3,12}\right) = C_{l_1 l_2 l_3,11} + C_{l_1 l_2 l_3,12} \end{aligned} \tag{6.147}$$

and

$$\begin{aligned} \frac{\partial M^{\gamma}_{l_1 l_2 l_3,2}}{\partial t} &+ \frac{\partial}{\partial x_1}\left(M^{\gamma+1}_{l_1+1\, l_2 l_3,2} + G_{1,l_1 l_2 l_3,21} + G_{1,l_1 l_2 l_3,22}\right) \\ &+ \frac{\partial}{\partial x_2}\left(M^{\gamma+1}_{l_1 l_2+1\, l_3,1} + G_{2,l_1 l_2 l_3,21} + G_{2,l_1 l_2 l_3,22}\right) \\ &+ \frac{\partial}{\partial x_3}\left(M^{\gamma+1}_{l_1 l_2 l_3+1,1} + G_{3,l_1 l_2 l_3,21} + G_{3,l_1 l_2 l_3,22}\right) = C_{l_1 l_2 l_3,21} + C_{l_1 l_2 l_3,22}. \end{aligned} \tag{6.148}$$

These equations have the same form as in the monodisperse case, expect that they contain additional terms due to collisions between unlike particles. Note that such collisions affect both the spatial flux (e.g. $\mathbf{G}_{l_1 l_2 l_3,12}$) and the collision source terms (e.g. $C_{l_1 l_2 l_3,12}$). Because Eq. (6.148) can be found from Eq. (6.147) by a permutation of the indices, hereafter we will consider only Eq. (6.147).

[15] As noted earlier, because the collisions are binary, the extension to polydisperse systems is straightforward and simply adds collision terms for every possible collision partner in the system.

The zeroth-order moment

For the zeroth-order moment, Eq. (6.147) reduces to

$$\frac{\partial M^0_{000,1}}{\partial t} + \frac{\partial}{\partial x_1}\left(M^1_{100,1}\right) + \frac{\partial}{\partial x_2}\left(M^1_{010,1}\right) + \frac{\partial}{\partial x_3}\left(M^1_{001,1}\right) = 0, \tag{6.149}$$

which can be rewritten in terms of the hydrodynamic variables as a mass continuity equation:[16]

$$\frac{\partial \rho_1 \alpha_1}{\partial t} + \nabla \cdot (\rho_1 \alpha_1 \mathbf{U}_1) = 0. \tag{6.150}$$

Note that this equation is closed in terms of the hydrodynamic variables α_1 and $\mathbf{U}_1$. However, if the mass-average velocity, defined by $\mathbf{U} = (\rho_1\alpha_1\mathbf{U}_1 + \rho_1\alpha_1\mathbf{U}_1)/\rho$ and $\rho = \rho_1\alpha_1 + \rho_2\alpha_2$, were used as the reference velocity then it would be necessary to provide a closure for $\mathbf{U}_1$ in terms of $\mathbf{U}$. This is typically accomplished by introducing a gradient-diffusion term involving the spatial gradient of α_1.

First-order moments

For the first-order moments, Eq. (6.147) reduces to

$$\begin{aligned}
\frac{\partial M^1_{100,1}}{\partial t} &+ \frac{\partial}{\partial x_1}\left(M^2_{200,1} + G_{1,100,11} + G_{1,100,12}\right) + \frac{\partial}{\partial x_2}\left(M^2_{110,1} + G_{2,100,11} + G_{2,100,12}\right) \\
&+ \frac{\partial}{\partial x_3}\left(M^2_{101,1} + G_{3,100,11} + G_{3,100,12}\right) = C_{100,12}, \\
\frac{\partial M^1_{010,1}}{\partial t} &+ \frac{\partial}{\partial x_1}\left(M^2_{110,1} + G_{1,010,11} + G_{1,010,12}\right) + \frac{\partial}{\partial x_2}\left(M^2_{020,1} + G_{2,010,11} + G_{2,010,12}\right) \\
&+ \frac{\partial}{\partial x_3}\left(M^2_{011,1} + G_{3,010,11} + G_{3,010,12}\right) = C_{010,12}, \\
\frac{\partial M^1_{001,1}}{\partial t} &+ \frac{\partial}{\partial x_1}\left(M^2_{101,1} + G_{1,001,11} + G_{1,001,12}\right) + \frac{\partial}{\partial x_2}\left(M^2_{011,1} + G_{2,001,11} + G_{2,001,12}\right) \\
&+ \frac{\partial}{\partial x_3}\left(M^2_{002,1} + G_{3,001,11} + G_{3,001,12}\right) = C_{001,12},
\end{aligned} \tag{6.151}$$

which can be rewritten in terms of the hydrodynamic variables as a mean momentum balance:

$$\frac{\partial \rho_1 \alpha_1 \mathbf{U}_1}{\partial t} + \nabla \cdot (\rho_1 \alpha_1 \mathbf{U}_1 \otimes \mathbf{U}_1 + \mathbf{P}_1) = \mathbf{C}_{12}. \tag{6.152}$$

The momentum exchange due to collisions $\mathbf{C}_{12}$ in Eq. (6.152) is defined by its components:

$$\begin{aligned}
C_{1,12} &\equiv \rho_1 \left(C^{(0)}_{100,12} + C^{(1)}_{100,12} + C^{(2)}_{100,12} + C^{(3)}_{100,12}\right), \\
C_{2,12} &\equiv \rho_1 \left(C^{(0)}_{010,12} + C^{(1)}_{010,12} + C^{(2)}_{010,12} + C^{(3)}_{010,12}\right), \\
C_{3,12} &\equiv \rho_1 \left(C^{(0)}_{001,12} + C^{(1)}_{001,12} + C^{(2)}_{001,12} + C^{(3)}_{001,12}\right).
\end{aligned} \tag{6.153}$$

[16] We introduce the particle density ρ_1, which is simply the particle mass m_1 divided by its volume $\pi d_1^3/6$, in order to conserve explicitly the mass and total mean momentum.

The symmetric second-order pressure tensor $\mathbf{P}_1$ is defined by its components:

$$\begin{aligned}
P_{11,1} &\equiv \rho_1 \left(\sigma_{11,1} + G^{(0)}_{1,100,11} + G^{(0)}_{1,100,12} + G^{(1)}_{1,100,12}\right),\\
P_{12,1} &\equiv \rho_1 \left(\sigma_{12,1} + G^{(0)}_{2,100,11} + G^{(0)}_{2,100,12} + G^{(1)}_{2,100,12}\right),\\
P_{13,1} &\equiv \rho_1 \left(\sigma_{13,1} + G^{(0)}_{3,100,11} + G^{(0)}_{3,100,12} + G^{(1)}_{3,100,12}\right),\\
P_{22,1} &\equiv \rho_1 \left(\sigma_{22,1} + G^{(0)}_{2,010,11} + G^{(0)}_{2,100,12} + G^{(1)}_{2,100,12}\right),\\
P_{23,1} &\equiv \rho_1 \left(\sigma_{23,1} + G^{(0)}_{3,010,11} + G^{(0)}_{3,100,12} + G^{(1)}_{3,100,12}\right),\\
P_{33,1} &\equiv \rho_1 \left(\sigma_{33,1} + G^{(0)}_{3,001,11} + G^{(0)}_{3,100,12} + G^{(1)}_{3,100,12}\right),
\end{aligned} \tag{6.154}$$

wherein the velocity covariance matrix is denoted by $\boldsymbol{\sigma}_1$. In order to show that $\mathbf{P}_1$ is indeed symmetric, it will be necessary to examine $\mathbf{G}^{(0)}_{l_1l_2l_3,12}$ and $\mathbf{G}^{(1)}_{l_1l_2l_3,12}$ for the first-order moments using their definitions from Section 6.2.

From the definition in Eq. (6.107) and the results in Table 6.4, we can represent the contribution $\mathbf{G}^{(0)}_{l_1l_2l_3,12}$ as a symmetric second-order tensor:

$$\mathbf{G}^{(0)}_{12} \equiv \frac{2\chi_{12}^3\mu_{12}(1+e_{12})g_{0,12}}{5\chi_{21}} \int_{\mathbb{R}^6} \left[\tfrac{1}{2}|\mathbf{v}_1 - \mathbf{v}_2|^2\mathbf{I} + (\mathbf{v}_1 - \mathbf{v}_2)\otimes(\mathbf{v}_1 - \mathbf{v}_2)\right] f_1(\mathbf{v}_1)f_2(\mathbf{v}_2)\mathrm{d}\mathbf{v}_1\,\mathrm{d}\mathbf{v}_2, \tag{6.155}$$

where $\mu_{12} = 2m_2/(m_1 + m_2)$. By separating out the mean velocities of each phase,

$$\begin{aligned}
&(\mathbf{v}_1 - \mathbf{v}_2)\otimes(\mathbf{v}_1 - \mathbf{v}_2)\\
&\quad = [(\mathbf{v}_1 - \mathbf{U}_1) - (\mathbf{v}_2 - \mathbf{U}_2) + (\mathbf{U}_1 - \mathbf{U}_2)] \otimes [(\mathbf{v}_1 - \mathbf{U}_1) - (\mathbf{v}_2 - \mathbf{U}_2) + (\mathbf{U}_1 - \mathbf{U}_2)],
\end{aligned}$$

and using the fact that

$$\int_{\mathbb{R}^6} (\mathbf{v}_1 - \mathbf{U}_1)\otimes(\mathbf{v}_2 - \mathbf{U}_2) f_1(\mathbf{v}_1)f_2(\mathbf{v}_2)\mathrm{d}\mathbf{v}_1\,\mathrm{d}\mathbf{v}_2 = \mathbf{0},$$

we can evaluate the integral in Eq. (6.155) to find

$$\mathbf{G}^{(0)}_{12} = \frac{2\alpha_1\alpha_2\chi_{12}^3\mu_{12}(1+e_{12})g_{0,12}}{5\chi_{21}} \left[\tfrac{1}{2}E_{12}\mathbf{I} + \boldsymbol{\sigma}_1 + \boldsymbol{\sigma}_2 + (\mathbf{U}_1 - \mathbf{U}_2)\otimes(\mathbf{U}_1 - \mathbf{U}_2)\right] \tag{6.156}$$

and, by a permutation of the indices,

$$\mathbf{G}^{(0)}_{21} = \frac{2\alpha_1\alpha_2\chi_{21}^3\mu_{21}(1+e_{12})g_{0,12}}{5\chi_{12}} \left[\tfrac{1}{2}E_{12}\mathbf{I} + \boldsymbol{\sigma}_1 + \boldsymbol{\sigma}_2 + (\mathbf{U}_1 - \mathbf{U}_2)\otimes(\mathbf{U}_1 - \mathbf{U}_2)\right], \tag{6.157}$$

which are closed in terms of the moments of up to second order. Thus, because $\rho_1\chi_{12}^3\mu_{12} = \rho_2\chi_{21}^3\mu_{21}$, we have

$$\rho_2\mathbf{G}^{(0)}_{21} = \frac{\chi_{21}}{\chi_{12}}\rho_1\mathbf{G}^{(0)}_{12},$$

so that in the limit where $d_1 \ll d_2$ ($\chi_{21} \to \infty$) the collisional-flux term $\rho_2\mathbf{G}^{(0)}_{21}$ will remain finite as $\rho_1\mathbf{G}^{(0)}_{12}$ goes to zero.[17] Likewise, from the definition of $\mathbf{G}^{(1)}_{12}$ in Eq. (6.108) and the

[17] In other words, in the fluid–particle limit where d_1 is the molecular diameter, the pressure in the fluid phase will be independent of the particle phase.

results in Table 6.10 it is possible to show that $\mathbf{G}_{12}^{(1)}$ has the same symmetry properties as $\mathbf{G}_{12}^{(0)}$. However, it is not possible to express $\mathbf{G}_{12}^{(1)}$ in terms of the moments in closed form.

Using the result in Eq. (6.156) and neglecting the contribution due to $\mathbf{G}_{12}^{(1)}$, we can then rewrite $\mathbf{P}_1$ as

$$\begin{aligned}\mathbf{P}_1 &= \rho_1\alpha_1\boldsymbol{\sigma}_1 + \tfrac{2}{5}\rho_1\alpha_1^2 g_{0,11}(1+e_{11})[3\Theta_1\mathbf{I} + 2\boldsymbol{\sigma}_1] \\ &\quad + \frac{2\alpha_1\alpha_2\rho_1\chi_{12}^3\mu_{12}(1+e_{12})g_{0,12}}{5\chi_{21}}\left[\tfrac{1}{2}E_{12}\mathbf{I} + \boldsymbol{\sigma}_1 + \boldsymbol{\sigma}_2 + (\mathbf{U}_1 - \mathbf{U}_2)\otimes(\mathbf{U}_1 - \mathbf{U}_2)\right].\end{aligned} \tag{6.158}$$

In terms of the decomposition $\mathbf{P}_1 = \boldsymbol{\tau}_1 + p_1\mathbf{I}$, the granular pressure for type-1 particles is

$$p_1 = \rho_1\alpha_1\Theta_1[1 + 2\alpha_1(1+e_{11})g_{0,11}] + \frac{\alpha_1\alpha_2\rho_1\chi_{12}^3\mu_{12}(1+e_{12})g_{0,12}E_{12}}{3\chi_{21}} \tag{6.159}$$

and the traceless granular stress tensor is

$$\begin{aligned}\boldsymbol{\tau}_1 &\equiv \rho_1\alpha_1\left[1 + \tfrac{4}{5}\alpha_1(1+e_{11})g_{0,11}\right](\boldsymbol{\sigma}_1 - \Theta_1\mathbf{I}) \\ &\quad + \frac{2\alpha_1\alpha_2\rho_1\chi_{12}^3\mu_{12}(1+e_{12})g_{0,12}}{5\chi_{21}}\left[\boldsymbol{\sigma}_1 + \boldsymbol{\sigma}_2 + (\mathbf{U}_1 - \mathbf{U}_2)\otimes(\mathbf{U}_1 - \mathbf{U}_2) - \tfrac{1}{3}E_{12}\mathbf{I}\right].\end{aligned} \tag{6.160}$$

As in the monodisperse case, these expressions for p_1 and $\boldsymbol{\tau}_1$ are exact and are closed in terms of the moments of up to second order. It is interesting to note that the granular pressure and granular stress tensor depend on the mean velocity difference (through E_{12}) in polydisperse granular flow, and, in the limit where the granular temperatures are zero (i.e. infinite particle Mach number) the granular pressure tensor will be nonzero if the mean phase velocities are different.[18] These facts follow directly from defining the hydrodynamic variables using the individual phase mean velocities (not the mass-average velocity). Only in the fluid–particle limit where $d_1 \ll d_2$ will p_1 and $\boldsymbol{\tau}_1$ be independent of the mean velocity difference.

For the collision source terms in Eq. (6.153), we can use the result in Eq. (6.155) to write $\rho_1[C_{100,12}^{(2)}, C_{010,12}^{(2)}, C_{001,12}^{(2)}]$ as a vector source term,

$$\mathbf{C}_{12}^{(2)} = \frac{4c_{12}\alpha_1\alpha_2\rho_1\chi_{12}^3\mu_{12}(1+e_{12})g_{0,12}}{5d_{12}E_{12}^{1/2}}\left[\left(\tfrac{1}{2}E_{12} + |\mathbf{U}_1 - \mathbf{U}_2|^2\right)\mathbf{I} + \boldsymbol{\sigma}_1 + \boldsymbol{\sigma}_2\right]\cdot(\mathbf{U}_2 - \mathbf{U}_1), \tag{6.161}$$

and it is straightforward to show that $\mathbf{C}_{21}^{(2)} = -\mathbf{C}_{12}^{(2)}$. This term has the form of a solid–solid drag (albeit with a second-order tensor drag coefficient) whose maximum value occurs when the granular temperatures are null. Note that for stability (i.e. the solid–solid drag should force $\mathbf{U}_1$ to relax to $\mathbf{U}_2$) the parameter $c_{12} = c_{21}$ must be greater than or equal to zero (i.e. $0 \le c_{12} \le 1$). We should emphasize that Eq. (6.161) is exact, following directly from the definition of the integrals for hard-sphere collisions. In the fluid–particle limit

[18] In the hydrodynamic limit, the mean velocity difference must be small compared with the granular temperatures of both particle types (Goldman & Sirovich (1967)). Thus, in the hydrodynamic limit $E_{12} \approx 3(\Theta_1 + \Theta_2)$ and the dependence of $\boldsymbol{\tau}_1$ on the mean velocity differences $(\mathbf{U}_1 - \mathbf{U}_2)$ is negligible. The moment equations derived in this section are not restricted to the hydrodynamic limit and can be applied even when the granular temperatures are null.

($d_1 \ll d_2$), Eq. (6.161) can be interpreted as a non-isotropic drag model. Nevertheless, for an inviscid fluid $\boldsymbol{\sigma}_1 = \Theta_1 \mathbf{I}$ with $|\boldsymbol{\sigma}_2| \ll \Theta_1$, and thus the drag coefficient found from Eq. (6.161) would be nearly isotropic.

Using the definition in Eq. (6.103), we can also write $\rho_1[C^{(0)}_{100,12}, C^{(0)}_{010,12}, C^{(0)}_{001,12}]$ as a vector source term:

$$\mathbf{C}^{(0)}_{12} = \frac{3\rho_1\chi^3_{12}\mu_{12}(1+e_{12})g_{0,12}}{2d_{12}} \int_{\mathbb{R}^6} |\mathbf{v}_1 - \mathbf{v}_2|(\mathbf{v}_2 - \mathbf{v}_1) f_1(\mathbf{v}_1) f_2(\mathbf{v}_2) \mathrm{d}\mathbf{v}_1 \, \mathrm{d}\mathbf{v}_2. \tag{6.162}$$

The integral on the right-hand side is not closed, but we can approximate it by

$$\int_{\mathbb{R}^6} |\mathbf{v}_1 - \mathbf{v}_2|(\mathbf{v}_2 - \mathbf{v}_1) f_1(\mathbf{v}_1) f_2(\mathbf{v}_2) \mathrm{d}\mathbf{v}_1 \, \mathrm{d}\mathbf{v}_2 \approx \frac{8c_0\alpha_1\alpha_2 E^{1/2}_{12}}{15} (\mathbf{U}_2 - \mathbf{U}_1), \tag{6.163}$$

where c_0 is a constant of order unity. The approximate value of $\mathbf{C}^{(0)}_{12}$ is then

$$\mathbf{C}^{(0)}_{12} \approx \frac{4c_0\alpha_1\alpha_2\rho_1\chi^3_{12}\mu_{12}(1+e_{12})g_{0,12}E^{1/2}_{12}}{5d_{12}} (\mathbf{U}_2 - \mathbf{U}_1). \tag{6.164}$$

This source term has the form of a solid–solid drag model with a scalar coefficient. The form is the same as in Eq. (6.161), indicating that $\mathbf{C}^{(0)}_{12}$ and $\mathbf{C}^{(2)}_{12}$ can be combined to yield the total solid–solid drag term with a tensor drag coefficient. Using quadrature-based moment methods, Eq. (6.162) can be evaluated using the quadrature weights and abscissas. Another alternative is to use a kinetic model, but this is essentially equivalent to getting an accurate estimate for c_0 in Eq. (6.163).

The other two collision source vectors, $\mathbf{C}^{(1)}_{12}$ and $\mathbf{C}^{(3)}_{12}$, can be evaluated using the definitions in Eqs. (6.104) and (6.106). As mentioned earlier, $\mathbf{C}^{(1)}_{12}$ will be closed in terms of the moments of order two and lower, and their gradients. In contrast, $\mathbf{C}^{(3)}_{12}$ will not be closed in terms of any finite set of moments. Nevertheless, it can be approximated using quadrature-based moment methods as described in Section 6.5. In the fluid–particle limit ($d_1 \ll d_2$), neither $\mathbf{C}^{(1)}_{12}$ nor $\mathbf{C}^{(3)}_{12}$ will contribute terms involving spatial gradients of the fluid properties (i.e. buoyancy, lift, etc.) to the fluid-phase momentum equation. As mentioned earlier, such terms result from the model for $g_{\alpha\beta}(\mathbf{x}_{12})$ and would appear, for example, on using the expression in Eq. (6.81). With the latter, Eq. (6.161) becomes

$$\mathbf{C}^{(2)}_{12} = -\tfrac{4}{5}\kappa_{12}\alpha_1\alpha_2\rho_1\chi^3_{12}\mu_{12}(1+e_{12})g_{0,12}\left[\left(\tfrac{1}{2}E_{12} + |\mathbf{U}_1 - \mathbf{U}_2|^2\right)\mathbf{I} + \boldsymbol{\sigma}_1 + \boldsymbol{\sigma}_2\right] \cdot \nabla_x \mathcal{H} \tag{6.165}$$

and $\kappa_{21} = -\kappa_{12}$ ensures that $\mathbf{C}^{(2)}_{21} = -\mathbf{C}^{(2)}_{12}$. In the fluid–particle limit,[19] the fluid pressure is $p_1 = \rho_1\alpha_1\Theta_1$, and, assuming that the mean velocity difference is negligible[20] compared with $\Theta^{1/2}_1$, Eq. (6.165) simplifies to $\mathbf{C}^{(2)}_{12} = -\kappa_{12}\alpha_2 p_1 \nabla_x \mathcal{H}$. Thus, for example, if $\mathcal{H} = -\ln p_1$ and $\kappa_{12} = 1$,[21] we find the buoyancy force on the fluid phase: $\mathbf{C}^{(2)}_{12} = \alpha_2 \nabla_x p_1$. Obviously, other functional forms for $\mathcal{H}$ could be proposed to model the phase-momentum exchange terms described in Chapter 5. Conversely, using the multi-scale approach described in Chapter 1, microscale simulations can be employed to develop

[19] In the fluid–particle limit, $\chi_{12} = 1/2$, $\mu_{12} = 2$, $e_{12} = 1$, and $g_{0,12} = 1$. Also, $\varrho_1 = \rho_1\alpha_1$ is the fluid density and α_1 is the *volume occupied by the molecules* (i.e. $\alpha_1 + \alpha_2 \neq 1$).

[20] For the buoyancy force, it suffices to consider the case in which the mean velocities are null.

[21] The realizability condition in this case is $|\kappa_{12}| < 2/|d_2 \nabla_x \ln p_1|$, and thus we must let $\kappa_{12} = \min(1, 2/|d_2 \nabla_x \ln p_1|)$. However, since under normal flow conditions the pressure is unlikely to change rapidly across the distance $d_2/2$, this condition will yield $\kappa_{12} = 1$.

physically accurate expressions for $\mathcal{H}$ that can subsequently be used in Eq. (6.165). In summary, using the kinetic-theory approach it is possible to derive from first principles a two-fluid model for gas–particle flows by considering the limit where $d_1 \ll d_2$. The important step in this process for capturing the phase-interaction terms is the definition of $g_{\alpha\beta}(\mathbf{x}_{12})$.

Second-order moments

As an example of the second-order moments, we will consider just one:

$$\frac{\partial M^2_{200,1}}{\partial t} + \frac{\partial}{\partial x_1}\left(M^3_{300,1} + G_{1,200,11} + G_{1,200,12}\right) + \frac{\partial}{\partial x_2}\left(M^3_{210,1} + G_{2,200,11} + G_{2,200,12}\right) + \frac{\partial}{\partial x_3}\left(M^3_{201,1} + G_{3,200,11} + G_{3,200,12}\right) = C_{200,11} + C_{200,12}. \tag{6.166}$$

As with the first-order moments, this expression has contributions due to the kinetic and collisional fluxes on the left-hand side, and due to the collision source terms on the right-hand side. The contributions due to like-particle collisions ($G_{1,200,11}$ and $C_{200,11}$) have the same forms as in the case of monodisperse particles described above. We will thus look briefly at the terms due to unlike-particle collisions.

The collisional-flux term $G_{m,200,12}$ has two contributions:

$$G^{(0)}_{m,200,12} = \frac{3\chi^2_{12} d_1 g_{0,\alpha\beta}}{d_2} \int_{\mathbb{R}^6} I^{(m)}_{200}(\omega_{12}, \mathbf{v}_1, \mathbf{v}_1 - \mathbf{v}_2) f_1(\mathbf{v}_1) f_2(\mathbf{v}_2) \mathrm{d}\mathbf{v}_1 \, \mathrm{d}\mathbf{v}_2 \tag{6.167}$$

and

$$G^{(1)}_{m,200,12} = \frac{3\chi^2_{12} d_1 g_{0,12} c_{12}}{d_2 E^{1/2}_{12}} \sum_{q=1}^{3} \int_{\mathbb{R}^6} \frac{U_{2,q} - U_{1,q}}{|\mathbf{v}_1 - \mathbf{v}_2|} I^{(m,q)}_{200}(\omega_{12}, \mathbf{v}_1, \mathbf{v}_1 - \mathbf{v}_2) f_1(\mathbf{v}_1) f_2(\mathbf{v}_2) \mathrm{d}\mathbf{v}_1 \, \mathrm{d}\mathbf{v}_2. \tag{6.168}$$

The term in Eq. (6.167) can be rewritten in closed form using integer moments of up to third order. On the other hand, the right-hand side of Eq. (6.168) cannot be written in terms of any finite set of integer moments and must therefore be closed (e.g. using quadrature).

The collision source term $C_{200,12}$ has four contributions:

$$C^{(0)}_{200,12} = \frac{6\chi^2_{12} g_{0,12}}{d_2} \int_{\mathbb{R}^6} I^{(0)}_{200}(\omega_{12}, \mathbf{v}_1, \mathbf{v}_1 - \mathbf{v}_2) |\mathbf{v}_1 - \mathbf{v}_2| f_1(\mathbf{v}_1) f_2(\mathbf{v}_2) \mathrm{d}\mathbf{v}_1 \, \mathrm{d}\mathbf{v}_1, \tag{6.169}$$

$$C^{(1)}_{200,12} = \frac{3\chi^2_{12} g_{0,12}}{d_2} \sum_{m=1}^{3} \int_{\mathbb{R}^6} I^{(m)}_{200}(\omega_{12}, \mathbf{v}_1, \mathbf{v}_1 - \mathbf{v}_2) \times \left(d_1 f_2(\mathbf{v}_2) \frac{\partial f_1(\mathbf{v}_1)}{\partial x_m} - d_2 f_1(\mathbf{v}_1) \frac{\partial f_2(\mathbf{v}_2)}{\partial x_m} \right) \mathrm{d}\mathbf{v}_1 \, \mathrm{d}\mathbf{v}_2, \tag{6.170}$$

$$C^{(2)}_{200,12} = \frac{6\chi^2_{12} g_{0,12} c_{12}}{d_2 E^{1/2}_{12}} \sum_{m=1}^{3} (U_{2,m} - U_{1,m}) \int_{\mathbb{R}^6} I^{(m)}_{200}(\omega_{12}, \mathbf{v}_1, \mathbf{v}_1 - \mathbf{v}_2) f_1(\mathbf{v}_1) f_2(\mathbf{v}_2) \mathrm{d}\mathbf{v}_1 \, \mathrm{d}\mathbf{v}_1, \tag{6.171}$$

$$C^{(3)}_{200,12} = \frac{3\chi^2_{12} g_{0,12} c_{12}}{d_2 E^{1/2}_{12}} \sum_{m,q=1}^{3} \int_{\mathbb{R}^6} \frac{U_{\beta,q} - U_{\alpha,q}}{|\mathbf{v}_\alpha - \mathbf{v}_\beta|} I^{(m,q)}_{200}(\omega_{12}, \mathbf{v}_1, \mathbf{v}_1 - \mathbf{v}_2) \times \left(d_1 f_2(\mathbf{v}_2) \frac{\partial f_1(\mathbf{v}_1)}{\partial x_m} - d_2 f_1(\mathbf{v}_1) \frac{\partial f_2(\mathbf{v}_2)}{\partial x_m} \right) \mathrm{d}\mathbf{v}_1 \, \mathrm{d}\mathbf{v}_2. \tag{6.172}$$

The integrals in Eqs. (6.170) and (6.171) can be evaluated explicitly in terms of the moments. In contrast, the integrals in Eqs. (6.169) and (6.172) are not closed in terms of a finite set of moments and must be closed (e.g. using quadrature).

6.5 Application of quadrature to collision terms

The moment-transport equations discussed above become more and more complicated as the order increases. Moreover, these equations are not closed. In quadrature-based moment methods, the velocity-distribution function is reconstructed from a finite set of moments, thereby providing a closure. In this section, we illustrate how the closure hypothesis is applied to solve the moment-transport equations with hard-sphere collisions. For clarity, we will consider the monodisperse case governed by Eq. (6.131). Formally, we can re-express this equation in conservative form:

$$\frac{\partial M^{\gamma}_{l_1l_2l_3}}{\partial t} + \nabla_{\mathbf{x}} \cdot \mathbf{F}^{\gamma}_{l_1l_2l_3} = S^{\gamma}_{l_1l_2l_3}, \tag{6.173}$$

where $\mathbf{F}^{\gamma}_{l_1l_2l_3}$ is the spatial flux for moment $M^{\gamma}_{l_1l_2l_3}$ and $S^{\gamma}_{l_1l_2l_3}$ is its source term. This type of equation can be efficiently solved using finite-volume methods (Leveque, 2002) if care is taken to ensure that the moments always remain realizable (Vikas *et al.*, 2009). (See Appendix B for more details.) The latter can be elegantly accomplished using kinetics-based fluxes defined in terms of the reconstructed velocity-distribution function. In the remainder of this section we will show how Eq. (6.173) is treated using quadrature-based moment methods.

6.5.1 Flux terms

Formally, the flux term in Eq. (6.173) can be expressed as

$$\mathbf{F}^{\gamma}_{l_1l_2l_3} = \begin{bmatrix} F^{\gamma}_{1,l_1l_2l_3} \\ F^{\gamma}_{2,l_1l_2l_3} \\ F^{\gamma}_{3,l_1l_2l_3} \end{bmatrix} = \begin{bmatrix} M^{\gamma+1}_{l_1+1l_2l_3} + G_{1,l_1l_2l_3} \\ M^{\gamma+1}_{l_1l_2+1l_3} + G_{2,l_1l_2l_3} \\ M^{\gamma+1}_{l_1l_2l_3+1} + G_{3,l_1l_2l_3} \end{bmatrix}, \tag{6.174}$$

where $\mathbf{G}_{l_1l_2l_3}$ is the collisional flux defined in Eq. (6.70). In a kinetics-based flux algorithm, the components of $\mathbf{F}^{\gamma}_{l_1l_2l_3}$ are split into two parts corresponding to the positive- and negative-velocity components in the direction of the flux. Consider, for example, the first component $F^{\gamma}_{1,l_1l_2l_3}$. Using the definitions of $M^{\gamma+1}_{l_1+1l_2l_3}$ and $G_{1,l_1l_2l_3}$, we can split the flux as

$$F^{\gamma}_{1,l_1l_2l_3} = F^{-}_{1,l_1l_2l_3} + F^{+}_{1,l_1l_2l_3}, \tag{6.175}$$

where

$$F^{-}_{1,l_1l_2l_3} = \int_{-\infty}^{0} \left[\int_{\mathbb{R}^2} \left(v_1^{l_1+1} v_2^{l_2} v_3^{l_3} + 3g_0 \int_{\mathbb{R}^3} I^{(1)}_{l_1l_2l_3}(\omega, \mathbf{v}, \mathbf{v}-\mathbf{v}^*) f(\mathbf{v}^*)\,\mathrm{d}\mathbf{v}^* \right) f(\mathbf{v})\mathrm{d}v_2\,\mathrm{d}v_3 \right] \mathrm{d}v_1 \tag{6.176}$$

and

$$F^{+}_{1,l_1l_2l_3} = \int_{0}^{\infty} \left[\int_{\mathbb{R}^2} \left(v_1^{l_1+1} v_2^{l_2} v_3^{l_3} + 3g_0 \int_{\mathbb{R}^3} I^{(1)}_{l_1l_2l_3}(\omega, \mathbf{v}, \mathbf{v}-\mathbf{v}^*) f(\mathbf{v}^*)\mathrm{d}\mathbf{v}^* \right) f(\mathbf{v})\mathrm{d}v_2\,\mathrm{d}v_3 \right] \mathrm{d}v_1 \tag{6.177}$$

are the flux components corresponding to the negative and positive v_1 velocities, respectively. Analogous definitions are found for the second and third components of $\mathbf{F}^{\gamma}_{l_1l_2l_3}$, so we will have a decomposition of the form $\mathbf{F}^{\gamma}_{l_1l_2l_3} = \mathbf{F}^{-}_{l_1l_2l_3} + \mathbf{F}^{+}_{l_1l_2l_3}$. The reader should note that, for each component, the outer integral in Eqs. (6.176) and (6.177) changes for each direction. In a finite-volume algorithm, the flux splitting in Eq. (6.175) is directly used to compute the upwind and downwind flux components relative to a given surface element (e.g. Desjardins *et al.*, 2008), so it is quite natural to have formulas for computing them separately.

The quadrature-based closure of Eqs. (6.176) and (6.177) then proceeds as follows. Let $f^*(t, \mathbf{x}, \mathbf{v})$ denote the velocity-distribution function reconstructed from the transported moments $M^{\gamma}_{l_1l_2l_3}$ (Eq. 6.173), and define the negative- and positive-integer moments for direction x_1 as[22]

$$\begin{aligned} M^{-}_{1,l_1l_2l_3} &= \int_{-\infty}^{0} \left(\int_{\mathbb{R}^2} v_1^{l_1} v_2^{l_2} v_3^{l_3} f^*(\mathbf{v}) \mathrm{d}v_2 \, \mathrm{d}v_3 \right) \mathrm{d}v_1, \\ M^{+}_{1,l_1l_2l_3} &= \int_{0}^{\infty} \left(\int_{\mathbb{R}^2} v_1^{l_1} v_2^{l_2} v_3^{l_3} f^*(\mathbf{v}) \mathrm{d}v_2 \, \mathrm{d}v_3 \right) \mathrm{d}v_1. \end{aligned} \tag{6.178}$$

Analogous moments can be defined for directions x_2 and x_3. Since f^* is a known function, these moments can be computed (usually analytically) for any choice of (l_1, l_2, l_3). Thus, using each set of moments, we can construct an arbitrarily high-order quadrature representation that can be used to evaluate the integrals in Eqs. (6.176) and (6.177). Formally, we can express the reconstruction procedure as

$$M^{\gamma}_{l_1l_2l_3} \Longrightarrow f^*(\mathbf{v}) \Longrightarrow \begin{cases} M^{-}_{i,l_1l_2l_3} \Longrightarrow n^{-}_{\alpha}, \mathbf{v}^{-}_{\alpha} & \text{for } \alpha = 1, \dots, N^{-}, \\ M^{+}_{i,l_1l_2l_3} \Longrightarrow n^{+}_{\alpha}, \mathbf{v}^{+}_{\alpha} & \text{for } \alpha = 1, \dots, N^{+}, \end{cases}$$

where $N^{\pm}$ are the numbers of quadrature nodes, $n^{\pm}_{\alpha}$ are the weights, and $\mathbf{v}^{\pm}_{\alpha}$ are the abscissas. The weights and abscissas can be computed for each moment set using, for example, CQMOM (Yuan & Fox, 2011). The reader should recognize that the moments computed for directions x_1, x_2, and x_3 will generate different quadratures, and hence that six sets of quadrature weights and abscissas will be constructed to evaluate the flux terms. If EQMOM is used to reconstruct f^*, the positive and negative abscissas can be found directly from the second (Gaussian) quadrature without computing the half-moments in Eq. (6.178). However, it is important (when using EQMOM without the half-moments) to employ a relatively large number of quadrature nodes to ensure that the half-moments are accurately reproduced.

Given the six sets of quadratures, it is now possible to evaluate the flux components. For example, from Eqs. (6.176) and (6.177) we find

$$F^{-}_{1,l_1l_2l_3} = \sum_{\alpha=1}^{N^-} n^{-}_{\alpha} \left((v^{-}_{1,\alpha})^{l_1+1} (v^{-}_{2,\alpha})^{l_2} (v^{-}_{3,\alpha})^{l_3} + 3g_0 \sum_{\beta=1}^{N^-+N^+} n_{\beta} I^{(1)}_{l_1l_2l_3}(\omega, \mathbf{v}^{-}_{\alpha}, \mathbf{v}^{-}_{\alpha} - \mathbf{v}_{\beta}) \right) \tag{6.179}$$

and

$$F^{+}_{1,l_1l_2l_3} = \sum_{\alpha=1}^{N^+} n^{+}_{\alpha} \left((v^{+}_{1,\alpha})^{l_1+1} (v^{+}_{2,\alpha})^{l_2} (v^{+}_{3,\alpha})^{l_3} + 3g_0 \sum_{\beta=1}^{N^-+N^+} n_{\beta} I^{(1)}_{l_1l_2l_3}(\omega, \mathbf{v}^{+}_{\alpha}, \mathbf{v}^{+}_{\alpha} - \mathbf{v}_{\beta}) \right), \tag{6.180}$$

[22]The reconstructed distribution function may be continuous (EQMOM) or discrete (QMOM), but we will assume that it is always realizable (i.e. nonnegative). For the case in which f^* is a set of weighted delta functions, the computation of the moments $M^{-}_{i,l_1l_2l_3}$ and $M^{+}_{i,l_1l_2l_3}$ is trivial. With EQMOM the integrals are evaluated using the Gaussian quadrature defined with respect to the kernel density functions.

where the summation over β is done for the entire set of $N^- + N^+$ weights and abscissas.[23] (By definition, if $N^- = 0$ ($N^+ = 0$), then $F^-_{1,l_1l_2l_3} = 0$ ($F^+_{1,l_1l_2l_3} = 0$).) In Eqs. (6.179) and (6.180), the collisional fluxes involving $I^{(1)}_{l_1l_2l_3}$ will usually require more quadrature nodes to compute accurately than are needed for the free-transport term. For this reason, EQMOM is especially useful because the number of quadrature nodes can be set independently from the number of transported moments.

An important point that is evident from the expressions in Eqs. (6.179) and (6.180) is that it suffices to know the functional form of $I^{(m)}_{l_1l_2l_3}$ in order to account for the effect of collisions on the spatial fluxes. As can be seen from Tables 6.4–6.9, these functional forms can be easily coded into a numerical algorithm and, hence, Eqs. (6.179) and (6.180) can be computed in a straightforward manner. This situation can be compared to the "traditional" closures proposed for the moment transport equations (e.g. Struchtrup, 2005), wherein the manipulations needed to derive closures become more and more complicated as the order of the moments increases. It is also important to note that Eqs. (6.179) and (6.180) are exact even in highly non-equilibrium cases for which traditional moment closures fail. For example, if the exact velocity-distribution function is monokinetic,

$$f(\mathbf{v}) = \rho\delta\left(\mathbf{v} - \mathbf{U}_\text{p}\right),$$

then $N^- + N^+ = 1$, so the contribution to the spatial fluxes due to collisions will be null, and the spatial fluxes will be computed exactly (i.e. only free transport with velocity $\mathbf{U}_\text{p}$). The ability to compute accurate spatial fluxes for highly non-equilibrium cases is one of the principal attractions of using quadrature-based moment methods.

6.5.2 Source terms

The source term in Eq. (6.173) is given by

$$S^\gamma_{l_1l_2l_3} = C^{(0)}_{l_1l_2l_3} + C^{(1)}_{l_1l_2l_3}, \tag{6.181}$$

where the two terms on the right-hand side are defined in Eq. (6.68) and Eq. (6.71), respectively. The first of these is evaluated using quadrature as

$$C^{(0)}_{l_1l_2l_3} = \frac{6g_0}{d_\text{p}} \sum_{\alpha=1}^{N}\sum_{\beta=1}^{N} n_\alpha n_\beta I^{(0)}_{l_1l_2l_3}(\omega, \mathbf{v}_\alpha, \mathbf{v}_\alpha - \mathbf{v}_\beta)|\mathbf{v}_\alpha - \mathbf{v}_\beta|, \tag{6.182}$$

where the N weights and N abscissas are found from the moments of the reconstructed velocity distribution function f^* (i.e. from its discrete quadrature representation). Note that these moments are related to those found from Eq. (6.178) by $M^*_{i,l_1l_2l_3} = M^-_{i,l_1l_2l_3} + M^+_{i,l_1l_2l_3}$ and hence do not need to be computed separately. As for the spatial fluxes, the computation of Eq. (6.182) is straightforward, requiring only the functional forms for $I^{(0)}_{l_1l_2l_3}$ (see Tables 6.1–6.3). Furthermore, Eq. (6.182) is exact even for highly non-equilibrium cases for which traditional moment closures fail.

The evaluation of $C^{(1)}_{l_1l_2l_3}$ is more complicated because it involves a spatial gradient,

$$C^{(1)}_{l_1l_2l_3} = 3g_0 \int_{\mathbb{R}^3}\left(\int_{\mathbb{R}^3} \mathbf{H}_{l_1l_2l_3}(\omega, \mathbf{v}_1, \mathbf{v}_2)\cdot\nabla_\mathbf{x} f(\mathbf{v}_1)\mathrm{d}\mathbf{v}_1\right) f(\mathbf{v}_2)\mathrm{d}\mathbf{v}_2, \tag{6.183}$$

[23] Alternatively, the summation over β can be defined using an N-node quadrature constructed from the moments $M^*_{i,l_1l_2l_3} = M^-_{i,l_1l_2l_3} + M^+_{i,l_1l_2l_3}$. If $N^- = N^+ = N$, the formal accuracy of both methods would be of order $2N - 1$ (Gautschi, 1997). With the quadrature found from $M^*_{i,l_1l_2l_3}$ the summations in the flux equations are shorter (N versus $2N$), but the additional computational cost of computing another quadrature may be higher.

where the components of the vector $\mathbf{H}_{l_1l_2l_3}$ are defined by Eq. (6.72). As noted earlier, $H_{m,l_1l_2l_3}$ is a homogeneous polynomial of order $\gamma + 1$ with respect to $\mathbf{v}_1$ and $\mathbf{v}_2$. Thus, formally, we can express $\mathbf{H}_{l_1l_2l_3}$ as

$$\mathbf{H}_{l_1l_2l_3} = \sum_{n=0}^{\gamma+1} a_n(\omega)\mathbf{v}_1^n \mathbf{v}_2^{\gamma+1-n},$$

so that[24]

$$C^{(1)}_{l_1l_2l_3} = 3g_0 \sum_{n=0}^{\gamma+1} a_n(\omega)(\nabla_{\mathbf{x}} \cdot M^n) M^{\gamma+1-n}, \tag{6.184}$$

where M^n denote the appropriate velocity moment of order n. In principle, the coefficients a_n can be found from the definitions of $I^{(m)}_{l_1l_2l_3}$, and hence Eq. (6.184) can be expanded out in terms of the velocity moments. As an example, the contribution to $C^{(1)}_{200}$ coming from $H_{1,200}$ is

$$\begin{aligned}
&-\frac{4\omega^2}{35}\Bigg[\left(\left(5M^3_{300} + 3M^3_{120} + 3M^3_{102}\right)\frac{\partial M^0_{000}}{\partial x_1} - M^0_{000}\frac{\partial}{\partial x_1}\left(5M^3_{300} + 3M^3_{120} + 3M^3_{102}\right)\right)\\
&\quad + 3\left(M^1_{100}\frac{\partial}{\partial x_1}\left(5M^2_{200} + M^2_{020} + M^2_{002}\right) - \left(5M^2_{200} + M^2_{020} + M^2_{002}\right)\frac{\partial M^1_{100}}{\partial x_1}\right)\\
&\quad + 6\left(M^1_{010}\frac{\partial M^2_{110}}{\partial x_1} - M^2_{110}\frac{\partial M^1_{010}}{\partial x_1}\right) + 6\left(M^1_{001}\frac{\partial M^2_{101}}{\partial x_1} - M^2_{101}\frac{\partial M^1_{001}}{\partial x_1}\right)\Bigg]\\
&\quad + \frac{4\omega}{15}\Bigg[\left(\left(3M^3_{300} + M^3_{120} + M^3_{102}\right)\frac{\partial M^0_{000}}{\partial x_1} - M^0_{000}\frac{\partial}{\partial x_1}\left(3M^3_{300} + M^3_{102} + M^3_{120}\right)\right)\\
&\qquad + \left(M^1_{100}\frac{\partial}{\partial x_1}\left(9M^2_{200} + M^2_{020} + M^2_{002}\right) - \left(9M^2_{200} + M^2_{020} + M^2_{002}\right)\frac{\partial M^1_{100}}{\partial x_1}\right)\\
&\qquad + 2\left(M^1_{010}\frac{\partial M^2_{110}}{\partial x_1} - M^2_{110}\frac{\partial M^1_{010}}{\partial x_1}\right) + 2\left(M^1_{001}\frac{\partial M^2_{101}}{\partial x_1} - M^2_{101}\frac{\partial M^1_{001}}{\partial x_1}\right)\Bigg],
\end{aligned}$$

while the contribution coming from $H_{2,200}$ is

$$\begin{aligned}
&-\frac{4\omega^2}{35}\Bigg[\left(\left(3M^3_{210} + M^3_{030} + M^3_{012}\right)\frac{\partial M^0_{000}}{\partial x_2} - M^0_{000}\frac{\partial}{\partial x_2}\left(3M^3_{210} + M^3_{030} + M^3_{012}\right)\right)\\
&\quad + \left(M^1_{010}\frac{\partial}{\partial x_2}\left(3M^2_{200} + 3M^2_{020} + M^2_{002}\right) - \left(3M^2_{200} + 3M^2_{020} + M^2_{002}\right)\frac{\partial M^1_{010}}{\partial x_2}\right)\\
&\quad + 6\left(M^1_{100}\frac{\partial M^2_{110}}{\partial x_2} - M^2_{110}\frac{\partial M^1_{100}}{\partial x_2}\right) + 2\left(M^1_{001}\frac{\partial M^2_{011}}{\partial x_2} - M^2_{011}\frac{\partial M^1_{001}}{\partial x_2}\right)\Bigg]\\
&\quad + \frac{8\omega}{15}\Bigg[\left(M^3_{210}\frac{\partial M^0_{000}}{\partial x_2} - M^0_{000}\frac{\partial M^3_{210}}{\partial x_2}\right) + \left(M^1_{010}\frac{\partial M^2_{200}}{\partial x_2} - M^2_{200}\frac{\partial M^1_{010}}{\partial x_2}\right)\\
&\qquad + 2\left(M^1_{100}\frac{\partial M^2_{110}}{\partial x_2} - M^2_{110}\frac{\partial M^1_{100}}{\partial x_2}\right)\Bigg].
\end{aligned}$$

---

[24]Here we abuse the notation for clarity, but the reader should recognize that $(\nabla_{\mathbf{x}} \cdot M^n)$ involves three different moments (one for each direction).

The contribution due to $H_{3,200}$ has a similar form, and thus the final expression for $C^{(1)}_{200}$ involves the spatial gradients of all 20 velocity moments of up to third order.[25] Using the N-node quadrature representation constructed from f^*, the moments appearing in Eq. (6.184) can be expressed, formally, as

$$M^n = \sum_{\alpha=1}^{N} n_\alpha \mathbf{v}_\alpha^n.$$

Given the moments, the gradients $(\nabla_{\mathbf{x}} \cdot M^n)$ can be computed using a consistent finite-volume representation. To the best of the authors' knowledge, this method for computing $C^{(1)}_{l_1 l_2 l_3}$ has not yet been implemented in any simulation code.

In summary, we have demonstrated in this section how quadrature-based moment methods can be used to evaluate the terms in the moment-transport equations arising from collisions. The principal observation is that it suffices to know the functional forms for the terms $I^{(m)}_{l_1 l_2 l_3}$, which are derived and tabulated in Section 6.1. We also observed that, unlike traditional moment closures, the closures developed in this section are applicable to highly non-equilibrium flows.

[25] Since $C^{(1)}_{000}$ and $C^{(1)}_{100}$ are zero, the nonzero contributions in $C^{(1)}_{200}$ represent source terms for the granular temperature due to density and velocity gradients. In the monokinetic case, these contributions will cause the granular temperature to increase to a nonzero value (thereby making the distribution no longer monokinetic). Physically, the terms in $C^{(1)}_{200}$ arise because particles with finite diameter d_p and finite separation can collide when they have different velocities (due to the mean velocity gradient) even when the granular temperature is null.

7

Solution methods for homogeneous systems

In this chapter we consider the solution of the generalized population-balance equation (GPBE) with a generic set of internal coordinates $\boldsymbol{\xi} = (\xi_1, \xi_2, ..., \xi_M)$ under the hypothesis of spatial homogeneity. Under this hypothesis, all spatial gradients are null and the GPBE depends only on time t and on $\boldsymbol{\xi}$. As discussed earlier, we refer here to the GPBE as a general equation describing the evolution of a number-density function (NDF) in space (neglected in this chapter), in time, and in the phase space generated by the internal coordinates. The extension to inhomogeneous systems is discussed in Chapter 8. It is also worth mentioning that the GPBE is given different names in different fields. It is called the population-balance equation (PBE) in crystallization, precipitation, and, in general, in the literature on particulate processes. In the simulation of aerosols, it is called the general aerosol dynamic equation (Friedlander, 2000), whereas in the simulation of sprays it is generally known as the Williams–Boltzmann equation (Williams, 1985) and, especially in this case, the droplet velocity is included as one of the internal coordinates. When dealing with solid particles, if the particle velocity is included as the only internal coordinate, the GPBE is called the Boltzmann equation. In general, when the particle velocity is the only internal coordinate, the GPBE is also called the kinetic equation (KE).[1] Many of the challenges associated with the KE occur with inhomogeneous systems, which are discussed in Chapter 8.

7.1 Overview of methods

Depending on the number of internal coordinates (univariate versus multivariate) and depending on whether the particle velocity is part of the internal-coordinate vector, very different solution methods have been developed. As a consequence, in this chapter the methods for cases with and without particle velocity are discussed separately. First, the methods developed for the solution of a univariate PBE (i.e. one internal coordinate) are discussed. Second, the approaches for the solution of bivariate and multivariate PBE (i.e.

[1] In the applied-mathematics literature, all balance equations with internal coordinates evolving in a phase space are often referred to as kinetic equations (KE). Using this classification, a GPBE or a PBE could simply be referred to as a KE. However, since the dynamics in physical and phase space lead to quite distinct differences, we prefer to refine the definitions and use KE to describe cases in which the only internal coordinates are velocity.

two or more internal coordinates) are presented. Finally, methods for the solution of the KE (i.e. the particle velocity is the only internal coordinate) are presented. Under the hypothesis of spatial homogeneity (i.e. the NDF and all of the terms in the GPBE do not depend on $\mathbf{x}$), the governing GPBE becomes[2]

$$\frac{\partial n}{\partial t} + \frac{\partial}{\partial \boldsymbol{\xi}} \cdot (\dot{\boldsymbol{\xi}} n) = \frac{\partial^2}{\partial \boldsymbol{\xi}\, \partial \boldsymbol{\xi}} : (\mathbf{D} n) + \mathcal{S}, \tag{7.1}$$

where $n(t, \boldsymbol{\xi})$ is the NDF that depends only on time and on the internal coordinates, $\dot{\boldsymbol{\xi}}(\boldsymbol{\xi})$ is the rate of change of the internal coordinates for continuous processes,[3] $\mathbf{D}(t, \boldsymbol{\xi})$ is the phase-space diffusion tensor, and $\mathcal{S}$ is the source term due to discontinuous jumps in phase space. As described in Chapter 5, the drift term involving $\dot{\boldsymbol{\xi}}$ represents the effect of continuous phenomena.[4] When the internal coordinate under study is the particle velocity, the drift term quantifies the effect of continuous forces such as body and fluid-particle forces (e.g. buoyancy, drag, and lift). The term on the right-hand side of Eq. (7.1) is often labeled as a discontinuous point process and represents the effect of discontinuous events. (See Chapter 5 for examples.) Let us now introduce the different approaches available for solving Eq. (7.1), which can be roughly classified into four groups: (i) class and sectional methods, (ii) the method of moments, (iii) quadrature-based moment methods and (iv) Monte Carlo methods.

Class and sectional methods

Class methods (CM) are based on the discretization of the internal coordinates so that the GPBE becomes a set of macroscopic balances. These methods have been developed and extensively applied in the case of univariate PBE by several authors in the field of the simulation of particulate systems, whereas in the case of bivariate and multivariate PBE their application is more difficult (Hounslow *et al.*, 1988; Kumar & Ramkrishna, 1996a, 1996b; Litster *et al.*, 1995; Marchal *et al.*, 1988; Vanni, 1999, 2000a; Wynn, 2004). Generally the application to multivariate systems is based on a combination of the reduction of the number of variables (Briesen, 2004; Puel *et al.*, 2003a, 2003b; Zhang & Doherty, 2004) and the extension of univariate methods (Kumar *et al.*, 2006, 2007, 2008b). The same basic idea of discretizing the internal coordinate has been used in other fields for the simulation of evaporating sprays, resulting in what is generally known as the spray multi-fluid model (Greenberg *et al.*, 1986, 1993; Laurent & Massot, 2001). The final equations are, however, very similar to the above-mentioned methods. The same approach, when employed for the solution of the KE, is referred to as a discrete-velocity model (DVM) and, together

---

[2]The GPBE is written in vector notation with dot products. In component form, where repeated Greek indices imply summation over the internal coordinates, the GPBE is

$$\frac{\partial n}{\partial t} + \frac{\partial}{\partial \xi_\alpha}(\dot{\xi}_\alpha n) = \frac{\partial^2}{\partial \xi_\alpha\, \partial \xi_\beta}(D_{\alpha\beta} n) + \mathcal{S}.$$

For homogeneous systems, the drift ($\dot{\xi}_\alpha$) and diffusion ($D_{\alpha\beta}$) components can be functions of $\boldsymbol{\xi}$ and t.

[3]The rate of continuous change can also be a function of time. However, in this chapter we will discuss only internal-coordinate dependences, since these are the ones which cause the most significant numerical issues.

[4]It is interesting to highlight that since the numerical solution of the drift term is problematic, especially when phase space is discretized, sometimes this perspective is inverted. In recent works, in fact, some continuous processes are described as if they were actually discontinuous processes. As we will see, this strategy solves some of the issues (i.e. numerical diffusion in phase space) but typically makes the problem very stiff, due to the different time scales governing the process. Readers interested in the details are referred to the work of Kumar *et al.* (2008b).

with standard approaches for PBE, will be presented and discussed in Section 7.2. As will become clearer in the next sections, with CM and DVM, the NDF is often approximated in each interval of the discretized internal coordinate by a low-order polynomial. Very often a polynomial of order zero (i.e. constant NDF within each single class) is used, resulting in zeroth-order methods.[5] It is important to mention that this choice is arbitrary and has some important implications for the performance of the method. Alternatively, it is possible to assume that all the particles belonging to a specific interval have the same value of the internal coordinate and therefore are concentrated at a single point, called the *pivotal point*. Higher-order methods are instead based on the approximation of the NDF in the interval of interest with higher-order polynomials or other basis functions. The latter numerical approaches fall into the category of methods of weighted residuals (MWR). Using a set of weighting functions, it is in fact possible to calculate the weighted sum of the residuals and subsequently to drive it to zero at selected points. When the local basis functions are orthogonal polynomials and the weighting functions are Dirac delta functions, the approach is known as a collocation technique (Crowley *et al.*, 2000; Gelbard & Seinfeld, 1978; Immanuel *et al.*, 2002; Immanuel & Doyle, 2003), whereas when the weighting functions are identical to the basis functions the method is known as a Galerkin method[6] (Chiu & Christofides, 1999; Godin *et al.*, 1999; Mahoney & Ramkrishna, 2002; Wulkow *et al.*, 2001). Since the sub-domain in which the internal coordinate is discretized is also known as a finite element, these methods are also labeled as finite-element methods (FEM). The FEM are indeed very powerful, but, due to their high computational costs, are not considered to be applicable for inhomogeneous systems, and therefore will not be discussed in detail. Another class of methods that will not be discussed is those based on a finite-volume discretization for phase space. These methods are very efficient in the solution of the drift and diffusion terms of Eq. (7.1), but the extension to problems involving discontinuous jumps (i.e. due, for example, to collisions) is not straightforward. Another important advantage lies in the fact that it is very easy to extend finite-volume methods to multivariate PBE (e.g. Gunawan *et al.*, 2004), but, even in the univariate case, the computational cost is too high for solution of a general inhomogeneous PBE.

The method of moments (MOM)

The MOM was introduced for particulate systems by Hulburt & Katz (1964). In their pioneering work these authors showed how it is possible to solve the PBE in terms of the moments of the NDF and to derive the corresponding transport equations. A similar approach can be used for the solution of the KE, and a detailed discussion on the derivation of the moment-transport equations can be found in the works of Struchtrup (2005) and Truesdell & Muncaster (1980). The main issue with this technique is in the closure problem, namely the impossibility of writing transport equations for the lower-order moments of the NDF involving only the lower-order moments. Since the work of Hulburt & Katz (1964) much progress has been made (Frenklach, 2002; Frenklach & Harris, 1987; Kazakov & Frenklach, 1998), and different numerical closures have been proposed (Alexiadis *et al.*, 2004; Kostoglou & Karabelas, 2004; Strumendo & Arastoopour, 2008). The basis

[5]This is a common but confusing notation. In fact, it is conceptually more correct to classify these methods according to the terms excluded by the approximation. With the latter approach, a method that approximates the NDF with a constant value within each interval can be denoted a first-order method, since when approximating the NDF as a Taylor series the first term excluded is the first-order derivative term.

[6]The software-package particle-size evaluator PARSIVAL is based on the Galerkin method.

of the MOM and the earlier strategies developed to overcome the closure problem (by employing different functional assumptions on the NDF) are discussed in Section 7.3.

Quadrature-based moment methods (QBMM)

In the context of moment methods, one particular closure proposed by McGraw (1997), which is based on a quadrature formula, is particularly interesting and generates an entirely new class of methods. These QBMM are discussed in detail in Section 7.4. As will become clearer, with QBMM the NDF is reconstructed as a summation of Dirac delta functions (and other basis functions), in such a way that both the weights and the locations in phase space of these basis functions are free to move and mimic the NDF, properly representing a large number of moments of the NDF. Technical details on Gaussian quadrature and moment-inversion algorithms are provided in Chapter 3. Thus, our focus in Section 7.4 is on the application of QBMM to the GPBE.

Monte Carlo methods

Monte Carlo (MC) methods are based on the solution of a set of stochastic differential equations that govern the behavior of the particulate system, and broadly MC methods are based on an artificial realization of the system behavior. Several approaches have been developed by various authors (Balthasar & Kraft, 2003; Lattuada *et al.*, 2004; Lee & Matsoukas, 2000; Lin *et al.*, 2002; Maisels *et al.*, 2004a, 2004b; Mishra, 2000; Rosner & Yu, 2001; Rosner *et al.*, 2003; Song & Qiu, 1999; Van Peborgh Gooch & Hounslow, 1996), and a short general description of MC methods is presented in Section 7.5. Since our focus in this chapter is on moment methods, the emphasis of our discussion in Section 7.5 is on introducing the reader to MC methods as a potential tool for the validation of moment closures.

7.2 Class and sectional methods

As introduced in the previous section, class and sectional methods are based on a discretization of the internal coordinate so that the GPBE becomes a set of macroscopic balances in state space. Indeed, the fineness of the discretization will be dictated by the accuracy needed in the approximation of the integrals and derivative terms appearing in the GPBE. As has already been anticipated, the methods differ according to the number of internal coordinates used in the description and depend on the nature of the internal coordinates. Therefore, in what follows, we will discuss separately the univariate, bivariate, and multivariate PBE, and the use of these methods for the solution of the KE.

7.2.1 Univariate PBE

Let ξ denote the unique internal coordinate and $n(t,\xi)$ the NDF. The univariate PBE is

$$\frac{\partial n}{\partial t} + \frac{\partial \dot{\xi} n}{\partial \xi} = \frac{\partial^2 D n}{\partial \xi^2} + S, \tag{7.2}$$

where $\dot{\xi}$ is the continuous rate of change of the internal coordinate, D is the diffusive change of the internal coordinate, and S refers to discontinuous jumps in the internal coordinate.

Expressing the drift, diffusion, and point processes as a unique source term (Ramkrishna, 2000) yields

$$\frac{\partial n}{\partial t} = S \equiv -\frac{\partial \dot{\xi} n}{\partial \xi} + \frac{\partial^2 D n}{\partial \xi^2} + \mathcal{S}, \tag{7.3}$$

where $S(t, \xi, n)$ is a functional that maps the NDF into a function of time and internal coordinate. We now define a finite interval I_i for the internal coordinate bounded by ξ_i and ξ_{i+1}. The macroscopic balance of elements in the interval I_i can be obtained by integrating the PBE over I_i:

$$\frac{\mathrm{d}}{\mathrm{d}t} \int_{\xi_i}^{\xi_{i+1}} n \, \mathrm{d}\xi = \int_{\xi_i}^{\xi_{i+1}} S \, \mathrm{d}\xi. \tag{7.4}$$

If we denote the number of particles in the I_i interval as

$$N_i \equiv \int_{\xi_i}^{\xi_{i+1}} n \, \mathrm{d}\xi, \tag{7.5}$$

we recognize that

$$\frac{\mathrm{d}N_i}{\mathrm{d}t} = \int_{\xi_i}^{\xi_{i+1}} S \, \mathrm{d}\xi, \tag{7.6}$$

where $i = 1, \ldots, M$, corresponding to the following partition P_M: $\xi_1 < \xi_2 < \xi_3 < \cdots < \xi_{M-1} < \xi_M < \xi_{M+1}$. The values of ξ_1 and ξ_M are chosen so that the excluded dynamics is negligible, whereas the number of intervals is fixed in order to achieve the desired accuracy. The resulting set of M ordinary differential equations (ODE) is unclosed. In fact, it is not possible to write the right-hand side of Eq. (7.6) in terms of N_i only, and it is said that this reflects a lack of autonomy. The use of discrete methods depends on the restoration of autonomy, which lies in our ability to express the right-hand side entirely in terms of the dependent variables N_i.

Typically, different assumptions regarding the functional form for the NDF within a single finite interval are employed, and, in general, these can be written as $n(t, \xi) = N_i(t) k_i(\xi)$ for $\xi_i < \xi < \xi_{i+1}$, where $i = 1, \ldots, M$, and $k_i(\xi)$ is a polynomial that has to fulfill the property

$$\int_{\xi_i}^{\xi_{i+1}} k_i(\xi) \mathrm{d}\xi = 1.$$

In the case of a zeroth-order polynomial, a (constant) approximation is employed within each class (Greenberg *et al.*, 1986, 1993):

$$n(t, \xi) = \frac{N_i(t)}{\xi_{i+1} - \xi_i} \quad \text{for } \xi_i < \xi < \xi_{i+1},$$

where $i = 1, \ldots, M$. An alternative assumption is that of pivotal points ζ_i where all the particles in the interval are concentrated:

$$n(t, \xi) = \sum_{i=1}^{M} N_i(t) \delta(\xi - \zeta_i). \tag{7.7}$$

Very often the discretization is written not in terms of the number of particles belonging to a single class, but rather in terms of the volume (or mass) of the particles belonging to a specific class. If ρ_p is the density of the particles and $V_\mathrm{p}(\xi)$ is the functional relationship between particle volume and internal coordinate,[7] the mass of the particles in the I_i interval is (Laurent & Massot, 2001)

$$M_i \equiv \int_{\xi_i}^{\xi_{i+1}} \rho_\mathrm{p} V_\mathrm{p}(\xi) n \, \mathrm{d}\xi, \tag{7.8}$$

whereas the corresponding fundamental assumption inside the I_i interval is $n(t,\xi) = M_i(t)k_i(\xi)$ for $\xi_i < \xi < \xi_{i+1}$, and

$$\rho_\mathrm{p} \int_{\xi_i}^{\xi_{i+1}} k_i(\xi) V_\mathrm{p}(\xi) \mathrm{d}\xi = 1.$$

Again, the case of constant approximation of the NDF within each interval results in

$$n(t,\xi) = \frac{M_i(t)}{\rho_\mathrm{p} \int_{\xi_i}^{\xi_{i+1}} V_\mathrm{p}(\xi)\mathrm{d}\xi} \quad \text{for } \xi_i < \xi < \xi_{i+1}.$$

In what follows some strategies to overcome and solve the problem of restoration of autonomy are discussed. Only the most popular methods are presented and analyzed separately for the discontinuous, drift, and diffusion processes.

Second- and first-order point processes

Let us start by discussing the implication related to the choice of a particular discretization and the restoration of autonomy for second-order point processes by using two different approaches: the fixed (and moving) pivot technique and the cell-average technique. If ξ represents an additive property of the particle (e.g. the particle mass), the source term for a second-order point process is

$$S = \frac{1}{2} \int_0^{\xi} \beta(\xi - \xi', \xi') n(\xi - \xi') n(\xi') \mathrm{d}\xi' - n(\xi) \int_0^{\infty} \beta(\xi, \xi') n(\xi') \mathrm{d}\xi', \tag{7.9}$$

where $\beta(\xi, \xi')$ is the aggregation kernel of two particles with masses equal to ξ and ξ', respectively. Substitution into Eq. (7.4) gives

$$\frac{\mathrm{d}N_i}{\mathrm{d}t} = \frac{1}{2} \int_{\xi_i}^{\xi_{i+1}} \int_0^{\xi} \beta(\xi - \xi', \xi') n(\xi - \xi') n(\xi') \mathrm{d}\xi' \, \mathrm{d}\xi - \int_{\xi_i}^{\xi_{i+1}} n(\xi) \int_0^{\infty} \beta(\xi, \xi') n(\xi') \mathrm{d}\xi' \, \mathrm{d}\xi. \tag{7.10}$$

Expressing the second integral as a summation of integrals between ξ_i and ξ_{i+1}, and rearranging the summation, yields

$$\begin{aligned} \frac{\mathrm{d}N_i}{\mathrm{d}t} = & \frac{1}{2} \sum_{j=1}^{i-1} \int_{\xi_i}^{\xi_{i+1}} \int_{\xi_j}^{\xi_{j+1}} \beta(\xi - \xi', \xi') n_\xi(\xi - \xi') n_\xi(\xi') \mathrm{d}\xi' \, \mathrm{d}\xi \\ & - \sum_{j=1}^{M} \int_{\xi_i}^{\xi_{i+1}} n_\xi(\xi) \int_{\xi_j}^{\xi_{j+1}} \beta(\xi, \xi') n_\xi(\xi') \mathrm{d}\xi' \, \mathrm{d}\xi. \end{aligned} \tag{7.11}$$

[7] Clearly this relationship depends on the nature of the internal coordinate. In the case of particle size (i.e. $\xi = L_\mathrm{p}$) $V_\mathrm{p} = k_\mathrm{V} L_\mathrm{p}^3$, where k_V is the volume shape factor, whereas in the case of particle surface area ($\xi = A_\mathrm{p}$) $V_\mathrm{p} = k_\mathrm{V} A_\mathrm{p}^{3/2} / k_\mathrm{A}^{3/2}$, where k_A is the surface shape factor.

The restoration of autonomy relies on our ability to write the integrals in Eq. (7.11) as functions of N_i. This can be done by using the mean-value theorem. For example, the second integral corresponds to the following equality:

$$\int_{\xi_i}^{\xi_{i+1}} n(\xi) \int_{\xi_j}^{\xi_{j+1}} \beta(\xi, \xi') n(\xi') \mathrm{d}\xi' \, \mathrm{d}\xi = N_i N_j \beta(\zeta_i, \zeta_j), \tag{7.12}$$

where ζ_i and ζ_j are the pivotal points or *pivots*. In the fixed-pivot technique, the pivots are fixed during the simulation (Kumar & Ramkrishna, 1996a), whereas in the moving-pivot technique (Kumar & Ramkrishna, 1996b) the pivots are free to move within the interval.[8] Applying the mean-value theorem and using the concept of pivots, Eq. (7.11) becomes

$$\frac{\mathrm{d}N_i}{\mathrm{d}t} = \frac{1}{2} \sum_{j=1}^{i-1} N_j \sum_{(\zeta_j+\zeta_k)\in I_i} \beta(\zeta_k, \zeta_j) N_k - N_i \sum_{j=1}^{M} \beta(\zeta_i, \zeta_j) N_j, \tag{7.13}$$

where the first term on the right-hand side represents the generation of new particles and the second summation is limited to the pairs of particles (i.e. ζ_j and ζ_k) whose internal coordinates fall into the interval I_i. The last term in Eq. (7.13) is the loss of particles.

The generation term in Eq. (7.13) causes most of the problems. In fact, if ξ is the particle mass, then, when a particle belonging to the interval I_j aggregates with a particle belonging to the interval I_k, a particle with mass equal to $\zeta_j + \zeta_k$ is formed. Unless a linear partition is used, it is very unlikely to have exactly a pivot at $\zeta_i = \zeta_j + \zeta_k$. This implies that the newly formed particle of mass $\zeta_j + \zeta_k$ belonging to the I_i interval might have a corresponding pivot ζ_i that is not equal to $\zeta_j + \zeta_k$. Therefore, the newly formed particle has to be shared between neighboring pivotal points in order to preserve some global properties of the distribution. This problem is extensively discussed in Ramkrishna (2000) and Kumar & Ramkrishna (1996a) by deriving the transport equation of the rth moment of the NDF. For a second-order point process, we obtain

$$\frac{\mathrm{d}m_r}{\mathrm{d}t} = \frac{1}{2} \sum_{i=1}^{M} \sum_{j=1}^{M} N_j \sum_{(\zeta_j+\zeta_k)\in I_i} \beta(\zeta_k, \zeta_j) N_k (\zeta_j + \zeta_k)^r - \sum_{i=1}^{M} N_i \zeta_i^r \sum_{j=1}^{M} \beta(\zeta_i, \zeta_j) N_j. \tag{7.14}$$

Comparison with Eq. (7.13) multiplied by ζ_i^r and summed over all possible values of the index i,

$$\frac{\mathrm{d}m_r}{\mathrm{d}t} = \frac{1}{2} \sum_{i=1}^{M} \sum_{j=1}^{M} N_j \sum_{(\zeta_j+\zeta_k)\in I_i} \beta(\zeta_k, \zeta_j) N_k \zeta_i^r - \sum_{i=1}^{M} N_i \zeta_i^r \sum_{j=1}^{M} \beta(\zeta_i, \zeta_j) N_j, \tag{7.15}$$

shows that Eq. (7.14) and Eq. (7.15) are identical, and therefore the method is internally consistent if and only if the chosen grid is such that a particle belonging to the I_j interval aggregating with a particle belonging to the I_k interval and generating a particle belonging to the I_i interval have pivots that respect the relationship $\zeta_i = \zeta_j + \zeta_k$. If this relationship is satisfied, the numerical scheme is internally consistent with respect to all the moments of the NDF. If the grid does not respect this relationship (which is usually the case), it is necessary to proceed to a reassignment of the newly generated particles to the neighboring

[8]This moving-pivot technique is different from QBMM, which will be introduced later in Section 7.4. In fact, in QBMM the pivots or nodes of the quadrature approximation are free to move not only in the interval I_i, but also in phase space, conferring greater flexibility on the method.

pivotal points. The reassignment is carried out in order to preserve some global properties of the distribution. For example, the preservation of the rth-order moment results in the reassignment equation $(\zeta_j + \zeta_k)^r = \gamma_i^{(i)} (\zeta_i)^r + \gamma_{i+1}^{(i)} (\zeta_{i+1})^r$, and, since two degrees of freedom are available, the values of $\gamma_i^{(i)}$ and $\gamma_{i+1}^{(i)}$ are chosen in order to preserve two different moments of the distribution. If, for example, the moment of order zero, m_0, and the moment of order one, m_1, are chosen to be correctly predicted, the reassignment of the particle belonging to the interval I_i to the pivotal points ζ_i and ζ_{i+1} will occur according to

$$\gamma_i^{(i)} = \frac{\zeta_{i+1} - (\zeta_j + \zeta_k)}{\zeta_{i+1} - \zeta_i}, \quad \gamma_{i+1}^{(i)} = \frac{(\zeta_j + \zeta_k) - \zeta_i}{\zeta_{i+1} - \zeta_i}. \tag{7.16}$$

It is clear that this approach appears internally consistent only for two chosen moments of the NDF, whereas for the other moments the method will behave inconsistently.[9] Naturally the definition of consistency can be extended to any specific property of the NDF and not only through their moments[10] (see, for example, Ramkrishna (2000)). This reassignment procedure is necessary any time a new particle appears in the system. Not only second-order, but also zeroth- and first-order point processes produce new particles and, in order to preserve some integral properties of the distribution, it is necessary to reassign particles in order to fit into the chosen grid. For example, in the source term for breakage the daughter particles generated by fragmentation of a particle belonging to the interval I_j might not fall into intervals with pivots that preserve the selected moments. It can be shown (Ramkrishna, 2000) that the condition for internal consistency for breakage is

$$\int_{\xi_i}^{\xi_{i+1}} \xi^r b(\xi|\xi_j) \mathrm{d}\xi = \zeta_i^r \int_{\xi_i}^{\xi_{i+1}} b(\xi|\xi_j) \mathrm{d}\xi \tag{7.17}$$

with $\zeta_i > \zeta_j$ and where $b(\xi|\xi_j)$ is the daughter distribution function. This condition is exactly respected only for $r = 0$ and in the other cases it is necessary to proceed to a particle-reassignment process similar to that for a second-order point process.

Let us analyze different methods resulting from different grids. If a linear grid is used, phase space is divided into a number of intervals delimited by a lower bound ξ_i and an upper bound ξ_{i+1}, and characterized by the pivotal point ζ_i according to the following scheme: $\xi_i = \frac{1}{2}ih$, $\xi_{i+1} = \frac{3}{2}ih$, $\zeta_i = ih$, for $i \in 1, \ldots, M$, where h is the size of the discretization. It is clear that this scheme is always consistent for aggregation. In fact, any time a particle belonging to the interval I_j with pivot equal to ζ_j aggregates with a particle belonging to the interval I_k with pivot equal to ζ_k the resulting particle belonging to the interval I_i has a pivot $\zeta_i = \zeta_j + \zeta_k$, which is the sum of the previous ones, as required by the internal consistency condition (see Eq. (7.16)). In the case of aggregation and breakage the evolution equation of the number of particles N_i with internal coordinate between ξ_i and ξ_{i+1} (or, in other words, belonging to the interval I_i) and characterized by a pivotal point equal to ζ_i becomes

$$\frac{\mathrm{d}N_i}{\mathrm{d}t} = \frac{1}{2} \sum_{j=1}^{i-1} \beta_{i-j,j} N_{i-j} N_j - N_i \sum_{j=1}^{M} \beta_{i,j} N_j + \sum_{j=i+1}^{M} a_j b_{i,j} N_j - a_i N_i, \tag{7.18}$$

[9] This fact implies that the moments computed from their definition will be different from the moments computed by their transport equations derived using Eq. (7.7).

[10] In general, any global property of the distribution $g(\xi)$ can be used in the reassignment process. However, in the cases analyzed in this book, the most important global properties of the NDF are the moments; therefore, we limit the discussion to the case in which $g(\xi) = \xi^r$.

where $\beta_{i,j} = \beta(\zeta_i, \zeta_j)$ is the aggregation kernel, $a_i = a(\zeta_i)$ is the breakage kernel, and $b_{i,j} = b(\zeta_i|\zeta_j)$ is the daughter distribution function.[11] Batterham *et al.* (1981) introduced a discretization that is based on a geometric grid with constant ratio equal to 2. In this geometric grid the internal coordinate is discretized into intervals I_i delimited by a lower bound ξ_i and an upper bound ξ_{i+1}, and characterized by the pivotal point ζ_i according to the following scheme: $\xi_i = \frac{3}{4}\zeta_i$, $\xi_{i+1} = \frac{3}{2}\zeta_i$, $\zeta_{i+1} = 2\zeta_i$, for $i \in 1, \ldots, M-1$. It is clear that the advantage of this geometric grid with respect to the linear one arises from its coarseness. For example, if with the linear grid 100, 1000, or 10 000 classes are necessary to cover some specific ranges, with the geometric grid only 6, 10, and 13 classes, respectively, are needed to cover the same ranges. This much lower computational cost is obtained at the expense of the internal consistency of the scheme. Therefore, in order to preserve some global properties, the reassignment process described above must be used. If the zeroth- and first-order moments are chosen, as described in Eq. (7.16), the following equation for the variation of the number of particles of the interval I_i undergoing pure aggregation is obtained:[12]

$$\frac{\mathrm{d}N_i}{\mathrm{d}t} = N_{i-1}\sum_{j=1}^{i-1} 2^{j-i+1}\beta_{i-1,j}N_j + \frac{1}{2}\beta_{i-1,j-1}N_{i-1}^2 - N_i\sum_{j=1}^{i-1} 2^{j-i}\beta_{i,j}N_j - N_i\sum_{j=1}^{M}\beta_{i,j}N_j. \quad (7.19)$$

By using the same approach, this method can be extended to breakage (Ramkrishna, 2000).

The fixed-pivot technique can, in general, be formulated for any grid while preserving two moments of the distribution of orders r_1 and r_2. The resulting equation for the evolution of the number of particles of the I_i interval is (Kumar & Ramkrishna, 1996a)

$$\begin{aligned}\frac{\mathrm{d}N_i}{\mathrm{d}t} = &\sum_{j=i}^{M} N_j a_j \pi_{i,j} - a_i N_i + \sum_{\zeta_{i-1}\le(\zeta_j+\zeta_k)<\zeta_i}^{j\ge k} \left(1 - \tfrac{1}{2}\delta_{jk}\right)\left[\gamma_i^{(i-1)}(\zeta_j+\zeta_k)\beta_{j,k}N_jN_k\right] \\ &+ \sum_{\zeta_i\le(\zeta_j+\zeta_k)<\zeta_{i+1}}^{j\ge k} \left(1 - \tfrac{1}{2}\delta_{jk}\right)\left[\gamma_i^{(i)}(\zeta_j+\zeta_k)\beta_{j,k}N_jN_k\right] - N_i\sum_{j=1}^{M}\beta_{i,j}N_j, \end{aligned} \quad (7.20)$$

where

$$\begin{aligned}\gamma_i^{(i)} &= \frac{(\zeta_j+\zeta_k)^{r_1}(\zeta_{i+1})^{r_2} - (\zeta_j+\zeta_k)^{r_2}(\zeta_{i+1})^{r_1}}{(\zeta_i)^{r_1}(\zeta_{i+1})^{r_2} - (\zeta_i)^{r_2}(\zeta_{i+1})^{r_1}}, \\ \gamma_i^{(i+1)} &= \frac{(\zeta_j+\zeta_k)^{r_2}(\zeta_i)^{r_1} - (\zeta_j+\zeta_k)^{r_1}(\zeta_i)^{r_2}}{(\zeta_i)^{r_1}(\zeta_{i+1})^{r_2} - (\zeta_i)^{r_2}(\zeta_{i+1})^{r_1}},\end{aligned} \quad (7.21)$$

and

$$\pi_{i,j} = \int_{\zeta_{i-1}}^{\zeta_i} \gamma_i^{(i-1)} b(\xi|\xi_j)\,\mathrm{d}\xi + \int_{\zeta_i}^{\zeta_{i+1}} \gamma_i^{(i-1)} b(\xi|\xi_j)\,\mathrm{d}\xi. \quad (7.22)$$

---

[11] It is interesting to note that in the case of aggregation and breakage of unbreakable monomers, Eq. (7.18) is the rigorous population balance equation. In fact, it describes the particulate system for all the possible classes (Vanni, 2000a). In this case the internal coordinate cannot vary continuously but instead can assume only discrete values depending on the number of monomers that constitute a single aggregate. For example, an aggregate constituted by four monomers will be characterized by a dimensionless mass of four, and this aggregate when aggregating with another particle, say with dimensionless mass five, will generate an aggregate with dimensionless mass equal to nine and similarly in the case of breakage. It is therefore clear that in the case with a linear grid and unbreakable monodisperse monomers the source term due to aggregation and breakage cannot generate particles that do not fall into a specific interval.

[12] Equation (7.19) was also obtained by Hounslow *et al.* (1988) after correcting the original equation proposed by Batterham *et al.* (1981). However, this equation is nothing other than the fixed-pivot technique applied for a geometric grid with ratio equal to two and after superimposing the conservation of the zeroth- and first-order moments (with particle mass as internal coordinate).

This is the most general formulation of the fixed-pivot technique that contains as special cases most of the previously developed class methods. For example, Marchal *et al.* (1988) developed a general method in which aggregation (and breakage) are considered as chemical reactions and stoichiometric coefficients are determined in order to preserve the first-order moment of the NDF. Litster *et al.* (1995) developed an approach that is based on a geometric grid in which the ratio is $2^{1/q}$, where q is an integer.[13] Among the other available methods, it is important to mention the scheme developed by Vanni (1999) for a generic grid for pure breakage. Although the fixed-pivot technique is the most popular discretizaton method, it has problems with the over-prediction of the number density, especially in the case of particle aggregation. To overcome this problem the moving-pivot technique, which offers the additional property of consistency for more moments than for the fixed-pivot technique, was formulated. We refer the reader to the original work of Kumar & Ramkrishna (1996b) for further details. It is instead interesting to devote a few more words to the cell-average technique (Kumar *et al.*, 2006, 2007, 2008b, 2009), which was developed to overcome some of the issues encountered with the fixed-pivot technique.

With the fixed-pivot technique, when a new particle is formed it is reassigned to the neighboring sections according to some reassignment factors $\gamma_i^{(i)}$ and $\gamma_{i+1}^{(i)}$ calculated *a priori* by imposing the conservation of two moments. It is clear, however, that this reassignment is totally arbitrary and is purely based on the grid structure. A simple way to improve the method is to calculate these reassignment factors adaptively, in terms of the net rate of change of two selected moments. When the moments of order zero and one are selected, the procedure is based on the evaluation of the rate of change of the moment of order zero, due to the total birth rate of particles in the I_i interval (Kumar & Ramkrishna, 1997),

$$B_i^{\text{agg}} = \sum_{\zeta_{i-1} \leq (\zeta_j + \zeta_k) < \zeta_i}^{j \geq k} \left(1 - \tfrac{1}{2}\delta_{jk}\right) \beta_{j,k} N_j N_k,$$

and on the evaluation of the rate of change of the moment of order one in the I_i interval,

$$M_i^{\text{agg}} = \sum_{\zeta_{i-1} \leq (\zeta_j + \zeta_k) < \zeta_i}^{j \geq k} \left(1 - \tfrac{1}{2}\delta_{jk}\right) \beta_{j,k} N_j N_k (\zeta_j + \zeta_k).$$

Consequently the average value of the internal coordinate for all newborn particles in the I_i interval can be evaluated as $\overline{\zeta}_i = M_i^{\text{agg}}/B_i^{\text{agg}}$, and the reassignment coefficients can therefore be calculated as $\gamma_i^{(i)} + \gamma_{i+1}^{(i)} = 1$ and $\gamma_i^{(i)}\zeta_i + \gamma_{i+1}^{(i)}\zeta_{i+1} = \overline{\zeta}_i$, resulting in $\gamma_i^{(i)} = (\overline{\zeta}_i - \zeta_{i+1})/(\zeta_i - \zeta_{i+1})$ and $\gamma_{i+1}^{(i)} = (\overline{\zeta}_i - \zeta_i)/(\zeta_{i+1} - \zeta_i)$. By summing up and collecting all contributions to the birth and the death of particles in the I_i interval, and by introducing a slightly different notation, the following global expression is obtained:

$$\begin{aligned} \frac{\mathrm{d}N_i}{\mathrm{d}t} &= B_{i-1}^{\text{agg}} \lambda_i^- \left(\overline{\zeta}_{i-1}\right) H\left(\overline{\zeta}_{i-1} - \zeta_{i-1}\right) + B_i^{\text{agg}} \lambda_i^- \left(\overline{\zeta}_i\right) H\left(\zeta_i - \overline{\zeta}_i\right) \\ &\quad + B_i^{\text{agg}} \lambda_i^+ \left(\overline{\zeta}_i\right) H\left(\overline{\zeta}_i - \zeta_i\right) + B_{i+1}^{\text{agg}} \lambda_i^+ \left(\overline{\zeta}_{i+1}\right) H\left(\zeta_{i+1} - \overline{\zeta}_{i+1}\right) - N_i \sum_{j=1}^{M} \beta_{ij} N_j, \end{aligned}$$

[13] It is evident that on increasing q the discretization becomes finer and finer, whereas for $q = 1$ the method becomes again the one described by Hounslow *et al.* (1988).

where the Heaviside step function is defined as

$$H(\xi) = \begin{cases} 1 & \text{if } \xi > 0, \\ 1/2 & \text{if } \xi = 0, \\ 0 & \text{if } \xi < 0, \end{cases}$$

and $\lambda^{\pm}(\xi) = (\xi - \zeta_{i\pm1})/(\zeta_i - \zeta_{i\pm1})$. Similarly, the net rate of change of the moment of order zero in the interval I_i due to breakage events can be written in discretized form as

$$B_i^{\text{break}} = \sum_{j \geq i} N_j a_j \int_{\zeta_i}^{\zeta_{i+1}} b(\xi|\xi_j)\,\mathrm{d}\xi,$$

whereas the net rate of change of the moment of order one into the interval for breakage events is

$$M_i^{\text{break}} = \sum_{j \geq i} N_j a_j \int_{\zeta_i}^{\zeta_{i+1}} b(\xi|\xi_j)\xi\,\mathrm{d}\xi,$$

from which the average value of the internal coordinate in the interval under scrutiny can be calculated as $\overline{\zeta_i} = M_i^{\text{break}}/B_i^{\text{break}}$, resulting in very similar equations for breakage. Readers interested in more details are referred to the work of Kumar *et al.* (2008b). Sectional methods that are based on the same concepts have been developed independently in the field of evaporating sprays (Greenberg *et al.*, 1993), resulting naturally in very similar equations. For an overview of this field, readers are referred to the work of Laurent *et al.* (2004).

Zeroth-order point processes

Let us consider as an example nucleation, which is a zeroth-order point process, since it introduces new particles into the system at a rate independent of the NDF. Zeroth-order point processes are usually treated in a very simple way by calculating the contribution to each class. For example, if just nucleation is occurring, the PBE becomes

$$\frac{\partial n}{\partial t} = J, \tag{7.23}$$

where J is the nucleation rate, namely the number of new particles produced per unit volume with internal coordinate between ξ and $\xi + \mathrm{d}\xi$. The source term in the discretized PBE simply becomes

$$\frac{\mathrm{d}N_i}{\mathrm{d}t} = \int_{\xi_i}^{\xi_{i+1}} J\,\mathrm{d}\xi = J_i, \tag{7.24}$$

where J_i represents the number of new particles introduced into the internal-coordinate interval I_i delimited by ξ_i and ξ_{i+1}.

The phase-space drift term

The drift term represents a continuous variation of the internal coordinate ξ due to a rate of change $\dot{\xi}$:

$$\frac{\partial n}{\partial t} = S = -\frac{\partial \dot{\xi} n}{\partial \xi}. \tag{7.25}$$

Therefore, after integration over the partition, we obtain

$$\frac{\mathrm{d}N_i}{\mathrm{d}t} = \int_{\xi_i}^{\xi_{i+1}} S \, \mathrm{d}\xi = (\dot{\xi}n)|_{\xi_i} - (\dot{\xi}n)|_{\xi_{i+1}}, \tag{7.26}$$

which states that the accumulation of particles belonging to class I_i equals the net flux of particles from the surrounding intervals I_{i-1} and I_{i+1}. It is important to note that the terms on the right-hand side of Eq. (7.26) involve the NDF evaluated at the points ξ_{i+1} and ξ_i. Unfortunately, the NDF is not known, but it can be approximated by a polynomial inside each interval. When the NDF is described by a zeroth-order polynomial in each interval (as illustrated in Eq. (7.7)), it is approximated by a constant value N_i within the I_i interval. Indeed, this is a discontinuous approximation, since, for example, $\xi = \xi_i$ corresponds to the boundary between two intervals, where the NDF is approximated by N_{i-1} in the I_{i-1} interval and by N_i in the I_i interval. When evaluating the fluxes of particles at the lower and upper boundaries of each single interval this distinction becomes very important. The simplest approach is the first-order upwind scheme, in which the value of $n(t, \xi_i)$ is taken equal to N_{i-1} for positive $\dot{\xi}(\xi_i)$ and equal to N_i for negative $\dot{\xi}(\xi_i)$. It can be shown that this method is of order one,[14] as demonstrated by Laurent *et al.* (2004). In the case of positive $\dot{\xi}(\xi_i)$, the equation for the number density of particles belonging to the I_i interval is

$$\frac{\mathrm{d}N_i}{\mathrm{d}t} = -\left[\frac{\dot{\xi}(\xi_i)N_i}{\xi_{i+1} - \xi_i}\right] + \left[\frac{\dot{\xi}(\xi_{i-1})N_{i-1}}{\xi_i - \xi_{i-1}}\right]. \tag{7.27}$$

In the case of a geometric grid, if s is the ratio of the geometric grid (i.e. $s = \xi_{i+1}/\xi_i$), we can write

$$\frac{\mathrm{d}N_i}{\mathrm{d}t} = \frac{1}{(s-1)\xi_i}\left[\dot{\xi}(\xi_{i-1})N_{i-1} - \dot{\xi}(\xi_i)N_i\right], \tag{7.28}$$

but this is a first-order approximation scheme that suffers from numerical diffusion. In fact, if an NDF exhibiting steeply increasing or decreasing internal-coordinate dependences were processed with this scheme, while growing (or shrinking according to the sign of $\dot{\xi}$) it would spread in phase space, as if an additional diffusion term were present. To overcome numerical diffusion, higher-order schemes can be developed, wherein N_i is written as a function of N_{i-1} and N_{i+1}:

$$\frac{\mathrm{d}N_i}{\mathrm{d}t} = \frac{1}{2\xi_i}\left[\frac{s}{s-1}\dot{\xi}(\xi_{i-1})N_{i-1} - \frac{1}{s^2 - s}\dot{\xi}(\xi_{i+1})N_{i+1}\right], \tag{7.29}$$

as described by Marchal *et al.* (1988) and by David *et al.* (1991). It is possible to derive an alternative second-order method by calculating the rate of variation of N_i as

$$\frac{\mathrm{d}N_i}{\mathrm{d}t} = \frac{1}{\xi_i}\left[a\dot{\xi}(\xi_{i-1})N_{i-1} + b\dot{\xi}(\xi_i)N_i + c\dot{\xi}(\xi_{i+1})N_{i+1}\right], \tag{7.30}$$

and choosing the constants a, b, and c in order to have three moments correctly predicted. If m_0, m_1, and m_3 are chosen, we obtain

$$a = \frac{2}{(1+s)(s^2-1)}, \qquad b = \frac{2}{1+s}, \qquad c = -\frac{2s}{(1+s)(s^2-1)}. \tag{7.31}$$

[14] In the sense that the first excluded term in the Taylor expansion approximating the NDF is the first-order derivative term.

Alternative higher-order methods are available in the literature, and most of them share the salient characteristics of the spatial discretization schemes used in finite-volume codes (such as, for example, QUICK and MUSCL). Some of them are based on polynomial reconstructions with some minor modifications in order to ensure positivity of the NDF, as explained in Laurent (2006), whereas others are based on the maximization of Shannon entropy (see, for example, Massot *et al.* (2010)).

The discretization in terms of particle mass, as described in Eq. (7.8), is particularly popular for the treatment of the drift term. In this case the rate of change of the mass of particles belonging to the I_i interval becomes

$$\frac{\mathrm{d}M_i}{\mathrm{d}t} = -\int_{\xi_i}^{\xi_{i+1}} \frac{\partial \dot{\xi} n}{\partial \xi} \rho_\mathrm{p} V_\mathrm{p}(\xi) \mathrm{d}\xi.$$

If the generic expression for particle volume $V_\mathrm{p}(\xi) = \alpha \xi^\gamma$ is used, and if the integral is solved by the rule of integration by parts, the following expression is obtained:

$$\frac{\mathrm{d}M_i}{\mathrm{d}t} = -\rho_\mathrm{p} \alpha \xi^\gamma \dot{\xi} n \Big|_{\xi_i}^{\xi_{i+1}} + \int_{\xi_i}^{\xi_{i+1}} \rho_\mathrm{p} \alpha \gamma \xi^{\gamma-1} \dot{\xi} n \, \mathrm{d}\xi.$$

If the NDF is approximated by a polynomial of order zero within the I_i interval, we obtain

$$n \approx \frac{M_i}{\rho_\mathrm{p} \alpha \gamma \left(\xi_{i+1}^{\gamma+1} - \xi_i^{\gamma+1} \right)}.$$

Then, if the first-order upwind scheme is employed, the final equation is

$$\frac{\mathrm{d}M_i}{\mathrm{d}t} = -M_i \frac{\xi_i^\gamma \dot{\xi}(\xi_i)}{\gamma \left(\xi_{i+1}^{\gamma+1} - \xi_i^{\gamma+1} \right)} + M_{i-1} \frac{\xi_{i-1}^\gamma \dot{\xi}(\xi_{i-1})}{\gamma \left(\xi_i^{\gamma+1} - \xi_{i-1}^{\gamma+1} \right)} + M_i \frac{\int_{\xi_i}^{\xi_{i+1}} \xi^{\gamma-1} \dot{\xi} \, \mathrm{d}\xi}{\xi_i^{\gamma+1} - \xi_{i-1}^{\gamma+1}}, \tag{7.32}$$

where the first term on the right-hand side represents the mass of the particles leaving the interval because they are growing into the next interval, the second term represents the mass of particles coming from the smaller size interval, and the last term represents the increase of mass within the interval due to the flux from the continuous phase. Once an expression for $\dot{\xi}(\xi)$ has been specified, the last term of Eq. (7.32) can be explicitly written, resulting in an evolution equation for M_i written as the summation of two terms involving M_i and one term involving M_{i-1}. Obviously, the fineness of the discretization determines the accuracy of the solution. As can easily be imagined, in the case of a very fine grid the different approaches give very similar performances, but some of them suffer from the possibility of negative (unrealizable) solutions for the discretized number or mass density. This is particularly severe when a problem with discontinuities is solved: unreasonable oscillations with excursions to negative values are observed, as explained by Kumar & Ramkrishna (1997). Moreover, class and sectional methods are numerically equivalent to finite-difference and finite-volume methods that suffer from numerical diffusion. The latter can be reduced by using higher-order schemes, but cannot be completely avoided.

The phase-space diffusion term

The diffusion term represents a continuous variation of the internal coordinate ξ with diffusivity D,

$$\frac{\partial n}{\partial t} = S = \frac{\partial^2 D n}{\partial \xi^2}, \tag{7.33}$$

and, therefore, after integration over the partition, we obtain

$$\frac{\mathrm{d}N_i}{\mathrm{d}t} = \int_{\xi_i}^{\xi_{i+1}} S \, \mathrm{d}\xi = \left.\frac{\partial Dn}{\partial \xi}\right|_{\xi_{i+1}} - \left.\frac{\partial Dn}{\partial \xi}\right|_{\xi_i}, \tag{7.34}$$

which states that the accumulation of particles belonging to class I_i equals the net diffusive flux of particles from the surrounding intervals I_{i-1} and I_{i+1}. It is important to note that the terms on the right-hand side of Eq. (7.34) involve the gradient of the NDF evaluated at the points ξ_{i+1} and ξ_i. Since the NDF is not known, usually its gradients are reconstructed from the available information, by employing for example the following approximation: $n_i = N_i/(\xi_i - \xi_{i-1})$. In the case of constant D the gradient results in

$$\left.\frac{\partial Dn}{\partial \xi}\right|_{\xi_i} = D\left(\frac{N_i}{(\xi_i - \xi_{i-1})^2} - \frac{N_{i-1}}{(\xi_i - \xi_{i-1})(\xi_{i-1} - \xi_{i-2})}\right), \tag{7.35}$$

resulting in a simple scheme for phase-space diffusion terms. Improved versions can be obtained with arguments very similar to those presented for the discretization of drift terms.

7.2.2 Bivariate and multivariate PBE

In the case of two internal coordinates with NDF $n(t, \xi_1, \xi_2)$, the PBE has the following form:

$$\frac{\partial n}{\partial t} + \frac{\partial \dot{\xi}_1 n}{\partial \xi_1} + \frac{\partial \dot{\xi}_2 n}{\partial \xi_2} = \frac{\partial^2 D_{11} n}{\partial \xi_1^2} + 2\frac{\partial^2 D_{12} n}{\partial \xi_1 \, \partial \xi_2} + \frac{\partial^2 D_{22} n}{\partial \xi_2^2} + \mathcal{S}, \tag{7.36}$$

where the second and third terms on the left-hand side represent advection in phase space, whereas the terms on the right-hand side are diffusion in phase space and the discontinuous point process. As before, it is possible to write the PBE using a generic source term:

$$S = -\frac{\partial \dot{\xi}_1 n}{\partial \xi_1} - \frac{\partial \dot{\xi}_2 n}{\partial \xi_2} + \frac{\partial^2 D_{11} n}{\partial \xi_1^2} + 2\frac{\partial^2 D_{12} n}{\partial \xi_1 \, \partial \xi_2} + \frac{\partial^2 D_{22} n}{\partial \xi_2^2} + \mathcal{S}. \tag{7.37}$$

Readers interested in a detailed discussion concerning the application of the different CM to this equation are referred to the specialized literature (Immanuel & Doyle, 2005; Laurenzi *et al.*, 2002; Qamar & Warnecke, 2007; Vale & McKenna, 2005; Xiong & Pratsinis, 1991, 1993). In what follows, only a brief summary is reported.[15] Let us now discuss a simple geometric extension to bivariate problems of the discretization presented for univariate PBE. Given the interval of the internal coordinate ξ_1 delimited by $\xi_1^{(i)}$ and $\xi_1^{(i+1)}$, denoted by $I_1^{(i)}$, and the interval of the second internal coordinate ξ_2 delimited by $\xi_2^{(j)}$ and

[15] Class methods can be applied to bivariate cases by following two alternative philosophies. The first is geometric in nature, where each internal coordinate is treated identically. The resulting two-dimensional Cartesian grid is formed by rectangles defined by the intersections of lines parallel to the internal coordinate axes (Kumar *et al.*, 2008a; Vale & McKenna, 2005). It is clear that in this case a particle formed by any event will have to be distributed over four different intervals, defined by four different pivots. For a general multidimensional PBE with n internal coordinates, this will result in distribution over 2^n intervals. The second approach is instead based on a new concept: a population of particles characterized by n internal coordinates is characterized by a minimum of $n + 1$ quantities that need to be preserved (in contrast with the 2^n with the previous method). In fact, as is well known, in a bi-dimensional space triangles are the simplest geometric object that can be used to represent a discretized form of phase space. Analogously, tetrahedrals can be used in three-dimensional spaces, and so on. These speculations suggest a minimal efficient way to discretize phase space, as highlighted by Chakraborty & Kumar (2007) and Nandanwar & Kumar (2008).

$\xi_2^{(j+1)}$, denoted by $I_2^{(j)}$, it is possible to define the number of particles belonging to the $I_1^{(i)}$ and the $I_2^{(j)}$ intervals as

$$N_{ij} \equiv \int_{\xi_1^{(i)}}^{\xi_1^{(i+1)}} \int_{\xi_2^{(j)}}^{\xi_2^{(j+1)}} n \, \mathrm{d}\boldsymbol{\xi}. \tag{7.38}$$

Also in this case it is possible to define pivotal points $\zeta_1^{(i)}$ and $\zeta_2^{(j)}$, and the resulting representation for the NDF is

$$n(t, \boldsymbol{\xi}) = \sum_{i=1}^{M_1} \sum_{j=1}^{M_2} N_{ij}(t) \delta\left(\xi_1 - \zeta_1^{(i)}\right) \delta\left(\xi_2 - \zeta_2^{(j)}\right), \tag{7.39}$$

where M_1 and M_2 are the number of sections or intervals used for the two internal coordinates. The evolution equation for N_{ij} results in

$$\frac{\mathrm{d}N_{ij}}{\mathrm{d}t} = \int_{\xi_1^{(i)}}^{\xi_1^{(i+1)}} \int_{\xi_2^{(j)}}^{\xi_2^{(j+1)}} S \, \mathrm{d}\boldsymbol{\xi} = S_{ij}. \tag{7.40}$$

Class methods have been applied to bivariate PBE only for simple problems, and generally their extension to multivariate problems is quite complicated. In what follows selected examples are discussed to illustrate these difficulties.

Point processes

In the case of two internal coordinates, the discontinuous point process can assume different forms, depending on the phenomenon considered (e.g. aggregation/coalescence versus breakage) and the nature of the internal coordinates. Let us illustrate the main issues through an example: aggregation. Let us, moreover, consider the problem of aggregative mixing, namely aggregation of two different types of particles. In this particular case the internal coordinates are additive during aggregation, as described by Matsoukas *et al.* (2006, 2009). Under these conditions the PBE is

$$\begin{aligned}\frac{\partial n}{\partial t} = &\frac{1}{2} \int_0^{\xi_1} \int_0^{\xi_2} \beta(\xi_1 - \xi_1', \xi_1', \xi_2 - \xi_2', \xi_2') n(\boldsymbol{\xi} - \boldsymbol{\xi}') n(\boldsymbol{\xi}') \mathrm{d}\boldsymbol{\xi}' \\ &- \iint_0^{\infty} \beta(\xi_1', \xi_1, \xi_2', \xi_2) n(\boldsymbol{\xi}') n(\boldsymbol{\xi}) \mathrm{d}\boldsymbol{\xi}'.\end{aligned}$$

By applying the integration over the $I_1^{(i)}$ and $I_2^{(j)}$ intervals for the two internal coordinates and by applying the definition in Eq. (7.40), we obtain

$$\begin{aligned}\frac{\mathrm{d}N_{ij}}{\mathrm{d}t} = &\frac{1}{2} \int_{\xi_1^{(i)}}^{\xi_1^{(i+1)}} \int_{\xi_2^{(i)}}^{\xi_2^{(i+1)}} \int_0^{\xi_1} \int_0^{\xi_2} \beta(\xi_1 - \xi_1', \xi_1', \xi_2 - \xi_2', \xi_2') n(\boldsymbol{\xi} - \boldsymbol{\xi}') n(\boldsymbol{\xi}') \mathrm{d}\boldsymbol{\xi}' \, \mathrm{d}\boldsymbol{\xi} \\ &- \int_{\xi_1^{(i)}}^{\xi_1^{(i+1)}} \int_{\xi_2^{(i)}}^{\xi_2^{(i+1)}} \iint_0^{\infty} \beta(\xi_1', \xi_1, \xi_2', \xi_2) n(\boldsymbol{\xi}') n(\boldsymbol{\xi}) \mathrm{d}\boldsymbol{\xi}' \, \mathrm{d}\boldsymbol{\xi}.\end{aligned}$$

As for the univariate case, when a particle is formed after an aggregation event, it has to be assigned to the neighboring cells (four in a bivariate case). When the fixed-pivot technique is used to assign the particles to cells, the following expression is obtained:

$$\frac{\mathrm{d}N_{ij}}{\mathrm{d}t} = -\sum_{k=1}^{M_1}\sum_{l=1}^{M_2}\beta_{ikjl}N_{ij}N_{kl} + \sum_{\xi_1^{(i-1)}<\zeta_1^{(k)}+\zeta_1^{(l)}<\xi_1^{(i+1)}}^{k\geq l}\ \sum_{\xi_2^{(j-1)}<\zeta_2^{(m)}+\zeta_2^{(n)}<\xi_2^{(j+1)}}^{m\geq n}\left(1-\tfrac{1}{2}\delta_{k,l}\delta_{m,n}\right)\eta_{ij} \times\left(\zeta_1^{(k)}+\zeta_1^{(l)},\zeta_2^{(m)}+\zeta_2^{(n)}\right)\beta_{klmn}N_{km}N_{ln},$$

where the second two summations are carried out over the pair of indices (k,l) and (m,n), respectively, M_1 and M_2 are the numbers of intervals used in the discretization of the two internal coordinates, and $\beta_{klmn} = \beta\left(\zeta_1^{(k)},\zeta_1^{(l)},\zeta_2^{(m)},\zeta_2^{(n)}\right)$. The reassignment coefficient η_{ij} is calculated by superimposing the conservation of some moments of the NDF. If, for example, the first three moments of the distribution (i.e. $m_{0,0}$, $m_{1,0}$, and $m_{0,1}$) are used, the following expression is obtained:

$$\eta_{ij}(\boldsymbol{\zeta}) = \begin{cases} \dfrac{\left(\zeta_1-\xi_1^{(i-1)}\right)\left(\zeta_2-\xi_2^{(j-1)}\right)}{\left(\xi_1^{(i)}-\xi_1^{(i-1)}\right)\left(\xi_2^{(j)}-\xi_2^{(j-1)}\right)} & \text{if } \xi_1^{(i-1)}\leq\zeta_1<\xi_1^{(i)} \text{ and } \xi_2^{(j-1)}\leq\zeta_2<\xi_2^{(j)}, \\ \dfrac{\left(\zeta_1-\xi_1^{(i-1)}\right)\left(\xi_2^{(j+1)}-\zeta_2\right)}{\left(\xi_1^{(i)}-\xi_1^{(i-1)}\right)\left(\xi_2^{(j+1)}-\xi_2^{(j)}\right)} & \text{if } \xi_1^{(i-1)}\leq\zeta_1<\xi_1^{(i)} \text{ and } \xi_2^{(j)}\leq\zeta_2<\xi_2^{(j+1)}, \\ \dfrac{\left(\xi_1^{(i+1)}-\zeta_1\right)\left(\xi_2^{(j+1)}-\zeta_2\right)}{\left(\xi_1^{(i+1)}-\xi_1^{(i)}\right)\left(\xi_2^{(j+1)}-\xi_2^{(j)}\right)} & \text{if } \xi_1^{(i)}\leq\zeta_1<\xi_1^{(i+1)} \text{ and } \xi_2^{(j)}\leq\zeta_2<\xi_2^{(j+1)}, \\ \dfrac{\left(\xi_1^{(i+1)}-\zeta_1\right)\left(\zeta_2-\xi_2^{(j-1)}\right)}{\left(\xi_1^{(i+1)}-\xi_1^{(i)}\right)\left(\xi_2^{(j)}-\xi_2^{(j-1)}\right)} & \text{if } \xi_1^{(i)}\leq\zeta_1<\xi_1^{(i+1)} \text{ and } \xi_2^{(j-1)}\leq\zeta_2<\xi_2^{(j)}, \end{cases} \tag{7.41}$$

where the moments are defined as $m_{k,l} = \iint_0^{\infty} \xi_1^k\xi_2^l n\,\mathrm{d}\boldsymbol{\xi}$. The coefficients in Eq. (7.41) were calculated by imposing the conservation of $m_{0,0}$, $m_{1,0}$, and $m_{0,1}$, but in theory any other set of moments could be accommodated. Analogously, by preserving the conservation of the same moment set, $m_{0,0}$, $m_{1,0}$, and $m_{0,1}$, the coefficient can also be calculated by adopting the cell-average technique:

$$\begin{aligned}\frac{\mathrm{d}N_{ij}}{\mathrm{d}t} = &\sum_{p=0}^{1}\sum_{q=0}^{1}\beta_{iqjp}N_{ij}N_{pq} \\ &+\sum_{p=0}^{1}\sum_{q=0}^{1}B_{i-p,j-q}\lambda_{i,j}^{-,-}\left(\zeta_1^{(i-p)},\zeta_2^{(j-q)}\right)H\left[(-1)^p\left(\zeta_1^{(i-p)}-\overline{\zeta}_{i-p}\right)\right]H\left[(-1)^q\left(\zeta_2^{(j-q)}-\overline{\zeta}_{j-q}\right)\right] \\ &+\sum_{p=0}^{1}\sum_{q=0}^{1}B_{i-p,j+q}\lambda_{i,j}^{-,+}\left(\zeta_1^{(i-p)},\zeta_2^{(j+q)}\right)H\left[(-1)^p\left(\zeta_1^{(i-p)}-\overline{\zeta}_{i-p}\right)\right]H\left[(-1)^{q+1}\left(\zeta_2^{(j+q)}-\overline{\zeta}_{j+q}\right)\right] \\ &+\sum_{p=0}^{1}\sum_{q=0}^{1}B_{i+p,j+q}\lambda_{i,j}^{+,+}\left(\zeta_1^{(i+p)},\zeta_2^{(j+q)}\right)H\left[(-1)^{p+1}\left(\zeta_1^{(i+p)}-\overline{\zeta}_{i+p}\right)\right]H\left[(-1)^{q+1}\left(\zeta_2^{(j+q)}-\overline{\zeta}_{j+q}\right)\right] \\ &+\sum_{p=0}^{1}\sum_{q=0}^{1}B_{i+p,j-q}\lambda_{i,j}^{+,-}\left(\zeta_1^{(i+p)},\zeta_2^{(j-q)}\right)H\left[(-1)^{p+1}\left(\zeta_1^{(i+p)}-\overline{\zeta}_{i+p}\right)\right]H\left[(-1)^q\left(\zeta_2^{(j-q)}-\overline{\zeta}_{j-q}\right)\right],\end{aligned}$$

where $H(\xi)$ is the Heaviside step function, the aggregation source term is

$$B_{ij} = \sum_{\xi_1^{(i-1)} \leq \left(\xi_1^{(k)} + \xi_1^{(l)}\right) \leq \xi_1^{(i+1)}}^{k \geq l} \quad \sum_{\xi_2^{(i-1)} \leq \left(\xi_2^{(m)} + \xi_2^{(n)}\right) \leq \xi_2^{(i+1)}}^{m \geq m} \left(1 - \tfrac{1}{2}\delta_{k,l}\delta_{m,n}\right)\beta_{klmn}N_{km}N_{ln},$$

and the other parameters are defined by $\overline{\zeta}_1^{(i)} = M_{ij,1}/B_{ij}$, $\overline{\zeta}_2^{(j)} = M_{ij,2}/B_{ij}$,

$$M_{ij,1} = \sum_{\xi_1^{(i-1)} \leq \left(\xi_1^{(k)} + \xi_1^{(l)}\right) \leq \xi_1^{(i+1)}}^{k \geq l} \quad \sum_{\xi_2^{(i-1)} \leq \left(\xi_2^{(m)} + \xi_2^{(n)}\right) \leq \xi_2^{(i+1)}}^{m \geq m} \left(1 - \tfrac{1}{2}\delta_{k,l}\delta_{m,n}\right)\beta_{klmn}N_{km}N_{ln}\left(\zeta_1^{(k)} + \zeta_1^{(l)}\right),$$

$$M_{ij,2} = \sum_{\xi_1^{(i-1)} \leq \left(\xi_1^{(k)} + \xi_1^{(l)}\right) \leq \xi_1^{(i+1)}}^{k \geq l} \quad \sum_{\xi_2^{(i-1)} \leq \left(\xi_2^{(m)} + \xi_2^{(n)}\right) \leq \xi_2^{(i+1)}}^{m \geq m} \left(1 - \tfrac{1}{2}\delta_{k,l}\delta_{m,n}\right)\beta_{klmn}N_{km}N_{ln}\left(\zeta_2^{(m)} + \zeta_2^{(n)}\right),$$

and

$$\lambda_{i,j}^{\pm,\pm}(\boldsymbol{\xi}) = \frac{\left(\xi_1 - \zeta_1^{(i\pm1)}\right)\left(\xi_2 - \zeta_2^{(j\pm1)}\right)}{\left(\zeta_1^{(i)} - \zeta_1^{(i\pm1)}\right)\left(\zeta_2^{(j)} - \zeta_2^{(j\pm1)}\right)}.$$

The method can be easily extended to the case of breakage, and we refer the reader to the specialized literature for details. It is important to report also the many efforts to reduce and optimize the computational costs by using, for example, expanding grids, as illustrated in the work of Chiney *et al.* (2010).

The phase-space drift term

In the case of pure drift the source term can be written as

$$S_{ij} = -\int_{\xi_1^{(i)}}^{\xi_1^{(i+1)}} \int_{\xi_2^{(j)}}^{\xi_2^{(j+1)}} \frac{\partial \dot{\xi}_1 n}{\partial \xi_1}\, \mathrm{d}\boldsymbol{\xi} - \int_{\xi_1^{(i)}}^{\xi_1^{(i+1)}} \int_{\xi_2^{(j)}}^{\xi_2^{(j+1)}} \frac{\partial \dot{\xi}_2 n}{\partial \xi_2}\, \mathrm{d}\boldsymbol{\xi}. \tag{7.42}$$

Similarly to what is done for one internal coordinate, this term can be written as

$$S_{ij} = f_1^{ij} + f_2^{ij}, \tag{7.43}$$

where f_1^{ij} and f_2^{ij} are the net fluxes of particles due to the continuous rate of change of the first and second internal coordinate, respectively. The analogy to finite-volume schemes is evident and, if a first-order upwind scheme is used, we can write

$$\begin{aligned} f_1^{ij} &= \left(a_i \dot{\xi}_1^{(i-1,j)} N_{i-1,j} - b_i \dot{\xi}_1^{(i,j)} N_{i,j}\right), \\ f_2^{ij} &= \left(c_i \dot{\xi}_1^{(i,j-1)} N_{i,j-1} - d_i \dot{\xi}_1^{(i,j)} N_{i,j}\right), \end{aligned} \tag{7.44}$$

where a_i, b_i, c_i, and d_i are coefficients calculated from the grid as explained by Puel *et al.* (2003a). Obviously this scheme is affected by numerical diffusion, and higher-order schemes could also be used.

The phase-space diffusion term

In the case of pure diffusion the source term can be written as

$$S_{ij} = \int_{\xi_1^{(i)}}^{\xi_1^{(i+1)}} \int_{\xi_2^{(j)}}^{\xi_2^{(j+1)}} \frac{\partial}{\partial \xi_1} \left(\frac{\partial D_{11} n}{\partial \xi_1} + \frac{\partial D_{12} n}{\partial \xi_2} \right) \mathrm{d}\boldsymbol{\xi} + \int_{\xi_1^{(i)}}^{\xi_1^{(i+1)}} \int_{\xi_2^{(j)}}^{\xi_2^{(j+1)}} \frac{\partial}{\partial \xi_2} \left(\frac{\partial D_{12} n}{\partial \xi_1} + \frac{\partial D_{22} n}{\partial \xi_2} \right) \mathrm{d}\boldsymbol{\xi}. \tag{7.45}$$

Similarly to what is done for one internal coordinate, this term can be written as

$$S_{ij} = d_1^{ij} + d_2^{ij}, \tag{7.46}$$

where d_1^{ij} and d_2^{ij} are the diffusive fluxes of particles that can be described with the standard methods used in finite-volume schemes.

7.2.3 Collisional KE

The kinetic equation (KE) for homogeneous systems can be written by grouping together all the source terms appearing in Eq. (7.1) on the right-hand side:

$$\frac{\partial n}{\partial t} = S = -\frac{\partial}{\partial \boldsymbol{\xi}} \cdot \left(\dot{\boldsymbol{\xi}} n \right) + \frac{\partial^2}{\partial \boldsymbol{\xi} \, \partial \boldsymbol{\xi}} : (\mathbf{D} n) + \mathbb{C}, \tag{7.47}$$

where the source term due to discontinuous events is denoted by $\mathbb{C}$ (see Chapter 6). Here, we will restrict our attention to systems without external forces and phase-space diffusion,[16] such that $\dot{\boldsymbol{\xi}}$ and $\mathbf{D}$ are null. The resulting collisional KE describes momentum exchange between particles with constant size and mass, for which the NDF is a function of the particle velocity $\boldsymbol{\xi} = (v_1, v_2, v_3)$. The Boltzmann hard-sphere collision integral is

$$\mathbb{C} = \int_{\mathbb{R}^3} \int_{\mathbb{S}^+} [n(\boldsymbol{\xi}')n(\boldsymbol{\xi}'_*) - n(\boldsymbol{\xi})n(\boldsymbol{\xi}_*)] B(\mathbf{g}, \mathbf{x}) \mathrm{d}\mathbf{x} \, \mathrm{d}\boldsymbol{\xi}_*, \tag{7.48}$$

where $\boldsymbol{\xi}$ and $\boldsymbol{\xi}_*$ are the pre-collision particle velocities of the test and field particles, $\boldsymbol{\xi}'$ and $\boldsymbol{\xi}'_*$ are the post-collision particle velocities, B is the collision kernel, $\mathbf{g} = \boldsymbol{\xi}_* - \boldsymbol{\xi}$ is the relative velocity of the field particle with respect to the test particle, and $\mathbf{x}$ is the unit vector along the direction connecting the center of the two particles during the instantaneous collision pointing from the field to the test particle. Conservation of linear momentum and kinetic energy of the colliding particles[17] results in the following relationships between the pre- and post-collision velocities:

$$\begin{aligned} \boldsymbol{\xi}' &= \boldsymbol{\xi} + (\mathbf{g} \cdot \mathbf{x}) \, \mathbf{x}, \\ \boldsymbol{\xi}'_* &= \boldsymbol{\xi} - (\mathbf{g} \cdot \mathbf{x}) \, \mathbf{x}. \end{aligned} \tag{7.49}$$

[16] For the KE, phase-space diffusion is usually associated with Brownian motion caused by random collisions with an external phase. In comparison, $\mathbb{C}$ is due to self-collisions between particles of the same type. The representation of Brownian motion as a diffusion term implies that collisions with the external phase occurs on a time scale that is much smaller than that for self-collisions.

[17] The kinetic energy is conserved during elastic collisions, whereas a fraction of it is irreversibly lost during inelastic collisions. The coefficient of restitution quantifies these two situations and more details can be found in Chapter 6.

The collision kernel depends upon the type of interactions between particles, and here we focus on the hard-sphere collision kernel: $B(\mathbf{g}, \mathbf{x}) = d_p^2 |\mathbf{g} \cdot \mathbf{x}|$. An alternative kernel is, for example, the Maxwell kernel described in Chapter 6.

Sectional and class methods for the solution of the collisional KE are generally called discrete-velocity methods (DVM). These methods are based on the simple idea of discretizing the velocity space into a grid constituted by a finite number of points. The existing methods are characterized by different grid structures (Aristov, 2001). For example, lattice Boltzmann methods discretize the velocity space into a regular cubic lattice with a constant lattice size (Li-Shi, 2000), whereas other methods employ different discretization schemes (Monaco & Preziosi, 1990). By using a similar approach to that used with PBE, it is possible to define N_i as the number density of the particles with velocity $\boldsymbol{\xi}_i$, and the discretized KE becomes

$$\frac{\mathrm{d}N_i}{\mathrm{d}t} = G_i - L_i, \tag{7.50}$$

where G_i and L_i represent the gain and loss rates, respectively, of particles with velocity equal to $\boldsymbol{\xi}_i$. In this discretized representation, a particle of velocity $\boldsymbol{\xi}_i$ is lost when it collides with another particle with velocity $\boldsymbol{\xi}_j$ and the collision leads to a different pair of post-collision velocities $(\boldsymbol{\xi}_k, \boldsymbol{\xi}_l)$. Analogously, collision of two particles of velocities $\boldsymbol{\xi}_k$ and $\boldsymbol{\xi}_l$ may generate a particle of velocity $\boldsymbol{\xi}_i$. These processes result in an expression for the discrete collision operator:

$$\frac{\mathrm{d}N_i}{\mathrm{d}t} = \sum_j \sum_{(k,l)} \left(A_{kl}^{ij} N_k N_l - A_{ij}^{kl} N_i N_j \right), \tag{7.51}$$

where A_{kl}^{ij} and A_{ij}^{kl} are collision frequencies or transition probabilities between particle-velocity classes, and the second summation is performed only over possible pairs (k, l). The discretized scheme must be consistent with the original KE. For example, since in a generic collision of two particles i and j, the particles k and l result, the four corresponding velocities are not arbitrary, due to conservation of momentum, $\boldsymbol{\xi}_i + \boldsymbol{\xi}_j = \boldsymbol{\xi}_k + \boldsymbol{\xi}_l$, and kinetic energy, $\boldsymbol{\xi}_i^2 + \boldsymbol{\xi}_j^2 = \boldsymbol{\xi}_k^2 + \boldsymbol{\xi}_l^2$. It is easy to show (see, for example, Chapman & Cowling (1961) for a graphical derivation) that, for collisions that respect these two conditions, the lengths of the pre- and post-collision relative velocity vectors are identical, $|\boldsymbol{\xi}_i - \boldsymbol{\xi}_j| = |\boldsymbol{\xi}_k - \boldsymbol{\xi}_l|$, and that, due to conservation of angular momentum, collisions of non-rotating particles must occur in a plane. Therefore, the structure of the discretization must be designed to result in transition probabilities that respect the properties of mass, momentum, and energy conservation. Moreover, it is highly desirable for the discretized method to converge to the continuum KE for finer and finer discretizations (Gorsch, 1999). This is particularly important when the full collision integral is included in the KE, whereas it is not an issue when the collision integral is replaced by a kinetic model (i.e. the BGK (Bhatnagar *et al.*, 1954) and ES-BGK (Holway, 1966) models). For example, when particles are represented as point masses with a finite radius, a particle belonging to class j is able to collide with a particle of class i only if the latter particle happens to be in a cylinder of cross-sectional area[18] S_c and height $|\boldsymbol{\xi}_i - \boldsymbol{\xi}_j|\,\mathrm{d}t$. The resulting transition probability is $A_{ij}^{kl} = S_c|\boldsymbol{\xi}_i - \boldsymbol{\xi}_j| a_{ij}^{kl}$. On assuming that in a collision all possible pairs are obtained with the same probability, it results that the transition probabilities are nonnegative constants (with the value zero corresponding to unrealizable collisions), which are perfectly symmetric with respect to the

[18]This cross-sectional area corresponds to the surface of a circle of radius equal to d_p in the case of hard spheres.

indices i, j and k, l: $A_{ij}^{kl} = A_{ji}^{kl} = A_{ij}^{lk} \geq 0$. As a consequence of the discrete formulation, we also ensure that the relation of micro-reversibility is respected, meaning that the collision of particles i and j to give particles k and l occurs with the same probability as the inverse collision. On using the relation of micro-reversibility and performing the summation over the indices j, k, and l instead of over pairs, the discretized kinetic equation becomes

$$\frac{\mathrm{d}N_i}{\mathrm{d}t} = \frac{1}{2} \sum_{j,k,l} \left(A_{kl}^{ij} N_k N_l - A_{ij}^{kl} N_i N_j \right), \tag{7.52}$$

where $i = 1, 2, \ldots, M$ and the coefficient $1/2$ has been added to avoid counting the same pair of particles twice.

As has already been anticipated, it is important to ensure that the discretization scheme respects some fundamental properties and it can be shown that the scheme in Eq. (7.52) possesses many properties of the original continuous KE, such as the conservation laws and an H-theorem with a Maxwellian equilibrium (Aristov, 2001). Comparison of Eq. (7.52) with Eq. (7.13) shows the similarities and differences between the two discretizations. For example, during particle aggregation a single particle is formed after collision, and therefore only three indices are involved (those of the two pre-collision particles and that of the daughter or post-collision particle). In contrast, during particle collisions (with exchange of just momentum and kinetic energy), the event leaves the number of particles unchanged and therefore four indices must be employed. Different techniques have been developed under different simplifying assumptions, and in what follows we will discuss two different cases, corresponding to the isotropic case and a generic non-isotropic case. In the parallelism with sectional methods for PBE, the methods illustrated for these two different cases share many similarities with the numerical methods developed for the solution of univariate and multivariate PBE.

The isotropic case

In many problems we are not interested in the distribution of particle-velocity vectors, but under certain conditions the distribution of the particle-velocity magnitude gives an adequate description of the system. One such situation is when the hypothesis of isotropy holds. Under this simplifying hypothesis, the direction of the velocity vector loses any importance and as a consequence $\boldsymbol{\xi}$ can be replaced by the scalar $\xi = |\boldsymbol{\xi}| = (v_1^2 + v_2^2 + v_3^2)^{1/2}$ corresponding to the particle-velocity magnitude (or particle speed). In general, if the NDF can be written in terms of the velocity vector, the velocity distribution function $n(\boldsymbol{\xi})$ is known. The NDF in terms of the velocity magnitude (denoted as $n_\xi(\xi)$) can be written explicitly by integrating over all possible directions:

$$n_\xi(\xi) = \int_0^{2\pi} \mathrm{d}\phi \int_0^{\pi} \mathrm{d}\theta \sin(\theta) \xi^2 n(\boldsymbol{\xi}), \tag{7.53}$$

where $\boldsymbol{\xi} = (\xi \sin\theta \cos\phi, \xi \sin\theta \sin\phi, \xi \cos\theta)$, and θ and ϕ are the polar and azimuthal angles, respectively, of the spherical coordinates. The latter are used to express the three-dimensional integrals appearing in the collision integrals (see, for example, Chapter 6). Equation (7.53) is always valid, but can be further simplified under the hypothesis of isotropy because $n(\boldsymbol{\xi})$ is constant on a sphere of constant radius ξ. This results in

$$n_\xi(\xi) = 4\pi\xi^2 n(\boldsymbol{\xi})_{|\boldsymbol{\xi}|=\xi}, \tag{7.54}$$

where $|\boldsymbol{\xi}|$ is constrained to equal ξ in the NDF since the direction of the velocity vector is unimportant. Henceforth we will abuse the notation by denoting $n(\xi) = n(\boldsymbol{\xi})_{|\boldsymbol{\xi}|=\xi}$. Equation (7.54) holds at the equilibrium and when the initial NDF is isotropic. In order to illustrate how the discretization for the kinetic equation is typically carried out, some additional simplifications must be introduced. It is convenient to write the collision integral as

$$\frac{\partial n}{\partial t} = \int_{\mathbb{R}^3} \mathrm{d}\boldsymbol{\xi}_* \int_0^{4\pi} \mathrm{d}\mathbf{x}[n(\boldsymbol{\xi}')n(\boldsymbol{\xi}'_*) - n(\boldsymbol{\xi})n(\boldsymbol{\xi}_*)]B(\mathbf{g},\mathbf{x})S(\mathbf{g}\cdot\mathbf{x}), \tag{7.55}$$

where the integral is now extended to the entire solid angle, but with the function S defined as

$$S(z) = \begin{cases} 1 & \text{if } z < 0, \\ 0 & \text{if } z \geq 0. \end{cases}$$

By writing the second integral with respect to the solid angle, we obtain

$$\frac{\partial n}{\partial t} = \int_0^{\infty} \mathrm{d}\xi_* \int_0^{4\pi} \mathrm{d}\mathbf{x}_* \int_0^{4\pi} \mathrm{d}\mathbf{x}[n(\boldsymbol{\xi}')n(\boldsymbol{\xi}'_*) - n(\boldsymbol{\xi})n(\boldsymbol{\xi}_*)]B(\mathbf{g},\mathbf{x})S(\mathbf{g}\cdot\mathbf{x})\xi_*^2. \tag{7.56}$$

In the case of hard-sphere collisions, the kernel is $B(\mathbf{g},\mathbf{x}) = d_\mathrm{p}^2|(\boldsymbol{\xi}_* - \boldsymbol{\xi})\cdot\mathbf{x}| = d_\mathrm{p}^2|\xi_*\cos(\alpha_y) - \xi\cos(\alpha_x)|$, where α_x is the angle formed by $\boldsymbol{\xi}$ and $\mathbf{x}$, and α_y is the angle formed by $\boldsymbol{\xi}_*$ and $\mathbf{x}$. If the solid angles are written in polar coordinates, $\mathrm{d}\mathbf{x} = \sin(\alpha_x)\mathrm{d}\alpha_x\,\mathrm{d}\beta_x$ and $\mathrm{d}\mathbf{x}_* = \sin(\alpha_y)\mathrm{d}\alpha_y\,\mathrm{d}\beta_y$, the collisional kinetic equation becomes

$$\begin{aligned}\frac{\partial n}{\partial t} = d_\mathrm{p}^2 \int_0^{\infty} \mathrm{d}\xi_* \int_0^{2\pi} \mathrm{d}\beta_x \int_0^{\pi} \sin(\alpha_x)\mathrm{d}\alpha_x \int_0^{2\pi} \mathrm{d}\beta_y \int_0^{\pi} \sin(\alpha_y)\mathrm{d}\alpha_y \\ \times [n(\boldsymbol{\xi}')n(\boldsymbol{\xi}'_*) - n(\boldsymbol{\xi})n(\boldsymbol{\xi}_*)]\left|\xi_*\cos(\alpha_y) - \xi\cos(\alpha_x)\right| S(\mathbf{g}\cdot\mathbf{x})\xi_*^2.\end{aligned} \tag{7.57}$$

On introducing the variables $x = \cos(\alpha_x)$ and $y = \cos(\alpha_y)$, and recalling that because of isotropy the NDF is not a function of α_x, β_x, α_y, and β_y, Eq. (7.57) simplifies to

$$\frac{\partial n}{\partial t} = 4\pi^2 d_\mathrm{p}^2 \int_0^{\infty} \mathrm{d}\xi_* \int_{-1}^{+1} \mathrm{d}x \int_{-1}^{+1} \mathrm{d}y[n(\boldsymbol{\xi}')n(\boldsymbol{\xi}'_*) - n(\boldsymbol{\xi})n(\boldsymbol{\xi}_*)]\,|\xi_* y - \xi x|\,S(\mathbf{g}\cdot\mathbf{x})\xi_*^2. \tag{7.58}$$

Moreover, because of isotropy, the collision operator is symmetric, resulting in

$$\frac{\partial n}{\partial t} = 2\pi^2 d_\mathrm{p}^2 \int_0^{\infty} \mathrm{d}\xi_* \int_{-1}^{+1} \mathrm{d}x \int_{-1}^{+1} \mathrm{d}y[n(\xi')n(\xi'_*) - n(\xi)n(\xi_*)]|\xi_* y - \xi x|\xi_*^2. \tag{7.59}$$

By separating the gain and loss terms, we obtain

$$\begin{aligned}\frac{\partial n}{\partial t} &= 2\pi^2 d_\mathrm{p}^2 \int_0^{\infty} \xi_*^2 \int_{-1}^{+1} \int_{-1}^{+1} n(\xi')n(\xi'_*)|\xi_* y - \xi x|\mathrm{d}x\,\mathrm{d}y\,\mathrm{d}\xi_* \\ &\quad - 2\pi^2 d_\mathrm{p}^2 \int_0^{\infty} \xi_*^2 \int_{-1}^{+1} \int_{-1}^{+1} n(\xi)n(\xi_*)|\xi_* y - \xi x|\mathrm{d}x\,\mathrm{d}y\,\mathrm{d}\xi_*,\end{aligned} \tag{7.60}$$

where the magnitudes of post-collision velocities (see Eq. (7.49)) are defined by

$$\begin{aligned}\xi' &= \left[\xi^2(1 - x^2) + \xi_* y^2\right]^{1/2}, \\ \xi'_* &= \left[\xi_*^2(1 - y^2) + \xi^2 x^2\right]^{1/2}.\end{aligned} \tag{7.61}$$

Details on this derivation can be found in Aristov (2001) for the hard-sphere case, whereas its extension to generic kernels is reported in Asinari (2010).

As with class and sectional methods for univariate PBE, the particle-velocity magnitude space ξ is uniformly discretized over M intervals of size $\Delta\xi$:[19] $\xi_1 < \xi_2 < \xi_3 < \cdots < \xi_{M-1} < \xi_M < \xi_{M+1}$. The following equation is obtained for the number density of particles with velocity magnitude equal to $N_i = n(\xi_i)$:

$$\begin{aligned}\frac{\mathrm{d}N_i}{\mathrm{d}t} &= 2\pi^2 d_\mathrm{p}^2 \sum_{j=1}^{M} \xi_j^2 \int_{-1}^{+1}\int_{-1}^{+1} n(\xi')n(\xi'_*)|\xi_j y - \xi_i x|\mathrm{d}x\,\mathrm{d}y\,\Delta\xi \\ &\quad - 2\pi^2 d_\mathrm{p}^2 \sum_{j=1}^{M} \xi_j^2 \int_{-1}^{+1}\int_{-1}^{+1} N_i N_j |\xi_j y - \xi_i x|\mathrm{d}x\,\mathrm{d}y\,\Delta\xi,\end{aligned} \tag{7.62}$$

where the integral with respect to the field particle has been replaced by a summation over the index j. In Eq. (7.62), ξ_i corresponds to the velocity of the test particle ξ, ξ_j corresponds to the velocity of the field particle ξ_*, and $n(\xi_i)$ and $n(\xi_j)$ have been readily replaced by N_i and N_j in the final term. It is immediately evident that, since N_i and N_j can be taken out of the final integral in Eq. (7.62), the loss term defined in Eq. (7.50) is

$$L_i = 2\pi^2 d_\mathrm{p}^2\,\Delta\xi \sum_{j=1}^{M} \xi_j^2 N_i N_j v_{ij}, \tag{7.63}$$

where

$$v_{ij} = \int_{-1}^{+1}\int_{-1}^{+1} |\xi_i x - \xi_j y|\mathrm{d}x\,\mathrm{d}y = \begin{cases} 2(3\xi_i^2 + \xi_j^2)/(3\xi_i) & \text{if } \xi_j \le \xi_i, \\ 2(3\xi_j^2 + \xi_i^2)/(3\xi_j) & \text{if } \xi_j > \xi_i. \end{cases} \tag{7.64}$$

The treatment of the gain term requires instead more steps. This term quantifies the rate of formation of a pair of particles with discretized velocities ξ_i and ξ_j, as a result of collisions of a different pair of particles with velocities ξ' and ξ'_*. Let us assign the indices k and l to indicate the latter particles using the following inequalities:

$$\begin{aligned}(k-1)^2\,\Delta\xi^2 &\le \xi_i^2(1-x^2) + \xi_j^2 y^2 \le k^2\,\Delta\xi^2, \\ (l-1)^2\,\Delta\xi^2 &\le \xi_j^2(1-y^2) + \xi_i^2 x^2 \le l^2\,\Delta\xi^2.\end{aligned} \tag{7.65}$$

For a given pair of discretized particles, there exist many pairs of particles that respect Eq. (7.65) for different values of the collision parameters $-1 \le x \le 1$ and $-1 \le y \le 1$. These pairs are aligned along a circle in the post-collision velocity plane of radius equal to $(\xi_i^2 + \xi_j^2)^{1/2}$, and, depending on the intersection between this circle and the grid generated by the linear discretization in the velocity plane, four different cases can be identified. On replacing $n(\xi')$ and $n(\xi'_*)$ with N_k and N_l, the gain term becomes

$$G_i = 2\pi^2 d_\mathrm{p}^2\,\Delta\xi \sum_{j=1}^{M} \xi_j^2 \sum_{(k,l)} \alpha_{ikl} N_k N_l, \tag{7.66}$$

[19] It is worth noticing that with this formulation the size of the particle-velocity interval, $\Delta\xi$, appears in the discretized equation, in contrast to what happens with population balances. This is simply because in the case of population balances, N_i was defined as the integral of $n(\xi)$ over the ith interval, representing therefore the finite number of particles with internal coordinate in between ξ and $\xi + \mathrm{d}\xi$: $N_i = \int_{\xi_i}^{\xi_{i+1}} n(\xi)\mathrm{d}\xi$. In the case of the KE, N_i is instead defined as the actual value of the distribution at the pivot: $N_i = n(\xi_i)$. Since the latter corresponds to taking $n(\xi)$ as a constant in the integral, the two approaches are essentially identical up to a factor of $\Delta\xi$.

where the summation over the indices k and l is defined by the circle in the plane of post-collision velocities and

$$\alpha_{ikl} = \int_{\Omega_{kl}} |\xi_l x - \xi_k y| \mathrm{d}x\,\mathrm{d}y. \tag{7.67}$$

Once the indices (i, j) and (k, l) are fixed, the integration domain Ω_{kl}, as well as the result of Eq. (7.67), can be calculated analytically by identifying six different possible cases (Aristov, 2001). The final discretized KE is

$$\frac{\mathrm{d}N_i}{\mathrm{d}t} = 2\pi^2 d_\mathrm{p}^2\, \Delta\xi \sum_{j=1}^{M} \xi_j^2 \left(\sum_{(k,l)} \alpha_{ikl} N_k N_l - N_i N_j \nu_{ij} \right), \tag{7.68}$$

and is valid when the NDF is isotropic.

The anisotropic case

When the hypothesis of isotropy is invalid, the discretization has to be realized for the three components of the velocity vector. An alternative and more convenient notation with respect to that in Eq. (7.50) makes use of an index vector $\mathbf{i} = (i_1, i_2, i_3)$ and defines the discrete velocity as

$$\boldsymbol{\xi}_\mathbf{i} = (i_1\, \Delta\xi, i_2\, \Delta\xi, i_3\, \Delta\xi), \tag{7.69}$$

where the index vector varies over positive and negative integer values, so that the particle velocity in each direction varies between a minimum (negative) value and a maximum (positive) value. The width of the interval $\Delta\xi$ is chosen so that the omitted part of the dynamics is smaller than the approximation of the numerical discretization. Note that the discretization in Eq. (7.69) corresponds to a linear grid in phase space and thus a large number of grid points may be needed for accurate solutions. As for the isotropic case, let us define the number density of particles with velocity vector equal to $\boldsymbol{\xi}_\mathbf{i}$ as $N_\mathbf{i} = n(\boldsymbol{\xi}_\mathbf{i})$. On applying the discretization to the kinetic equation and replacing the integration with respect to the field particle with a summation over the vector index $\mathbf{j} = (j_1, j_2, j_3)$, the rate of change of particles belonging to the $\mathbf{i}$th class becomes

$$\frac{\mathrm{d}N_\mathbf{i}}{\mathrm{d}t} = \frac{6\,\Delta\xi^3}{\pi d_\mathrm{p}} \sum_\mathbf{j} \int_{\mathbb{S}^+} n(\boldsymbol{\xi}')n(\boldsymbol{\xi}'_*)|\mathbf{g}\cdot\mathbf{x}|\mathrm{d}\mathbf{x} - \frac{6\,\Delta\xi^3}{\pi d_\mathrm{p}} \sum_\mathbf{j} \int_{\mathbb{S}^+} n(\boldsymbol{\xi}_\mathbf{i})n(\boldsymbol{\xi}_\mathbf{j})|\mathbf{g}\cdot\mathbf{x}|\mathrm{d}\mathbf{x}, \tag{7.70}$$

where the velocity difference vector $\mathbf{g}$ now indicates the difference between $\boldsymbol{\xi}_\mathbf{i}$ and $\boldsymbol{\xi}_\mathbf{j}$ and the summation over the vector index $\mathbf{j}$ corresponds to a triple summation over the three indices j_1, j_2, and j_3. By recognizing that $N_\mathbf{i} = n(\boldsymbol{\xi}_\mathbf{i})$ and $N_\mathbf{j} = n(\boldsymbol{\xi}_\mathbf{j})$ can be extracted from the final integral in Eq. (7.70), the loss term becomes

$$L_\mathbf{i} = \frac{6\,\Delta\xi^3}{\pi d_\mathrm{p}} \sum_\mathbf{j} N_\mathbf{i} N_\mathbf{j} \int_{\mathbb{S}^+} |\mathbf{g}\cdot\mathbf{x}|\mathrm{d}\mathbf{x}, \tag{7.71}$$

where the surface integral in the case of hard spheres is $\int_{\mathbb{S}^+} |\mathbf{g}\cdot\mathbf{x}|\mathrm{d}\mathbf{x} = \pi d_\mathrm{p}^2 |\boldsymbol{\xi}_\mathbf{i} - \boldsymbol{\xi}_\mathbf{j}|$. The gain term (also known as the integral of inverse collisions) can be further simplified by recalling that the relationship between the post- and pre-collision velocities can be written as $\boldsymbol{\xi}' = \frac{1}{2}(\boldsymbol{\xi} + \boldsymbol{\xi}_*) + \frac{1}{2}(\mathbf{g}\cdot\mathbf{x})\mathbf{x}$ and $\boldsymbol{\xi}'_* = \frac{1}{2}(\boldsymbol{\xi} + \boldsymbol{\xi}_*) - \frac{1}{2}(\mathbf{g}\cdot\mathbf{x})\mathbf{x}$. If these relationships are discretized and written in terms of the corresponding indices, we obtain $\mathbf{k} = \frac{1}{2}(\mathbf{i}+\mathbf{j}) + \mathbf{r}$

and $\mathbf{l} = \frac{1}{2}(\mathbf{i}+\mathbf{j})-\mathbf{r}$, where $\mathbf{r}$ is an integer or half-integer index. The approximate expression for the gain term is then

$$G_{\mathbf{i}} = \frac{6\,\Delta\xi^3}{\pi d_{\mathrm{p}}} \sum_{\mathbf{j}} \sum_{\mathbf{k},\mathbf{l}} N_{\mathbf{k}} N_{\mathbf{l}} \int_{\mathbb{S}^+} |\mathbf{g}\cdot\mathbf{x}| \mathrm{d}\mathbf{x}. \tag{7.72}$$

The remaining part of the collision integral can be calculated analytically by identifying the different cases, thereby reducing the discretized KE to a form similar to Eq. (7.52). A detailed description can be found in Aristov (2001).

7.3 The method of moments

The basic idea behind the MOM is to solve the evolution equations for the moments of the NDF instead of determining the evolution of the NDF by solving the actual GPBE. The general evolution equation for the $\mathbf{k}$th moment ($\mathbf{k} = (k_1, \ldots, k_M)$ is the moment exponent vector containing the order of the moments with respect to each of the components of $\boldsymbol{\xi}$) for a homogeneous system can easily be obtained by applying the moment transform to Eq. (7.1):

$$\frac{\mathrm{d}m_{\mathbf{k}}}{\mathrm{d}t} = \overline{S}_{\mathbf{k}}, \tag{7.73}$$

where (see Eq. (2.2)) the moments are defined by $m_{\mathbf{k}} \equiv \int \xi_1^{k_1} \ldots \xi_M^{k_M} n \,\mathrm{d}\boldsymbol{\xi}$ and

$$\overline{S}_{\mathbf{k}} = \int \boldsymbol{\xi}^{\mathbf{k}} \left[-\frac{\partial}{\partial \boldsymbol{\xi}} \cdot \left(\dot{\boldsymbol{\xi}} n\right) + \frac{\partial^2}{\partial \boldsymbol{\xi}\,\partial \boldsymbol{\xi}} : (\mathbf{D}n) + \mathcal{S} \right] \mathrm{d}\boldsymbol{\xi}. \tag{7.74}$$

Alternative, but equivalent, notations for the moments are $m_{k_1,\ldots,k_M} = m_{\mathbf{k}} = m(k_1, \ldots, k_M) = m(\mathbf{k})$. The MOM presents a number of advantages that make it interesting. The first advantage arises from the reduction of the dimensionality of the problem. In fact, by solving equations for the moments rather than for the NDF, the internal coordinates are integrated out. We will see that this is typically done at the expense of a crucial loss of information that must be retrieved (by reconstructing the NDF, for example). Another important advantage is that the moments correspond to quantities that have meaningful physical interpretations and are therefore, at least in principle, directly measurable. This is a crucial point, since in many applications the NDF is not directly measured, but is inferred from measurements of integral quantities. These inference methods solve the ill-conditioning problem underlying the inversion, which typically has multiple solutions, and introduce non-negligible uncertainty into the determination of the NDF. Since the measured integral properties are instead directly and uniquely related to the moments of the NDF, in many applications formulating and solving the problem in terms of the moments themselves turns out to be a more consistent and convenient approach. Two main issues arise when using the MOM. The first one is related to the number of moments to be tracked. Theoretically the knowledge of an infinite set of moments is equivalent to accessing the underlying NDF, and therefore truncating and projecting the original infinite set onto a finite set implies a loss of information. The crucial point is to evaluate whether this loss is acceptable or not. This point is strictly related to the second issue, namely the impossibility of writing the source term as a function of the moments only. Different methods to overcome this closure problem have been developed and generally their accuracy depends on the number of moments available. Therefore the issues of the number of moments to be tracked and the closure adopted are typically addressed together. In the following, these issues will be discussed separately for the univariate and bivariate PBE, and for the KE.

7.3.1 Univariate PBE

Let us consider just one internal coordinate. The PBE for the homogeneous system is Eq. (7.3), and the resulting evolution equation for the moment of order k is

$$\frac{\mathrm{d}m_k}{\mathrm{d}t} = \overline{S}_k, \tag{7.75}$$

where

$$\overline{S}_k = \int \xi^k \left[-\frac{\partial \dot{\xi} n}{\partial \xi} + \frac{\partial^2 D n}{\partial \xi^2} + S \right] \mathrm{d}\xi \tag{7.76}$$

and $m_k \equiv \int \xi^k n \, \mathrm{d}\xi$, with the integral extending over the phase space. For example, in the case of the particle size, mass, volume, or surface area the definition of the moment is $m_k \equiv \int_0^\infty \xi^k n \, \mathrm{d}\xi$, and in the case of the particle velocity the integral is $m_k \equiv \int_{-\infty}^{+\infty} \xi^k n \, \mathrm{d}\xi$. In the case of univariate distributions the moment of order zero, m_0, is the total particle number density, representing the number of particles per unit volume. If the internal coordinate is, for example, the particle size, then m_1 is the total particle length density, or, in other words, the total length of the particles per unit volume. It is therefore clear that the ratio of these two quantities represents the number-average mean particle size: $d_{10} = m_1/m_0$. The moment of order two, m_2, is proportional to the total particle surface area per unit volume, through the surface shape factor k_A. The moment of order three, m_3, is related through the volume shape factor k_V to the total particle volume per unit volume, or, in other words, the total disperse-phase volume fraction. Another very common mean particle size is the surface-area-average particle size, often called the Sauter mean diameter, $d_{32} = m_3/m_2$, whereas $d_{43} = m_4/m_3$ quantifies the mean particle size weighted on particle volume rather than particle surface area. If the internal coordinate is instead the particle volume, then m_0 is again the total particle number density, whereas m_1 is the total disperse-phase volume fraction.

Moment equations with phase-space drift

In the case of drift, the source term of Eq. (7.75) can be calculated by using the rule of integration by parts:

$$\frac{\mathrm{d}m_k}{\mathrm{d}t} = -\left. \xi^k n(t,\xi)\dot{\xi}(\xi) \right|_0^\infty + k \int_0^\infty \xi^{k-1} \dot{\xi} n \, \mathrm{d}\xi. \tag{7.77}$$

The first term on the right-hand side of Eq. (7.77) is always null for $\xi \to \infty$; in fact, in this limit $n(t,\xi)$ tends to zero faster than $\xi^k \dot{\xi}(\xi)$ tends to infinity. For $\xi = 0$ the first term is null when $k > 0$, whereas for $k = 0$ we obtain

$$\frac{\mathrm{d}m_0}{\mathrm{d}t} = n(t,0)\dot{\xi}(0), \tag{7.78}$$

which represents the net flux of particles through the boundary at $\xi = 0$. For particle growth where $\dot{\xi}(\xi) > 0$, this term can be easily calculated and is usually related to the nucleation rate:[20]

$$\frac{\mathrm{d}m_0}{\mathrm{d}t} = J(t,0). \tag{7.79}$$

[20] Very often nucleation is assumed to occur only at $\xi = 0$ and therefore $J(t,0)$ is the overall nucleation rate.

More complex is the situation when $\dot{\xi}(0) < 0$, for example during particle dissolution or droplet evaporation. In this case the first term in Eq. (7.78) is not related to the nucleation rate but to the pointwise value of the NDF, thus $n(t, 0)$ must be inferred from the known moments of the distribution.[21] In the case of positive drift (i.e. growth) with a rate independent of the internal coordinate, it is easy to show that

$$\frac{\mathrm{d}m_k}{\mathrm{d}t} = k\dot{\xi}m_{k-1}, \tag{7.80}$$

whereas if the growth rate follows a power-law relationship with respect to the internal coordinate, $\dot{\xi} = \dot{\xi}_0\xi^\alpha$, we find

$$\frac{\mathrm{d}m_k}{\mathrm{d}t} = k\dot{\xi}_0 m_{k+\alpha-1}, \tag{7.81}$$

leading to an unclosed set of equations for particular values of α. For more complex growth rates, no analytical solutions are available and numerical closures must be sought.

Moment equations with phase-space diffusion

Diffusion in phase space is mainly associated with the velocity, so for this term we will integrate over the interval $(-\infty, +\infty)$. In the case of pure diffusion the source term of Eq. (7.75) can be calculated by using integration by parts twice:

$$\begin{aligned}\frac{\mathrm{d}m_k}{\mathrm{d}t} &= -\xi^k \frac{\partial Dn}{\partial \xi}\bigg|_{\xi=-\infty}^{\xi=+\infty} + k\int_{-\infty}^{+\infty} \xi^{k-1}\frac{\partial Dn}{\partial \xi}\,\mathrm{d}\xi \\ &= -\xi^k \frac{\partial Dn}{\partial \xi}\bigg|_{\xi=-\infty}^{\xi=+\infty} - k\,\xi^{k-1}Dn\Big|_{\xi=-\infty}^{\xi=+\infty} + k(k-1)\int_{-\infty}^{+\infty} \xi^{k-2}Dn\,\mathrm{d}\xi. \end{aligned} \tag{7.82}$$

The first two terms on the right-hand side of Eq. (7.82) are always null for $|\xi| \to \infty$; in fact, in this limit the NDF tends to zero faster than $\xi^k D$ tends to infinity. We thus obtain

$$\frac{\mathrm{d}m_k}{\mathrm{d}t} = k(k-1)\int_{-\infty}^{+\infty} \xi^{k-2}Dn\,\mathrm{d}\xi, \tag{7.83}$$

which acts a source term for $k \geq 2$. For the case in which $D = D_0 + D_1\xi^2$ (recall that $D \geq 0$ for all ξ), we find a closed expression for the diffusion term:

$$\frac{\mathrm{d}m_k}{\mathrm{d}t} = k(k-1)(D_0 m_{k-2} + D_1 m_k). \tag{7.84}$$

Otherwise, a moment closure must be provided for Eq. (7.83).

Moment equations with point processes

In the case of aggregation, when the internal coordinate ξ is purely additive during one aggregation event, the PBE has the source term given in Eq. (7.9) and the evolution equation for the moment of order k is

$$\frac{\mathrm{d}m_k}{\mathrm{d}t} = \frac{1}{2}\iint_0^\infty \beta(\xi,\xi')(\xi+\xi')^k n(\xi)n(\xi')\mathrm{d}\xi\,\mathrm{d}\xi' - \int_0^\infty \xi^k n(\xi)\int_0^\infty \beta(\xi,\xi')n(\xi')\mathrm{d}\xi'\,\mathrm{d}\xi. \tag{7.85}$$

[21] This is particularly important in the case of evaporating sprays. For the related theoretical and numerical issues, as well as for a thorough discussion of the topic, we refer the reader to the specialized literature.

In the case of a constant aggregation kernel $\beta(\xi, \xi') = \beta_0$, it is possible to obtain

$$\frac{\mathrm{d}m_k}{\mathrm{d}t} = \frac{\beta_0}{2} \iint_0^\infty (\xi + \xi')^k n(\xi) n(\xi') \mathrm{d}\xi \, \mathrm{d}\xi' - \beta_0 m_0 m_k. \tag{7.86}$$

Therefore, for $k = 0$,

$$\frac{\mathrm{d}m_0}{\mathrm{d}t} = -\frac{1}{2}\beta_0 m_0^2, \tag{7.87}$$

and for $k = 1$,

$$\frac{\mathrm{d}m_1}{\mathrm{d}t} = 0, \tag{7.88}$$

which states the conservation of the first-order moment during aggregation for a purely additive internal coordinate. Higher-order moments are instead governed by

$$\frac{\mathrm{d}m_k}{\mathrm{d}t} = \frac{1}{2}\beta_0 \sum_{i=1}^{k-1} \binom{k}{i} m_i m_{k-i}. \tag{7.89}$$

In the case of pure breakage the PBE is

$$\frac{\partial n}{\partial t} = \int_\xi^\infty a(\xi') b(\xi'|\xi) n(\xi') \mathrm{d}\xi' - a(\xi) n(\xi), \tag{7.90}$$

and the evolution equation for the moment of order k is

$$\frac{\mathrm{d}m_k}{\mathrm{d}t} = \int_0^\infty \xi^k \int_0^\infty a(\xi') b(\xi'|\xi) n(\xi') \mathrm{d}\xi' \, \mathrm{d}\xi - \int_0^\infty \xi^k a(\xi) n(\xi) \mathrm{d}\xi. \tag{7.91}$$

Analytical solutions are available for some particular cases for a constant breakage kernel, $a(\xi) = a_0$. For example, in the case of symmetric fragmentation with daughter distribution

$$b(\xi'|\xi) = 2\delta(\xi - \xi'/2), \tag{7.92}$$

the evolution equation for m_k is

$$\frac{\mathrm{d}m_k}{\mathrm{d}t} = a_0 2^{1-k} m_k - a_0 m_k, \tag{7.93}$$

which again respects the conservation constraint for the moment of order one. In the case of breakage into two fractional fragments characterized by x and $1 - x$,

$$b(\xi'|\xi) = \delta\left(\xi - \frac{\xi'}{x}\right) + \delta\left(\xi - \frac{\xi'}{1 - x}\right), \tag{7.94}$$

the resulting evolution equation for m_k is

$$\frac{\mathrm{d}m_k}{\mathrm{d}t} = a_0 \left[x^k + \left(1 - x^k\right)\right] m_k - a_0 m_k. \tag{7.95}$$

In the case of size-dependent kernels, no analytical solutions are available and numerical closures are needed.

Moment equations for PBE

The general evolution equation for the homogeneous system undergoing nucleation, growth, diffusion, aggregation, and breakage is

$$\begin{aligned}\frac{\mathrm{d}m_k}{\mathrm{d}t} &= \int J\xi^k \,\mathrm{d}\xi + k\int \dot{\xi}\xi^{k-1} n \,\mathrm{d}\xi + k(k-1)\int \xi^{k-2} D n \,\mathrm{d}\xi \\ &+ \frac{1}{2}\int \beta(\xi,\xi')(\xi+\xi')^k n(\xi)n(\xi')\mathrm{d}\xi\,\mathrm{d}\xi' - \int \xi^k n(\xi)\int_0^\infty \beta(\xi,\xi')n(\xi')\mathrm{d}\xi'\,\mathrm{d}\xi \\ &+ \int \xi^k \int_0^\infty a(\xi')b(\xi'|\xi)n(\xi')\mathrm{d}\xi'\,\mathrm{d}\xi - \int \xi^k a(\xi) n \,\mathrm{d}\xi. \end{aligned} \quad (7.96)$$

A moment closure is an algorithm for writing the source terms on the right-hand side of Eq. (7.96), explicitly or implicitly, in terms of the moments:

$$\begin{aligned}\frac{\mathrm{d}m_0}{\mathrm{d}t} &= f_0(m_0, m_1, \ldots, m_k), \\ \frac{\mathrm{d}m_1}{\mathrm{d}t} &= f_1(m_0, m_1, \ldots, m_k), \\ &\vdots \\ \frac{\mathrm{d}m_k}{\mathrm{d}t} &= f_k(m_0, m_1, \ldots, m_k). \end{aligned} \quad (7.97)$$

We have shown how this set of differential equations is closed in some simple cases (i.e. for nucleation at zero size, with constant growth rate, and constant aggregation and breakage kernels). In general, when the system is not closed, closure must be sought through a stable numerical method.

Closures for moment equations

Over the years several methods have been developed, and we begin a brief review of these methods by starting with the method of moments with interpolative closure (MOMIC) (Frenklach, 2002; Frenklach & Harris, 1987). This method is based on the very simple and effective idea that unknown moments can be interpolated (or extrapolated) from the known set of moments by using, for example, Lagrange (or spline) interpolation. This concept is clearly illustrated by plotting the moments of a distribution versus the moment order. As can be seen from Figure 7.1, knowledge of the integer moments suffices for the estimation (through interpolation) of the unknown non-integer moments. The method is effective only for those cases in which the source terms for the moments can be reduced in terms of some (known or unknown) moments. This is almost always the case for simple problems, but it turns out to be impossible in a number of (not very rare) cases. We now illustrate these concepts with two examples that are based on aggregation problems. The main difficulty with aggregation is the dependence of the aggregation kernel on the internal coordinate and, in particular, its non-additive character. For example, in the continuum regime[22] the aggregation kernel when ξ represents particle mass is

$$\beta(\xi,\xi') = \beta_0\left(\frac{1}{\xi^{1/3}} + \frac{1}{\xi'^{1/3}}\right)\left(\xi^{1/3} + \xi'^{1/3}\right). \quad (7.98)$$

[22] The different aggregation/coagulation regimes and the corresponding kernels are discussed in Chapter 5.

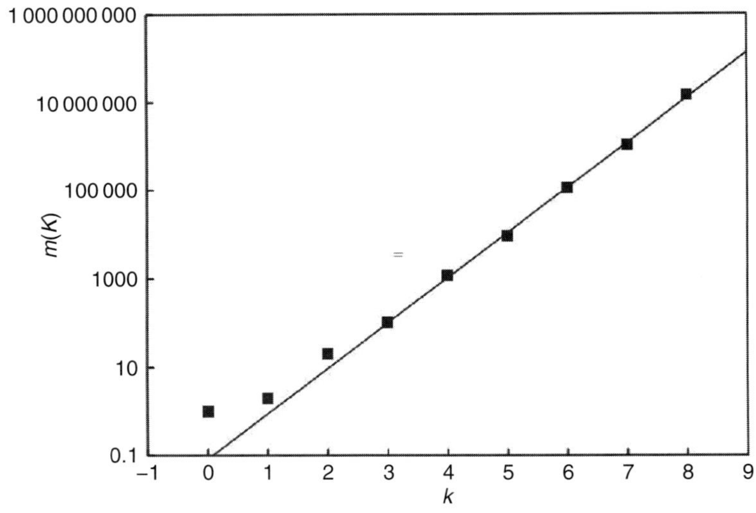

Figure 7.1. Moments of a normal distribution versus moment order. If just integer moments are known (symbols), simple interpolation (continuous line) allows one to estimate non-integer moments.

The source terms for aggregation can be written as

$$\begin{aligned}
\frac{\mathrm{d}m_0}{\mathrm{d}t} &= -\frac{1}{2}\iint_0^\infty \beta(\xi,\xi')n(\xi)n(\xi')\mathrm{d}\xi\,\mathrm{d}\xi', \\
\frac{\mathrm{d}m_1}{\mathrm{d}t} &= 0, \\
\frac{\mathrm{d}m_k}{\mathrm{d}t} &= \frac{1}{2}\sum_{i=1}^{k-1}\binom{k}{i}\left(\iint_0^\infty \beta(\xi,\xi')\xi^i\xi'^{k-i}n(\xi)n(\xi')\mathrm{d}\xi\,\mathrm{d}\xi'\right)
\end{aligned} \tag{7.99}$$

with $k = 2, 3, \ldots$. If Eq. (7.98) is substituted into Eq. (7.99), we obtain

$$\frac{\mathrm{d}m_0}{\mathrm{d}t} = -\beta_0\left(m_0^2 + m_{1/3}m_{-1/3}\right) \tag{7.100}$$

and

$$\frac{\mathrm{d}m_k}{\mathrm{d}t} = \frac{1}{2}\beta_0\sum_{i=1}^{k-1}\binom{k}{i}\left(2m_im_{k-i} + m_{i+1/3}m_{k-i-1/3} + m_{i-1/3}m_{k-i+1/3}\right) \tag{7.101}$$

with $k = 2, 3, \ldots$ resulting in an unclosed set of equations. To obtain closure the fractional-order moments appearing in the source terms must be closed. In the MOMIC this is accomplished by interpolation among integer moments. Interpolation is more efficient when carried out on the logarithm of reduced moments, $\log(\mu_k) = \log(m_k) - \log(m_0)$, using, for example, Lagrange polynomials, and on the entire set of moments available for positive fractional moments, $\log(\mu_p) = L_p(\log(\mu_0), \log(\mu_1), \log(\mu_2), \ldots, \log(\mu_N))$ for $p > 0$, where N is the maximum number of moments tracked, and by extrapolating for negative fractional moments only from the first three moments, $\log(\mu_p) = L_p(\log(\mu_0), \log(\mu_1), \log(\mu_2))$ for $p < 0$. The MOMIC becomes very cumbersome when it is impossible to write the source terms as explicit functions of some known or unknown moments. For example, for aggregation in the free-molecular regime the aggregation kernel is

$$\beta(\xi, \xi') = \beta_0 \left(\frac{1}{\xi} + \frac{1}{\xi'} \right)^{1/2} \left(\xi^{1/3} + \xi'^{1/3} \right)^2 . \qquad (7.102)$$

It is immediately evident that the presence of the exponent $1/2$ does not allow us to write the source term as an explicit summation. An interesting alternative in this case would be to estimate the resulting summations appearing in the source terms for the following kernel:

$$\beta(\xi, \xi') = \beta_0 \left(\frac{1}{\xi} + \frac{1}{\xi'} \right)^{l} \left(\xi^{1/3} + \xi'^{1/3} \right)^2 \qquad (7.103)$$

for $l = 0, 1, 2, \ldots,$ and then interpolate again to obtain the solution for $l = 1/2$. As anticipated, in these cases the method loses its efficiency and simplicity, whereas it is not applicable in those cases (which are not very rare, unfortunately!) in which the kernel (or the continuous rate of change) contains exponential dependences on the internal coordinate, or any other type of functional dependence that cannot be accurately approximated by a polynomial in terms of the internal coordinate (e.g. exponential breakage kernel, exponential growth rate, etc.). We refer readers interested in the mathematical details concerning the validity of the interpolation to Frenklach (2002). See Alexiadis *et al.* (2004), Barrett & Jheeta (1996), and Diemer & Olson (2002b), for the description of related interpolation methods.

A second class of methods for overcoming the closure problem is to make a functional assumption regarding the NDF. The simplest is to assume that the NDF is composed of a delta function centered on the mean value of the internal coordinate (e.g. m_1/m_0), or, in other words, assume that the population of particles is monodisperse. On resorting to this approach the missing moments can be readily calculated (e.g. $m_k = m_0^{1-k} m_1^k$) as illustrated by Kruis *et al.* (1993). A very popular alternative is to use a less approximate functional form for the NDF, as, for example, in the case of the log-normal distribution. Knowledge of three lower-order moments suffices for the calculation of the three parameters of the log-normal distribution. (See Chapter 3 for details.) The closed NDF can be used to compute all the missing moments and the source and sink terms (Bensberg *et al.*, 1999; Friedlander & Wang, 1966; Lee, 1983; Swift & Friedlander, 1964). For problems that clearly present bimodal (or multimodal) distributions, the population of particles can be divided into different groups (such as seeds and growing crystals or nuclei and aggregates, etc.) described by delta functions centered on different mean values of the internal coordinate (e.g. the number average and volume average) as explained by Wu *et al.* (1988), or by log-normal distributions (Megaridis & Dobbins, 1990). A very efficient technique is the finite-size domain complete set of trial functions method of moments (FCMOM) (Strumendo & Arastoopour, 2008; Strumendo & Arastoopour, 2009, 2010), in which the NDF is reconstructed from the moments by using a series expansion by a complete system of orthonormal functions. Readers interested in the details of the different closures, as well as in the problem of distribution reconstruction, should consult Diemer & Olson (2002a, 2002b). It should be highlighted, however, that most of the proposed methods for univariate PBE suffer from some important drawbacks, mainly related to the extent of their applicability and to the numerical accuracy. McGraw (1997) proposed a method that appears to overcome these limitations and is based on a quadrature approximation. This method, called the quadrature method of moments (QMOM), is described in Section 7.4.

7.3.2 Bivariate and multivariate PBE

The MOM is also applicable for bivariate PBEs, as explained by Hulburt & Katz (1964). In this case the moments of the NDF are defined as $m_{\mathbf{k}} \equiv \int \xi_1^{k_1}\xi_2^{k_2} n\, \mathrm{d}\boldsymbol{\xi}$, where ξ_1 and ξ_2 are the two internal coordinates and $\mathbf{k} = (k_1, k_2)$ is the exponent vector containing the order of the moment with respect to the two internal coordinates. Moments can have a particular physical meaning and, as for the univariate case, $m_{0,0}$ is the total particle number density. Likewise, the ratio of different moments can be used to characterize integral properties of the population, for example $m_{1,0}/m_{0,0}$ is the number-average value of the first internal coordinate ξ_1, whereas $m_{0,1}/m_{0,0}$ is the number-average value of the second one ξ_2. For details and examples on the relationship between bivariate moments and measurable quantities, readers are referred, for example, to the work of Rosner *et al.* (2003). As for the univariate case, one has to solve the evolution equation for the moments,

$$\frac{\mathrm{d}m_{k_1,k_2}}{\mathrm{d}t} = \overline{S}_{k_1,k_2} = \int \xi_1^{k_1}\xi_2^{k_2}\left[-\frac{\partial}{\partial \boldsymbol{\xi}} \cdot \left(\dot{\boldsymbol{\xi}} n\right) + \frac{\partial^2}{\partial \boldsymbol{\xi}\, \partial \boldsymbol{\xi}} : (\mathbf{D}n) + \mathcal{S}\right] \mathrm{d}\boldsymbol{\xi}, \tag{7.104}$$

belonging to the moment set that is tracked during a MOM simulation. The existence of the two internal coordinates complicates both the choice of the number of moments and the nature of the moments to be included in the moment set. In fact, if for univariate cases the most natural choice would be to select a complete set of moments starting from m_0 up to a certain order,[23] for bivariate cases no immediate and trivial choice exists. One possible strategy could be to define the moment set by classifying moments through their global order, $\gamma = k_1 + k_2$. According to this classification, there is only one moment of global order zero ($\gamma = 0$), namely $m_{0,0}$, but there are two first-order moments ($\gamma = 1$), namely $m_{1,0}$ and $m_{0,1}$, three second-order moments ($\gamma = 2$), $m_{2,0}$, $m_{1,1}$, and $m_{0,2}$, and so on for higher-order moments. In order to rationalize the process of selecting the moments to be included in the set, very often they are organized in triangular tables, as illustrated in Table 3.6 of Section 3.2.1.

Insofar as the possibility of directly using Eq. (7.104) is concerned, as for univariate cases, the main limitation resides in the difficulty of finding analytical solutions and developing simple and stable closures. Analytical solutions are very rare and exist only for rates of continuous change and for kernels independent of the internal coordinates. Under these simplification hypotheses, for the case of nucleation, growth, diffusion,[24] aggregation, and breakage (with additive internal coordinates), the bivariate PBE becomes

$$\begin{aligned}\frac{\partial n}{\partial t} &= J - \dot{\xi}_1 \frac{\partial n}{\partial \xi_1} - \dot{\xi}_2 \frac{\partial n}{\partial \xi_2} + D_{11}\frac{\partial^2 n}{\partial \xi_1^2} + 2D_{12}\frac{\partial^2 n}{\partial \xi_1\, \partial \xi_2} + D_{22}\frac{\partial^2 n}{\partial \xi_2^2} \\ &+ \frac{\beta_0}{2}\int n(\boldsymbol{\xi}')n(\boldsymbol{\xi} - \boldsymbol{\xi}')\mathrm{d}\boldsymbol{\xi}' - \beta_0 n \int n(\boldsymbol{\xi}')\mathrm{d}\boldsymbol{\xi}' + a_0 \int b(\boldsymbol{\xi}|\boldsymbol{\xi}')n(\boldsymbol{\xi}')\mathrm{d}\boldsymbol{\xi}' - a_0 n,\end{aligned} \tag{7.105}$$

[23]For simple applications one could choose (m_0, m_1, m_2, m_3), but for complex problems more moments might be needed, and therefore the highest order could be fixed equal to five or six. This also has some important implications for NDF reconstruction; as a matter of fact, the accurate reconstruction of a complex NDF requires knowledge of many moments.

[24]We remind the reader that diffusion is usually associated with velocity phase space on an unbounded domain. If diffusion is used on a bounded domain, proper boundary conditions for the diffusive fluxes must be provided. In Eq. (7.105) we assume that the NDF does not change due to diffusion through the boundaries of phase space.

and therefore the transport equation for the mixed moment of order $\mathbf{k}$ is

$$\begin{aligned}\frac{\mathrm{d}m_{\mathbf{k}}}{\mathrm{d}t} &= \overline{J}_{\mathbf{k}} + k_1\dot{\xi}_1 m_{k_1-1,k_2} + k_2\dot{\xi}_2 m_{k_1,k_2-1} \\ &+ k_1(k_1-1)D_{11}m_{k_1-2,k_2} + 2k_1k_2D_{12}m_{k_1-1,k_2-1} + k_2(k_2-1)D_{22}m_{k_1,k_2-2} \\ &+ \frac{\beta_0}{2}\int\left[(\xi_1+\xi_1')^{k_1}(\xi_2+\xi_2')^{k_2} - \xi_1^{k_1}\xi_2^{k_2} - (\xi'_1)^{k_1}(\xi'_2)^{k_2}\right]n(\boldsymbol{\xi}')n(\boldsymbol{\xi})\mathrm{d}\boldsymbol{\xi}'\,\mathrm{d}\boldsymbol{\xi} \\ &+ a_0\int \xi_1^{k_1}\xi_2^{k_2} b(\boldsymbol{\xi}|\boldsymbol{\xi}')n(\boldsymbol{\xi}')n(\boldsymbol{\xi})\mathrm{d}\boldsymbol{\xi}'\,\mathrm{d}\boldsymbol{\xi} - a_0 m_{\mathbf{k}}.\end{aligned} \tag{7.106}$$

As for the univariate case for constant kernels and symmetric breakage (see Eq. (7.92)), under the assumption of additive internal coordinates for breakage,

$$b(\boldsymbol{\xi}|\boldsymbol{\xi}') = 2\delta(\xi_1 - \xi_1'/2)\delta(\xi_2 - \xi_2'/2), \tag{7.107}$$

the set of moment evolution equations (see Eq. (7.106)) is closed and some analytical solutions exist. For the lower-order moments we can write

$$\begin{aligned}\frac{\mathrm{d}m_{0,0}}{\mathrm{d}t} &= \overline{J}_{0,0} - \frac{\beta_0}{2}m_{0,0}^2 + a_0 m_{0,0}, \\ \frac{\mathrm{d}m_{1,0}}{\mathrm{d}t} &= \dot{\xi}_1 m_{0,0}, \\ \frac{\mathrm{d}m_{0,1}}{\mathrm{d}t} &= \dot{\xi}_2 m_{0,0},\end{aligned} \tag{7.108}$$

and similar equations can be written for higher-order moments. In all other cases, closure must be restored using the known moments. With a similar formalism these concepts can be extended to multivariate systems; however, as for bivariate systems no general procedure for the solution of the moment-closure problem exists. Moreover, as the number of internal coordinates increases, it becomes increasingly difficult to guarantee that a given moment closure is realizable (see Chapter 3 for more details).

7.3.3 Collisional KE

In the case of velocity distributions, the moments of the NDF represent important quantities. For example, the moment of order zero, $m(\mathbf{0}) = m(0,0,0) = \int n\,\mathrm{d}\boldsymbol{\xi}$, represents the particle number density N. If this quantity is multiplied by the mass of a single particle M_p, the disperse-phase mass density ϱ_p is obtained. The moment of order one with respect to the particle velocity is related to the disperse-phase momentum density: $\varrho_\mathrm{p}\mathbf{U}_\mathrm{p} = M_\mathrm{p}\int \boldsymbol{\xi} n\,\mathrm{d}\boldsymbol{\xi}$. Second-order moments of the NDF are related to the total particle stress tensor: $\mathbf{P} = M_\mathrm{p}\int(\boldsymbol{\xi}-\mathbf{U}_\mathrm{p})\otimes(\boldsymbol{\xi}-\mathbf{U}_\mathrm{p})n\,\mathrm{d}\boldsymbol{\xi}$. It is well known that the trace of this tensor is related to the total disperse-phase kinetic energy, which in turn can be written as $\varrho_\mathrm{p}e_\mathrm{p} = \frac{1}{2}M_\mathrm{p}\int \xi^2 n\,\mathrm{d}\boldsymbol{\xi}$, where e_p is the disperse-phase specific energy. Since $\xi^2 = \xi_1^2 + \xi_2^2 + \xi_3^2$, the previous equation can be written as $\varrho_\mathrm{p}e_\mathrm{p} = \frac{1}{2}M_\mathrm{p}[m(2,0,0) + m(0,2,0) + m(0,0,2)]$. When solving problems related to the collisional KE using the MOM, the set of moments to be tracked has to be defined. In the case of lower-order moments the selection is quite straightforward, since all lower-order moments have clear and important physical meanings, and should be included in any description. When the method is extended to higher-order moments, with no direct physical meaning, the question of which moments one should consider for an accurate description of the particulate system arises.[25] One answer to this

[25]The other important question related to the number of moments will be addressed later, as well as having been touched on in Chapter 3.

question is to include all the moments up to a specific order. For example, if just first-order moments are considered, then four moments will be used (i.e. $m_{0,0,0}$, $m_{1,0,0}$, $m_{0,1,0}$, and $m_{0,0,1}$). If second-order moments are considered then six moments should be added to the set, resulting in a total of ten. On increasing the order of the moments included in the moment set, 20, 35, 56, 84, 120, 165, 220, 286, 364, ... moments will be used. An alternative answer could be based on the selection of an extended set of moments (Grad, 1949a, 1958).

The kinetic equation for homogeneous systems is given by Eq. (7.47). The evolution equation for the zeroth-order moment of the NDF is null, which is due to the fact that the collision integral does not change the number of particles, or, more explicitly, $\int \mathbb{C}\,\mathrm{d}\boldsymbol{\xi} = 0$. If the rate of change of the particle velocity (i.e. particle acceleration) is a linear function of the particle velocity (i.e. $\dot{\boldsymbol{\xi}} = \mathbf{a} + b\boldsymbol{\xi}$), then the evolution equation for the first-order moments are

$$\begin{aligned}\frac{\mathrm{d}m_{1,0,0}}{\mathrm{d}t} &= a_1 m_{0,0,0} + b m_{1,0,0},\\ \frac{\mathrm{d}m_{0,1,0}}{\mathrm{d}t} &= a_2 m_{0,0,0} + b m_{0,1,0},\\ \frac{\mathrm{d}m_{0,0,1}}{\mathrm{d}t} &= a_3 m_{0,0,0} + b m_{0,0,1},\end{aligned} \tag{7.109}$$

where we have used the fact that the first-order moments of the collision integral are identically null: $\int \boldsymbol{\xi}\mathbb{C}\,\mathrm{d}\boldsymbol{\xi} = \mathbf{0}$. The evolution equations for higher-order moments are complicated by the fact that now the moments of the collision integral are not null. For example, the evolution equation for the generic moment of the KE with isotropic diffusion[26] is

$$\begin{aligned}\frac{\mathrm{d}m_{i,j,k}}{\mathrm{d}t} = {} & i a_1 m_{i-1,j,k} + j a_2 m_{i,j-1,k} + k a_3 m_{i,j,k-1} + (i+j+k) b m_{i,j,k}\\ & + i(i-1) D m_{i-2,j,k} + j(j-1) D m_{i,j-2,k} + k(k-1) D m_{i,j,k-2} + \Pi_{i,j,k},\end{aligned} \tag{7.110}$$

where $\Pi_{i,j,k} = \int \xi_1^i \xi_2^j \xi_3^k \mathbb{C}\,\mathrm{d}\boldsymbol{\xi}$. As with the application of the MOM to PBE, Eq. (7.110) is generally unclosed, because the moments of the collision integral cannot be written in terms of the moments themselves.[27] It is important to remind readers that Eq. (7.110) is the spatially homogeneous version of Eq. (6.131) (plus the drift term due to continuous rate of change of the particle velocity). For the KE, simple moment closures are available for some limiting cases. For example, in the case of Maxwell particles the moment of order $\gamma = i + j + k$ of the collision integral can be written explicitly in terms of the moments up to order γ, as illustrated in Eq. (6.109) of Chapter 6. Also, when the collision integral is linearized by using the BGK approximation the closure problem related to the collision integral is overcome (see Eq. (6.120) and Eq. (6.121)). In all the other cases some form of closure is needed. For a general and thorough discussion of the system of equations generated by applying the MOM to collisional KE, readers are referred to the classical book of Truesdell & Muncaster (1980). In what follows the closure offered by the method of Grad will be briefly discussed.

[26] An isotropic diffusion model that conserves the total energy can be constructed by setting $b = -1/\tau$ and $D = [m(2,0,0) + m(0,2,0) + m(0,0,2)]/[\tau m(0,0,0)]$, where τ is a relaxation time scale.

[27] In addition, when dealing with inhomogeneous systems, the presence of the convective term generates some additional closure problems. These issues will be discussed in Chapter 8.

Grad's 13-moment closure

Grad's 13-moment theory is based on the solution of the evolution equations for the disperse-phase mass density $\varrho_{\rm p}$, for the three components of the disperse-phase momentum density $\varrho_{\rm p}\mathbf{U}_{\rm p}$, for the six components of the total disperse-phase stress tensor $\mathbf{P}$, and for the three components of the disperse-phase energy flux $\mathbf{q}$, which are defined by $\mathbf{q} = \frac{1}{2}M_{\rm p}\int|\boldsymbol{\xi}-\mathbf{U}_{\rm p}|^2(\boldsymbol{\xi}-\mathbf{U}_{\rm p})n\,{\rm d}\boldsymbol{\xi}$. The resulting closure for the NDF is an expansion of the Maxwellian distribution f^*: $n^{[13]} = [\lambda^0+\lambda_i^0(\xi_i-U_{{\rm p}i})+\lambda^1|\boldsymbol{\xi}-\mathbf{U}_{\rm p}|^2+\lambda_{ij}^0(\xi_i-U_{{\rm p}i})(\xi_j-U_{{\rm p}j})+\lambda_i^1|\boldsymbol{\xi}-\mathbf{U}_{\rm p}|^2(\xi_i-U_{{\rm p}i})]f^*$, where repeated indices imply summation and the coefficients are chosen so that $n^{[13]}$ reproduces the 13-moment set. Note that this velocity distribution function is defined in terms of velocity fluctuations about the mean velocity ($\boldsymbol{\xi}' = \boldsymbol{\xi}-\mathbf{U}_{\rm p}$). This choice ensures that, like the original KE and the moment equations, $n^{[13]}$ and the resulting closures will be Galilean invariant (i.e. the values of the coefficients do not depend on the mean velocity). We refer the reader to Struchtrup (2005) for details on this method and for a description of its extension to 26 moments. One important limitation of Grad's method is that far away from equilibrium, even though the evolution of the moments could be decently described, the corresponding assumption for the NDF often turns out to be unphysical, in that it results in a negative NDF (as illustrated in Icardi *et al.* (2012) for the homogeneous isotropic Boltzmann equation). This problem is related to the fact that the coefficients in the definition of $n^{[13]}$ appear in a linear form. When the 13 moments computed from $n^{[13]}$ are set equal to the known moments a linear system results, and there is no guarantee that the resulting coefficients will keep $n^{[13]} \geq 0$ for all $\boldsymbol{\xi}$. The same problem occurs with any moment closure that is based on linear expansions (e.g. Strumendo & Arastoopour, 2008, 2009, 2010). In comparison, the methods described in Section 7.4 use nonlinear reconstructions for which it is possible to guarantee a realizable NDF. (See Chapter 3 for more details.) Another crucial point with Grad's method is the number of moments that should be tracked. Ideally, when solving a KE (as well as in the context of the GPBE) it would be nice to know this number in advance, namely before running a specific simulation for the multiphase process. It is, moreover, desirable to have this number defined as a function of the dimensionless numbers involved (see Chapter 1). Unfortunately such a simple criterion is quite difficult to define, therefore the best option is to simulate the process with an increasing number of moments and then to stop when a further increase in the number of moments does not produce any significant difference in the final results.

The lattice Boltzmann method

The lattice Boltzmann method (LBM) can be formulated as a multivariate MOM (Lallemand & Luo, 2003) similar in spirit to Grad's method. In the LBM, the lattice velocities $\mathbf{c}_\alpha$ are fixed such that on a uniform lattice the vector $\mathbf{c}_i\,\Delta t$, where Δt is the time step, corresponds to a lattice vector. While this point is unimportant for homogeneous systems, it is crucial for the numerical algorithm used for spatial transport in inhomogeneous systems. Each lattice velocity has a nonnegative weight f_α, and the assumed form for the NDF is $n(t,\boldsymbol{\xi}) = \sum_\alpha f_\alpha(t)\delta(\boldsymbol{\xi}-\mathbf{c}_\alpha)$, where the multidimensional Dirac delta function is defined as $\delta(\boldsymbol{\xi}-\mathbf{c}_\alpha) = \prod_{i=1}^3\delta(\xi_i-c_{i,\alpha})$. As in Grad's method, the N weights in the LBM are determined by enforcing consistency with the KE for N moments given by $m(k_1,k_2,k_3) = \sum_\alpha c_{1\alpha}^{k_1}c_{2\alpha}^{k_2}c_{3\alpha}^{k_3}f_\alpha$. Because the components $c_{1\alpha}$, $c_{2\alpha}$, and $c_{3\alpha}$ are fixed, the moments can be written as a linear system for the unknown weights:

$$\mathbf{A}(\mathbf{c}_\alpha)\begin{bmatrix} f_1 \\ \vdots \\ f_N \end{bmatrix} = \begin{bmatrix} m(0,0,0) \\ m(1,0,0) \\ \vdots \end{bmatrix}, \tag{7.111}$$

where the coefficient matrix **A** depends on the components of $\mathbf{c}_\alpha$. The set of N moments used on the right-hand side of Eq. (7.111) includes all of the moments up to second order and selected higher-order moments (Lallemand & Luo, 2003), and is chosen such that **A** is invertible. The equilibrium weights f_α^* could be determined from Eq. (7.111) using Gaussian moments, but in practice a linearized form of the equilibrium distribution is employed (Lallemand & Luo, 2003). As with the coefficients in Grad's method, the weights in the LBM are linear functions of the moments (i.e. $\mathbf{f} = \mathbf{A}^{-1}\mathbf{m}$) and, therefore, some of them will be negative for realizable moment sets **m**. It is also noteworthy that the weights will depend on the value of the mean velocity, and hence the LBM is not intrinsically Galilean invariant (Lallemand & Luo, 2003). In fact, the LBM can be used only when the magnitude of the mean velocity is small relative to the magnitude of the lattice velocities (i.e. for low Mach numbers). In practice, the LBM is used for low-Mach-number inhomogeneous flows wherein the weights f_α remain close to f_α^*. Indeed, as mentioned earlier, the real advantage of the LBM lies in the efficient advection scheme that results from the choice of the lattice velocities, but not in its ability to capture non-equilibrium solutions to the KE.

7.4 Quadrature-based moment methods

Quadrature-based moment methods (QBMM) constitute a particular class of very successful moment methods that overcome the closure problem by using quadrature approximations. A general (multivariate) quadrature of N nodes for M internal coordinates requires knowledge of N weights, w_α, and N nodes (or abscissas), $\boldsymbol{\xi}_\alpha = (\xi_{1,\alpha}, \xi_{2,\alpha}, \ldots, \xi_{M,\alpha})$, with $1 \leq \alpha \leq N$. Note that each node $\boldsymbol{\xi}_\alpha$ is a vector of length M. In our notation the first index of the node defines the internal coordinate, whereas the second index defines the number of the node. The N-point, M-dimensional quadrature overcomes the closure problem (which appears always in an integral form) with the following equation: $\int g(\boldsymbol{\xi})n\,\mathrm{d}\boldsymbol{\xi} \approx \sum_{\alpha=1}^{N} w_\alpha g(\boldsymbol{\xi}_\alpha)$. The reader will recognize this expression as a classical quadrature formula for approximating integrals. There are two main approaches available to simulate the evolution of a specific set of moments (henceforth referred to as a *moment set*). In the first approach (called the quadrature method of moments (QMOM)) the evolution equations for the moments are solved, with the estimation of the source terms (appearing on the right-hand side of Eq. (7.73) and detailed in Eq. (7.74)) through the quadrature approximation. The quadrature approximation is in turn determined from the available moments belonging to the moment set by employing moment-inversion algorithms. (The different moment-inversion algorithms available are discussed in Chapter 3.) In the second approach (called the direct quadrature method of moments (DQMOM)) the evolution of the quadrature approximation itself is solved by ensuring that the moments belonging to the moment set obey the proper evolution equations. The basic idea of the QMOM/DQMOM is that the NDF appearing in the GPBE is represented by a sum of N multidimensional Dirac delta functions: $n(t,\boldsymbol{\xi}) = \sum_{\alpha=1}^{N} w_\alpha(t)\delta(\boldsymbol{\xi} - \boldsymbol{\xi}_\alpha(t))$. This equation defines the main assumption regarding the functional form of the NDF, which is at the foundation of QBMM. The reader may have remarked that the QMOM is very similar to the LBM. However, there are some very important differences. First, the nodes in the QMOM are not constant, but rather are allowed to vary. Second, the moment evolution equations in the QMOM depend nonlinearly on the abscissas and, therefore, the moment-inversion problem is much more difficult than that for the LBM. As we shall see below, this additional complexity has a silver lining (see Chapter 3 for details): all realizable moment sets will lead to nonnegative weights and, hence, the QMOM will always represent a realizable NDF. The QBMM will be discussed below first for a univariate PBE, then for bivariate and multivariate PBE, and finally for the collisional KE. As described in Chapter 3, it is possible to expand the accuracy of QBMM by using kernel density functions (i.e. with finite spread) in the definition of $n(t, \boldsymbol{\xi})$. However, for simplicity, we will assume Dirac delta functions in the discussion below.

7.4.1 Univariate PBE

The closure problem in QBMM is overcome by using a quadrature approximation that presents the great advantage of being very accurate for calculating integrals. In fact, they typically exhibit a degree of accuracy much higher than that of other interpolation schemes of the same order. Quadrature approximations are built by determining the sequence of polynomials orthogonal to the NDF. The main result of this theory (see Chapter 3 for details) states that, if the nodes of the quadrature approximation are the roots of the polynomial of order N orthogonal to the NDF, the quadrature approximation given by $\int f(\xi) n \, \mathrm{d}\xi \approx \sum_{\alpha=1}^{N} f(\xi_\alpha) w_\alpha$ is exact if $f(\xi)$ is a polynomial of order $2N - 1$ or smaller (Dette & Studden, 1997). In all other cases, the approximation is more accurate the closer $f(\xi)$ is to a polynomial. Before explaining how the weights and nodes of the quadrature approximation can be calculated from the moments of the NDF and the similarities and differences between the QMOM and DQMOM, let us discuss the estimation of the source term $\overline{S}_k$ in Eq. (7.76). This term is estimated with a certain error that depends on the order of the quadrature approximation. Intuitively, one could think that on increasing the number of nodes the error decreases to very small values; however, the dependence of the error on the number of nodes is difficult to predict *a priori*. Moreover, it is useful to remind the reader that increasing the number of nodes means that more moments will have to be tracked. In general, for univariate systems a quadrature of order N can be constructed by tracking $2N$ moments. This generally impacts both the computational time (which of course increases consequently since more equations have to be solved) and the stability of the method (due to issues related to the moment-inversion algorithm).

Because the accuracy of the quadrature approximation strongly depends on the application, let us discuss some of these issues in detail for specific examples. In the case of standard nucleation, (positive) growth, aggregation, and breakage, application of the QBMM to the source term in Eq. (7.96) yields

$$\overline{S}_k = \overline{J}_k + k \sum_{\alpha=1}^{N} \xi_\alpha^{k-1} \dot{\xi}_\alpha w_\alpha + k(k-1) \sum_{\alpha=1}^{N} \xi_\alpha^{k-2} D_\alpha w_\alpha$$
$$+ \frac{1}{2} \sum_{\alpha=1}^{N} \sum_{\gamma=1}^{N} \left[(\xi_\alpha + \xi_\gamma)^k - \xi_\alpha^k - \xi_\gamma^k \right] \beta_{\alpha,\gamma} w_\alpha w_\gamma + \sum_{\alpha=1}^{N} a_\alpha \bar{b}_\alpha^k w_\alpha - \sum_{\alpha=1}^{N} \xi_\alpha^k a_\alpha w_\alpha, \quad (7.112)$$

where $\dot{\xi}_\alpha = \dot{\xi}(\xi_\alpha)$, $D_\alpha = D(\xi_\alpha)$, $\beta_{\alpha,\gamma} = \beta(\xi_\alpha, \xi_\gamma)$, and $a_\alpha = a(\xi_\alpha)$, and the moments of the daughter distribution function are $\bar{b}_\alpha^k = \int \xi^k b(\xi|\xi_\alpha) \mathrm{d}\xi$. Table 7.1 reports a few examples for the daughter distribution function and their corresponding moment transforms. When breakage produces two fragments, the distribution function is composed by two delta

Table 7.1. Common daughter distribution functions and their moment transforms.

Mechanism	$b(\xi\|\xi')$	$\overline{b}_{\alpha}^{k}$
Symmetric fragmentation	2 if $\xi = \xi'/2$ 0 otherwise	$2^{(1-k)}\xi_{\alpha}^{k}$
Erosion	1 if $\xi = \xi_0$ 1 if $\xi = (\xi' - \xi_0)$ 0 otherwise	$\xi_0 + (\xi_{\alpha} - \xi_0)^k$
Internal-coordinate ratio 1:4	1 if $\xi = \xi'\,(1/5)$ 1 if $\xi = \xi'\,(4/5)$ 0 otherwise	$\xi_{\alpha}^{k}(4^k + 1)/5^k$
Uniform	$2/\xi'$ if $0 < \xi < \xi'$ 0 otherwise	$2\xi_{\alpha}^{k}/(k+1)$

functions centered at the values of the internal coordinate characterizing the fragments. If breakage results in the formation of two equal fragments the two delta functions overlap, whereas in the case of erosion one delta function is centered at $\xi = \xi_0$ (i.e. monomer), with the other one at $\xi' - \xi_0$. In the case of formation of two fragments with ratio 1 to 4 for the internal coordinate, the distribution function is composed by two delta functions centered at $\xi'(1/5)^{1/3}$ and $\xi'(4/5)^{1/3}$. For simple aggregation and breakage problems the QMBB results in excellent accuracy (Marchisio *et al.*, 2003b, 2003c; Marchisio & Fox, 2005). In these cases the use of a small number of nodes (i.e. $2 \leq N \leq 5$), and therefore the solution of a small number of equations for the evolution of the moments belonging to the moment set, is sufficient to keep the overall error acceptable. Typically the errors remain very small not only for the tracked moments, but also for higher-order moments. In fact, the quadrature approximation can be successfully used to estimate moments not included in the original moment set, by extrapolating the quadrature approximation. This is illustrated in Figure 7.2, where the predictions obtained with the QBMM are compared with the exact solution. The reported predictions correspond to the steady-state solution of case 4 in Marchisio *et al.* (2003c). The accuracy is analyzed in terms of the global error function, which quantifies the error committed in predicting a moment of order k, for different moment orders, by using QBMM with $N = 2$, 3 and 4 (and therefore by tracking four, six, and eight moments). As can be seen from Figure 7.2, not only is the error committed in predicting the moments involved in the construction of the quadrature acceptable, but also the error committed in predicting higher-order moments is reasonably small.

When more nodes are employed, the source term is approximated with higher accuracy and the NDF is represented more realistically. This is illustrated in Figure 7.3, where the NDF predicted with a sectional method is compared with the position of the nodes of the quadrature approximation, schematically represented by vertical bars. It can be seen that the source term is driving the NDF to the shape described by the sectional method. When just two nodes are used, they are positioned in order to capture the main features of the NDF and to represent four moments of the NDF. When more nodes are added, they are used to represent more moments, corresponding to one tail of the distribution. Readers interested in the details of this example are referred to the original work (Marchisio *et al.*, 2006). As a general rule it can safely be stated that, for normal kernels and rates of continuous change,

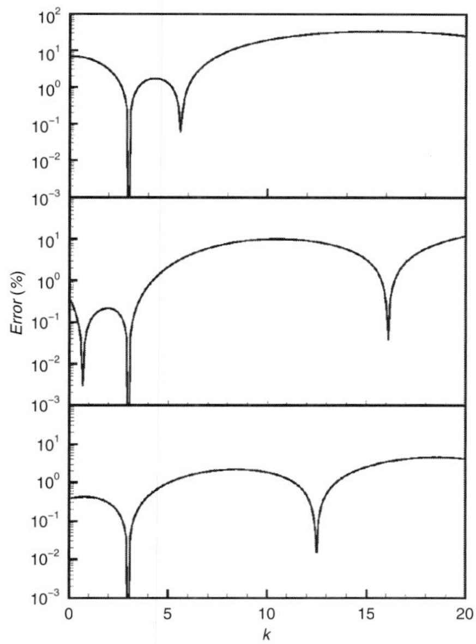

Figure 7.2. The global error function, $E(k)$, versus the moment order, k, for predictions concerning a particulate system undergoing aggregation and breakage obtained by using the QBMM (from top to bottom: $N = 2$, 3 and 4) against the exact solution. Adapted from Marchisio *et al.* (2003c).

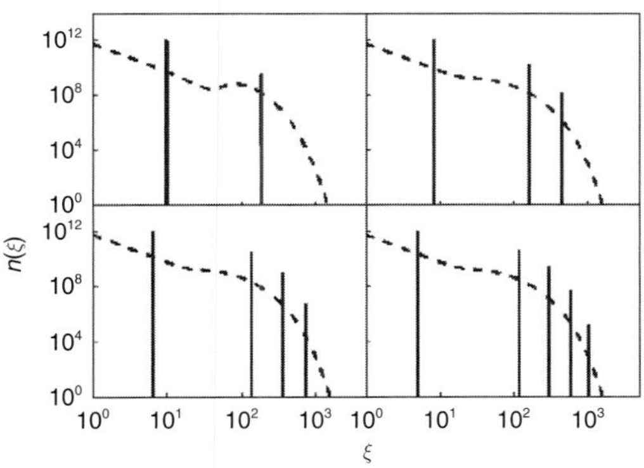

Figure 7.3. The NDF predicted by a sectional method (dashed line) and by a QBMM (vertical bars) with (from left to right and top to bottom) $N = 2$, 3, 4, and 5 for a standard aggregation and breakage problem. Adapted from Marchisio *et al.* (2006).

the resulting source term is a smooth function that can be approximated with an accuracy that increases constantly as the number of nodes of the quadrature approximation increases. The number of nodes of the quadrature approximation (N), and therefore the corresponding number of moments tracked ($2N$), cannot be increased indefinitely because of the resulting computational time and because of stability issues. In fact, as N is increased, a larger set of moments has to be tracked, resulting in longer computational times. Moreover, as the number of moments is increased, numerical issues arise. These issues are related to the moment-inversion algorithm, which sometimes fails to calculate the quadrature approximation for large N values. For most of the cases which have been investigated reasonable accuracy is achieved with $2 \leq N \leq 5$. For additional analyses and discussions on particulate systems undergoing aggregation and breakage problems, readers are referred to the work of Alopaeus *et al.* (2006), Diemer & Ehrman (2005), Gerber & Mousavi (2007), Prat & Ducoste (2006), Silva *et al.* (2010), Upadhyay & Ezekoye (2003, 2005), and Wang *et al.* (2005a, 2005b). The same conclusions can be drawn in the case of standard rates of continuous change of the internal coordinates. Most of the functional dependences of these rates with respect of the internal coordinate can be handled easily by the quadrature approximation, as has been demonstrated in a number of works (see, for example, McGraw (1997) and Wright *et al.* (2000, 2001a)).

Very critical are those cases in which the source term is highly localized in phase space (see, for example, Grosch *et al.* (2007)). In fact, when dealing with highly localized phenomena, it is possible to run the risk of not having enough nodes in the regions of phase space where they are needed. A typical example is that of particle disappearance (due to fine dissolution loops), for which a negative sink term is formulated for a specific (and typically limited) range of values for the phase space. If no nodes are localized in this window the phenomenon is ignored. Moreover, since the position of new nodes, when their number is increased, is not easy to predict, the behavior of the error with which the source term is approximated when the number of nodes is varied is unpredictable. In such cases, simply adding more nodes (and solving more moment-evolution equations) is not a good strategy. Instead, using extended Gaussian quadrature with kernel density functions as described in Chapter 3 may produce better results because the order of the quadrature can be changed independently from the number of transported moments. Another relevant case, for which QBMM tend not to perform as well as in other cases, is when one of the two limits of integration appearing in the definition of the moment has a nonzero NDF. A typical example is when the NDF is defined for values of the internal coordinate in between zero and infinity (and consequently also the moments are defined between zero and infinity) and particles are subjected to a negative rate of change for the internal coordinate. Starting from a given initial condition, as the particles decrease their internal-coordinate value, the NDF moves towards zero, which is one of the two limits of integration. When the NDF at zero, $n(0)$, is not null, some particles escape the phase space, disappear, and generate a nonzero flux at one of the two boundaries (see Eq. (7.77)). This is what happens during particle dissolution or droplet evaporation: when particles reach a finite size they leave phase space, causing a change in the moment of order zero of the NDF. In order to correctly predict these phenomena, QBMM have to be coupled with reconstruction methods capable of estimating $n(0)$ with good accuracy. Some interesting solutions are those proposed by Strumendo & Arastoopour (2008, 2010), Massot *et al.* (2010), and Yuan *et al.* (2012), among which the last is the more computationally efficient and robust. In what follows the two most popular versions of the QBMM are discussed in detail. Readers interested in recent elaborations and new developments are referred to the specialized literature (Aamir *et al.*, 2009, 2010; Attarakih *et al.*, 2009; Bove *et al.*, 2005; Gimbun *et al.*, 2009; Nagy, 2009; Qamar *et al.*, 2011a; Su *et al.*, 2007, 2008).

The QMOM

When this approach is employed, evolution equations for the moments m_k with $k = 0, 1, 2, \ldots, 2N - 1$ are solved, and the source terms are evaluated at each time step by resorting to the quadrature approximation. The number of moments to be included can theoretically be any integer number; however, usually only the first $2N - 1$ moments of the NDF are tracked, since these suffice for the calculation of a quadrature approximation of order N. Of course, an initial condition is needed, and this is readily calculated from the initial condition for the NDF (which is generally known):[28] $m_k(0) = \int \xi^k n(0, \xi) \mathrm{d}\xi$. Integration of the moment-evolution equations requires the evaluation at each time step of the source terms, and to do this a moment-inversion algorithm capable of providing the weights w_α and nodes ξ_α of the quadrature approximation is needed. A direct way to calculate the quadrature approximation is by means of its definition for the moments: $m_k = \sum_{\alpha=1}^{N} w_\alpha \xi_\alpha^k$ with $k = 0, 1, 2, \ldots, 2N - 1$. In fact, the quadrature approximation of order N is defined by its N weights, w_α, and N abscissas, ξ_α, and can be calculated by taking its first $2N$ moments, $m_0, \ldots, m_{2N-1}$. However, this is not the best way, since a very good initial guess is needed. A smarter way is to write the recursive relationship for the polynomials orthogonal to the NDF in terms of the moments m_k. Once this relationship has been written in matrix form it is easy to show that the roots of the polynomials correspond to the eigenvalues of a Jacobi matrix. (For a complete discussion of univariate moment-inversion algorithms see Sections 3.1.2 and 3.1.3.) Moment-inversion algorithms require that the moment set be realizable. Thus, the ODE solver used to solve the moment-evolution equations must ensure that, given realizable moments $m_k(0)$, the moment set $m_k(t)$ is always realizable. In practice, for homogeneous systems the source terms generate realizable moments with most ODE solvers. However, this is not the case for inhomogeneous systems, for which the advection terms can be problematic. We will return to this topic in Chapter 8.

The DQMOM

Instead of solving evolution equations for the moments, the evolution of the weights and nodes in the quadrature approximation can be directly tracked (Marchisio & Fox, 2005; McGraw & Wright, 2003). The evolution equations for weights and nodes can be derived by formally substituting the delta-function representation of the NDF into Eq. (7.3). If the weights and nodes are continuous functions of time, this procedure yields

$$\sum_{\alpha=1}^{N} \delta(\xi - \xi_\alpha) \left(\frac{\mathrm{d}w_\alpha}{\mathrm{d}t} \right) - \sum_{\alpha=1}^{N} \delta'(\xi - \xi_\alpha) \left(w_\alpha \frac{\mathrm{d}\xi_\alpha}{\mathrm{d}t} \right) = S(\xi), \tag{7.113}$$

where $\delta'(\xi)$ is the first derivative of the generalized function $\delta(\xi)$, as explained by Arfken (1985). If the weighted nodes (or weighted abscissas) $\varsigma_\alpha = w_\alpha \xi_\alpha$ are introduced, then, after some manipulations, Eq. (7.113) can be rewritten as

$$\sum_{\alpha=1}^{N} \delta(\xi - \xi_\alpha) \left(\frac{\mathrm{d}w_\alpha}{\mathrm{d}t} \right) - \sum_{\alpha=1}^{N} \delta'(\xi - \xi_\alpha) \left(-\xi_\alpha \frac{\mathrm{d}w_\alpha}{\mathrm{d}t} + \frac{\mathrm{d}\varsigma_\alpha}{\mathrm{d}t} \right) = S(\xi). \tag{7.114}$$

[28] For example, in a typical gas- or liquid-phase particle-formation process all the moments are identically null at $t = 0$. In these cases, in fact, generally no particles are present at the beginning of the simulation, since they are produced through nucleation or inception mechanisms. In other cases (for example in seeded crystallization processes or in industrial coagulators) the initial moments correspond to the NDF of the seeds/monomers/particles initially present.

We now define a_α and b_α to be the source terms for the weights w_α and weighted abscissas ς_α, respectively:

$$\frac{\mathrm{d}w_\alpha}{\mathrm{d}t} = a_\alpha, \quad \frac{\mathrm{d}\varsigma_\alpha}{\mathrm{d}t} = b_\alpha. \tag{7.115}$$

Using these definitions, Eq. (7.114) can be rewritten in a simpler form:

$$\sum_{\alpha=1}^{N} [\delta(\xi - \xi_\alpha) + \delta'(\xi - \xi_\alpha)\xi_\alpha] a_\alpha - \sum_{\alpha=1}^{N} \delta'(\xi - \xi_\alpha) b_\alpha = S(\xi). \tag{7.116}$$

This equation can now be used to determine the unknown functions a_α and b_α by applying the moment transformation. In general, the DQMOM can be applied for any independent set of moments that one wishes to control, with the only necessary condition being that the number of moments has to be equal to the number of unknown functions (i.e. a_α and b_α, or a total of $2N$ unknowns). Given that $\int \xi^k \delta(\xi - \xi_\alpha)\,\mathrm{d}\xi = \xi_\alpha^k$ and $\int \xi^k \delta'(\xi - \xi_\alpha)\mathrm{d}\xi = -k\xi_\alpha^{k-1}$, the moment transform of Eq. (7.116) yields

$$(1-k)\sum_{\alpha=1}^{N} \xi_\alpha^k a_\alpha + k \sum_{\alpha=1}^{N} \xi_\alpha^{k-1} b_\alpha = \overline{S}_k \tag{7.117}$$

with $k = k_1, k_2, \ldots, k_{2N}$. As can be seen from Eq. (7.117), the source terms of the transport equations of the N weights w_α and the N weighted abscissas ς_α can be defined through a linear system involving some $2N$ integer moments. The linear system in Eq. (7.117) can be written in matrix form, $\mathbf{A}\boldsymbol{\alpha} = \mathbf{d}$, where $\boldsymbol{\alpha} = [a_1 \cdots a_N\, b_1 \cdots b_N]^{\mathrm{T}} = [\mathbf{a}^{\mathrm{T}}\, \mathbf{b}^{\mathrm{T}}]^{\mathrm{T}}$, $\mathbf{d} = [\overline{S}_{k_1} \cdots \overline{S}_{k_{2N-1}}]^{\mathrm{T}}$, and the components of the matrix $\mathbf{A}$ are

$$a_{ij} = \begin{cases} (1-k_i)\xi_j^{k_i} & \text{if } 1 \le j \le N, \\ k_i \xi_j^{k_i - 1} & \text{if } N+1 \le j \le 2N. \end{cases} \tag{7.118}$$

If (as in the QMOM) the first $2N$ integer moments are chosen (i.e. $k = 0, \ldots, 2N-1$), the matrix of the linear system is

$$\mathbf{A} = \begin{bmatrix} 1 & \cdots & 1 & 0 & \cdots & 0 \\ 0 & \cdots & 0 & 1 & \cdots & 1 \\ -\xi_1^2 & \cdots & -\xi_N^2 & 2\xi_1 & \cdots & 2\xi_N \\ \vdots & & \vdots & \vdots & & \vdots \\ 2(1-N)\xi_1^{2N-1} & \cdots & 2(1-N)\xi_N^{2N-1} & (2N-1)\xi_1^{2N-2} & \cdots & (2N-1)\xi_N^{2N-2} \end{bmatrix}. \tag{7.119}$$

Note that $\mathbf{A}$ does not depend on the weights w_α. Moreover, if the abscissas ξ_α are unique, then $\mathbf{A}$ will be of full rank for arbitrary N. For this case, the source terms for the transport equations of the weights w_α and weighted abscissas ς_α are simply $\boldsymbol{\alpha} = \mathbf{A}^{-1}\mathbf{d}$. One can summarize the procedure for the DQMOM as follows. The evolution equations for the weights and nodes of the quadrature approximation, reported in Eq. (7.115), are solved by calculating the source terms inverting the linear system. The equations are integrated by using the following initial conditions: $w_\alpha(0) = w_\alpha^0$, $\varsigma_\alpha(0) = w_\alpha^0 \xi_\alpha^0$ for $k = 1, \ldots, N$. The values for w_α^0 and ξ_α^0 with $1 \le \alpha \le N$ are calculated from the initial condition for the moments belonging to the moment set $m_k(0)$, which in turn are calculated from the NDF. This is generally done by applying the PD or Wheeler algorithm, in which case the first $2N$ moments are chosen (as described in Chapter 3), and by directly solving the non-linear system when another non-sequential moment set is chosen. Many practical cases are characterized by zero initial conditions for the moments of the NDF and as a consequence for the weights w_α and weighted abscissas ς_α. Thus, in the first time step, the local abscissas ξ_α are undefined, the matrix $\mathbf{A}$ is not invertible, and therefore the source terms $\mathbf{a}$ and $\mathbf{b}$ cannot be found. A possible solution is to introduce "seeds" in the initial conditions in order to start the simulation with very small values of the weights and weighted abscissas. Although this approach could seem questionable due to the arbitrary choice of the initial values, it has been found to be very stable and efficient. Another possibility is to solve the moment-evolution equations for one very small time step δt to find $m_k(\delta t)$ (since $m_k(0) = 0$ in this case, $m_k(\delta t)$ will be determined by nucleation), and then to apply the moment-inversion algorithm to $m_k(\delta t)$. It is worth discussing in more detail the calculation of the source term in the case of continuous changes in the internal coordinate (i.e. the drift term) because, through some mathematical passages, it can be further simplified. In this case the source term reduces to $S(\xi) = -\partial\dot{\xi}n/\partial\xi$ and $\overline{S}_k$ can be expressed as

$$\overline{S}_k = -\int \xi^k \frac{\partial \dot{\xi} n}{\partial \xi} \, d\xi = k \int \xi^{k-1} \dot{\xi} n \, d\xi. \tag{7.120}$$

On representing the NDF as a summation of delta functions centered on the nodes, we obtain $\overline{S}_k = k \sum_{\alpha=1}^{N} \xi_\alpha^{k-1} \dot{\xi}_\alpha w_\alpha$, which is the N-point quadrature approximation of the source term for integer moment m_k. In this case, the linear system that defines the source terms in the DQMOM transport equations reduces to

$$\mathbf{A} \begin{bmatrix} \mathbf{a} \\ \mathbf{b}^* \end{bmatrix} = \mathbf{0}, \tag{7.121}$$

where $\mathbf{b}^* = \mathbf{b} - \mathbf{P}\mathbf{g}$. The diagonal weight matrix is defined by $\mathbf{P} = \text{diag}(w_1, \ldots, w_N)$, and $\mathbf{g} = [\dot{\xi}_1 \cdots \dot{\xi}_N]^{\mathrm{T}}$. The solution to Eq. (7.121) is $\mathbf{a} = \mathbf{0}$ and $\mathbf{b} = \mathbf{P}\mathbf{g}$ so that the source term for the weighted abscissas is found directly from the drift term, while the weights remain constant. For the univariate case, the QMOM and DQMOM yield exactly the same results (if the weights and nodes are continuous functions of time), and thus they can be used interchangeably. The principal motivation for developing the DQMOM was the treatment of multivariate PBE for which the QMOM is problematic due to the lack of a robust moment-inversion algorithm.

7.4.2 Bivariate and multivariate PBE

By extending the concepts of the previous section, it is possible to derive an approximation (which is no longer a real Gaussian quadrature) for the bivariate system. The resulting N-point quadrature approximation transforms the definition of the general bivariate moment into (Wright *et al.*, 2001b) $m_{\mathbf{k}} = \sum_{\alpha=1}^{N} w_\alpha \xi_{1,\alpha}^{k_1} \xi_{2,\alpha}^{k_2}$, where, as for the univariate case, w_α is the weight of the αth node, $\xi_{1,\alpha}$ is the position of the αth node for the first internal coordinate, and $\xi_{2,\alpha}$ is the position of the αth node for the second internal coordinate. Using this quadrature approximation, the source term of the moment of order $\mathbf{k} = (k_1, k_2)$ can be estimated with good accuracy. As an example, the approximated expression of the source term in the case of nucleation, molecular growth, diffusion, aggregation, and breakage (in the general case of drift rates and kernels depending on internal coordinates) is

$$
\begin{aligned}
\overline{S}_{\mathbf{k}} = \overline{J}_{\mathbf{k}} &+ k_1 \sum_{\alpha=1}^{N} \xi_{1,\alpha}^{k_1-1} \xi_{2,\alpha}^{k_2} \dot{\xi}_1(\xi_{1,\alpha}, \xi_{2,\alpha}) w_\alpha + k_2 \sum_{\alpha=1}^{N} \xi_{1,\alpha}^{k_1} \xi_{2,\alpha}^{k_2-1} \dot{\xi}_2(\xi_{1,\alpha}, \xi_{2,\alpha}) w_\alpha \\
&+ k_1(k_1 - 1) \sum_{\alpha=1}^{N} \xi_{1,\alpha}^{k_1-2} \xi_{2,\alpha}^{k_2} D_{11}(\xi_{1,\alpha}, \xi_{2,\alpha}) w_\alpha \\
&+ 2k_1 k_2 \sum_{\alpha=1}^{N} \xi_{1,\alpha}^{k_1-1} \xi_{2,\alpha}^{k_2-1} D_{12}(\xi_{1,\alpha}, \xi_{2,\alpha}) w_\alpha \\
&+ k_2(k_2 - 1) \sum_{\alpha=1}^{N} \xi_{1,\alpha}^{k_1} \xi_{2,\alpha}^{k_2-2} D_{22}(\xi_{1,\alpha}, \xi_{2,\alpha}) w_\alpha \\
&+ \frac{1}{2} \sum_{\alpha=1}^{N} \sum_{\gamma=1}^{N} \left[(\xi_{1,\alpha} + \xi_{1,\gamma})^{k_1} (\xi_{2,\alpha} + \xi_{2,\gamma})^{k_2} - \xi_{1,\alpha}^{k_1} \xi_{2,\alpha}^{k_2} - \xi_{1,\gamma}^{k_1} \xi_{2,\gamma}^{k_2} \right] \beta_{\alpha,\gamma} w_\alpha w_\gamma \\
&+ \sum_{\alpha=1}^{N} \left[\overline{b}_\alpha^{\mathbf{k}} - \xi_{1,\alpha}^{k_1} \xi_{2,\alpha}^{k_2} \right] a_\alpha w_\alpha,
\end{aligned} \tag{7.122}
$$

where the aggregation kernel is evaluated for the values of the internal coordinates in the αth and γth nodes, $\beta_{\alpha,\gamma} = \beta(\xi_{1,\alpha}, \xi_{1,\gamma}; \xi_{2,\alpha}, \xi_{2,\gamma})$, and the breakage kernel $a_\alpha = (\xi_{1,\alpha}, \xi_{2,\alpha})$ is evaluated at the αth node. For the moments of the daughter distribution function the following general expression is obtained: $\overline{b}_\alpha^{\mathbf{k}} = \int \xi_1^{k_1} \xi_2^{k_2} b(\xi_{1,\alpha}, \xi_{2,\alpha} | \boldsymbol{\xi}) \mathrm{d}\boldsymbol{\xi}$. In the case of symmetric breakage, which has already been discussed in Eq. (7.107), the expression is $\overline{b}_\alpha^{\mathbf{k}} = 2^{1-k_1-k_2} \xi_{1,\alpha}^{k_1} \xi_{2,\alpha}^{k_2}$. More details on the use of the QBMM for bivariate particulate systems can be found in Buffo *et al.* (2012) and Zucca *et al.* (2007). As with univariate systems, there are two ways to approach the bivariate quadrature problem: (i) by solving the equations for a set of bivariate moments (QMOM) and calculating the quadrature approximation with a moment-inversion algorithm,[29] or (ii) by directly tracking the evolution of the quadrature by solving the equations for the weights and weighted nodes/abscissas of the approximation (DQMOM). The main difference from univariate problems arises from the higher number of degrees of freedom in the choice of the moments to be represented and tracked. In fact, if with univariate distributions the choice of the moments is automatically defined (in order to have a true quadrature approximation of order N we must use the first $2N$ moments), with bivariate and multivariate distributions theoretically any set of moments could be considered, as shown for example in the work of McGraw *et al.* (1998), McGraw & Wright (2003), Rosner *et al.* (2003), and Yoon & McGraw (2004a, 2004b). Although the choice of moments is affected by the method employed (i.e. QMOM versus DQMOM) and, when using the QMOM, by the moment-inversion algorithm adopted (i.e. brute-force QMOM, tensor-product QMOM, CQMOM), some general comments can be made. Some of these comments are reported in the next section, when discussing bivariate and multivariate applications of QMOM and DQMOM, whereas the opportunity to define and use an optimal moment set is discussed in Section 3.2.1.

[29] The moment-inversion algorithms for bivariate and multivariate problems are discussed in Section 3.2.

The QMOM

The extension of the QMOM to bivariate and multivariate systems is relatively straightforward. The moment set $m_{\mathbf{k}}$, defined in Eq. (2.2), whose evolution is dictated by Eq. (7.73), can be solved by estimating the source term in Eq. (7.74) with an N-point quadrature approximation. An example of how this works for a simple bivariate system undergoing continuous changes of the internal coordinates, as well as aggregation, breakage, and nucleation, is shown in Eq. (7.122). Of course, an initial condition for the moments is needed, and this can be calculated from knowledge of the initial NDF, as in univariate cases. We should also note that the QMOM is valid even for source terms that are discontinuous in time (which is rare), while the DQMOM cannot be used for such cases. With the QMOM closure is achieved by employing the quadrature approximation found from the moment-inversion algorithms discussed in detail in Chapter 3.

The DQMOM

The extension of the DQMOM to bivariate and multivariate systems uses a quadrature approximation of order N. The simple case of a bivariate distribution uses

$$n(\xi_1, \xi_2) = \sum_{\alpha=1}^{N} w_\alpha \delta(\xi_1 - \xi_{1,\alpha})\delta(\xi_2 - \xi_{2,\alpha}). \tag{7.123}$$

The DQMOM equations are obtained by substituting Eq. (7.123) into Eq. (7.104). If the weights and nodes are continuous functions of time,[30] we obtain

$$\begin{aligned}\sum_{\alpha=1}^{N} \delta(\xi_1 - \xi_{1,\alpha})\delta(\xi_2 - \xi_{2,\alpha})a_\alpha - \sum_{\alpha=1}^{N} \delta'(\xi_1 - \xi_{1,\alpha})\delta(\xi_2 - \xi_{2,\alpha})\,(b_{1,\alpha} - \xi_{1,\alpha}a_\alpha) \\ - \sum_{\alpha=1}^{N} \delta(\xi_1 - \xi_{1,\alpha})\delta'(\xi_2 - \xi_{2,\alpha})\,(b_{2,\alpha} - \xi_{2,\alpha}a_\alpha) = S(\xi_1, \xi_2).\end{aligned} \tag{7.124}$$

The unknown functions $\mathbf{a} = (a_1, \ldots, a_N)$, $\mathbf{b}_1 = (b_{1,1}, \ldots, b_{1,N})$, and $\mathbf{b}_2 = (b_{2,1}, \ldots, b_{2,N})$ are the source terms of DQMOM equations:

$$\frac{\mathrm{d}w_\alpha}{\mathrm{d}t} = a_\alpha, \quad \frac{\mathrm{d}\varsigma_{1,\alpha}}{\mathrm{d}t} = b_{1,\alpha}, \quad \frac{\mathrm{d}\varsigma_{2,\alpha}}{\mathrm{d}t} = b_{2,\alpha}, \tag{7.125}$$

for $1 \leq \alpha \leq N$, where the weighted abscissas are defined by $\varsigma_{1,\alpha} = w_\alpha \xi_{1,\alpha}$ and $\varsigma_{2,\alpha} = w_\alpha \xi_{2,\alpha}$. The DQMOM source terms can be found using the moment transform in terms of the moments of order $\mathbf{k} = \{k_1, k_2\}$: $m_{k_1,k_2} = \int \xi_1^{k_1} \xi_2^{k_2} n \, \mathrm{d}\boldsymbol{\xi}$. If the moment transform is applied to Eq. (7.124), the right-hand sides of Eq. (7.125) are found by solving the linear system

$$\sum_{\alpha=1}^{N} (1 - k_1 - k_2)\xi_{1,\alpha}^{k_1} \xi_{2,\alpha}^{k_2} a_\alpha + \sum_{\alpha=1}^{N} \left[k_1 \xi_{1,\alpha}^{k_1-1} \xi_{2,\alpha}^{k_2} b_{1,\alpha} + k_2 \xi_{1,\alpha}^{k_1} \xi_{2,\alpha}^{k_2-1} b_{2,\alpha} \right] = \overline{S}_{k_1,k_2}. \tag{7.126}$$

Here the source term $\overline{S}_{k_1,k_2}$ is defined by the right-hand side of Eq. (7.104): $\overline{S}_{k_1,k_2} = \int \xi_1^{k_1} \xi_2^{k_2} S \, \mathrm{d}\boldsymbol{\xi}$. It is important to stress again here that the source term is evaluated by using

[30] It is possible for the moments to be continuous in time, while the weights and/or nodes are not. However, discontinuous solutions are most likely to occur for inhomogeneous systems for which the transport equations are hyperbolic.

an N-point quadrature scheme and therefore sometimes it is reported with the superscript $^{(N)}$. For the bivariate case there are $3N$ unknowns $\{a_\alpha, b_{1,\alpha}, b_{2,\alpha}\}$, and thus an equal number of moments must be used with Eq. (7.126). The resulting linear system of $3N$ equations can be written in matrix form: $\mathbf{A}\boldsymbol{\alpha} = \mathbf{d}$, where $\boldsymbol{\alpha} = [a_1 \cdots a_N\, b_{1,1} \cdots b_{1,N}\, b_{2,1} \cdots b_{2,N}]^{\mathrm{T}} = [\mathbf{a}^{\mathrm{T}}\, \mathbf{b}_1^{\mathrm{T}}\, \mathbf{b}_2^{\mathrm{T}}]^{\mathrm{T}}$. It is interesting to notice here that the coefficient matrix $\mathbf{A}$ of the linear system is the same as that used with brute-force QMOM and is defined for a general multivariate case in Eq. (3.44). For a bivariate system (i.e. $M = 2$) and for $N = 1$, the natural choice of mixed moments is $\{k_1, k_2\} = \{0,0\}, \{1,0\}, \{0,1\}$. As a consequence, the DQMOM equations reduce to

$$\frac{\mathrm{d}w_1}{\mathrm{d}t} = \overline{S}^{(1)}_{0,0}, \quad \frac{\mathrm{d}\varsigma_{1,1}}{\mathrm{d}t} = \overline{S}^{(1)}_{1,0}, \quad \frac{\mathrm{d}\varsigma_{2,1}}{\mathrm{d}t} = \overline{S}^{(1)}_{0,1}. \tag{7.127}$$

As in the monovariate case, the one-point quadratures used to approximate the source terms on the right-hand side are usually inaccurate. Thus, only cases with $N > 1$ will normally be of interest. For $N = 2$, the NDF is described by two quadrature points. Since each point α in phase space is determined by three parameters $\{w_\alpha, \xi_{1,\alpha}, \xi_{2,\alpha}\}$, six mixed moments are required in order to close the problem. There is no unique way to choose the moments. If the six lowest-order moments, $\{k_1, k_2\} = \{0,0\}, \{1,0\}, \{0,1\}, \{2,0\}, \{1,1\}$, and $\{0,2\}$, are chosen, then the coefficient matrix is

$$\mathbf{A} = \begin{bmatrix} 1 & 1 & 0 & 0 & 0 & 0 \\ 0 & 0 & 1 & 1 & 0 & 0 \\ 0 & 0 & 0 & 0 & 1 & 1 \\ -(\xi_{1,1})^2 & -(\xi_{1,2})^2 & 2\xi_{1,1} & 2\xi_{1,2} & 0 & 0 \\ -\xi_{1,1}\xi_{2,1} & -\xi_{1,2}\xi_{2,2} & \xi_{2,1} & \xi_{2,2} & \xi_{1,1} & \xi_{1,2} \\ -(\xi_{2,1})^2 & -(\xi_{2,2})^2 & 0 & 0 & 2\xi_{2,1} & 2\xi_{2,2} \end{bmatrix}. \tag{7.128}$$

However, this choice is a poor one. For any values of $\xi_{1,\alpha}$ and $\xi_{2,\alpha}$ the matrix in Eq. (7.128) is singular. This is due to the fact that the covariance (where $p_\alpha = w_\alpha/(w_1 + w_2)$), $\sigma_{12} = p_1\xi_{1,1}\xi_{2,1} + p_2\xi_{1,2}\xi_{2,2} - (p_1\xi_{1,1} + p_2\xi_{1,2})(p_1\xi_{2,1} + p_2\xi_{2,2})$, is related to the variances, $\sigma_1^2 = p_1\xi_{1,1}^2 + p_2\xi_{1,2}^2 - (p_1\xi_{1,1} + p_2\xi_{1,2})^2$ and $\sigma_2^2 = p_1\xi_{2,1}^2 + p_2\xi_{2,2}^2 - (p_1\xi_{2,1} + p_2\xi_{2,2})^2$, by $\sigma_{12}^2 = \sigma_1^2\sigma_2^2$. Geometrically, this equality results from the fact that two points in phase space are necessarily collinear. Thus, in order to control the covariance, $N \geq 3$ is required. The issue of the choice of the moments for bivariate and multivariate systems is further discussed below and in Chapter 3. We will close the discussion on the DQMOM by noting that the multivariate version of Eq. (7.126) follows by inspection:

$$\begin{aligned} \sum_{\alpha=1}^{N} \left[\left(1 - \sum_{n=1}^{M} k_n \right) \prod_{n=1}^{M} \xi_{n,\alpha}^{k_n} \right] a_\alpha + \sum_{\alpha=1}^{N} \sum_{n=1}^{M} \frac{\partial}{\partial \xi_{n,\alpha}} \left(\prod_{m=1}^{M} \xi_{m,\alpha}^{k_m} \right) b_{n,\alpha} \\ = \sum_{\alpha=1}^{N} \sum_{n=1}^{M} \sum_{m=1}^{M} \frac{\partial^2}{\partial \xi_{n,\alpha}\, \partial \xi_{m,\alpha}} \left(\prod_{p=1}^{M} \xi_{p,\alpha}^{k_p} \right) + \overline{S}_{k_1,\ldots,k_M}, \end{aligned} \tag{7.129}$$

where the mixed-moment source term is defined by $\overline{S}_{k_1,\ldots,k_M} = \int S \prod_{n=1}^{M} \xi_n^{k_n}\, \mathrm{d}\boldsymbol{\xi}$ and the resulting coefficient matrix for the linear system is Eq. (3.44). On choosing an independent set of low-order moments $\{k_1, \ldots, k_M\}$, Eq. (7.129) generates a linear system that can be used to find the $N(M+1)$ unknown source terms $\{a_\alpha, b_{1,\alpha}, \ldots, b_{M,\alpha}\}$ appearing in the DQMOM equations for $\{w_\alpha, \varsigma_{1,\alpha}, \ldots, \varsigma_{M,\alpha}\}$ (cf. Eq. (7.125)). A brief summary of all the relevant issues concerning the choice of the moment set is reported below, whereas a detailed discussion can be found in Chapter 3. Last, but not least, for a multivariate NDF defined on a finite domain in phase space, the question of realizability must be carefully examined (see, for example, the work of Wang & Fox (2003)).

Choice of moment set

We now briefly discuss the issue of choosing the moments of the NDF for a generic application of the QBMM. First of all, for a specific problem, it is necessary to establish the total number of moments within the set and the highest order of the moment set. This is generally determined by the order of the quadrature approximation to be used. The higher the number of nodes of the quadrature approximation, N, the more moments will be included in the moment set, $N(M + 1)$, and as a consequence the higher the order of the moments included in the set. There exists no *a priori* criterion for choosing N, and, therefore, depending on the nature of the source term, the order of the quadrature approximation must be determined by comparison with alternative, more sophisticated, methods. The choice of the moment set also influences the overall stability of the method and therefore this aspect too should be carefully considered. We will first analyze the bivariate case and then extend the results to multivariate cases. The discussion will be based on the impact that the choice of the moment set has on the coefficient matrix $\mathbf{A}$, which, as has already been mentioned, plays an important role both in the QMOM and in the DQMOM. Moreover, although with some of the moment-inversion algorithms the moment set is automatically defined by fixing the number of nodes of the quadrature approximation, it is worth discussing what moment set the different methods require, and what impact this has in terms of the final coefficient matrix. As a general rule we can infer from Eq. (7.128) that in order to have a correct representation of the NDF at the second-order moment level the number of nodes N has to be greater than the number of internal coordinates M. When working with a univariate NDF ($M = 1$), N has to be at least equal to two in order to predict the variance. For a bivariate NDF ($M = 2$), N has to be at least equal to three in order to represent correctly the second-order moments (i.e. the variances and covariance). Note, however, that it is possible to use $N = 2$ for the bivariate case if $(k_1, k_2) = (1, 1)$ is replaced in Eq. (7.128) by a higher-order moment (e.g. $(2, 2)$). In fact, as a general rule, even for the multivariate case in which $M > 2$ it is usually possible to find a moment set for any value of $N \leq M$ for which $\mathbf{A}$ is of full rank. If we decide to describe the bivariate distribution by two nodes (i.e. $N = 2$) the problem is defined in terms of six variables, namely two weights and two abscissas for each internal coordinate, and as a consequence six moments must be chosen to generate the linear system. If we consider all the possible integer moments up to the fifth order, we will have 21 moments and therefore up to 54 264 possible choices of six moments to be tracked. Of course, not all of these choices are valid; 373 of them result in a singular matrix. The number of possible combinations of moments is greatly reduced if we take into consideration the physical meanings of the first moments that we wish to evaluate with high accuracy. The first three mixed moments are in fact related to the total number of particles ($m_{0,0}$) and to the mean values of the internal coordinates ($m_{1,0}$ and $m_{0,1}$), properties that are usually of great interest and need to be known as accurately as possible.

If we restrict our analysis to moment sets containing the first three moments, we just need to choose the remaining three moments in a group of 18, and the choices leading to a non-singular system reduce to nine. All of them can be obtained by choosing any group of pure moments of a single internal coordinate. It is sufficient to choose mixed moments or pure moments of both the internal coordinates in order to have a non-singular matrix, $\mathbf{A}$. However, one may want to know with good accuracy also the values of the pure moments of order two (i.e. $m_{2,0}$ and $m_{0,2}$), which are related to the variances of the two internal coordinates.[31] In this case, we just need to choose one of the remaining 16 moments of the original set of 21 moments of order between zero and five. The first possible moment that can be selected is $m_{3,0}$. With this set of moments the method can be applied and the numerical stability is satisfactory. However, it is important to notice that using only pure moments in the formulation of the problem presents some drawbacks, because the correlation between the internal coordinates is not properly taken into account. This can be important if one is interested in gaining information on the shape of the bivariate distribution function. Since the two-node formulation is not satisfactory for bivariate problems, let us consider the issues involved when using three nodes, $N = 3$. In this case a set of nine integer moments, which are generally chosen from among the moments up to global order five, must be tracked.[32] There are in this case 293 930 possible choices, only 36 of which lead to a non-singular matrix, **A**. If we again restrict our analysis to the sets including the first five pure moments ($(k_1, k_2) = (0, 0), (1, 0), (0, 1), (2, 0)$, and (0,2)), we obtain non-singular matrices only by including in the moment set $m_{1,1}$ and two moments of global order three and one moment of global order four, resulting in $m_{3,0}$, $m_{2,1}$, $m_{4,0}$ or $m_{0,3}$, $m_{1,2}$, and $m_{0,4}$, or alternatively $m_{2,1}$, $m_{1,2}$, and $m_{2,2}$. It is worth noticing that some regular structures can be found in the sets of moments. These structures are particularly evident when the moments of a bivariate distribution are displayed in a triangular table, as illustrated by Zucca (2005) (see, for example, Table 7.2).

We have discussed the selection of a set of moments leading to an invertible matrix **A**. This is a necessary condition, and further refinement in the choice of the moments can be driven by the conditioning and the numerical stability. If the problem becomes ill-conditioned, the accuracy in the computation can be dramatically reduced, thus leading to numerical instability. The more the matrix **A** is kept far from singularity during the computation, the more accurate the predictions of both tracked and non-tracked moments will be. The determinant becomes close to zero if the nodes approach each other. As the number of nodes is increased, the probability of finding two nodes approaching each other increases. Another consideration concerns the order of the mixed moments used to compute the source terms for weights and abscissas. From this point of view, including higher-order moments that do not need to be known should be avoided, and it is interesting to evaluate the possibility of using fractional-order moments, to reduce the degree of the determinant (see, for example, the work of Zucca (2005)). It is worth noticing as well that, when the

Table 7.2. A triangular table reporting the different moments of a bivariate NDF up to global order five

$m(5, 0)$					
$m(4, 0)$	$m(4, 1)$				
$m(3, 0)$	$m(3, 1)$	$m(3, 2)$			
$m(2, 0)$	$m(2, 1)$	$m(2, 2)$	$m(2, 3)$		
$m(1, 0)$	$m(1, 1)$	$m(1, 2)$	$m(1, 3)$	$m(1, 4)$	
$m(0, 0)$	$m(0, 1)$	$m(0, 2)$	$m(0, 3)$	$m(0, 4)$	$m(0, 5)$

[31] If the internal coordinate is the particle size, the second-order moment can be related to the mean particle surface area.

[32] The global order of the moment is defined as the summation of the order with respect to each internal coordinate. For example, $m_{0,1}$ is of global order one, whereas $m_{2,3}$ is of global order five.

two internal coordinates differ by orders of magnitude, the condition number can be higher than in the case of internal coordinates of the same order of magnitude. As a general rule, we can say that internal coordinates characterized by great differences in magnitude lead to bad scaling of the matrix, and their use should be avoided. If necessary, this problem can be overcome by rescaling and normalization. All the conclusions formulated thus far are based on the analysis of numerical issues, but it should be kept in mind that, as has already been stated, the accuracy of the prediction for the tracked moments is usually much better than that for the other moments not included in the set. Therefore, important moments should always, as far as possible, be included in the set of selected moments. This is illustrated in Figure 7.4, where a simple aggregation and breakage problem is considered. Under the operating conditions considered, a steady-state solution for the NDF is reached. Figure 7.4 compares the predictions for the third-order moment $m_{3,0}$ obtained with a Monte Carlo method with the predictions obtained with the DQMOM for $N = 3$ and for different moment sets. All the sets considered contain $m_{0,0}$, $m_{1,0}$, $m_{0,1}$, $m_{2,0}$, $m_{1,1}$, and $m_{0,2}$, and saturate the degrees of freedom with three moments of global order three, four, and five. As can be seen in Figure 7.4, only by including moments of global order three in the moment set is $m_{3,0}$ correctly predicted. The extension to multivariate cases is complicated by a number of factors, and, especially for high-dimensional cases, it is necessary to identify a reliable procedure. Applications in different fields have shown that simulations should be carried out by tracking the evolution of the moments belonging to the *optimal-moment-set DQMOM* described in Section 3.2.1. In summary, when using the optimal moment set, typically simulations run smoothly and without many problems. When homogeneous systems are considered, the evolution of the optimal moment set can be tracked by solving equations either for the moments (QMOM) or for the quadrature approximation (DQMOM).

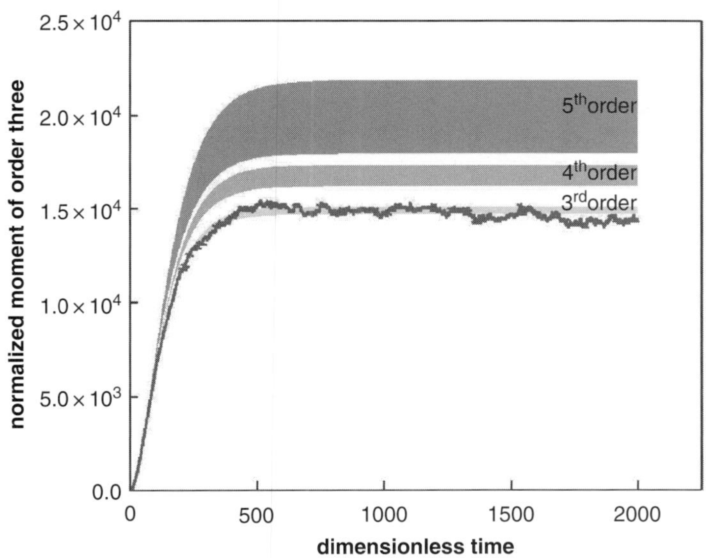

Figure 7.4. Predictions of the evolution of $m_{3,0}$ (normalized with its initial value) by the QBMM with $N = 3$ for a standard aggregation-and-breakage problem with different choices for the last three moments from among moments of global order three, global order four, and global order five, given the first six ($m_{0,0}$, $m_{1,0}$, $m_{0,1}$, $m_{2,0}$, $m_{1,1}$, and $m_{0,2}$). Thin lines, QBMM; thick line: Monte Carlo method. Adapted from Zucca *et al.* (2007).

When the first procedure is adopted, the CQMOM should be used as the moment-inversion algorithm (Yuan & Fox, 2011). When the second procedure is adopted, the evolution of the M-dimensional, N-point quadrature approximation is directly tracked. In the case of spatially homogeneous systems the two methods are equivalent.

7.4.3 Collisional KE

The generic application of QBMM to the homogeneous collisional KE is straightforward. Recall that the generic definition of the moment of the velocity NDF is $m_{ijk} = m(i, j, k) = \int \xi_1^i \xi_2^j \xi_3^k n \, \mathrm{d}\boldsymbol{\xi}$, and that its evolution obeys the following equation:

$$\frac{\mathrm{d}m_{ijk}}{\mathrm{d}t} = \overline{S}_{ijk} = \int \xi_1^i \xi_2^j \xi_3^k S \, \mathrm{d}\boldsymbol{\xi}. \tag{7.130}$$

On highlighting the presence of the drift and discontinuous-event terms, the following equation is obtained:

$$\frac{\mathrm{d}m_{ijk}}{\mathrm{d}t} = i \int \xi_1^{i-1} \xi_2^j \xi_3^k \dot{\xi}_1 n \, \mathrm{d}\boldsymbol{\xi} + j \int \xi_1^i \xi_2^{j-1} \xi_3^k \dot{\xi}_2 n \, \mathrm{d}\boldsymbol{\xi} + k \int \xi_1^i \xi_2^j \xi_3^{k-1} \dot{\xi}_3 n \, \mathrm{d}\boldsymbol{\xi} + \Pi_{ijk}, \tag{7.131}$$

where Π_{ijk} is the moment transform of the collision integral. The procedure for a KE is identical to that for a PBE. First a moment set is defined. Then, from the moments, it is possible (using any of the moment-inversion algorithms discussed in Chapter 3) to determine the quadrature approximation of order N that represents the moments of the NDF: $m_{ijk} = \sum_{\alpha=1}^{N} w_\alpha \xi_{1,\alpha}^i \xi_{2,\alpha}^j \xi_{3,\alpha}^k$. By applying the quadrature approximation to the source term of Eq. (7.131) the following evolution equation is obtained:

$$\frac{\mathrm{d}m_{ijk}}{\mathrm{d}t} = i \sum_{\alpha=1}^{N} w_\alpha \xi_{1,\alpha}^{i-1} \xi_{2,\alpha}^j \xi_{3,\alpha}^k + j \sum_{\alpha=1}^{N} w_\alpha \xi_{1,\alpha}^i \xi_{2,\alpha}^{j-1} \xi_{3,\alpha}^k + k \sum_{\alpha=1}^{N} w_\alpha \xi_{1,\alpha}^i \xi_{2,\alpha}^j \xi_{3,\alpha}^{k-1} + \Pi_{ijk}, \tag{7.132}$$

resulting in a closed set of differential equations that represents the evolution of the moment set. The collision source term, which is typically expressed as the summation of two terms (see, for example, Eq. (6.181)), for the spatially homogeneous case is simplified into a single term:

$$\Pi_{ijk} = \frac{6}{d_\mathrm{p}} \sum_{\alpha=1}^{N} \sum_{\beta=1}^{N} w_\alpha w_\beta I_{ijk}^{(0)}(1, \boldsymbol{\xi}_\alpha, \boldsymbol{\xi}_\alpha - \boldsymbol{\xi}_\beta) |\boldsymbol{\xi}_\alpha - \boldsymbol{\xi}_\beta|. \tag{7.133}$$

The function $I_{ijk}^{(0)}$ is derived analytically from the hard-sphere-collision integral, and readers interested in the exact forms are referred to Tables 6.1–6.3 of Chapter 6.[33] One crucial issue is the description of the equilibrium distribution with QBMM. In fact, since the nonlinear collision source terms that drive the NDF and its moments to the Maxwellian equilibrium are approximated, the equilibrium is generally not perfectly described. The error involved is generally very small, and is reduced when the number of nodes is increased, but can be easily overcome by using some simple corrections. Details on these corrections for the isotropic Boltzmann equation test case are reported in Icardi *et al.* (2012).

[33] The representation of the collision integral with simpler approaches, such as Maxwell particles or BGK, is not discussed here, because, as has already been stated, no closure problem arises in these cases.

7.5 Monte Carlo methods

The moment methods discussed thus far in this chapter almost always involve some type of closure. Thus, in order to validate the accuracy of the closures, it is necessary to compare the predictions for the moments with more sophisticated solution methods. For a univariate PBE, the class and sectional methods described in Section 7.2 are an excellent choice for validation of moment methods. However, we have seen that for a multivariate PBE the implementation of sectional methods becomes quite tedious and must be done with great care in order to avoid errors. For multivariate systems, an attractive alternative to sectional methods is a class of simulation techniques known as Monte Carlo (MC) methods. Monte Carlo (and direct-simulation Monte Carlo (DSMC)) methods are based on *artificial realizations* of the behavior of the multiphase system under investigation (Kendal, 1950). As has already been mentioned, we are interested in predicting the time evolution of a multiphase system constituted by a population of particles, generally also spatially distributed, governed by some continuous and discontinuous events. In MC methods the evolution of each particle is tracked. Continuous events are represented deterministically as a continuous change (with a specific rate, $\dot{\boldsymbol{\xi}}$) of the internal-coordinate vector, $\boldsymbol{\xi}$, whereas discontinuous events are represented by random events occurring with specific and prescribed probabilities. The connection with the deterministic approach used in the methods previously presented is clear (Ramkrishna, 1981): the underlying multiparticle joint PDF (see Chapter 4) or a similar master density equation (see Chapter 2 of Ramkrishna (2000)) is not directly averaged (leading to the GPBE) but is directly solved. This solution is obtained by adopting appropriate methods for stochastic differential equations, leading to numerous complex solutions, each representing a single realization. These realizations represent different *sample paths* that must be averaged out, thereby eliminating low-probability events and eventually resulting in averaged values that can be compared with results obtained using the other deterministic methods (e.g. CM, MOM, QMOM, DQMOM, etc.). This is generally done by repeating the same simulations many times (for time-dependent problems) or by averaging of a single simulation at different times (for steady-state problems).

In this section the different MC methods that can be used to validate moment closures will be briefly described, focusing the discussion on spatially homogeneous systems, although the extension to inhomogeneous systems is straightforward. This will be rapidly analyzed at the end of the section, but more details are available in the literature. All of these methods are based on knowledge of the initial states of the particles. This is known in a statistical sense, meaning that the initial NDF of the multiphase system is specified. A group of N particles is then randomly created (representing the given NDF) in the computational domain or box, and its evolution is directly tracked. The main distinction between the different methods is related to the way in which time advances during the simulation. In time-driven simulations a time step Δt is chosen, and then the number and the type of event acting on the population of particles and realized within the time interval are defined (Collins & Knudsen, 1970; Liffman, 1992; Spielman & Levenspiel, 1965; Zeitlin & Tavlarides, 1972a, 1972b). In event-driven simulations an event is selected to occur and then time is advanced accordingly (Shah *et al.*, 1976, 1977). This method does not require explicit time discretization, and the time step, which is calculated during the simulation, adjusts itself to the rates of the various processes. Another important distinction between methods is related to what happens to the total number of particles N belonging to the computational domain as time advances. In fact, some of the discontinuous events (e.g.

aggregation and breakage) change the number of particles in the domain. After an aggregation event, for example, $N-1$ particles remain in the domain, whereas after a breakage event $N+1$ particles have to be tracked. As the simulation proceeds the total number of particles changes, whereas the total volume (or total mass) of particles in the computational domain remains constant. For this reason these methods are called constant-volume Monte Carlo. However, the simulation cannot be run indefinitely with this approach because after some time, if breakage prevails, the number of particles to be tracked is too large, whereas if aggregation prevails the number of particles is too low to guarantee acceptable statistical accuracy. In order to overcome these limitations, at every time step for which the particle number in the computational domain has increased or decreased by a factor of two of its initial value, N is halved or doubled, respectively (Kruis *et al.*, 2000; Liffman, 1992; Maisels *et al.*, 2004a, 2004b). The logical extension of this concept is to introduce continuous refilling of the computational domain, by selecting a particle from the domain at random and placing its copy in the position vacated by aggregation, and by overwriting old particles with newly formed particles in the case of breakage (Khalili *et al.*, 2010; Lee *et al.*, 2008; Lee & Matsoukas, 2000; Lin *et al.*, 2002; Matsoukas *et al.*, 2006, 2009; Smith & Matsoukas, 1998). The different methods have been compared in the case of aggregation, breakage, growth, and settling, showing that, when they are run with comparable number of particles, all methods compute the evolution of the NDF with comparable levels of error. However, although the computational costs and overall accuracy of the methods strongly depend on the details, event-driven methods generally perform faster and are characterized by higher accuracy (Zhao *et al.*, 2007).

Let us very briefly discuss the theoretical foundation of these methods. Since real systems are constituted by very large particle number densities (e.g. from $<10^9$ per cubic meter in the case of millimeter and micrometer particles to $<10^{18}$ per cubic meter in the case of nanoparticles), in MC simulations only a subsystem of test or *notional* particles (e.g. 10^4–10^6) is considered, corresponding to a small fraction of the real domain. When particle movement within the computational domain is also considered, periodic boundary conditions are implemented, so that a particle escaping the domain is immediately replaced with an identical particle. The underlying implicit assumption is that the behavior of the subsystem duplicates the system as a whole. If $n_0(\boldsymbol{\xi}) = n(0, \boldsymbol{\xi})$ is the initial NDF, the initial states of the population of N particles can be readily determined by using the corresponding normalized cumulative distribution: $N_0(\boldsymbol{\xi}) = (1/m(\mathbf{0})) \int_0^{\xi_1} \dots \int_0^{\xi_M} n_0 \, \mathrm{d}\boldsymbol{\xi}$, where M is the dimension of the internal-coordinate vector. In fact, by generating random numbers, uniformly distributed between 0 and 1, and then identifying the set of internal coordinates resulting in a normalized cumulative distribution value equal to that specific random number, one particle belonging to the distribution is generated. The procedure can then be implemented N times, to produce the N particles belonging to the initial distribution. At each instant the state of the system is identified by the values of the internal-coordinate vector, $\boldsymbol{\xi}_i(t)$, characterizing each individual particle (with $i = 1, 2, \dots, N$). We now assume that the population of particles changes its state continuously between two subsequent discontinuous events. This time interval during which no discontinuous event occurs is called an *interval of quiescence*, T. A population of particles starting from time instant t will evolve during quiescence until the time instant $t+T$ is reached, or, in other words, until a new discontinuous event is encountered, according to the following set of ordinary differential equations:

$$\frac{\mathrm{d}\boldsymbol{\xi}_i}{\mathrm{d}t'} = \dot{\boldsymbol{\xi}}(\boldsymbol{\xi}_i), \quad i = 1, 2, \dots, N, \quad t' > t; \tag{7.134}$$

assuming that the initial condition for $\boldsymbol{\xi}_i(t)$ is known. The interval of quiescence is a random variable and, before proceeding any further, the corresponding probability has to be quantified. In what follows only one first-order (or single-particle) event will be considered (i.e. breakage), together with one second-order point process (e.g. collision and/or aggregation). The extension to any other type of particulate processes is straightforward, so let us start by considering only breakage. The breakage frequency is defined by the breakage kernel. From this quantity it is possible to estimate the probability that one of the particles breaks up during the infinitesimal time interval $\mathrm{d}\tau$, resulting in

$$\Pr\{\text{one particle breaks in interval } \mathrm{d}\tau\} = \sum_{i=1}^{N} b(\boldsymbol{\xi}_i(t+\tau))\mathrm{d}\tau. \tag{7.135}$$

Equation (7.135) allows us to derive the following expression for the cumulative distribution for the probability of the interval of quiescence:

$$F_{\mathrm{T}}(\tau) = 1 - \exp\left(-\sum_{i=1}^{N}\int_0^{\tau} b(\boldsymbol{\xi}_i(t+\tau'))\mathrm{d}\tau'\right). \tag{7.136}$$

The cumulative distribution $F_{\mathrm{T}}(\tau)$ represents the probability that the quiescence time T is smaller than τ. By assuming that the effect of the continuous change of the internal-coordinate vector is negligible, for the calculation of this probability, and by including also collision and aggregation as an additional discontinuous event, the following expression is obtained:

$$F_{\mathrm{T}}(\tau) = 1 - \exp\left(-\left\{\sum_{i=1}^{N} b(\boldsymbol{\xi}_i(t)) + \sum_{i=1}^{N-1}\sum_{j=i+1}^{N} \beta(\boldsymbol{\xi}_i(t), \boldsymbol{\xi}_j(t))\right\}\tau\right). \tag{7.137}$$

The corresponding expected interval of quiescence conditioned on the initial states of the particles can readily be found by calculating first the probability distribution $f_{\mathrm{T}} = \mathrm{d}F_{\mathrm{T}}/\mathrm{d}\tau$ and then its first moment with respect to τ, resulting in

$$\langle\tau\rangle = \frac{1}{\sum_{i=1}^{N} b(\boldsymbol{\xi}_i(t)) + \sum_{i=1}^{N-1}\sum_{j=i+1}^{N} \beta^V(\boldsymbol{\xi}_i(t), \boldsymbol{\xi}_j(t))}, \tag{7.138}$$

where β^V is the collision rate in the volume constituted by the computational box.

In time-driven MC simulations a particular time interval, Δt, is prescribed. This time interval has to be smaller than $\langle\tau\rangle$ and is generally different from one time step to another. Once the time step has been arbitrarily fixed, each particle is analyzed in turn and it is decided whether that particle will undergo a discontinuous event or not. Slightly different strategies are used to describe collisions, aggregation, and breakage, so let us discuss them separately. The probability of a single particle (identified by the index i) breaking during a time interval Δt is

$$F_{\mathrm{T}}^{(i)}(\Delta t) = 1 - \exp\left(-\frac{\Delta t}{\langle\tau_i\rangle}\right), \tag{7.139}$$

where $\langle\tau_i\rangle = 1/b(\boldsymbol{\xi}_i(t))$. Once a random number uniformly distributed between 0 and 1 has been selected, the particles breaking are those characterized by the probability calculated in Eq. (7.139) being greater than the random number. Then the selected particles will break up, forming new fragments that will have to be accommodated in the simulation box.[34] In

[34] This procedure is very easy in the case of simple daughter distribution functions (e.g. symmetric binary breakage), whereas additional random numbers might have to be selected for more complex daughter distribution functions (e.g. beta, uniform, etc.)

the case of collision or aggregation, things are a bit more complicated since, when the particles that aggregate are identified, the particles they interact with must be also identified. In the case of collision or aggregation, the discontinuous event occurs only for those particles that exhibit a probability calculated with Eq. (7.139) that is greater than the generated random number. For this calculation the characteristic time is $\langle \tau_i \rangle = 1/\sum_{j=i+1}^{N} \beta(\boldsymbol{\xi}_i(t), \boldsymbol{\xi}_j(t))$. Once the index of the colliding or aggregating particle i has been found, the index of the other particle must be identified. This is done by defining the generic probability of the encounter of particles i and j as

$$P_{ij} = \frac{\beta(\boldsymbol{\xi}_i(t), \boldsymbol{\xi}_j(t))}{\sum_{j=i+1}^{N} \beta(\boldsymbol{\xi}_i(t), \boldsymbol{\xi}_j(t))}. \tag{7.140}$$

Another random number, r, uniformly distributed between 0 and 1, is now generated, and the encounter is selected from among all possible encounters by satisfying the following relationship:

$$\sum_{j=1}^{k-1} P_{ij} \leq r \leq \sum_{j=1}^{k} P_{ij}, \tag{7.141}$$

which in this case implies that particle k will encounter particle i. In the case of aggregation there will be a single resulting particle with an internal-coordinate vector satisfying the basic conservation laws (i.e. conservation of mass and momentum), whereas in the case of particle collision two particles will be generated, with final velocities calculated on the basis of conservation of linear and angular momentum, as well as kinetic energy. More details concerning in particular the latter case (of particles colliding and not aggregating) and its relationship to the Boltzmann and kinetic equations can be found in the specialized literature (Bird, 1963, 1994).

In event-driven MC simulations the amplitude of the time step, $\Delta t = \tau$, is decided by sampling from a random variable with a cumulative probability distribution given by Eq. (7.137). When the time at which the new discontinuous event occurs has been determined, we have to choose the event from among all the possible ones with a procedure very similar to the one just described. The selection of the event is governed by the inversion method: the possible events are ordered, a random number is picked from a uniform distribution between 0 and 1, and the event that has the closest probability to the random number is eventually selected. Starting from the first applications of MC methods to generic aggregation and breakage problems (Hsu & Tsao, 1992; Laurenzi & Diamond, 1999; Tandon & Rosner, 1999; Van Peborgh Gooch & Hounslow, 1996) some specific problems were investigated, such as liquid–liquid dispersions (Balmelli & Steiner, 2000; Ghalehchian, 2002; Goodson & Kraft, 2004), polymerization processes (Saliakas *et al.*, 2008; Tong, 2001), nanoparticle formation in flames (Balthasar & Kraft, 2003; Celnik *et al.*, 2007), and grinding (Mishra, 2000). Particular attention was paid to the use of MC methods for the simulation of granulation (Braumann *et al.*, 2010; Cameron *et al.*, 2005; Gantt & Gatzke, 2006; Oullion *et al.*, 2009; Wildeboer *et al.*, 2005), spray drying (Terrazas-Velarde *et al.*, 2009), and aerosol processes (Kalani & Christofides, 2002; Rosner *et al.*, 2003). Crystallization and precipitation processes (Haseltine *et al.*, 2005), self-assembly of colloidal systems (Ethayaraja *et al.*, 2006), and electrostatic precipitation (Zhao & Zheng, 2008) processes were also investigated by means of MC methods.

As far as the development of new methodologies is concerned, it is worth mentioning the research efforts for the estimation of the potential source of errors (Li *et al.*, 2000) and for the reduction of the statistical noise (Zhao *et al.*, 2009). It is also important to cite the extension of constant-number Monte Carlo methods to breakage (Friesen & Dabros, 2003) as well as for the acceleration of the overall algorithm by Irizarry (2008a, 2008b). Particular attention was also paid to the improvement of performance in the simulation of multicomponent systems (Laurenzi *et al.*, 2002; Zhao *et al.*, 2010) and to the solution of other issues (Vikhansky & Kraft, 2004; Zhao *et al.*, 2005a, 2005b; Zhao & Zheng, 2009a, 2009b). It is also important to mention that because of the large amount of information produced and available in MC simulations, and because of the ease of extension to multivariate problems, MC methods are often employed for accurate solution of the GPBE for testing, comparing, and validating other methods (see, for example, the work of Alexopoulos & Kiparissides (2007) and Meimaroglou & Kiparissides (2007)). It should be mentioned, however, that since higher-order moments of the NDF are very much affected by the tails of the distribution a very large number of particles is needed in order to estimate with great accuracy their evolution. In this context, it is particularly interesting to cite a comparison between MC simulations and QBMM (Marchisio *et al.*, 2003b) that showed how QMOM and DQMOM calculations with quadratures of only three and four nodes are as accurate as MC simulations carried out with a very large number of particles (i.e. $N > 20\,000$ particles). The extension to spatially distributed systems is very intuitive. The spatially distributed computational domain is divided into several boxes, with many particles in each box. Movement of the particles from box to box due to advection and diffusion is now represented as an additional random event, with a prescribed probability dictated by the convective and diffusive fluxes.

7.6 Example homogeneous PBE

We conclude the chapter with a discussion of two example PBE involving (i) a univariate homogeneous continuous system and (ii) a bivariate homogeneous batch system. For both cases a comparison between two solution methods is presented, namely between QMOM and DQMOM for the first system and between CQMOM and Monte Carlo methods for the second. Note, however, that these examples could be computed with even greater accuracy (Yuan *et al.*, 2012) using EQMOM and ECQMOM as described in Section 3.3. We begin with a discussion of the validity of assuming homogeneity for a PBE.

7.6.1 A few words on the spatially homogeneous PBE

Before looking at example homogeneous systems, it is useful to analyze when the hypothesis of spatially homogeneity can be used for a PBE. For a PBE, the internal coordinates do not include the particle velocity. For example, consider the following PBE:

$$\begin{aligned}\frac{\partial n}{\partial t} + \frac{\partial}{\partial \mathbf{x}} \cdot (\mathbf{U}_{\mathrm{f}} n) + \frac{\partial}{\partial \boldsymbol{\xi}} \cdot (\dot{\boldsymbol{\xi}} n) = \frac{\partial^2}{\partial \boldsymbol{\xi}\, \partial \boldsymbol{\xi}} : (\mathbf{D} n) &+ \int \beta(\boldsymbol{\xi}', \tilde{\boldsymbol{\xi}}) n(\boldsymbol{\xi}') n(\tilde{\boldsymbol{\xi}}) J(\tilde{\boldsymbol{\xi}}, \boldsymbol{\xi}) \mathrm{d}\boldsymbol{\xi}' \\ &- \int \beta(\boldsymbol{\xi}', \boldsymbol{\xi}) n(\boldsymbol{\xi}') n(\boldsymbol{\xi}) \mathrm{d}\boldsymbol{\xi}' \\ &+ \int b(\boldsymbol{\xi}') N(\boldsymbol{\xi}|\boldsymbol{\xi}') n(\boldsymbol{\xi}') \mathrm{d}\boldsymbol{\xi}' - b(\boldsymbol{\xi}) n + J_{\mathrm{f}}. \end{aligned} \tag{7.142}$$

This PBE is written in a general form and contains the terms representing accumulation, real-space advection, phase-space advection, phase-space diffusion, and second-, first-, and zeroth-order point processes. (See Chapter 5 for more details on these processes.) Let us now assume that the PBE admits a steady-state solution; on separating out the term for real-space advection, we have (with the notation employing indices)

$$\frac{\partial U_{\mathrm{f}i} n}{\partial x_i} = -\frac{\partial \dot{\xi}_i n}{\partial \xi_i} - \frac{\partial^2 D_{ij} n}{\partial \xi_i \partial \xi_j} + \int \beta(\boldsymbol{\xi}', \tilde{\boldsymbol{\xi}}) n(\boldsymbol{\xi}') n(\tilde{\boldsymbol{\xi}}) J(\tilde{\boldsymbol{\xi}}, \boldsymbol{\xi}) \mathrm{d}\boldsymbol{\xi}'$$
$$- \int \beta(\boldsymbol{\xi}', \boldsymbol{\xi}) n(\boldsymbol{\xi}') n(\boldsymbol{\xi}) \mathrm{d}\boldsymbol{\xi}' + \int b(\boldsymbol{\xi}') N(\boldsymbol{\xi}|\boldsymbol{\xi}') n(\boldsymbol{\xi}') \mathrm{d}\boldsymbol{\xi}' - b(\boldsymbol{\xi}) n + J_{\mathrm{f}}. \quad (7.143)$$

Equation (7.143) can be made dimensionless by a proper choice of characteristic length and time scales. It is easy for a specific system to identify a characteristic length L and a characteristic velocity $\mathcal{U}$ resulting in a characteristic advection time $\tau_{\mathrm{M}} = L/\mathcal{U}$. Once the steady state has been reached, it is straightforward to calculate average values for the internal coordinates:

$$\bar{\boldsymbol{\xi}} = \frac{1}{\int n \, \mathrm{d}\boldsymbol{\xi}} \int \boldsymbol{\xi} n \, \mathrm{d}\boldsymbol{\xi} = \frac{1}{m_{0,\dots,0}} \left(\int \xi_1 n \, \mathrm{d}\boldsymbol{\xi}, \dots, \int \xi_M n \, \mathrm{d}\boldsymbol{\xi} \right). \quad (7.144)$$

By using these values, it is possible to estimate the characteristic rate of change $\dot{\boldsymbol{\xi}}(\bar{\boldsymbol{\xi}})$ for the internal coordinates, which can in turn be used to define characteristic time scales for phase-space advection for each internal coordinate: $\boldsymbol{\tau}_{\dot{\xi}} = [\bar{\xi}_1/\dot{\xi}_1(\bar{\xi}_1), \dots, \bar{\xi}_M/\dot{\xi}_M(\bar{\xi}_M)]$. Analogously for diffusion a characteristic time scale is easily defined: $\tau_{\mathrm{D},ij} = \bar{\xi}_i \bar{\xi}_j / D_{ij}$. Also for the source term for point processes some characteristic time scales can be defined. For the rate of formation of the disperse phase the time scale is defined as $\tau_J = m_{0,\dots,0}/J_{\mathrm{f}}(\bar{\boldsymbol{\xi}})$. For second-order point processes the time scale is $\tau_{\mathrm{II}} = 1/[\beta(\bar{\boldsymbol{\xi}}, \bar{\boldsymbol{\xi}}) m_{0,\dots,0}]$, whereas $\tau_{\mathrm{I}} = 1/b(\bar{\boldsymbol{\xi}})$ is used for first-order processes. By introducing the normalized quantities $\mathbf{X} = \mathbf{x}/L$, $\boldsymbol{\zeta} = \boldsymbol{\xi}/\bar{\boldsymbol{\xi}}$, $\mathbf{V} = \mathbf{U}_{\mathrm{f}}/\mathcal{U}$, $F(\boldsymbol{\zeta}) = n(\boldsymbol{\xi}) \bar{\xi}_1 \bar{\xi}_2 \dots \bar{\xi}_M / m_{0,\dots,0}$, $\mathbf{D}^* = \mathbf{D}/\mathbf{D}(\bar{\boldsymbol{\xi}})$, $\dot{\boldsymbol{\zeta}} = \dot{\boldsymbol{\xi}}/\dot{\boldsymbol{\xi}}(\bar{\boldsymbol{\xi}})$, $J^* = J_{\mathrm{f}}/J_{\mathrm{f}}(\bar{\boldsymbol{\xi}})$, $\beta^* = \beta/\beta(\bar{\boldsymbol{\xi}}, \bar{\boldsymbol{\xi}})$, and $b^* = b/b(\bar{\boldsymbol{\xi}})$, and inserting them into Eq. (7.143), we obtain

$$\frac{1}{\tau_{\mathrm{M}}} \frac{\partial V_i F}{\partial X_i} = -\frac{1}{\tau_{\dot{\xi}_i}} \frac{\partial \dot{\zeta}_i F}{\partial \zeta_i} - \frac{1}{\tau_{\mathrm{D},ij}} \frac{\partial^2 D^*_{ij} F}{\partial \zeta_i \partial \zeta_j} + \frac{J^*}{\tau_J} + \frac{1}{\tau_I} \left(\int b^* N(\boldsymbol{\zeta}|\boldsymbol{\zeta}') F(\boldsymbol{\zeta}') \mathrm{d}\boldsymbol{\zeta}' - b^* F \right)$$
$$+ \frac{1}{\tau_{II}} \left(\int F(\boldsymbol{\zeta}') F(\tilde{\boldsymbol{\zeta}}) \beta^* J(\tilde{\boldsymbol{\zeta}}, \boldsymbol{\zeta}) \mathrm{d}\boldsymbol{\zeta}' - \int_{\Omega_{\boldsymbol{\zeta}'}} F(\boldsymbol{\zeta}') F(\boldsymbol{\zeta}) \beta^* \, \mathrm{d}\boldsymbol{\zeta}' \right). \quad (7.145)$$

It is important to stress here that in Eq. (7.145) there are explicit dependences on spatial coordinates in the normalized NDF F and fluid velocity $\mathbf{V}$, but also implicit dependences on the rates of change of internal coordinates $\dot{\boldsymbol{\zeta}}$, the rate of formation of the disperse phase J^*, and the kernels β^* and b^*. In fact, as described in Chapter 5, these rates and the kernels depend on the flow properties, which change from point to point in the system.

We begin our analysis of Eq. (7.145) by noticing that, since all the terms are normalized, the importance of the different terms is dictated by the values of the time scales. If, for example, τ_{M} is much greater than $\tau_{\dot{\xi}_i}$, $\tau_{\mathrm{D},ij}$, τ_J, τ_{II}, and τ_{I}, then particle advection is much slower than the other particulate processes, and the term on the left-hand side is negligible with respect to the terms on the right-hand side. In contrast, when τ_{M} is much smaller than $\tau_{\dot{\xi}_i}$, $\tau_{\mathrm{D},ij}$, τ_J, τ_{II}, and τ_{I}, particle advection is much faster than the particulate processes involved, and the term on the left-hand side is more important than the terms on the right-hand side.

It is interesting to remind readers that, if, for example, Eq. (7.145) describes these specific particulate processes for particles dispersed in a liquid and suspended in a stirred tank, then their rates will significantly change from point to point. In fact, the aggregation and breakage rates and the corresponding kernels for particles suspended in a stirred tank will be very high around the impeller and very low far away from it. As τ_M tends to zero, advection becomes faster, and by multiplying Eq. (7.145) by τ_M it is easy to show that spatial gradients of the NDF must tend to zero. Therefore our first conclusion is that, as τ_M tends to zero, the NDF becomes more uniform throughout the computational domain. In contrast, when the other particulate time scales $\tau_{\dot{\xi}_i}$, $\tau_{D,ij}$, τ_J, τ_{II}, and τ_I decrease the corresponding particulate processes become faster. Moreover, if one of these characteristic time scales tends to zero, and if the rates exhibit strong dependences on the spatial coordinates, the resulting NDF will exhibit the same spatial dependence. Therefore our second conclusion is that if $\tau_{\dot{\xi}_i}$, $\tau_{D,ij}$, τ_J, τ_{II} or τ_I tends to zero the NDF will become less and less uniform in the domain. Our interpretation of Eq. (7.145) is that there are two groups of processes with opposite effects. On the one hand, advection mixes particles and smoothes down all the gradients of the NDF, making it more and more spatially homogeneous. On the other hand, particulate processes, whose rates depend strongly on the spatial coordinates, create and sustain these gradients in the NDF. One typical example is particle aggregation and breakage in stirred tanks (Marchisio *et al.*, 2006). In this case very often a steady state in which particles keep breaking in regions of high shear rate and aggregating in more stagnant regions is reached. When the characteristic mixing time for particles is much shorter than the characteristic time for aggregation and breakage, particle advection is faster than aggregation and breakage, with the result that the NDF is quite uniform throughout the vessel. In contrast, when the characteristic time scales for aggregation and breakage are much faster than the mixing time, mixing is not fast enough to homogenize and smooth down the gradients generated, and therefore the resulting NDF is very different from point to point in the vessel. This provides a criterion for establishing whether a system can be analyzed as a spatially uniform system. Nevertheless, while the homogeneity of the NDF allows us to describe the system in terms of a single quantity, it does not imply that the rates are constant throughout the vessel![35]

If we end up in the first situation, we can simplify Eq. (7.142) by using the volume-average NDF: $\hat{n} = (1/V)\int n\,\mathrm{d}\mathbf{x}$, where V is the total volume of the system under investigation. The situation of perfect mixing implies that the spatial variations of the NDF are so small that $n \approx \hat{n}$ throughout the entire vessel. By applying the volume-averaging procedure to Eq. (7.142) under the condition of perfect mixing we obtain

$$\frac{\partial \hat{n}}{\partial t} + \frac{\partial}{\partial \boldsymbol{\xi}} \cdot (\hat{\dot{\boldsymbol{\xi}}}\hat{n}) = \frac{\partial^2}{\partial \boldsymbol{\xi}\,\partial \boldsymbol{\xi}} : (\hat{\mathbf{D}}\hat{n}) + \hat{J}_{\mathrm{f}} + \int \hat{b}(\boldsymbol{\xi}')N(\boldsymbol{\xi}|\boldsymbol{\xi}')\hat{n}(\boldsymbol{\xi}')\mathrm{d}\boldsymbol{\xi}' - \hat{b}(\boldsymbol{\xi})\hat{n}$$
$$+ \int \hat{\beta}(\boldsymbol{\xi}', \tilde{\boldsymbol{\xi}})\hat{n}(\boldsymbol{\xi}')\hat{n}(\tilde{\boldsymbol{\xi}})J(\tilde{\boldsymbol{\xi}}, \boldsymbol{\xi})\mathrm{d}\boldsymbol{\xi}' - \int \hat{\beta}(\boldsymbol{\xi}', \boldsymbol{\xi})\hat{n}(\boldsymbol{\xi}')\hat{n}(\boldsymbol{\xi})\mathrm{d}\boldsymbol{\xi}', \qquad (7.146)$$

where $\hat{\dot{\boldsymbol{\xi}}}$, $\hat{\mathbf{D}}$, $\hat{J}_{\mathrm{f}}$, $\hat{\beta}$, and $\hat{b}$ are the volume-average rates of change of the internal coordinate, the rate of formation of the disperse phase, and the kernels for the second- and first-order processes, respectively. Equation (7.146) is the volume-average PBE that has to be solved when describing a system as spatially homogeneous. This equation can be solved only if $\hat{\dot{\boldsymbol{\xi}}}$, $\hat{\mathbf{D}}$, $\hat{J}_{\mathrm{f}}$, $\hat{\beta}$, and $\hat{b}$ are known. Note that these quantities are difficult to estimate because they contain hidden spatial dependences. Referring again to aggregation and breakage in stirred

[35] This misconception is particularly common in crystallization. The hypothesis of a perfectly mixed system is, for crystallization and precipitation processes, labeled as mixed-suspension, mixed-product removal (MSMPR). With this model the crystallizer is modeled with a spatially homogeneous NDF, generally called the crystal-size distribution (CSD). However, the fact that the CSD is constant through the vessel does not mean that the rates of crystal nucleation, molecular growth, aggregation, and breakage are constant.

tanks, the mesoscale aggregation and breakage rates vary spatially and must be accounted for in the volume-average rates of Eq. (7.146). For example, as highlighted in Chapter 5, $\dot{\xi}$ and J_f may depend on supersaturation and on the local flow and turbulence, which can vary quite significantly from point to point in the vessel. One way to estimate their volume-average values is to account explicitly for the spatial dependence. For example, in the case of mass transfer the parameter that makes $\dot{\xi}$ strongly dependent on the position could be the turbulence dissipation rate $\varepsilon(t, \mathbf{x})$ and therefore $\hat{\dot{\xi}} = (1/V) \int \dot{\xi} \, d\mathbf{x} = \int \dot{\xi} f_\varepsilon(\varepsilon) d\varepsilon$, where $f(t, \varepsilon)$ contains information on the spatial distribution of the turbulence dissipation rate. Here, $f_\varepsilon(\varepsilon) d\varepsilon$ represents the volume fraction of the vessel characterized by a turbulence dissipation rate of ε and the volume distribution function integrates to unity, $\int f_\varepsilon(\varepsilon) d\varepsilon \equiv 1$, whereas its first moment represents the volume-average turbulence dissipation rate in the vessel, $\hat{\varepsilon} = \int \varepsilon f_\varepsilon(\varepsilon) d\varepsilon$. Since the relationship between $\hat{\dot{\xi}}$ and ε is nonlinear, the inequality $\hat{\dot{\xi}} \neq \dot{\xi}(\hat{\varepsilon})$ holds most of the time. Analogously, for the volume-average kernel for the second-order point process $\hat{\beta} = (1/V) \int \beta \, d\mathbf{x}$ and for the volume-average kernel for the first-order process $\hat{b} = (1/V) \int b \, d\mathbf{x}$, and similarly for all the other rates and kernels appearing in the volume-average PBE. For example, in the case of aggregation and breakage the key parameter that could contain significant spatial dependence is the shear rate G_f, as evidenced by Eqs. (5.170), (5.136), and (5.136). In these cases, once the volume-shear-rate distribution in the vessel f_{G_f} has been determined, the volume-average aggregation and breakage kernels can be calculated as $\hat{\beta} = \int \beta(G_f) f_{G_f} \, dG_f$ and $\hat{b} = \int b(G) f_{G_f} \, dG_f$. Also for the other variables involved, once the parameters that contain the dependence over spatial coordinates have been identified, the volume distribution of this parameter can be determined and used to calculate the corresponding volume-average rates and kernels.

In summary, what has been discussed above has important implications for multiphase processes governed by a PBE.

1. The fact that the NDF is nearly constant throughout the domain generally means that the system can be described with a volume-average NDF by solving the volume-average PBE reported in Eq. (7.146). However, it does not mean that the rates appearing in the equation are known.

2. If the rate of change of internal coordinates, the rate of formation of the disperse phase, and the kernels for second- and first-order processes do not exhibit a strong variation according to the spatial coordinates, then their volume-average values appearing in Eq. (7.146) can be easily estimated.

3. In all the other cases the parameters appearing in the rates and kernels responsible for spatial dependences must be identified and their variation throughout the vessel must be quantified, by identifying their volume distribution. The latter quantity can be typically measured or estimated through simulations. Once these distributions are known, they can be used to estimate the volume-average rates and kernels that appear in Eq. (7.146).

4. In addition, when the inverse problem is solved, namely the volume-average rates and kernels are identified by comparison between experiments and model predictions under the assumption of perfect mixing (or, in other words, by using Eq. (7.146)), one must remember that the volume-average rates and kernels are not truly "kinetic" parameters but contain some fluid-dynamic factors in them. This is why, on transferring these rates and kernels to different operating conditions or from one system to another, very poor performances can be observed.

7.6.2 Comparison between the QMOM and the DQMOM

Consider a simple homogeneous system with a continuous inlet and outlet. The corresponding univariate PBE is

$$\frac{\partial n}{\partial t} = \frac{1}{\tau_R}(n^{in} - n) + S, \tag{7.147}$$

where $n(t,\xi)$ is the NDF, ξ is the internal coordinate (e.g. the particle mass), $n^{in}(t,\xi)$ is the inlet NDF, τ_R is the mean residence time of the particles, and $S(\xi)$ is the source term representing the different particulate processes. When the moment transform is applied, the equation for the moments $m_k(t)$ is obtained:

$$\frac{dm_k}{dt} = \frac{1}{\tau_R}(m_k^{in} - m_k) + \overline{S}_k. \tag{7.148}$$

The moments can be found from Eq. (7.148) using the QMOM with $k = 0, \ldots, 2N - 1$ and a given initial condition $m_k(0) = m_k^o$. The source term is closed with a quadrature approximation using N nodes ξ_α and N weights w_α. The latter are calculated with a univariate moment-inversion algorithm (i.e. the PD or Wheeler algorithm). Let us define a new source term grouping together the entire right-hand side of Eq. (7.148):

$$\overline{S}_k^* = \frac{1}{\tau_R}(m_k^{in} - m_k) + \overline{S}_k. \tag{7.149}$$

If we solve the problem using the DQMOM, the transport equations for the N weights w_α and N weighted nodes $\zeta_\alpha = w_\alpha \xi_\alpha$ are

$$\frac{dw_\alpha}{dt} = a_\alpha^*,$$
$$\frac{d\zeta_\alpha}{dt} = b_\alpha^*,$$

with $\alpha = 1, \ldots, N$ and initial conditions $w_\alpha(0)$ and $\xi_\alpha(0)$ calculated from m_k^o by applying the univariate moment-inversion algorithm. The source terms a_α^* and b_α^* are calculated from a linear system $\mathbf{A}\boldsymbol{\alpha}^* = \mathbf{d}^*$, where $\boldsymbol{\alpha}^* = [a_1^*, \ldots, a_N^*, b_1^*, \ldots, b_N^*]^T$ and $\mathbf{d}^* = [\overline{S}_0^*, \ldots, \overline{S}_{2N-1}^*]^T$. If we now write out the source terms as indicated in Eq. (7.149) and make use of the properties of linear systems, we obtain $\boldsymbol{\alpha}^* = \mathbf{A}^{-1}(\mathbf{d}^{in} - \mathbf{d}^{out} + \mathbf{d}) = \boldsymbol{\alpha}^{in} - \boldsymbol{\alpha}^{out} + \boldsymbol{\alpha}$, where $\boldsymbol{\alpha}^{in} = [a_1^{in}, \ldots, a_N^{in}, b_1^{in}, \ldots, b_N^{in}]^T$ is the part of the source terms representing particles entering the system, $\boldsymbol{\alpha}^{out} = [a_1^{out}, \ldots, a_N^{out}, b_1^{out}, \ldots, b_N^{out}]^T$ is the part representing particles exiting the system, and $\boldsymbol{\alpha} = [a_1, \ldots, a_N, b_1, \ldots, b_N]^T$ is the part representing the particulate processes. As indicated in Eq. (7.149), $\mathbf{d}^{in} = [m_0^{in}/\tau_R, \ldots, m_{2N-1}^{in}/\tau_R]^T$, $\mathbf{d}^{out} = [m_0^{out}/\tau_R, \ldots, m_{2N-1}^{out}/\tau_R]^T$, and $\mathbf{d} = [\overline{S}_0, \ldots, \overline{S}_{2N-1}]^T$. By making use of the fact that the matrix $\mathbf{A}$ depends only on the values of the nodes, a further simplification is possible:

$$\begin{aligned} \frac{dw_\alpha}{dt} &= a_\alpha^{in} - \frac{w_\alpha}{\tau_R} + a_\alpha, \\ \frac{d\zeta_\alpha}{dt} &= b_\alpha^{in} - \frac{\zeta_\alpha}{\tau_R} + b_\alpha, \end{aligned} \tag{7.150}$$

with $\alpha = 1, \ldots, N$. On solving the problem using either Eq. (7.148) or Eq. (7.150), the same solution should be obtained.[36]

[36] It is interesting to point out that if the source terms $\boldsymbol{\alpha}^{in}$ were calculated by using the matrix $\mathbf{A}$ determined not with the nodes ξ_α, but with the nodes ξ_α^{in} corresponding to the inlet moments (or in other words determined with

As an example, consider a monodisperse population of particles characterized by mass as internal coordinate and moments $m_k^{\text{in}} = m_k(0) = 1$ with $k = 0, \ldots, 2N-1$. This population of particles is continuously fed to a system wherein particles undergo aggregation and symmetric binary breakage with constant kernels. The equations describing the evolution of the moments are

$$\frac{\mathrm{d}m_k}{\mathrm{d}t} = \frac{1}{\tau_\mathrm{R}}(m_k^{\text{in}} - m_k) + \frac{\beta_\mathrm{o}}{2}\sum_{\alpha=1}^{N}\sum_{\gamma=1}^{N}[(\xi_\alpha + \xi_\gamma)^k - \xi_\alpha^k - \xi_\gamma^k]w_\alpha w_\gamma + a_\mathrm{o}\sum_{\alpha=1}^{N}(2^{1-k} - 1)\xi_\alpha^k w_\alpha,$$

with $k = 0, \ldots, 2N-1$ and where $\beta_\mathrm{o} = 1$ and $a_\mathrm{o} = 0.02$ are the constant aggregation and breakage kernels. Binary symmetric breakage is used in this example in analogy with case 1 of Marchisio *et al.* (2003c). If the QMOM is used with $N = 2$, four equations are solved, corresponding to the first four moments of the NDF shown in Figure 7.5. It can be seen that this is a case dominated by aggregation; in fact, m_0 decreases whereas m_2 and m_3 increase, and m_1 stays constant since particle mass is conserved during aggregation and breakage. The corresponding evolution of the two nodes and the two weights is shown in Figure 7.6. If the DQMOM is used[37] then Eq. (7.150) is solved, resulting in the very same evolution as seen in Figures 7.5 and 7.6.

7.6.3 Comparison between the CQMOM and Monte Carlo

Let us now consider a population of particles characterized by two internal coordinates, namely size ξ_1 and composition ξ_2. The composition is interpreted as the mass of one component within the particles. During aggregation, if ξ_1' and $\tilde{\xi}_1$ are the sizes of the aggregating particles resulting in a particle of size ξ_1, and ξ_1' and $\tilde{\xi}_1$ are the compositions of the aggregating particles resulting in a particle of composition ξ_1, the following relations hold: $\xi_1^3 = \xi_1'^3 + \tilde{\xi}_1^3$ and $\xi_2 = \xi_2' + \tilde{\xi}_2$. The particles are confined in a homogeneous domain and undergo aggregation and breakage. The evolution of the NDF $n(t, \xi_1, \xi_2)$ is described by

$$\begin{aligned}\frac{\partial n}{\partial t} &= \frac{\xi_1^2}{2}\int_0^{\xi_1}\int_0^{\xi_2}\frac{\beta(\xi_1', \xi_2', \tilde{\xi}_1, \tilde{\xi}_2)}{\tilde{\xi}_1^2}n(\xi_1', \xi_2')n(\tilde{\xi}_1, \tilde{\xi}_2)\mathrm{d}\xi_1'\,\mathrm{d}\xi_2' \\ &\quad - \int_0^{\xi_1}\int_0^{\xi_2} n(\xi_1, \xi_2)n(\xi_1', \xi_2')\beta(\xi_1, \xi_2, \xi_1', \xi_2')\mathrm{d}\xi_1'\mathrm{d}\xi_2' \\ &\quad + \int_{\xi_1}^{\infty}\int_{\xi_2}^{\infty} N(\xi_1, \xi_2|\xi_1', \xi_2')n(\xi_1', \xi_2')a(\xi_1', \xi_2')\mathrm{d}\xi_1'\mathrm{d}\xi_2' - a(\xi_1, \xi_2)n, \end{aligned} \quad (7.151)$$

the moment-inversion algorithm applied to the inlet moment set m_k^{in}), the following set of equations would have been obtained:

$$\frac{\mathrm{d}w_\alpha}{\mathrm{d}t} = \frac{1}{\tau_\mathrm{R}}(w_\alpha^{\text{in}} - w_\alpha) + a_\alpha,$$
$$\frac{\mathrm{d}\zeta_\alpha}{\mathrm{d}t} = \frac{1}{\tau_\mathrm{R}}(\zeta_\alpha^{\text{in}} - \zeta_\alpha) + b_\alpha,$$

where w_α^{in} and $\zeta_\alpha^{\text{in}} = w_\alpha^{\text{in}}\xi_\alpha^{\text{in}}$ are calculated from inlet moment set m_k^{in} by applying the moment-inversion algorithm. However, this set of equations does not result in the correct moment evolution, since only m_0 and m_1 will be correctly predicted, whereas all the other moments (i.e. m_k with $k \geq 2$) will inevitably be wrong.

[37] When using the DQMOM, we solve equations for nodes and weights, and in order to calculate the source terms a linear system has to be solved. This linear system is singular if two or more nodes are identical, and, especially for $N \geq 3$, starting the simulation with a perfectly monodisperse distribution can lead to problems. For example for the case under investigation the original moment set is $m_k(0) = 1$ for $k = 1, \ldots, 3$, but for the DQMOM the following moment set is used instead: $m_0 = m_1 = 1$, $m_2 = 1.009$, and $m_3 = 1.0027$. These moments represent a slightly polydisperse population of particles, practically equivalent to the perfectly monodisperse one, but resulting in a much more stable calculation.

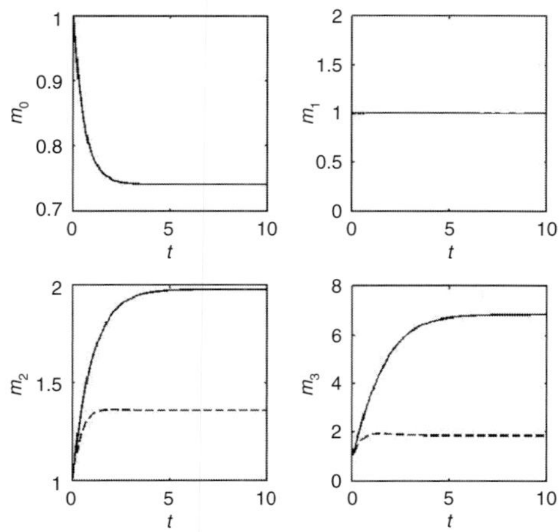

Figure 7.5. Evolution of the first four moments for a continuous homogeneous system undergoing aggregation and breakage with the QMOM and DQMOM.

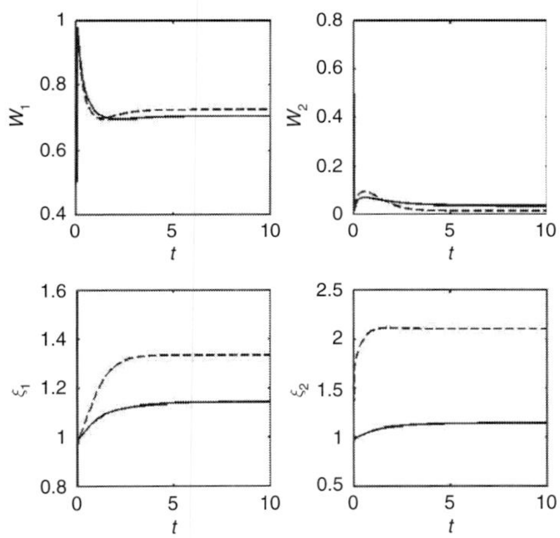

Figure 7.6. Evolution of the weights and nodes for a continuous homogeneous system undergoing aggregation and breakage with the QMOM and DQMOM.

where all the terms have been derived with the rules explained in the previous chapters. From Eq. (7.151) the evolution equation for the mixed moment $m_{k,l}$ can be easily derived:

$$\begin{aligned}\frac{\mathrm{d}m_{k,l}}{\mathrm{d}t} = &\frac{1}{2}\int\left[\left(\xi_1^3 + {\xi_1'}^3\right)^{k/3}(\xi_2 + \xi_2')^l - \xi_1^k\xi_2^l - {\xi_1'}^k{\xi_2'}^l\right] \\ &\times \beta(\xi_1,\xi_2,\xi_1',\xi_2')n(\xi_1,\xi_2)n(\xi_1',\xi_2')\mathrm{d}\xi_1\,\mathrm{d}\xi_1'\,\mathrm{d}\xi_2\,\mathrm{d}\xi_2' \\ &+ \int\left[\left(\int N(\xi_1,\xi_2|\xi_1',\xi_2')\xi_1^k\xi_2^l\,\mathrm{d}\xi_1\,\mathrm{d}\xi_2\right) - {\xi_1'}^k{\xi_2'}^l\right]a(\xi_1',\xi_2')n(\xi_1',\xi_2')\mathrm{d}\xi_1'\,\mathrm{d}\xi_2'. \quad (7.152)\end{aligned}$$

When Eq. (7.152) is solved with the CQMOM, the NDF is approximated by

$$n(\xi_1,\xi_2) = \sum_{\alpha=1}^{N_1} w_\alpha\delta(\xi_1 - \xi_{1,\alpha})\sum_{\gamma=1}^{N_2} w_{\alpha,\gamma}\delta(\xi_2 - \xi_{2;\alpha,\gamma}),$$

where N_1 is the number of nodes chosen for the particle size and the N_2 is the number of nodes for the particle composition, resulting in a quadrature approximation of $N_1 N_2$ nodes. Applying this approximation to Eq. (7.152) yields

$$\frac{\mathrm{d}m_{k,l}}{\mathrm{d}t} = \sum_{\alpha=1}^{N_1}\sum_{\gamma=1}^{N_2} w_\alpha w_{\alpha,\gamma} a_{\alpha,\gamma}[\overline{N}_{k,l}^{(\alpha,\gamma)} - \xi_{1,\alpha}^k \xi_{2;\alpha,\gamma}^l] + \frac{1}{2}\sum_{\alpha,\delta=1}^{N_1}\sum_{\gamma,\sigma=1}^{N_2} w_\alpha w_\delta w_{\alpha,\gamma} w_{\delta,\sigma} \beta_{\alpha,\gamma,\delta,\sigma}[(\xi_{1,\alpha}^3 + \xi_{1,\delta}^3)^{k/3}(\xi_{2;\alpha,\gamma} + \xi_{2;\delta,\sigma})^l - \xi_{1,\alpha}^k \xi_{2;\alpha,\gamma}^l - \xi_{1,\delta}^k \xi_{2;\delta,\sigma}^l], \tag{7.153}$$

where $\beta_{\alpha,\gamma,\delta,\sigma} = \beta(\xi_{1,\alpha}, \xi_{2;\alpha,\gamma}, \xi_{1,\delta}, \xi_{2;\delta,\sigma})$, $a_{\alpha,\gamma} = a(\xi_{1,\alpha}, \xi_{2;\alpha,\gamma})$, and the moments of the daughter distribution function are

$$\overline{N}_{k,l}^{(\alpha,\gamma)} = \int N(\xi_1, \xi_2|\xi_{1,\alpha}, \xi_{2;\alpha,\gamma})\xi_1^k \xi_2^l \,\mathrm{d}\xi_1\,\mathrm{d}\xi_2.$$

For this specific example, consider the case of constant aggregation and breakage kernels, namely $\beta_{\alpha,\gamma,\delta,\sigma} = \beta_0$, $a_{\alpha,\gamma} = a_0$, and a beta daughter distribution function

$$N(\xi_1, \xi_2|\xi_1', \xi_2') = 180\frac{\xi_1^2}{\xi_1'^3}\left(\frac{\xi_1^3}{\xi_1'^3}\right)^2\left(1 - \frac{\xi_1^3}{\xi_1'^3}\right)^2 \delta\left(\xi_2 - \frac{\xi_1^3}{\xi_1'^3}\xi_2'\right),$$

resulting in

$$N_{k,l}^{(\alpha,\gamma)} = \frac{3240}{(3l+k+9)(3l+k+12)(3l+k+15)}\xi_{1,\alpha}^k \xi_{2;\alpha,\gamma}^l.$$

As the initial condition, we take the bivariate Gaussian distribution of Section 3.2.3 with $\mu_1 = 4 \times 10^{-3}$ m, $\mu_2 = 2.4 \times 10^{-7}$ mol, $\sigma_1 = 0.15$, and $\sigma_2 = 0.15$.

Let us now investigate the evolution of the initial distribution undergoing aggregation and breakage with the CQMOM and with the MC method. The MC simulation was carried out using the constant-number MC method reported in Section 7.5 with 10 000 notional

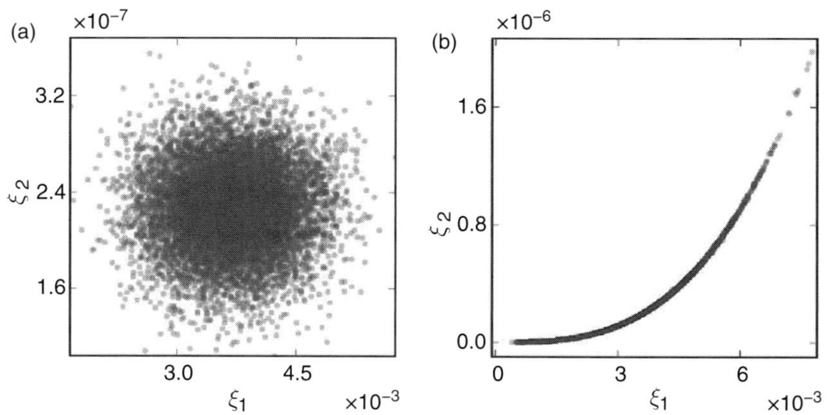

Figure 7.7. A bivariate distribution (a) as represented in a Monte Carlo simulation at the beginning of the simulation and (b) when a steady state is reached.

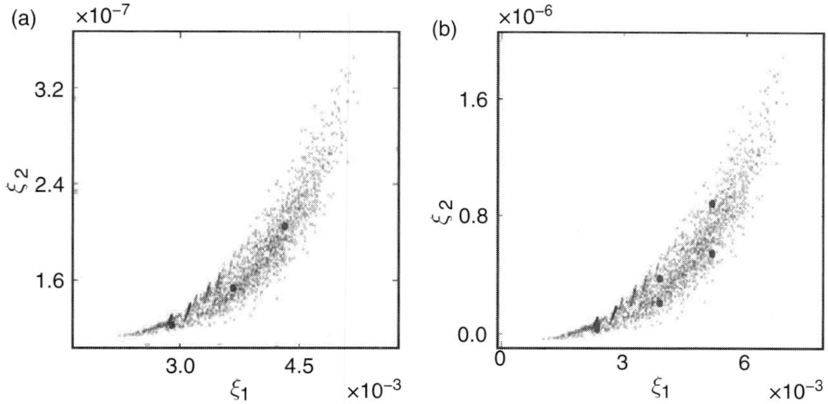

Figure 7.8. A comparison of the bivariate distribution, as represented in a Monte Carlo simulation, and the representation with the CQMOM with (a) $N_1 = 3$, $N_2 = 1$ and (b) $N_1 = 3$, $N_2 = 2$ at an intermediate instant of the simulation.

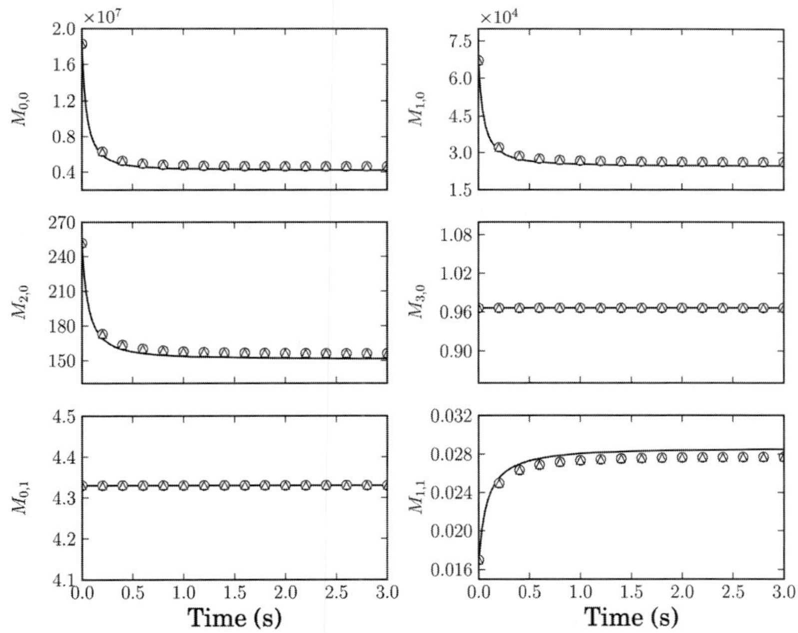

Figure 7.9. A comparison of the evolution for the lower-order moments as predicted by the Monte Carlo simulation and by the CQMOM: solid black line, MC method; white circles, CQMOM with $N_1 = 3$ and $N_2 = 1$; and white triangles, CQMOM with $N_1 = 3$ and $N_2 = 2$ (in units of a generic moment $M_{k,l}$: $\mathrm{m}^{k-3}\ \mathrm{mol}^l$).

particles. More details can be found in the work of Buffo *et al.* (2013). Figure 7.7 reports the bivariate distribution (as represented in the MC simulation) at the beginning of the simulation and when the steady state is reached. It can be seen that at the beginning the distribution is spread along both internal coordinates. However, as soon as aggregation and breakage start, the population of particles is homogenized and assumes a cubic shape, representing the relationship between composition and size (i.e. $\xi_2 \propto \xi_1^3$). It is interesting to see how the distribution is represented with the CQMOM. Figure 7.8 reports the bivariate distribution at an intermediate time from the MC simulation and compares it with the implicit assumption in the CQMOM. It can be seen that when $N_1 = 3$ and $N_2 = 1$ the population of particles is represented in three groups, each characterized by distinct node values for ξ_1 and ξ_2. In contrast, when $N_1 = 3$ and $N_2 = 2$ for each node for ξ_1 two nodes for ξ_2 are identified. The comparison for the evolution of the moments predicted by the MC simulation and CQMOM is reported in Figure 7.9. It can be seen that $m_{3,0}$ and $m_{0,1}$ are constant, because these are conserved quantities during aggregation and breakage, whereas all the other moments evolve as expected. It is interesting to observe that the discrepancy between MC and the CQMOM is very small, both when using $N_1 = 3$ and $N_2 = 1$ for a total of three nodes and when using $N_1 = 3$ and $N_2 = 2$ for a total of six. In both cases, in fact, the quadrature provided by the CQMOM seems to evaluate the source terms for the tracked moments quite accurately. A crucial role is played also by the numerical scheme employed for the time integration of the set of ODE reported in Eq. (7.153). This numerical scheme, in fact, must not introduce additional errors that can corrupt the moment set, transforming a realizable moment set into an unrealizable one. This is typically not an issue with the numerical integration of the ODE involved in the simulation of spatially homogeneous systems, whereas the problem becomes serious when dealing with the inhomogeneous systems described in Chapter 8. Readers interested in the problem of moment realizability should consult Chapter 3.

A separate issue is that of the realizability of the quadrature approximation. For example, since for the case under investigation both ξ_1 and ξ_2 (representing the size and the chemical composition of the particles) are positive, it must be ensured that a *realizable quadrature* is employed to sample the phase space. A realizable quadrature in this case has positive nodes that result in sensible values. For this specific problem, it turns out that the portion of phase space explored is fully realizable only if some mixed moments, such as $m_{2,1}$, are included in the tracked moment set. This means that the option $N_1 = 2$ and $N_2 = 1$ that does not include $m_{2,1}$ can potentially lead to problems (for example when mass transfer and chemical reactions are involved), whereas the option $N_1 = 3$ and $N_2 = 1$ is a better choice.

8

Moment methods for inhomogeneous systems

In this chapter we discuss issues specific to applying moment methods with spatially inhomogeneous systems. In particular, we focus on the spatial transport of moment sets by advection, diffusion, and free transport. In Chapter 7, issues related to transport in phase space are thoroughly treated and, here, we will discuss such terms only inasmuch as they affect spatial transport. In Section 8.1, the principal modeling issues that arise with spatially inhomogeneous systems are briefly reviewed. In the sections that follow, we discuss separately moment methods for (i) the inhomogeneous population-balance equation (PBE) (i.e. where the internal coordinates do not include or affect the velocity) in Section 8.3, (ii) the inhomogeneous kinetic equation (KE) (i.e. where the only internal coordinate is velocity) in Section 8.4, and finally (iii) the full inhomogeneous generalized population-balance equation (GPBE) in Section 8.5. Concrete examples, and the corresponding discretized formulas, are provided for each type of system in order for the reader to understand fully the issues that arise when simulating inhomogeneous systems. An important theme running through the entire chapter is the issue of realizable moment sets, and how realizability is affected by spatial transport. Thus, in order to have explicit examples of the numerical issues, we introduce kinetics-based finite-volume methods (KBFVM) for moment sets in Section 8.2. Nevertheless, the reader should keep in mind that these numerical issues are generic to moment transport and will arise with all spatial-discretization methods. Finally, we concentrate our discussion in this chapter on spatial transport of the moment sets needed for quadrature-based moment methods (QBMM). As seen in Chapter 7, QBMM offer many advantages for treating the phase-space transport terms in the GPBE. As we shall discover in this chapter, QBMM are also an effective tool for addressing realizability issues stemming from spatial transport.

8.1 Overview of spatial modeling issues

When using moment methods for inhomogeneous systems, the moment set is transported in physical space due to advection, diffusion, and free transport. Since the moment-transport equations are derived from a transport equation for the NDF, the problem of moment transport is closely related to the problem of transporting the NDF. Denoting the NDF by $n(t, \mathbf{x}, \boldsymbol{\xi})$, the process of spatial transport involves changes in $n(t, \mathbf{x}, \boldsymbol{\xi})$ for *fixed* values of the internal coordinates $\boldsymbol{\xi}$. For example, advection with a constant velocity $\mathbf{u}$, expressed mathematically as

$$\frac{\partial n}{\partial t} + \mathbf{u} \cdot \frac{\partial n}{\partial \mathbf{x}} = 0, \tag{8.1}$$

simply moves the NDF from its original value $n(0, \mathbf{x}, \boldsymbol{\xi})$ to its new value $n(t, \mathbf{x}, \boldsymbol{\xi})$ at which the exact solution is $n(t, \mathbf{x}, \boldsymbol{\xi}) = n(0, \mathbf{x} - \mathbf{u}t, \boldsymbol{\xi})$. Note that the exact solution implies that the NDF moves in physical space without changing its shape in phase space. Furthermore, since the moments $m_{\mathbf{k}}(t, \mathbf{x})$ are a linear transformation of the NDF, we find from Eq. (8.1) that the moments are advected by

$$\frac{\partial m_{\mathbf{k}}}{\partial t} + \mathbf{u} \cdot \frac{\partial m_{\mathbf{k}}}{\partial \mathbf{x}} = 0, \tag{8.2}$$

whose exact solution is $m_{\mathbf{k}}(t, \mathbf{x}) = m_{\mathbf{k}}(0, \mathbf{x} - \mathbf{u}t)$, which is again a linear transformation of the exact NDF. Unfortunately, the *numerical* solutions to Eqs. (8.1) and (8.2) will never be exact, and it is precisely these numerical errors that lead to serious problems with moment methods. It is worth emphasizing that these errors have nothing whatsoever to do with the derivation of Eq. (8.2) from Eq. (8.1), or with the closures used in QBMM to close the phase-space terms. Instead, they are a direct result of the numerical approximation for advection. The reader familiar with numerical advection algorithms will be aware of some of the most relevant issues such as numerical diffusion (in first-order schemes) and loss of maximum/minimum bounds (in high-order schemes) due to oscillations. Unlike with classical CFD algorithms for advection of *independent* scalars, the scalars (i.e. the moments) in a moment set are *correlated* and thus cannot be truncated (or limited) independently without running the risk that the moment set will no longer be realizable.

In the remainder of this section, we introduce the principal modeling issues related to spatial transport using moment methods. First, we discuss the realizability of the NDF and of moment sets (which are related to the numerical errors discussed above). Second, we introduce the phenomenon of particle trajectory crossing (PTC) that occurs with the inhomogeneous KE (and is exactly captured by the NDF), and describe how it leads to a closure problem in the moment-transport equations. Next, we look at issues related to coupling between spatial and phase-space transport in the GPBE (i.e. due to correlations between velocity and internal coordinates such as particle volume). Finally, we introduce KBFVM for solving the moment-transport equations in connection with QBMM, and briefly discuss how they can be used to ensure realizability as well as to capture PTC and to treat coupled moments.

8.1.1 Realizability

As the discussion above suggests, there are two types of realizability that concern us: (i) realizability of the NDF and (ii) realizability of the moments. In theory, since a realizable NDF generates realizable moments, we could limit our discussion to type (i) realizability. However, in moment methods we will transport moments without knowledge of the true NDF, so type (ii) realizability will be the most important issue. Some of these issues have already been discussed in Section 3.1.4, with particular attention to algorithms for assessing and for restoring the realizability of a moment set. By definition, the NDF is realizable if it is nonnegative for all possible values of the internal coordinates: $n(t, \mathbf{x}, \boldsymbol{\xi}) \geq 0$ for all $\boldsymbol{\xi} \in \Omega_{\boldsymbol{\xi}}$. Thus, the principal issue with the spatial transport of the NDF is that the discretized transport algorithm must never yield a negative NDF: if $n(t, \mathbf{x}, \boldsymbol{\xi}) < 0$ for any $\boldsymbol{\xi}$, then the NDF is unrealizable. For classical advection schemes with first-order accuracy, it is relatively straightforward to ensure that the NDF remains realizable; however, this is not the case for high-order schemes. Because first-order schemes exhibit excessive numerical diffusion, only *realizable* high-order advection schemes are of interest for computing the NDF. As a practical matter, due to the number of internal coordinates and the resulting high dimensionality of the NDF, the direct numerical solution of the GPBE is usually not feasible. Thus, the main concern when developing numerical schemes will be the realizability of the corresponding moment set.

As explained in previous chapters, a moment set is a vector of moments. For example, for univariate QMOM we use the moment set $\mathbf{M} = (m_0, m_1, \ldots, m_{2N-1})$, where N is the number of quadrature nodes. If the components of the moment set are used as the coordinates of a $2N$-dimensional *moment space*, we can plot the set of all points in moment space corresponding to a particular NDF. As a general rule, the moment space is *convex*, and hence if $\mathbf{M}_1$ and $\mathbf{M}_2$ are two realizable "points" in moment space, then the "point" defined by $\mathbf{M}_3 = \alpha\mathbf{M}_1 + (1 - \alpha)\mathbf{M}_2$ for $0 \leq \alpha \leq 1$ is a realizable moment set. More generally, any convex combination[1] of realizable moments sets will be realizable. Thus, one convenient way to demonstrate that a numerical algorithm generates only realizable moment sets is to show that it updates the moments using convex combinations of moment sets (e.g. moment sets at different spatial locations) from the previous time step. As a general rule, the boundaries of moment space can be quite complex, making it difficult to check whether a particular multivariate moment set $\mathbf{M}$ is realizable or not. For the special case of univariate moments for an NDF on a bounded interval, it is possible to define a *canonical* moment set $\mathbf{P}$ (Dette & Studden, 1997), using a nonlinear transformation of $\mathbf{M}$, for which the moment space is the unit cube: $\mathbf{P} \in [0, 1]^{2N}$. (The canonical moments are related to the Hankel–Hadamard determinants defined in Eq. (3.4).) Thus, for univariate moment sets, it is possible to formulate realizable spatial-transport schemes by formulating them in terms of canonical moments (Kah *et al.*, 2012). Another possibility for univariate moment sets is to use the moment-correction algorithms described in Section 3.1.4. For multivariate moment sets, the range of known possibilities is much more limited, and most of them rely on the convex moment combinations mentioned above.

To describe advection for a univariate PBE, it is necessary to advect using Eq. (8.2) a moment set $\mathbf{M}$ of length $2N$, and this moment set will be realizable for all t and $\mathbf{x}$ if it corresponds to a nonnegative NDF $n(t, \mathbf{x}, \xi)$. For a multivariate PBE, exactly the same condition applies, but now the dimensionality of the moment set used in QBMM is much larger. In practice, the reason why realizability is so important when solving a PBE using QBMM is that only realizable moment sets can be used with the moment-inversion algorithms described in Chapter 3. Thus, if spatial transport generates an unrealizable moment set at some spatial location $\mathbf{x}$, it will not be possible to apply QBMM to evaluate the moment source terms and the computation will abort. ***In short, the entire procedure of using QBMM to solve moment-transport equations will be robust if and only if the discretized spatial-transport scheme is guaranteed to produce realizable moment sets!*** For this reason, we take the position that, while it may sometimes be possible to make *ad hoc* corrections to recover realizable moments, the best approach is to employ only numerical algorithms for which it can be proven mathematically that the moment set is always realizable. In addition, for spatial transport we will need high-order advection schemes in order to minimize numerical diffusion. Hereafter, we will refer to algorithms with these properties as *realizable, high-order, spatial-transport schemes*.

[1] A convex combination is a linear combination with coefficients $\alpha_n \geq 0$ that satisfy $\sum_n \alpha_n = 1$.

Another type of realizability problem, not directly related to spatial transport, is the construction of a *realizable quadrature*. For a univariate NDF defined on a convex domain Ω_ξ, the Gaussian quadrature algorithm is guaranteed to return a set of N abscissas with the property $\xi_\alpha \in \Omega_\xi$ for all $\alpha \in N$. In other words, the abscissas correspond to realizable samples from the NDF. For multivariate quadrature, there is no guarantee that all of the abscissas will lie in Ω_ξ (unless, of course, Ω_ξ is the entire real line), even if the moment set used to construct the quadrature is realizable. In other words, realizable multivariate moment sets do not always generate realizable abscissas. This point is very important when Ω_ξ is a finite domain because points outside of Ω_ξ are often not physically meaningful. For example, if the internal coordinates are chemical mass fractions, then they must be nonnegative and sum to unity. Employing a negative mass fraction or a mass fraction greater than unity to evaluate the chemical reaction rate would yield a nonsensical value. As a consequence, using a non-realizable abscissa to evaluate a phase-space source term (as described in Chapter 7) could lead to large quadrature errors and hence compromise the simulation. For this reason, the moment-inversion algorithm used to construct multivariate quadratures from realizable moment sets must verify that the resulting quadrature is realizable. This is of course related to the moment set used to build it (as discussed in Section 7.6.3). We reiterate, however, that spatial transport is not a factor in determining the moment-inversion algorithm and thus our focus in this chapter will be on realizable moment sets.

8.1.2 Particle trajectory crossing

The occurrence of particle trajectory crossing (PTC) is associated with the free-transport term in the collisionless KE (and, by extension, the GPBE), and leads to a multi-velocity state that is difficult to capture with Eulerian solvers (Sachdev *et al.*, 2007; Saurel *et al.*, 1994). In a 1D velocity phase space, the simplest KE for the NDF $n(t, x, v)$ is

$$\frac{\partial n}{\partial t} + v\frac{\partial n}{\partial x} = 0, \tag{8.3}$$

where the second term is referred to as *free transport*. The analytical solution to this KE is $n(t, x, v) = n(0, x - vt, v)$ and the crucial difference from Eq. (8.1) is that the velocity v is both the internal coordinate *and* the transport velocity. This implies that the NDF will not simply be transported in physical space without changing shape, but rather that free transport will change the spatial location of the NDF and its shape. This leads to the counterintuitive phenomena of PTC wherein "packets" of density can cross at a single point in physical space.

For example, consider the initial NDF $n(0, x, v) = \delta(x + 1)\delta(v - 1) + \delta(x - 1)\delta(v + 1)$, which corresponds to two packets, one located at $x = -1$, with velocity $v = 1$, and the other located at $x = 1$, with velocity $v = -1$. The initial values for the moments are $m_k(0, x) = \delta(x+1) + (-1)^k\delta(x-1)$. Using the analytical solution to Eq. (8.3), we then have

$$\begin{aligned} n(t, x, v) &= \delta(x - vt + 1)\delta(v - 1) + \delta(x - vt - 1)\delta(v + 1) \\ &= \delta(x - t + 1)\delta(v - 1) + \delta(x + t - 1)\delta(v + 1), \end{aligned} \tag{8.4}$$

where the final expression follows on applying the Dirac delta functions for the velocity v. This leads to the following analytical solution for the moments: $m_k(t, x) = \delta(x - t + 1) + (-1)^k\delta(x + t - 1)$. From the solution in Eq. (8.4), we can observe that at $t = 1$ there occurs a PTC at which $n(1, x, v) = \delta(x)[\delta(v - 1) + \delta(v + 1)]$ and $m_k(1, x) = [1 + (-1)^k]\delta(x)$. In words,

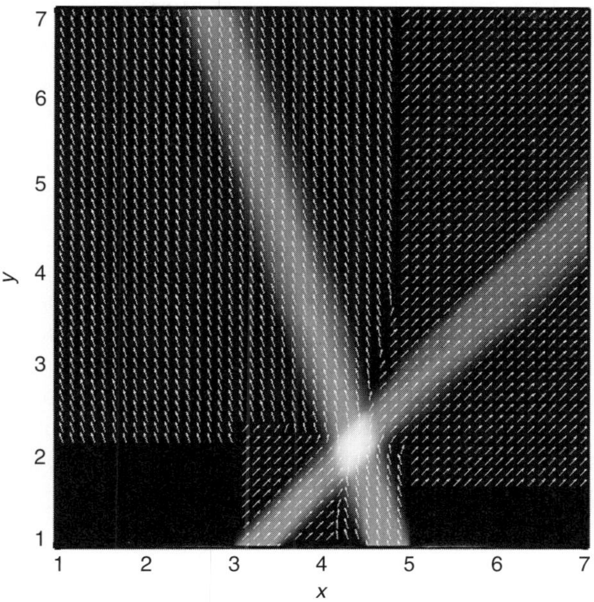

Figure 8.1. An example of particle trajectory crossing in a 2D domain. The fully developed particle number density m_{00} is shown together with the mean particle velocity vector $(m_{10}/m_{00}, m_{01}/m_{00})$. This example was computed using QBMM with a KBFVM (Yuan & Fox, 2011) and realizable second-order fluxes (Vikas *et al.*, 2011a). Note that the mean particle velocity is not defined in regions where the density is exactly zero. Owing to a small amount of numerical diffusion, the density is very small, but nonzero, outside of the two jets.

at $t = 1$ the NDF is concentrated at $x = 0$ and has two equal parts, one moving to the right with velocity $v = 1$ and one moving to the left with velocity $v = -1$. Likewise, at $t = 1$ the moments are nonzero only at $x = 0$. In essence, the two packets cross at $t = 1$, and the exact solution to Eq. (8.3) reproduces this behavior.

The extension of PTC to higher spatial dimensions is straightforward and many examples can be found in the literature (see, for example, Desjardins *et al.* (2008)). In Figure 8.1 we show a 2D example with two jets crossing. An important question that arises with PTC is whether or not the moment-transport equations can predict it. To answer this question, we return to Eq. (8.3) and integrate over phase space to find the moment-transport equation:

$$\frac{\partial m_k}{\partial t} + \frac{\partial m_{k+1}}{\partial x} = 0. \tag{8.5}$$

We immediately observe that the free-transport term in the KE leads to a closure problem in the moment-transport equation: if we truncate the moment set at m_{2N-1}, then, in order to solve Eq. (8.5), we must provide a moment closure for m_{2N}. For the example given in Eq. (8.4), it is straightforward to verify that the analytical expression for the moments satisfies Eq. (8.5). In fact, for this example, we can observe that a two-point quadrature with $w_1(t,x) = \delta(x - t + 1)$, $w_2(t,x) = \delta(x + t + 1)$, $\xi_1(t,x) = 1$, and $\xi_2(t,x) = -1$ exactly reproduces the moment $m_k(t,x)$ for arbitrary k. Thus, starting from the moment set (m_0, m_1, m_2, m_3), the two-point quadrature is the optimal closure for m_4. The moment-transport equations needed for a two-point quadrature are

$$\begin{aligned}\frac{\partial m_0}{\partial t} + \frac{\partial m_1}{\partial x} &= 0,\\ \frac{\partial m_1}{\partial t} + \frac{\partial m_2}{\partial x} &= 0,\\ \frac{\partial m_2}{\partial t} + \frac{\partial m_3}{\partial x} &= 0,\\ \frac{\partial m_3}{\partial t} + \frac{\partial m_4}{\partial x} &= 0,\end{aligned} \tag{8.6}$$

where the optimal choice for the unclosed moment is $m_4 = w_1\xi_1^4 + w_2\xi_2^4$. As in this example, in 1D problems the unknown moment is m_{2N}, but in higher dimensions other unknown moments will appear in the spatial fluxes and are closed using quadrature. In summary, using QBMM the free-transport term in the KE is closed by first constructing an N-node quadrature from the set of transported moments, and then using the N weights and N abscissas to estimate the unknown moments appearing in the spatial fluxes.

The conservative-flux form of the system in Eq. (8.6),

$$\frac{\partial \mathbf{M}}{\partial t} + \frac{\partial \mathbf{F}(\mathbf{M})}{\partial x} = 0, \tag{8.7}$$

where $\mathbf{F}$ is the flux function, naturally leads to the question of whether the system is hyperbolic. In other words, if we consider the Jacobian of the flux function, $\partial \mathbf{F}/\partial \mathbf{M}$, are its eigenvalues all real? This question has been investigated by Chalons *et al.* (2012), where it was shown that the system is weakly hyperbolic, i.e. it has N pairs of repeated (real) eigenvalues that correspond to the abscissas. The classical weakly hyperbolic system is the pressure-less gas-dynamic equation:

$$\begin{aligned}\frac{\partial \rho}{\partial t} + \frac{\partial \rho u}{\partial x} &= 0,\\ \frac{\partial \rho u}{\partial t} + \frac{\partial \rho u^2}{\partial x} &= 0,\end{aligned} \tag{8.8}$$

which is known to produce delta shocks in the density ρ (Bouchut *et al.*, 2003). It is interesting to note that Eq. (8.8) is the QBMM approximation to free transport with $N = 1$ (i.e. $m_0 = \rho$, $m_1 = \rho u$, $w_1 = \rho$, and $\xi_1 = u$ so that $m_2 = w_1 u^2$). If we were to attempt to solve the PTC problem in Eq. (8.4) using $N = 1$, we would find that for all $1 \le t$ the moments would remain "stuck" at $x = 0$: $m_0(t \ge 1, x) = 2\delta(x)$, $m_1(t \ge 1, x) = 0 \Rightarrow \rho(t \ge 1, x) = 2\delta(x)$, and $u(t \ge 1, x) = 0$, which corresponds to the definition of a delta shock in the density. However, we have seen above that, with $N = 2$, no delta shock occurs and the same will be true for $N \ge 2$ for this example. In general, the solution to Eq. (8.7) will exhibit delta shocks in some of the weights $w_\alpha(t, x)$ whenever the number of PTC at a given point x is greater than the number of available abscissas N.

In summary, although the weakly hyperbolic nature of Eq. (8.7) has been shown rigorously only for 1D phase space (i.e. one velocity component in the KE), experience strongly suggests that the full 3D system is also weakly hyperbolic. This observation implies that the numerical schemes used to solve the moment-transport equations closed with QBMM must be able to handle local delta shocks in the moments. One such class of numerical schemes consists of the kinetics-based finite-volume solvers presented in Section 8.2. As a final note, we should mention that the work of Chalons *et al.* (2012) using extended Gaussian quadrature (see Section 3.3.2) and kinetics-based finite-volume solvers to close Eq. (8.7) suggests that the system with $2N + 1$ moments is fully hyperbolic and thus does not exhibit delta shocks. As we shall see below (and as we saw in Figure 8.1 above), kinetics-based solvers can be designed with realizable, high-order, spatial-transport schemes, thereby both ensuring the optimal QBMM closure of the free-transport term and guaranteeing that the transported moments are always realizable. Moreover, such methods can be extended to the GPBE involving both active (e.g. velocity or internal coordinates that affect the advection velocity) and passive internal coordinates.[2]

8.1.3 Coupling between active and passive internal coordinates

In the GPBE the internal coordinates include both active and passive variables. For example, consider the 1D NDF for the velocity (v) and surface area (s) of a particle: $n(t, x, v, s)$, where the two internal coordinates are $\boldsymbol{\xi} = (v, s)$. The free-transport term for the NDF is the same as in Eq. (8.3). However, we now have bivariate moments defined by $m_{k,l} = \int v^k s^l n \, \mathrm{d}v \, \mathrm{d}s$, so that the moment transport equation is

$$\frac{\partial m_{k,l}}{\partial t} + \frac{\partial m_{k+1,l}}{\partial x} = 0. \tag{8.9}$$

As before, the exact solution to Eq. (8.3) is $n(t, x, v, s) = n(0, x - vt, v, s)$, from which we can observe that s is indeed passive because it is simply carried along with velocity v. Or, in other words, the free-transport term in the GPBE can be solved separately for each value of s. However, the moment-transport equations are coupled because they depend on both k and l. Nevertheless, we can observe that if we consider a set of moments with k fixed, but l free, it is possible to use QBMM to represent the unclosed moments. For example, with $N = 2$ the moment-transport equations are

$$\begin{aligned}
\frac{\partial m_{0,l}}{\partial t} + \frac{\partial m_{1,l}}{\partial x} &= 0, \\
\frac{\partial m_{1,l}}{\partial t} + \frac{\partial m_{2,l}}{\partial x} &= 0, \\
\frac{\partial m_{2,l}}{\partial t} + \frac{\partial m_{3,l}}{\partial x} &= 0, \\
\frac{\partial m_{3,l}}{\partial t} + \frac{\partial m_{4,l}}{\partial x} &= 0,
\end{aligned} \tag{8.10}$$

and, for each l, we can use the moment set $(m_{0,l}, m_{1,l}, m_{2,l}, m_{3,l})$ to find two weights and two abscissas ($w_{1,l}, w_{2,l}, \xi_{1,l}$, and $\xi_{2,l}$), from which the unknown moment can be closed: $m_{4,l} = w_{1,l}\,(\xi_{1,l})^4 + w_{2,l}\,(\xi_{2,l})^4$. Again, this decomposition of the solution algorithm is possible because of the passive nature of the size moments in Eq. (8.10). However, in general, the right-hand side of the GBPE will include unclosed source terms that will require a 2D quadrature approximation for closure.

For example, if a 2D quadrature with $N = 4$ nodes is constructed, the 12 mixed moments in Table 3.5 must be transported. The free-transport term will correspond to Eq. (8.10) with $l = 0, 1, 2, 3$. However, for $l = 2, 3$, only the four moments with $k = 0, 1$ are transported (i.e. $m_{0,2}$, $m_{1,2}$, $m_{0,3}$, and $m_{1,3}$), and hence we will not be able to find two weights and

[2] We shall use the term "active" to refer to the velocity internal coordinate because its value directly affects the spatial transport. Other internal coordinates (e.g. size, mass, etc.) are referred to as "passive" if they are simply advected in space by the free-transport term. However, if the advection velocity is modeled as a function of internal coordinates such as size, then such coordinates are also referred to as active.

two abscissas for $l = 2, 3$. Instead, we have essentially two options: (1) compute a 2D, four-node quadrature using the brute-force QMOM described in Section 3.2, or (2) work with the conditional moments of velocity given the surface area to construct a CQMOM quadrature. As shown in Section 3.2, the CQMOM for this case requires the ten mixed moments $m_{0,0}$, $m_{0,1}$, $m_{0,2}$, $m_{0,3}$, $m_{1,0}$, $m_{1,1}$, $m_{2,0}$, $m_{2,1}$, $m_{3,0}$, and $m_{3,1}$, for which the free-transport equations are

$$\begin{aligned}
&\frac{\partial m_{0,0}}{\partial t} + \frac{\partial m_{1,0}}{\partial x} = 0, \quad &&\frac{\partial m_{0,1}}{\partial t} + \frac{\partial m_{1,1}}{\partial x} = 0, \quad &&\frac{\partial m_{0,2}}{\partial t} + \frac{\partial m_{1,2}}{\partial x} = 0,\\
&\frac{\partial m_{1,0}}{\partial t} + \frac{\partial m_{2,0}}{\partial x} = 0, &&\frac{\partial m_{1,1}}{\partial t} + \frac{\partial m_{2,1}}{\partial x} = 0, &&\frac{\partial m_{0,3}}{\partial t} + \frac{\partial m_{1,3}}{\partial x} = 0,\\
&\frac{\partial m_{2,0}}{\partial t} + \frac{\partial m_{3,0}}{\partial x} = 0, &&\frac{\partial m_{2,1}}{\partial t} + \frac{\partial m_{3,1}}{\partial x} = 0, &&\\
&\frac{\partial m_{3,0}}{\partial t} + \frac{\partial m_{4,0}}{\partial x} = 0, &&\frac{\partial m_{3,1}}{\partial t} + \frac{\partial m_{4,1}}{\partial x} = 0. &&
\end{aligned} \tag{8.11}$$

The reader can verify that the system in Eq. (8.11) has four unclosed moments, $m_{1,2}$, $m_{1,3}$, $m_{4,0}$, and $m_{4,1}$. The CQMOM closure for the moments has the form $m_{k,l} = w_1(w_{11}v_{11}^k + w_{12}v_{12}^k)s_1^l + w_2(w_{21}v_{21}^k + w_{22}v_{22}^k)s_2^l$, where the weights and abscissas are determined from the ten transported moments. (By definition, $w_{11} + w_{12} = 1$ and $w_{21} + w_{22} = 1$, so there are exactly ten unknowns.) Thus, for example, the closure for $m_{4,0}$ is $m_{4,0} = w_1(w_{11}v_{11}^4 + w_{12}v_{12}^4) + w_2(w_{21}v_{21}^4 + w_{22}v_{22}^4)$, which, in general, is different than the closure which would be found using the moment set $(m_{0,0}, m_{1,0}, m_{2,0}, m_{3,0})$ directly. Only in cases in which the NDF is separable, $n(v, s) = n(v)n(s)$, will the two quadrature approximations for $m_{4,0}$ be the same. In other words, only when the active and passive internal coordinates are *uncoupled* will all possible quadrature reconstructions be the same.

As an example, let us consider a generalization of the NDF in Eq. (8.4):

$$n(t, x, v, s) = \delta(x - t + 1)\delta(v - 1)\delta(s - 1) + \delta(x + t - 1)\delta(v + 1)\delta(s - 2), \tag{8.12}$$

which can be exactly represented by a two-point quadrature with $w_1(t, x) = \delta(x - t + 1)$, $v_1(t, x) = 1$, $s_1(t, x) = 1$, $w_2(t, x) = \delta(x + t + 1)$, $v_2(t, x) = -1$, and $s_2(t, x) = 2$. The moment set for arbitrary l is thus

$$\begin{aligned}
m_{0,l} &= \delta(x - t + 1) + (2)^l\delta(x + t + 1),\\
m_{1,l} &= \delta(x - t + 1) - (2)^l\delta(x + t + 1),\\
m_{2,l} &= \delta(x - t + 1) + (2)^l\delta(x + t + 1),\\
m_{3,l} &= \delta(x - t + 1) - (2)^l\delta(x + t + 1),
\end{aligned} \tag{8.13}$$

which yield the weights and abscissas $w_{1,l} = \delta(x - t + 1)$, $w_{2,l} = \delta(x + t + 1)$, $\xi_{1,l} = 1$, and $\xi_{2,l} = 2$. Thus, the unknown moment $m_{4,l}(t, x) = \delta(x - t + 1) + (2)^l\delta(x + t + 1)$ is closed exactly. For the CQMOM, we start with the four moments

$$\begin{aligned}
m_{0,0} &= \delta(x - t + 1) + \delta(x + t + 1),\\
m_{0,1} &= \delta(x - t + 1) + 2\delta(x + t + 1),\\
m_{0,2} &= \delta(x - t + 1) + 4\delta(x + t + 1),\\
m_{0,3} &= \delta(x - t + 1) + 8\delta(x + t + 1),
\end{aligned} \tag{8.14}$$

and calculate the weights and abscissas $w_1 = \delta(x - t + 1)$, $w_2 = \delta(x + t + 1)$, $s_1 = 1$, and $s_2 = 2$, which, again, are exact. In order to calculate the conditional weights and abscissas, we use the six moments

$$\begin{aligned}
m_{1,0} &= \delta(x-t+1) - \delta(x+t+1), \\
m_{2,0} &= \delta(x-t+1) + \delta(x+t+1), \\
m_{3,0} &= \delta(x-t+1) - \delta(x+t+1), \\
m_{1,1} &= \delta(x-t+1) - 2\delta(x+t+1), \\
m_{2,1} &= \delta(x-t+1) + 2\delta(x+t+1), \\
m_{3,1} &= \delta(x-t+1) - 2\delta(x+t+1).
\end{aligned} \tag{8.15}$$

The conditional moments $\langle v^k \rangle_\alpha$ for $k = 1, 2, 3$ are found by solving the linear system

$$\begin{bmatrix} 1 & 1 \\ s_1 & s_2 \end{bmatrix} \begin{bmatrix} w_1 & 0 \\ 0 & w_2 \end{bmatrix} \begin{bmatrix} \langle v \rangle_1 & \langle v^2 \rangle_1 & \langle v^3 \rangle_1 \\ \langle v \rangle_2 & \langle v^2 \rangle_2 & \langle v^3 \rangle_2 \end{bmatrix} = \begin{bmatrix} m_{1,0} & m_{2,0} & m_{3,0} \\ m_{1,1} & m_{2,1} & m_{3,1} \end{bmatrix}. \tag{8.16}$$

This yields $\langle v \rangle_1 = \langle v^2 \rangle_1 = \langle v^3 \rangle_1 = 1$ and $\langle v \rangle_2 = -1$, $\langle v^2 \rangle_2 = 1$, and $\langle v^3 \rangle_2 = -1$, and hence the conditional weights and abscissas are $w_{11} = w_{12} = w_{21} = w_{22} = 1/2$, $v_{11} = v_{12} = 1$, and $v_{21} = v_{22} = -1$. This result can then be used to estimate the unknown moments in the transport equations (Eq. (8.11)). For example, the CQMOM closure for $m_{4,0}$ is $m_{4,0} = w_1 v_{11}^4 + w_2 v_{21}^4 = \delta(x-t+1) + \delta(x+t+1)$, which is exact. We should note that, if the adaptive Wheeler algorithm described in Yuan & Fox (2011) were used with the conditional moments, then only one node would be returned for the conditional weights and abscissas: $w_{11} = 1$, $v_{11} = 1$ and $w_{21} = 1$, $v_{21} = -1$. In other words, as shown in Yuan & Fox (2011), the numerical algorithm used in the CQMOM will find the exact NDF when the latter is composed of weighted Dirac delta functions (i.e. an N-point distribution).

In summary, the CQMOM quadrature is exact for this example because the NDF can be represented by two multivariate delta functions (Eq. (8.12)). For cases in which the NDF is continuous, the CQMOM quadrature will provide an approximation of the NDF that can be used to estimate the unknown moments in the moment-transport equations. In practice, due to the phase-space source terms, the active and passive internal coordinates will be strongly coupled. For example, in a fluid–particle flow the particle velocity will change due to the fluid-drag force, and the magnitude of the drag force is inversely proportional to the particle surface area. Thus, particles with small s will have a very different velocity v than that of particles with large s. Because it is exactly such correlations between the internal coordinates that we wish to capture by using a GPBE, it is imperative that the solution methods used to handle the spatial transport in the moment-transport equations be able to work with sets of coupled mixed moments such as in Eq. (8.11). As we have shown in this section, the CQMOM is an efficient tool for constructing quadratures for highly correlated mixed moment sets. In addition, the CQMOM can be easily combined with kinetics-based methods (KBFVM) to treat the spatial-transport terms in the moment-transport equations while maintaining realizable moment sets. In Section 8.5 we show that the EQMOM can also be used for bivariate quadrature with size-dependent spatial fluxes.

8.1.4 The QMOM versus the DQMOM

The quadrature method of moments (QMOM) and the direct quadrature method of moments (DQMOM) were introduced in Chapter 3 as equivalent methods for solving a homogeneous GPBE. In fact, the DQMOM was derived by Marchisio & Fox (2005) primarily for the purpose of solving spatially inhomogeneous *multivariate* moment-transport equations. Unlike for the univariate case, where the moment-inversion algorithm is uniquely defined for a given set of moments, the QMOM in the multivariate case is much more difficult to apply. With the DQMOM, once the moment set has been chosen (e.g. the optimal moment set in Fox (2009b)), and the weights and abscissas initialized, it is (in principle) no longer necessary to apply the moment-inversion algorithm since the weights and abscissas are solved for directly. In practice, the DQMOM works well under the condition that the abscissas remain continuous and distinct at every point in space and time during the simulation. Indeed, the derivation of the DQMOM equations (see Marchisio & Fox (2005) and Fox *et al.* (2008)) requires that the space and time derivatives of the abscissas be finite (in order for the derivatives of the Dirac delta function to be defined) and that the coefficient matrix **A** (e.g. see Eq. (7.118)) be non-singular. If either of these conditions fails, then the DQMOM simulation will be compromised. It is thus instructive to consider cases for which the DQMOM is likely to fail.

Before looking at specific examples for which the DQMOM is likely to fail, we should note that the direct solution of the moment-transport equations using the QMOM is very robust. In the context of conservation equations, the moment-transport equations (and, for that matter, the NDF) are written in a conservative form:

$$\frac{\partial \mathbf{M}}{\partial t} + \frac{\partial}{\partial \mathbf{x}} \cdot \mathbf{F}(\mathbf{M}) = \mathbf{S}(\mathbf{M}), \tag{8.17}$$

where **F** is a flux function and **S** is a source term.[3] The moments are thus *conserved variables* (when $\mathbf{S} = \mathbf{0}$) in the parlance of transport theory. In contrast, the weights and abscissas are *primitive variables* for which the transport equations (i.e. the DQMOM) do not appear in conservative form. As is well known from gas dynamics, equations of the form of Eq. (8.17) can exhibit shocks in the primitive variables (e.g. in the Euler equation the gas velocity can be discontinuous in space). We can thus expect that, depending on the properties of **F** in Eq. (8.17), the weights and abscissas in the DQMOM may be discontinuous, even though the moment set **M** is well defined. As is well known in the CFD community, finite-volume methods (FVM) are well suited to solve Eq. (8.17). However, as discussed below in Section 8.2, when solving for moment sets it is very important to use *realizable* FVM.

With this background information on the DQMOM and on conservative versus primitive variables in mind, we can now discuss cases for which the DQMOM is likely to fail.

(a) If the moment-transport equation (Eq. (8.17)) is purely hyperbolic (i.e. pure advection of the NDF or the KE), the DQMOM will fail when there are shocks in the abscissas, across which the abscissas (and maybe also the weights) are discontinuous. As described in Appendix C, this occurs when characteristics (which correspond to the abscissas) cross in physical space, and thus are not unique. For such cases, using QMOM with a multivariate moment-inversion algorithm appears at present to be the only viable solution. This situation occurs most commonly with the KE or the GPBE, wherein velocity is an internal coordinate.

(b) Despite the fact that the advection terms in the moment-transport equations are purely hyperbolic, even if there are no shocks and the moments are continuous in space, when solved with a standard FVM the non-conservative form of the transport equations used in the DQMOM may still fail to conserve the moments due to the inevitable presence of numerical errors and numerical diffusion. In fact, only

[3] Strictly speaking, the conservative form refers to conservation equations without a source term. When one is solving moment-transport equations, operator splitting wherein the transport step is solved with $\mathbf{S} = \mathbf{0}$ is often employed.

the weights and abscissas, and linear combinations of these variables (i.e. moments of order zero and one), will be conserved with a finite-volume solution of the DQMOM equations. All higher-order moments, which are nonlinear combinations of the weights and abscissas, will not be rigorously conserved by the numerical flux functions. Such numerical errors can generate unrealizable moment sets, especially when high-order spatial reconstructions are used in the numerical flux functions. One way to handle this problem is to treat moment advection in physical space directly in terms of realizable numerical flux functions written in terms of the moments (which are conserved quantities), instead of using the numerical flux functions for the DQMOM equations. This variation of the DQMOM, which is referred to as fully conservative DQMOM (DQMOM-FC), is discussed in detail in Appendix D.

(c) If the moment-transport equation contains spatial-diffusion terms, but the diffusivity is very small (and hence spatial gradients are very difficult to resolve), the DQMOM correction terms for diffusion (see Marchisio & Fox (2005) for details) can be inaccurate. For such cases, the DQMOM will not be exactly the same as the QMOM unless the spatial gradients in the abscissas are fully resolved on the computational grid. This situation is likely to occur when one is using an FVM for solving the DQMOM equations and the numerical diffusion is of the same order of magnitude as (or larger than!) the true diffusivity. For this reason, the classical DQMOM equations are more suitable for use with turbulence models that provide significant "turbulent diffusivity" in the model equations, whereas cases with very small spatial diffusion should be treated with DQMOM-FC as explained in the previous point.

(d) If the initial conditions for the abscissas are discontinuous inside the flow domain or at an inlet, then the DQMOM can fail if the transport equations are purely hyperbolic (see Appendix C). This situation commonly occurs when, for example, the top half of the domain has one NDF and the bottom half has a different NDF. Note that, in this case, the moments **M** are not continuous throughout the domain, but this does not impede our getting the correction solution by solving Eq. (8.17) using a realizable FVM. Obviously, the same problem will occur when the inflow NDF is different than the initial NDF in the solution domain. In general, the interior of the flow domain must be initialized with the same abscissas as used to represent the inflow(s) in order for the DQMOM to work correctly. Many of the reported difficulties with the DQMOM in the literature are due to discontinuous initial/boundary conditions, resulting in crossing characteristics.

(e) Another, more subtle, problem can occur for degenerate moment sets. By the latter, we mean moment sets for which the number of abscissas required to represent all of the moments is finite (or changes in space or time). The most obvious example of this problem is when the initial conditions correspond to an NDF composed of a single weighted Dirac delta function. For this case, the moments **M** are well defined (as is Eq. (8.17)), but degenerate in the sense that only two of them are needed in order to reconstruct the NDF. If, for example, due to the source term the moments become non-degenerate at some points in the flow, then the QMOM with the adaptive Wheeler algorithm (Yuan & Fox, 2011) will automatically adapt the value of N in such a way as to increase the number of abscissas. In contrast, the DQMOM was formulated with a fixed value of N, so something "clever" must be done to handle such cases (e.g. make N adaptive).

(f) Another, even more subtle, problem is the so-called "X-problem" arising in cases with spatial diffusion of the moments. Consider a univariate PBE with zero advection, but nonzero diffusion:

$$\frac{\partial n}{\partial t} = \frac{\partial^2 n}{\partial x^2} \quad \Longrightarrow \quad \frac{\partial m_k}{\partial t} = \frac{\partial^2 m_k}{\partial x^2}.$$

For simplicity, let $m_0(t,x) = 1$, $m_1(t,x) = 0$, $m_2(0,x) = x^2$, and $m_3(t,x) = 0$, so that the variance at $x = 0$ is initially zero. The exact solution for the second-order moment is $m_2(t,x) = 2t + x^2$. The corresponding weights and abscissas are $w_1(t,x) = w_2(t,x) = 1/2$ and $\xi_1(t,x) = -\xi_2(t,x) = \sqrt{2t + x^2}$ for $0 \leq t$. Thus, at $t = 0$, the abscissas form an "X" that separates into two parts for $0 < t$. Obviously, the same thing could occur for higher-order moments, and the DQMOM must be modified to treat such cases. In practice, this diffusion-induced change in the number of degenerate moments occurs for problems in which one of the internal coordinates is generated by a source term, for example, for products of a chemical reaction.

There are undoubtedly other examples for which the DQMOM fails but the QMOM works. However, as a general rule, the more dominant the diffusion component of the spatial fluxes, the more likely it is that the abscissas will exhibit smooth variations in space and the DQMOM will be applicable. In cases for which the DQMOM is applicable, the problem of realizable moment sets is fairly simple to address because the only requirement is that the weights be nonnegative. Nevertheless, it is still necessary in multivariate cases to verify that the abscissas are physically realizable, i.e. they must take on values in the phase space of the internal coordinates (see Section 8.1.1). With the DQMOM (as with the QMOM), there is no guarantee that the solutions will represent realizable quadratures for a multivariate NDF.

8.2 Kinetics-based finite-volume methods

In this section, we introduce KBFVM for the moment-transport equations and show how they can be formulated to guarantee that the updated moment set is realizable. A more detailed presentation can be found in Appendix B. For clarity, we begin with the PBE and then discuss the additional complications that arise with the KE and, finally, the GPBE. Because the advection term is the principal cause of unrealizable moment sets, we limit our discussion to this term and focus on the high-order spatial reconstruction needed in order to reduce numerical diffusion. The material in this section follows roughly the work of Vikas *et al.* (2011a, 2012), and the reader is referred there and to Appendix B for more details.

The main idea behind KBFVM for moment-transport equations is that behind every moment set there exists a corresponding NDF. Thus, the moments are not treated as independent scalars, but rather as linear transformations of a single NDF. With this idea in mind, we can freely write the same scheme in terms of either the moments or the NDF. Then, in order to guarantee that the moment set is realizable, we enforce a *realizability condition* on the NDF, which states that the "numerical" NDF must be nonnegative for the parameters used in the scheme (e.g. the time step and the grid spacing). For an arbitrary NDF, only a first-order spatial reconstruction (i.e. the NDF is constant in each grid cell) will guarantee realizability of the moment set. However, we shall see that with QBMM the NDF has a weighted Dirac-delta-function representation, making it possible to use high-order spatial reconstruction for the weights while still guaranteeing that the numerical NDF is nonnegative and thus that the moment set is realizable. In the following, we demonstrate the basic ideas behind KBFVM using simple cases involving 1D spatial advection or free transport. More details can be found in Appendix B.

8.2.1 Application to PBE

For simplicity, only the 1D case is discussed here, and we denote the univariate NDF by $n(t, x, \xi)$, where ξ is a passive internal coordinate. The PBE with only the advection term is

$$\frac{\partial n}{\partial t} + \frac{\partial un}{\partial x} = 0, \tag{8.18}$$

where $u(t, x)$ is a known velocity. The moment-transport equation found from Eq. (8.18) is

$$\frac{\partial m_k}{\partial t} + \frac{\partial u m_k}{\partial x} = 0. \tag{8.19}$$

Denoting the moment set used in QBMM[4] as $\mathbf{M} = [m_0\ m_1\ m_2\ \ldots\ m_{2N-1}]^{\mathrm{T}}$ and the moment fluxes as $\mathbf{F}(u, \mathbf{M}) = u\mathbf{M}$, the moment-transport equations can be written as

$$\frac{\partial \mathbf{M}}{\partial t} + \frac{\partial \mathbf{F}(u, \mathbf{M})}{\partial x} = 0. \tag{8.20}$$

The moment set and moment fluxes can be written in the general forms $\mathbf{M} = \int \mathbf{K} n\, \mathrm{d}\xi$ and $\mathbf{F} = u \int \mathbf{K} n\, \mathrm{d}\xi$, where $\mathbf{K} = [1\ \xi\ \xi^2\ \cdots\ \xi^{2N-1}]^{\mathrm{T}}$. Starting from Eq. (8.20), we now develop a realizable, high-order scheme to update the moment set at time $t + \Delta t$ given a realizable moment set at time t.

On a uniform grid with spacing Δx, we define the volume-average moment set at time t as

$$\mathbf{M}_i^m \equiv \frac{1}{\Delta x} \int_{x_i - \Delta x/2}^{x_i + \Delta x/2} \mathbf{M}(m\, \Delta t, x) \mathrm{d}x, \tag{8.21}$$

where x_i is the cell center and $t = m\, \Delta t$ is the current time. If a single-stage Euler explicit time-integration scheme is used, the updated moment set can be written as

$$\begin{aligned} \mathbf{M}_i^{m+1} = \mathbf{M}_i^m - \lambda \Big[& \mathbf{G}\left(u_{i+1/2,l}^m, \mathbf{M}_{i+1/2,l}^m, u_{i+1/2,r}^m, \mathbf{M}_{i+1/2,r}^m\right) \\ & - \mathbf{G}\left(u_{i-1/2,l}^m, \mathbf{M}_{i-1/2,l}^m, u_{i-1/2,r}^m, \mathbf{M}_{i-1/2,r}^m\right)\Big], \end{aligned} \tag{8.22}$$

where $\lambda = \Delta t / \Delta x$, and $\mathbf{G}$ is the numerical flux function defined as

$$\mathbf{G}(u_l, \mathbf{M}_l, u_r, \mathbf{M}_r) = u_l^+ \int \mathbf{K} n_l\, \mathrm{d}\xi + u_r^- \int \mathbf{K} n_r\, \mathrm{d}\xi \tag{8.23}$$

with $u_l^+ = \frac{1}{2}(u_l + |u_l|)$ and $u_r^- = \frac{1}{2}(u_r - |u_r|)$. The subscript $_{i-1/2}$ ($_{i+1/2}$) denotes the interface between the grid cells $i - 1$ and i (i and $i + 1$). The subscript $_l$ ($_r$) denotes the limit on approaching the interface from the left (right). For example, the point $x_{i-1/2,l}$ is located in grid cell $i - 1$ but near the interface with grid cell i, and the point $x_{i-1/2,r}$ is located near the same interface but in grid cell i. Likewise, $u_{i-1/2,l}^m$ denotes the velocity at time t in grid

[4]For simplicity, we employ the QMOM to illustrate the numerical methods. In Section 8.3 examples using the EQMOM are provided.

cell $i-1$, but located at the interface between grid cells $i-1$ and i. The reader should note in Eq. (8.23) that the numerical flux function requires us to estimate the NDF n_l and n_r located to the left and right of the cell interfaces. These estimates must be constructed from the known moment sets in all of the grid cells, i.e. from $\mathbf{M}_i^m$.

On introducing the numerical NDF $h_i(\xi)$, the updated moment set in Eq. (8.22) can be written as $\mathbf{M}_i^{m+1} = \int \mathbf{K} h_i \, \mathrm{d}\xi$, where

$$\begin{aligned} h_i(\xi) = n_i^m(\xi) - \lambda u_{i+1/2,r}^{m-} n_{i+1/2,r}^m(\xi) + \lambda u_{i-1/2,l}^{m+} n_{i-1/2,l}^m(\xi) - \lambda u_{i+1/2,l}^{m+} n_{i+1/2,l}^m(\xi) \\ + \lambda u_{i-1/2,r}^{m-} n_{i-1/2,r}^m(\xi). \end{aligned} \tag{8.24}$$

The reader should note that the ξ dependence in this expression comes from the reconstructed NDF: e.g. $n_i^m(\xi) = n(m\,\Delta t, x_i, \xi)$ is reconstructed from $\mathbf{M}_i^m$. Assuming that the moments at time level m correspond to a nonnegative density function (i.e. n_i^m is nonnegative for all i), the first three terms on the right-hand side of Eq. (8.24) are nonnegative and the last two terms are non-positive. For the moments at time level $m+1$ to correspond to a nonnegative density function, $h_i(\xi)$ must be nonnegative for all ξ. Using a first-order spatial reconstruction for the NDF, the NDF at the cell interfaces are the same as those at the cell centers, $n_i^m(\xi) = n_{i+1/2,l}^m(\xi) = n_{i-1/2,r}^m(\xi)$, for which Eq. (8.24) reduces to

$$h_i(\xi) = \left(1 - \lambda u_{i+1/2,l}^{m+} + \lambda u_{i-1/2,r}^{m-}\right) n_i^m(\xi) - \lambda u_{i+1/2,r}^{m-} n_{i+1}^m(\xi) + \lambda u_{i-1/2,l}^{m+} n_{i-1}^m(\xi), \tag{8.25}$$

and hence the updated moment set is

$$\mathbf{M}_i^{m+1} = \left(1 - \lambda u_{i+1/2,l}^{m+} + \lambda u_{i-1/2,r}^{m-}\right) \mathbf{M}_i^m - \lambda u_{i+1/2,r}^{m-} \mathbf{M}_{i+1}^m + \lambda u_{i-1/2,l}^{m+} \mathbf{M}_{i-1}^m. \tag{8.26}$$

The nonnegativity of $h_i(\xi)$ is guaranteed under the following *realizability condition*:

$$\left(1 - \lambda u_{i+1/2,l}^{m+} + \lambda u_{i-1/2,r}^{m-}\right) \geq 0 \quad \Rightarrow \quad 0 < \Delta t \leq \frac{\Delta x}{\max_i \left(u_{i+1/2,l}^{m+} - u_{i-1/2,r}^{m-}\right)}. \tag{8.27}$$

The reader should notice that this realizability condition on the time step Δt applies even when a high-order spatial reconstruction is used for the velocity. It is also worth noting that Eq. (8.26) applies even when the velocity is discontinuous (e.g. at a shock).

For second-order and, in general, any high-order spatial reconstruction of the NDF in Eq. (8.24), nonnegativity of $h_i(\xi)$ cannot be guaranteed. However, if QBMM are used along with the *special* reconstruction discussed below, nonnegativity of $h_i(\xi)$ can be guaranteed. Using QBMM with an N-node quadrature,[5] the NDF is

$$n(t, x, \xi) = \sum_{\alpha=1}^{N} w_\alpha(t, x) \delta(\xi - \xi_\alpha(t, x)), \tag{8.28}$$

where w_α are the weights and ξ_α are the abscissas. For this reconstructed NDF, the following theorem due to Vikas *et al.* (2011a) applies.

Theorem 8.1 *Let $N, p \in \mathbb{N}$ and $\alpha \in \{1, 2, \ldots, N\}$. For the volume-average moment sets $\mathbf{M}_i^m$, let the reconstructed values of the weights satisfy $w_{i,\alpha}^m > 0$ and $w_{i+1/2,\alpha,l}^m, w_{i-1/2,\alpha,r}^m \geq 0$ $\forall i, \alpha$. If a finite-volume scheme using a single-stage Euler explicit time-integration scheme is devised that uses a pth-order spatial reconstruction for the weights and first-order spatial reconstruction for the abscissas, the nonnegativity of the numerical NDF in Eq. (8.24) can always be guaranteed under an explicit constraint on the time step ($\Delta t \in \mathbb{R}^+$).*

[5]Both the QMOM and the EQMOM can be written in this discrete form by using the quadrature nodes.

The proof is as follows. Using Eq. (8.24), the numerical density function, regardless of the scheme used, can be written as

$$h_i = n_i^m - \lambda u_{i+1/2,l}^{m+} n_{i+1/2,l}^m + \lambda u_{i-1/2,r}^{m-} n_{i-1/2,r}^m + \varepsilon^+. \tag{8.29}$$

In the above expression, the nonnegative terms in Eq. (8.24) have been grouped under ε^+. For an N-node quadrature, using Eq. (8.28), the expression for h_i^m becomes

$$h_i = \sum_{\alpha=1}^{N} \left[w_{i,\alpha}^m \delta\left(\xi - \xi_{i,\alpha}^m\right) - \lambda u_{i+1/2,l}^{m+} w_{i+1/2,\alpha,l}^m \delta\left(\xi - \xi_{i+1/2,\alpha,l}^m\right) \right.$$
$$\left. + \lambda u_{i-1/2,r}^{m-} w_{i-1/2,\alpha,r}^m \delta\left(\xi - \xi_{i-1/2,\alpha,r}^m\right) \right] + \varepsilon^+. \tag{8.30}$$

If a first-order spatial reconstruction is used for the abscissas, then the interface values of the abscissas will be the same as the "volume-average" values:[6] $\xi_{i,\alpha}^m = \xi_{i+1/2,\alpha,l}^m = \xi_{i-1/2,\alpha,r}^m$. On putting this into Eq. (8.30), the numerical NDF becomes

$$h_i = \sum_{\alpha=1}^{N} \left(w_{i,\alpha}^m - \lambda u_{i+1/2,l}^{m+} w_{i+1/2,\alpha,l}^m + \lambda u_{i-1/2,r}^{m-} w_{i-1/2,\alpha,r}^m \right) \delta\left(\xi - \xi_{i,\alpha}^m\right) + \varepsilon^+. \tag{8.31}$$

For $\Delta t \in \mathbb{R}^+$ satisfying the realizability condition,

$$\Delta t \le \min_{i,\alpha} \left(\frac{w_{i,\alpha}^m \, \Delta x}{w_{i+1/2,\alpha,l}^m u_{i+1/2,l}^{m+} - w_{i-1/2,\alpha,r}^m u_{i-1/2,r}^{m-}} \right), \tag{8.32}$$

h_i is nonnegative for all ξ. This concludes the proof.

When high-order spatial reconstruction is used for the weights, the updated moment set is

$$\mathbf{M}_i^{m+1} = \sum_{\alpha=1}^{N} \left(w_{i,\alpha}^m - \lambda u_{i+1/2,l}^{m+} w_{i+1/2,\alpha,l}^m + \lambda u_{i-1/2,r}^{m-} w_{i-1/2,\alpha,r}^m \right) \mathbf{K}_{i,\alpha}^m - \lambda u_{i+1/2,r}^{m-} \mathbf{M}_{i+1}^m$$
$$+ \lambda u_{i-1/2,l}^{m+} \mathbf{M}_{i-1}^m, \tag{8.33}$$

where $\mathbf{K}_{i,\alpha}^m = [1 \; \xi_{i,\alpha}^m \; (\xi_{i,\alpha}^m)^2 \; \cdots \; (\xi_{i,\alpha}^m)^{2N-1}]^{\mathrm{T}}$. For example, with a second-order spatial reconstruction for grid cell i, we have $w_\alpha(t, x) = w_{i,\alpha}^m + a_{i,\alpha}(x - x_i)$ for $x_i - \Delta x/2 \le x \le x_i + \Delta x/2$,

---

[6] Denoting the weights and abscissas as "volume averages" is technically incorrect because only the moments are volume-averaged. From the definition of the volume-average moments, we can write for each moment

$$(m_k)_i^m = \sum_{\alpha=1}^{N} \frac{1}{\Delta x} \int_{x_i-\Delta x/2}^{x_i+\Delta x/2} w_\alpha(m\,\Delta t, x)[\xi_\alpha(m\,\Delta t, x)]^k \, \mathrm{d}x.$$

Owing to the nonlinear dependence on the abscissas, the right-hand side of this expression cannot, in general, be replaced by the true volume-average weights and abscissas:

$$\frac{1}{\Delta x} \int_{x_i-\Delta x/2}^{x_i+\Delta x/2} w_\alpha(m\,\Delta t, x)[\xi_\alpha(m\,\Delta t, x)]^k \, \mathrm{d}x \ne \frac{1}{\Delta x} \int_{x_i-\Delta x/2}^{x_i+\Delta x/2} w_\alpha(m\,\Delta t, x)\mathrm{d}x \left[\frac{1}{\Delta x} \int_{x_i-\Delta x/2}^{x_i+\Delta x/2} \xi_\alpha(m\,\Delta t, x)\mathrm{d}x \right]^k.$$

However, for the special reconstruction for which $\xi_\alpha(m\,\Delta t, x) = \xi_{i,\alpha}^m$ is constant in each grid cell, it is correct to refer to the weights and abscissas as the volume-average values. Nevertheless, the nonlinear dependence between the moments and the abscissas makes it challenging to use high-order spatial reconstructions for the abscissas, while preserving the values of the volume-average moments.

where $w_{i,\alpha}^m$ and $\xi_{i,\alpha}^m$ are found from the volume-average moment set $\mathbf{M}_i^m$. The reconstructed moments in cell i are thus

$$m_k(m\,\Delta t, x) = \sum_{\alpha=1}^{N} \left[w_{i,\alpha}^m + a_{i,\alpha}(x - x_i)\right] \left(\xi_{i,\alpha}^m\right)^k \quad \text{for } x_i - \Delta x/2 \leq x \leq x_i + \Delta x/2. \tag{8.34}$$

Note that, because the abscissas are constant in the interval $x_i - \Delta x/2 \leq x \leq x_i + \Delta x/2$, the reconstructed moment sets automatically satisfy Eq. (8.21) for any values of $a_{i,\alpha}$. In practice, the $a_{i,\alpha}$ are estimated using the values of $w_{i,\alpha}^m$ in neighboring cells with a flux limiter (Vikas *et al.*, 2011a). The extension of the formulas in Eq. (8.33) to 2D and 3D spatial domains, and to unstructured grids, is straightforward, as described in Vikas *et al.* (2011a, 2012).

In order to improve the time accuracy, a second-order, two-stage, strong stability-preserving Runge–Kutta (RK2SSP) scheme can be used (Vikas *et al.*, 2011a). The updated moment sets are found from

$$\mathbf{M}_i^{m+1} = \frac{1}{2}(\mathbf{M}_i^m + \mathbf{M}_i^*), \tag{8.35}$$

where the first-stage moment sets are defined by

$$\mathbf{M}_i^{(1)} = \sum_{\alpha=1}^{N} \left(w_{i,\alpha}^m - \lambda u_{i+1/2,l}^{m+} w_{i+1/2,\alpha,l}^m + \lambda u_{i-1/2,r}^{m-} w_{i-1/2,\alpha,r}^m\right) \mathbf{K}_{i,\alpha}^m - \lambda u_{i+1/2,r}^{m-} \mathbf{M}_{i+1}^m + \lambda u_{i-1/2,l}^{m+} \mathbf{M}_{i-1}^m, \tag{8.36}$$

and the second-stage moment sets by

$$\mathbf{M}_i^* = \sum_{\alpha=1}^{N} \left(w_{i,\alpha}^{(1)} - \lambda u_{i+1/2,l}^{m+1+} w_{i+1/2,\alpha,l}^{(1)} + \lambda u_{i-1/2,r}^{m+1-} w_{i-1/2,\alpha,r}^{(1)}\right) \mathbf{K}_{i,\alpha}^{(1)} - \lambda u_{i+1/2,r}^{m+1-} \mathbf{M}_{i+1}^{(1)} + \lambda u_{i-1/2,l}^{m+1+} \mathbf{M}_{i-1}^{(1)}. \tag{8.37}$$

The weights and abscissas appearing in the second stage are computed from $\mathbf{M}_i^{(1)}$. The reader can note that the formulas used in each stage are exactly the same as Eq. (8.33), and hence are realizable. Surprisingly, the standard Runge–Kutta schemes are not realizable (Vikas *et al.*, 2011a), and therefore care must be taken when devising higher-order time-stepping schemes.

The extension of Eq. (8.33) (or Eq. (8.35)) to a multivariate PBE $n(t, x, \boldsymbol{\xi})$ is fairly straightforward (see Section B.3 of Appendix B). First, a multivariate quadrature is needed:

$$n(t, x, \boldsymbol{\xi}) = \sum_{\alpha=1}^{N} w_\alpha(t, x)\delta(\boldsymbol{\xi} - \boldsymbol{\xi}_\alpha(t, x)), \tag{8.38}$$

and the volume-average moment set $\mathbf{M}_i^m$ now contains the multivariate moments needed to find $w_{i,\alpha}^m$ and $\boldsymbol{\xi}_{i,\alpha}^m$. As before, high-order spatial reconstruction is employed only for the weights, so that Eq. (8.33) remains the same except for the definition of $\mathbf{K}_{i,\alpha}^m$, which now includes all of the mixed moments used to construct the multivariate quadrature. As described in Chapter 3, a convenient choice for the transported moment set is the optimal moment set. If the CQMOM is used to construct the quadrature, the weights and abscissas will depend on the order of the variables used in conditioning (i.e. on the permutation). In order to use high-order spatial reconstruction for the weights, the same permutation must be used in all grid cells (see Yuan & Fox (2011) and Section B.3 of Appendix B for details). Moreover, each permutation of the CQMOM, when applied in Eq. (8.33), will yield a different updated moment set $(\mathbf{M}_i^{m+1})_p$, where $p \in 1, \ldots, P$, and P is the number of permutations (e.g. $P = 2$ for a 2D phase space, $P = 6$ for a 3D phase space). The final updated moment set is then defined by

$$\mathbf{M}_i^{m+1} = \frac{1}{P}\sum_{p=1}^{P}(\mathbf{M}_i^{m+1})_p. \tag{8.39}$$

The reader seeking more details is referred to Yuan & Fox (2011) and Section B.3 of Appendix B.

8.2.2 Application to KE

We again consider only the 1D KE, and denote the univariate NDF by $n(t, x, v)$, where v is the velocity. The KE with only the free-transport term is

$$\frac{\partial n}{\partial t} + \frac{\partial v n}{\partial x} = 0, \tag{8.40}$$

and the moment-transport equation is

$$\frac{\partial m_k}{\partial t} + \frac{\partial m_{k+1}}{\partial x} = 0. \tag{8.41}$$

Denoting the moment fluxes as $\mathbf{F}(n)$, where n is the NDF reconstructed from the transported moment set $\mathbf{M}$, the moment-transport equations can be written as

$$\frac{\partial \mathbf{M}}{\partial t} + \frac{\partial \mathbf{F}(n)}{\partial x} = 0. \tag{8.42}$$

For convenience, the moment fluxes are written in a split form, namely $\mathbf{F}(n) = \mathbf{Q}^-(n) + \mathbf{Q}^+(n)$, where

$$\mathbf{Q}^-(n) \equiv \int_{-\infty}^{0} v\mathbf{K}n \, \mathrm{d}v, \quad \mathbf{Q}^+(n) \equiv \int_{0}^{\infty} v\mathbf{K}n \, \mathrm{d}v, \tag{8.43}$$

and $\mathbf{K} = [1 \ v \ v^2 \ \cdots \ v^{2N-1}]^{\mathrm{T}}$. Finally, in order to accommodate arbitrary reconstructions of the NDF (e.g. Gaussian EQMOM), we define two new half-moment sets:

$$\mathbf{M}^- \equiv \int_{-\infty}^{0} \mathbf{K}n \, \mathrm{d}v, \quad \mathbf{M}^+ \equiv \int_{0}^{\infty} \mathbf{K}n \, \mathrm{d}v, \tag{8.44}$$

where, by definition, $\mathbf{M} = \mathbf{M}^- + \mathbf{M}^+$.[7] The half-moment sets can be used with QBMM to construct two sets of weights and abscissas: $\mathbf{M}^- \Rightarrow \{w_\alpha^-, v_\alpha^-\}$ and $\mathbf{M}^+ \Rightarrow \{w_\alpha^+, v_\alpha^+\}$ for $\alpha = 1, \ldots, M$, in which $v_\alpha^- \le 0$ and $0 \le v_\alpha^+$.[8] Note that in these quadratures M can take on any positive value and, hence, if n is reconstructed as N Dirac delta functions, then an adaptive quadrature algorithm should be used to set the unused weights to zero (see Yuan & Fox (2011) for details). Starting from Eq. (8.42), we can develop a realizable, high-order scheme to update the transported moment set at time $t + \Delta t$ given a realizable moment set at time t.

If a single-stage Euler explicit time-integration scheme is used, the updated moment set can be written as

$$\mathbf{M}_i^{m+1} = \mathbf{M}_i^{m-} + \mathbf{M}_i^{m+} - \lambda \left[\mathbf{G}\left(n_{i+1/2,l}^m, n_{i+1/2,r}^m\right) - \mathbf{G}\left(n_{i-1/2,l}^m, n_{i-1/2,r}^m\right)\right], \tag{8.45}$$

[7] In general, the transported moment set $\mathbf{M}$ could contain fewer moments than the half-moment sets $\mathbf{M}^-$ and $\mathbf{M}^+$. In practice, the lengths of $\mathbf{M}^-$ and $\mathbf{M}^+$ are chosen such that the Gaussian quadratures constructed from the half-moment sets include the highest-order moment appearing in the transport equation for $\mathbf{M}$, e.g. m_{2N}. This implies that $M = N + 1$.

[8] In Section 8.4, an alternative definition using the EQMOM directly with the moments is employed to determine the weights and abscissas. In both formulations, it is always possible to split the set of abscissas into the positive and negative velocities.

where the numerical flux function is defined by $\mathbf{G}(n_l, n_r) = \mathbf{Q}^-(n_r) + \mathbf{Q}^+(n_l)$. Then, using the definitions in Eq. (8.43), we can write

$$\mathbf{Q}^-(n_r) = \sum_{\alpha=1}^{M} w_{\alpha,r}^- v_{\alpha,r}^- \mathbf{K}_{\alpha,r}^-, \quad \mathbf{Q}^+(n_l) = \sum_{\alpha=1}^{M} w_{\alpha,l}^+ v_{\alpha,l}^+ \mathbf{K}_{\alpha,l}^+, \tag{8.46}$$

where $\mathbf{K}_{\alpha,r}^-$ ($\mathbf{K}_{\alpha,l}^+$) is defined the same as $\mathbf{K}$ but evaluated using the abscissa $v_{\alpha,r}^-$ ($v_{\alpha,l}^+$). Similarly, we have[9]

$$\mathbf{M}^- = \sum_{\alpha=1}^{M} w_\alpha^- \mathbf{K}_\alpha^-, \quad \mathbf{M}^+ = \sum_{\alpha=1}^{M} w_\alpha^+ \mathbf{K}_\alpha^+, \tag{8.47}$$

and thus Eq. (8.45) can be rewritten as

$$\begin{aligned} \mathbf{M}_i^{m+1} = & \sum_{\alpha=1}^{M} \left(w_{\alpha,i}^{m-} \mathbf{K}_{\alpha,i}^{m-} + \lambda w_{\alpha,i-1/2,r}^{m-} v_{\alpha,i-1/2,r}^{m-} \mathbf{K}_{\alpha,i-1/2,r}^{m-} - \lambda w_{\alpha,i+1/2,r}^{m-} v_{\alpha,i+1/2,r}^{m-} \mathbf{K}_{\alpha,i+1/2,r}^{m-} \right) \\ & + \sum_{\alpha=1}^{M} \left(w_{\alpha,i}^{m+} \mathbf{K}_{\alpha,i}^{m+} - \lambda w_{\alpha,i+1/2,l}^{m+} v_{\alpha,i+1/2,l}^{m+} \mathbf{K}_{\alpha,i+1/2,l}^{m+} + \lambda w_{\alpha,i-1/2,l}^{m+} v_{\alpha,i-1/2,l}^{m+} \mathbf{K}_{\alpha,i-1/2,l}^{m+} \right). \end{aligned} \tag{8.48}$$

The summations on the right-hand side of Eq. (8.48) can be written in terms of the following two numerical NDF:

$$\begin{aligned} h_i^-(v) &= n_i^m(v) - \lambda|v| n_{i-1/2,r}^m(v) + \lambda|v| n_{i+1/2,r}^m(v) \qquad \text{for } v \le 0, \\ h_i^+(v) &= n_i^m(v) - \lambda v n_{i+1/2,l}^m(v) + \lambda v n_{i-1/2,l}^m(v) \qquad \text{for } 0 \le v. \end{aligned} \tag{8.49}$$

According to the realizability condition, these NDF must both be nonnegative for all values of v.

As before, the numerical NDF in Eq. (8.49) will be nonnegative only if we use QBMM and impose the following realizability conditions on the abscissas:

$$v_{\alpha,i+1/2,l}^{m-} = v_{\alpha,i-1/2,r}^{m-} = v_{\alpha,i}^{m-}, \quad v_{\alpha,i+1/2,l}^{m+} = v_{\alpha,i-1/2,r}^{m+} = v_{\alpha,i}^{m+}. \tag{8.50}$$

We then have

$$\begin{aligned} h_i^-(v) &= \sum_{\alpha=1}^{M} \left[\left(w_{\alpha,i}^{m-} - \lambda|v| w_{\alpha,i-1/2,r}^{m-} \right) \delta\left(v - v_{\alpha,i}^{m-}\right) + \lambda|v| w_{\alpha,i+1/2,r}^{m-} \delta\left(v - v_{\alpha,i+1}^{m-}\right) \right], \\ h_i^+(v) &= \sum_{\alpha=1}^{M} \left[\left(w_{\alpha,i}^{m+} - \lambda v w_{\alpha,i+1/2,l}^{m+} \right) \delta\left(v - v_{\alpha,i}^{m+}\right) + \lambda v w_{\alpha,i-1/2,l}^{m} \delta\left(v - v_{\alpha,i-1}^{m+}\right) \right], \end{aligned} \tag{8.51}$$

and thus the realizability condition on Δt is

$$\left. \begin{aligned} w_{\alpha,i}^{m-} - \lambda |v_{\alpha,i}^{m-}| w_{\alpha,i-1/2,r}^{m-} \ge 0 \\ w_{\alpha,i}^{m+} - \lambda v_{\alpha,i}^{m+} w_{\alpha,i+1/2,l}^{m+} \ge 0 \end{aligned} \right\} \Rightarrow \Delta t \le \min_{\alpha,i} \left(\frac{w_{\alpha,i}^{m-} \Delta x}{w_{\alpha,i-1/2,r}^{m-} |v_{\alpha,i}^{m-}|}, \frac{w_{\alpha,i}^{m+} \Delta x}{w_{\alpha,i+1/2,l}^{m+} v_{\alpha,i}^{m+}} \right). \tag{8.52}$$

This result generalizes Theorem 1 in Vikas *et al.* (2011a) to an arbitrary reconstruction of the NDF used in the definition of $\mathbf{M}^-$ and $\mathbf{M}^+$. Note that, if a first-order spatial reconstruction is used for the weights so that $w_{\alpha,i}^{m+} = w_{\alpha,i+1/2,l}^{m+}$ and $w_{\alpha,i}^{m-} = w_{\alpha,i-1/2,r}^{m-}$, then Eq. (8.52)

[9] These equations are identities because $M \ge N$, and thus the half-moment sets up to order $2M - 1$ are exact.

reduces to a CFL condition that is based on the largest magnitude of the velocity abscissas. A subtle, but very important, point introduced in the discussion above is that, even though the reconstructed NDF may be continuous, we still use quadrature with the half-moment sets $\mathbf{M}^-$ and $\mathbf{M}^+$ to compute the weights and abscissas appearing in Eq. (8.48). This step ensures that the realizability condition in Eq. (8.52) is well defined for a finite value of Δt.

In summary, the overall time-stepping procedure starting from the transported moment set $\mathbf{M}^m$ is as follows:

$$\mathbf{M}^m \overset{\text{(i)}}{\Longrightarrow} n(v) \overset{\text{(ii)}}{\Longrightarrow} \mathbf{M}^-, \mathbf{M}^+ \overset{\text{(iii)}}{\Longrightarrow} \{w_\alpha^-, v_\alpha^-\}, \{w_\alpha^+, v_\alpha^+\} \overset{\text{(iv)}}{\Longrightarrow} \mathbf{M}^{m+1}.$$

(i) Use the transported moment set to reconstruct the NDF (e.g. Gaussian EQMOM).

(ii) Use the reconstructed NDF to compute the half-moment sets analytically.

(iii) Use half-moment sets to construct two Gaussian quadratures with M nodes.

(iv) Advance the transported moment set using weights and abscissas computed from half-moment sets.

The updated moment set in step (iv) is found from[10]

$$\begin{aligned}\mathbf{M}_i^{m+1} = &\sum_{\alpha=1}^{M} \left[\left(w_{\alpha,i}^{m-} + \lambda w_{\alpha,i-1/2,r}^{m-} v_{\alpha,i}^{m-} \right) \mathbf{K}_{\alpha,i}^{m-} - \lambda w_{\alpha,i+1/2,r}^{m-} v_{\alpha,i+1}^{m-} \mathbf{K}_{\alpha,i+1}^{m-} \right] \\ &+ \sum_{\alpha=1}^{M} \left[\left(w_{\alpha,i}^{m+} - \lambda w_{\alpha,i+1/2,l}^{m+} v_{\alpha,i}^{m+} \right) \mathbf{K}_{\alpha,i}^{m+} + \lambda w_{\alpha,i-1/2,l}^{m+} v_{\alpha,i-1}^{m+} \mathbf{K}_{\alpha,i-1}^{m+} \right].\end{aligned} \tag{8.53}$$

If the realizability condition in Eq. (8.52) is satisfied, $\mathbf{M}_i^{m+1}$ found from Eq. (8.53) is guaranteed to be realizable. Using the method described above for the PBE (Eq. (8.35)), it is straightforward to extend Eq. (8.53) to second-order time-stepping. The extension to multiple velocity components $\mathbf{v}$ in multiple spatial dimensions is a bit more complicated. As described in Yuan & Fox (2011) and in Section B.3 of Appendix B, when the CQMOM is used to construct the multivariate quadratures all permutations must be used in a consistent manner in order to get the correct kinetic energy fluxes. Nevertheless, for each permutation of the CQMOM, the basic time-stepping formula in Eq. (8.53) is used to update the transported moment set by modifying the definition of $\mathbf{K}_{\alpha,i}^m$ to include the multivariate moments in the optimal-moment set. Readers interested in more details on multidimensional free transport should consult Yuan & Fox (2011) and Section B.3 of Appendix B.

8.2.3 Application to GPBE

The application of KBFVM to GPBE is very similar to the KE case discussed above, but now the NDF contains either an active scalar $n(t, x, \xi)$ with known (or computed) advection

[10] As discussed in Yuan & Fox (2011) and Section B.3 of Appendix B, in the multivariate case Eq. (8.53) is not the same as

$$\mathbf{M}_i^{m+1} = \mathbf{M}_i^m - \lambda \sum_{\alpha=1}^{M} \left(w_{\alpha,i+1/2,l}^{m+} v_{\alpha,i}^{m+} \mathbf{K}_{\alpha,i}^{m+} + w_{\alpha,i+1/2,r}^{m-} v_{\alpha,i+1}^{m-} \mathbf{K}_{\alpha,i+1}^{m-} - w_{\alpha,i-1/2,l}^{m+} v_{\alpha,i-1}^{m+} \mathbf{K}_{\alpha,i-1}^{m+} - w_{\alpha,i-1/2,r}^{m-} v_{\alpha,i}^{m-} \mathbf{K}_{\alpha,i}^{m-} \right)$$

because the CQMOM does not control all of the optimal moments appearing in $\mathbf{M}_i^m$. Or, in other words, only the form given in Eq. (8.53) is guaranteed to be realizable. For this reason, it is preferable to leave the right-hand side of Eq. (8.53) in terms of the weights and abscissas even for the univariate case.

velocity $u(t, x, \xi)$, or it is bivariate with velocity v and a passive scalar ξ. We start with the case of an active scalar, for which the GPBE has the form of Eq. (8.18). The single-step Euler explicit time-integration scheme is the same as Eq. (8.22), but the numerical flux function is now

$$\mathbf{G}(u_l, \mathbf{M}_l, u_r, \mathbf{M}_r) = \int u_l^+ \mathbf{K} n_l \, \mathrm{d}\xi + \int u_r^- \mathbf{K} n_r \, \mathrm{d}\xi \tag{8.54}$$

with $u_l^+(\xi) = \frac{1}{2}(u_l(\xi) + |u_l(\xi)|)$ and $u_r^-(\xi) = \frac{1}{2}(u_r(\xi) - |u_r(\xi)|)$. Note that the advection velocity is defined separately for each value of ξ, but otherwise the notation is the same as in Eq. (8.23). Using high-order spatial reconstruction for the weights and for the x-dependence of the advection velocity $u(m\,\Delta t, x, \xi)$, the updated moment set is

$$\begin{aligned}\mathbf{M}_i^{m+1} = & \sum_{\alpha=1}^{N} \left(w_{i,\alpha}^m - \lambda u_{i+1/2,l}^{m+}(\xi_{i,\alpha}^m) w_{i+1/2,\alpha,l}^m + \lambda u_{i-1/2,r}^{m-}(\xi_{i,\alpha}^m) w_{i-1/2,\alpha,r}^m \right) \mathbf{K}_{i,\alpha}^m \\ & - \lambda u_{i+1/2,r}^{m-}(\xi_{i+1,\alpha}^m) \mathbf{M}_{i+1}^m + \lambda u_{i-1/2,l}^{m+}(\xi_{i-1,\alpha}^m) \mathbf{M}_{i-1}^m,\end{aligned} \tag{8.55}$$

where $\mathbf{K}_{i,\alpha}^m$ is defined as in Eq. (8.33). The reader will note that this expression for $\mathbf{M}_i^{m+1}$ is very similar to the one for free transport in Eq. (8.53), but with the velocities $u_{i+1/2,l}^{m+}(\xi_{i,\alpha}^m)$ and $u_{i-1/2,r}^{m-}(\xi_{i,\alpha}^m)$ in place of the velocity abscissas. The realizability condition for Eq. (8.55) is then

$$\Delta t \leq \min_{i,\alpha} \left(\frac{w_{i,\alpha}^m \, \Delta x}{w_{i+1/2,\alpha,l}^m u_{i+1/2,l}^{m+}(\xi_{i,\alpha}^m) - w_{i-1/2,\alpha,r}^m u_{i-1/2,r}^{m-}(\xi_{i,\alpha}^m)} \right), \tag{8.56}$$

which is analogous to Eq. (8.32) except that now the advection velocity depends on the abscissas.

We next consider the case of free transport in a 1D domain with $n(t, x, v, s)$, where s is a passive internal coordinate (e.g. surface area). More general cases with the GPBE are treated in Appendix B. The moment-transport equation with free transport is given in Eq. (8.9). The transported bivariate moment set will again be denoted by $\mathbf{M}$ and the compact moment-transport equation by Eq. (8.42) with the moment fluxes as in Eq. (8.43), but with n denoting the reconstructed bivariate NDF. The split form of the moment fluxes is again given by Eq. (8.43) but the vector $\mathbf{K}$ must now be defined with the bivariate internal coordinates: $\mathbf{K} = [1 \; v \; s \; v^2 \; vs \; s^2 \; \cdots \; v^k s^l \; \cdots]^{\mathrm{T}}$, where the length of $\mathbf{K}$ is determined by the optimal-moment set. The half-moment sets are defined using the velocity:

$$\mathbf{M}^- \equiv \int_{-\infty}^{0} \int_{0}^{\infty} \mathbf{K} n \, \mathrm{d}s \, \mathrm{d}v, \quad \mathbf{M}^+ \equiv \int_{0}^{\infty} \int_{0}^{\infty} \mathbf{K} n \, \mathrm{d}s \, \mathrm{d}v, \tag{8.57}$$

where, by definition, $\mathbf{M} = \mathbf{M}^- + \mathbf{M}^+$. As with the KE, the half-moment sets can be used with QBMM to construct two sets of weights and abscissas:[11]

$$\mathbf{M}^- \Rightarrow \{w_\alpha^-, v_\alpha^-, s_\alpha^-\}, \quad \mathbf{M}^+ \Rightarrow \{w_\alpha^+, v_\alpha^+, s_\alpha^+\} \quad \text{for } \alpha = 1, \dots, M; \tag{8.58}$$

from which the velocity abscissas $v_\alpha^- \leq 0$ and $0 \leq v_\alpha^+$ are used to define the spatial fluxes. The single-stage Euler explicit time-integration scheme proceeds exactly as in Eq. (8.45) and results in Eq. (8.48). However, in this case, the two bivariate numerical NDF are

$$\begin{aligned} h_i^-(v, s) &= n_i^m(v, s) - \lambda |v| n_{i-1/2,r}^m(v, s) + \lambda |v| n_{i+1/2,r}^m(v, s) && \text{for } v \leq 0, \\ h_i^+(v, s) &= n_i^m(v, s) - \lambda v n_{i+1/2,l}^m(v, s) + \lambda v n_{i-1/2,l}^m(v, s) && \text{for } 0 \leq v, \end{aligned} \tag{8.59}$$

[11] As was done in Section 8.4, it is also possible to get the weights and abscissas directly from the EQMOM quadrature.

and they must both be nonnegative for all values of v and s. Using the QBMM reconstruction of the NDF, and imposing the realizability condition in Eq. (8.50) along with the realizability conditions $s^{m-}_{\alpha,i+1/2,l} = s^{m-}_{\alpha,i-1/2,r} = s^{m-}_{\alpha,i}$ and $s^{m+}_{\alpha,i+1/2,l} = s^{m+}_{\alpha,i-1/2,r} = s^{m+}_{\alpha,i}$, we find

$$\begin{aligned} h^-_i(v,s) = \sum_{\alpha=1}^{M} \Big[&\left(w^{m-}_{\alpha,i} - \lambda|v|w^{m-}_{\alpha,i-1/2,r}\right)\delta\left(v - v^{m-}_{\alpha,i}\right)\delta\left(s - s^{m-}_{\alpha,i}\right) \\ &+ \lambda|v|w^{m-}_{\alpha,i+1/2,r}\delta\left(v - v^{m-}_{\alpha,i+1}\right)\delta\left(s - s^{m-}_{\alpha,i+1}\right)\Big], \\ h^+_i(v,s) = \sum_{\alpha=1}^{M} \Big[&\left(w^{m+}_{\alpha,i} - \lambda v w^{m+}_{\alpha,i+1/2,l}\right)\delta\left(v - v^{m+}_{\alpha,i}\right)\delta\left(s - s^{m+}_{\alpha,i}\right) \\ &+ \lambda v w^{m}_{\alpha,i-1/2,l}\delta\left(v - v^{m+}_{\alpha,i-1}\right)\delta\left(s - s^{m+}_{\alpha,i-1}\right)\Big]. \end{aligned} \tag{8.60}$$

The realizability condition on Δt is thus the same as in Eq. (8.52). In summary, the four-stage time-stepping procedure starting from the transported moment set $\mathbf{M}^m$ is as follows:

$$\mathbf{M}^m \overset{\text{(i)}}{\Longrightarrow} n(v,s) \overset{\text{(ii)}}{\Longrightarrow} \mathbf{M}^-, \mathbf{M}^+ \overset{\text{(iii)}}{\Longrightarrow} \{w^-_\alpha, v^-_\alpha, s^-_\alpha\}, \{w^+_\alpha, v^+_\alpha, s^+_\alpha\} \overset{\text{(iv)}}{\Longrightarrow} \mathbf{M}^{m+1},$$

where the updated moment set $\mathbf{M}^{m+1}$ is found from Eq. (8.53) using the definition of $\mathbf{K}$ given above. The comments on the extension to higher-order time stepping and multivariate active and/or passive internal coordinates appearing after Eq. (8.53) also apply for the GPBE.

8.3 Inhomogeneous PBE

In this section, we consider the advection and diffusion of a univariate NDF $n(t,x,\xi)$, where $\xi \in \mathbb{R}^+$ is a passive scalar (i.e. the velocity $u(t,x)$ does not depend on ξ). For example, ξ could denote the particle mass for fine particles in a liquid with very small Stokes number. The diffusion coefficient, $\Gamma(\xi)$, is assumed to be a function of ξ. For example, using the Stokes–Einstein formula for diffusivity reported in Eq. (5.116) on page 187, $\Gamma(\xi) = \Gamma_0/\xi$, where ξ is the particle mass. The PBE for this case reads

$$\frac{\partial n}{\partial t} + \frac{\partial u n}{\partial x} = \frac{\partial}{\partial x}\left(\Gamma(\xi)\frac{\partial n}{\partial x}\right), \tag{8.61}$$

where the velocity u and the initial NDF $n(0,x,\xi)$ are known. For the examples considered in this section, we do not add terms in Eq. (8.61) for phase-space processes such as growth and aggregation. When operator splitting is employed, such terms are treated separately from the spatial-transport terms, as described in Chapter 7. Nevertheless, in order to have realistic examples, we will assume that the initial NDF is sufficiently complex to be representative of a system with aggregation and breakage.

8.3.1 Moment-transport equations

On defining the integer moments of the NDF as $m_k(t,x) = \int \xi^k n(t,x,\xi)\mathrm{d}\xi$, the moment-transport equation found from Eq. (8.61) is

$$\frac{\partial m_k}{\partial t} + \frac{\partial u m_k}{\partial x} = \frac{\partial}{\partial x}\left(\int_0^\infty \xi^k \Gamma(\xi)\frac{\partial n}{\partial x}\,\mathrm{d}\xi\right), \tag{8.62}$$

which must be solved with initial conditions $m_k(0,x) = \int \xi^k n(0,x,\xi)\mathrm{d}\xi$. The reader can note that, because of the ξ-dependent diffusion coefficient, Eq. (8.62) is not closed. If $\Gamma(\xi)$ does not depend on t or x (e.g. due to spatial gradients of temperature), we can rewrite the diffusion term in Eq. (8.62) using

$$\int_0^\infty \xi^k \Gamma(\xi) \frac{\partial n}{\partial x} \,\mathrm{d}\xi = \frac{\partial}{\partial x}\left(\int_0^\infty \xi^k \Gamma(\xi) n \,\mathrm{d}\xi \right).$$

Moreover, employing the Stokes–Einstein formula, we can write $\int \xi^k \Gamma(\xi) n \,\mathrm{d}\xi = \Gamma_0 m_{k-1}$, which is essentially the form[12] that will be used in the numerical examples in Section 8.3.4. In summary, we will consider two variations of the moment-transport equations in the numerical examples in Section 8.3.4. The first example will use a closed moment system wherein the diffusivity does not depend on ξ:

$$\frac{\partial m_k}{\partial t} + \frac{\partial u m_k}{\partial x} = \frac{\partial}{\partial x}\left(\Gamma \frac{\partial m_k}{\partial x} \right), \tag{8.63}$$

and u and Γ are known constants. The second example will consider the unclosed moment system

$$\frac{\partial m_k}{\partial t} + \frac{\partial u m_k}{\partial x} = \frac{\partial}{\partial x}\left(\Gamma_0 \frac{\partial m_{k-1}}{\partial x} \right), \tag{8.64}$$

where u and Γ_0 are known constants. Because it is closed, Eq. (8.63) can be solved using standard FVM. Thus, in Section 8.3.2, we use Eq. (8.63) to illustrate the fact that high-order schemes can generate unrealizable moments due both to the advection and to the diffusion terms. In Section 8.3.3, we implement realizable high-order schemes both for Eq. (8.63) and for Eq. (8.64) using gamma EQMOM to reconstruct the NDF from the moments.

8.3.2 Standard finite-volume schemes for moments

In this section, we develop a high-order scheme to solve Eq. (8.63) on a uniform mesh with spacing Δx. Denoting the mesh points by x_i, the volume-average moments are

$$m_{k,i}(t) = \frac{1}{\Delta x} \int_{x_{i-1/2}}^{x_{i+1/2}} m_k(t,x)\mathrm{d}x,$$

where $x_{i-1/2}$ and $x_{i+1/2}$ denote the locations of the left and right interfaces, respectively, of the grid cell centered at x_i. Volume-averaging of Eq. (8.63) leads to

$$\frac{\mathrm{d}m_{k,i}}{\mathrm{d}t} = -\frac{1}{\Delta x}\left[u m_{k,i+1/2}(t) - u m_{k,i-1/2}(t) - \Gamma \left.\frac{\partial m_k}{\partial x}\right|_{i+1/2} + \Gamma \left.\frac{\partial m_k}{\partial x}\right|_{i-1/2} \right], \tag{8.65}$$

where $m_{k,i\pm1/2}(t)$ are the reconstructed moments at the interfaces, and $\Gamma\, \partial m_k/\partial x|_{i+1/2}$ and $\Gamma\, \partial m_k/\partial x|_{i-1/2}$ are the reconstructed gradients. In standard schemes, the reconstructed values are found from the volume-average moments in neighboring grid cells. Below we consider schemes with first-order, second-order, and fourth-order accuracy.

[12] Since the Stokes–Einstein formula is valid for finite-size particles, the NDF should be nonzero only for $\xi_0 \le \xi < \infty$. For convenience, we will allow $0 \le \xi < \infty$ and add ξ_0 to the denominator of the Stokes–Einstein formula in order to ensure that the diffusivity remains finite with $\xi = 0$.

The first-order scheme

Using a first-order upwind reconstruction for the advection term in Eq. (8.65), we have

$$m_{k,i+1/2}(t) = c^+ m_{k,i}(t) + c^- m_{k,i+1}(t) = \begin{cases} m_{k,i}(t) & \text{if } u > 0, \\ m_{k,i+1}(t) & \text{if } u < 0, \end{cases}$$

where

$$c^- = \frac{1}{2}\left(1 - \frac{u}{|u|}\right) \quad \text{and} \quad c^+ = \frac{1}{2}\left(1 + \frac{u}{|u|}\right). \tag{8.66}$$

Using central differences, the gradient terms are approximated by

$$\left.\frac{\partial m_k}{\partial x}\right|_{i+1/2} = \frac{m_{k,i+1}(t) - m_{k,i}(t)}{\Delta x}.$$

Implementing these approximations in Eq. (8.65) yields

$$\frac{\mathrm{d}m_{k,i}}{\mathrm{d}t} = \frac{|u|}{\Delta x}\left[c^- m_{k,i+1}(t) - m_{k,i}(t) + c^+ m_{k,i-1}(t)\right] + \frac{\Gamma}{\Delta x^2}\left[m_{k,i+1}(t) - 2m_{k,i}(t) + m_{k,i-1}(t)\right]. \tag{8.67}$$

This system of coupled equations must be solved by integrating forward in time starting from the initial conditions $m_{k,i}(0)$. For stability, the time integration of the diffusion term can be treated implicitly using, for example, the Crank–Nicolson (CN) scheme. If we denote the volume-average moments at time $t = n\,\Delta t$ by $m^n_{k,i}$, a semi-implicit scheme for solving Eq. (8.67) is

$$\begin{aligned} m^{n+1}_{k,i} - \alpha\lambda_1\left[m^{n+1}_{k,i+1} - 2m^{n+1}_{k,i} + m^{n+1}_{k,i-1}\right] &= m^n_{k,i} + (1-\alpha)\lambda_1\left[m^n_{k,i+1} - 2m^n_{k,i} + m^n_{k,i-1}\right] \\ &\quad + \lambda_2\left[c^- m^n_{k,i+1} - m^n_{k,i} + c^+ m^n_{k,i-1}\right], \end{aligned} \tag{8.68}$$

where $\lambda_1 = \Gamma\,\Delta t/(\Delta x^2)$ and $\lambda_2 = |u|\Delta t/\Delta x$. The scheme in Eq. (8.68) with $\alpha = 1$ is fully implicit for diffusion, and with $\alpha = 1/2$ is the CN scheme. With $\alpha = 0$, the overall scheme is explicit Euler. The system in Eq. (8.68) can be rewritten as a tridiagonal linear system for $m^{n+1}_{k,i}$:

$$-\alpha\lambda_1 m^{n+1}_{k,i+1} + (1 + 2\alpha\lambda_1)m^{n+1}_{k,i} - \alpha\lambda_1 m^{n+1}_{k,i-1} = a_u m^n_{k,i+1} + a_d m^n_{k,i} + a_l m^n_{k,i-1}, \tag{8.69}$$

where the coefficients are defined by

$$\begin{aligned} a_u &= (1-\alpha)\lambda_1 + c^-\lambda_2, \\ a_d &= 1 - 2(1-\alpha)\lambda_1 - \lambda_2, \\ a_l &= (1-\alpha)\lambda_1 + c^+\lambda_2. \end{aligned} \tag{8.70}$$

Writing Eq. (8.69) as $\mathbf{A}\mathbf{m}^{n+1} = \mathbf{B}\mathbf{m}^n$, the formal solution for the updated moments is thus $\mathbf{m}^{n+1} = \mathbf{A}^{-1}\mathbf{B}\mathbf{m}^n$. Defining the matrix $\mathbf{P} = \mathbf{A}^{-1}\mathbf{B}$, the well-known stability condition for Eq. (8.69) is that all eigenvalues of $\mathbf{P}$ have magnitude less than unity. For moment methods, an equally important condition is realizability of the moment set. A *sufficient* condition[13] for the realizability of Eq. (8.69) is that all components of $\mathbf{P}$ be nonnegative. Because

[13] The moments at time $n + 1$ are weighted sums of the moments at time n, where the coefficients are the rows of $\mathbf{P}$. If all components are positive, then the moments are guaranteed to be realizable since moment space is convex.

the matrix $\mathbf{A}^{-1}$ has nonnegative components, it will suffice that the components of $\mathbf{B}$ are nonnegative. Thus, a *sufficient condition* for realizable moments is $a_d > 0$, which yields

$$\Delta t < \left(\frac{2(1-\alpha)\Gamma}{\Delta x^2} + \frac{|u|}{\Delta x}\right)^{-1}. \tag{8.71}$$

When the first-order scheme is semi-implicit with $0 < \alpha < 1$, this realizability condition is a mix of those found for pure advection and pure diffusion.

The second-order scheme

Using a second-order upwind reconstruction for the advection term in Eq. (8.65), we have

$$m_{k,i+1/2}(t) = c^+\left[m_{k,i}(t) + \frac{1}{2}S_{k,i}(t)\right] + c^-\left[m_{k,i+1}(t) - \frac{1}{2}S_{k,i+1}(t)\right].$$

The reconstructed slopes are defined by

$$S_{k,i}(t) = \text{minmod}[m_{k,i+1}(t) - m_{k,i}(t), m_{k,i}(t) - m_{k,i-1}(t)],$$

where the minmod slope limiter is

$$\text{minmod}(a,b) = \begin{cases} \text{sign}(a)\min(|a|,|b|) & \text{if } ab > 0, \\ 0 & \text{otherwise.} \end{cases}$$

In the diffusion term, the gradients are again approximated by

$$\left.\frac{\partial m_k}{\partial x}\right|_{i+1/2} = \frac{m_{k,i+1}(t) - m_{k,i}(t)}{\Delta x}$$

and

$$\left.\frac{\partial m_k}{\partial x}\right|_{i-1/2} = \frac{m_{k,i}(t) - m_{k,i-1}(t)}{\Delta x}.$$

Implementing these approximations in Eq. (8.65) yields

$$\begin{aligned}\frac{\mathrm{d}m_{k,i}}{\mathrm{d}t} &= \frac{|u|}{\Delta x}\left[c^- m_{k,i+1}(t) - m_{k,i}(t) + c^+ m_{k,i-1}(t)\right] + \frac{|u|}{2\Delta x}\left[c^- S_{k,i+1}(t) - S_{k,i}(t) + c^+ S_{k,i-1}(t)\right] \\ &\quad + \frac{\Gamma}{\Delta x^2}\left[m_{k,i+1}(t) - 2m_{k,i}(t) + m_{k,i-1}(t)\right].\end{aligned} \tag{8.72}$$

This system of coupled equations must be solved by integrating forward in time starting from the initial conditions $m_{k,i}(0)$. Using an explicit time-stepping algorithm, Eq. (8.72) leads to

$$\begin{aligned}m_{k,i}^{n+1} &= m_{k,i}^n + \lambda_1\left[m_{k,i+1}^n - 2m_{k,i}^n + m_{k,i-1}^n\right] \\ &\quad + \lambda_2\left[c^-\left(m_{k,i+1}^n - \frac{1}{2}S_{k,i+1}^n\right) - \left(m_{k,i}^n + \frac{1}{2}(c^+ - c^-)S_{k,i}^n\right) + c^+\left(m_{k,i-1}^n + \frac{1}{2}S_{k,i-1}^n\right)\right].\end{aligned} \tag{8.73}$$

In order to check for realizability, we can rearrange the terms as

$$\begin{aligned}m_{k,i}^{n+1} &= (1 - 2\lambda_1 - \lambda_2)m_{k,i}^n + (\lambda_1 + c^-\lambda_2)m_{k,i+1}^n + (\lambda_1 + c^+\lambda_2)m_{k,i-1}^n \\ &\quad - \frac{\lambda_2}{2}\left[c^+\left(S_{k,i}^n - S_{k,i-1}^n\right) + c^-\left(S_{k,i+1}^n - S_{k,i}^n\right)\right].\end{aligned}$$

The first term on the right-hand side will be nonnegative when Δt satisfies the realizability condition

$$\Delta t < \left(\frac{2\Gamma}{\Delta x^2} + \frac{|u|}{\Delta x} \right)^{-1},$$

and the next two terms are always nonnegative. However, the last term can be positive or negative, depending on the slopes, and thus we cannot guarantee that Eq. (8.73) will produce realizable moments. In fact, experience has shown that standard schemes of order higher than first order will almost always produce unrealizable moments at some point in the computational domain. In fact, even the diffusion term will yield unrealizable moments if a fourth-order scheme is employed.[14] Because unrealizable moments cannot be used in moment-inversion algorithms to find the weights and abscissas that are needed to evaluate source terms, standard schemes are poor choices for solving moment-transport equations.

8.3.3 Realizable finite-volume schemes for moments

In this section, we discuss realizable schemes for Eq. (8.61) that are based on the KBFVM presented in Section 8.2 and Appendix B. In order to implement these methods, we must reconstruct the NDF from the moments. For this purpose, we will employ gamma EQMOM (Eq. (3.84)) with $N = 2$ nodes, with which the reconstructed moments are given by

$$m_{k,i}^n = \sum_{\alpha=1}^{2} w_{\alpha,i}^n \sum_{\beta=1}^{M} w_{\alpha\beta,i}^n \left(\xi_{\alpha\beta,i}^n \right)^k, \tag{8.74}$$

where $\xi_{\alpha\beta,i}^n = \sigma_i^n t_{\alpha\beta,i}^n$. In this expression, for each α the M (normalized) weights $w_{\alpha\beta,i}^n$ and M abscissas $t_{\alpha\beta,i}^n$ correspond to the Gauss–Laguerre quadrature using the associated Laguerre polynomials $P^{(a)}(t)$ with $a = (\xi_{\alpha,i}^n/\sigma_i^n) - 1$. The two weights $w_{\alpha,i}^n$, two abscissas $\xi_{\alpha,i}^n$, and σ_i^n are found using the moment-inversion algorithm for gamma EQMOM: $\{m_{0,i}^n, m_{1,i}^n, m_{2,i}^n, m_{3,i}^n, m_{4,i}^n\} \Longrightarrow \{w_{1,i}^n, w_{2,i}^n, \xi_{1,i}^n, \xi_{2,i}^n, \sigma_i^n\}$. The Gauss–Laguerre quadrature is defined such that $\lim_{\sigma \to 0} \xi_{\alpha\beta} = \xi_\alpha$. Furthermore, the second summation in Eq. (8.74) corresponds to the moments of the gamma PDF,

$$m_k^{(\alpha)} = \sum_{\beta=1}^{M} w_{\alpha\beta} \left(\xi_{\alpha\beta} \right)^k,$$

and is exact for moments $m_k^{(\alpha)}$ up to order $2M - 1$. Thus, in order for Eq. (8.74) to be exact for moments up to $m_{4,i}^n$, we must have $M \geq 3$. For the examples below with constant diffusivity, we will set $M = 3$, whereas with size-dependent diffusivity we use $M = 20$. In summary, the realizable schemes will be written for the moment set $\{m_{0,i}^n, m_{1,i}^n, m_{2,i}^n, m_{3,i}^n, m_{4,i}^n\}$. At each time step n and in each grid cell i, we employ gamma EQMOM to find $(w_{1,i}^n, w_{2,i}^n, \xi_{1,i}^n, \xi_{2,i}^n, \sigma_i^n)$, from which we define the moment reconstruction in Eq. (8.74). The latter will be used to define the spatial fluxes and, in particular, to develop quasi-high-order flux reconstructions at the interfaces between grid cells. Finally, for convenience, we let $W_{\alpha\beta} = w_\alpha w_{\alpha\beta}$ in the following.

The first-order scheme

The first-order scheme for convection is exactly the same as the one described in Section 8.3.2. The only difference (which is not strictly necessary) is that we will replace the

[14] In a fourth-order diffusion scheme, the moments at $i + 2$ and $i - 2$ appear with negative coefficients.

moments by Eq. (8.74). The second-order scheme[15] for diffusion is derived in Section B.6 of Appendix B. Using an explicit time-stepping scheme and combining the schemes for advection and diffusion yields

$$\begin{aligned} m_{k,i}^{n+1} = & \sum_{\alpha=1}^{2}\sum_{\beta=1}^{M} W_{\alpha\beta,i}^{n}\left(1 - 2c_{\alpha\beta,i}^{n}\lambda_1 - \lambda_2\right)\left(\xi_{\alpha\beta,i}^{n}\right)^k + \sum_{\alpha=1}^{2}\sum_{\beta=1}^{M} W_{\alpha\beta,i+1}^{n}\left(c_{\alpha\beta,i+1}^{n}\lambda_1 + c^{-}\lambda_2\right)\left(\xi_{\alpha\beta,i+1}^{n}\right)^k \\ & + \sum_{\alpha=1}^{2}\sum_{\beta=1}^{M} W_{\alpha\beta,i-1}^{n}\left(c_{\alpha\beta,i-1}^{n}\lambda_1 + c^{+}\lambda_2\right)\left(\xi_{\alpha\beta,i-1}^{n}\right)^k, \end{aligned} \tag{8.75}$$

where

$$c_{\alpha\beta,i}^{n} = \begin{cases} 1 & \text{for constant diffusivity,} \\ \xi_0/(\xi_{\alpha\beta,i}^{n} + \xi_0) & \text{for Stokes–Einstein diffusivity.} \end{cases} \tag{8.76}$$

The constant ξ_0 in the Stokes–Einstein diffusivity is added to treat the limit of vanishing size so that the diffusivity coefficient remains finite.[16] The realizability condition for Eq. (8.75) is

$$1 - 2c_{\alpha\beta,i}^{n}\lambda_1 - \lambda_2 > 0 \quad \Longrightarrow \quad \Delta t < \min_{\alpha,\beta,i}\left[\left(\frac{2c_{\alpha\beta,i}^{n}\Gamma}{\Delta x^2} + \frac{|u|}{\Delta x}\right)^{-1}\right].$$

Note that it is not possible to write an implicit form for the diffusion term with Stokes–Einstein diffusivity because the weights and abscissas at time $n + 1$ depend in a highly nonlinear manner on $m_{k,i}^{n+1}$. Also, we should note that a more accurate representation of the Stokes–Einstein (or any ξ-dependent) diffusion coefficient can be attained by increasing M. The reader should keep in mind that increasing M does not change the number of moments that need to be solved for, and thus the computational cost is relatively insensitive to the value of M.

The quasi-second-order scheme

The quasi-second-order scheme for advection reconstructs the interface fluxes using constant abscissas and slope estimators with limiters for the weights. The reconstructed weights are then given by

$$\begin{aligned} W_{\alpha\beta,i+1/2,l}^{n} &= W_{\alpha\beta,i}^{n} + \frac{1}{2}S_{\alpha\beta,i}^{n}, \\ W_{\alpha\beta,i-1/2,r}^{n} &= W_{\alpha\beta,i}^{n} - \frac{1}{2}S_{\alpha\beta,i}^{n}. \end{aligned} \tag{8.77}$$

The reconstructed slopes are defined in terms of $W_{\alpha\beta}$:

$$S_{\alpha\beta,i}^{n} = \text{minmod}\left(W_{\alpha\beta,i+1}^{n} - W_{\alpha\beta,i}^{n}, W_{\alpha\beta,i}^{n} - W_{\alpha\beta,i-1}^{n}\right).$$

Owing to the properties of the slope limiter, the reconstructed weights obey a maximum/minimum principle and cannot be greater/less than their values in cells $i - 1, i$, and

[15] On a uniform grid, the diffusion term is always of at least second order.

[16] Since $0 < \xi_0$ can be set very small, introducing it simply allows us to define the NDF on the interval $[0, \infty)$ instead of $[\xi_0, \infty)$.

$i + 1$. Thus, we are guaranteed that the reconstructed weights in Eq. (8.77) are never negative. The second-order finite-volume approximation for the diffusion term is the same as in Eq. (8.75). The realizable quasi-second-order scheme is thus

$$\begin{aligned} m_{k,i}^{n+1} = & \sum_{\alpha=1}^{2} \sum_{\beta=1}^{M} \left(W_{\alpha\beta,i}^{n} - 2W_{\alpha\beta,i}^{n} c_{\alpha\beta,i}^{n} \lambda_1 - W_{\alpha\beta,i+1/2,l}^{n} c^{+} \lambda_2 - W_{\alpha\beta,i-1/2,r}^{n} c^{-} \lambda_2 \right) \left(\xi_{\alpha\beta,i}^{n} \right)^{k} \\ & + \sum_{\alpha=1}^{2} \sum_{\beta=1}^{M} \left(W_{\alpha\beta,i+1}^{n} c_{\alpha\beta,i+1}^{n} \lambda_1 + W_{\alpha\beta,i+1/2,r}^{n} c^{-} \lambda_2 \right) \left(\xi_{\alpha\beta,i+1}^{n} \right)^{k} \\ & + \sum_{\alpha=1}^{2} \sum_{\beta=1}^{M} \left(W_{\alpha\beta,i-1}^{n} c_{\alpha\beta,i-1}^{n} \lambda_1 + W_{\alpha\beta,i-1/2,l}^{n} c^{+} \lambda_2 \right) \left(\xi_{\alpha\beta,i-1}^{n} \right)^{k}. \end{aligned} \tag{8.78}$$

The realizability condition for Eq. (8.78) is that all of the coefficients involving the weights must be nonnegative. This condition yields

$$0 \le \min_{\alpha,\beta,i} \left[W_{\alpha\beta,i}^{n} \left(\frac{1}{\Delta t} - \frac{2c_{\alpha\beta,i}^{n} \Gamma}{\Delta x^2} - \frac{|u|}{\Delta x} \right) - (c^{+} - c^{-}) S_{\alpha\beta,i}^{n} \frac{|u|}{2\,\Delta x} \right]. \tag{8.79}$$

Note that only the first term on the right-hand side of Eq. (8.79) is positive and can be made arbitrarily large by decreasing Δt.

The quasi-third-order scheme

The quasi-third-order MUSCL scheme for advection reconstructs the interface fluxes using constant abscissas and slope estimators with limiters for the weights. The reconstructed weights are defined by

$$\begin{aligned} W_{\alpha\beta,i+1/2,l}^{n} &= W_{\alpha\beta,i}^{n} + \frac{1}{6} S_{\alpha\beta,i-1/2}^{n} + \frac{1}{3} S_{\alpha\beta,i+1/2}^{n}, \\ W_{\alpha\beta,i-1/2,r}^{n} &= W_{\alpha\beta,i}^{n} - \frac{1}{6} S_{\alpha\beta,i+1/2}^{n} - \frac{1}{3} S_{\alpha\beta,i-1/2}^{n}. \end{aligned} \tag{8.80}$$

The definitions of the slope functions are

$$\begin{aligned} S_{\alpha\beta,i+1/2}^{n} &= \phi^{+} \left(\Delta W_{\alpha\beta,i-1/2}^{n}, \Delta W_{\alpha\beta,i+1/2}^{n} \right), \\ S_{\alpha\beta,i-1/2}^{n} &= \phi^{-} \left(\Delta W_{\alpha\beta,i-1/2}^{n}, \Delta W_{\alpha\beta,i+1/2}^{n} \right), \end{aligned} \tag{8.81}$$

where the differences are defined by

$$\begin{aligned} \Delta W_{\alpha\beta,i+1/2}^{n} &= W_{\alpha\beta,i+1}^{n} - W_{\alpha\beta,i}^{n}, \\ \Delta W_{\alpha\beta,i-1/2}^{n} &= W_{\alpha\beta,i}^{n} - W_{\alpha\beta,i-1}^{n}. \end{aligned} \tag{8.82}$$

For a third-order scheme, the functions $\phi^{-}(a, b)$ and $\phi^{+}(a, b)$ in Eq. (8.81) are defined for four distinct cases:

(i) If $a, b \ge 0$:

$$\phi^{-}(a, b) = \begin{cases} b & \text{if } b \le 4a, \\ 4a & \text{if } b > 4a, \end{cases} \qquad \phi^{+}(a, b) = \begin{cases} a & \text{if } a \le 4b, \\ 4b & \text{if } a > 4b. \end{cases}$$

(ii) If $a \geq 0$ and $b < 0$:

$$\phi^-(a,b) = \begin{cases} b & \text{if } b \geq -2a, \\ -2a & \text{if } b < -2a, \end{cases} \quad \phi^+(a,b) = \begin{cases} a & \text{if } a \leq -2b, \\ -2b & \text{if } a > -2b. \end{cases}$$

(iii) If $a, b < 0$:

$$\phi^-(a,b) = \begin{cases} b & \text{if } b \geq 4a, \\ 4a & \text{if } b < 4a, \end{cases} \quad \phi^+(a,b) = \begin{cases} a & \text{if } a \geq 4b, \\ 4b & \text{if } a < 4b. \end{cases}$$

(iv) If $a < 0$ and $b \geq 0$:

$$\phi^-(a,b) = \begin{cases} b & \text{if } b \leq -2a, \\ -2a & \text{if } b > -2a, \end{cases} \quad \phi^+(a,b) = \begin{cases} a & \text{if } a \geq -2b, \\ -2b & \text{if } a < -2b. \end{cases}$$

The update due to the advection term can then be expressed as

$$\begin{aligned} m_{k,i}^{n+1} = & \sum_{\alpha=1}^{2}\sum_{\beta=1}^{M}\left(W_{\alpha\beta,i}^{n} - W_{\alpha\beta,i+1/2,l}^{n}c^{+}\lambda_2 - W_{\alpha\beta,i-1/2,r}^{n}c^{-}\lambda_2\right)\left(\xi_{\alpha\beta,i}^{n}\right)^k \\ & + \sum_{\alpha=1}^{2}\sum_{\beta=1}^{M} W_{\alpha\beta,i+1/2,r}^{n}c^{-}\lambda_2\left(\xi_{\alpha\beta,i+1}^{n}\right)^k + \sum_{\alpha=1}^{2}\sum_{\beta=1}^{M} W_{\alpha\beta,i-1/2,l}^{n}c^{+}\lambda_2\left(\xi_{\alpha\beta,i-1}^{n}\right)^k, \end{aligned} \tag{8.83}$$

and will be realizable as long as none of the coefficients involving the weights are negative. In practice, it is necessary to lower the order of the advection scheme for a given cell i when the realizability condition with the quasi-third-order scheme fails. The quasi-fourth-order scheme for the diffusion term is described in Section B.6 of Appendix B. The update due to diffusion has the form

$$\begin{aligned} m_{k,i}^{n+1} = & \sum_{\alpha=1}^{2}\sum_{\beta=1}^{M}\left[W_{\alpha\beta,i}^{n} - \left(W_{\alpha\beta,i+1/2}^{n} - \frac{1}{2}D_{\alpha\beta,i+1/2}^{n} + W_{\alpha\beta,i-1/2}^{n} + \frac{1}{2}D_{\alpha\beta,i-1/2}^{n}\right)c_{\alpha\beta,i}^{n}\lambda_1\right]\left(\xi_{\alpha\beta,i}^{n}\right)^k \\ & + \sum_{\alpha=1}^{2}\sum_{\beta=1}^{M}\left(W_{\alpha\beta,i+1/2}^{n} + \frac{1}{2}D_{\alpha\beta,i+1/2}^{n}\right)c_{\alpha\beta,i+1}^{n}\lambda_1\left(\xi_{\alpha\beta,i+1}^{n}\right)^k \\ & + \sum_{\alpha=1}^{2}\sum_{\beta=1}^{M}\left(W_{\alpha\beta,i-1/2}^{n} - \frac{1}{2}D_{\alpha\beta,i-1/2}^{n}\right)c_{\alpha\beta,i-1}^{n}\lambda_1\left(\xi_{\alpha\beta,i-1}^{n}\right)^k, \end{aligned} \tag{8.84}$$

where

$$W_{\alpha\beta,i+1/2}^{n} = -\frac{1}{12}W_{\alpha\beta,i-1}^{n} + \frac{7}{12}W_{\alpha\beta,i}^{n} + \frac{7}{12}W_{\alpha\beta,i+1}^{n} - \frac{1}{12}W_{\alpha\beta,i+2}^{n} \tag{8.85}$$

and

$$D_{\alpha\beta,i+1/2}^{n} = \frac{1}{12}W_{\alpha\beta,i-1}^{n} - \frac{5}{4}W_{\alpha\beta,i}^{n} + \frac{5}{4}W_{\alpha\beta,i+1}^{n} - \frac{1}{12}W_{\alpha\beta,i+2}^{n}. \tag{8.86}$$

The diffusion update in Eq. (8.84) will be realizable as long as all of the coefficients are non-negative. This leads to two cases.

1. If either $W_{\alpha\beta,i+1/2}^{n} + 0.5D_{\alpha\beta,i+1/2}^{n}$ or $W_{\alpha\beta,i-1/2}^{n} - 0.5D_{\alpha\beta,i-1/2}^{n}$ is negative, then revert to the second-order scheme for diffusion as in Eq. (8.78).

2. Otherwise, fix the simulation parameters such that the first coefficient in Eq. (8.84) is positive.[17]

In practice, the first case rarely occurs, and thus the quasi-fourth-order scheme in Eq. (8.84) is almost always used for diffusion. By combining the schemes in Eqs. (8.83) and (8.84), we find the realizable quasi-third-order scheme:

$$\begin{aligned} m_{k,i}^{n+1} = \sum_{\alpha=1}^{2}\sum_{\beta=1}^{M} \Bigg[& W_{\alpha\beta,i}^{n} - \left(W_{\alpha\beta,i+1/2}^{n} - \frac{1}{2}D_{\alpha\beta,i+1/2}^{n}\right) c_{\alpha\beta,i}^{n}\lambda_1 - W_{\alpha\beta,i+1/2,l}^{n} c^{+}\lambda_2 \\ & - \left(W_{\alpha\beta,i-1/2}^{n} + \frac{1}{2}D_{\alpha\beta,i-1/2}^{n}\right) c_{\alpha\beta,i}^{n}\lambda_1 - W_{\alpha\beta,i-1/2,r}^{n} c^{-}\lambda_2 \Bigg] \left(\xi_{\alpha\beta,i}^{n}\right)^{k} \\ & + \sum_{\alpha=1}^{2}\sum_{\beta=1}^{M} \left[\left(W_{\alpha\beta,i+1/2}^{n} + \frac{1}{2}D_{\alpha\beta,i+1/2}^{n}\right) c_{\alpha\beta,i+1}^{n}\lambda_1 + W_{\alpha\beta,i+1/2,r}^{n} c^{-}\lambda_2\right] \left(\xi_{\alpha\beta,i+1}^{n}\right)^{k} \\ & + \sum_{\alpha=1}^{2}\sum_{\beta=1}^{M} \left[\left(W_{\alpha\beta,i-1/2}^{n} - \frac{1}{2}D_{\alpha\beta,i-1/2}^{n}\right) c_{\alpha\beta,i-1}^{n}\lambda_1 + W_{\alpha\beta,i-1/2,l}^{n} c^{+}\lambda_2\right] \left(\xi_{\alpha\beta,i-1}^{n}\right)^{k}. \end{aligned} \tag{8.87}$$

The realizability condition for Eq. (8.87) is that all of the coefficients in square brackets must be nonnegative. This again leads to two cases.

1. If either $W_{\alpha\beta,i+1/2}^{n} + 0.5D_{\alpha\beta,i+1/2}^{n}$ or $W_{\alpha\beta,i-1/2}^{n} - 0.5D_{\alpha\beta,i-1/2}^{n}$ is negative, then in cell i revert to the second-order scheme for diffusion as in Eq. (8.78).
2. Otherwise, check whether

$$\begin{aligned} \min_{\alpha,\beta,i} \Bigg[& W_{\alpha\beta,i}^{n} - \left(W_{\alpha\beta,i+1/2}^{n} - \frac{1}{2}D_{\alpha\beta,i+1/2}^{n}\right) c_{\alpha\beta,i}^{n}\lambda_1 - W_{\alpha\beta,i+1/2,l}^{n} c^{+}\lambda_2 \\ & - \left(W_{\alpha\beta,i-1/2}^{n} + \frac{1}{2}D_{\alpha\beta,i-1/2}^{n}\right) c_{\alpha\beta,i}^{n}\lambda_1 - W_{\alpha\beta,i-1/2,r}^{n} c^{-}\lambda_2 \Bigg] > 0, \end{aligned} \tag{8.88}$$

and, if not, lower the order of the advection and diffusion schemes in cell i.

In practice, the first case occurs only at a few isolated points, and thus the quasi-fourth-order diffusion scheme is almost always used.

The overall finite-volume scheme

All of the finite-volume schemes discussed above can be written in the standard flux form:

$$m_{k,i}^{n+1} = \sum_{\alpha=1}^{2}\sum_{\beta=1}^{M} \left[W_{\alpha\beta,i}^{n} \left(\xi_{\alpha\beta,i}^{n}\right)^{k} + \lambda_1 \left(H_{\alpha\beta,i+1/2}^{k,n} - H_{\alpha\beta,i-1/2}^{k,n}\right) - \lambda_2 \left(G_{\alpha\beta,i+1/2}^{k,n} - G_{\alpha\beta,i-1/2}^{k,n}\right)\right], \tag{8.89}$$

where $H_{\alpha\beta,i+1/2}^{k,n}$ and $G_{\alpha\beta,i+1/2}^{k,n}$ are, respectively, the diffusive and advective numerical flux functions. These functions can be written in the following forms:

$$\begin{aligned} H_{\alpha\beta,i+1/2}^{k,n} &= W_{\alpha\beta,i+1/2}^{n} \left[c_{\alpha\beta,i+1}^{n} \left(\xi_{\alpha\beta,i+1}^{n}\right)^{k} - c_{\alpha\beta,i}^{n} \left(\xi_{\alpha\beta,i}^{n}\right)^{k} \right] \\ &\quad + \frac{1}{2} D_{\alpha\beta,i+1/2}^{n} \left[c_{\alpha\beta,i+1}^{n} \left(\xi_{\alpha\beta,i+1}^{n}\right)^{k} + c_{\alpha\beta,i}^{n} \left(\xi_{\alpha\beta,i}^{n}\right)^{k} \right], \\ G_{\alpha\beta,i+1/2}^{k,n} &= W_{\alpha\beta,i+1/2,l}^{n} c^{+} \left(\xi_{\alpha\beta,i}^{n}\right)^{k} - W_{\alpha\beta,i+1/2,r}^{n} c^{-} \left(\xi_{\alpha\beta,i+1}^{n}\right)^{k}. \end{aligned} \tag{8.90}$$

[17] As discussed below, it is usually preferable to fix Δt using the stability condition and to lower the order of the scheme to ensure realizability.

The definitions of the terms on the right-hand sides depend on the orders of the schemes used for diffusion and advection. For diffusion we have

$$W^n_{\alpha\beta,i+1/2} = \begin{cases} \frac{1}{2}\left(W^n_{\alpha\beta,i+1} + W^n_{\alpha\beta,i+1}\right) & \text{for second-order diffusion,} \\ \text{Eq. (8.85)} & \text{for quasi-fourth-order diffusion,} \end{cases} \tag{8.91}$$

and

$$D^n_{\alpha\beta,i+1/2} = \begin{cases} W^n_{\alpha\beta,i+1} - W^n_{\alpha\beta,i} & \text{for second-order diffusion,} \\ \text{Eq. (8.86)} & \text{for quasi-fourth-order diffusion} \end{cases} \tag{8.92}$$

while for advection we have

$$W^n_{\alpha\beta,i+1/2,l} = \begin{cases} W^n_{\alpha\beta,i} & \text{for first-order advection,} \\ \text{Eq. (8.77)} & \text{for quasi-second-order advection,} \\ \text{Eq. (8.80)} & \text{for quasi-third-order advection,} \end{cases} \tag{8.93}$$

and

$$W^n_{\alpha\beta,i+1/2,r} = \begin{cases} W^n_{\alpha\beta,i+1} & \text{for first-order advection,} \\ \text{Eq. (8.77)} & \text{for quasi-second-order advection,} \\ \text{Eq. (8.80)} & \text{for quasi-third-order advection.} \end{cases} \tag{8.94}$$

The time step Δt and the order of the methods used are determined as follows. An overall CFL condition is applied to fix the time step on a given grid:

$$\Delta t = \text{CFL}\left(\frac{2\Gamma}{\Delta x^2} + \frac{|u|}{\Delta x}\right)^{-1},$$

where CFL $= 1/4$ is chosen to ensure that the quasi-third-order advection is stable. Note that, with this choice for Δt, the first-order advection and second-order diffusion schemes are always stable and realizable. Next, $\lambda_1 = \Gamma\,\Delta t/(\Delta x^2)$ and $\lambda_2 = |u|\Delta t/\Delta x$ are computed. For the quasi-fourth-order diffusion scheme, we then check whether the realizability conditions

$$\min_{\alpha\beta}\left(W^n_{\alpha\beta,i+1/2} + 0.5D^n_{\alpha\beta,i+1/2}\right) > 0 \qquad \text{and} \qquad \min_{\alpha\beta}\left(W^n_{\alpha\beta,i-1/2} - 0.5D^n_{\alpha\beta,i-1/2}\right) > 0$$

hold for each i, and, if not, the second-order diffusion scheme is used in cell i. Next, the realizability condition in Eq. (8.88) is checked for each i, and if it fails the advection scheme is lowered to quasi-second order in cell i. Finally, the quasi-second-order cells are rechecked for the realizability condition in Eq. (8.88), and if it again fails the advection and diffusion schemes are set at their lowest orders (which, as noted earlier, are always realizable). In this manner, the time step is held constant and the orders of the finite-volume schemes are changed locally to ensure realizability of the moments. In practice, the highest-order schemes are used in most of the cells at each time step, especially after diffusion has significantly smoothed the profiles.

8.3.4 Example results for an inhomogeneous PBE

The realizable schemes developed in the preceding section are applied to the moment distributions shown in Figure 8.2. The normalized moments are defined by dividing their values by the initial values at $x = 0$: (1, 10, 200, 6000, 240 000). For the cases with constant diffusivity, we set $M = 20$, while for those with Stokes–Einstein diffusivity we use $\xi_0 = 1$, and set $M = 100$ to improve the representation of the volume-dependent diffusive fluxes.

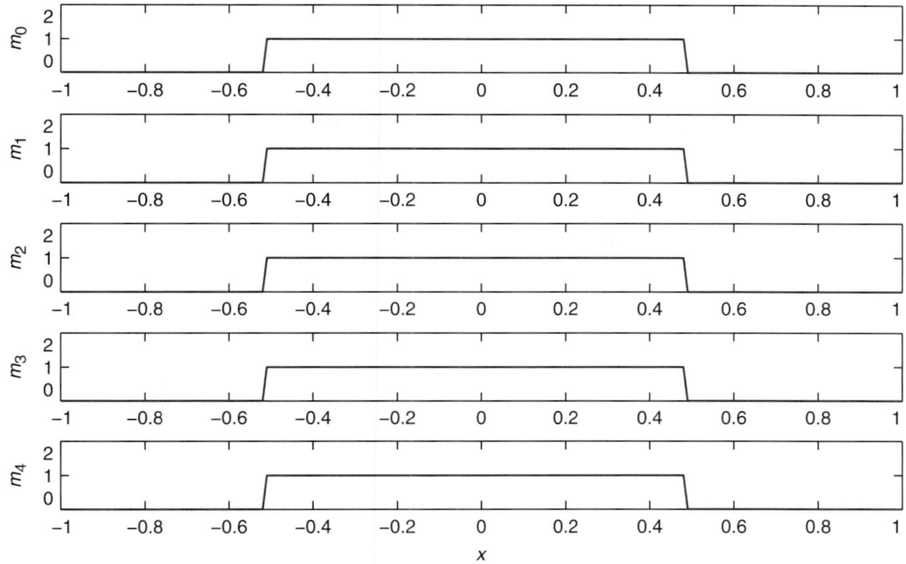

Figure 8.2. Normalized moments of PBE at t = 0.

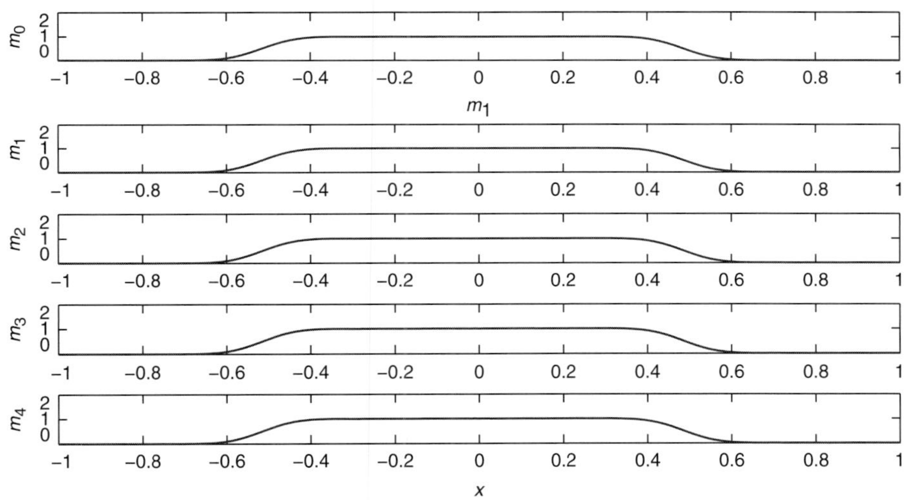

Figure 8.3. Normalized moments of PBE at t = 2 found with $u = 0$ and constant diffusivity $\Gamma = 10^{-3}$.

Constant diffusivity

Results for pure diffusion with $u = 0$ and $\Gamma = 10^{-3}$ at time $t = 2$ are shown in Figure 8.3. Without advection, we can observe that all the moments diffuse at the same rate, starting from the square-wave initial conditions. For this case all of the abscissas used to represent the NDF are independent of x and t. The fourth-order diffusion scheme is used almost everywhere in the domain for $t > 0$. Results for advection and diffusion with $u = 1$ and $\Gamma = 10^{-3}$ at time $t = 2$ are shown in Figure 8.4. Since the boundary conditions are periodic, $t = 2$ corresponds to the time needed for the square wave to return to its original position.

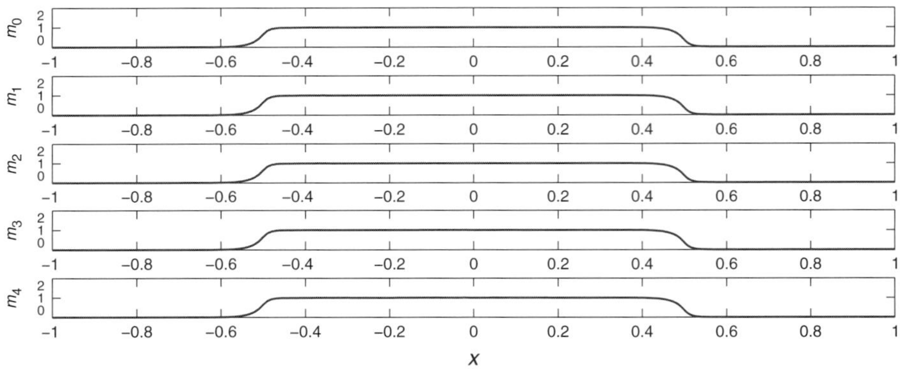

Figure 8.4. Normalized moments of PBE at $t = 2$ found with $u = 1$ and constant diffusivity $\Gamma = 10^{-3}$.

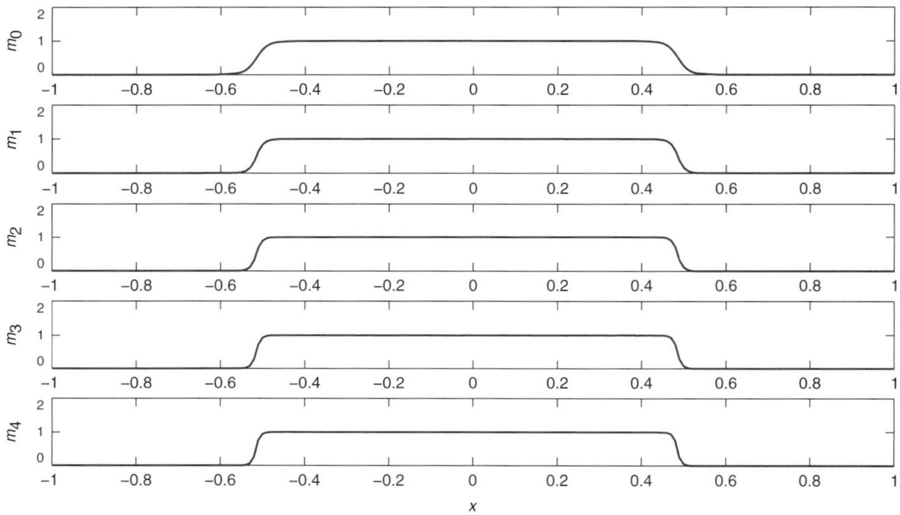

Figure 8.5. Normalized moments of PBE at $t = 2$ found with $u = 0$ and Stokes–Einstein diffusivity $\Gamma = 10^{-3}$.

As in the previous case, all of the abscissas used to represent the NDF are independent of x and t. The high-order advection and diffusion schemes are used almost everywhere in the domain for $t > 0$.

Stokes–Einstein diffusivity

Results for Stokes–Einstein diffusion with $u = 0$ and $\Gamma = 10^{-3}$ at time $t = 2$ are shown in Figure 8.5. We can observe that lower-order moments diffuse faster than the higher-order moments. For this case, the abscissas used to represent the NDF are dependent on x and t. Indeed, to the right of the square wave the abscissas for $t > 0$ are smaller because the small particles diffuse faster than the large particles. Conversely, just inside the square wave the abscissas are larger. Results for Stokes–Einstein diffusion with $u = 1$ and $\Gamma = 10^{-3}$ at time $t = 2$ are shown in Figure 8.6. With advection, the shapes of the moment profiles

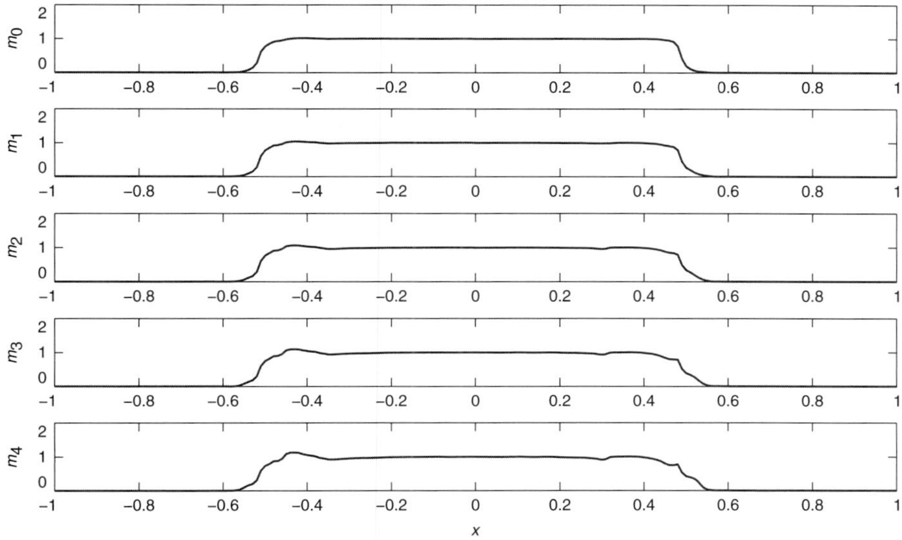

Figure 8.6. Normalized moments of PBE at $t = 2$ found with $u = 1$ and Stokes–Einstein diffusivity $\Gamma = 10^{-3}$.

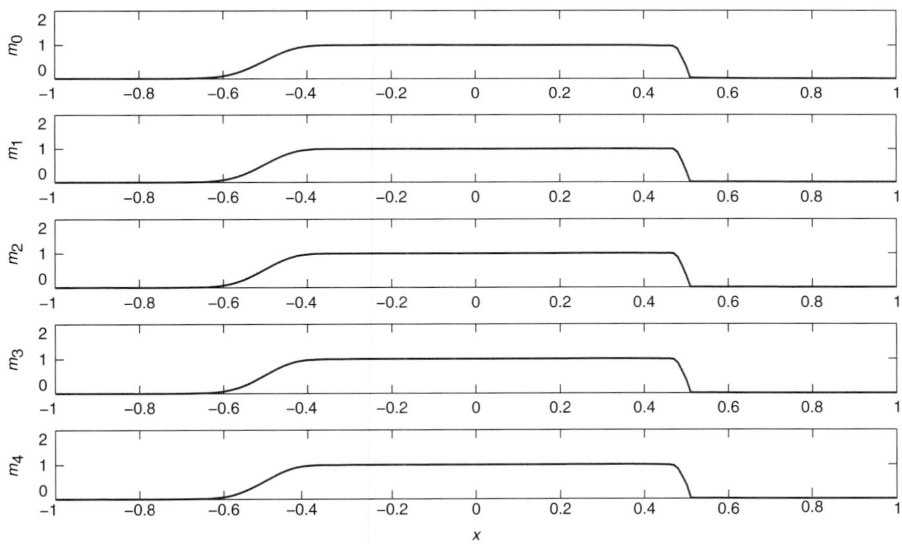

Figure 8.7. Normalized moments of PBE at $t = 2$ found with $u = 1$ and Stokes–Einstein diffusivity $\Gamma = 10^{-3}$. The CFL number is set to 1.

are sensitive to the EQMOM reconstruction of the NDF, which is used with the second quadrature to estimate the size-dependent diffusion coefficient. In general, better results can be found by increasing n (the number of EQMOM nodes) at the expense of transporting more moments. In Figure 8.7 we repeat the same simulation as in Figure 8.6, but with the CFL number set to 1. It can be seen that the front of the square wave is less diffusive than the back side, which is a numerical artifact that results from using the high-order advection scheme with a CFL number above its stability limit. This can be verified by switching

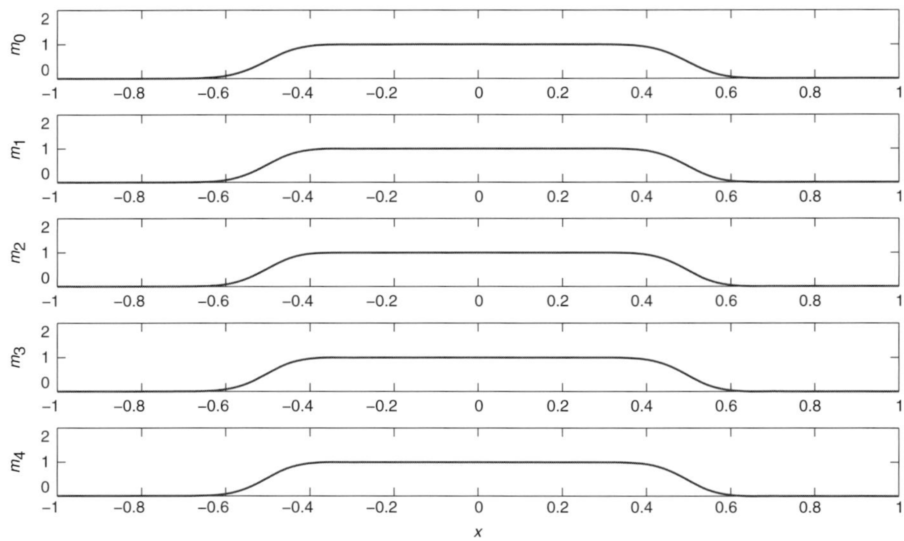

Figure 8.8. Normalized moments of PBE at $t = 2$ found with $u = 1$ and Stokes–Einstein diffusivity $\Gamma = 10^{-3}$. The first-order advection scheme with a CFL number of 1 is used.

to the first-order advection scheme, as can be seen in Figure 8.8. Note that in all cases the moments remain realizable, as is guaranteed by the choice of the time step and the realizable scheme.

8.4 Inhomogeneous KE

In this section, we look at a kinetic equation for the velocity NDF $n(t, x, \mathbf{v})$, where $\mathbf{v} = (u, v)$ is a two-component velocity vector (i.e. the velocity phase space is two-dimensional). In order to show the dynamics for different amounts of particle–particle collisions, we will use the BGK collision model. (See Chapter 6 for more details on collision models.) The inhomogeneous kinetic equation for this case is

$$\frac{\partial n}{\partial t} + \frac{\partial v n}{\partial x} = \frac{m_0 \Theta^{1/2}}{\tau}(n^* - n), \tag{8.95}$$

where τ is a collision time constant (proportional to the Knudsen number), and Θ is the granular temperature defined in terms of the bivariate velocity moments by

$$\Theta = \frac{1}{3}\left[\frac{m_{2,0}}{m_{0,0}} - \frac{m_{1,0}^2}{m_{0,0}^2} + 2\left(\frac{m_{0,2}}{m_{0,0}} - \frac{m_{0,1}^2}{m_{0,0}^2}\right)\right].$$

Here we have used the statistical symmetry between the second and third directions in velocity phase space to express the granular temperature in terms of the bivariate moments $m_{j,k}$. The NDF n^* is the equilibrium (Maxwellian) distribution with the conservation properties: $m_{0,0}^* = m_{0,0}$, $m_{1,0}^* = m_{1,0}$, $m_{0,1}^* = m_{0,1}$, and $\Theta^* = \Theta$. Physically, these equalities result from the fact that the mass, momentum, and energy are conserved during collisions.

8.4.1 The moment-transport equation

The transport equation for the bivariate moments found from Eq. (8.95) is

$$\frac{\partial m_{j,k}}{\partial t} + \frac{\partial m_{j+1,k}}{\partial x} = \frac{m_{0,0}\Theta^{1/2}}{\tau}\left(m^*_{j,k} - m_{j,k}\right), \tag{8.96}$$

where the equilibrium moments $m^*_{j,k}$ are known functions of the conserved quantities $m_{0,0}$, $m_{1,0}$, $m_{0,1}$, and Θ. The collision term in Eq. (8.96) is closed, and thus the only unclosed term is the spatial flux involving $m_{j+1,k}$. In order to close this term, we will use the multivariate EQMOM described in Section 3.3.4 with Gaussian functions (see Eq. (3.134) for details):

$$n(t, x, \mathbf{v}) = \sum_{\alpha=1}^{2}\sum_{\beta=1}^{2} w_{\alpha\beta}\delta_{\sigma_1}(u, u_\alpha)\delta_{\sigma_{2\alpha}}\left(v, v_{\alpha\beta}\right), \tag{8.97}$$

where $\delta_\sigma(x, y)$ is defined by Eq. (3.88). (See also Eq. (3.141) on page 98.) The moment-inversion algorithm for Eq. (8.97) is described after Eq. (3.140) on page 98 and requires the following set of 13 moments:

$$\begin{bmatrix} m_{0,0} & m_{0,1} & m_{0,2} & m_{0,3} & m_{0,4} \\ m_{1,0} & m_{1,1} & m_{1,2} & m_{1,3} & m_{1,4} \\ m_{2,0} & & & & \\ m_{3,0} & & & & \\ m_{4,0} & & & & \end{bmatrix}.$$

However, by symmetry, the four moments $m_{0,1}$, $m_{1,1}$, $m_{0,3}$, and $m_{1,3}$ are null if the initial NDF is Gaussian, in which case only nine moments are required in order to solve Eq. (8.96). Nevertheless, it is convenient to solve for all 13 moments (referred to hereafter as the transported moment set) and use the known zero moments to check for the numerical accuracy of the algorithm.

8.4.2 Operator splitting for moment equations

Because the collision time τ is arbitrary and can lead to stiff equations, it is best to solve Eq. (8.96) using operator splitting, i.e. solve first

$$\frac{\partial m_{j,k}}{\partial t} + \frac{\partial m_{j+1,k}}{\partial x} = 0, \tag{8.98}$$

followed by

$$\frac{\partial m_{j,k}}{\partial t} = \frac{m_{0,0}\Theta^{1/2}}{\tau}\left(m^*_{j,k} - m_{j,k}\right). \tag{8.99}$$

The collision term in Eq. (8.99) can be solved semi-analytically by setting

$$\kappa(t, x) = m_{0,0}(t, x)\Theta^{1/2}(t, x)/\tau.$$

This leads to the approximation

$$m_{j,k}(t + \Delta t, x) = m^*_{j,k}(t, x)\left(1 - \mathrm{e}^{-\kappa(t,x)\Delta t}\right) + m_{j,k}(t, x)\mathrm{e}^{-\kappa(x,t)\Delta t}.$$

Written in terms of the finite-volume moments on a uniform mesh, this formula becomes

$$m^{n+1}_{j,k;i} = m^*_{j,k;i}\left(1 - \mathrm{e}^{-\kappa^n_i\Delta t}\right) + m^\dagger_{j,k;i}\mathrm{e}^{-\kappa^n_i\Delta t}, \tag{8.100}$$

where $\kappa_i^n = m_{0,0;i}^n (\Theta_i^n)^{1/2}/\tau$ and

$$m_{j,k;i}^n = \frac{1}{\Delta x}\int_{x_i-1/2}^{x_i+1/2} m_{j,k}(n\,\Delta t, x)\mathrm{d}x.$$

The moments $m_{j,k;i}^\dagger$ result from solving Eq. (8.98) starting from $m_{j,k;i}^n$, and the equilibrium moments $m_{j,k;i}^*$ are known functions of $m_{j,k;i}^\dagger$ (with $j,k = 0,1,2$).[18] Because Eq. (8.100) is a convex sum of realizable moments, it is always realizable for $\Delta t \geq 0$. Thus, in practice, the time step Δt can be chosen solely on the basis of the realizability condition for the spatial fluxes. It remains to describe how the spatial fluxes in Eq. (8.98) are treated in a realizable finite-volume scheme for the velocity moments.

Before leaving the collision term, we should point out to the reader that the right-hand side of Eq. (8.99) can be replaced with a more sophisticated collision model, such as the Boltzmann hard-sphere model described in Chapter 6. In general, the collision model will not be closed in terms of the transported moments, and hence the moment update will be more complicated than in Eq. (8.100). For example, with the Boltzmann collision term the right-hand side will involve the double integral over the product NDF $n(t, x, \mathbf{v}_1)n(t, x, \mathbf{v}_2)$ (see Eq. (6.9) on page 217). By using the multi-Gaussian closure in Eq. (8.96), the double integral can be written as a double summation over Gauss–Hermite quadrature nodes (see Section 3.3.1 for more details). For the case under consideration here, the Gauss–Hermite quadratures would be used to represent the integrals in the bivariate moment source terms given in Eq. (6.68) on page 232.[19] Note that the expressions $I_{l_1 l_2 l_3}^{(0)}$ (with $\omega = 1$) needed to evaluate the hard-sphere collision source terms are given in Tables 6.1–6.3.

8.4.3 A realizable finite-volume scheme for bivariate velocity moments

The realizable scheme for Eq. (8.98) can be written in the standard flux form:

$$m_{j,k;i}^\dagger = \sum_{\alpha=1}^{4}\sum_{\beta_1=1}^{M_1}\sum_{\beta_2=1}^{M_2}\left[W_{\alpha\beta_1\beta_2,i}^n \left(U_{\alpha\beta_1,i}^n\right)^j \left(V_{\alpha\beta_2,i}^n\right)^k - \lambda_1\left(G_{\alpha\beta_1\beta_2,i+1/2}^{j,k;n} - G_{\alpha\beta_1\beta_2,i-1/2}^{j,k;n}\right)\right], \quad (8.101)$$

where $\lambda_1 = \Delta t/\Delta x$ and $G_{\alpha\beta_1\beta_2,i+1/2}^{j,k;n}$ is the numerical flux function. The quadrature weights $W_{\alpha\beta_1\beta_2,i}^n$ and abscissas $U_{\alpha\beta_1,i}^n$ and $V_{\alpha\beta_2,i}^n$ are found from the moments $m_{j,k;i}^n$ using the bivariate Gaussian ECQMOM described in Section 3.3.4. Concretely, this process involves the following steps for each grid cell i at each time step n.

(i) Starting from the transported moment set $m_{j,k;i}^n$, apply the moment-inversion algorithm to find the parameters of the four Gaussian distributions: w_α, u_α, v_α, $\sigma_{1\alpha}$, and $\sigma_{2\alpha}$. Note that here we employ a single summation over $\alpha = 1, \ldots, 4$, in place of the double summation used in Eq. (3.134) on page 96. Thus, for example, $\sigma_{1\alpha}$ is the same for all α, and $u_1 = u_2$, $u_3 = u_4$, $\sigma_{21} = \sigma_{22}$, and $\sigma_{23} = \sigma_{24}$, since velocity component v is conditioned on the value of u. The exact steps needed to implement the moment-inversion algorithm are described after Eq. (3.140) on page 98.

[18] Recall that the collision term has conservation properties so that $m_{0,0;i}^\dagger$, $m_{1,0;i}^\dagger$, $m_{0,1;i}^\dagger$, and $m_{2,0;i}^\dagger + 2m_{0,2;i}^\dagger$ do not change during the collision step. Thus, in the absence of the spatial fluxes, $\kappa(t, x)$ is constant.

[19] The Gauss–Hermite quadrature in the third direction of velocity phase space is the same as for velocity component v, and thus can be formed as a tensor product of the quadrature weights and abscissas in the second direction. However, in practice, it is more computationally efficient to rewrite the terms in Eq. (6.68) using symmetry to eliminate the dependence on the third velocity component.

(ii) Given M_1 and M_2, find the corresponding Gauss–Hermite quadrature weights and abscissas $(w^{(1)}_{\beta_1}, t^{(1)}_{\beta_1})$ and $(w^{(2)}_{\beta_2}, t^{(2)}_{\beta_2})$, which we will assume are standardized such that the weights sum to unity and the variance is one.

(iii) For each α, the quadrature weights and abscissas are then defined by

$$\begin{aligned} W^n_{\alpha\beta_1\beta_2,i} &= w_\alpha w^{(1)}_{\beta_1} w^{(2)}_{\beta_2}, \\ U^n_{\alpha\beta_1,i} &= u_\alpha + \sigma_{1\alpha} t^{(1)}_{\beta_1}, \\ V^n_{\alpha\beta_2,i} &= v_\alpha + \sigma_{2\alpha} t^{(2)}_{\beta_2}. \end{aligned} \tag{8.102}$$

Note that, because the Gauss–Hermite quadrature weights and abscissas depend only on the number of points, they need be computed only once if M_1 and M_2 are constant (which will usually be the case).

In Section B.4 of Appendix B we present an alternative method for computing the quadrature for the fluxes by using the half-moments defined in Eq. (B.37) on page 431 with a second application of the CQMOM. In practice, there exists a trade-off between the accuracy of the CQMOM and the number of quadrature nodes that can be determined without excess round-off error. Thus, if many flux nodes are needed in order to compute the spatial fluxes accurately, it may be difficult to find them using the CQMOM to the desired accuracy, even if the half-moments are exact. In contrast, with the Gauss–Hermite quadrature, M_1 and M_2 can be made very large without losing any accuracy since they do not require a moment-inversion algorithm. In practice, this means that we should be able to achieve a very accurate representation of the fluxes without passing by the half-moments. For example, with $M = 14$ the smallest Gauss–Hermite weights are of the order of 10^{-8}, and adding more nodes has hardly any effect on the numerical flux function. For the cases considered here, we find that the transported moment set can be accurately simulated using $M_1 = 14$ and $M_2 = 4$. However, in general, the number of nodes required will depend on the highest order of the moments in the transported moment set.

For the free-transport problem in Eq. (8.98), the realizable numerical flux function is

$$\begin{aligned} G^{j,k;n}_{\alpha\beta_1\beta_2,i+1/2} = {} & W^n_{\alpha\beta_1\beta_2,i+1/2,l} \max\left(U^n_{\alpha\beta_1,i}, 0\right)\left(U^n_{\alpha\beta_1,i}\right)^j \left(V^n_{\alpha\beta_2,i}\right)^k \\ & + W^n_{\alpha\beta_1\beta_2,i+1/2,r} \min\left(U^n_{\alpha\beta_1,i+1}, 0\right)\left(U^n_{\alpha\beta_1,i+1}\right)^j \left(V^n_{\alpha\beta_2,i+1}\right)^k. \end{aligned} \tag{8.103}$$

The weights $W^n_{\alpha\beta_1\beta_2,i+1/2,l}$ and $W^n_{\alpha\beta_1\beta_2,i+1/2,r}$ depend on the order of the scheme. For example, the first-order scheme uses

$$\begin{aligned} W^n_{\alpha\beta_1\beta_2,i+1/2,l} &= W^n_{\alpha\beta_1\beta_2,i}, \\ W^n_{\alpha\beta_1\beta_2,i+1/2,r} &= W^n_{\alpha\beta_1\beta_2,i+1}, \end{aligned} \tag{8.104}$$

while the quasi-second- and quasi-third-order schemes use forms analogous to Eqs. (8.77) and (8.80), respectively. The realizability condition that results from Eq. (8.101) is

$$W^n_{\alpha\beta_1\beta_2,i} \geq \lambda_1 W^n_{\alpha\beta_1\beta_2,i+1/2,l} \max\left(U^n_{\alpha\beta_1,i}, 0\right) - \lambda_1 W^n_{\alpha\beta_1\beta_2,i-1/2,r} \min\left(U^n_{\alpha\beta_1,i}, 0\right) \tag{8.105}$$

for each value of $(\alpha, \beta_1, \beta_2, i)$. In practice, the time step Δt is determined using a CFL condition that is based on the first-order scheme for nonzero weights (i.e. when $0 < W^n_{\alpha\beta_1\beta_2,i}$):

$$\Delta t = \text{CFL} \min_{\alpha,\beta_1,i} \left(\frac{\Delta x}{|U^n_{\alpha\beta_1,i}|} \right) \tag{8.106}$$

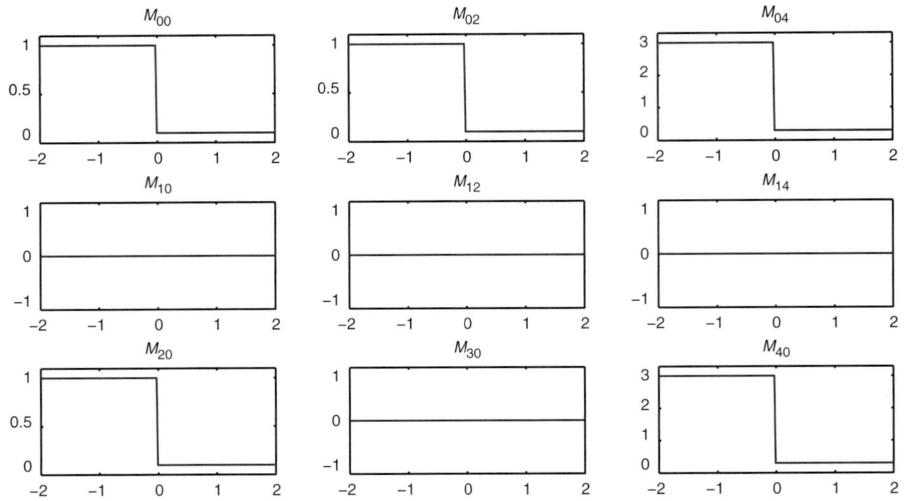

Figure 8.9. Nonzero initial moments for Riemann shock.

with CFL < 1. The quasi-third/second-order convection scheme is used when Eq. (8.105) is satisfied with fixed $\lambda_1 = \Delta t/\Delta x$, otherwise a lower-order reconstruction scheme is employed for given values of $(\alpha, \beta_1, \beta_2, i)$. For the first-order scheme, Eq. (8.106) suffices to guarantee realizability of the updated moment set $m^{\dagger}_{j,k;i}$. In summary, the realizable scheme starts with $m^{n}_{j,k;i}$ and computes $m^{\dagger}_{j,k;i}$ using Eq. (8.101). This result is used with Eq. (8.100) to find the updated moments $m^{n+1}_{j,k;i}$. As long as Eqs. (8.105) and (8.106) are satisfied, the updated moments are guaranteed to be realizable.

8.4.4 Example results for an inhomogeneous KE

Here we apply the finite-volume scheme to simulate two different examples of inhomogeneous kinetic equations. The first example is a non-equilibrium Riemann shock problem with different values of the collision time τ. The second example is two 1D crossing jets with different collision times. In reality, the collision time is controlled by the number density M_{00}, which we normalize with respect to unity in these examples. Thus, the reader can interpret the different values of τ as different values of the unnormalized number density. As noted above, for the multi-Gaussian quadrature we compute the spatial fluxes using $M_1 = 14$ and $M_2 = 4$ with a CFL number of unity.

Riemann shock

The initial conditions for the nonzero moments in the Riemann shock problem are shown in Figure 8.9. On the left half of the domain the normalized number density is $M_{00} = 1$, and on the right half it is 0.1. Initially the RMS velocity is 1 on both sides of the domain and the mean velocity is null. The moments are initialized as Maxwellian, so that the initial conditions are at local equilibrium. As time increases, a shock wave starting at $x = 0$ moves to the right and a deflation wave moves to the left. In the limit of $\tau = 0$, the solution is the same as for the Euler equation of gas dynamics. Sample results for the moments with $\tau = 100$ at time $t = 0.5$ are shown in Figure 8.10. For this case, the collisions are very weak, and thus there is little transfer of kinetic energy from the u-component to the

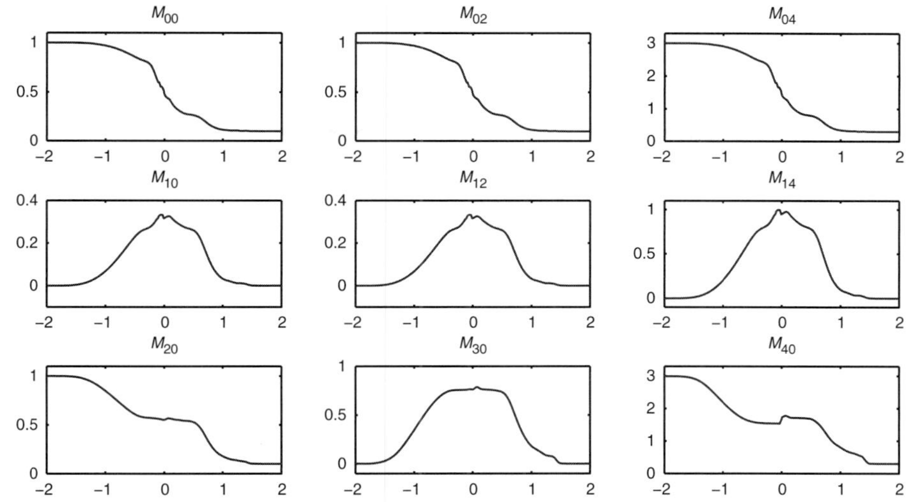

Figure 8.10. Nonzero moments for Riemann shock with $\tau = 100$ at $t = 0.5$.

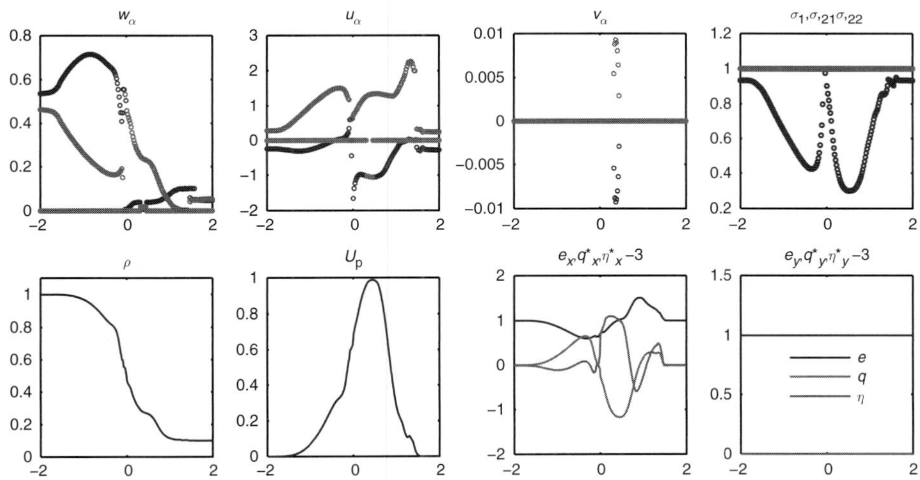

Figure 8.11. Statistics for Riemann shock with $\tau = 100$ at $t = 0.5$. The key indicates the lines plotting e_x and e_y as e, and similarly for q and η in the key.

v-component of the velocity fluctuations. This can be seen in Figure 8.11, where the v statistics remain near the equilibrium values. For example $\sigma_{21} = \sigma_{22} \approx 1$, indicating that all of the v velocity fluctuations are contained in the Gaussian velocity distribution. However, σ_1 is much less than unity, indicating that the u-component of velocity is very much out of equilibrium. Likewise, the energy in the v-component is $e_y \approx 1$ for all x, and the v skewness is $q_y \approx 0$, and the v kurtosis is $\eta_y \approx 3$. In contrast, the values for the u-component of velocity are far from equilibrium. As expected for a Riemann-shock problem, the mean u velocity U_{p} is nonzero. Note that the weights and the u-velocity abscissas are far from the equilibrium values. In comparison, the v-velocity abscissas are very close to zero, as would be expected for a Gaussian velocity distribution. Sample results for the moments with $\tau = 1$ at time $t = 0.5$ are shown in Figure 8.12. For this case, the collisions are still

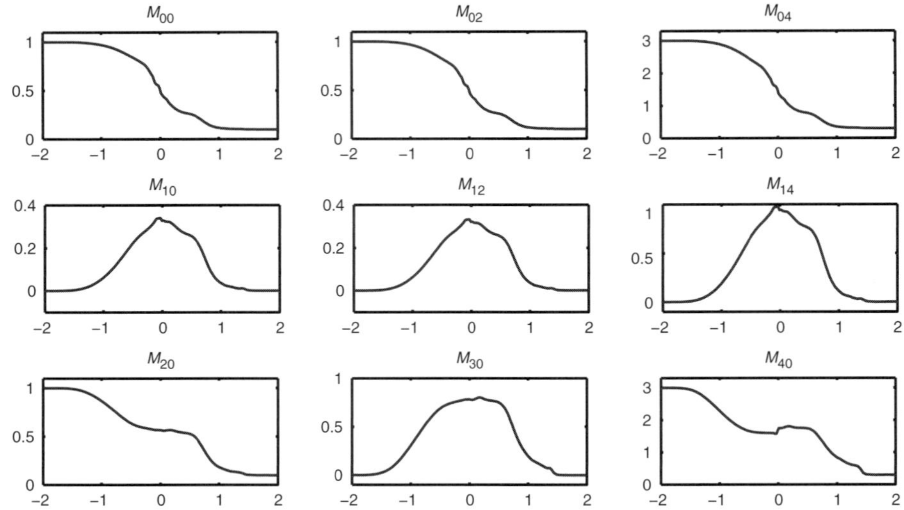

Figure 8.12. Nonzero moments for Riemann shock with $\tau = 1$ at $t = 0.5$.

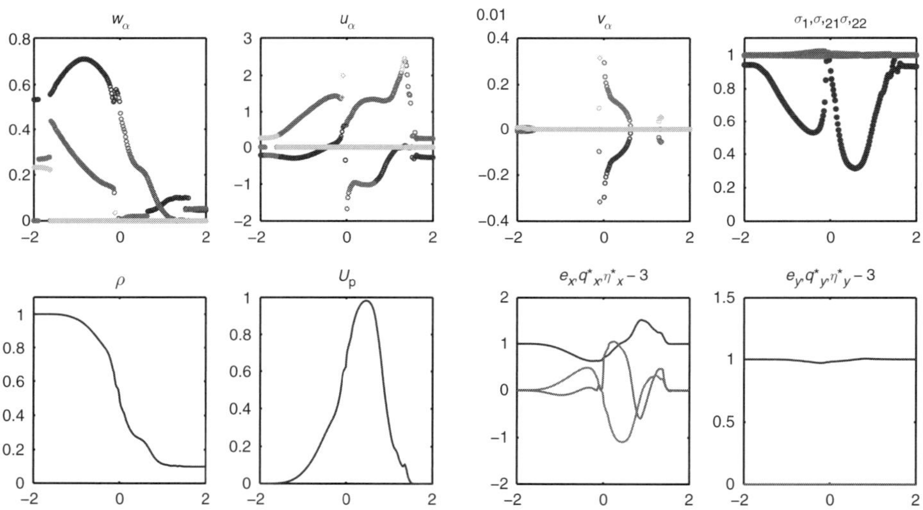

Figure 8.13. Statistics for Riemann shock with $\tau = 1$ at $t = 0.5$. See the key for the different lines in Figure 8.11.

relatively weak, and thus there is not much transfer of kinetic energy from the u-component to the v-component of the velocity fluctuations. This can be seen in Figure 8.13, where the v statistics deviate only slightly from the equilibrium values. Also σ_{21} and σ_{22} are close to unity, indicating that the v-velocity fluctuations are near a Gaussian velocity distribution. Again, σ_1 is significantly less than unity, indicating that the u-component of velocity is out of equilibrium. The mean u velocity U_p is nonzero, as expected, and the weights and the u-velocity abscissas are far from the equilibrium values. However, now the v-velocity abscissas are farther from zero, indicating deviations from a Gaussian velocity distribution. Sample results for the moments with $\tau = 0.01$ at time $t = 0.5$ are shown in Figure 8.14. For

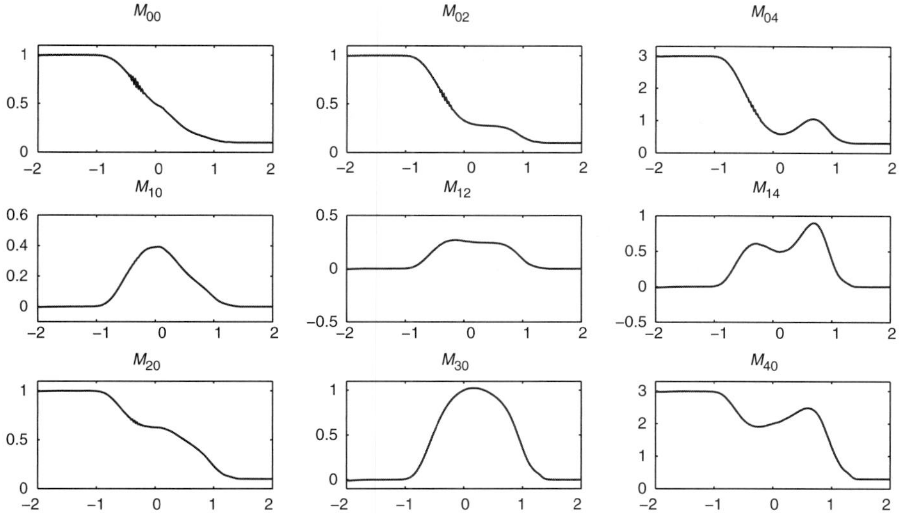

Figure 8.14. Nonzero moments for Riemann shock with $\tau = 0.01$ at $t = 0.5$.

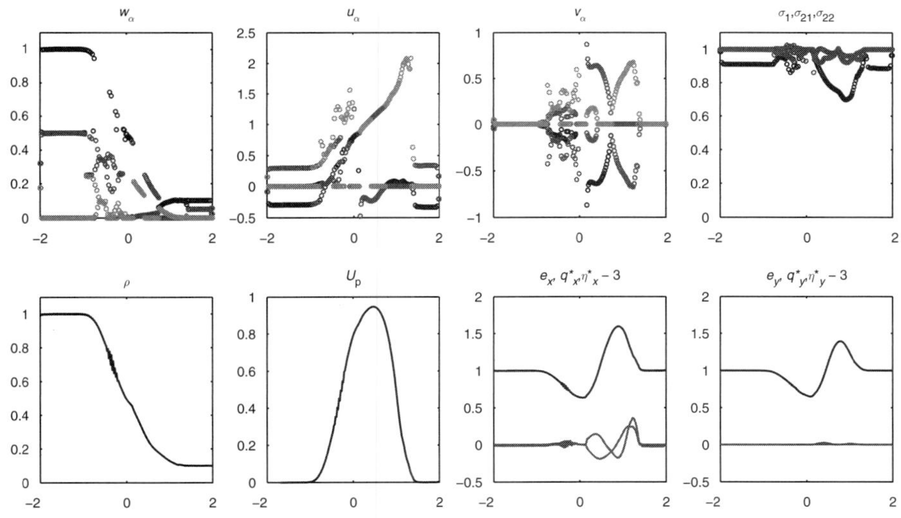

Figure 8.15. Statistics for Riemann shock with $\tau = 0.01$ at $t = 0.5$. See the key for the different lines in Figure 8.11.

this case, the collisions are relatively strong, and thus there is significant transfer of kinetic energy from the u-component to the v-component of the velocity fluctuations. This can be seen in Figure 8.15, where the v statistics approach the values seen for the u-component. In this case, σ_{21} and σ_{22} are close to unity, but have a complex dependence on x. The same is true for σ_1. In the limit where $\tau = 0$, all components of σ would equal unity. The mean u velocity U_p is again non-zero, and the weights and abscissas exhibit a complex dependence on x. Overall, the realizable scheme is able to handle the entire range of τ values while ensuring that the moments are always realizable.

Crossing jets

The initial conditions for the nonzero moments in the crossing-jets problem are shown in Figure 8.16. The initial number density is uniform with $M_{00} = 1$, but on the left half of the domain the mean velocity is $U_p = 1$, while on the right half it is -1. Initially the RMS velocity is 0.1 on both sides of the domain and the moments are initialized as Maxwellian. As time increases, the density on the left half side moves to the right and that on the right half side moves to the left, causing a 1D crossing jet at $x = 0$. Depending on the value of τ, the two jets will relax towards an equilibrium velocity-distribution function. Results for $\tau = 100$ at time $t = 1$ are shown in Figures 8.17 and 8.18. For this value of τ, the collisions are very weak, so the relaxation to equilibrium is very slow. This is evident

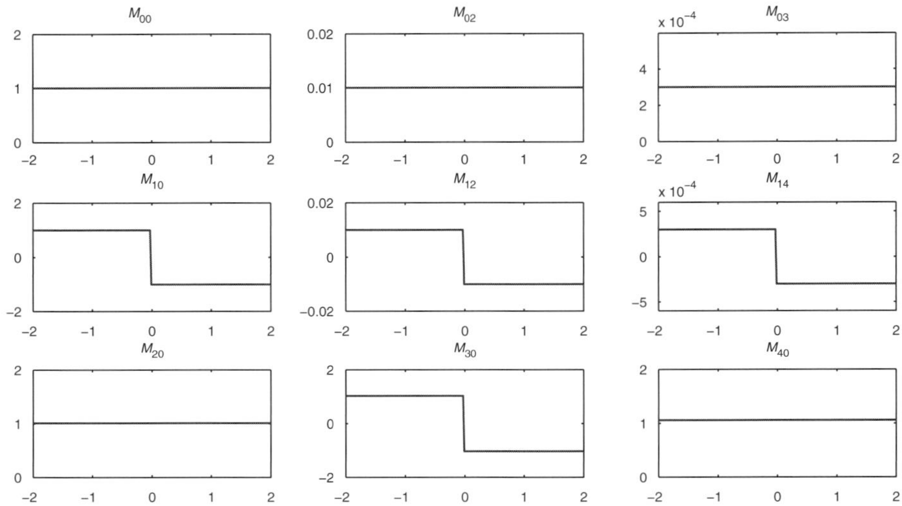

Figure 8.16. Nonzero initial moments for crossing jets.

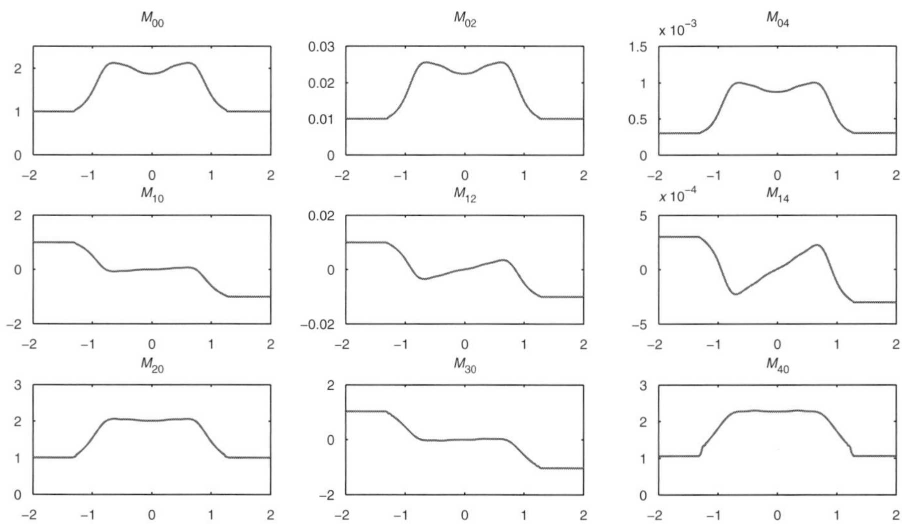

Figure 8.17. Nonzero moments for crossing jets with $\tau = 100$ at $t = 1$.

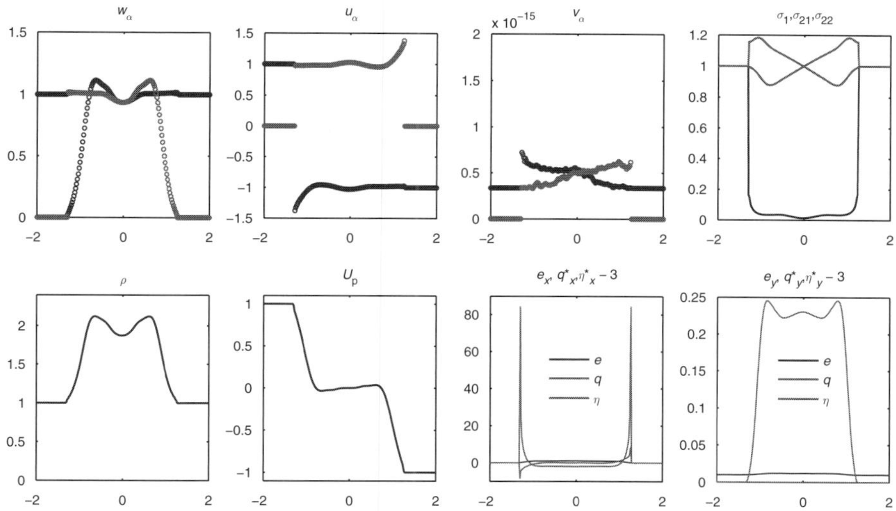

Figure 8.18. Statistics for crossing jets with $\tau = 100$ at $t = 1$.

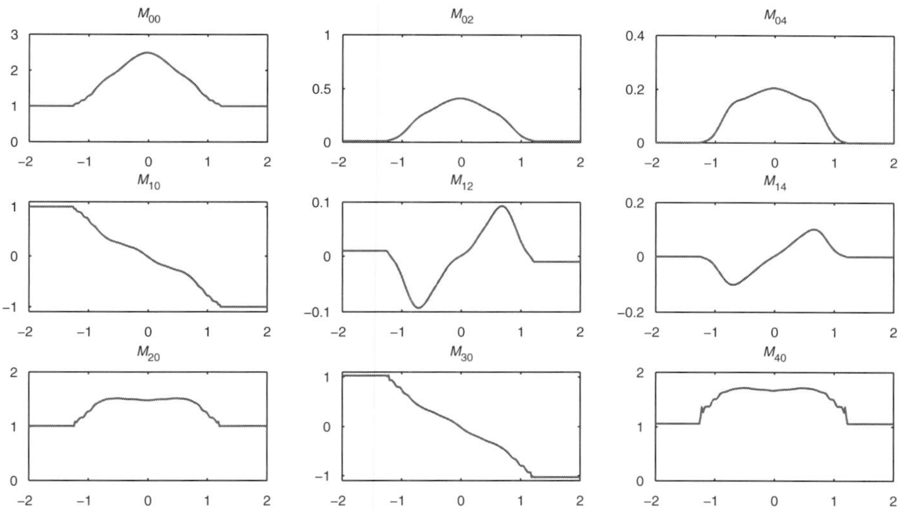

Figure 8.19. Nonzero moments for crossing jets with $\tau = 1$ at $t = 1$.

from the v-component statistics for energy e_y, skewness q_y, and kurtosis η_y, which are very close to their initial values. In comparison, the same statistics for the u-component exhibit strong deviations from equilibrium (as is expected for crossing jets). The behavior of the components of σ is also interesting. In the center of the domain, σ_1 is nearly zero, showing that the u-component of velocity has almost a double-Dirac distribution centered about the two velocity abscissas u_α. In contrast, the components σ_{21} and σ_{22} deviate from each other and from unity, as is expected when the v-component of velocity is not in equilibrium. The mean density ρ is higher in the center of the domain due to the jet crossing, and the mean velocity U_p exhibits a complex spatial dependence. Results for $\tau = 1$ at time $t = 1$ are shown in Figures 8.19 and 8.20. For this value of τ, the collisions are significant, so the relaxation to equilibrium occurs fairly rapidly. The v-component statistics for energy

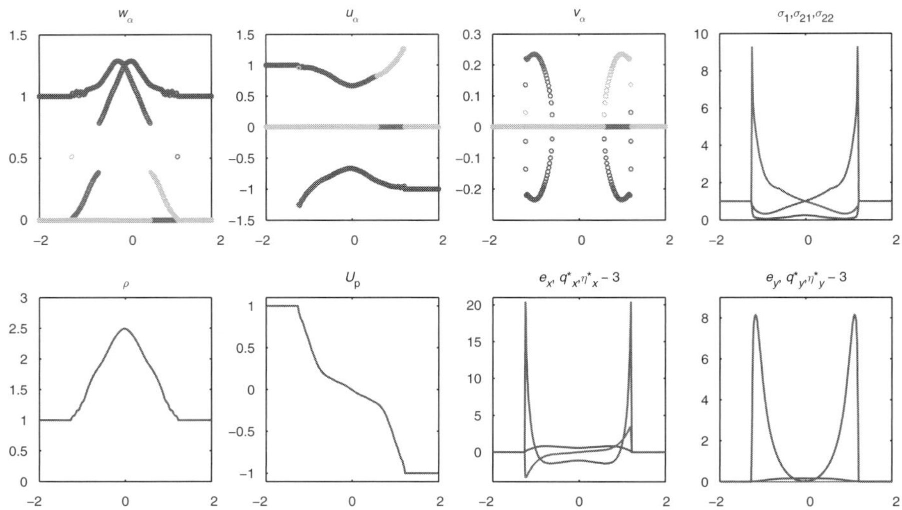

Figure 8.20. Statistics for crossing jets with $\tau = 1$ at $t = 1$. See the key for the different lines in Figure 8.18.

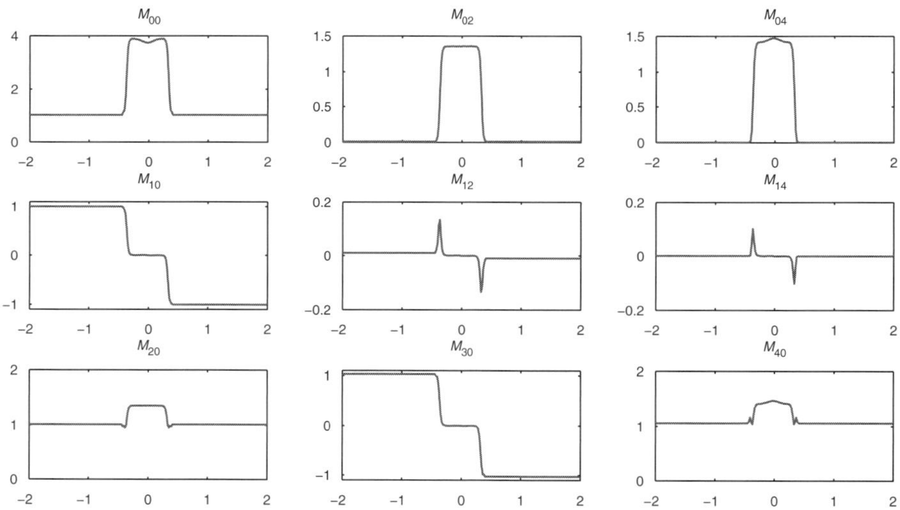

Figure 8.21. Nonzero moments for crossing jets with $\tau = 0.01$ at $t = 1$.

e_y, skewness q_y, and kurtosis η_y deviate significantly from the equilibrium values, while for the u-component these statistics again exhibit strong deviations from equilibrium. In the center of the domain, σ_1 is again close to zero, but now the components σ_{21} and σ_{22} deviate very strongly from the equilibrium values. Also note that the abscissas v_α are now far from zero in a significant part of the domain. The mean density ρ is, as expected, higher in the center of the domain, and the mean velocity U_p decreases monotonically from left to right. Results for $\tau = 0.01$ at time $t = 1$ are shown in Figures 8.21 and 8.22. For this value of τ, the collisions are very strong, so the relaxation to equilibrium is fast. The v-component statistics for energy e_y, skewness q_y, and kurtosis η_y are near the equilibrium values everywhere, expect at the points where the two jets cross. For the u-component,

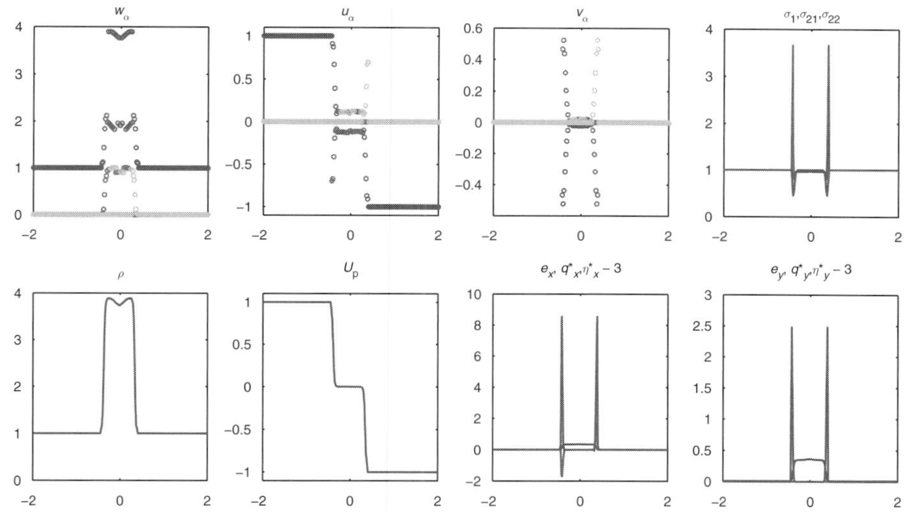

Figure 8.22. Statistics for crossing jets with $\tau = 0.01$ at $t = 1$. See the key for the different lines in Figure 8.18.

these statistics are also near equilibrium with the v-component. The abscissas v_α are near zero in most of the domain, except at the jet-crossing points. The mean density ρ is close to a square wave in the center of the domain, while the mean velocity U_p is near zero in the center of the domain. Overall, we can conclude from the crossing-jet example that the realizable scheme coupled to the collision term can capture the complex physics in velocity phase space without any difficulties, while always yielding realizable moment sets.

8.5 Inhomogeneous GPBE

8.5.1 Classes of GPBE

In the generalized PBE (GPBE), the phase-space variables include both the particle velocity and particle properties (such as size or volume). Generally speaking, there are two main classes of GPBE, which depend on how the particle velocity is represented:

(1) particle velocity is given by a known scalar-dependent advection velocity $u(\xi)$, and

(2) particle velocity is found from the joint velocity-scalar NDF.

The first class is a generalization of the advection–diffusion problem discussed in Section 8.3, and much of the material developed there can be reformulated to develop a realizable scheme for the GPBE. The second class is a generalization of the kinetic equation considered in Section 8.4, and has been referred to in the literature as the *semi-kinetic model* (Laurent *et al.*, 2004; Laurent & Massot, 2001). In the following, we will treat each class separately, although the reader will undoubtedly note many similarities between the numerical schemes. A third class of GPBE, lying between the two listed above, can also be identified wherein the scalar-dependent convection velocity has a parametric form such as

$$u(t, \mathbf{x}, \xi) = u_0(t, \mathbf{x}) + u_1(t, \mathbf{x})\xi + u_2(t, \mathbf{x})\xi^2. \quad (8.107)$$

In particular, the left-hand side can be interpreted as the particle velocity conditioned on the particle size, and then the right-hand side is a local (in real space) second-order approximation of the conditional velocity. However, in order to accurately reproduce the moment fluxes, the approximation in Eq. (8.107) must be close to the true velocity. Or, to put it a different way, the conditional variance of the particle velocity must be small so that the conditional velocity distribution function is tightly centered at $u(t, \mathbf{x}, \xi)$. This will be true, for example, when the particle Stokes number is below the critical value at which PTC begin to occur (i.e. particles with very small inertia).

Determination of velocity parameters

The flow-dependent parameters u_0, u_1, and u_2 in Eq. (8.107) can be determined by introducing the cross moments between the velocity v and size ξ. This is done by noting that the parametric form for the convection velocity is valid when the joint velocity-scalar NDF is

$$n(v, \xi) = \delta\left(v - u_0 - u_1\xi - u_2\xi^2\right) n(\xi), \tag{8.108}$$

where $n(\xi)$ is the size NDF. Thus, the joint integer moments are given by

$$m_{j,k} = \int_0^\infty \left(u_0 + u_1\xi + u_2\xi^2\right)^j \xi^k n(\xi) \mathrm{d}\xi, \tag{8.109}$$

and the first three cross moments are

$$\begin{aligned} m_{1,0} &= u_0 m_{0,0} + u_1 m_{0,1} + u_2 m_{0,2}, \\ m_{1,1} &= u_0 m_{0,1} + u_1 m_{0,2} + u_2 m_{0,3}, \\ m_{1,2} &= u_0 m_{0,2} + u_1 m_{0,3} + u_2 m_{0,4}. \end{aligned} \tag{8.110}$$

By transporting three additional moments, $m_{1,0}$, $m_{1,1}$, and $m_{1,2}$, this linear system can be solved for the three unknown parameters appearing in Eq. (8.107):

$$\begin{bmatrix} u_0 \\ u_1 \\ u_2 \end{bmatrix} = \begin{bmatrix} m_{0,0} & m_{0,1} & m_{0,2} \\ m_{0,1} & m_{0,2} & m_{0,3} \\ m_{0,2} & m_{0,3} & m_{0,4} \end{bmatrix}^{-1} \begin{bmatrix} m_{1,0} \\ m_{1,1} \\ m_{1,2} \end{bmatrix}. \tag{8.111}$$

The moments appearing in the inverse matrix on the right-hand side are the pure moments for the size NDF. Thus, as long as the size NDF is not composed of one or two delta functions, the inverse matrix will be well defined. Otherwise, in cases where the inverse is poorly conditioned, the order of the approximation in Eq. (8.107) can simply be reduced by setting u_2 (and, if necessary, u_0) to zero, and solving for a reduced set of parameters.[20] Note that this is done locally in space and time since the parameters evolve with the mixed moments. It then suffices to derive for the expanded moment set $\{m_{0,0}, m_{0,1}, m_{0,2}, m_{0,3}, m_{0,4}, m_{1,0}, m_{1,1}, m_{1,2}\}$ a realizable scheme that is consistent with the GPBE. We consider an example system of particles with Stokes drag in Section 8.5.4.

[20] More precisely, defining

$$\mathbf{A} = \frac{1}{m_{0,0}} \begin{bmatrix} m_{0,0} & m_{0,1} & m_{0,2} \\ m_{0,1} & m_{0,2} & m_{0,3} \\ m_{0,2} & m_{0,3} & m_{0,4} \end{bmatrix},$$

if the condition number of $\mathbf{A}$ is small enough for accurate inversion, we use Eq. (8.111). Otherwise we set $u_2 = 0$, redefine the coefficient matrix as

$$\mathbf{A} = \frac{1}{m_{0,0}} \begin{bmatrix} m_{0,0} & m_{0,1} \\ m_{0,1} & m_{0,2} \end{bmatrix},$$

and solve the reduced system for u_0 and u_1. However, if this new $\mathbf{A}$ is also poorly conditioned (i.e. the variance of ξ is negligible), then we set $u_0 = 0$ and $u_1 = m_{1,1}/m_{0,2}$.

Determination of velocity parameters with the EQMOM

If the EQMOM is used to reconstruct the size NDF, the joint velocity-scalar NDF for $\xi \in [0,1]$ can be written is a piecewise linear form:[21]

$$n(v,\xi) = \sum_{\alpha=1}^{2} w_\alpha \delta(v - U(\xi))\delta_\sigma(\xi,\xi_\alpha), \tag{8.112}$$

where the parameters w_α, σ, and ξ_α are defined in Section 3.3, the conditional velocity is modeled as $U(\xi) = u_0 g_0(\xi) + u_1 g_1(\xi) + u_2 g_2(\xi) + u_3 g_3(\xi)$, and the functions g_n are defined by

$$\begin{aligned}
g_0(\xi) &= \left(1 - \frac{\xi}{\xi_1}\right) I_{[0,\xi_1]}(\xi), \\
g_1(\xi) &= \left(\frac{\xi}{\xi_1}\right) I_{[0,\xi_1)}(\xi) + \left(\frac{\xi_2 - \xi}{\xi_2 - \xi_1}\right) I_{[\xi_1,\xi_2]}(\xi), \\
g_2(\xi) &= \left(\frac{\xi - \xi_1}{\xi_2 - \xi_1}\right) I_{[\xi_1,\xi_2)}(\xi) + \left(\frac{1 - \xi}{1 - \xi_2}\right) I_{[\xi_2,1]}(\xi), \\
g_3(\xi) &= \left(\frac{\xi - \xi_2}{1 - \xi_2}\right) I_{[\xi_2,1]}(\xi).
\end{aligned} \tag{8.113}$$

The indicator function $I_{[x_1,x_2]}(x)$ is unity if $x \in [x_1, x_2]$, and zero otherwise. Note that $g_0(0) = 1$, $g_1(\xi_1) = 1$, $g_2(\xi_2) = 1$, and $g_3(1) = 1$. In general, $0 < \xi_1 < \xi_2 < 1$ with beta EQMOM; however, if $\xi_1 = \xi_2$ then $u_1 = u_2$. In the degenerate case where $\sigma = 0$, only g_1 and g_2 are nonzero, so u_0 and u_3 are not defined. The four velocity parameters $(u_0, u_1, u_2,$ and $u_3)$ are determined from four mixed moments. The mixed integer moments are given by

$$m_{j,k} = \sum_{\alpha=1}^{2} w_\alpha \int_0^\infty \left[u_0 g_0(\xi) + u_1 g_1(\xi) + u_2 g_2(\xi) + u_3 g_3(\xi)\right]^j \xi^k \delta_\sigma(\xi,\xi_\alpha) \mathrm{d}\xi. \tag{8.114}$$

The velocity parameters are found using $j = 1$ and $k = 0, 1, 2, 3$. These moments can be written as

$$m_{1,k} = u_0 G_0^k + u_1 G_1^k + u_2 G_2^k + u_3 G_3^k, \tag{8.115}$$

where

$$G_n^k = \sum_{\alpha=1}^{2} w_\alpha \int_0^\infty g_n(\xi) \xi^k \delta_\sigma(\xi,\xi_\alpha) \mathrm{d}\xi = \sum_{\alpha=1}^{2} \sum_{\beta=1}^{M} w_{\alpha\beta} g_n(\xi_{\alpha\beta}) \left(\xi_{\alpha\beta}\right)^k, \tag{8.116}$$

and $w_{\alpha\beta}$ and $\xi_{\alpha\beta}$ are the weights and abscissas from the second Gauss–Jacobi quadrature. Note that M must be chosen large enough that the G_n^k are computed accurately. The linear system for the velocity parameters is

$$\begin{bmatrix} G_0^0 & G_1^0 & G_2^0 & G_3^0 \\ G_0^1 & G_1^1 & G_2^1 & G_3^1 \\ G_0^2 & G_1^2 & G_2^2 & G_3^2 \\ G_0^3 & G_1^3 & G_2^3 & G_3^3 \end{bmatrix} \begin{bmatrix} u_0 \\ u_1 \\ u_2 \\ u_3 \end{bmatrix} = \begin{bmatrix} m_{1,0} \\ m_{1,1} \\ m_{1,2} \\ m_{1,3} \end{bmatrix}. \tag{8.117}$$

[21] For clarity, we consider the case with two-node beta EQMOM defined for $\xi \in [0,1]$. More general forms are undoubtedly possible.

The coefficient matrix on the left-hand side will be invertible if $\xi_1 < \xi_2$ and $\sigma > 0$. In the limiting case where $\sigma = 0$, the coefficient matrix is

$$m_{0,0}\begin{bmatrix} 0 & 1 & 1 & 0 \\ 0 & \xi_1 & \xi_2 & 0 \\ 0 & \xi_1^2 & \xi_2^2 & 0 \\ 0 & \xi_1^3 & \xi_2^3 & 0 \end{bmatrix},$$

which is of rank 2 as long as $\xi_1 < \xi_2$; otherwise it is of rank 1. In the case of rank 1, only one velocity parameter (u_1) can be determined, whereas in the case of rank 2, both u_1 and u_2 can be found. These can be found (for $\sigma > 0$) by removing the last two rows of the system in Eq. (8.117). Thus, Eq. (8.117) can be solved to find the four velocity parameters. It then suffices to derive for the expanded moment set $\{m_{0,0}, m_{0,1}, m_{0,2}, m_{0,3}, m_{0,4}, m_{1,0}, m_{1,1}, m_{1,2}, m_{1,3}\}$ a realizable scheme that is consistent with the GPBE. It is interesting to note that, with the EQMOM reconstruction, more moments are transported than with Eq. (8.108), which would suggest that a better representation can be obtained using Eq. (8.112). However, due to the need to adjust the algorithm to account for the possible rank deficiency of Eq. (8.117), its numerical implementation is more difficult.

8.5.2 Spatial transport with known scalar-dependent velocity

Let ξ denote a nonnegative scalar with NDF $n(t, x, \xi)$. Generalizing Eq. (8.61), we will assume that the GPBE is

$$\frac{\partial n}{\partial t} + \frac{\partial u(\xi) n}{\partial x} = \frac{\partial}{\partial x}\left(\Gamma(\xi)\frac{\partial n}{\partial x}\right), \tag{8.118}$$

where the advection velocity $u(\xi)$ is a known function of ξ. The moment-transport equation for this case is

$$\frac{\partial m_k}{\partial t} + \frac{\partial}{\partial x}\left(\int_0^\infty \xi^k u(\xi) n \, \mathrm{d}\xi\right) = \frac{\partial}{\partial x}\left(\int_0^\infty \xi^k \Gamma(\xi)\frac{\partial n}{\partial x} \, \mathrm{d}\xi\right), \tag{8.119}$$

and it is clear that a closure will be needed for the spatial fluxes. As done in Section 8.3.3, a quadrature-based closure can be used for this purpose:

$$\int_0^\infty \xi^k u(\xi) n \, \mathrm{d}\xi = \sum_{\alpha=1}^{2}\sum_{\beta=1}^{M} W_{\alpha\beta} u\left(\xi_{\alpha\beta}\right)\left(\xi_{\alpha\beta}\right)^k,$$

where $W_{\alpha\beta}$ and $\xi_{\alpha\beta}$ depend on the Gauss–Laguerre quadrature nodes. For this case, the value of M must be chosen large enough to represent the flux integral to the desired accuracy. In practice, this will depend on the exact functional form of $u(\xi)$. The realizable scheme for solving Eq. (8.119) is the same as in Eq. (8.89), but with a slightly different definition for $G^{k,n}_{\alpha\beta,i+1/2}$:

$$G^{k,n}_{\alpha\beta,i+1/2} = W^n_{\alpha\beta,i+1/2,l} \max\left[u\left(\xi^n_{\alpha\beta,i}\right), 0\right]\left(\xi^n_{\alpha\beta,i}\right)^k + W^n_{\alpha\beta,i+1/2,r} \min\left[u\left(\xi^n_{\alpha\beta,i+1}\right), 0\right]\left(\xi^n_{\alpha\beta,i+1}\right)^k. \tag{8.120}$$

The formulas for the weights $W^n_{\alpha\beta,i+1/2,l}$ and $W^n_{\alpha\beta,i+1/2,r}$ are the same as those given in Eqs. (8.93) and (8.94). The realizability condition in Eq. (8.88) must be modified to read

$$\min_{\alpha,\beta,i}\left\{W^n_{\alpha\beta,i}-\left(W^n_{\alpha\beta,i+1/2}-\frac{1}{2}D^n_{\alpha\beta,i+1/2}\right)c^n_{\alpha\beta,i}\lambda_1-W^n_{\alpha\beta,i+1/2,l}\max\left[u\left(\xi^n_{\alpha\beta,i}\right),0\right]\lambda_2\right.$$
$$\left.-\left(W^n_{\alpha\beta,i-1/2}+\frac{1}{2}D^n_{\alpha\beta,i-1/2}\right)c^n_{\alpha\beta,i}\lambda_1+W^n_{\alpha\beta,i-1/2,r}\min\left[u\left(\xi^n_{\alpha\beta,i}\right),0\right]\lambda_2\right\}>0. \quad (8.121)$$

As is done for advection–diffusion cases, the time step is fixed by setting an overall CFL condition:

$$\Delta t=\mathrm{CFL}\left(\frac{2\Gamma}{\Delta x^2}+\max_{\alpha,\beta,i}\left[\frac{\left|u\left(\xi^n_{\alpha\beta,i}\right)\right|}{\Delta x}\right]\right)^{-1}.$$

With this value for Δt, the order of the convection scheme in grid cell i is determined from Eq. (8.121).

8.5.3 Example results with known scalar-dependent velocity

As examples of scalar-dependent velocity, we will consider two cases in which the velocity is a linear function of ξ. For simplicity, we will assume that at time zero ξ can be represented by a beta distribution on interval [0, 1] with the spatial distribution of the moments shown in Figure 8.23. As discussed in Appendix C, the evolution of the moments depends on whether the velocity is an increasing or decreasing function of size. Here we consider a segregating system with velocity $u(\xi) = -1 + 2\xi$, defined such that for $\xi > 0.5$ the particles rise, while for $\xi < 0.5$ the particles fall; and a mixing system with $u(\xi) = 1 - 2\xi$. From the initial conditions in Figure 8.23, we can observe that initially the smaller particles are located below the point $x = 0$, while the larger particles are located above this point. Thus, for the segregating system the particles will move away from $x = 0$, creating a void; whereas for the mixing system the particles on the top and bottom halves of the domain will move towards each other, causing particle mixing. For these simulations, we will use beta EQMOM with $n = 2$ to reconstruct the NDF, and set the diffusivity Γ equal to zero. Thus, as in Appendix C, the GPBE is purely hyperbolic with crossing characteristics in

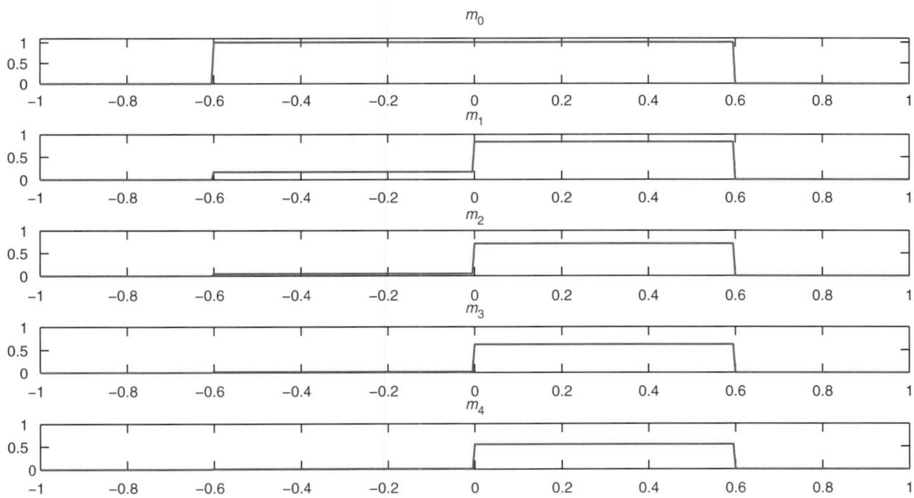

Figure 8.23. Initial moments for scalar-dependent-velocity cases.

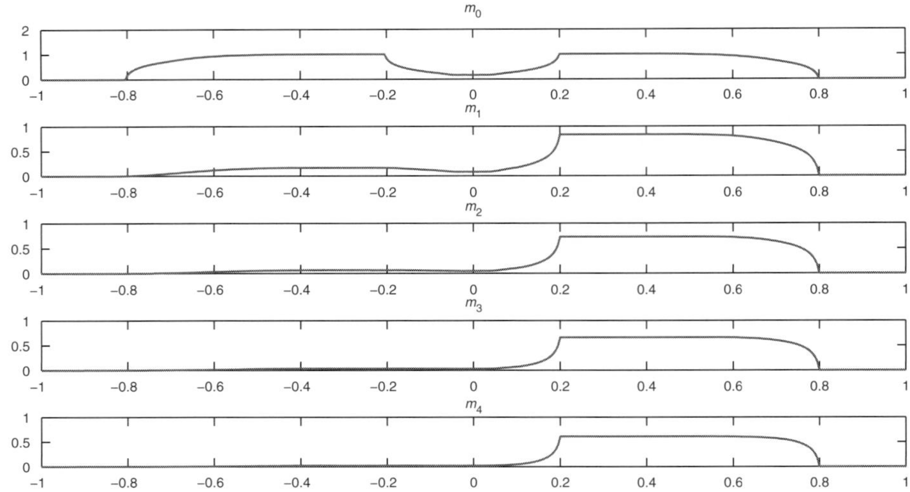

Figure 8.24. Moments for the segregating system at $t = 0.2$.

the mixing case. In order to achieve highly accurate spatial fluxes, we use $M = 100$ in the second Gaussian quadrature and set the CFL number to unity. Thus, all of the quadrature error is due to the first beta EQMOM quadrature. Note that unlike in Appendix C, where the initial NDF is cleanly separated into two parts (i.e. above and below $\xi = 0.5$), here the EQMOM representation uses beta distributions on the entire interval [0, 1], which implies that above and below $x = 0$ there will initially be particles of all possible sizes (albeit with very different mean sizes, as shown in Figure 8.23).

The segregating system

Sample results for the segregating system at time $t = 0.2$ are shown in Figures 8.24 and 8.25. As expected, m_0 has a minimum at $x = 0$ as particles are advected away from the midpoint. The hyperbolic nature of the underlying GPBE is clearly evident in the moment profiles in Figure 8.24, where we see that particles have risen/fallen by exactly 0.2 spatial units relative to their initial locations in Figure 8.23. Since there are no crossing characteristics for the segregating system, the moment profiles behave as would be expected, with the smaller particles falling and the larger particles rising. Nevertheless, the weights and abscissas for the beta EQMOM shown in Figure 8.25 exhibit a complex spatial dependence, as does the spread parameter σ. Note that the mean particle size m_1/m_0 increases monotonically with x from zero to unity, and thus that the beta EQMOM is able to handle the two extreme cases in which the NDF is nearly a delta function at the boundary of the interval. Finally, in Figure 8.25, the spatial distributions of the first five flux abscissas $\xi_{\alpha\beta}$ and flux weights $W_{\alpha\beta}$ for $\alpha = 1$ are shown. From the flux abscissas, it can be seen that the size phase space is well sampled by using the second Gaussian quadrature with $M = 100$ for the fluxes.

The mixing system

Sample results for the mixing system at time $t = 0.2$ are shown in Figures 8.26 and 8.27. As expected, m_0 has a maximum at $x = 0$ as particles are advected towards from the

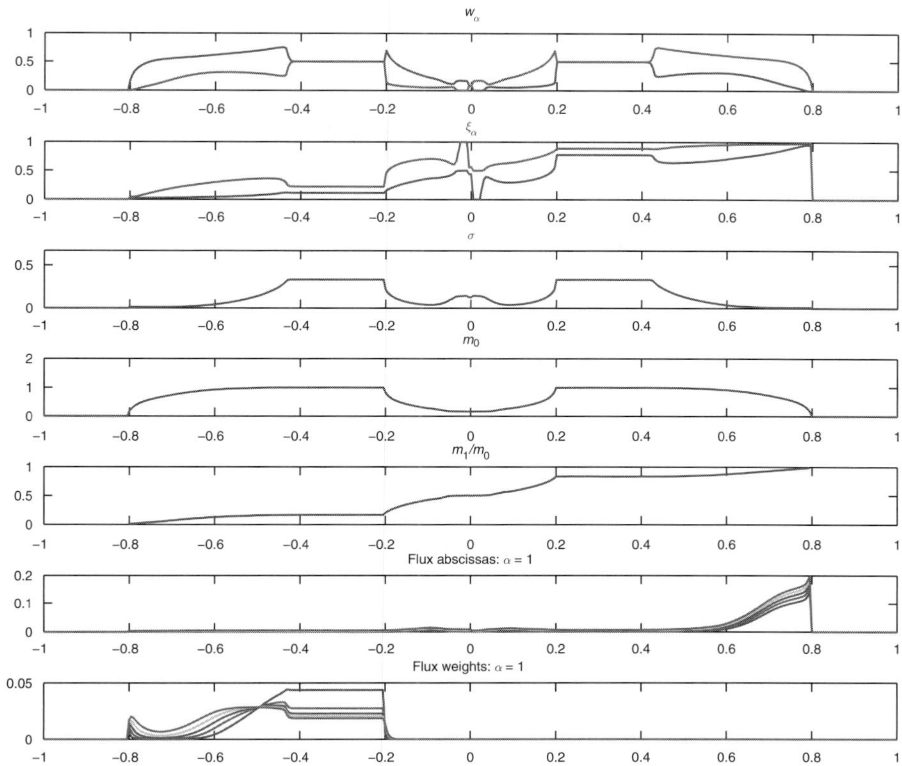

Figure 8.25. Statistics for the segregating system at $t = 0.2$.

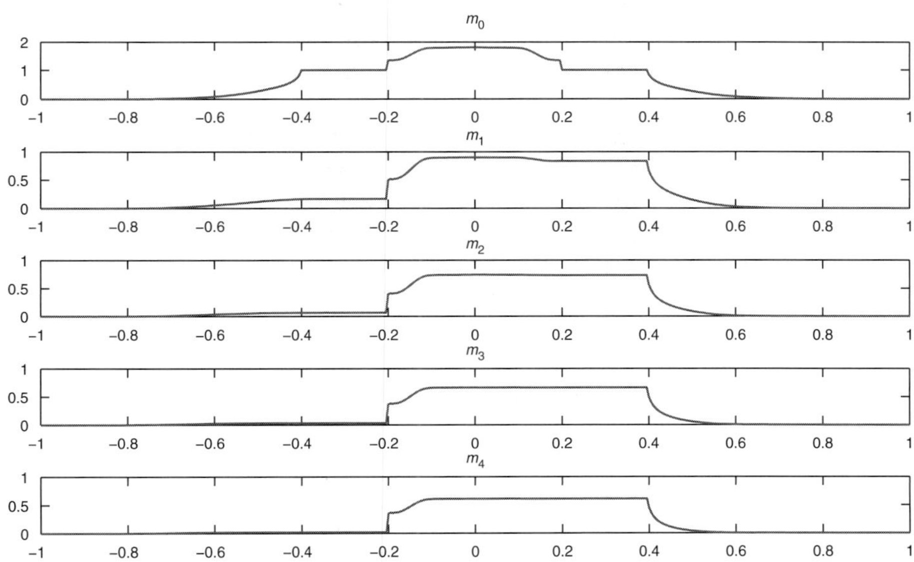

Figure 8.26. Moments for the mixing system at $t = 0.2$.

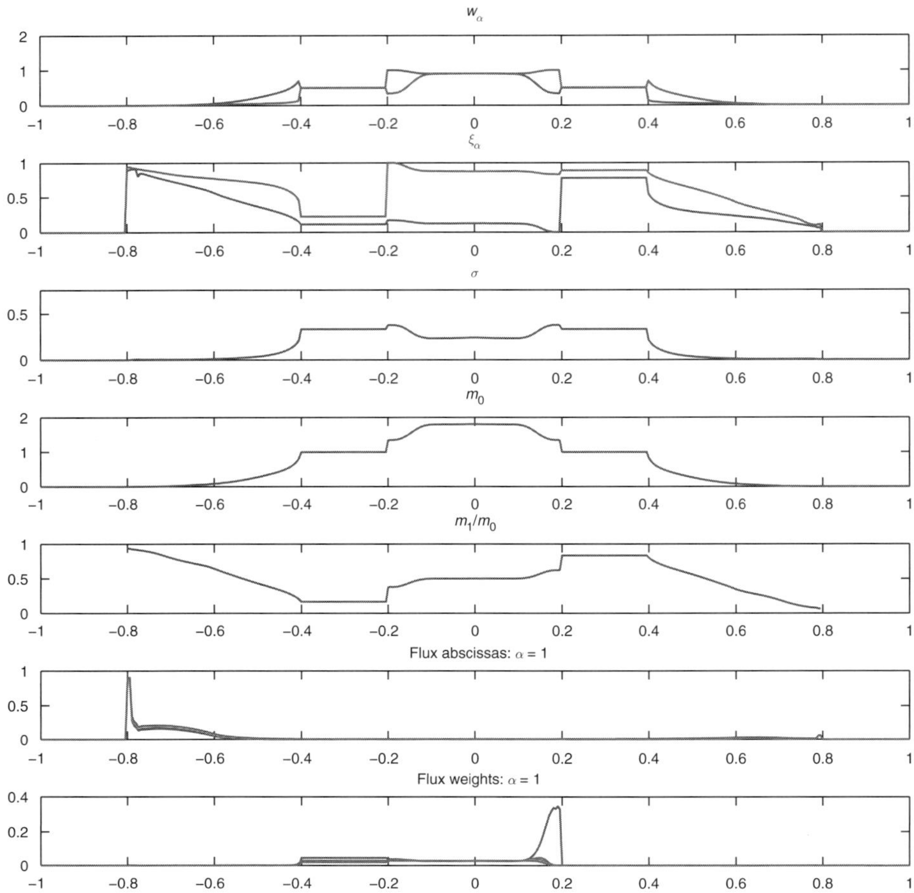

Figure 8.27. Statistics for the mixing system at $t = 0.2$.

midpoint. The hyperbolic nature of the underlying GPBE is again clearly evident in the moment profiles in Figure 8.26, where we see that particles have risen/fallen by exactly 0.2 spatial units relative to their initial locations in Figure 8.23. In this example, crossing characteristics are an important feature and the moment profiles exhibit mixing behavior as the larger/smaller particles overlap in space. Here again the weights and abscissas for the beta EQMOM shown in Figure 8.27 exhibit a complex spatial dependence, as does the spread parameter σ. Moreover, the mean particle size m_1/m_0 first decreases from unity and then increases towards 0.5 for $x < 0$ due to the mixing behavior of the characteristics. In Figure 8.27, the spatial distributions of the first five flux abscissas $\xi_{\alpha\beta}$ and flux weights $W_{\alpha\beta}$ for $\alpha = 1$ exhibit a nontrivial dependence on x. Overall, the realizable scheme has no difficulties dealing with characteristics that cross due to the initial conditions and the dependence of velocity on the particle size. As mentioned earlier, the accuracy of the solution to the moment-transport equations is controlled by the beta EQMOM quadrature and could be improved by using a larger value for n (i.e. the number of EQMOM nodes). Nevertheless, despite their complex spatial profiles, the moment sets are always realizable.

8.5.4 Spatial transport with scalar-conditioned velocity

In this section, we consider the intermediate case in which the particle velocity is given by Eq. (8.107) with flow-dependent parameters u_0, u_1, and u_2, or by the EQMOM representation given in Eq. (8.112). As a starting point, we will take the following GPBE for the joint velocity-size NDF $n(t, x, v, \xi)$:

$$\frac{\partial n}{\partial t} + \frac{\partial vn}{\partial x} + \frac{1}{\tau(\xi)}\frac{\partial vn}{\partial v} = 0, \tag{8.122}$$

where $\tau(\xi) = \tau_0 \xi^{2/3}$ is the characteristic time scale for drag for a particle with normalized volume $\xi = 1$. In this GPBE, ξ is the particle volume and the drag term implicitly assumes that the fluid velocity is null. Thus, the drag term will have the effect of damping out the initial particle velocity at a size-dependent decay rate. The moment-transport equation corresponding to Eq. (8.122) is

$$\frac{\partial m_{j,k}}{\partial t} + \frac{\partial m_{j+1,k}}{\partial x} = -\frac{j m_{j,k-2/3}}{\tau_0}. \tag{8.123}$$

Thus, the expanded moment set is governed by[22]

$$\begin{aligned}
\frac{\partial m_{0,0}}{\partial t} + \frac{\partial m_{1,0}}{\partial x} &= 0,\\
\frac{\partial m_{0,1}}{\partial t} + \frac{\partial m_{1,1}}{\partial x} &= 0,\\
\frac{\partial m_{0,2}}{\partial t} + \frac{\partial m_{1,2}}{\partial x} &= 0,\\
\frac{\partial m_{0,3}}{\partial t} + \frac{\partial m_{1,3}}{\partial x} &= 0,\\
\frac{\partial m_{0,4}}{\partial t} + \frac{\partial m_{1,4}}{\partial x} &= 0,\\
\frac{\partial m_{1,0}}{\partial t} + \frac{\partial m_{2,0}}{\partial x} &= -\frac{1}{\tau_0} m_{1,-2/3},\\
\frac{\partial m_{1,1}}{\partial t} + \frac{\partial m_{2,1}}{\partial x} &= -\frac{1}{\tau_0} m_{1,1/3},\\
\frac{\partial m_{1,2}}{\partial t} + \frac{\partial m_{2,2}}{\partial x} &= -\frac{1}{\tau_0} m_{1,4/3},\\
\frac{\partial m_{1,3}}{\partial t} + \frac{\partial m_{2,3}}{\partial x} &= -\frac{1}{\tau_0} m_{1,7/3}.
\end{aligned}$$

Note that both the spatial-flux terms and the drag terms require moment closures. These closures will depend on whether the NDF is represented by Eq. (8.108) or Eq. (8.112). When the NDF is represented by Eq. (8.108), the transport equation for $m_{1,3}$ is not used.

[22] The choice of the moment set is not unique. However, the moment $m_{1,1}$ corresponds to the momentum density, which is an overall conserved quantity, and thus is a natural choice to include in the expanded moment set.

Velocity parameters from Eq. (8.108)

As an example, we will employ a two-node beta EQMOM to represent $n(\xi)$ and, thus, the mixed moments can be expressed in closed form as

$$m_{j,k} = \sum_{\alpha=1}^{2} \sum_{\beta=1}^{M} W_{\alpha\beta} \left[u_0(t,x) + u_1(t,x)\xi_{\alpha\beta} + u_2(t,x)\xi_{\alpha\beta}^2 \right]^j \xi_{\alpha\beta}^k, \tag{8.124}$$

where $W_{\alpha\beta}$ and $\xi_{\alpha\beta}$ depend on the Gauss–Jacobi quadrature nodes. The functions $u_0(t,x)$, $u_1(t,x)$, and $u_2(t,x)$ are found from the transported moment set using Eq. (8.111). The fractional moments appearing in the drag terms in Eq. (8.123) can be evaluated using Eq. (8.124) with k set to the appropriate fraction. In cases in which the NDF has a nonzero value at $\xi = 0$, the moment $m_{1,-2/3}$ can be infinite, and thus the numerical method must account for this possibility. The system in Eq. (8.123) is conveniently solved using operator splitting to separate the flux terms from the drag terms. In the first step, the flux terms are updated by solving

$$\frac{\partial m_{j,k}}{\partial t} + \frac{\partial m_{j+1,k}}{\partial x} = 0. \tag{8.125}$$

The realizable scheme for solving Eq. (8.125) is the same as in Eq. (8.89) (without diffusive fluxes), but with a slightly different definition for the numerical fluxes:

$$\begin{aligned} G^{j,k;n}_{\alpha\beta,i+1/2} = {} & W^n_{\alpha\beta,i+1/2,l} \max\left[u^n_i\left(\xi^n_{\alpha\beta,i}\right), 0\right]\left[u^n_i\left(\xi^n_{\alpha\beta,i}\right)\right]^j \left(\xi^n_{\alpha\beta,i}\right)^k \\ & + W^n_{\alpha\beta,i+1/2,r} \min\left[u^n_{i+1}\left(\xi^n_{\alpha\beta,i+1}\right), 0\right]\left[u^n_{i+1}\left(\xi^n_{\alpha\beta,i+1}\right)\right]^j \left(\xi^n_{\alpha\beta,i+1}\right)^k. \end{aligned} \tag{8.126}$$

The formulas for the weights $W^n_{\alpha\beta,i+1/2,l}$ and $W^n_{\alpha\beta,i+1/2,r}$ are the same as those given in Eqs. (8.93) and (8.94). The function $u^n_i(\xi)$ is defined as $u^n_i(\xi) = u^n_{0;i} + u^n_{1;i}\xi + u^n_{2;i}\xi^2$ by assuming that the parameters in Eq. (8.110) are piecewise constant in each finite-volume cell.[23] This leads to the linear system

$$\begin{bmatrix} u^n_{0;i} \\ u^n_{1;i} \\ u^n_{2;i} \end{bmatrix} = \begin{bmatrix} m^n_{0,0;i} & m^n_{0,1;i} & m^n_{0,2;i} \\ m^n_{0,1;i} & m^n_{0,2;i} & m^n_{0,3;i} \\ m^n_{0,2;i} & m^n_{0,3;i} & m^n_{0,4;i} \end{bmatrix}^{-1} \begin{bmatrix} m^n_{1,0;i} \\ m^n_{1,1;i} \\ m^n_{1,2;i} \end{bmatrix}, \tag{8.127}$$

where only the volume-average transported moments appear on the right-hand side. Note that this specification of the parameters is consistent with the realizable schemes wherein the velocity abscissas are constant in each grid cell. The realizability condition in Eq. (8.88) must be modified to read

$$\min_{\alpha,\beta,i}\left\{W^n_{\alpha\beta,i} - W^n_{\alpha\beta,i+1/2,l}\max\left[u^n_i\left(\xi^n_{\alpha\beta,i}\right),0\right]\lambda_1 + W^n_{\alpha\beta,i-1/2,r}\min\left[u^n_i\left(\xi^n_{\alpha\beta,i}\right),0\right]\lambda_1\right\} > 0, \tag{8.128}$$

where $\lambda_1 = \Delta t/\Delta x$. The time step is fixed by setting an overall CFL condition:

$$\Delta t = \text{CFL} \min_{\alpha,\beta,i}\left[\frac{\Delta x}{\left|u^n_i\left(\xi^n_{\alpha\beta,i}\right)\right|}\right]$$

[23] The finite-volume representation of the mixed moments in each cell is found by computing the volume average of Eq. (8.110). On the right-hand side the volume averages of products such as $u_0 m_{0,0}$ appear. Thus, if u_0 is piecewise constant in the cell, the volume-average product depends on the volume-average moments, which are the conserved quantities in the finite-volume scheme.

for CFL ≤ 1. With this value for Δt, the order of the convection scheme in grid cell i is determined from Eq. (8.128). At the end of the advection step, the updated moment set is denoted by $m^*_{j,k;i}$, and is guaranteed to be realizable.

Next, the drag terms are solved using an equivalent ODE system for the velocity parameters with fixed weights and abscissas. The ODE system is found starting from

$$\frac{\mathrm{d}m_{1,k;i}}{\mathrm{d}t} = -\frac{1}{\tau_0} m_{1,k-2/3;i},$$

which must be solved with initial conditions $m_{1,k;i}(0) = m^*_{1,k;i}$. By substituting for the moments written in terms of the weights and abscissas the expression

$$m_{1,k;i}(t) = \sum_{\alpha=1}^{2} \sum_{\beta=1}^{M} W_{\alpha\beta,i} \left[u_{0;i}(t) + u_{1;i}(t)\xi_{\alpha\beta,i} + u_{2;i}(t)\xi^2_{\alpha\beta,i} \right] \xi^k_{\alpha\beta,i},$$

we find an ODE system for each grid cell i:

$$\sum_{\alpha=1}^{2} \sum_{\beta=1}^{M} W_{\alpha\beta;i} \left(\frac{\mathrm{d}u_{0;i}}{\mathrm{d}t} + \frac{\mathrm{d}u_{1;i}}{\mathrm{d}t}\xi_{\alpha\beta,i} + \frac{\mathrm{d}u_{2;i}}{\mathrm{d}t}\xi^2_{\alpha\beta,i} \right) \xi^k_{\alpha\beta,i}$$
$$= -\frac{1}{\tau_0} \sum_{\alpha=1}^{2} \sum_{\beta=1}^{M} W_{\alpha\beta,i} \left(u_{0;i} + u_{1;i}\xi_{\alpha\beta,i} + u_{2;i}\xi^2_{\alpha\beta,i} \right) \xi^{k-2/3}_{\alpha\beta,i}, \quad (8.129)$$

which is the same as

$$m_{0,k;i}\frac{\mathrm{d}u_{0;i}}{\mathrm{d}t} + m_{0,k+1;i}\frac{\mathrm{d}u_{1;i}}{\mathrm{d}t} + m_{0,k+2;i}\frac{\mathrm{d}u_{2;i}}{\mathrm{d}t} = -\frac{1}{\tau_0}\left(m_{0,k-2/3;i}u_{0;i} + m_{0,k+1/3;i}u_{1;i} + m_{0,k+4/3;i}u_{2;i}\right). \quad (8.130)$$

The pure size moments $m_{0,k;i}$ are time-independent during the drag step, so one can solve this constant-coefficient, linear ODE system for the parameters $u_{0;i}(t)$, $u_{1;i}(t)$, and $u_{2;i}(t)$ given their initial conditions found from Eq. (8.127) and the moments $m^*_{j,k;i}$. With $k = 0, 1, 2$, the linear ODE system resulting from Eq. (8.130) can be written as (dropping the i dependence for clarity)[24]

$$\mathbf{A}\frac{\mathrm{d}}{\mathrm{d}t}\begin{bmatrix} u_0 \\ u_1 \\ u_2 \end{bmatrix} = -\frac{1}{\tau_0}\mathbf{B}\begin{bmatrix} u_0 \\ u_1 \\ u_2 \end{bmatrix}, \quad (8.131)$$

where the coefficient matrices are defined in terms of the size moments by

$$\mathbf{A} = \begin{bmatrix} m_{0,0} & m_{0,1} & m_{0,2} \\ m_{0,1} & m_{0,2} & m_{0,3} \\ m_{0,2} & m_{0,3} & m_{0,4} \end{bmatrix}, \quad \mathbf{B} = \begin{bmatrix} m_{0,-2/3} & m_{0,1/3} & m_{0,4/3} \\ m_{0,1/3} & m_{0,4/3} & m_{0,7/3} \\ m_{0,4/3} & m_{0,7/3} & m_{0,10/3} \end{bmatrix}. \quad (8.132)$$

Note that, during the drag evaluation, the size moments $m_{0,k}$ are constants,[25] and thus **A** and **B** are constant matrices. Moreover, these coefficient matrices are both symmetric and positive definite, implying that the eigenvalues for the first-order ODE system in

[24] If the fluid velocity were nonzero, the right-hand side of Eq. (8.131) would have an additional inhomogeneous term depending on the fluid velocity. If the fluid velocity u_f were time-independent, an equivalent representation would be found by simply replacing u_0 by $u_0 - u_f$ on the right-hand side.

[25] The size moments are constant because the weights and abscissas are constant, which is consistent with the assumption used to derive Eq. (8.130).

Eq. (8.131) are real-valued and negative. Thus, the parameters $u_0(t), u_1(t)$, and $u_2(t)$ will decay exponentially towards zero with increasing t.[26]

Using an implicit Euler scheme to update the parameter vector (and the definition in Eq. (8.111) to rewrite the right-hand side), we find for grid cell i:

$$\begin{bmatrix} u_{0;i}^{n+1} \\ u_{1;i}^{n+1} \\ u_{2;i}^{n+1} \end{bmatrix} = \left(\mathbf{A}_i^* + \frac{\Delta t}{\tau_0} \mathbf{B}_i^* \right)^{-1} \begin{bmatrix} m_{1,0;i}^* \\ m_{1,1;i}^* \\ m_{1,2;i}^* \end{bmatrix}, \tag{8.133}$$

where $m_{j,k;i}^*$ are the volume-average moments in cell i after the convection step. The matrices $\mathbf{A}_i^*$ and $\mathbf{B}_i^*$ are evaluated with $W_{\alpha\beta;i}^*$ and $\xi_{\alpha\beta;i}^*$, which are the weights and abscissas found using the EQMOM with the moments $m_{0,k;i}^*$. The time step Δt is the same as the one used in the convection step. An important special case occurs when $m_{0,-2/3} \to \infty$, i.e. the NDF has droplets approaching zero volume due, for example, to evaporation. In this case, the inverse matrix on the right-hand side of Eq. (8.133) reduces to[27]

$$\lim_{m_{0,-2/3} \to \infty} (\mathbf{A} + \tau \mathbf{B})^{-1} = \begin{bmatrix} 0 & 0 & 0 \\ 0 & m_{0,4} + m_{0,10/3}\tau & -m_{0,3} - m_{0,7/3}\tau \\ 0 & -m_{0,3} - m_{0,7/3}\tau & m_{0,2} + m_{0,4/3}\tau \end{bmatrix}, \tag{8.134}$$

where $\tau = \Delta t / \tau_0$ and[28] $D = m_{0,2}m_{0,4} - m_{0,3}^2 + (m_{0,4/3}m_{0,4} - 2m_{0,7/3}m_{0,3} + m_{0,10/3}m_{0,2})\tau + (m_{0,4/3}m_{0,10/3} - m_{0,7/3}^2)\tau^2$. Here, $D = 0$ implies that the NDF is a delta function at $\xi = 0$, otherwise $D > 0$. From this expression, we can clearly see that $u_{0;i}^* = 0$ when $m_{0,-2/3} = \infty$. In practice, Eq. (8.134) could be used in Eq. (8.133) whenever $m_{0,-2/3}$ is very large relative to $m_{0,0}$. Because $m_{0,-2/3}$ is evaluated from the Gaussian quadrature weights and abscissas ($W_{\alpha\beta;i}^*$ and $\xi_{\alpha\beta;i}^*$), it will never be infinite. The matrix exponential in the exact solution to Eq. (8.131) may be difficult to implement, especially in cases for which $\mathbf{B}$ is ill-conditioned.

An alternative formulation for the drag force that is physically consistent and numerically robust can be implemented as follows.

(i) After the advection step, use the moments $m_{0,k;i}^*$ to compute the weights $W_{\alpha\beta,i}^*$ and abscissas $\xi_{\alpha\beta,i}^*$.

(ii) Construct the matrix $\mathbf{A}$ in Eq. (8.127) using the moments $m_{0,k;i}^*$, and compute the velocity parameters resulting from advection:

$$\begin{bmatrix} u_{0;i}^* \\ u_{1;i}^* \\ u_{2;i}^* \end{bmatrix} = \mathbf{A}^{-1} \begin{bmatrix} m_{1,0;i}^* \\ m_{1,1;i}^* \\ m_{1,2;i}^* \end{bmatrix}.$$

[26] The exact solution is $\mathbf{u}(t) = \exp\left(-\mathbf{A}^{-1}\mathbf{B}t/\tau_0\right)\mathbf{u}(0) = \exp\left(-\mathbf{A}^{-1}\mathbf{B}t/\tau_0\right)\mathbf{A}^{-1}\mathbf{m}(0)$, where the matrix $\mathbf{C} = \mathbf{A}^{-1}\mathbf{B}$ is positive definite. Because $m_{0,-2/3}$ is evaluated from the quadrature (as are all of the fractional moments appearing in $\mathbf{B}$), it should always be bounded, although it may be very large. Thus, in principle, the exact solution can be used to update the velocity coefficients.

[27] If a linear function is used to represent $u(\xi)$, the corresponding matrices are

$$\mathbf{A} = \begin{bmatrix} m_{0,0} & m_{0,1} \\ m_{0,1} & m_{0,2} \end{bmatrix}, \quad \mathbf{B} = \begin{bmatrix} m_{0,-2/3} & m_{0,1/3} \\ m_{0,1/3} & m_{0,4/3} \end{bmatrix},$$

and the limiting expression is

$$\lim_{m_{0,-2/3} \to \infty} (\mathbf{A} + \tau \mathbf{B})^{-1} = \frac{1}{D} \begin{bmatrix} 0 & 0 \\ 0 & 1/(m_{0,2} + m_{0,4/3}\tau) \end{bmatrix}.$$

In this case, $u_{0;i}^{n+1} = 0$ and $u_{1;i}^{n+1} = m_{1,1;i}^*/(m_{0,2;i}^* + m_{0,4/3;i}^*\tau)$.

[28] From the definition of τ_0, τ has units of length squared and thus the components of $\mathbf{A}$ and $\mathbf{B}$ have consistent units.

Recall that **A** may be ill-conditioned (or even singular), so care must be taken when treating the degenerate cases during this step.

(iii) Use the velocity parameters to create an ensemble of sample velocities corresponding to each $W^*_{\alpha\beta,i}$ and $\xi^*_{\alpha\beta,i}$: $V^*_{\alpha\beta;i} = u^*_{0;i} + u^*_{1;i}\xi^*_{\alpha\beta,i} + u^*_{2;i}(\xi^*_{\alpha\beta,i})^2$ for $\alpha = 1, 2$ and $\beta = 1, 2, \ldots, M$.

(iv) Update the sample velocities according to the drag law:

$$\frac{\mathrm{d}u}{\mathrm{d}t} = -\frac{u}{\tau_0 \xi^{2/3}} \quad \Longrightarrow \quad V^\dagger_{\alpha\beta;i} = \exp\left(-\frac{\Delta t}{\tau_0 \left(\xi^*_{\alpha\beta,i}\right)^{2/3}}\right) V^*_{\alpha\beta;i}.$$

The scalar exponential in this step is always well defined, even in the limit where $\xi^*_{\alpha\beta,i} \to 0$.

(v) Use the new sample velocities to compute the sample mixed moments:

$$m^\dagger_{1,k;i} = \sum_{\alpha=1}^{2} \sum_{\beta=1}^{M} W^*_{\alpha\beta;i} V^\dagger_{\alpha\beta;i} \left(\xi^*_{\alpha\beta;i}\right)^k.$$

(vi) Update the velocity parameters using the sample mixed moments and **A** from step (ii):

$$\begin{bmatrix} u^{n+1}_{0;i} \\ u^{n+1}_{1;i} \\ u^{n+1}_{2;i} \end{bmatrix} = \mathbf{A}^{-1} \begin{bmatrix} m^\dagger_{1,0;i} \\ m^\dagger_{1,1;i} \\ m^\dagger_{1,2;i} \end{bmatrix}.$$

This formulation is also convenient for situations in which other forces (e.g. buoyancy, virtual mass, lift, etc.) are present. For such cases, it suffices to add the additional forces into the update of the sample velocities in step (iv). After updating the velocity parameters, the transported moments at the end of the drag step are found from

$$m^{n+1}_{j,k;i} = \begin{cases} m^*_{0,k;i} & \text{if } j = 0, \\ \sum_{\alpha=1}^{2} \sum_{\beta=1}^{M} W^*_{\alpha\beta;i} \left[u^{n+1}_{0;i} + u^{n+1}_{1;i} \xi^*_{\alpha\beta;i} + u^{n+1}_{2;i} \left(\xi^*_{\alpha\beta;i}\right)^2\right] \left(\xi^*_{\alpha\beta;i}\right)^k & \text{if } j = 1. \end{cases} \tag{8.135}$$

Because the weights are constant during the drag step, and always nonnegative after the advection step, the updated moment set in Eq. (8.135) is guaranteed to be realizable. In Section 8.5.5, we apply the algorithm described above to a case in which τ_0 is small enough to avoid PTC.

Velocity parameters from Eq. (8.112)

We will again use a two-node beta EQMOM to represent $n(\xi)$, and thus the mixed moments can be expressed in closed form as

$$m_{j,k} = \sum_{\alpha=1}^{2} \sum_{\beta=1}^{M} W_{\alpha\beta} \left[U(t, x, \xi_{\alpha\beta})\right]^j \xi^k_{\alpha\beta}, \tag{8.136}$$

with the conditional velocity defined by $U(t, x, \xi) = u_0(t, x)g_0(\xi) + u_1(t, x)g_1(\xi) + u_2(t, x)g_2(\xi) + u_3(t, x)g_3(\xi)$ and where $W_{\alpha\beta}$ and $\xi_{\alpha\beta}$ depend on the Gauss–Jacobi quadrature nodes. The functions $u_k(t,x)$ are found from the transported moment set using Eq. (8.117). The functions $g_k(\xi)$ depend on t and x through the EQMOM abscissas $\xi_\alpha(t,x)$. To avoid cumbersome notation, we will assume that the t and x dependence of $g_k(\xi)$ is the same as its argument. For example, $g_k(\xi^n_{\alpha\beta,i})$ is evaluated at time step n and grid cell i. The system in Eq. (8.123) is conveniently solved using operator splitting to separate the flux terms from the drag terms. In the first step, the flux terms are updated by solving Eq. (8.125). The realizable scheme for solving Eq. (8.125) is the same as in Eq. (8.89) (without diffusive fluxes), but with a slightly different definition for the numerical fluxes $G^{j,k;n}_{\alpha\beta,i+1/2}$:

$$G^{j,k;n}_{\alpha\beta,i+1/2} = W^n_{\alpha\beta,i+1/2,l} \max\left(U^n_{\alpha\beta,i}, 0\right)\left(U^n_{\alpha\beta,i}\right)^j \left(\xi^n_{\alpha\beta,i}\right)^k + W^n_{\alpha\beta,i+1/2,r} \min\left(U^n_{\alpha\beta,i+1}, 0\right)\left(U^n_{\alpha\beta,i+1}\right)^j \left(\xi^n_{\alpha\beta,i+1}\right)^k, \tag{8.137}$$

where the conditional velocity is defined by $U^n_{\alpha\beta,i} = u^n_{0;i}g_0(\xi^n_{\alpha\beta,i}) + u^n_{1;i}g_1(\xi^n_{\alpha\beta,i}) + u^n_{2;i}g_2(\xi^n_{\alpha\beta,i}) + u^n_{3;i}g_3(\xi^n_{\alpha\beta,i})$. The formulas for the weights $W^n_{\alpha\beta,i+1/2,l}$ and $W^n_{\alpha\beta,i+1/2,r}$ are the same as those given in Eqs. (8.93) and (8.94). The velocity parameters $u^n_{k;i}$ are assumed to be piecewise constant in each finite-volume cell. Starting from Eq. (8.136), this leads to

$$m^n_{1,k;i} = u^n_{0;i}G^{k;n}_{0;i} + u^n_{1;i}G^{k;n}_{1;i} + u^n_{2;i}G^{k;n}_{2;i} + u^n_{3;i}G^{k;n}_{3;i}, \tag{8.138}$$

where, following Eq. (8.116),

$$G^{k;n}_{m;i} = \sum_{\alpha=1}^{2}\sum_{\beta=1}^{M} W^n_{\alpha\beta,i} g_m(\xi^n_{\alpha\beta,i})\left(\xi^n_{\alpha\beta,i}\right)^k. \tag{8.139}$$

Taking $k = 0, 1, 2, 3$ in Eq. (8.138) leads to a linear system:[29]

$$\begin{bmatrix} u^n_{0;i} \\ u^n_{1;i} \\ u^n_{2;i} \\ u^n_{3;i} \end{bmatrix} = \begin{bmatrix} G^{0;n}_{0;i} & G^{0;n}_{1;i} & G^{0;n}_{2;i} & G^{0;n}_{3;i} \\ G^{1;n}_{0;i} & G^{1;n}_{1;i} & G^{1;n}_{2;i} & G^{1;n}_{3;i} \\ G^{2;n}_{0;i} & G^{2;n}_{1;i} & G^{2;n}_{2;i} & G^{2;n}_{3;i} \\ G^{3;n}_{0;i} & G^{3;n}_{1;i} & G^{3;n}_{2;i} & G^{3;n}_{3;i} \end{bmatrix}^{-1} \begin{bmatrix} m^n_{1,0;i} \\ m^n_{1,1;i} \\ m^n_{1,2;i} \\ m^n_{1,3;i} \end{bmatrix}, \tag{8.140}$$

where only components found from the transported moments appear on the right-hand side. Note that this specification of the velocity parameters is consistent with the realizable schemes wherein the velocity abscissas are constant in each grid cell. The realizability condition in Eq. (8.88) must be modified to read

$$\min_{\alpha,\beta,i}\left\{W^n_{\alpha\beta,i} - W^n_{\alpha\beta,i+1/2,l}\max\left(U^n_{\alpha\beta,i}, 0\right)\lambda_1 + W^n_{\alpha\beta,i-1/2,r}\min\left(U^n_{\alpha\beta,i}, 0\right)\lambda_1\right\} > 0, \tag{8.141}$$

where $\lambda_1 = \Delta t/\Delta x$. The time step is fixed by setting an overall CFL condition:

$$\Delta t = \text{CFL} \min_{\alpha,\beta,i}\left(\frac{\Delta x}{\left|U^n_{\alpha\beta,i}\right|}\right)$$

for CFL ≤ 1. With this value for Δt, the order of the convection scheme in grid cell i is determined from Eq. (8.141). At the end of the advection step, the updated moment set is denoted by $m^*_{j,k;i}$, and is guaranteed to be realizable.

Next, the drag term is evaluated as follows.

[29] The properties of the coefficient matrix are discussed after Eq. (8.117), and the numerical method must account for the possible singular cases.

(i) Use the moments $m^*_{0,k;i}$ to compute the weights $W^*_{\alpha\beta,i}$ and abscissas $\xi^*_{\alpha\beta,i}$.

(ii) Solve the linear system in Eq. (8.140) for the velocity parameters resulting from advection:

$$\begin{bmatrix} u^*_{0;i} \\ u^*_{1;i} \\ u^*_{2;i} \\ u^*_{3;i} \end{bmatrix} = \mathbf{A}^{-1} \begin{bmatrix} m^*_{1,0;i} \\ m^*_{1,1;i} \\ m^*_{1,2;i} \\ m^*_{1,3;i} \end{bmatrix},$$

where

$$\mathbf{A} = \begin{bmatrix} G^{0;*}_{0;i} & G^{0;*}_{1;i} & G^{0;*}_{2;i} & G^{0;*}_{3;i} \\ G^{1;*}_{0;i} & G^{1;*}_{1;i} & G^{1;*}_{2;i} & G^{1;*}_{3;i} \\ G^{2;*}_{0;i} & G^{2;*}_{1;i} & G^{2;*}_{2;i} & G^{2;*}_{3;i} \\ G^{3;*}_{0;i} & G^{3;*}_{1;i} & G^{3;*}_{2;i} & G^{3;*}_{3;i} \end{bmatrix}$$

and the components $G^{k;*}_{m;i}$ are computed from Eq. (8.139) using $W^*_{\alpha\beta,i}$ and $\xi^*_{\alpha\beta,i}$. Recall that $\mathbf{A}$ may be ill-conditioned (i.e. when the abscissas approach each other, or the bounds of the domain), so care must be taken when treating the degenerate cases during this step.

(iii) Use the velocity parameters to create an ensemble of sample velocities corresponding to each $W^*_{\alpha\beta,i}$ and $\xi^*_{\alpha\beta,i}$: $V^*_{\alpha\beta;i} = u^*_{0;i}g_0(\xi^*_{\alpha\beta,i}) + u^*_{1;i}g_1(\xi^*_{\alpha\beta,i}) + u^*_{2;i}g_2(\xi^*_{\alpha\beta,i}) + u^*_{3;i}g_3(\xi^*_{\alpha\beta,i})$ for $\alpha = 1, 2$ and $\beta = 1, 2, \dots, M$.

(iv) Update the sample velocities according to the drag law:

$$\frac{\mathrm{d}u}{\mathrm{d}t} = -\frac{u}{\tau_0 \xi^{2/3}} \quad \Longrightarrow \quad V^\dagger_{\alpha\beta;i} = \exp\left(-\frac{\Delta t}{\tau_0 \left(\xi^*_{\alpha\beta,i}\right)^{2/3}} \right) V^*_{\alpha\beta;i}.$$

The scalar exponential in this step is always well defined, even in the limit where $\xi^*_{\alpha\beta,i} \to 0$.

(v) Use the new sample velocities to compute the sample mixed moments:

$$m^\dagger_{1,k;i} = \sum_{\alpha=1}^{2} \sum_{\beta=1}^{M} W^*_{\alpha\beta;i} V^\dagger_{\alpha\beta;i} \left(\xi^*_{\alpha\beta;i}\right)^k.$$

(vi) Update the velocity parameters using the sample mixed moments and $\mathbf{A}$ from step (ii):

$$\begin{bmatrix} u^{n+1}_{0;i} \\ u^{n+1}_{1;i} \\ u^{n+1}_{2;i} \\ u^{n+1}_{3;i} \end{bmatrix} = \mathbf{A}^{-1} \begin{bmatrix} m^\dagger_{1,0;i} \\ m^\dagger_{1,1;i} \\ m^\dagger_{1,2;i} \\ m^\dagger_{1,3;i} \end{bmatrix}.$$

This formulation is also convenient for situations in which other forces (e.g. buoyancy, virtual mass, lift, etc.) are present. For such cases, it suffices to add the additional forces into the update of the sample velocities in step (iv). Because the functions g_k are constant during the drag step (i.e. they depend only on the size moments), it is also possible to construct a linear ODE system such as Eq. (8.131). However, in practice, the algorithm described above is more convenient and offers more flexibility to incorporate other forces. After updating the velocity parameters, the transported moments at the end of the drag step are found from

$$m_{j,k;i}^{n+1} = \begin{cases} m_{0,k;i}^{*} & \text{if } j = 0, \\ \sum_{\alpha=1}^{2} \sum_{\beta=1}^{M} W_{\alpha\beta;i}^{*} U_{\alpha\beta,i}^{n+1} \left(\xi_{\alpha\beta;i}^{*}\right)^{k} & \text{if } j = 1, \end{cases} \tag{8.142}$$

where $U_{\alpha\beta,i}^{n+1} = u_{0;i}^{n+1} g_0(\xi_{\alpha\beta,i}^{*}) + u_{1;i}^{n+1} g_1(\xi_{\alpha\beta,i}^{*}) + u_{2;i}^{n+1} g_2(\xi_{\alpha\beta,i}^{*}) + u_{3;i}^{n+1} g_3(\xi_{\alpha\beta,i}^{*})$. Because the weights are constant during the drag step, and always nonnegative after the advection step, the updated moment set in Eq. (8.142) is guaranteed to be realizable.

8.5.5 Example results with scalar-conditioned velocity

In this section, we present results found with the scalar-conditioned velocity algorithm introduced in the previous section using the NDF representation in Eq. (8.108). All of these calculations have been done with beta EQMOM using $n = 2$ nodes for the first quadrature and $M = 50$ nodes for the second. The initial NDF, where $\xi \in [0, 1]$ is the particle volume, has one of two forms. In the first case, referred to as "small spread," the NDF is represented by two beta functions, one near $\xi = 1/3$ and one near $\xi = 2/3$, with a small spread parameter σ. The initial moments for this case are shown in Figure 8.28. As can be observed from this figure, the moments are nonzero only within the spatial domain $x \in (-0.8, -0.2)$. Some statistics computed from the initial NDF are given in Figure 8.29. Note that the initial conditional velocity $u(0, x, \xi) = 1$ is independent of the particle size, and the mean and standard deviation of the volume do not depend on x.

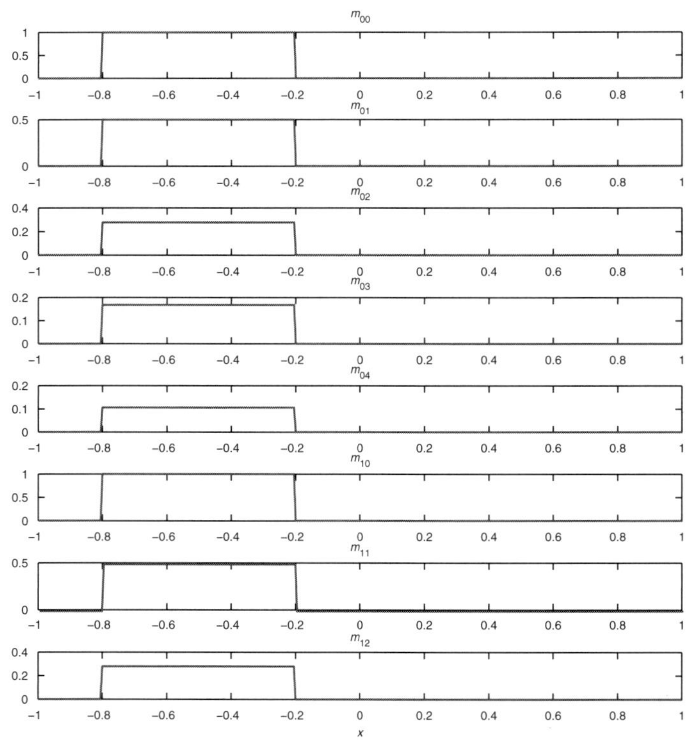

Figure 8.28. Initial moments for the scalar-conditioned-velocity case with small spread.

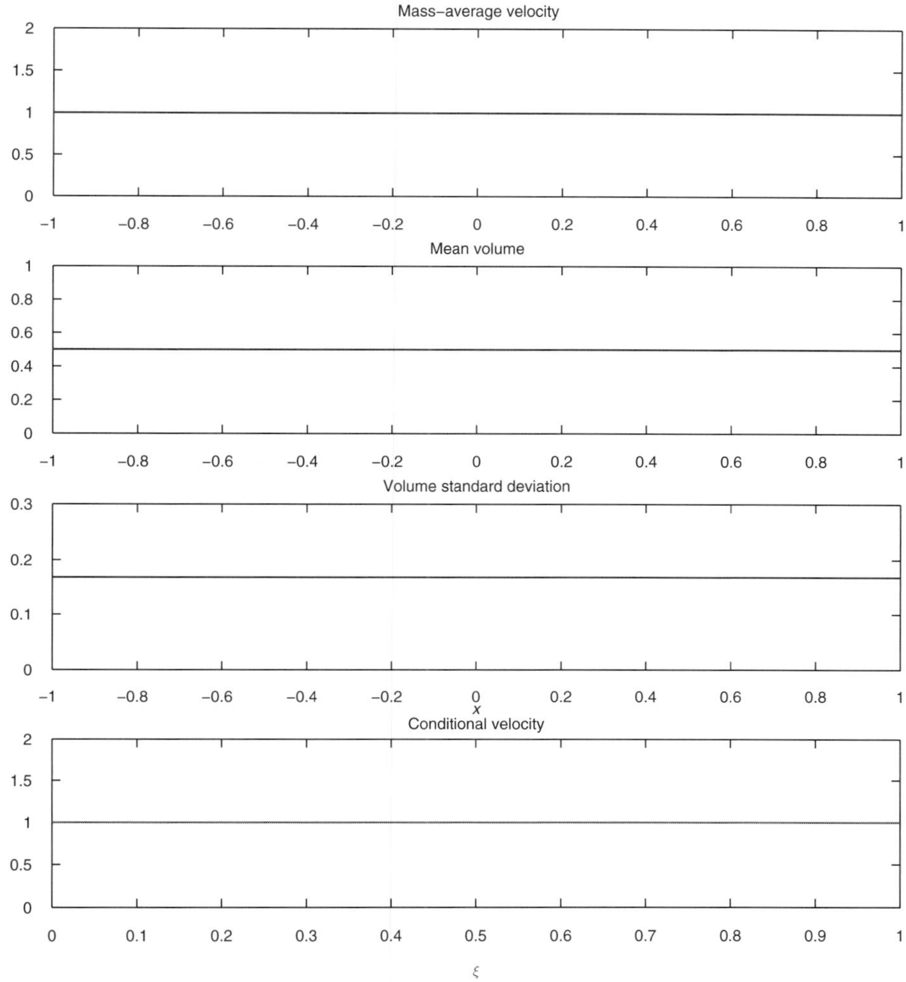

Figure 8.29. Initial statistics for the scalar-conditioned-velocity case with small spread.

For this case, the NDF at $\xi = 0$ is null. The second case, referred to as "large spread," has a uniform initial NDF on the interval $\xi \in [0, 1]$. For this case, the NDF at $\xi = 0$ is unity. The initial spatial distributions of the moments are shown in Figure 8.30, and selected statistics in Figure 8.31. The principal difference between the cases with small and large spread is the volume standard deviation. By definition, the beta EQMOM with $n = 2$ can exactly reconstruct both of the initial NDF. In addition to these cases, we have also successfully tested the algorithm with even larger spreads for which the exact NDF at $\xi = 0$ is infinite (as is the exact value of $m_{0,-2/3}$). As example results, we consider three values of τ_0: 10, 1, and 0.1. Note that, since the initial mean particle velocity is unity (and the fluid velocity is null), the initial Stokes number is equal to τ_0. As noted earlier, the scalar-conditioned-velocity model is valid only for small Stokes numbers. However, if we avoid PTC, the model can in principle be used for larger Stokes numbers. Since our simulation domain has periodic boundary conditions, we can avoid PTC by stopping the simulation before particles can exit the domain (i.e. $t = 1.2$ for $\tau_0 = \infty$). Moreover, since the particle velocity decreases due to drag, the Stokes number will decrease with

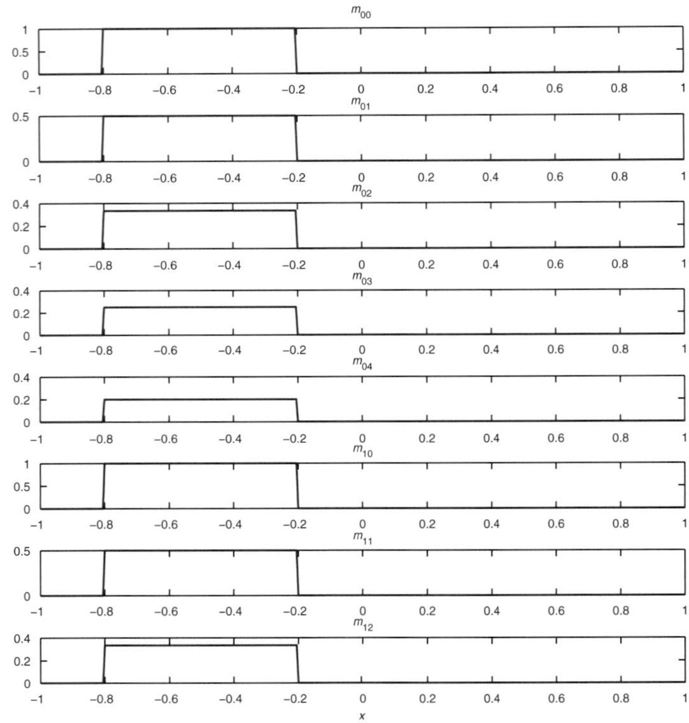

Figure 8.30. Initial moments for the scalar-conditioned-velocity case with large spread.

time. In fact, for $\tau_0 \leq 1$ the particles will stop due to drag before the end of the domain. Nevertheless, we shall see that the cases with $\tau_0 > 0.1$ can lead to other problems unrelated to PTC.

Small spread

We begin by looking at the case with $\tau_0 = 10$ (i.e. very weak drag), for which the moments at $t = 1$ are shown in Figure 8.32 and the sample statistics in Figure 8.33. From the moment plots, it is obvious that the initial NDF is mainly just transported in space with an average velocity near unity, which is consistent with the initial conditions. From the higher-order moments, we can see that the left side of the spatial distribution has more small particles, while the right side has more larger particles. This is also evident from the plot of the mean volume in Figure 8.33. The mass-average velocity, defined by $m_{1,1}/m_{0,1}$, is slightly lower on the left side of the distribution and higher on the right. From the distribution of the standard deviation of the volume, we can observe that the differences in velocity as a function of particle volume cause the large and small particles to segregate, with the small particles trailing behind the leading large particles. Finally, the conditional velocity $u(\xi) = u_0 + u_1\xi + u_2\xi^2$ is shown in Figure 8.33 for five values of x. Because the spread of the distribution is small, there are essentially no particles with volumes $\xi = 0$ or $\xi = 1$ for this case. Nevertheless, the function $u(\xi)$ can be extrapolated to these points. Note that, because the drag is very weak (high Stokes number), the conditional velocity moves very slowly towards zero (i.e. towards the fluid velocity). The results for $\tau_0 = 1$ (i.e. moderate drag) at $t = 3$ are shown in Figures 8.34 and 8.35. As noted above, by this time the NDF is

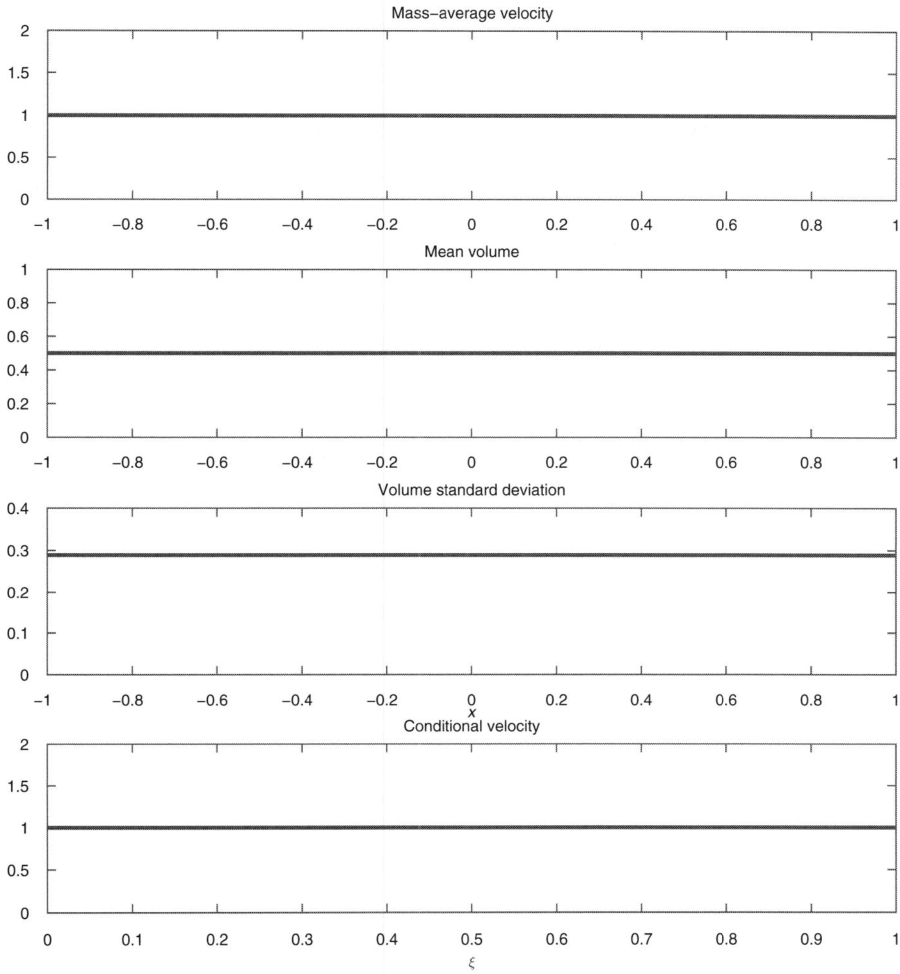

Figure 8.31. Initial statistics for the scalar-conditioned-velocity case with large spread.

essentially frozen in space because the velocity is near zero. The spatial distributions of the moments show very strong spatial segregation of the large and small particles, as do the plots for the mean and standard deviation of the particle volume in Figure 8.35. Although it is small, the mass-average velocity has a shape that follows that of the mean volume. Also, note that the maximum and minimum of the mean volume curve correspond roughly to the peaks of the two beta functions used in the small-spread case, and thus the particles are almost completely segregated at $x = -0.45$ and at $x = 0.65$. The conditional velocity curves in Figure 8.35 show a distinct dependence on x. In fact, for $x = 0.8$ the conditional velocity is negative at $\xi = 0$, which is not physically consistent with the fact that the drag should only make particles slow down monotonically towards zero. Nevertheless, the curve for $u(\xi)$ is consistent with the moments $m_{j,k}$ used to find it from Eq. (8.127). Indeed, the reconstructed velocity in the form of Eq. (8.107) does not impose realizability constraints to limit the maximum/minimum values of $u(\xi)$. When the Stokes number is not small, this can lead to large values of velocity, even though the expression reproduces the correct velocity-scalar moments. In practice, since large velocities lead to a smaller time step,

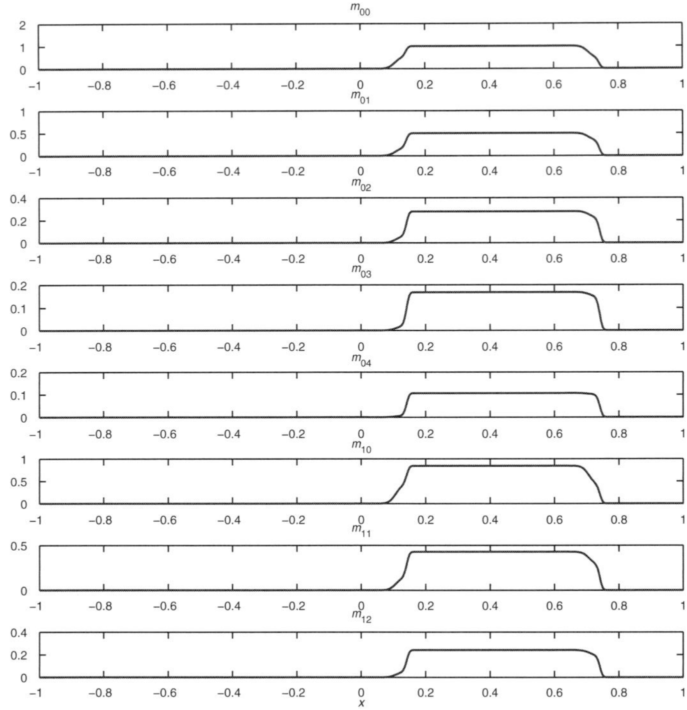

Figure 8.32. Moments for the scalar-conditioned-velocity case with small spread, $\tau_0 = 10$ at time $t = 1$.

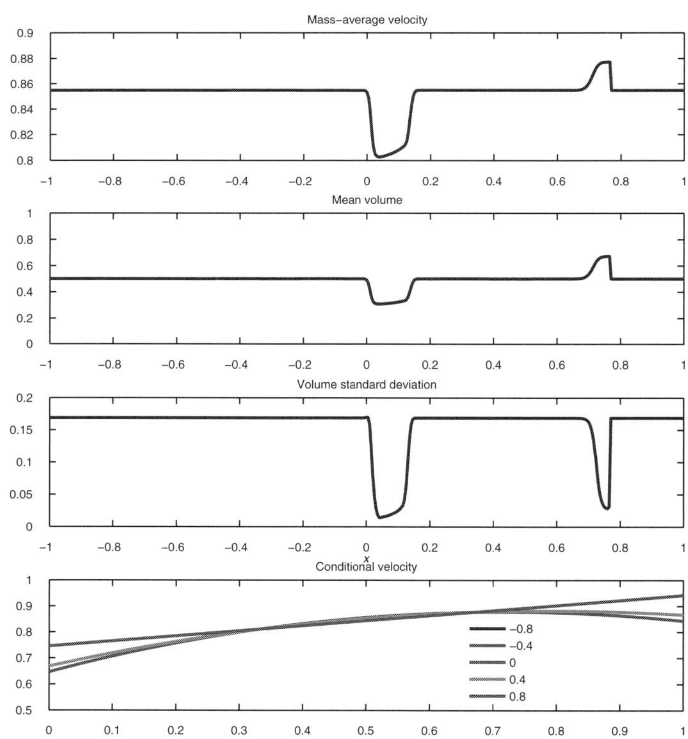

Figure 8.33. Statistics for the scalar-conditioned-velocity case with small spread, $\tau_0 = 10$ at time $t = 1$. The conditional velocities are shown for five spatial locations: $x = -0.8$, -0.4, 0, 0.4, and 0.8.

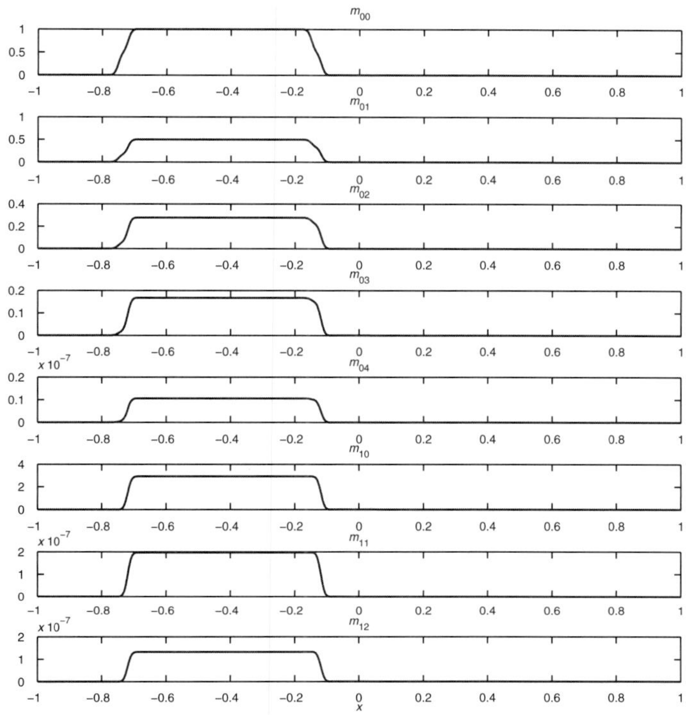

Figure 8.34. Moments for the scalar-conditioned-velocity case with small spread, $\tau_0 = 1$ at time $t = 3$.

such realizability errors can have a negative impact on the simulations. For small Stokes numbers (i.e. within the range of validity of the model), the deviation of $u(\xi)$ from the fluid velocity is small, and thus the simulation time step is fixed by the fluid CFL number. The results for $\tau_0 = 0.1$ (i.e. strong drag) at $t = 1$ are presented in Figures 8.36 and 8.37. For this case, the distributions become frozen very quickly due to the strong drag. However, from the statistics we can still observe many of the same trends as in previous cases. Note that this case is within the expected range of Stokes numbers for which the conditional velocity model will be valid. For this reason, the conditional velocity curves are tightly coupled to the fluid velocity, and thus unrealizable velocities are very small in Figure 8.37.

Large spread

We next look at the cases with large spread. For $\tau_0 = 10$ (i.e. very weak drag) at $t = 1$, the results are shown in Figures 8.38 and 8.39. In comparison with the case with small spread, we now observe from the mean volume plot that after segregation the remaining particles have mean volume (and volume standard deviation) near zero. The corresponding conditional velocity at $x = -0.4$ in Figure 8.39 exhibits wide deviations from zero velocity. This is because the NDF at $x = -0.4$ is entirely concentrated at values of ξ near zero, and hence the moments used to compute $u(\xi)$ contain information solely about this region of phase space. For larger values of ξ, the curve simply extrapolates using the quadratic model for $u(\xi)$. Another telltale sign of unrealizable velocities is the fact that the statistics in Figure 8.39 change past $x = 0.8$ (i.e. across the periodic boundary). Since the largest initial velocity was unity, information in the real system cannot travel at a rate faster than unity. As

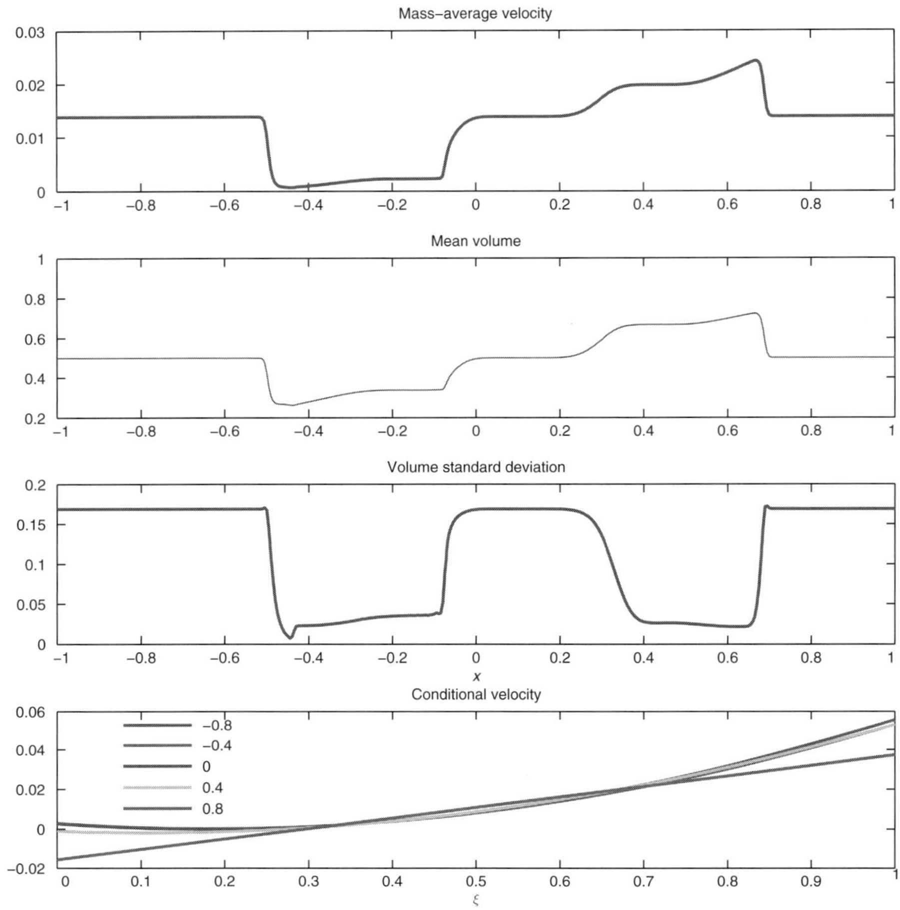

Figure 8.35. Statistics for the scalar-conditioned-velocity case with small spread, $\tau_0 = 1$ at time $t = 3$. The conditional velocities are shown for five spatial locations: $x = -0.8, -0.4, 0, 0.4$, and 0.8.

discussed above, this unrealizable result occurs because the conditional-velocity model is not applicable for large Stokes numbers. In fact, on analyzing the simulations more closely, the unrealizable velocities are found to arise during the spatial-transport step (i.e. due to Eqs. (8.125) and (8.127)). In theory, it might be possible to develop consistent limiters for the coefficients in Eq. (8.127). However, the real difficulty is that the model does not capture the correct physics unless the Stokes number is small. The results for $\tau_0 = 1$ (i.e. moderate drag) at $t = 3$ are shown in Figures 8.40 and 8.41. Here, due to the large spread of the NDF, the size segregation is very pronounced. Again, by the end of the simulation the NDF is nearly frozen in place due to the effect of drag. However, during the initial steps of the simulation the wide spread and the unrealizable reconstruction of $u(\xi)$ lead to advection velocities significantly larger than unity, which results in relatively small time steps. The case with small Stokes number is shown in Figures 8.42 and 8.43. Again, although the distributions are quickly frozen in place, the size segregation is very significant. Overall, with a small Stokes number, velocity realizability is a minor issue when reconstructing $u(\xi)$ from the moments. This would suggest that, when used in its range of validity, the

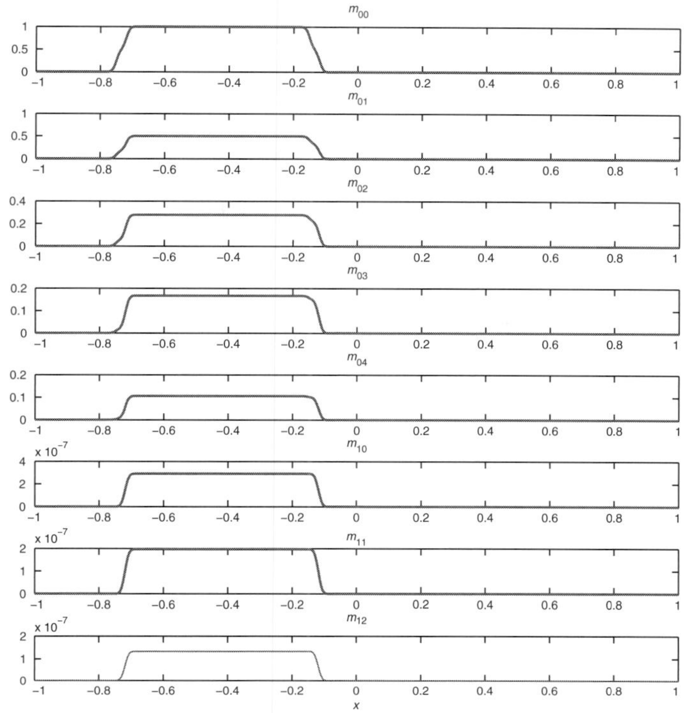

Figure 8.36. Moments for the scalar-conditioned-velocity case with small spread, $\tau_0 = 0.1$ at time $t = 1$.

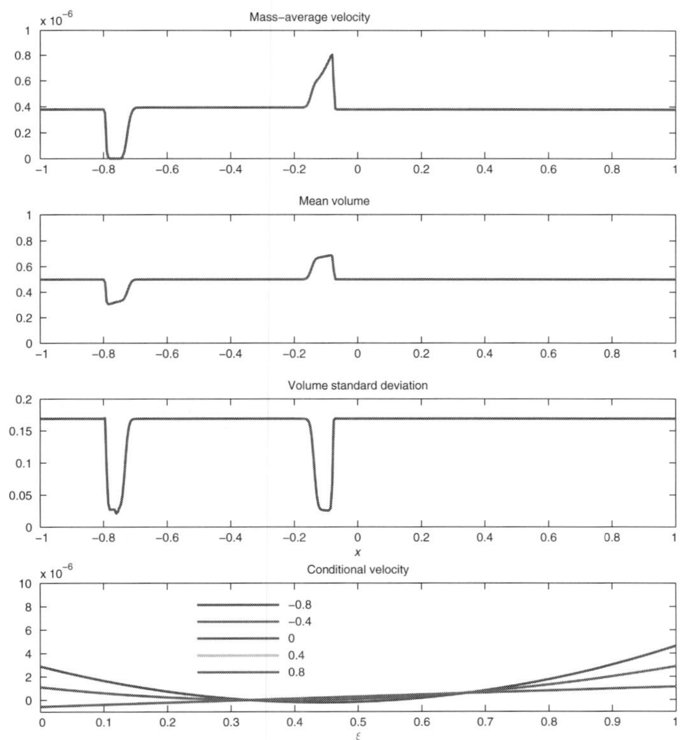

Figure 8.37. Statistics for the scalar-conditioned-velocity case with small spread, $\tau_0 = 0.1$ at time $t = 1$. The conditional velocities are shown for five spatial locations: $x = -0.8$, -0.4, 0, 0.4, and 0.8.

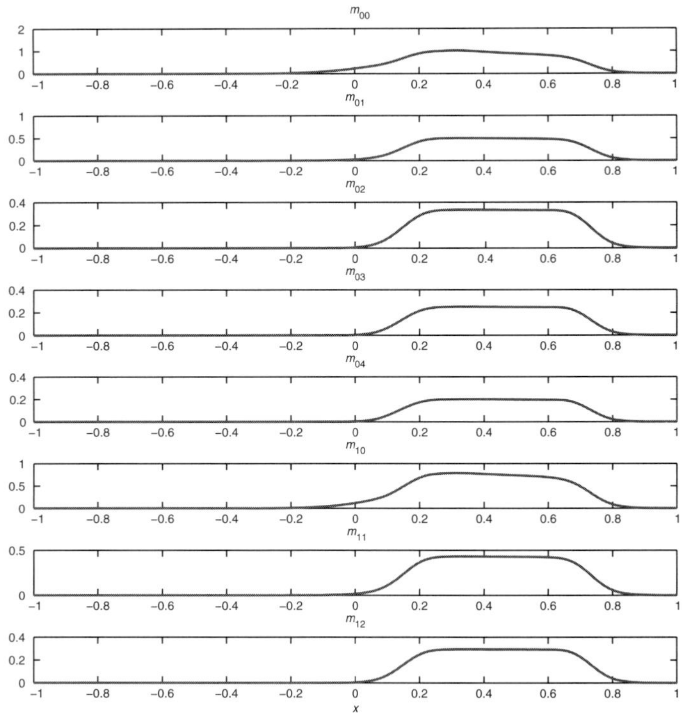

Figure 8.38. Moments for the scalar-conditioned-velocity case with large spread, $\tau_0 = 10$ at time $t = 1$.

scalar-conditioned velocity is a numerically efficient choice for simulating polydisperse multiphase flows by adding only a small set of velocity moments needed to compute $u(\xi)$. For instance, disperse bubbly flows typically have very small Stokes numbers and thus could be handled in this manner. For larger Stokes numbers, a better solution is to use the velocity-scalar NDF method described next.

8.5.6 Spatial transport of the velocity-scalar NDF

As a final case, we will again consider the joint velocity-scalar NDF governed by Eq. (8.122), but without specifying a functional form for the scalar-conditioned velocity. The moment-transport equation is again given by Eq. (8.122). However, we will now use a scalar-conditioned multivariate EQMOM to reconstruct the joint NDF:

$$n(v,\xi) = \sum_{\alpha=1}^{2} \sum_{\beta=1}^{2} w_{\alpha\beta}\delta_{\sigma_1}(\xi,\xi_\alpha)\delta_{\sigma_{2\alpha}}\left(v, v_{\alpha\beta}\right). \tag{8.143}$$

For the scalar, the kernel density function will be defined by

$$\delta_{\sigma_1}(\xi,\xi_\alpha) = \frac{1}{\sigma_1 \Gamma\left(\xi_\alpha/\sigma_1\right)} \left(\frac{\xi}{\sigma_1}\right)^{\xi_\alpha/\sigma_1 - 1} \mathrm{e}^{-\xi/\sigma_1}, \tag{8.144}$$

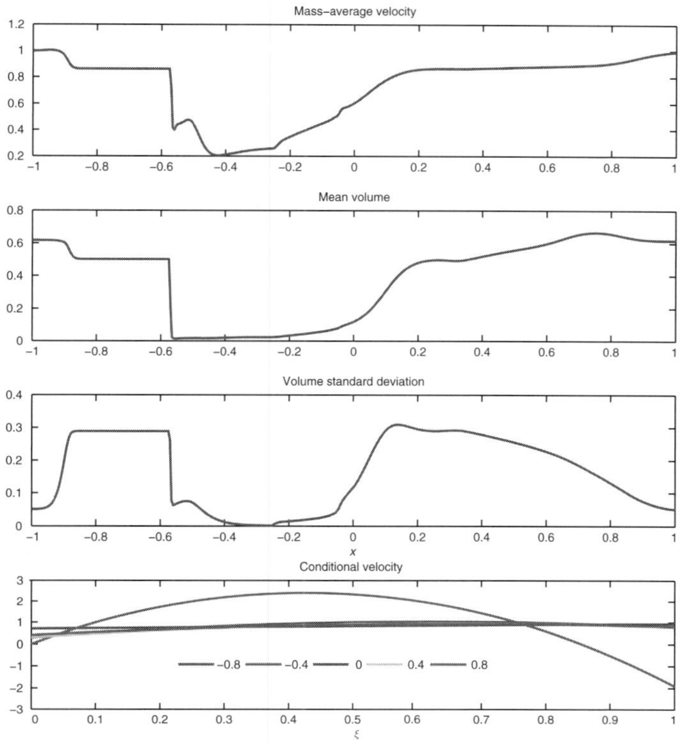

Figure 8.39. Statistics for the scalar-conditioned-velocity case with large spread, $\tau_0 = 10$ at time $t = 1$. The conditional velocities are shown for five spatial locations: $x = -0.8$, -0.4, 0, 0.4, and 0.8.

which corresponds to gamma EQMOM. For the velocity, a Gaussian kernel density function will be used:

$$\delta_{\sigma_{2\alpha}}\left(v, v_{\alpha\beta}\right) = \frac{1}{\sqrt{2\pi}\sigma_{2\alpha}} \exp\left(-\frac{(v - v_{\alpha\beta})^2}{2\sigma_{2\alpha}^2}\right). \tag{8.145}$$

As for the bivariate velocity moments in Section 8.4, 13 bivariate moments must be transported to find the parameters in Eq. (8.143): $\{m_{0,0}, m_{0,1}, m_{0,2}, m_{0,3}, m_{0,4}, m_{1,0}, m_{1,1}, m_{2,0}, m_{2,1}, m_{3,0}, m_{3,1}, m_{4,0}, m_{4,1}\}$, where the moments are defined by $m_{j,k} = \int v^j \xi^k n(v, \xi) \mathrm{d}v \, \mathrm{d}\xi$. The weights ($w_{\alpha\beta}$), abscissas ($\xi_\alpha, v_{\alpha\beta}$), and width parameters ($\sigma_1, \sigma_{1\alpha}$) are determined from the 13 bivariate moments using the ECQMOM as described in Section 3.3.4. For clarity, the steps required for bivariate ECQMOM for the case under consideration are as follows.

(i) Use gamma EQMOM with the size moments ($m_{0,0}, m_{0,1}, m_{0,2}, m_{0,3}, m_{0,4}$) to find the quadrature w_α, ξ_α, and σ_1 for $\alpha = 1, 2$.

(ii) Find the scalar-conditioned velocity moments μ_α^k by solving the linear system

$$\begin{bmatrix} 1 & 1 \\ \xi_1 & \xi_2 \end{bmatrix} \begin{bmatrix} w_1 & \\ & w_2 \end{bmatrix} \begin{bmatrix} \mu_1^1 & \mu_1^2 & \mu_1^3 & \mu_1^4 \\ \mu_2^1 & \mu_2^2 & \mu_2^3 & \mu_2^4 \end{bmatrix} = \begin{bmatrix} m_{1,0} & m_{2,0} & m_{3,0} & m_{4,0} \\ m_{1,1} & m_{2,1} & m_{3,1} & m_{4,1} \end{bmatrix}.$$

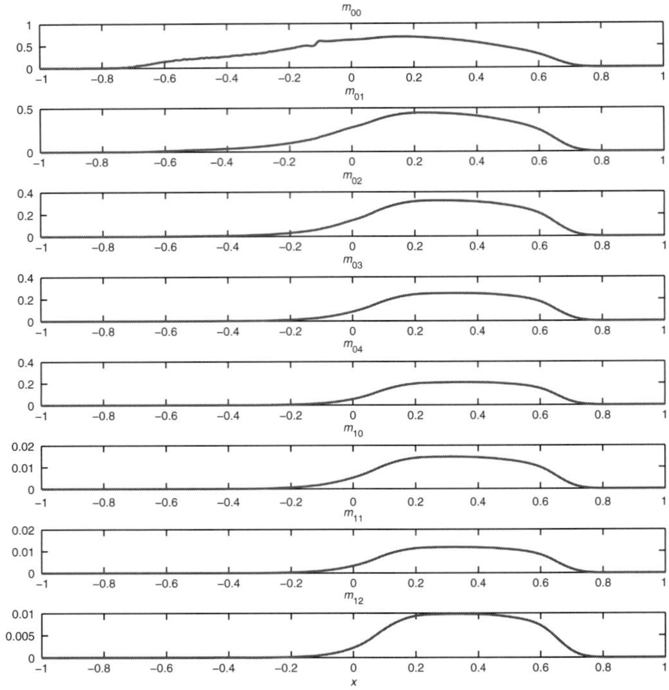

Figure 8.40. Moments for the scalar-conditioned-velocity case with small spread, $\tau_0 = 1$ at time $t = 3$.

(iii) For each $\alpha = 1, 2$, use Gaussian EQMOM with the conditional velocity moments $(1, \mu_\alpha^1, \mu_\alpha^2, \mu_\alpha^3, \mu_\alpha^4)$ to find the quadrature $w^*_{\alpha\beta}$, $v_{\alpha\beta}$, and $\sigma_{2\alpha}$ for $\beta = 1, 2$.

(iv) The final quadrature weights are then given by $w_{\alpha\beta} = w_\alpha w^*_{\alpha\beta}$.

As was done with the bivariate velocity moments, we will simplify the notation by replacing the double sum in Eq. (8.143) by a single sum with index $\alpha = 1, \ldots, 4$. We then denote the ECQMOM quadrature by w_α, ξ_α, v_α, $\sigma_{1\alpha}$, and $\sigma_{2\alpha}$. As a final quadrature step, we introduce a Gauss–Laguerre quadrature for the size variable, and a Gauss–Hermite quadrature for the velocity variable. Letting $m^n_{j,k;i}$ denote the volume-average bivariate moments in grid cell i at time step n, the dual-quadrature representation of the moments is

$$m^n_{j,k;i} = \sum_{\alpha=1}^{4} \sum_{\beta_1=1}^{M_1} \sum_{\beta_2=1}^{M_2} W^n_{\alpha\beta_1\beta_2,i} \left(V^n_{\alpha\beta_1,i}\right)^j \left(\xi^n_{\alpha\beta_2,i}\right)^k .$$

As in Eq. (8.102), the weights and abscissas in this expression come from (i) the ECQMOM, (ii) Gauss–Laguerre quadrature, and (iii) Gauss–Hermite quadrature. Likewise, M_1 and M_2 can be chosen as large as needed in order to minimize the quadrature error in the spatial fluxes and drag terms.

The numerical solution of the moment-transport equation is again done by using operator splitting to update first for the spatial fluxes and then for the drag. The realizable scheme

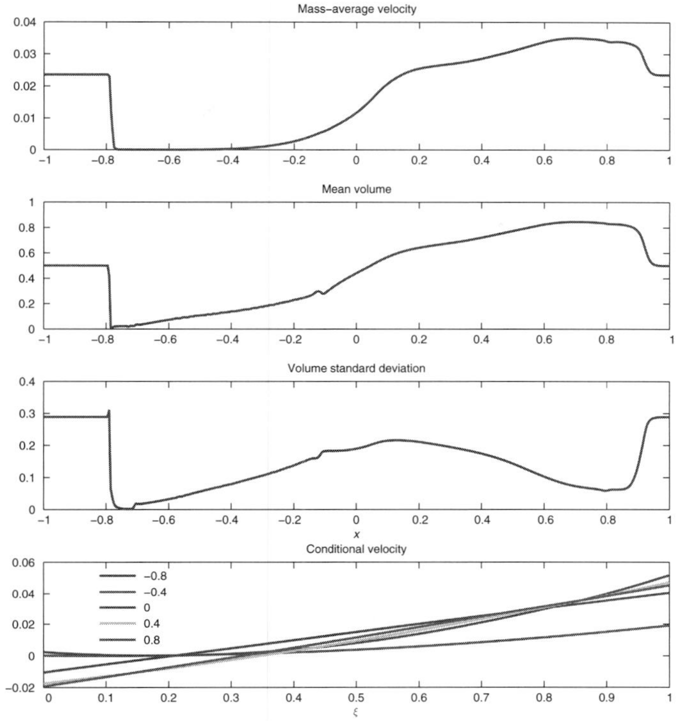

Figure 8.41. Statistics for the scalar-conditioned-velocity case with large spread, $\tau_0 = 1$ at time $t = 3$. The conditional velocities are shown for five spatial locations: $x = -0.8, -0.4, 0, 0.4$, and 0.8.

for solving Eq. (8.125) is the same as in Eq. (8.89) (without diffusive fluxes), but with the following definition for the numerical fluxes:

$$\begin{aligned} G^{j,k;n}_{\alpha\beta_1\beta_2,i+1/2} = {} & W^n_{\alpha\beta_1\beta_2,i+1/2,l} \max\left(V^n_{\alpha\beta_1,i}, 0\right)\left(V^n_{\alpha\beta_1,i}\right)^j \left(\xi^n_{\alpha\beta_2,i}\right)^k \\ & + W^n_{\alpha\beta_1\beta_2,i+1/2,r} \min\left(V^n_{\alpha\beta_1,i+1}, 0\right)\left(V^n_{\alpha\beta_1,i+1}\right)^j \left(\xi^n_{\alpha\beta_2,i+1}\right)^k. \end{aligned} \tag{8.146}$$

The formulas for the weights $W^n_{\alpha\beta_1\beta_2,i+1/2,l}$ and $W^n_{\alpha\beta_1\beta_2,i+1/2,r}$ are the same as those described after Eqs. (8.103). The realizability condition for this case is

$$W^n_{\alpha\beta_1\beta_2,i} \geq \lambda_1 W^n_{\alpha\beta_1\beta_2,i+1/2,l} \max\left(V^n_{\alpha\beta_1,i}, 0\right) - \lambda_1 W^n_{\alpha\beta_1\beta_2,i-1/2,r} \min\left(V^n_{\alpha\beta_1,i}, 0\right). \tag{8.147}$$

The time step is fixed as in Eq. (8.106) using

$$\Delta t = \text{CFL} \min_{\alpha,\beta_1,i}\left(\frac{\Delta x}{|V^n_{\alpha\beta_1,i}|}\right). \tag{8.148}$$

The bivariate moments at the end of the convection step are denoted by $m^*_{j,k;i}$, and are guaranteed to be realizable if Eq. (8.147) is satisfied. After the convection step, a new dual-quadrature representation is constructed starting from $m^*_{j,k;i}$. This new quadrature, with parameters $W^*_{\alpha\beta_1\beta_2,i}$, $V^*_{\alpha\beta_1,i}$, and $\xi^*_{\alpha\beta_2,i}$, is used in the drag step. Note that the weights and size abscissas are invariant with respect to drag, and thus $W^*_{\alpha\beta_1\beta_2,i}$ and $\xi^*_{\alpha\beta_2,i}$ do not change.

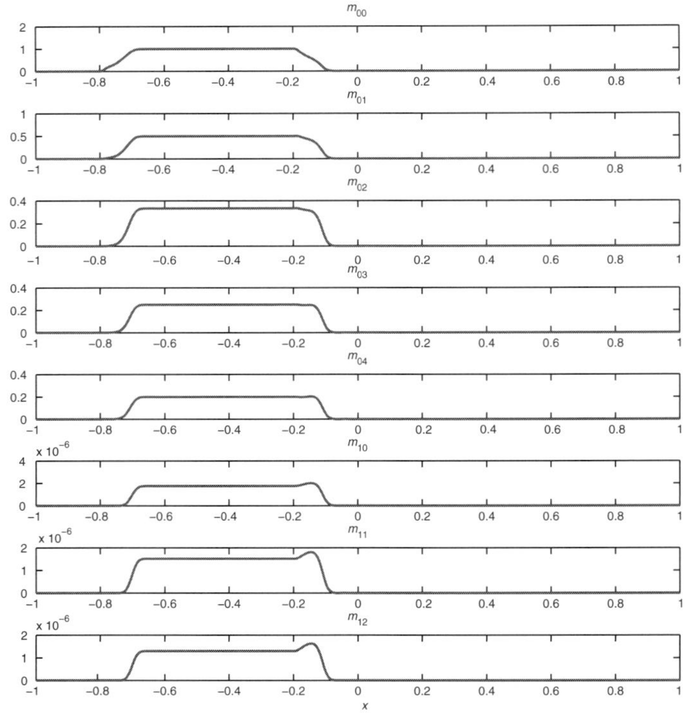

Figure 8.42. Moments for the scalar-conditioned-velocity case with large spread, $\tau_0 = 0.1$ at time $t = 1$.

In contrast, the velocity abscissas in each grid cell i are found by solving an ODE:

$$\frac{\mathrm{d}V_{\alpha\beta_1,i}}{\mathrm{d}t} = -\frac{1}{\tau_0}\left[\sum_{\beta_2=1}^{M_2} w^*_{\alpha\beta_2,i}\left(\xi^*_{\alpha\beta_2,i}\right)^{-2}\right] V_{\alpha\beta_1,i}$$

with initial condition $V^*_{\alpha\beta_1,i}$. The weights $w^*_{\alpha\beta_2,i}$ are found from the Gauss–Laguerre quadrature and sum to unity with respect to β_2 for all values of α and i. The term in square brackets is thus the expected value of $1/\xi^2$ for each of the gamma kernel density functions used to represent $n(\xi)$. Using an implicit Euler scheme, the updated velocity abscissas are

$$V^\dagger_{\alpha\beta_1,i} = \left(1 + \frac{\Delta t}{\tau_0}\sum_{\beta_2=1}^{M_2} w^*_{\alpha\beta_2,i}\left(\xi^*_{\alpha\beta_2,i}\right)^{-2}\right)^{-1} V^*_{\alpha\beta_1,i}, \tag{8.149}$$

with the same Δt as in the advection step. Finally, the updated moment set is found from

$$m^{n+1}_{j,k;i} = \sum_{\alpha=1}^{4}\sum_{\beta_1=1}^{M_1}\sum_{\beta_2=1}^{M_2} W^*_{\alpha\beta_1\beta_2,i}\left(V^\dagger_{\alpha\beta_1,i}\right)^j\left(\xi^*_{\alpha\beta_2,i}\right)^k. \tag{8.150}$$

By construction, this moment set is guaranteed to be realizable.

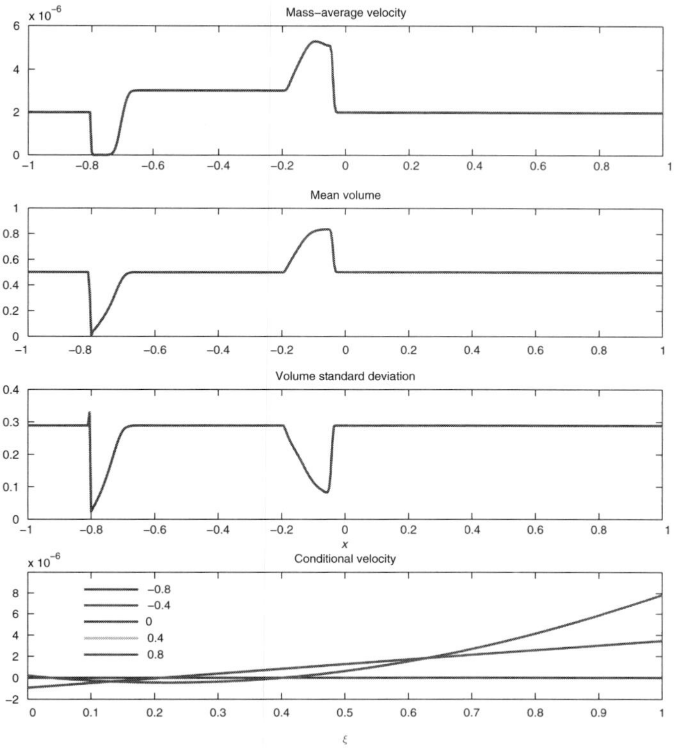

Figure 8.43. Statistics for the scalar-conditioned-velocity case with large spread, $\tau_0 = 0.1$ at time $t = 1$. The conditional velocities are shown for five spatial locations: $x = -0.8$, -0.4, 0, 0.4, and 0.8.

8.6 Concluding remarks

In this chapter we have described the numerical methods and algorithms needed to solve inhomogeneous kinetic equations with multiple phase-space variables using moment methods. A key point when using moment methods is the issue of realizability of the moment set. We have seen that certain numerical schemes (such as high-order spatial reconstruction of the moments) will inevitably lead to unrealizable moments. Thus, we have emphasized realizable finite-volume schemes for which we can explicitly demonstrate that the moment set will be realizable subject to a realizability constraint. Because research on the development of realizable schemes is relatively recent, the reader can expect that further improvements will be made in the near future, and thus that some of the methods presented in this chapter may be replaced by more accurate schemes. A second key point when working with the moment methods described in this book is the algorithm for reconstructing the NDF starting from a realizable moment set. In general, the accuracy of the solution will depend strongly on the quality of the NDF reconstruction, especially for inhomogeneous systems wherein the numerical fluxes are closely related to the reconstructed NDF. Because of their flexibility, robustness, and computational efficiency, in this book we have focused on quadrature-based moment methods for NDF reconstruction. For example, with univariate distributions, the reconstructed NDF found using the EQMOM can usually be made sufficiently accurate by increasing the number of nodes in the first quadrature. Moreover, we have seen that the second quadrature in the EQMOM allows great flexibility in developing numerical schemes due to the duality between the continuous NDF and the discrete representation in terms of the Gaussian quadrature nodes. However, when working with a multivariate NDF, the task of reconstructing the NDF is much more difficult. In this book, we have focused on the existing multivariate quadrature methods, but here again the reader can expect that further improvements will result from the increasing research interest in multivariate quadrature.

Appendix A

Moment-inversion algorithms

In this appendix MATLAB scripts for the moment-inversion algorithms discussed in Chapter 3 are reported.

A.1 Univariate quadrature

A.1.1 The PD algorithm

Below a MATLAB script calculating the nodes `xi` and the weights `w` of a quadrature approximation starting from the moments of the NDF is reported. The script uses the vector `m` containing the `N` moments to calculate the matrix `P`, the vector `zeta`, and the coefficients of the Jacobi matrix `jacobi`. The eigenvalues and eigenvectors (`eval`, `evect`) are then calculated by using the MATLAB function `eig`, and eventually nodes and weights are calculated and returned.

```
function [xi,w]=PD(m,N)
%
% compute quadrature approximation with the Product-Difference algorithm
%
% INPUT:
%   N = number of nodes of the quadrature approximation
%   m = moments from 0 to 2N-1 [mom(1), ..., mom(2N)]
%
% OUTPUT:
%   xi = abscissas
%   w  = weights
%
% compute matrix P
%
P(1,1)= 1.0;
for i=2:2*N+1
    P(i,1)=0.0;
end
for i=1:2*N
    P(i,2)=(-1)^(i-1)*m(i);
end
for j=3:2*N+1
    for i=1:2*N+2-j
        P(i,j)=P(1,j-1)*P(i+1,j-2)-P(1,j-2)*P(i+1,j-1);
    end
end
%
% compute coefficients of continued fraction
%
```

```
zeta(1)=0.0;
for i=2:2*N
    if P(1,i)*P(1,i-1)>0.0
       zeta(i)=P(1,i+1)/(P(1,i)*P(1,i-1));
    else
       display('m1 is equal to zero!')
       zeta(i)=0.0;
    end
end
%
% compute coefficients for Jacobi matrix
%
for i=1:N
    a(i)=zeta(2*i)+zeta(2*i-1);
end
for i=1:N-1
    b(i) = zeta(2*i+1)*zeta(2*i);
end
%
% compute Jacobi matrix
%
for i=1:N
    jacobi(i,i)=a(i);
end
for i=1:N-1
    jacobi(i,i+1) = -(b(i))^0.5;
    jacobi(i+1,i) = -(b(i))^0.5;
end
%
% compute eigenvalues and eigenvectors
%
[evec,eval]=eig(jacobi);
%
% return weights
%
for i=1:N
    w(i)=evec(1,i)^2*m(1);
end
%
% return abscissas
%
for i=1:N
    xi(i)=eval(i,i);
end
```

A.1.2 The adaptive Wheeler algorithm

Below a MATLAB script for the calculation of a quadrature approximation of order `N` from a known set of moments `m` using the Wheeler algorithm is reported. The script computes the intermediate coefficients `sigma` and the `jacobi` matrix, and, as for the PD algorithm, determines the nodes and weights of the quadrature approximation from the eigenvalues and eigenvectors of the matrix.

```
function [xi,w]=WH(m,N)
%
% compute quadrature approximation with the Wheeler algorithm
%
% INPUT:
%   N = number of nodes of the quadrature approximation
%   m = moments from 0 to 2N-1 [mom(1), ..., mom(2N)]
%
% OUTPUT:
%   xi = abscissas
%   w  = weights
%
% compute intermediate coefficients
%
for l=1:2*N-2
```

```
    sigma(1,l+1) = 0.0;
end
%
for l=0:2*N-1
    sigma(2,l+1) = m(l+1);
end
%
% compute coefficients for Jacobi matrix
%
a(1) = m(2)/m(1);
b(1) = 0.0;
%
for k=1:N-1
    for l=k:2*N-k-1
        sigma(k+2,l+1) = sigma(k+1,l+2)-a(k)*sigma(k+1,l+1)-b(k)*sigma(k,l+1);
        a(k+1) = -sigma(k+1,k+1)/sigma(k+1,k)+sigma(k+2,k+2)/sigma(k+2,k+1);
        b(k+1) = sigma(k+2,k+1)/sigma(k+1,k);
    end
end
%
% compute Jacobi matrix
%
for i=1:N
    jacobi(i,i)=a(i);
end
%
for i=1:N-1
    jacobi(i,i+1) = -(abs(b(i+1)))^0.5;
    jacobi(i+1,i) = -(abs(b(i+1)))^0.5;
end
%
% compute eigenvalues and eigenvectors
%
[evec,eval]=eig(jacobi);
%
% return weights
%
for i=1:N
    w(i)=evec(1,i)^2*m(1);
end
%
% return abscissas
%
for i=1:N
    xi(i)=eval(i,i);
end
```

A.2 Moment-correction algorithms

A.2.1 The correction algorithm of McGraw

Below a MATLAB script with the implementation of the algorithm for the correction of an unrealizable moment set is reported. The script first analyzes the moment set and then, if it is unrealizable, the script identifies the moment that has to be changed the least to make the set realizable. The procedure is iterated until the moment set becomes realizable or the maximum iteration number is reached. Realizability is verified by ensuring positiveness of the second-order differences; however, other more stringent conditions could more effectively be used.

```
function [m,iter]=CORR(m,n)
%
% correct a moment set
%
% INPUT:
```

```
%   m = moments
%   n = number of moments
%
% OUTPUT:
%   m    = corrected moment set
%   iter = number of iteration required
%
% define maximum number of iteration
%
iter_max = 100.0;
%
% define the unitary correction matrix
%
for i=1:n
    for j=1:n
        bkk(i,j) = 0.0;
    end
end
%
for k=1:n
    for i=1:n
        if (i==k)
            bkk(i,1) = 1.0;
        else
            bkk(i,1) = 0.0;
        end
    end
%
    for j=2:4
        for i=1:n-j+1
            bkk(i,j) = bkk(i+1,j-1)-bkk(i,j-1);
        end
    end

    for i=1:n
        bk(k,i)=bkk(i,4);
    end

end
%
check_corr = 1.0;
iter =0.0;
%
% start the correction loop
%
while (check_corr==1.0 && iter<iter_max)
%
    check_corr = 0.0;
%
    for i=1:n
        d(i,1) = log(m(i));
    end
    for j=2:n+1
        for i=1:n-j+1
            d(i,j) = d(i+1,j-1)-d(i,j-1);
            if (j==3 && d(i,j)<0)
                check_corr = 1.0;
            end
        end
    end
%
    if (check_corr == 1.0)
%
        iter   = iter + 1.0;
        k_star = 1.0;
%
        for k=1:n
            cos_quad_alfa(k) = ((bk(k,:)*d(:,4))/((bk(k,:)*bk(k,:)')^0.5*(d(:,4)'
             *d(:,4))^0.5))^2.0;
            if (cos_quad_alfa(k) >= cos_quad_alfa(k_star))
                k_star=k;
            end
        end
```

```
%
        lnck = -(bk(k_star,:)*d(:,4))/(bk(k_star,:)*bk(k_star,:)');
        m(k_star) = exp(lnck)*m(k_star);
%
    end
%
end
```

A.2.2 The correction algorithm of Wright

Below a simple MATLAB script implementing the Wright version of the correction algorithm is reported.

```
function [m]=LOGNORM(m,n)
%
% correct a moment set
%
% INPUT:
%   m = moments
%   n = number of moments
%
% OUTPUT:
%   m = corrected moment set
%
% select the two additional moments to be used for the first log-normal
%
i = 2;
j = 3;
%
% calculate the corresponding mean and variance
%
mu        = (j/(i*j-i^2))*log(m(i+1)/m(1)) + (i/(i*j-j^2))*log(m(j+1)/m(1));
%
sigma_var = ((2.0/(j^2))*log(m(j+1)/m(1)) - (2.0/(i*j))*log(m(i+1)/m(1)))/(1.0-i/j);
%
if (sigma_var < 0)
    sigma_var = 0.0;
end
%
% calculate the new moments for the first distribution
%
for k=1:n
    m1(k) = m(1)*exp((k-1.0)*mu + ((k-1.0)^2*sigma_var)/2.0);
end
%
% select the two additional moments to be used for the first log-normal
%
i=1;
j=3;
%
% calculate the corresponding mean and variance
%
mu        = (j/(i*j-i^2))*log(m(i+1)/m(1)) + (i/(i*j-j^2))*log(m(j+1)/m(1));
%
sigma_var = ((2.0/(j^2))*log(m(j+1)/m(1)) - (2.0/(i*j))*log(m(i+1)/m(1)))/(1.0-i/j);
%
if (sigma_var < 0)
    sigma_var = 0.0;
end
%
% calculate the new moments for the second distribution
%
for k=1:n
    m2(k) =  m(1)*exp((k-1.0)*mu + ((k-1.0)^2*sigma_var)/2.0);
end
%
% calculate the new set of moments from the arithmetic mean of the previous
% two sets of moments
```

```
%
for k=1:n
    m(k) = (m1(k)+m2(k))/2.0;
end
%
end
```

A.3 Multivariate quadrature

A.3.1 Brute-force QMOM

Below a MATLAB script implementing the brute-force QMOM for a bivariate problem ($M = 2$) is reported. The script requires as input the total number of nodes used in the quadrature approximation `N`. Moreover, since with this method theoretically any set of moments of size $N(M + 1)$ can be used, the set chosen for the inversion must be specified. This is done through the exponent matrix `k`, which has $M = 2$ columns and $N(M + 1)$ rows, reporting the indices with respect to the two internal coordinates of the orders of the moments employed. The $N(M + 1)$ moments are then passed through the vector `m` containing in the different positions the different moments, respecting the order indicated in the exponent matrix `k`. As anticipated, a good initial guess is necessary to start the search. In the matrix variable `xi_i` the initial guess for the nodes is organized into two different rows (corresponding to the two internal coordinates) and `N` columns (corresponding to the different nodes), whereas `w_i` is a vector containing the `N` initial guesses for the weights. The script calculates the quadrature and gives back the `N` weights `w`, the `N` nodes, again organized in the matrix `xi`, and the number of iterations (`iter`) required to converge to the final solution. The script employs the iterative Newton–Raphson method to solve the nonlinear system and the convergence criterion requires the overall normalized residuals to be smaller than `tol` with a default value of 1×10^{-6}. To help convergence, the under-relaxation factor `u_rel` can be taken much smaller than unity; typical values range from 1 to 0.001. The iteration loop is closed upon convergence or when the maximum number of iterations `iter_max` is reached. The extension to $M > 2$ is straightforward, but convergence is generally much more difficult to achieve.

```
function [w,xi,iter]=BF(N,k,m,xi_i,w_i,u_rel)
%
% compute bivariate quadrature approximation with brute-force QMOM
%
% INPUT:
%
%   N     = number of nodes of the quadrature approximation
%   k     = exponent matrix
%   m     = moments of the NDF
%   xi_i  = initial guess for nodes
%   w_i   = initial guess for weights
%   u_rel = under-relaxation factor for Newton-Raphson
%
% OUTPUT:
%
%   xi   = abscissas
%   w    = weights
%
% set the number of internal coordinates
M = 2;
%
% set maximum number of iteration
iter_max = 10000;
%
```

```
% set overall tolerance
tol = 1.0e-6;
%
error = 1.0;
iter  = 0;
%
% copy the initial guess
for i=1:N
    w(i) = w_i(i);
end
%
for i=1:M
    for j=1:N
        xi(i,j) = xi_i(i,j);
    end
end
%
% define unknown vector
for i=1:N
    Z(i) = w(i);
end
for i=1:N
    for j=1:M
        Z(i+j*N) = w(i)*xi(j,i);
    end
end
Z = Z';

% start iteration loop
while (error > tol && iter < iter_max)
%
    iter = iter + 1;
%
    % construct the A matrix
    for i=1:(M+1)*N
        for j=1:(M+1)*N
            if (j<=N)
                a(i,j) = 1.0;
                for q=1:M
                    a(i,j) = a(i,j) - k(i,q);
                end
                for q=1:M
                    a(i,j) = a(i,j)*xi(q,j)^k(i,q);
                end
            elseif (j<=2*N)
                a(i,j) = k(i,1) / xi(1,j-N);
                for q=1:M
                    a(i,j) = a(i,j)*xi(q,j-N)^k(i,q);
                end
            elseif (j<=(M+1)*N)
                a(i,j) = k(i,2) / xi(2,j-2*N);
                for q=1:M
                    a(i,j) = a(i,j)*xi(q,j-2*N)^k(i,q);
                end
            end
        end
    end
%
    % Newton-Raphson method
    for i=1:N*(M+1)
        sum = 0.0;
        for alpha=1:N
            prod = 1.0;
            for beta=1:M
                prod = prod*xi(beta,alpha)^k(i,beta);
            end
            sum = sum + w(alpha)*prod;
        end
        F(i) = sum - m(i);
    end
%
    if abs(det(a)) > 0
        Z = Z - u_rel*inv(a)*F';
```

```
    else
        'Matrix A is singular: change the moment set!'
        iter = iter_max;
    end
%
    % recalculate weights and nodes
    error  = 0.0;
%
    for i=1:N
        w_new(i) = Z(i);
        if w_new(i) > 0
            error = error + abs((w(i)-w_new(i))/w(i));
            w(i) = w_new(i);
        else
            error = error + 1.0;
            w(i) = w(i);
        end
    end
    for i=1:N
        for j=1:M
            if w_new(i)>0.0
                xi_new(j,i) = Z(i+j*N)/w_new(i);
                error = error + abs((xi(j,i)-xi_new(j,i))/xi(j,i));
                xi(j,i) = xi_new(j,i);
            else
                error = error + 1.0;
                xi(j,i) = xi(j,i);
            end
        end
    end
end
```

A.3.2 Tensor-product QMOM

Below a MATLAB script implementing the tensor-product QMOM for a simple bivariate case described in this section is reported. The required inputs are the number of nodes for the first (N1) and for the second (N2) internal coordinates. Since in the formulation described above the moments used for the calculation of the quadrature approximation are defined by the method itself, no exponent matrix is needed. The moments used are passed though a matrix variable m, whose elements are defined by two indices. The first one indicates the order of the moments with respect to the first internal coordinates (index 1 for moment 0, index 2 for moment order 1, etc.), whereas the second one is for the order of the moments with respect to the second internal coordinate. The final matrix is very similar to that reported in Table 3.8. The script returns the quadrature approximation in the usual form: the weights are stored in the weight vector w of size $N = N_1N_2$, whereas the nodes are stored in a matrix with two rows (corresponding to the first and second internal coordinate) and $N = N_1N_2$ columns (corresponding to the different nodes).

```
function [w,xi]=TP(N1,N2,m)
%
% compute bivariate quadrature approximation with the Tensor-Product method
%
%
% INPUT:
%
%   N1    = number of nodes for the first internal coordinate
%   N2    = number of nodes for the second internal coordinate
%   m     = moments of the NDF
%
% OUTPUT:
%
%   xi    = abscissas
%   w     = weights
```

```
%
%
% define a small cutoff value
%
cutoff = 1.0e-6;
%
% define a flag variable
%
flag = 0;
%
% construct the quadrature in the first direction
%
[xi1,w1] = PD(m(:,1)',N1);
%
% construct the quadrature in the second direction
%
[xi2,w2] = PD(m(1,:)',N2);
%
for i=1:N1
    if (abs(w1(i))/m(1,1) < cutoff)
        'One of the weights is null! Reduce the number of nodes in direction 1'
        flag = 1;
    end
end
%
for i=1:N1
    if (abs(w2(i))/m(1,1) < cutoff)
        'One of the weights is null! Reduce the number of nodes in direction 2'
        flag = 1;
    end
end
%
if (flag == 0)
%
% assign the different nodes
%
    index = 0.0;
    for i=1:N1
        for j=1:N2
            index = index + 1.0;
            xi(1,index) = xi1(i);
            xi(2,index) = xi2(j);
        end
    end
 %
 % create the linear system to calculate the final weights
 %
    index = 0.0;
    for i=1:N1
        for j=1:N2
            index = index + 1.0;
            b(index) = m(i,j);
            for index2=1:N1*N2
                A(index,index2) = xi(1,index2)^(i-1)*xi(2,index2)^(j-1);
            end
        end
    end
%
    w = inv(A)*b';
%
elseif (flag ==1)
%
    index = 0.0;
    for i=1:N1
        for j=1:N2
            index = index + 1.0;
            xi(1,index) = 0.0;
            xi(2,index) = 0.0;
            w(index) = 0.0;
        end
    end
%
end
```

A.3.3 The CQMOM

Below a minimal MATLAB script for the determination of the CQMOM approximation (Yuan & Fox, 2011) for a bivariate case is reported. The script requires the specification of the number of nodes desired for the first (`N1`) and second (`N2`) internal coordinate. The moments used in the calculation must be provided in matrix form. The matrix containing the moments `m` is defined by two indices; the first one indicates the order of the moments with respect to the first internal coordinate (index 1 for moment 0, index 2 for moment order 1, etc.), whereas the second one is for the order of the moments with respect to the second internal coordinate. The structure of the data resembles that of Tables 3.9–3.11. The script calculates the $N = N_1 N_2$ weights and nodes and stores them in the vector `w` and in the matrix `xi`, which have the same structure as in the previous script. The procedure is based on the calculation of the moments with respect to the second internal coordinate, conditioned on the value of the first internal coordinate. The calculation of the quadrature constructed on moments conditioned on the second internal coordinate can be simply carried out by providing the script with the transpose of the moment matrix.

```
function [w,xi]=CQMOM(N1,N2,m)
%
% compute bivariate quadrature approximation with the Conditional
% Quadrature Method Of Moments
%
% INPUT:
%
%    N1    = number of nodes for the first internal coordinate
%    N2    = number of nodes for the second internal coordinate
%    m     = moments of the NDF
%
% OUTPUT:
%
%    xi   = abscissas
%    w    = weights
%
% define a small cutoff value
%
cutoff = 1.0e-6;
%
% define a flag variable
%
flag = 0;
%
% construct the quadrature in the first direction
%
[xi1,w1] = PD(m(:,1)',N1);
%
for i=1:N1
    if (abs(w1(i))/m(1,1) < cutoff)
        'One of the weights is null! Reduce the number of nodes in direction 1'
        flag = 1;
    end
end
%
if (flag == 0)
%
% define the Vandermonde matrix for conditional moments
%
    for i=1:N1
        for j=1:N1
            V1(i,j) = xi1(j)^(i-1.0);
        end
    end
%
% define the R1 matrix
%
```

```
    for i=1:N1
        R1(i,i) = w1(i);
    end
%
% calculate the conditional moments
%
    for i=1:2*N2
        c_m(:,i) = inv(V1*R1)*m(1:N1,i);
    end
%
    for i=1:N1
        [xi2(:,i),w2(:,i)] = PD(c_m(i,:),N2);
    end
%
    index = 0.0;
    for i=1:N1
        for j=1:N2
            index = index + 1.0;
            w(index) = w1(i)*w2(j,i);
            xi(1,index) = xi1(i);
            xi(2,index) = xi2(j,i);
        end
    end
elseif (flag == 1)
    index = 0.0;
    for i=1:N1
        for j=1:N2
            index = index + 1.0;
            w(index) = 0.0;
            xi(1,index) = 0.0;
            xi(2,index) = 0.0;
        end
    end
end
```

A.4 The EQMOM

In this section we provide example codes for computing extended quadratures as described in Section 3.3.2. These codes are meant to illustrate the general structure of the iterative method needed to find σ, and can undoubtedly be improved.

A.4.1 Beta EQMOM

```
function [w,x,sig,nout,flag] = beta_Extended(m, rmax, eabs, nnodes, b)
% b is the upper limit of beta distribution, lower limit is set as 0
%
n = nnodes ;
nmom = 2*n+1 ;
w = zeros(n,1) ;
x = zeros(n,1) ;
% Use an iterative method to find sig (based on 1D bounded secant method)
itmax = 100 ;
mtol = 1.d-8 ;
ftol = 1.d-14/m(nmom) ;
stol = 1.d-14 ;
% check realizable density
if m(1) <= 0
    sig = 0 ;
    flag = 0 ;
    display('zero or negative density!')
    return
end
% Find maximum value of sigma = total variance (upper bound)
sigmax = abs(m(1)*m(3) - m(2)^2)/(m(1)*(m(3) - m(2))) ; % variance like condition
sigmaxx = abs(m(3)/(m(3) - m(2))) ;   % ms positive condition for ms(3:5)
sigmax = max(sigmax, sigmaxx) ;
```

```
% Before iterating, try sigmax
sig = sigmax ;
ms = beta_m2ms(m, sig, nmom-1, b) ;
[w,x,nout,~] = Wheeler_moments_adaptive(ms(1:2*n), n, rmax, eabs) ;
f = beta_targetfun(m, nmom, n, b, rmax, eabs, sig) ;
if abs(f) < ftol
    mre = beta_moments_1D_NVsig(w, x, sig, nmom, b) ;
    mdiff = (m(1:nmom) - mre(1:nmom))./m(1:nmom) ;
    if norm(mdiff) > mtol
        format longg
        disp(norm(mdiff))
        disp(mdiff)
        disp(m)
        disp(sig)
        disp(f)
        disp('moments not conserved in EQMOM when using sigmax')
    end
    flag = 0 ;
    return
end
sold = 0 ;
sig = sold ;
%
% First compute f(sig) = fold
ms = beta_m2ms(m, sig, nmom-1, b) ;
[w,x,nout,~] = Wheeler_moments_adaptive(ms(1:2*n), n, rmax, eabs) ;
fold = beta_targetfun(m, nmom, n, b, rmax, eabs, sig) ;
flag = 0 ;
% check for convergence, and stop if converged
if abs(fold) >= ftol
    % 1 Newton step (with approximate slope df) to get things started
    switch nnodes
        case 1
            df = -ms(2)/m(3) ;
        case 2
            df = -6*ms(4)/m(5) ;
        case 3
            df = -15*ms(6)/m(7) ;
        case 4
            df = -28*ms(8)/m(9) ;
        otherwise
            disp('wrong nnodes')
            error('beta_Extended')
    end
    ds = -fold/df ;
    snew = sold + ds ;
    sig = min(snew, sigmax) ;
    sig = max(sig, 0) ;
    snew = sig ;
    fnew = beta_targetfun(m, nmom, n, b, rmax, eabs, sig) ;
    flag = 0 ;
    method = 0 ;
    lambda = 0.1 ;
    dsmax = sigmaxx ;
    % check for convergence, and stop if converged
    if abs(fnew) >= ftol
        % Now try bounded secant method when f > 0 , otherwise bisection
        df = (fnew - fold)/(snew - sold) ;
        ds = -fnew/df ;
        sig = snew + ds ;
        if fnew < 0
            sigmax = snew ;
            sig = snew/2 ;
            snew = 0 ;
        end
        sig = min(sig, sigmax) ;
        sig = max(sig, 0) ;
        ds = sig - snew ;
        for j = 1:itmax
            % compute new value of f
            f = beta_targetfun(m, nmom, n, b, rmax, eabs, sig) ;
            % new guess
            df = f - fold ;
```

```
        % check for convergence
        if abs(f) <= ftol || abs(ds) <= stol || abs(df) <= ftol
            ms = beta_m2ms(m, sig, nmom-1, b) ;
            [w,x,nout,~] = Wheeler_moments_adaptive(ms(1:2*n), n, rmax, eabs) ;
            mre = beta_moments_1D_NVsig (w, x, sig, nmom, b) ;
            mdiff = mre(1:nmom) - m(1:nmom) ;
            if norm(mdiff) > mtol
                disp(m)
                disp(mdiff)
                disp(norm(mdiff)) ;
                disp(f)
                disp(ds)
                disp(df)
                disp('moments not conserved')
            end
            return
        end
        if dsmax == 0 && f > 0
            display('function has no zero!')
            display(j)
            display(f)
            ms = moments_1D(w,x,nmom) ;
            ms(nmom) = 0 ;
            for i=1:nout
                ms(nmom) = ms(nmom) + w(i)*x(i)^(nmom-1) ;
            end
            normdiff = max(m - beta_ms2m(ms,sig,nmom,b)) ;
            if normdiff > ftol
                display('normdiff larger then tol')
                display(normdiff)
            end
            return
        end
        % new guess
        if f > 0 && method == 0 % secant
            fnew = f ;
            snew = sig ;
            dfds = df/(snew - sold) ;
            ds = -fnew/dfds ;
            sig1 = snew + lambda*ds ;
            sig1 = min(sig1, sigmax) ;
            sig1 = max(sig1, 0) ;
            dsmax = sigmaxx - sig1 ;
            kk = 0 ;
            while dsmax == 0 && kk < itmax
                kk = kk + 1 ;
                ds = lambda*min(ds,sigmaxx-snew)/2 ;
                sig1 = snew + ds ;
                sig1 = min(sig1, sigmax) ;
                sig1 = max(sig1, 0) ;
                dsmax = sigmaxx - sig1 ;
            end
            sig = sig1 ;
            fold = fnew ;
            sold = snew ;
            if j > 10
                lambda = 1 ;
            end
        else % bisection
            if f > 0
                fold = f ;
                sold = sig ;
                ds = (snew - sold)/2 ;
            else
                fnew = f ;
                snew = sig ;
                ds = (snew - sold)/2 ;
                sigmax = sig ;
            end
            sig = sold + ds ;
            sig = min(sig, sigmax) ;
            sig = max(sig, 0) ;
            dsmax = sigmaxx - sig ;
```

```
                method = 1 ;
            end
            flag = j ;
            if j >= itmax
                ms = beta_m2ms(m, sig, nmom-1, b) ;
                [w,x,nout,~] = Wheeler_moments_adaptive(ms(1:2*n), n, rmax, eabs) ;
                format long
                display('too many iterations! : increase itmax')
                display(fnew)
                display(fold)
                display(ds)
                display(method)
                display(dsmax)
                ms = moments_1D(w,x,nmom) ;
                ms(nmom) = 0 ;
                for i=1:nout
                    ms(nmom) = ms(nmom) + w(i)*x(i)^(nmom-1) ;
                end
                normdiff = max(m - beta_ms2m(ms,sig,nmom,b)) ;
                if normdiff > ftol
                    display('normdiff larger then tol')
                    display(normdiff)
                end
                format short
                return
            end
        end
    end
end
end
```

A.4.2 Gamma EQMOM

```
function [w,x,sig,nout,flag] = gamma_Extended(m, rmax, eabs, nodes)
% Routine for 1D extended gamma EQMOM
%
n = 4 ;
nmom = 2*n+1 ;
w = zeros(n,1) ;
x = zeros(n,1) ;
% Use an iterative method to find sig (based on 1D bounded secant method)
itmax = 100 ;
ftol = 1.d-15 ;
stol = 1.d-15 ;
% check realizable density
if m(1) <= 0
    sig = 0 ;
    flag = 0 ;
    display('zero or negative density!')
    return
end
% Find maximum value of sigma = total variance (upper bound)
sigmax = abs(m(1)*m(3) - m(2)^2 )/(m(1)*m(2)) ; % variance like condition
sigmaxx = sigmax ;
sigmax = abs(m(3)/m(2)) ;   % ms positive condition for ms(3:5)
sigmaxx = max(sigmax, sigmaxx) ;
sigmax = sqrt(abs(m(4)/(2*m(2)))) ;
sigmaxx = max(sigmax, sigmaxx) ;
sigmax = (m(5)/(6*m(2)))^(1/3) ;
sigmaxx = max(sigmax, sigmaxx) ;
sigmax = sigmaxx ;
% Before iterating, try sigmax
sig = sigmax ;
ms = m2ms(m, sig, nmom-1) ;
[w,x,nout,~] = Wheeler_moments_adaptive(ms(1:2*n), n, rmax, eabs) ;
ms(nmom) = m(nmom) ;
for i=1:nout
    ms(nmom) = ms(nmom) - w(i)*x(i)^(nmom-1) ;
end
f = ms(9) - (28*sig*ms(8)+322*sig^2*ms(7)+1960*sig^3*ms(6)+6769*sig^4*ms(5) ...
    +13132*sig^5*ms(4)+13068*sig^6*ms(3)+5040*sig^7*ms(2)) ;
```

```
if abs(f) < ftol
    flag = 1 ;
    return
elseif f > 0  && nout == nodes
    display('fmax > 0 : no minimum available!')
    display(f)
    display(sig)
    display(nout)
end
sold = 0 ;
sig = sold ;
%
% First compute f(sig) = fold
ms = m2ms(m, sig, nmom-1) ;
[w,x,nout,~] = Wheeler_moments_adaptive(ms(1:2*n), n, rmax, eabs) ;
ms(nmom) = m(nmom) ;
for i=1:nout
    ms(nmom) = ms(nmom) - w(i)*x(i)^(nmom-1) ;
end
fold = ms(nmom) ;
flag = 1 ;
% check for convergence, and stop if converged
if abs(fold) >= ftol
    % 1 Newton step (with approximate slope df) to get things started
    df = -28*ms(8) ;
    ds = -fold/df ;
    snew = sold + ds ;
    sig = min(snew, sigmax) ;
    sig = max(sig, 0) ;
    snew = sig ;
    ms = m2ms(m, sig, nmom-1) ;
    [w,x,nout,~] = Wheeler_moments_adaptive(ms(1:2*n), n, rmax, eabs) ;
    ms(nmom) = m(nmom) ;
    for i=1:nout
        ms(nmom) = ms(nmom) - w(i)*x(i)^(nmom-1) ;
    end
    fnew  = ms(9) - (28*sig*ms(8)+322*sig^2*ms(7)+1960*sig^3*ms(6) ...
            +6769*sig^4*ms(5)+13132*sig^5*ms(4)+13068*sig^6*ms(3)+5040*sig^7*ms(2)) ;
    flag = 2 ;
    method = 0 ;
    lambda = 1 ;
    dsmax = sigmaxx ;
    % check for convergence, and stop if converged
    if abs(fnew) >= ftol
        % Now try bounded secant method when f > 0 , otherwise bisection
        df = (fnew - fold)/(snew - sold) ;
        ds = -fnew/df ;
        sig = snew + ds ;
        if fnew < 0
            sigmax = snew ;
            sig = snew/2 ;
            snew = 0 ;
        end
        sig = min(sig, sigmax) ;
        sig = max(sig, 0) ;
        ds = sig - snew ;
        for j = 1:itmax
            % compute new value of f
            ms = m2ms(m, sig, nmom-1) ;
            [w,x,nout,~] = Wheeler_moments_adaptive(ms(1:2*n), n, rmax, eabs) ;
            ms(nmom) = m(nmom) ;
            for i=1:nout
                ms(nmom) = ms(nmom) - w(i)*x(i)^(nmom-1) ;
            end
            f = ms(9) - (28*sig*ms(8)+322*sig^2*ms(7)+1960*sig^3*ms(6)+6769*sig^4*ms(5) ...
                +13132*sig^5*ms(4)+13068*sig^6*ms(3)+5040*sig^7*ms(2)) ;
            % new guess
            df = f - fold ;
            % check for convergence
            if abs(f) <= ftol || abs(ds) <= stol || abs(df) <= ftol
                return
            end
            if dsmax == 0 && f > 0
```

```
                display('function has no zero!')
                display(j)
                display(f)
                ms = moments_1D(w, x, nmom) ;
                ms(nmom) = 0 ;
                for i=1:nout
                    ms(nmom) = ms(nmom) + w(i)*x(i)^(nmom-1) ;
                end
                normdiff = max(m - ms2m(ms, sig, nmom)) ;
                if normdiff > ftol
                    display('normdiff larger then tol')
                    display(normdiff)
                end
                return
            end
        fnew = f ;
        snew = sig ;
        dfds = df/(snew - sold) ;
        ds = -fnew/dfds ;
        sig1 = snew + lambda*ds ;
        sig1 = min(sig1, sigmax) ;
        sig1 = max(sig1,0) ;
        dsmax = sigmaxx - sig1 ;
        kk = 0 ;
        while dsmax == 0 && kk < itmax
            kk = kk + 1 ;
            ds = lambda*min(ds, sigmaxx - snew)/2 ;
            sig1 = snew + ds ;
            sig1 = min(sig1, sigmax) ;
            sig1 = max(sig1, 0) ;
            dsmax = sigmaxx - sig1 ;
        end
        sig = sig1 ;
        fold = fnew ;
        sold = snew ;
        if j > 10
            lambda = min(1.9, 1.01*lambda) ;
        end
            flag = j ;
            if j >= itmax
                format long
                display('too many iterations! : increase itmax')
                display(fnew)
                display(fold)
                display(ds)
                display(method)
                display(dsmax)
                ms = moments_1D(w, x, nmom) ;
                ms(nmom) = 0 ;
                for i=1:nout
                    ms(nmom) = ms(nmom) + w(i)*x(i)^(nmom-1) ;
                end
                normdiff = max(m - ms2m(ms, sig, nmom)) ;
                if normdiff > ftol
                    display('normdiff larger then tol')
                    display(normdiff)
                end
                format short
                return
            end
        end
    end
end
end
```

A.4.3 Gaussian EQMOM

```
function [w, x, sig, nout, flag] = gaussian_Extended(m)
% Routine for 1D extended three-node Gaussian EQMOM
%
n = 3 ;
```

```
nmom = 2*n+1 ;
% check realizable density
if m(1) <= 0
    w = zeros(n,1) ;
    x = w ;
    sig = 0 ;
    flag = 0 ;
    display('zero or negative density!')
    return
end
% set quadrature parameters
rmax(1) = 1.d-10 ; % smallest weight with more than 1 node
rmax(2) = 1.d-6 ; % minimum weight ratio
rmax(3) = 1.d-4 ;
eabs    = 1.d-6 ; % minimum distance between distinct absicssas
%
% Find maximum value of sigma = total variance (upper bound)
sigmax = (m(3) - m(2)^2/m(1))/m(1) ;
% Use an iterative method to find sig (based on 1D bounded secant method)
itmax = 100 ;
ftol = 1.d-15 ;
stol = 1.d-15 ;
% initial guess sigma: sig = 0 (lower bound)
sold = 0 ;
sig = sold ;
%
% First compute f(sig) = fold
ms = m2ms(m, sig, nmom-1) ;
[w,x,nout,~] = Wheeler_moments_adaptive(ms(1:2*n), n, rmax, eabs) ;
ms(nmom) = m(7) ;
for i=1:nout
    ms(nmom) = ms(nmom) - w(i)*x(i)^6 ;
end
fold = ms(7) ;
flag = 1 ;
% check for convergence, and stop if converged
if abs(fold) >= ftol
    % 1 Newton step (with approximate slope df) to get things started
    df = -15*ms(5) ;
    ds = -fold/df ;
    snew = sold + ds ;
    sig = min(snew, sigmax) ;
    sig = max(sig, 0) ;
    snew = sig ;
    ms = m2ms(m, sig, nmom-1) ;
    [w,x,nout,~] = Wheeler_moments_adaptive(ms(1:2*n), n, rmax, eabs) ;
    ms(nmom) = m(7) ;
    for i=1:nout
        ms(nmom) = ms(nmom) - w(i)*x(i)^6 ;
    end
    fnew = ms(7) - (15*ms(5) + (45*ms(3) + 15*ms(1)*sig)*sig)*sig ;
    flag = 2 ;
    % check for convergence, and stop if converged
    if abs(fnew) >= ftol
        % Now try bounded secant method
        df = (fnew - fold)/(snew - sold) ;
        ds = -fnew/df ;
        sig = snew + ds ;
        sig = min(sig, sigmax) ;
        sig = max(sig, 0) ;
        ds = sig - snew ;
        for j = 1:itmax
            % compute new value of f
            ms = m2ms(m, sig, nmom-1) ;
            [w,x,nout,~] = Wheeler_moments_adaptive(ms(1:2*n), n, rmax, eabs) ;
            ms(nmom) = m(7) ;
            for i=1:nout
                ms(nmom) = ms(nmom) - w(i)*x(i)^6 ;
            end
            f = ms(7) - (15*ms(5) + (45*ms(3) + 15*ms(1)*sig)*sig)*sig ;
            % reset end points
            fold = fnew ;
            sold = snew ;
```

```
            fnew = f ;
            snew = sig ;
            % new guess
            df = fnew - fold ;
            % check for convergence
            if abs(f) <= ftol || abs(ds) <= stol || abs(df) <= ftol
                return
            end
            % new guess
            dfds = df/(snew - sold) ;
            ds = -fnew/dfds ;
            sig = snew + ds ;
            sig = min(sig, sigmax) ;
            sig = max(sig,0) ;
            ds = sig - snew ;
            flag = j ;
            if j >= itmax
                display('too many iterations! : increase itmax')
            end
        end
    end
end
end
```

Appendix B

Kinetics-based finite-volume methods

In this appendix we discuss the solution of the moment-transport equations found from the GPBE using kinetics-based finite-volume methods (KBFVM) in multiple spatial dimensions. As in Section 8.2, the discussion focuses on time advancement using a single-stage Euler explicit time-integration scheme. Readers interested in more details on finite-volume methods are referred to Leveque (2002).

B.1 Spatial dependence of GPBE

In the following, we denote the NDF by $n(t, \mathbf{x}, \mathbf{v}, \boldsymbol{\xi})$, where the velocity internal coordinates are $\mathbf{v}$ and the remaining internal coordinates are $\boldsymbol{\xi}$. The general form of the GPBE to be considered is

$$\frac{\partial n}{\partial t} + \frac{\partial}{\partial \mathbf{x}} \cdot (\mathbf{u} n) + \mathbf{v} \cdot \frac{\partial n}{\partial \mathbf{x}} = \frac{\partial^2}{\partial \mathbf{x}\, \partial \mathbf{v}} : (\boldsymbol{\lambda} n) + \frac{\partial}{\partial \mathbf{x}} \cdot \left(\boldsymbol{\Gamma} \cdot \frac{\partial n}{\partial \mathbf{x}} \right) + S, \tag{B.1}$$

where the second and third terms on the left-hand side are multidimensional advection with a known velocity $\mathbf{u}(t, \mathbf{x}, \boldsymbol{\xi})$ and free transport, respectively, and the final term on the right-hand side, $S(t, \mathbf{x}, \mathbf{v}, \boldsymbol{\xi})$, is the phase-space source term. The term involving the nonnegative second-order tensor $\boldsymbol{\lambda}(t, \mathbf{x}, \mathbf{v}, \boldsymbol{\xi})$ is the mixed advection term[1] describing how gradients in velocity phase space lead to fluxes in real space. This term appears in turbulence models for gas–particle flows to describe the effect of unresolved gas-velocity fluctuations on the particle velocity-distribution function. The term involving the nonnegative, second-order tensor $\boldsymbol{\Gamma}(t, \mathbf{x}, \mathbf{v}, \boldsymbol{\xi})$ models multidimensional diffusion in real space due, for example, to Brownian motion. For convenience, we have included all possible terms in Eq. (B.1), but in applications it is almost always the case that only a subset of the terms will be present.

[1] Mixed advection is usually written in the form (Reeks, 1983, 1991, 1992; Zaichik, 1997; Zaichik *et al.*, 2008)

$$\frac{\partial}{\partial \mathbf{v}} \cdot \left(\boldsymbol{\lambda} \cdot \frac{\partial n}{\partial \mathbf{x}} + \boldsymbol{\mu} \cdot \frac{\partial n}{\partial \mathbf{v}} \right) = \frac{\partial^2}{\partial \mathbf{x}\, \partial \mathbf{v}} : (\boldsymbol{\lambda} n) - \frac{\partial}{\partial \mathbf{v}} \cdot \left[\left(\frac{\partial}{\partial \mathbf{x}} \cdot \boldsymbol{\lambda} \right) n \right] + \frac{\partial}{\partial \mathbf{v}} \cdot \left(\boldsymbol{\mu} \cdot \frac{\partial n}{\partial \mathbf{v}} \right),$$

where $\boldsymbol{\mu}$ is a nonnegative, second-order tensor. The second and third contributions on the right-hand side of this expression can be added to the phase-space source term S, and thus Eq. (B.1) is an equivalent form.

For example, in a PBE the velocity internal coordinate is absent, so only advection, diffusion, and the source term will appear. Likewise, for a KE only the free-transport, mixed advection, and source terms will appear.

For clarity and simplicity, we discuss the finite-volume method (FVM) for a 3D structured Cartesian grid with uniform spacing in each direction: $(\Delta x, \Delta y, \Delta z)$. Moreover, we consider only isotropic diffusion and isotropic mixed advection so that $\mathbf{\Gamma}(t, \mathbf{x}, \mathbf{v}, \boldsymbol{\xi}) = \Gamma(t, \mathbf{x}, \mathbf{v}, \boldsymbol{\xi})\delta_{ij}$ and $\boldsymbol{\lambda}(t, \mathbf{x}, \mathbf{v}, \boldsymbol{\xi}) = \lambda(t, \mathbf{x}, \mathbf{v}, \boldsymbol{\xi})\delta_{ij}$. The extension of the realizable high-order FVM to unstructured grids is described in Vikas *et al.* (2011a). Here our focus is on ensuring realizability of the transported moment set when high-order spatial reconstructions are employed and thus the simpler formulas for structured grids suffice. Furthermore, by using operator splitting each term in Eq. (B.1) can be advanced in time separately, so it suffices to describe the realizable FVM for each term separately. Thus, in Section B.3, we describe the realizable FVM for multidimensional advection:

$$\frac{\partial n}{\partial t} = -\frac{\partial}{\partial x_i}(u_i n) \tag{B.2}$$

for each direction $i = 1, 2, 3$. In Section B.4 we describe the realizable FVM for free transport:

$$\frac{\partial n}{\partial t} = -\frac{\partial}{\partial x_i}(v_i n), \tag{B.3}$$

in Section B.5 we do the same for mixed advection:

$$\frac{\partial n}{\partial t} = \frac{\partial^2}{\partial x_i \partial v_i}(\lambda n), \tag{B.4}$$

and in Section B.6 we do it for diffusion:

$$\frac{\partial n}{\partial t} = \frac{\partial}{\partial x_i}\left(\Gamma \frac{\partial n}{\partial x_i}\right). \tag{B.5}$$

The reader should note that in Eqs. (B.2)–(B.5) the spatial derivative appears on the right-hand side, and therefore it will be necessary to define a *realizable high-order* FVM for each case. In contrast, the source term S in Eq. (B.1) contains no spatial derivatives[2] and hence is *local* in each finite-volume grid cell. In other words, with operator splitting the source term leads to a (stiff) ordinary differential equation (ODE) for each grid cell.

As described in Chapter 7, the source term can be treated using QBMM, where the weights and abscissas are computed from the multivariate moments. It will thus be necessary to design a *realizable* ODE solver for use with the source term in Eq. (B.1). In the remainder of this appendix, we assume that such an ODE solver exists, and thus we focus our attention on the spatial-transport solver. Nevertheless, the reader should consult Yuan & Fox (2011) for details on how to use the CQMOM with multiple permutations to evaluate the source term in a statistically consistent manner. Briefly, when using the CQMOM

[2]The source term can contain derivatives in the internal coordinates. For example, diffusion in velocity phase space is written as

$$\frac{\partial n}{\partial t} = \frac{\partial^2}{\partial v_i \partial v_j}\left(B_{ij} n\right),$$

where $\mathbf{B}(t, \mathbf{x}, \mathbf{v})$ is a nonnegative diffusion tensor. Since this term does not involve spatial derivatives, it is included in the source term S.

as described in Section 3.2.3, the user can choose which internal coordinate comes first, second, etc. when defining the conditional PDF, and we refer to each unique choice as a *permutation* of the CQMOM. Unless the underlying NDF is exactly reproduced by all permutations of the CQMOM, the resulting sets of weights and abscissas will be different. *Nevertheless, each permutation generates a valid reconstruction of the NDF!* Thus, if we assume that there are P permutations, then the source term will be closed differently for each one. Let S_p denote the source-term closure for the pth permutation of the CQMOM. The change in the transported moment set $\mathbf{M}$ due to the source terms S_p can then be expressed as

$$\frac{\mathrm{d}\mathbf{M}}{\mathrm{d}t} = \frac{1}{P}\sum_{p=1}^{P}\int_{\Omega_{\xi}}\int_{\mathbb{R}^3}\mathbf{K}S_p\,\mathrm{d}\mathbf{v}\,\mathrm{d}\boldsymbol{\xi}, \tag{B.6}$$

where $\mathbf{K}$ is the moment vector (see Eq. (B.9) for an example) corresponding to the transported moment set. In other words, all P permutations are treated equally when defining the closure for the (homogeneous) moment equations. For completeness, we should note that, if the brute-force QMOM described in Section 3.2.1 could be applied, then $P = 1$ since the maximum possible number of moments is controlled by the quadrature reconstruction. Unfortunately, brute-force QMOM are not guaranteed to converge and thus are ill-suited for use with FVM.

B.2 Realizable FVM

On a structured Cartesian grid, the grid cell centered at the point $(i\,\Delta x,\ j\,\Delta y,\ k\,\Delta z)$ can be denoted by Ω_{ijk} and its volume by $|\Omega_{ijk}| = \Delta x\,\Delta y\,\Delta z$. The volume-average NDF in Ω_{ijk} at time $t = m\,\Delta t$,[3] denoted by $n^m_{ijk}(\mathbf{v},\boldsymbol{\xi})$, is defined by (Leveque, 2002)

$$n^m_{ijk} \equiv \frac{1}{|\Omega_{ijk}|}\int_{\Omega_{ijk}} n(m\,\Delta t,\mathbf{x},\mathbf{v},\boldsymbol{\xi})\mathrm{d}\mathbf{x}. \tag{B.7}$$

The transported moment set $\mathbf{M}(t,\mathbf{x})$ is related to the NDF by

$$\mathbf{M} = \int_{\Omega_{\xi}}\int_{\mathbb{R}^3}\mathbf{K}n\,\mathrm{d}\mathbf{v}\,\mathrm{d}\boldsymbol{\xi}, \tag{B.8}$$

where $\mathbf{K}$ is the moment vector defined by

$$\mathbf{K} = \begin{bmatrix}1 & v_1 & v_2 & v_3 & \xi_1 & \xi_2 & \cdots & v_1^{k_1}v_2^{k_2}v_3^{k_3}\xi_1^{k_4}\xi_2^{k_5}\cdots & \cdots\end{bmatrix}^{\mathrm{T}} \tag{B.9}$$

and the length of $\mathbf{K}$ is determined by the multivariate transported moment set. Multiplying Eq. (B.7) by $\mathbf{K}$ and integrating over the internal coordinates yields the volume-average transported moment set:

$$\mathbf{M}^m_{ijk} \equiv \frac{1}{|\Omega_{ijk}|}\int_{\Omega_{ijk}}\mathbf{M}(m\,\Delta t,\mathbf{x})\mathrm{d}\mathbf{x} = \int_{\Omega_{\xi}}\int_{\mathbb{R}^3}\mathbf{K}n^m_{ijk}\,\mathrm{d}\mathbf{v}\,\mathrm{d}\boldsymbol{\xi}. \tag{B.10}$$

[3] Writing $t = m\,\Delta t$ is correct only if the time step is constant. In general, this will not be the case. Nevertheless, we will abuse the notation by writing $t = m\,\Delta t$ in all cases with the implicit understanding that the formula for t is more complicated when Δt varies on each time step.

In general, the volume-average transported moment set is related to a volume-average NDF. Thus, in order to show that $\mathbf{M}^m_{ijk}$ is a realizable moment set, it suffices to show that n^m_{ijk} is nonnegative. This idea of using the NDF to show the realizability of the moment set is the basis of *kinetics-based* FVM.

Given n^0_{ijk}, the value of n^m_{ijk} for $0 < m$ is determined from the finite-volume solution to Eq. (B.1). Thus, as a first step, we need to consider the volume-average form of Eq. (B.1) or, equivalently, the volume-average forms of the individual terms in Eqs. (B.2)–(B.5). Using a single-stage Euler explicit time-integration scheme (Leveque, 2002), the finite-volume expression for the updated NDF has the form[4]

$$\begin{aligned} n^{m+1}_{ijk} = n^m_{ijk} &- \frac{\Delta t}{\Delta x}\left[G\left(n^m_{i+1/2\,jk,l}, n^m_{i+1/2\,jk,r}\right) - G\left(n^m_{i-1/2\,jk,l}, n^m_{i-1/2\,jk,r}\right)\right] \\ &- \frac{\Delta t}{\Delta y}\left[G\left(n^m_{ij+1/2\,k,l}, n^m_{ij+1/2\,k,r}\right) - G\left(n^m_{ij-1/2\,k,l}, n^m_{ij-1/2\,k,r}\right)\right] \\ &- \frac{\Delta t}{\Delta z}\left[G\left(n^m_{ijk+1/2,l}, n^m_{ijk+1/2,r}\right) - G\left(n^m_{ijk-1/2,l}, n^m_{ijk-1/2,r}\right)\right], \end{aligned} \tag{B.11}$$

where $G(n_l, n_r)$ is a numerical flux function that must be constructed for each spatial-transport term. An important point to observe in Eq. (B.11) is that the NDF appearing in the numerical flux function corresponds to the cell faces. For example, $n^m_{i+1/2\,jk,l}$ is the NDF to the left[5] of the cell face located at $(i + 1/2)\Delta x$, and $n^m_{i-1/2\,jk,r}$ is the NDF to the right of the cell face located at $(i - 1/2)\Delta x$. In the FVM, only the volume-average values n^m_{ijk} are known and, hence, spatial reconstruction is used to find the face values. For example, with a first-order spatial reconstruction, we set all values in the cell equal to n^m_{ijk}. On the other hand, with a second-order spatial reconstruction, we use a linear approximation:

$$n^m(\mathbf{x}, \mathbf{v}, \boldsymbol{\xi}) = n^m_{ijk}(\mathbf{v}, \boldsymbol{\xi}) + \mathbf{a}^m_{ijk}(\mathbf{v}, \boldsymbol{\xi}) \cdot \left(\mathbf{x} - \mathbf{x}_{ijk}\right) \qquad \text{for } \mathbf{x} \in \Omega_{ijk}, \tag{B.12}$$

where $\mathbf{x}_{ijk}$ denotes the cell center and the slope vector $\mathbf{a}^m_{ijk}$ is computed from the values of n^m_{ijk} in neighboring cells using flux limiters to ensure that $n^m(\mathbf{x}, \mathbf{v}, \boldsymbol{\xi})$ is always nonnegative for $\mathbf{x} \in \Omega_{ijk}$. In a realizable, high-order FVM the numerical flux functions must be defined in such a way as to guarantee that $n^{m+1}_{ijk}(\mathbf{v}, \boldsymbol{\xi})$ computed from Eq. (B.11) is always nonnegative. In general, with a first-order spatial reconstruction it is straightforward to construct realizable numerical flux functions. However, with high-order spatial reconstruction this is usually not the case.

In QBMM, the volume-average NDF n^m_{ijk} is reconstructed from the volume-average transported moment set $\mathbf{M}^m_{ijk}$ using a moment-inversion algorithm (see Chapter 3 for details). The resulting NDF has the form[6]

$$n^m_{ijk}(\mathbf{v}, \boldsymbol{\xi}) = \sum_{\alpha=1}^{N} w^m_{ijk,\alpha} \delta\left(\mathbf{v} - \mathbf{v}^m_{ijk,\alpha}\right) \delta\left(\boldsymbol{\xi} - \boldsymbol{\xi}^m_{ijk,\alpha}\right) \qquad \text{for } \mathbf{x} \in \Omega_{ijk}, \tag{B.13}$$

[4]The extension to unstructured grids is presented in Vikas *et al.* (2011a). The key point both for structured and for unstructured grids is that it will be necessary to reconstruct the NDF at cell faces given the volume-average NDF in the grid cells.

[5]We use the same convention as in 1D to define left and right: increasing values of x_i correspond to right, and decreasing values correspond to left.

[6]Even when multivariate EQMOM is used with kernel density functions, a Dirac delta function (dual-quadrature) representation is employed to close the terms in the GPBE. See Section 3.3.4 for more details.

where $w^m_{ijk,\alpha}$ are the weights and $(\mathbf{v}^m_{ijk,\alpha}, \boldsymbol{\xi}^m_{ijk,\alpha})$ are the abscissas corresponding to the volume-average moments in grid cell Ω_{ijk}. Hereafter, to avoid confusion, we will refer to this step as *quadrature* reconstruction. The quadrature reconstruction is done separately in each grid cell, while the spatial reconstruction uses information from (potentially) all grid cells. As shown in Vikas *et al.* (2011a), if the NDF at the cell faces is found by using quadrature reconstruction of the moments at the cell faces found by spatial reconstruction of the volume-average moments, it is not possible to guarantee that the NDF at the cell faces will be realizable. Likewise, if a spatial reconstruction of the weights and abscissas is used with Eq. (B.13) to approximate the NDF at the cell faces, then the latter will be unrealizable when high-order spatial reconstructions are used (see Section 8.2.1 for an example).

If the CQMOM is used for the multivariate quadrature reconstruction, another important technical point to keep in mind is that Eq. (B.13) corresponds to *one* permutation of the CQMOM. In other words, each of the P permutations of the CQMOM will generate a different representation of $n^m_{ijk}(\mathbf{v}, \boldsymbol{\xi})$. Nevertheless, the formulas given below for computing $n^{m+1}_{ijk}(\mathbf{v}, \boldsymbol{\xi})$ starting from $\mathbf{M}^m_{ijk}$ can be applied for each permutation, so that at the end of one time step each permutation will yield a different updated moment set, denoted by $(\mathbf{M}^{m+1}_{ijk})_p$. We can then define the permutation-average updated moment set as

$$\mathbf{M}^{m+1}_{ijk} = \frac{1}{P} \sum_{p=1}^{P} \left(\mathbf{M}^{m+1}_{ijk}\right)_p, \tag{B.14}$$

which is then the starting point for the next time iteration. As we shall see in Section B.4, the use of CQMOM permutations is slightly different for the velocity internal coordinates. Nevertheless, the general idea of using all valid CQMOM permutations remains unchanged.

Following Vikas *et al.* (2011a), in order to ensure that the spatial reconstruction of the NDF is realizable, we will use first-order spatial reconstruction for the quadrature abscissas and high-order spatial reconstruction for the quadrature weights. Formally, this can be written as

$$n^m(\mathbf{x}, \mathbf{v}, \boldsymbol{\xi}) = \sum_{\alpha=1}^{N} w^m_\alpha(\mathbf{x}) \delta\left(\mathbf{v} - \mathbf{v}^m_{ijk,\alpha}\right) \delta\left(\boldsymbol{\xi} - \boldsymbol{\xi}^m_{ijk,\alpha}\right) \quad \text{for } \mathbf{x} \in \Omega_{ijk}, \tag{B.15}$$

where the weights $0 \leq w^m_\alpha(\mathbf{x})$ are reconstructed with flux limiters using the known values of $w^m_{ijk,\alpha}$ in all grid cells. For example, with first-order spatial reconstruction the NDF inside grid cell Ω_{ijk} is represented by

$$n^m(\mathbf{x}, \mathbf{v}, \boldsymbol{\xi}) = \sum_{\alpha=1}^{N} \left[w^m_{ijk,\alpha} + \mathbf{a}_{ijk,\alpha} \cdot \left(\mathbf{x} - \mathbf{x}_{ijk}\right)\right] \delta\left(\mathbf{v} - \mathbf{v}^m_{ijk,\alpha}\right) \delta\left(\boldsymbol{\xi} - \boldsymbol{\xi}^m_{ijk,\alpha}\right) \quad \text{for } \mathbf{x} \in \Omega_{ijk}, \tag{B.16}$$

where the slope vector $\mathbf{a}^m_{ijk}$ is computed from the values of w^m_{ijk} in neighboring cells using flux limiters to ensure that $n^m(\mathbf{x}, \mathbf{v}, \boldsymbol{\xi})$ is always nonnegative. Note that Eqs. (B.15) and (B.16) are valid for arbitrary cells. For the particular case of Cartesian cells, the 12 NDF located at cell faces that appear on the right-hand side of Eq. (B.11) become

$$
\begin{aligned}
n^m_{i+1/2\,jk,l} &= \sum_{\alpha=1}^{N} \left(w^m_{ijk,\alpha} + \tfrac{1}{2} a_{1,ijk,\alpha}\, \Delta x \right) \delta\left(\mathbf{v} - \mathbf{v}^m_{ijk,\alpha} \right) \delta\left(\boldsymbol{\xi} - \boldsymbol{\xi}^m_{ijk,\alpha} \right), \\
n^m_{i+1/2\,jk,r} &= \sum_{\alpha=1}^{N} \left(w^m_{i+1\,jk,\alpha} - \tfrac{1}{2} a_{1,i+1\,jk,\alpha}\, \Delta x \right) \delta\left(\mathbf{v} - \mathbf{v}^m_{i+1\,jk,\alpha} \right) \delta\left(\boldsymbol{\xi} - \boldsymbol{\xi}^m_{i+1\,jk,\alpha} \right), \\
n^m_{i-1/2\,jk,l} &= \sum_{\alpha=1}^{N} \left(w^m_{i-1\,jk,\alpha} + \tfrac{1}{2} a_{1,i-1\,jk,\alpha}\, \Delta x \right) \delta\left(\mathbf{v} - \mathbf{v}^m_{i-1\,jk,\alpha} \right) \delta\left(\boldsymbol{\xi} - \boldsymbol{\xi}^m_{i-1\,jk,\alpha} \right), \\
n^m_{i-1/2\,jk,r} &= \sum_{\alpha=1}^{N} \left(w^m_{ijk,\alpha} - \tfrac{1}{2} a_{1,ijk,\alpha}\, \Delta x \right) \delta\left(\mathbf{v} - \mathbf{v}^m_{ijk,\alpha} \right) \delta\left(\boldsymbol{\xi} - \boldsymbol{\xi}^m_{ijk,\alpha} \right), \\
n^m_{ij+1/2\,k,l} &= \sum_{\alpha=1}^{N} \left(w^m_{ijk,\alpha} + \tfrac{1}{2} a_{2,ijk,\alpha}\, \Delta y \right) \delta\left(\mathbf{v} - \mathbf{v}^m_{ijk,\alpha} \right) \delta\left(\boldsymbol{\xi} - \boldsymbol{\xi}^m_{ijk,\alpha} \right), \\
n^m_{ij+1/2\,k,r} &= \sum_{\alpha=1}^{N} \left(w^m_{ij+1\,k,\alpha} - \tfrac{1}{2} a_{2,ij+1\,k,\alpha}\, \Delta y \right) \delta\left(\mathbf{v} - \mathbf{v}^m_{ij+1\,k,\alpha} \right) \delta\left(\boldsymbol{\xi} - \boldsymbol{\xi}^m_{ij+1\,k,\alpha} \right), \\
n^m_{ij-1/2\,k,l} &= \sum_{\alpha=1}^{N} \left(w^m_{ij-1\,k,\alpha} + \tfrac{1}{2} a_{2,ij-1\,k,\alpha}\, \Delta y \right) \delta\left(\mathbf{v} - \mathbf{v}^m_{ij-1\,k,\alpha} \right) \delta\left(\boldsymbol{\xi} - \boldsymbol{\xi}^m_{ij-1\,k,\alpha} \right), \\
n^m_{ij-1/2\,k,r} &= \sum_{\alpha=1}^{N} \left(w^m_{ijk,\alpha} - \tfrac{1}{2} a_{2,ijk,\alpha}\, \Delta y \right) \delta\left(\mathbf{v} - \mathbf{v}^m_{ijk,\alpha} \right) \delta\left(\boldsymbol{\xi} - \boldsymbol{\xi}^m_{ijk,\alpha} \right), \\
n^m_{ijk+1/2,l} &= \sum_{\alpha=1}^{N} \left(w^m_{ijk,\alpha} + \tfrac{1}{2} a_{3,ijk,\alpha}\, \Delta z \right) \delta\left(\mathbf{v} - \mathbf{v}^m_{ijk,\alpha} \right) \delta\left(\boldsymbol{\xi} - \boldsymbol{\xi}^m_{ijk,\alpha} \right), \\
n^m_{ijk+1/2,r} &= \sum_{\alpha=1}^{N} \left(w^m_{ijk+1,\alpha} - \tfrac{1}{2} a_{3,ijk+1,\alpha}\, \Delta z \right) \delta\left(\mathbf{v} - \mathbf{v}^m_{ijk+1,\alpha} \right) \delta\left(\boldsymbol{\xi} - \boldsymbol{\xi}^m_{ijk+1,\alpha} \right), \\
n^m_{ijk-1/2,l} &= \sum_{\alpha=1}^{N} \left(w^m_{ijk-1,\alpha} + \tfrac{1}{2} a_{3,ijk-1,\alpha}\, \Delta z \right) \delta\left(\mathbf{v} - \mathbf{v}^m_{ijk-1,\alpha} \right) \delta\left(\boldsymbol{\xi} - \boldsymbol{\xi}^m_{ijk-1,\alpha} \right), \\
n^m_{ijk-1/2,r} &= \sum_{\alpha=1}^{N} \left(w^m_{ijk,\alpha} - \tfrac{1}{2} a_{3,ijk,\alpha}\, \Delta z \right) \delta\left(\mathbf{v} - \mathbf{v}^m_{ijk,\alpha} \right) \delta\left(\boldsymbol{\xi} - \boldsymbol{\xi}^m_{ijk,\alpha} \right),
\end{aligned}
\tag{B.17}
$$

wherein the components of the slope vector are written as $\mathbf{a} = [a_1\ a_2\ a_3]^{\mathrm{T}}$. By construction, all of the weights appearing in Eq. (B.17) are nonnegative, and hence the reconstructed NDF at the cell faces are all realizable.

In summary, the realizable FVM solves for the volume-average transported moment set at each time step and grid cell, $\mathbf{M}^m_{ijk}$, and uses these moment sets at each time step to reconstruct the NDF on cell faces:

$$
\mathbf{M}^m_{ijk} \xrightarrow[\text{reconstruction}]{\text{quadrature}} \left\{ w^m_{ijk,\alpha}, \mathbf{v}^m_{ijk,\alpha}, \boldsymbol{\xi}^m_{ijk,\alpha} \right\} \xrightarrow[\text{reconstruction}]{\text{spatial}} \left\{ n^m_{i+1/2\,jk,l}, \ldots, n^m_{ijk-1/2,r} \right\}.
$$

In the following sections we describe how the reconstructed NDF at the cell faces can be used to define a numerical flux function $G(n_l, n_r)$ for each of the spatial-transport terms in Eqs. (B.2)–(B.5) that guarantees that the updated NDF n^{m+1}_{ijk} found from Eq. (B.11) (and thus the volume-average transported moment set $\mathbf{M}^{m+1}_{ijk}$) is realizable. The extension of realizable FVM to second order in time using RK2SSP (Vikas *et al.*, 2011a) is described in Section 8.2.

B.3 Advection

For clarity, we will describe the advection term in Eq. (B.2) using operator splitting in each direction. Formally, this procedure can be represented as

$$\mathbf{M}_{ijk}^{m} \xrightarrow{x \text{ direction}} \mathbf{M}_{ijk}^{(1)} \xrightarrow{y \text{ direction}} \mathbf{M}_{ijk}^{(2)} \xrightarrow{z \text{ direction}} \mathbf{M}_{ijk}^{m+1},$$

where we successively solve the advection problem for each direction using a single global time step Δt. Taking as an example the x direction with a $\boldsymbol{\xi}$-independent velocity, the advection equations for the NDF and transported moment set are

$$\frac{\partial n}{\partial t} = -\frac{\partial}{\partial x}(u_1 n) \quad \text{and} \quad \frac{\partial \mathbf{M}}{\partial t} = -\frac{\partial}{\partial x}(u_1 \mathbf{M}). \tag{B.18}$$

Since the moment-transport equation is closed, one might be tempted to try to design a high-order scheme directly for $\mathbf{M}$. However, for such a scheme it would be difficult to ensure realizability, and thus a better approach is to work directly with the NDF. Moreover, when the advection velocity depends on $\boldsymbol{\xi}$ the moment-transport equation is not closed, thus working with the NDF will result in more general formulas.

Consider now the case in which the known advection velocity $\mathbf{u}(t, \mathbf{x}, \boldsymbol{\xi})$ depends on $\boldsymbol{\xi}$.[7] For this case, the moment-transport equation is not closed, but the NDF transport equation is still given by Eq. (B.18). For the x direction, the finite-volume formula for the NDF is

$$n_{ijk}^{(1)} = n_{ijk}^{m} - \lambda_x \left[G\left(n_{i+1/2jk,l}^{m}, n_{i+1/2jk,r}^{m}\right) - G\left(n_{i-1/2jk,l}^{m}, n_{i-1/2jk,r}^{m}\right) \right], \tag{B.19}$$

where $\lambda_x = \Delta t/\Delta x$ and the numerical-flux function is defined by

$$G(n_l, n_r) = u_{1,l}^{+} n_l + u_{1,r}^{-} n_r \tag{B.20}$$

with

$$u_{1,l}^{+}(\boldsymbol{\xi}) = \frac{1}{2}\left[u_{1,l}(\boldsymbol{\xi}) + |u_{1,l}(\boldsymbol{\xi})|\right] \quad \text{and} \quad u_{1,r}^{-}(\boldsymbol{\xi}) = \frac{1}{2}\left[u_{1,r}(\boldsymbol{\xi}) - |u_{1,r}(\boldsymbol{\xi})|\right]. \tag{B.21}$$

Note that, if the velocity $\mathbf{u}(t, \mathbf{x}, \boldsymbol{\xi})$ is known for all $\mathbf{x}$, then $u_{1,l}$ and $u_{1,r}$ correspond to velocity component $u_1(t, \mathbf{x}, \boldsymbol{\xi})$ evaluated at the cell faces for fixed values of $\boldsymbol{\xi}$. Otherwise, $u_{1,l}$ and $u_{1,r}$ must be reconstructed from the volume-average values of the velocity in the corresponding grid cells. Throughout this appendix, we denote $\mathbf{x}$-dependent variables, such as $u_1(t, \mathbf{x}, \boldsymbol{\xi})$, evaluated at the cell faces (at time step m) by, for example, $u_{1,i+1/2jk}^{m}$ and $u_{1,i-1/2jk}^{m}$. An additional subscript $_l$ or $_r$ is used to indicate which side of the cell face is of interest.

[7] As described in Section 5.2, a typical dependence on $\boldsymbol{\xi}$ can be written as

$$\mathbf{u}(t, \mathbf{x}, \boldsymbol{\xi}) = \mathbf{u}_{\mathrm{f}}(t, \mathbf{x}) + g(\boldsymbol{\xi})\mathbf{u}^{*}(t, \mathbf{x}),$$

where $\mathbf{u}_{\mathrm{f}}$, $\mathbf{u}^{*}$, and g are known functions. For example in a fluid–particle system, $\mathbf{u}_{\mathrm{f}}$ could be the fluid velocity and g could be the Stokes number, so that $\mathbf{u}^{*}$ represents the velocity correction valid for very small Stokes numbers.

Using the reconstructed NDF from Eq. (B.15), the updated NDF in Eq. (B.19) becomes[8]

$$
\begin{aligned}
n_{ijk}^{(1)} = \sum_{\alpha=1}^{N} \left[w_{ijk,\alpha}^{m} - \lambda_x u_{1,i+1/2jk,l}^{+}(\boldsymbol{\xi}) w_{i+1/2jk,\alpha,l}^{m} + \lambda_x u_{1,i-1/2jk,r}^{-}(\boldsymbol{\xi}) w_{i-1/2jk,\alpha,r}^{m} \right] \delta\left(\boldsymbol{\xi} - \boldsymbol{\xi}_{ijk,\alpha}^{m}\right) \\
- \sum_{\alpha=1}^{N} \lambda_x u_{1,i+1/2jk,r}^{-}(\boldsymbol{\xi}) w_{i+1/2jk,\alpha,r}^{m} \delta\left(\boldsymbol{\xi} - \boldsymbol{\xi}_{i+1jk,\alpha}^{m}\right) \\
+ \sum_{\alpha=1}^{N} \lambda_x u_{1,i-1/2jk,l}^{+}(\boldsymbol{\xi}) w_{i-1/2jk,\alpha,l}^{m} \delta\left(\boldsymbol{\xi} - \boldsymbol{\xi}_{i-1jk,\alpha}^{m}\right).
\end{aligned} \tag{B.22}
$$

The last two terms in Eq. (B.22) are nonnegative, thus $n_{ijk}^{(1)}$ will be realizable if

$$
0 \le w_{ijk,\alpha}^{m} - \lambda_x u_{1,i+1/2jk,l}^{+}(\boldsymbol{\xi}_{ijk,\alpha}^{m}) w_{i+1/2jk,\alpha,l}^{m} + \lambda_x u_{1,i-1/2jk,r}^{-}(\boldsymbol{\xi}_{ijk,\alpha}^{m}) w_{i-1/2jk,\alpha,r}^{m}
$$

for all values of ijk and α, which yields the first realizability condition for Δt:

$$
\Delta t \le \Delta t_1 = \min_{ijk,\alpha} \left(\frac{w_{ijk,\alpha}^{m} \, \Delta x}{w_{i+1/2jk,\alpha,l}^{m} u_{1,i+1/2jk,l}^{+}(\boldsymbol{\xi}_{ijk,\alpha}^{m}) - w_{i-1/2jk,\alpha,r}^{m} u_{1,i-1/2jk,r}^{-}(\boldsymbol{\xi}_{ijk,\alpha}^{m})} \right). \tag{B.23}
$$

Note that if $w_{ijk,\alpha}^{m} = 0$, then $w_{i+1/2jk,\alpha,l}^{m} = w_{i-1/2jk,\alpha,r}^{m} = 0$ and the realizability condition is satisfied for all $0 \le \Delta t$.

If the realizability condition in Eq. (B.23) is satisfied, then the moment set $\mathbf{M}_{ijk}^{(1)}$ can be found from Eq. (B.22) using the definition[9]

$$
\mathbf{M}_{ijk}^{(1)} \equiv \int_{\Omega_{\xi}} \mathbf{K} n_{ijk}^{(1)} \, \mathrm{d}\boldsymbol{\xi}, \tag{B.24}
$$

where (as in Eq. (B.9)) $\mathbf{K}$ is the moment vector determined by the transported moment set. The quadrature reconstruction using $\mathbf{M}_{ijk}^{(1)}$ then yields the weights $w_{ijk,\alpha}^{(1)}$ and abscissas $\boldsymbol{\xi}_{ijk,\alpha}^{(1)}$, from which the NDF $n_{ijk}^{(2)}$ can be computed:

$$
\begin{aligned}
n_{ijk}^{(2)} = \sum_{\alpha=1}^{N} \left(w_{ijk,\alpha}^{(1)} - \lambda_y u_{2,ij+1/2k,l}^{+}(\boldsymbol{\xi}) w_{ij+1/2k,\alpha,l}^{(1)} + \lambda_y u_{2,ij-1/2k,r}^{-}(\boldsymbol{\xi}) w_{ij-1/2k,\alpha,r}^{(1)} \right) \delta\left(\boldsymbol{\xi} - \boldsymbol{\xi}_{ijk,\alpha}^{(1)}\right) \\
- \sum_{\alpha=1}^{N} \lambda_y u_{2,ij+1/2k,r}^{-}(\boldsymbol{\xi}) w_{ij+1/2k,\alpha,r}^{(1)} \delta\left(\boldsymbol{\xi} - \boldsymbol{\xi}_{ij+1k,\alpha}^{(1)}\right) \\
+ \sum_{\alpha=1}^{N} \lambda_y u_{2,ij-1/2k,l}^{+}(\boldsymbol{\xi}) w_{ij-1/2k,\alpha,l}^{(1)} \delta\left(\boldsymbol{\xi} - \boldsymbol{\xi}_{ij-1k,\alpha}^{(1)}\right).
\end{aligned} \tag{B.25}
$$

The second realizability condition for Δt is thus

$$
\Delta t \le \Delta t_2 = \min_{ijk,\alpha} \left(\frac{w_{ijk,\alpha}^{(1)} \, \Delta y}{w_{ij+1/2k,\alpha,l}^{(1)} u_{2,ij+1/2k,l}^{+}(\boldsymbol{\xi}_{ijk,\alpha}^{(1)}) - w_{ij-1/2k,\alpha,r}^{(1)} u_{2,ij-1/2k,r}^{-}(\boldsymbol{\xi}_{ijk,\alpha}^{(1)})} \right). \tag{B.26}
$$

[8] With advection, the velocity internal coordinate is not required, so we represent all internal coordinates by $\boldsymbol{\xi}$.

[9] Because $n_{ijk}^{(1)}(\boldsymbol{\xi})$ involves weighted Dirac delta functions, the moment set $\mathbf{M}_{ijk}^{(1)}$ involves summations and can be computed analytically.

If this condition holds, then $n_{ijk}^{(2)}$ is realizable.

Analogous formulas are found for the z direction:

$$\mathbf{M}_{ijk}^{(2)} \equiv \int_{\Omega_\xi} \mathbf{K} n_{ijk}^{(2)} \, \mathrm{d}\boldsymbol{\xi}, \tag{B.27}$$

from which we find the weights $w_{ijk,\alpha}^{(2)}$ and abscissas $\boldsymbol{\xi}_{ijk,\alpha}^{(2)}$, and then

$$\begin{aligned} n_{ijk}^{m+1} &= \sum_{\alpha=1}^{N} \left(w_{ijk,\alpha}^{(2)} - \lambda_z u_{3,ijk+1/2,l}^{+}(\boldsymbol{\xi}) w_{ijk+1/2,\alpha,l}^{(2)} + \lambda_z u_{3,ijk-1/2,r}^{-}(\boldsymbol{\xi}) w_{ijk-1/2,\alpha,r}^{(2)} \right) \delta\left(\boldsymbol{\xi} - \boldsymbol{\xi}_{ijk,\alpha}^{(2)}\right) \\ &\quad - \sum_{\alpha=1}^{N} \lambda_z u_{3,ijk+1/2,r}^{-}(\boldsymbol{\xi}) w_{ijk+1/2,\alpha,r}^{(2)} \delta\left(\boldsymbol{\xi} - \boldsymbol{\xi}_{ijk+1,\alpha}^{(2)}\right) \\ &\quad + \sum_{\alpha=1}^{N} \lambda_z u_{3,ijk-1/2,l}^{+}(\boldsymbol{\xi}) w_{ijk-1/2,\alpha,l}^{(2)} \delta\left(\boldsymbol{\xi} - \boldsymbol{\xi}_{ijk-1,\alpha}^{(2)}\right). \end{aligned} \tag{B.28}$$

The third realizability condition is thus

$$\Delta t \leq \Delta t_3 = \min_{ijk,\alpha} \left(\frac{w_{ijk,\alpha}^{(2)} \Delta z}{w_{ijk+1/2,\alpha,l}^{(2)} u_{3,ijk+1/2,l}^{+}(\boldsymbol{\xi}_{ijk,\alpha}^{(2)}) - w_{ijk-1/2,\alpha,r}^{(2)} u_{3,ijk-1/2,r}^{-}(\boldsymbol{\xi}_{ijk,\alpha}^{(2)})} \right). \tag{B.29}$$

The *global* realizability condition (which determines the time step) is

$$\Delta t \leq \min(\Delta t_1, \Delta t_2, \Delta t_3). \tag{B.30}$$

If this condition holds, then the realizable updated moment set is found from

$$\mathbf{M}_{ijk}^{m+1} \equiv \int_{\Omega_\xi} \mathbf{K} n_{ijk}^{m+1} \, \mathrm{d}\boldsymbol{\xi}, \tag{B.31}$$

which completes one time step of multidimensional advection. In the case in which the CQMOM is used for the quadrature reconstructions, the steps leading up to Eq. (B.31) are followed for *each* permutation to find $(\mathbf{M}_{ijk}^{m+1})_p$, followed by application of Eq. (B.14) to find $\mathbf{M}_{ijk}^{m+1}$. An equally valid alternative would be to define the intermediate moment set $\mathbf{M}_{ijk}^{(1)}$ using the sum over all CQMOM permutations, and then do the same for $\mathbf{M}_{ijk}^{(2)}$ and $\mathbf{M}_{ijk}^{m+1}$.

In conclusion, it is important to understand that Eq. (B.22) is key to the construction of a realizable, high-order scheme for multidimensional advection. The resulting scheme is referred to as *kinetics-based* because the numerical-flux function and the realizability condition are formulated in terms of the NDF, even though we are ultimately interested in knowing only the value of the transported moment set. Thus, while working with the NDF directly may seem overly restrictive in the context of moment methods, it allows us to design a high-order scheme for multidimensional advection that will be realizable for all possible transported moment sets.

B.4 Free transport

For clarity, we will again describe the free-transport term in Eq. (B.3) using operator splitting in each direction. Taking as an example the x direction, the free-transport equation for the NDF is

$$\frac{\partial n}{\partial t} = -\frac{\partial}{\partial x}(v_1 n), \tag{B.32}$$

where v_1 is an internal coordinate.[10] For simplicity, in this section we will assume that the NDF contains only the velocity: $n(t, \mathbf{x}, \mathbf{v})$. Nevertheless, since the other internal coordinates are passive with respect to free transport, it will be relatively straightforward to include additional internal coordinates. The finite-volume formula for the NDF in Eq. (B.18) is the same as in Eq. (B.19) but with the numerical-flux function defined by

$$G(n_l, n_r) = v_1^+ n_l + v_1^- n_r \tag{B.33}$$

with

$$v_1^+ = \frac{1}{2}(v_1 + |v_1|) \quad \text{and} \quad v_1^- = \frac{1}{2}(v_1 - |v_1|). \tag{B.34}$$

Note that the numerical NDF $n_{ijk}^m(\mathbf{v})$ used to reconstruct n_l and n_r now depends on the free-transport velocity component v_1.

Once again, we can use the quadrature reconstruction of $n_{ijk}^m(\mathbf{v})$ to rewrite the scheme in Eq. (B.19) as

$$\begin{aligned} n_{ijk}^{(1)} = {} & \sum_{\alpha=1}^{N} \left(w_{ijk,\alpha}^m - \lambda_x v_{1,ijk,\alpha}^+ w_{i+1/2jk,\alpha,l}^m + \lambda_x v_{1,ijk,\alpha}^- w_{i-1/2jk,\alpha,r}^m \right) \delta\left(\mathbf{v} - \mathbf{v}_{ijk,\alpha}^m\right) \\ & - \sum_{\alpha=1}^{N} \lambda_x v_{1,i+1jk,\alpha}^- w_{i+1/2jk,\alpha,r}^m \delta\left(\mathbf{v} - \mathbf{v}_{i+1jk,\alpha}^m\right) \\ & + \sum_{\alpha=1}^{N} \lambda_x v_{1,i-1jk,\alpha}^+ w_{i-1/2jk,\alpha,l}^m \delta\left(\mathbf{v} - \mathbf{v}_{i-1jk,\alpha}^m\right) \end{aligned} \tag{B.35}$$

with weights $w_{ijk,\alpha}^m$ and velocity abscissas $\mathbf{v}_{ijk,\alpha}^m$ found using quadrature reconstruction from the volume-average moment set $\mathbf{M}_{ijk}^m$ and

$$v_{1,ijk,\alpha}^+ = \frac{1}{2}\left(v_{1,ijk,\alpha}^m + \left|v_{1,ijk,\alpha}^m\right|\right) \quad \text{and} \quad v_{1,ijk,\alpha}^- = \frac{1}{2}\left(v_{1,ijk,\alpha}^m - \left|v_{1,ijk,\alpha}^m\right|\right). \tag{B.36}$$

The resulting realizability condition is analogous to Eq. (B.23), but with the velocity abscissas $v_{1,ijk,\alpha}^+$ and $v_{1,ijk,\alpha}^-$ appearing in place of $u_{1,i+1/2jk,l}^+$ and $u_{1,i-1/2jk,r}^-$, respectively. In fact, the net effect of Eq. (B.36) is to separate the v_1 abscissas into two groups: (1) N_{ijk}^+ abscissas with $0 \le v_{1,ijk,\alpha}^m$ and (2) N_{ijk}^- abscissas with $v_{1,ijk,\alpha}^m < 0$, where $N = N_{ijk}^- + N_{ijk}^+$ and the numbers of positive (negative) abscissas can vary in space and time. This decomposition of the spatial flux is exact in the case in which the true NDF is an N-point distribution function.[11]

In many systems (e.g. those involving to collisions), the velocity NDF will be continuous and thus will not be well represented by an N-point distribution function. In such cases, in order to have a more accurate representation of the spatial fluxes, the multivariate EQMOM described in Section 3.3.4 can be used to reconstruct the NDF $n_{ijk}^m(\mathbf{v})$ from the transported moment set $\mathbf{M}_{ijk}^m$. Note that, unlike with "regular" quadrature,[12] this NDF will be a continuous (known) function of $\mathbf{v}$. Nevertheless, for evaluating the spatial fluxes in Eq. (B.35), it will still be advantageous to construct a regular quadrature using the moments of $n_{ijk}^m(\mathbf{v})$. As

[10] Free transport can be seen as a special case of multidimensional advection with $\mathbf{u}(t, \mathbf{x}, \boldsymbol{\xi}) = \boldsymbol{\xi}$. However, the absence of a $\boldsymbol{\xi}$-independent velocity component leads us to consider it separately.

[11] An N-point distribution function has the form of N weighted Dirac delta functions located at distinct points in phase space with positive weights.

[12] By "regular" we mean a quadrature for which the NDF can be represented by a sum of Dirac delta functions.

discussed in Section 3.3.4, the order of the quadrature that is constructed from the moments of $n^m_{ijk}(\mathbf{v})$ is arbitrary. In other words, it can be different than the order of the regular quadrature that can be constructed from the transported moment set $\mathbf{M}^m_{ijk}$. This additional degree of freedom allows us to achieve higher accuracy for the spatial fluxes without increasing the size of the transported moment set. Naturally, in the limit where $n^m_{ijk}(\mathbf{v})$ is an N-point distribution function, the multivariate EQMOM reduces to the regular quadrature (i.e. the CQMOM), in which case Eq. (B.35) remains unchanged.

With the EQMOM, we can represent the velocity moments needed to compute the spatial fluxes as the sum of two contributions:

$$\begin{aligned}\mathbf{M}^+_{1,ijk} &= \int_0^{+\infty} \int_{\mathbb{R}^2} \mathbf{K}^* n^m_{ijk} \, \mathrm{d}v_3 \, \mathrm{d}v_2 \, \mathrm{d}v_1, \\ \mathbf{M}^-_{1,ijk} &= \int_{-\infty}^{0} \int_{\mathbb{R}^2} \mathbf{K}^* n^m_{ijk} \, \mathrm{d}v_3 \, \mathrm{d}v_2 \, \mathrm{d}v_1,\end{aligned} \tag{B.37}$$

where $\mathbf{K}^*$ is the moment vector needed to compute an N^*-node regular multivariate quadrature and $N \leq N^*$ (i.e. the vector $\mathbf{K}^*$ is longer than $\mathbf{K}$, but all of the components of $\mathbf{K}$ are in $\mathbf{K}^*$). In Eq. (B.37), $n^m_{ijk}(\mathbf{v})$ is reconstructed using multivariate EQMOM from the transported moment set $\mathbf{M}^m_{ijk}$. As discussed in Section 3.3.4, since $n^m_{ijk}(\mathbf{v})$ has a known functional form (e.g. multivariate Gaussian) the half-moment sets in Eq. (B.37) can be evaluated analytically. Once these half-moment sets are known, the N^*-node multivariate quadrature is applied to find the two sets of weights and abscissas. The overall procedure is as follows:[13]

$$\mathbf{M}^m_{ijk} \xrightarrow[\text{with } N \text{ nodes}]{\text{extended quadrature}} n^m_{ijk} \xrightarrow[\text{analytically}]{\text{evaluate moments}} \begin{matrix}\mathbf{M}^+_{1,ijk} \\ \mathbf{M}^-_{1,ijk}\end{matrix} \xrightarrow[\text{with } N^* \text{ nodes}]{\text{regular quadrature}} \begin{matrix}\left\{w^+_{ijk,\beta}, \mathbf{v}^+_{ijk,\beta}\right\} \\ \left\{w^-_{ijk,\beta}, \mathbf{v}^-_{ijk,\beta}\right\}\end{matrix},$$

where $\beta = 1, 2, \ldots, N^*$. By construction, the x-component of the vectors $\mathbf{v}^+_{ijk,\beta}$ (or $\mathbf{v}^-_{ijk,\beta}$) is positive (or negative). We can now re-express Eq. (B.35) as the sum of two contributions:

$$n^{(1)}_{ijk} = n^+_{ijk} + n^-_{ijk}, \tag{B.38}$$

where

$$n^+_{ijk} = \sum_{\beta=1}^{N^*} \left(w^+_{ijk,\beta} - \lambda_x v^+_{1,ijk,\beta} w^+_{i+1/2jk,\beta,l} \right) \delta\left(\mathbf{v} - \mathbf{v}^+_{ijk,\beta}\right) + \sum_{\beta=1}^{N^*} \lambda_x v^+_{1,i-1jk,\beta} w^+_{i-1/2jk,\beta,l} \delta\left(\mathbf{v} - \mathbf{v}^+_{i-1jk,\beta}\right) \tag{B.39}$$

and

$$n^-_{ijk} = \sum_{\beta=1}^{N^*} \left(w^-_{ijk,\beta} + \lambda_x v^-_{1,ijk,\beta} w^-_{i-1/2jk,\beta,r} \right) \delta\left(\mathbf{v} - \mathbf{v}^-_{ijk,\beta}\right) - \sum_{\beta=1}^{N^*} \lambda_x v^-_{1,i+1jk,\beta} w^-_{i+1/2jk,\beta,r} \delta\left(\mathbf{v} - \mathbf{v}^-_{i+1jk,\beta}\right). \tag{B.40}$$

By construction, the final terms in Eqs. (B.39) and (B.40) are nonnegative. Thus, the first realizability condition for Δt is

$$\Delta t \leq \Delta t_1 = \min_{ijk,\beta} \left(\frac{w^+_{ijk,\beta} \, \Delta x}{w^+_{i+1/2jk,\beta,l} v^+_{1,ijk,\beta}}, \frac{w^-_{ijk,\beta} \, \Delta x}{w^-_{i-1/2jk,\beta,r} |v^-_{1,ijk,\beta}|} \right). \tag{B.41}$$

[13] As was done in Section 8.5, the split into positive and negative velocities can also be done directly with the EQMOM quadrature.

If this condition is satisfied, then $n^{(1)}_{ijk}$ is realizable and hence the moment set $\mathbf{M}^{(1)}_{ijk}$ is also realizable.[14]

Combining the intermediate steps, the procedure for computing $\mathbf{M}^{(1)}_{ijk}$ from $\mathbf{M}^{m}_{ijk}$ for free transport is

$$\mathbf{M}^{m}_{ijk} \xrightarrow[\text{with } N \text{ nodes}]{\text{extended quadrature}} \begin{matrix}\mathbf{M}^{+}_{1,ijk}\\ \mathbf{M}^{-}_{1,ijk}\end{matrix} \xrightarrow[\text{with } N^{*} \text{ nodes}]{\text{regular quadrature}} \mathbf{M}^{(1)}_{ijk}.$$

At this point, one technical question remains: how will the multivariate quadratures be defined? If a brute-force QMOM were used, there would be only one choice since all of the transported moments are reproduced with such methods. However, since brute-force QMOM are numerically ill-conditioned, a more likely choice will be to use the CQMOM (see Section 3.2.3 for details). As described in Yuan & Fox (2011), when the CQMOM is used for spatial transport care must be taken when choosing the permutation. For example, with $\mathbf{v}$ (i.e. a 3D phase space) there are $P = 6$ permutations of the CQMOM, but only two of them give the correct spatial flux in the x direction (i.e. permutations 123 and 132).[15] In order to accommodate the two CQMOM permutations, the update procedure described above is modified as follows:

$$\mathbf{M}^{m}_{ijk} \rightarrow \left\{ \begin{matrix} \xrightarrow[\text{123 permutation}]{\text{CQMOM}} \left(\mathbf{M}^{(1)}_{ijk}\right)_1 \\ \\ \xrightarrow[\text{132 permutation}]{\text{CQMOM}} \left(\mathbf{M}^{(1)}_{ijk}\right)_2 \end{matrix} \right\} \rightarrow \mathbf{M}^{(1)}_{ijk} = \frac{1}{2}\left(\mathbf{M}^{(1)}_{ijk}\right)_1 + \frac{1}{2}\left(\mathbf{M}^{(1)}_{ijk}\right)_2 .$$

Note that the same CQMOM permutation is used both for the regular quadrature and for the extended quadrature when finding $\mathbf{M}^{(1)}_{ijk}$.

The formulas for the y direction are analogous to those given above for the x direction. The only difference is that now we must compute the half-moment sets in the y direction defined by

$$\begin{aligned} \mathbf{M}^{+}_{2,ijk} &= \int_0^{+\infty} \int_{\mathbb{R}^2} \mathbf{K}^{*} n^{m}_{ijk} \,\mathrm{d}v_1 \,\mathrm{d}v_3 \,\mathrm{d}v_2, \\ \mathbf{M}^{-}_{2,ijk} &= \int_{-\infty}^{0} \int_{\mathbb{R}^2} \mathbf{K}^{*} n^{m}_{ijk} \,\mathrm{d}v_1 \,\mathrm{d}v_3 \,\mathrm{d}v_2. \end{aligned} \tag{B.42}$$

The procedure for computing $\mathbf{M}^{(2)}_{ijk}$ can be summarized as follows:

$$\mathbf{M}^{(1)}_{ijk} \xrightarrow[\text{with } N \text{ nodes}]{\text{extended quadrature}} \begin{matrix}\mathbf{M}^{+}_{2,ijk}\\ \mathbf{M}^{-}_{2,ijk}\end{matrix} \xrightarrow[\text{with } N^{*} \text{ nodes}]{\text{regular quadrature}} \mathbf{M}^{(2)}_{ijk},$$

and the second realizability condition for Δt is

$$\Delta t \leq \Delta t_2 = \min_{ijk,\beta} \left(\frac{w^{+}_{ijk,\beta} \Delta y}{w^{+}_{ij+1/2k,\beta,l} v^{+}_{2,ijk,\beta}}, \frac{w^{-}_{ijk,\beta} \Delta y}{w^{-}_{ij-1/2k,\beta,r} |v^{-}_{2,ijk,\beta}|} \right). \tag{B.43}$$

[14]Formally, the moment set $\mathbf{M}^{(1)}_{ijk}$ is found from the definition in Eq. (B.24). However, given that $n^{(1)}_{ijk}$ is a $2N^{*}$-point distribution function, the integrals in Eq. (B.24) become summations.

[15]The permutations are numbered according to the conditioning variables. For example, permutation 123 starts the CQMOM with the v_1 direction, followed by the v_2 direction, and finally the v_3 direction. In other words, v_2 is conditioned on v_1, and v_3 on v_2 and v_1.

When the CQMOM is used for the multivariate quadrature, we must accommodate the two CQMOM permutations:

$$\mathbf{M}_{ijk}^{(1)} \rightarrow \left\{ \begin{aligned} &\xrightarrow[\text{213 permutation}]{\text{CQMOM}} \left(\mathbf{M}_{ijk}^{(2)}\right)_1 \\ &\xrightarrow[\text{231 permutation}]{\text{CQMOM}} \left(\mathbf{M}_{ijk}^{(2)}\right)_2 \end{aligned} \right\} \rightarrow \mathbf{M}_{ijk}^{(2)} = \frac{1}{2}\left(\mathbf{M}_{ijk}^{(2)}\right)_1 + \frac{1}{2}\left(\mathbf{M}_{ijk}^{(2)}\right)_2 .$$

If the second realizability condition holds, then $\mathbf{M}_{ijk}^{(2)}$ is realizable.

The treatment of the z direction proceeds in the same manner, starting from $\mathbf{M}_{ijk}^{(2)}$. The third realizability condition for Δt is

$$\Delta t \leq \Delta t_3 = \min_{ijk,\beta} \left(\frac{w_{ijk,\beta}^{+} \Delta z}{w_{ijk+1/2,\beta,l}^{+} v_{3,ijk,\beta}^{+}}, \frac{w_{ijk,\beta}^{-} \Delta z}{w_{ijk-1/2,\beta,r}^{-} |v_{3,ijk,\beta}^{-}|} \right), \tag{B.44}$$

and thus the global time step must satisfy $\Delta t \leq \min(\Delta t_1, \Delta t_2, \Delta t_3)$. When the CQMOM is used for the multivariate quadrature, we must again accommodate the two CQMOM permutations:

$$\mathbf{M}_{ijk}^{(2)} \rightarrow \left\{ \begin{aligned} &\xrightarrow[\text{321 permutation}]{\text{CQMOM}} \left(\mathbf{M}_{ijk}^{m+1}\right)_1 \\ &\xrightarrow[\text{312 permutation}]{\text{CQMOM}} \left(\mathbf{M}_{ijk}^{m+1}\right)_2 \end{aligned} \right\} \rightarrow \mathbf{M}_{ijk}^{m+1} = \frac{1}{2}\left(\mathbf{M}_{ijk}^{m+1}\right)_1 + \frac{1}{2}\left(\mathbf{M}_{ijk}^{m+1}\right)_2 .$$

The updated transported moment set $\mathbf{M}_{ijk}^{m+1}$ will be realizable provided that the global realizability condition is satisfied for all permutations of the CQMOM.[16]

To conclude this section, we discuss three technical points that arise when applying the realizable high-order scheme for free transport. The first point is how to choose the value of N^* given that $N = n^3$ is used for multivariate EQMOM. To answer this question, we first note that the highest-order transported moment in any one direction is of order $2n$ (which corresponds to $2n+1$ pure moments in any one direction). The "extra" (even-order) moment is used in the EQMOM to determine the spread of the kernel density functions (see Section 3.3.2 for details). When free transport is applied to the moment of order $2n$, the flux function involves the moment of order $2n + 1$. Therefore, in order to exactly predict the flux function for free transport using the half-moment sets, we must have $N^* \geq (n+1)^3$. The obvious choice for N^* is thus $N^* = (n + 1)^3$.

The second point concerns the high-order spatial reconstruction of the N^* weights. Since the CQMOM uses an adaptive 1D quadrature routine to control for quadrature errors (see Yuan & Fox (2011) for details), it is likely that the CQMOM reconstructions in neighboring cells will sometimes be different in the sense that not all cells will have N^* nonzero weights. The use of high-order spatial reconstruction for the weights requires that the CQMOM reconstructions in neighboring cells be done in exactly the same manner. If this is not done, only a first-order spatial reconstruction can be used. Thus, in order to use high-order spatial reconstruction, an index must be added to the CQMOM routine that uniquely defines the CQMOM reconstruction (see Yuan & Fox (2011) for examples). By comparing indices in neighboring cells, the correct *local* order for the spatial reconstruction can be determined.

[16]In other words, the realizability condition for a given direction must be satisfied for both permutations of the CQMOM used in that direction.

The final point deals with the addition of passive internal coordinates so that the NDF is $n(t, \mathbf{x}, \mathbf{v}, \boldsymbol{\xi})$. When using the CQMOM to reconstruct the multivariate quadrature approximation, there is (in principle) no limitation on adding passive internal coordinates. However, in practice, the number of transported moments scales like $(1 + 3 + M)n^{3+M}$, where M is the length of ξ, and thus constructing quadratures for even moderate value of M requires a large number of moments. Nevertheless, even if $M = 1$, the question of how to choose the CQMOM permutation when evaluating the free-transport numerical fluxes arises. For example, in the x direction we could start the CQMOM with ξ_1, following this choice by choosing either 123 or 132 for the velocity variables, or we could put ξ_1 somewhere else in the conditioning order. At present, this question has not been clearly resolved, but, from our observations in Section 8.1.3, it would appear that, in order to correctly predict the spatial fluxes of $\boldsymbol{\xi}$, it would be best to put the passive variables at the *beginning* of the CQMOM conditioning sequence when computing the free-transport spatial fluxes.

B.5 Mixed advection

The mixed advection term in Eq. (B.4) involves both a spatial and a phase-space gradient of the NDF, and thus it is more complicated than the pure advection term described in Section B.3 of this appendix. Nevertheless, by working with the transport equation for the moments, we can show that it can be treated like an advection term with "special" properties.

Denoting the NDF by $n(t, \mathbf{x}, \mathbf{v}, \boldsymbol{\xi})$, with mixed advection the NDF transport equation for the x direction is

$$\frac{\partial n}{\partial t} = \frac{\partial^2 \lambda n}{\partial x \, \partial v_1}, \tag{B.45}$$

where $\lambda(t, \mathbf{x}, \mathbf{v}, \boldsymbol{\xi})$ is nonnegative. The corresponding moment-transport equation is

$$\frac{\partial \mathbf{M}}{\partial t} = \frac{\partial}{\partial x}\left(\int \mathbf{K} \frac{\partial \lambda n}{\partial v_1} \, \mathrm{d}\mathbf{v} \, \mathrm{d}\boldsymbol{\xi} \right) = -\frac{\partial}{\partial x}\left(\int \lambda n \frac{\partial \mathbf{K}}{\partial v_1} \, \mathrm{d}\mathbf{v} \, \mathrm{d}\boldsymbol{\xi} \right), \tag{B.46}$$

where the final form on the right-hand side is found using integration by parts. To go further, we note that the moment vector $\mathbf{K}$ has components of the form $v_1^{k_1} v_2^{k_2} v_3^{k_3} \boldsymbol{\xi}^{\mathbf{k}_\xi}$, where the exponent vector $\mathbf{k} = (k_1, k_2, k_3, \mathbf{k}_\xi)$ contains nonnegative integers that correspond to the transported moment set $\mathbf{M}$. Thus, the v_1 derivative of $\mathbf{K}$ appearing in Eq. (B.46) has the form

$$k_1 v_1^{k_1 - 1} v_2^{k_2} v_3^{k_3} \boldsymbol{\xi}^{\mathbf{k}_\xi} \quad \Longrightarrow \quad \frac{\partial \mathbf{K}}{\partial v_1} = \frac{k_1}{v_1} \mathbf{K}.$$

By removing the phase-space integration, Eq. (B.46) can be rewritten as a k_1-dependent advection term for the NDF:

$$\frac{\partial n}{\partial t} = -\frac{\partial}{\partial x}\left(\frac{k_1}{v_1} \lambda n \right). \tag{B.47}$$

Formally, we can define the advection velocity as $u_1(t, \mathbf{x}, \mathbf{v}, \boldsymbol{\xi}) = k_1 \lambda / v_1$, so that Eq. (B.47) reduces to an advection equation for the NDF:

$$\frac{\partial n}{\partial t} = -\frac{\partial}{\partial x} (u_1 n) . \tag{B.48}$$

Notice that the advection velocity depends linearly on k_1, so the highest velocity will be associated with the highest-order moments appearing in $\mathbf{M}$. Let $k_1^* = \max(k_1)$ be the largest value of k_1 appearing in the transported moment set (i.e. $k_1^* = 2n - 1$ for regular quadrature and $2n$ for extended quadrature). We can then define the maximum advection velocity as $u_1^* = k_1^* \lambda / v_1$, and it is this velocity that will appear in the realizability condition. Note that, because λ is nonnegative, the sign of u_1 is determined by v_1. Also, because u_1 is inversely proportional to v_1, the advection velocity will be infinite for $v_1 = 0$, unless λ is also zero. In general, the functional form of λ is very important for determining the behavior of the algorithm, and we will return to this subject below after introducing the scheme for mixed advection.

The advection problem in Eq. (B.48) has exactly the same form as the one discussed in relationship to Eq. (B.3), but with a particular definition of the advection velocity. For example, corresponding to Eq. (B.21), we have

$$u_{1,l}^{+} = \frac{k_1 \lambda_l}{2 v_1} \left(1 + \frac{v_1}{|v_1|} \right) \quad \text{and} \quad u_{1,r}^{-} = \frac{k_1 \lambda_r}{2 v_1} \left(1 - \frac{v_1}{|v_1|} \right). \tag{B.49}$$

The updated numerical NDF $n_{ijk}^{(1)}$ can then be written as the sum of two contributions as shown in Eq. (B.38), where

$$\begin{aligned} n_{ijk}^{+} = & \sum_{\beta=1}^{N} \left(w_{ijk,\beta}^{+} - k_1 \lambda_x v_1^{-1} \lambda_{i+1/2jk,l}^{m}(\mathbf{v}, \boldsymbol{\xi}) w_{i+1/2jk,\beta,l}^{+} \right) \delta\left(\mathbf{v} - \mathbf{v}_{ijk,\beta}^{+}\right) \delta\left(\boldsymbol{\xi} - \boldsymbol{\xi}_{ijk,\beta}^{+}\right) \\ & + \sum_{\beta=1}^{N} k_1 \lambda_x v_1^{-1} \lambda_{i-1/2jk,l}^{m}(\mathbf{v}, \boldsymbol{\xi}) w_{i-1/2jk,\alpha,l}^{+} \delta\left(\mathbf{v} - \mathbf{v}_{i-1jk,\beta}^{+}\right) \delta\left(\boldsymbol{\xi} - \boldsymbol{\xi}_{i-1jk,\beta}^{+}\right) \end{aligned} \tag{B.50}$$

and

$$\begin{aligned} n_{ijk}^{-} = & \sum_{\beta=1}^{N} \left(w_{ijk,\beta}^{-} + k_1 \lambda_x v_1^{-1} \lambda_{i-1/2jk,r}^{m}(\mathbf{v}, \boldsymbol{\xi}) w_{i-1/2jk,\beta,r}^{-} \right) \delta\left(\mathbf{v} - \mathbf{v}_{ijk,\beta}^{-}\right) \delta\left(\boldsymbol{\xi} - \boldsymbol{\xi}_{ijk,\beta}^{-}\right) \\ & - \sum_{\beta=1}^{N} k_1 \lambda_x v_1^{-1} \lambda_{i+1/2jk,r}^{m}(\mathbf{v}, \boldsymbol{\xi}) w_{i+1/2jk,\alpha,r}^{-} \delta\left(\mathbf{v} - \mathbf{v}_{i+1jk,\beta}^{-}\right) \delta\left(\boldsymbol{\xi} - \boldsymbol{\xi}_{i+1jk,\beta}^{-}\right). \end{aligned} \tag{B.51}$$

The final terms in Eqs. (B.50) and Eq. (B.51) are nonnegative, thus $n_{ijk}^{(1)}$ will be realizable if the following realizability condition for Δt holds:

$$\Delta t \leq \min_{ijk,\beta} \left(\frac{w_{ijk,\beta}^{+} v_{1,ijk,\beta}^{+} \Delta x}{k_1^* w_{i+1/2jk,\beta,l}^{+} \lambda_{i+1/2jk,l}^{m}(\mathbf{v}_{ijk,\beta}^{+}, \boldsymbol{\xi}_{ijk,\beta}^{+})}, \frac{w_{ijk,\beta}^{-} |v_{1,ijk,\beta}^{-}| \Delta x}{k_1^* w_{i-1/2jk,\beta,r}^{-} \lambda_{i-1/2jk,r}^{m}(\mathbf{v}_{ijk,\beta}^{-}, \boldsymbol{\xi}_{ijk,\beta}^{-})} \right). \tag{B.52}$$

It then follows that the updated moment set, defined by

$$\mathbf{M}_{ijk}^{(1)} = \int_{\Omega_\xi} \int_{\mathbb{R}^3} \mathbf{K} n_{ijk}^{(1)} \, \mathrm{d}\mathbf{v} \, \mathrm{d}\boldsymbol{\xi}, \tag{B.53}$$

is also realizable. As discussed earlier, the positive integer k_1^* appears in the realizability condition, causing the time step to be smaller when higher-order moments are included in the transported moment set. Likewise, we can observe that the time step will be controlled by the $|v_1|$ abscissa *closest* to zero.

Similarly to what we did in the case of free transport, with mixed advection we define the half-moment sets needed to compute the weights and abscissas used in Eqs. (B.50) and (B.51) as

$$\mathbf{M}^{+}_{1,ijk} = \int_{0}^{+\infty} \int_{\mathbb{R}^2} \int_{\Omega_\xi} \mathbf{K} n^{m}_{ijk} \, \mathrm{d}\boldsymbol{\xi} \, \mathrm{d}v_3 \, \mathrm{d}v_2 \, \mathrm{d}v_1,$$
$$\mathbf{M}^{-}_{1,ijk} = \int_{-\infty}^{0} \int_{\mathbb{R}^2} \int_{\Omega_\xi} \mathbf{K} n^{m}_{ijk} \, \mathrm{d}\boldsymbol{\xi} \, \mathrm{d}v_3 \, \mathrm{d}v_2 \, \mathrm{d}v_1. \tag{B.54}$$

In this definition, $n^{m}_{ijk}(\mathbf{v}, \boldsymbol{\xi})$ is reconstructed using extended multivariate quadrature from the transported moment set $\mathbf{M}^{m}_{ijk}$. Since $n^{m}_{ijk}(\mathbf{v}, \boldsymbol{\xi})$ has a known functional form, the half-moment sets in Eq. (B.54) can be evaluated analytically. Once these half-moment sets are known, the N-node multivariate quadrature is applied to find the two set of weights and abscissas. The overall procedure is as follows:

$$\mathbf{M}^{m}_{ijk} \xrightarrow[\text{with } N \text{ nodes}]{\text{extended quadrature}} n^{m}_{ijk} \xrightarrow[\text{analytically}]{\text{evaluate moments}} \begin{matrix} \mathbf{M}^{+}_{1,ijk} \\ \mathbf{M}^{-}_{1,ijk} \end{matrix} \xrightarrow[\text{with } N \text{ nodes}]{\text{regular quadrature}} \begin{matrix} \left\{ w^{+}_{ijk,\beta}, \mathbf{v}^{+}_{ijk,\beta}, \boldsymbol{\xi}^{+}_{ijk,\beta} \right\} \\ \left\{ w^{-}_{ijk,\beta}, \mathbf{v}^{-}_{ijk,\beta}, \boldsymbol{\xi}^{-}_{ijk,\beta} \right\} \end{matrix}.$$

Notice that for mixed advection both the extended quadrature and the regular quadrature use N nodes. This is possible because with mixed advection the highest-order moment appearing in the spatial fluxes has a *lower* order due to the division by v_1 (i.e. instead of multiplication by v_1 for free transport). However, this observation only holds when $\boldsymbol{\lambda}$ does not depend on $\mathbf{v}$. The mixed advection formulas in the y and z directions are analogous to those derived above for the x direction. Likewise, the comments in Section B.4 of this appendix concerning the use of the CQMOM for free transport apply also for mixed advection.

A key technical point for mixed advection is the dependence of $\boldsymbol{\lambda}$ on $\mathbf{v}$, in particular when any one of the components of $\mathbf{v}$ tends towards zero. For example, $\boldsymbol{\lambda}(\mathbf{v})$ in Eq. (B.52) has three possible limiting behaviors:

$$\lim_{v_1 \to 0} \frac{\lambda}{|v_1|} = \begin{cases} 0 & \text{free transport,} \\ \text{constant} & \text{advection,} \\ \infty & \text{effective (elliptic) pressure.} \end{cases}$$

The first two correspond to free transport and advection wherein the time step is controlled by the velocity abscissa with the largest magnitude. On the other hand, the third case is controlled by the velocity abscissa with the *smallest* magnitude. A simple form for the second-order tensor $\boldsymbol{\lambda}$ that exhibits these behaviors is

$$\boldsymbol{\lambda}(t, \mathbf{x}, \mathbf{v}, \boldsymbol{\xi}) = \lambda_0(\boldsymbol{\xi}) \begin{bmatrix} |v_1|^{\nu} & 0 & 0 \\ 0 & |v_2|^{\nu} & 0 \\ 0 & 0 & |v_3|^{\nu} \end{bmatrix}, \tag{B.55}$$

where $0 \leq \nu \leq 2$ and $0 \leq \lambda_0(\boldsymbol{\xi})$ has dimensions of $(\mathrm{m/s})^{2-\nu}$ and depends, for example, on the particle Stokes number.

In proposed applications of mixed advection to gas–particle flows (Pandya & Mashayek, 2003b; Reeks, 1977, 1983, 1993; Zaichik, 1997, 1999; Zaichik *et al.*, 2008), $\boldsymbol{\lambda}$ is taken to be independent of $\mathbf{v}$, and hence it behaves as an effective pressure. Therefore, in order for the time step computed from Eq. (B.52) to be nonzero, it will be necessary for $\boldsymbol{\lambda}$ to be zero at spatial locations where either $v^{-}_{1,ijk,\beta}$ or $v^{+}_{1,ijk,\beta}$ is zero. Since these abscissas are computed from the half-moment sets $\mathbf{M}^{-}_{1,ijk}$ and $\mathbf{M}^{+}_{1,ijk}$, respectively, they can be zero only if the kernel density functions used in the EQMOM have zero width. (See Section 3.3.2

for more details on the EQMOM.) As an example, let us consider kernel density functions that are Gaussian with width σ and mean u_1. For computing v_1^+, we need the moments in the positive tail defined by

$$\mu_k = \frac{\int_0^\infty x^k \exp[-(x-u_1)^2/(2\sigma^2)]\mathrm{d}x}{\int_0^\infty \exp[-(x-u_1)^2/(2\sigma^2)]\mathrm{d}x},$$

and the worse-case scenario will occur when $u_1 \ll 0$ (i.e. for regular quadrature the Dirac delta functions would be located at negative v_1 velocities). In this limit, we find that

$$v_1^+ \propto \mu_1 \propto \frac{\sigma^2}{|u_1|},$$

and thus σ^2 controls the distance of the velocity abscissa from the origin when $u_1 \ll 0$. The same will be true for v_1^- in the limit $0 \ll u_1$. For multivariate EQMOM, the matrix $\boldsymbol{\Sigma}$ plays the role of σ^2. This observation would suggest that a good model for λ would be

$$\lambda(t, \mathbf{x}, \boldsymbol{\xi}) = h(\boldsymbol{\xi})\boldsymbol{\Sigma}(t, \mathbf{x}). \tag{B.56}$$

Note that, since $\boldsymbol{\Sigma}$ has units $(\mathrm{m/s})^2$, the nonnegative function $h(\boldsymbol{\xi})$ would be dimensionless. With this model for λ the realizability condition in Eq. (B.52) would always yield a nonzero upper bound on Δt when $h(\boldsymbol{\xi})$ is finite.[17] Physically, $\boldsymbol{\Sigma}$ is null in the limit of pure particle trajectory crossing where the true NDF is a sum of Dirac delta functions. On the other hand, when $\boldsymbol{\Sigma}$ reaches its maximum value, the NDF is Gaussian. Thus, since mixed advection is associated with "random" particle motion, the model in Eq. (B.56) also makes physical sense. Nonetheless, the potential for singular behavior in the update formula makes the treatment of mixed advection problematic.

In summary, in this section we have shown that the realizable scheme for mixed advection is very similar to those introduced for advection and free transport. However, the numerical properties of the scheme are very dependent on the functional form used for λ. In order for the "standard" models for mixed advection to be well conditioned, the functional form for λ must compensate for velocity abscissas located too close to the origin. With multivariate EQMOM, the model given in Eq. (B.56) can be used for this purpose. The potential for singular behavior with mixed advection makes it nevertheless problematic, and alternative schemes may be required to properly handle such behavior.

B.6 Diffusion

For multidimensional diffusion with diffusion coefficient $\Gamma(t, \mathbf{x}, \boldsymbol{\xi})$, we will again use operator splitting in each spatial direction (Vikas *et al.*, 2012). In the x direction, the diffusion equation is

$$\frac{\partial n}{\partial t} = \frac{\partial}{\partial x}\left(\Gamma \frac{\partial n}{\partial x}\right). \tag{B.57}$$

[17] In gas–particle flows, h depends on the particle Stokes number as

$$h \propto \frac{1}{\mathrm{St}(\mathrm{St}+1)}.$$

Thus, h will be finite when the Stokes number is nonzero.

Formally, the corresponding finite-volume formula is

$$n^{(1)}_{ijk} = n^m_{ijk} + \lambda_x \left(\Gamma^m_{i+1/2jk} \frac{\partial n^m_{i+1/2jk}}{\partial x} - \Gamma^m_{i-1/2jk} \frac{\partial n^m_{i-1/2jk}}{\partial x} \right) \tag{B.58}$$

wherein the spatial gradients of the NDF must be defined in terms of the quadrature representation:

$$\frac{\partial n^m_{i\pm1/2jk}}{\partial x} = \sum_{\alpha=1}^{N} \left[w^m_{i\pm1/2jk,\alpha} \frac{\partial \delta(\boldsymbol{\xi} - \boldsymbol{\xi}^m_{i\pm1/2jk,\alpha})}{\partial x} + \frac{\partial w^m_{i\pm1/2jk,\alpha}}{\partial x} \delta(\boldsymbol{\xi} - \boldsymbol{\xi}^m_{i\pm1/2jk,\alpha}) \right]. \tag{B.59}$$

For each order of the spatial reconstruction, it is then necessary to define the spatial gradient terms appearing in Eq. (B.59).

For second-order spatial reconstruction, the weights at the interfaces are defined by

$$\begin{aligned} w^m_{i+1/2jk,\alpha} &= \frac{1}{2} w^m_{ijk,\alpha} + \frac{1}{2} w^m_{i+1jk,\alpha}, \\ w^m_{i-1/2jk,\alpha} &= \frac{1}{2} w^m_{i-1jk,\alpha} + \frac{1}{2} w^m_{ijk,\alpha}, \end{aligned} \tag{B.60}$$

and the delta functions at the interfaces by

$$\begin{aligned} \delta(\boldsymbol{\xi} - \boldsymbol{\xi}^m_{i+1/2jk,\alpha}) &= \frac{1}{2}\delta(\boldsymbol{\xi} - \boldsymbol{\xi}^m_{ijk,\alpha}) + \frac{1}{2}\delta(\boldsymbol{\xi} - \boldsymbol{\xi}^m_{i+1jk,\alpha}), \\ \delta(\boldsymbol{\xi} - \boldsymbol{\xi}^m_{i-1/2jk,\alpha}) &= \frac{1}{2}\delta(\boldsymbol{\xi} - \boldsymbol{\xi}^m_{i-1jk,\alpha}) + \frac{1}{2}\delta(\boldsymbol{\xi} - \boldsymbol{\xi}^m_{ijk,\alpha}). \end{aligned} \tag{B.61}$$

The gradients are defined by

$$\begin{aligned} \frac{\partial w^m_{i+1/2jk,\alpha}}{\partial x} &= \frac{1}{\Delta x} \left(w^m_{i+1jk,\alpha} - w^m_{ijk,\alpha} \right), \\ \frac{\partial w^m_{i-1/2jk,\alpha}}{\partial x} &= \frac{1}{\Delta x} \left(w^m_{ijk,\alpha} - w^m_{i-1jk,\alpha} \right), \end{aligned} \tag{B.62}$$

and

$$\begin{aligned} \frac{\partial \delta(\boldsymbol{\xi} - \boldsymbol{\xi}^m_{i+1/2jk,\alpha})}{\partial x} &= \frac{1}{\Delta x} \left[\delta(\boldsymbol{\xi} - \boldsymbol{\xi}^m_{i+1jk,\alpha}) - \delta(\boldsymbol{\xi} - \boldsymbol{\xi}^m_{ijk,\alpha}) \right], \\ \frac{\partial \delta(\boldsymbol{\xi} - \boldsymbol{\xi}^m_{i-1/2jk,\alpha})}{\partial x} &= \frac{1}{\Delta x} \left[\delta(\boldsymbol{\xi} - \boldsymbol{\xi}^m_{ijk,\alpha}) - \delta(\boldsymbol{\xi} - \boldsymbol{\xi}^m_{i-1jk,\alpha}) \right]. \end{aligned} \tag{B.63}$$

Using these formulas in Eq. (B.58) leads to

$$\begin{aligned} n^{(1)}_{ijk} = &\sum_{\alpha=1}^{N} \left[1 - \gamma_x \left(\Gamma^m_{i+1/2jk}(\boldsymbol{\xi}) + \Gamma^m_{i-1/2jk}(\boldsymbol{\xi}) \right) \right] w^m_{ijk,\alpha} \delta(\boldsymbol{\xi} - \boldsymbol{\xi}^m_{ijk,\alpha}) \\ &+ \sum_{\alpha=1}^{N} \gamma_x \Gamma^m_{i+1/2jk}(\boldsymbol{\xi}) w^m_{i+1jk,\alpha} \delta(\boldsymbol{\xi} - \boldsymbol{\xi}^m_{i+1jk,\alpha}) \\ &+ \sum_{\alpha=1}^{N} \gamma_x \Gamma^m_{i-1/2jk}(\boldsymbol{\xi}) w^m_{i-1jk,\alpha} \delta(\boldsymbol{\xi} - \boldsymbol{\xi}^m_{i-1jk,\alpha}), \end{aligned} \tag{B.64}$$

where $\gamma_x = \Delta t/(\Delta x)^2$. This NDF is realizable provided that

$$\Delta t \leq \min_{ijk,\alpha} \left(\frac{(\Delta x)^2}{\Gamma^m_{i+1/2jk}(\boldsymbol{\xi}^m_{ijk,\alpha}) + \Gamma^m_{i-1/2jk}(\boldsymbol{\xi}^m_{ijk,\alpha})} \right), \tag{B.65}$$

and the expressions for the y and z directions are analogous to Eq. (B.64).

The reader can note that, when $\Gamma(t, \mathbf{x})$ does not depend on $\boldsymbol{\xi}$, Eq. (B.64) leads to a closed, realizable formula for the updated moment set:

$$\mathbf{M}^{(1)}_{ijk} = \left[1 - \gamma_x \left(\Gamma^m_{i+1/2jk} + \Gamma^m_{i-1/2jk}\right)\right] \mathbf{M}^m_{ijk} + \gamma_x \Gamma^m_{i+1/2jk} \mathbf{M}^m_{i+1jk} + \gamma_x \Gamma^m_{i-1/2jk} \mathbf{M}^m_{i-1jk}. \tag{B.66}$$

In fact, this expression is the same as the second-order expression found when working directly with the diffusion term written using the transported moment set. In general, the second-order spatial reconstruction on unstructured grids will also be realizable. Perhaps for this reason, the diffusion term is not usually identified as a potential source of unrealizable moment sets.

For higher-order spatial reconstruction, the NDF will be realizable only if the abscissas in neighboring cells are employed to construct the scheme. In other words, we must continue to use Eqs. (B.61) and (B.63). We can then approximate the spatial derivatives of the weights using a higher-order scheme. For example, with a fourth-order scheme, we will use

$$\begin{aligned} w^m_{i+1/2jk,\alpha} &= -\frac{1}{12} w^m_{i-1jk,\alpha} + \frac{7}{12} w^m_{ijk,\alpha} + \frac{7}{12} w^m_{i+1jk,\alpha} - \frac{1}{12} w^m_{i+2jk,\alpha}, \\ w^m_{i-1/2jk,\alpha} &= -\frac{1}{12} w^m_{i-2jk,\alpha} + \frac{7}{12} w^m_{i-1jk,\alpha} + \frac{7}{12} w^m_{ijk,\alpha} - \frac{1}{12} w^m_{i+1jk,\alpha}, \end{aligned} \tag{B.67}$$

and

$$\begin{aligned} \frac{\partial w^m_{i+1/2jk,\alpha}}{\partial x} &= \frac{1}{\Delta x} \left(-\frac{1}{12} w^m_{i+2jk,\alpha} + \frac{5}{4} w^m_{i+1jk,\alpha} - \frac{5}{4} w^m_{ijk,\alpha} + \frac{1}{12} w^m_{i-1jk,\alpha} \right), \\ \frac{\partial w^m_{i-1/2jk,\alpha}}{\partial x} &= \frac{1}{\Delta x} \left(-\frac{1}{12} w^m_{i+1jk,\alpha} + \frac{5}{4} w^m_{ijk,\alpha} - \frac{5}{4} w^m_{i-1jk,\alpha} + \frac{1}{12} w^m_{i-2jk,\alpha} \right). \end{aligned} \tag{B.68}$$

Note that if the abscissas are constant in all cells, then the overall scheme will have the same order as the spatial reconstruction of the weights. On the other hand, if the weights are constant, then the overall scheme for the abscissas is of second order.

Applying the formulas above, the updated numerical NDF can be written as

$$\begin{aligned} n^{(1)}_{ijk} = \sum_{\alpha=1}^{N} \Bigg[& w^m_{ijk,\alpha} - \gamma_x \left(\Gamma^m_{i+1/2jk}(\boldsymbol{\xi}) w^m_{i+1/2jk,\alpha} + \Gamma^m_{i-1/2jk}(\boldsymbol{\xi}) w^m_{i-1/2jk,\alpha} \right) \\ & - \frac{\lambda_x}{2} \left(\Gamma^m_{i+1/2jk}(\boldsymbol{\xi}) \frac{\partial w^m_{i+1/2jk,\alpha}}{\partial x} - \Gamma^m_{i-1/2jk}(\boldsymbol{\xi}) \frac{\partial w^m_{i-1/2jk,\alpha}}{\partial x} \right) \Bigg] \delta(\boldsymbol{\xi} - \boldsymbol{\xi}^m_{ijk,\alpha}) \\ & + \sum_{\alpha=1}^{N} \Gamma^m_{i+1/2jk}(\boldsymbol{\xi}) \left(\gamma_x w^m_{i+1/2jk,\alpha} + \frac{\lambda_x}{2} \frac{\partial w^m_{i+1/2jk,\alpha}}{\partial x} \right) \delta(\boldsymbol{\xi} - \boldsymbol{\xi}^m_{i+1jk,\alpha}) \\ & + \sum_{\alpha=1}^{N} \Gamma^m_{i-1/2jk}(\boldsymbol{\xi}) \left(\gamma_x w^m_{i-1/2jk,\alpha} - \frac{\lambda_x}{2} \frac{\partial w^m_{i-1/2jk,\alpha}}{\partial x} \right) \delta(\boldsymbol{\xi} - \boldsymbol{\xi}^m_{i-1jk,\alpha}). \end{aligned} \tag{B.69}$$

This NDF is realizable provided that each coefficient multiplying a delta function is nonnegative for all grid cells and all nodes. As before, the expressions for the y and z directions are analogous to Eq. (B.69).

The final equation for the change in the moment set due to diffusion in the x direction is

$$\begin{aligned}
\mathbf{M}_{ijk}^{(1)} = \sum_{\alpha=1}^{N} \Big[w_{ijk,\alpha}^{m} &- \gamma_x \left(\Gamma_{i+1/2jk}^{m}(\xi_{ijk,\alpha}^{m}) w_{i+1/2jk,\alpha}^{m} + \Gamma_{i-1/2jk}^{m}(\xi_{ijk,\alpha}^{m}) w_{i-1/2jk,\alpha}^{m} \right) \\
&- \frac{\lambda_x}{2} \left(\Gamma_{i+1/2jk}^{m}(\xi_{ijk,\alpha}^{m}) \frac{\partial w_{i+1/2jk,\alpha}^{m}}{\partial x} - \Gamma_{i-1/2jk}^{m}(\xi_{ijk,\alpha}^{m}) \frac{\partial w_{i-1/2jk,\alpha}^{m}}{\partial x} \right) \Big] \mathbf{K}_{ijk,\alpha}^{m} \\
&+ \sum_{\alpha=1}^{N} \Gamma_{i+1/2jk}^{m}(\xi_{i+1jk,\alpha}^{m}) \left(\gamma_x w_{i+1/2jk,\alpha}^{m} + \frac{\lambda_x}{2} \frac{\partial w_{i+1/2jk,\alpha}^{m}}{\partial x} \right) \mathbf{K}_{i+1jk,\alpha}^{m} \\
&+ \sum_{\alpha=1}^{N} \Gamma_{i-1/2jk}^{m}(\xi_{i-1jk,\alpha}^{m}) \left(\gamma_x w_{i-1/2jk,\alpha}^{m} - \frac{\lambda_x}{2} \frac{\partial w_{i-1/2jk,\alpha}^{m}}{\partial x} \right) \mathbf{K}_{i-1jk,\alpha}^{m},
\end{aligned} \tag{B.70}$$

with analogous expressions for the y and z directions. As described earlier, if the CQMOM is employed for multivariate quadrature reconstruction, each of the CQMOM permutations should be used to update the moment set, and the results then averaged to find $\mathbf{M}_{ijk}^{m+1}$.

Appendix C

Moment methods with hyperbolic equations

In the literature (Mazzei, 2011) it has been argued that quadrature-based moment methods (QBMM) cannot be used to represent solutions to hyperbolic kinetic equations without adding "diffusion" terms. The argument goes that diffusion terms are necessary to represent "micromixing" between two NDFs originating from spatially distinct regions of the system. The purpose of this appendix is to present an analytical counterexample that clearly shows that the exact solution to the hyperbolic kinetic equation (i.e. without diffusion terms) can result in a "completely mixed" (or, more precisely, a homogeneous) NDF at large times. As we shall see, this solution arises entirely due to the size dependence of the advection velocity. We also use the analytical solution to illustrate the issues that arise when using the DQMOM and QMOM to solve kinetic equations with spatially non-smooth solutions (i.e. when the spatial derivatives of the moments are not continuous). Finally, it is noteworthy that the analytical solutions derived in this appendix could be useful for comparison with numerical solutions of the model kinetic equation in order to verify the accuracy of a numerical algorithm.

C.1 A model kinetic equation

In order to clearly illustrate the relevant issues, we consider an NDF with one spatial dimension, advected by a velocity that depends on a single internal coordinate ξ (which we will refer to as "size"). The corresponding kinetic equation is

$$\frac{\partial f}{\partial t} + \frac{\partial}{\partial x}[u(\xi)f] = 0, \tag{C.1}$$

where $f(t, x, \xi)$ is the NDF, and the advection velocity $u(\xi)$ is assumed to be known. Note that, while it is possible to generalize $u(\xi)$ to depend on t and x without altering the basic mathematical properties of Eq. (C.1), doing so would greatly complicate the analytical solution, and thus serve only to obscure its essential properties. In applications to multiphase flows, $u(\xi)$ would correspond to the average velocity of all particles with size ξ (i.e. a conditional velocity), and thus by this definition $u(\xi)$ is single-valued (i.e. for each ξ there is a single velocity). This in no way excludes the possibility of particle trajectory crossings because the latter is a property of the instantaneous velocity, not of its conditional average.

C.2 Analytical solution for segregated initial conditions

Given the initial NDF $f(0, x, \xi)$, Eq. (C.1) has the analytical solution

$$f(t, x, \xi) = f(0, x - u(\xi)t, \xi), \tag{C.2}$$

which can be verified by substituting Eq. (C.2) into Eq. (C.1). This solution is exact for an arbitrary initial NDF. Here we shall consider a spatially segregated initial NDF:

$$f_0(x, \xi) = \begin{cases} f_1(\xi) & \text{for } x < 0 \text{ and } 0 \leq \xi < \xi^*, \\ f_2(\xi) & \text{for } x \geq 0 \text{ and } \xi^* \leq \xi < 1, \\ 0 & \text{otherwise,} \end{cases} \tag{C.3}$$

where $0 < \xi^* < 1$ denotes the cut-off particle size. In words, all of the smaller particles are initially located below $x = 0$, and the larger particles above them. The NDFs f_1 and f_2 are defined for the ranges of ξ indicated in Eq. (C.3), but otherwise have arbitrary shapes (i.e. they could be smooth functions or even delta functions). Using Eqs. (C.2) and (C.3), the analytical solution with segregated initial conditions can be written as

$$f(t, x, \xi) = f_1(\xi) I_{[0,\xi^*)}(\xi) I_{(-\infty,u(\xi)t)}(x) + f_2(\xi) I_{[\xi^*,1]}(\xi) I_{[u(\xi)t,\infty)}(x), \tag{C.4}$$

where $I_S(x)$ is the indicator function for the interval S:

$$I_S(x) = \begin{cases} 1 & \text{for } x \in S, \\ 0 & \text{otherwise.} \end{cases} \tag{C.5}$$

Note that the analytical solution in Eq. (C.4) is discontinuous along the surface in (t, x, ξ)-space defined by $x = u(\xi)t$. Thus, as we shall see next, the functional form of $u(\xi)$ will determine whether the NDF mixes or segregates further for increasing t.

C.2.1 Segregating solution

We first consider the case where $u(\xi)$ is a monotonically increasing function of ξ. Specifically, we let $u(\xi) = \xi - \xi^*$. The characteristics of Eq. (C.1) are then defined by $x(t, \xi) = x(0) + (\xi - \xi^*)t$, and thus, using Eq. (C.3), we see that

$$x(t, \xi) = \begin{cases} -|x(0)| - |\xi^* - \xi|t & \text{for } x(0) < 0, \\ x(0) + |\xi - \xi^*|t & \text{for } x(0) \geq 0. \end{cases} \tag{C.6}$$

In words, the characteristics starting at negative x become more negative with increasing t, while those starting at positive x become more positive. In all cases, we can see from Eq. (C.6) that the characteristics never cross each other (i.e. if $\xi_1 \neq \xi_2$, then $x(t, \xi_1) \neq x(t, \xi_2)$ for any $t > 0$). The characteristic starting at $x(0) = 0$ with $\xi = \xi^*$ has the property $x(t, \xi^*) = 0$ and corresponds to a contact point (or discontinuity) between the two initial solutions with left and right NDF $f_1(\xi)$ and $f_2(\xi)$, respectively. Because the characteristics do not cross, the analytical solution can be written in terms of the two half intervals in space:

$$f(t, x, \xi) = \begin{cases} f_1(\xi) I_{[\max(0,\xi^*+x/t),\xi^*)}(\xi) & \text{for } -\infty < x/t < 0, \\ f_2(\xi) I_{[\xi^*,\min(1,\xi^*+x/t)]}(\xi) & \text{for } 0 \leq x/t < \infty. \end{cases} \tag{C.7}$$

As a consequence, the NDF remains "unmixed" for all times, and changes discontinuously at the line $x = (\xi - \xi^*)t$. Note that, for large t, only the characteristics starting at ξ^* (i.e. with zero velocity) will remain in the system. As a result, the initial NDF becomes more and more segregated according to size as time increases. Finally, note that as long as $u(\xi)$ is monotonically *increasing*, changing the functional dependence on ξ will not change the qualitative behavior of Eq. (C.6) or Eq. (C.7).

C.2.2 Mixing solution

We next consider the case in which $u(\xi)$ is a monotonically decreasing function of ξ. Specifically, we let $u(\xi) = \xi^* - \xi$. The characteristics of Eq. (C.1) are defined by $x(t,\xi) = x(0) + (\xi^* - \xi)t$, and thus, using Eq. (C.3), we see that

$$x(t,\xi) = \begin{cases} -|x(0)| + |\xi^* - \xi|t & \text{for } x(0) < 0, \\ x(0) - |\xi - \xi^*|t & \text{for } x(0) \geq 0. \end{cases} \tag{C.8}$$

In words, the characteristics starting at negative x eventually become positive with increasing t, while those starting at positive x eventually become negative. Hence, we can conclude from Eq. (C.8) that the characteristics corresponding to $\xi < \xi^*$ will cross all of the characteristics corresponding to $\xi^* < \xi$ at some value of t.[1] The characteristic starting at $x(0) = 0$ with $\xi = \xi^*$ has the property $x(t, \xi^*) = 0$, and again corresponds to a contact point between the two initial NDF. However, there are now two discontinuities corresponding the two characteristics starting at $x(0) = 0$ with $\xi = 0$ and $\xi = 1$, respectively. The first represents the penetration of the NDF f_1 into the spatial region originally occupied by f_2, and vice versa for the second. By accounting for the crossing characteristics, the analytical solution can be written as

$$f(t,x,\xi) = \begin{cases} f_1(\xi) I_{[0,\min(\xi^*,\xi^*-x/t))}(\xi) & \text{for } -\infty < x/t < \xi^* - 1, \\ f_1(\xi) I_{[0,\min(\xi^*,\xi^*-x/t)}(\xi) & \\ \quad + f_2(\xi) I_{[\max(\xi^*,\xi^*-x/t),1]}(\xi) & \text{for } \xi^* - 1 \leq x/t < \xi^*, \\ f_2(\xi) I_{[\max(\xi^*,\xi^*-x/t),1]}(\xi) & \text{for } \xi^* \leq x/t < \infty. \end{cases} \tag{C.9}$$

From this analytical solution, it can clearly be seen that the NDF for $\xi^* - 1 < x/t < \xi^*$ is a combination of the two original NDFs. Moreover, as t increases to infinity, we find

$$\lim_{t\to\infty} f(t,x,\xi) = \left[f_1(\xi) I_{[0,\xi^*)}(\xi) + f_2(\xi) I_{[\xi^*,1]}(\xi) \right] I_{(-\infty,\infty)}(x), \tag{C.10}$$

which corresponds to the initial NDF spread over the entire real line (i.e. a "well-mixed" system). Note again that, as long as $u(\xi)$ is monotonically decreasing, changing the functional dependence on ξ will not change the qualitative behavior of Eq. (C.8) or Eq. (C.10). As stated earlier, the mixing solution is obtained from the fully hyperbolic kinetic equation without adding diffusion. Indeed, the key technical point that leads to mixing is the crossing of characteristics in solutions to hyperbolic equations.

[1] The exact crossing time is $t_c = (x_2(0) - x_1(0))/(\xi_2 - \xi_1)$, where $\xi_1 < \xi^* < \xi_2$.

C.3 Moments and the quadrature approximation

Given the analytical solution, it is straightforward to compute the integer moments using the definition

$$m_k(t, x) = \int_0^1 \xi^k f(t, x, \xi) \mathrm{d}\xi. \tag{C.11}$$

Likewise, the quadrature approximation can be found from these moments. For example, for a one-node quadrature the weights and abscissas are defined by

$$w_1(t, x) = m_0(t, x) \qquad \text{and} \qquad \xi_1(t, x) = \frac{m_1(t, x)}{m_0(t, x)}. \tag{C.12}$$

Note that due to the hyperbolic form of the kinetic equation, the moments, weights, and abscissas need not be smooth functions of x. For simplicity, we will consider the moments for two cases. The first uses a continuous initial NDF defined by $f_1 = f_2 = I_{[0,1]}(\xi)$, and the second a discrete NDF defined by $f_1 = \delta(\xi - \xi^*/2)$ and $f_2 = \delta(\xi - 3\xi^*/2)$.

C.3.1 Moments of segregating solution

Continuous NDF

The moments of the segregating solution with the continuous initial NDF can be computed analytically from Eq. (C.7):

$$m_k(t, x) = \begin{cases} [1/(k+1)]\left[(\xi^*)^{k+1} - \max(0, \xi^* + x/t)^{k+1}\right] & \text{for } -\infty < x < 0, \\ [1/(k+1)]\left[\min(1, \xi^* + x/t)^{k+1} - (\xi^*)^{k+1}\right] & \text{for } 0 \leq x < \infty. \end{cases} \tag{C.13}$$

Examples are shown in Figure C.1. The one-node quadrature approximation is given by

$$w_1(t, x) = \begin{cases} \xi^* - \max(0, \xi^* + x/t) & \text{for } -\infty < x < 0, \\ \min(1, \xi^* + x/t) - \xi^* & \text{for } 0 \leq x < \infty, \end{cases}$$

$$\xi_1(t, x) = \begin{cases} [(\xi^*)^2 - \max(0, \xi^* + x/t)^2]/\{2[\xi^* - \max(0, \xi^* + x/t)]\} & \text{for } -\infty < x < 0, \\ [\min(1, \xi^* + x/t)^2 - (\xi^*)^2]/\{2[\min(1, \xi^* + x/t) - \xi^*]\} & \text{for } 0 \leq x < \infty, \end{cases} \tag{C.14}$$

and a two-node quadrature is shown in Figure C.2. For this case, the limit from above and below at $x = 0$ for $t > 0$ yields $\xi_1(t, 0) = \xi^*$, and thus the abscissas are continuous in all x. Note, however, that $m_k(t, 0) = 0$ for $t > 0$ and the spatial gradients of the moments are discontinuous at $x = 0$. When multiple abscissas are used to represent the moments for $t > 0$, all of the weights will be zero at $x = 0$. However, since the true NDF approaches a delta function centered at $\xi = \xi^*$ for x close to zero, all abscissas smoothly approach ξ^* as $x \to 0$, as can be seen in Figure C.2.

Discrete NDF

For the discrete initial conditions, the moments are

$$m_k(t, x) = \begin{cases} (\xi^*/2)^k & \text{for } -\infty < x/t \leq -\xi^*/2, \\ 0 & \text{for } -\xi^*/2 < x/t \leq \xi^*/2, \\ (3\xi^*/2)^k & \text{for } \xi^*/2 \leq x/t < \infty. \end{cases} \tag{C.15}$$

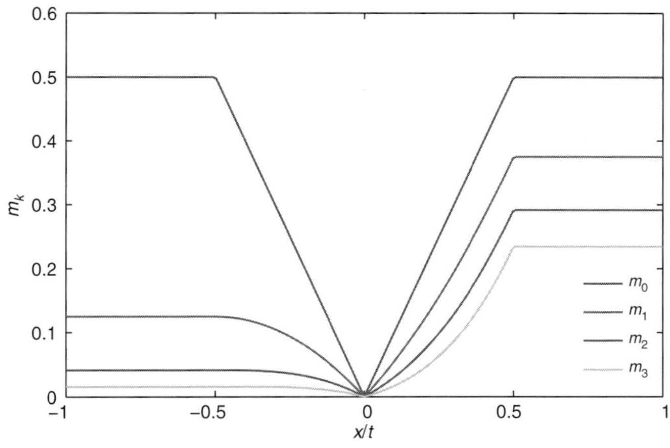

Figure C.1. Examples of moments found for segregating solution in Eq. (C.13) with $\xi^* = 1/2$.

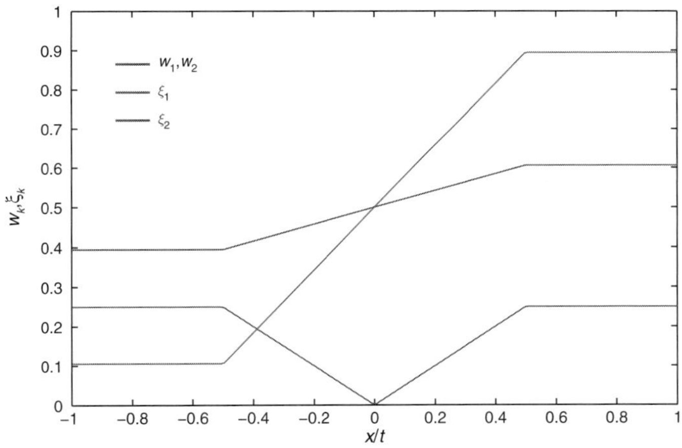

Figure C.2. Weights and abscissas found from four moments in Figure C.1.

Readers can note that the upper and lower bounds on the nonzero region correspond to the characteristics computed from Eq. (C.6) starting at $x(0) = 0$. The weights and abscissas for this case have a simple form:

$$\begin{aligned} &w_1 = 1,\ \xi_1 = \xi^*/2 && \text{for } -\infty < x/t \leq -\xi^*/2, \\ &w_1 = 0,\ \xi_1 \text{ not defined} && \text{for } -\xi^*/2 < x/t \leq \xi^*/2, \\ &w_1 = 1,\ \xi_1 = 3\xi^*/2 && \text{for } \xi^*/2 \leq x/t < \infty. \end{aligned} \tag{C.16}$$

Owing to segregation, this system creates a vacuum zone in which the abscissa cannot be defined. Outside this region the weights and abscissas are constant.

C.3.2 Moments of mixing solution

Continuous NDF

The moments of the segregating solution can be computed analytically from Eq. (C.10):

$$m_k(t,x) = \begin{cases} [1/(k+1)] \min(\xi^*, \xi^* - x/t)^{k+1} & \text{for } -\infty < x/t < \xi^* - 1, \\ [1/(k+1)] \left[1 + \min(\xi^*, \xi^* - x/t)^{k+1} - \max(\xi^*, \xi^* - x/t)^{k+1}\right] & \text{for } \xi^* - 1 \le x/t < \xi^*, \\ [1/(k+1)] \left[1 - \max(\xi^*, \xi^* - x/t)^{k+1}\right] & \text{for } \xi^* \le x/t < \infty. \end{cases} \tag{C.17}$$

Examples are shown in Figure C.3, and the corresponding quadrature weights and abscissas are shown in Figure C.4. For this case, the moments at $x = 0$ are just the sum of the moments found from f_1 and f_2 for all $t > 0$. It can also be seen that, although the moments

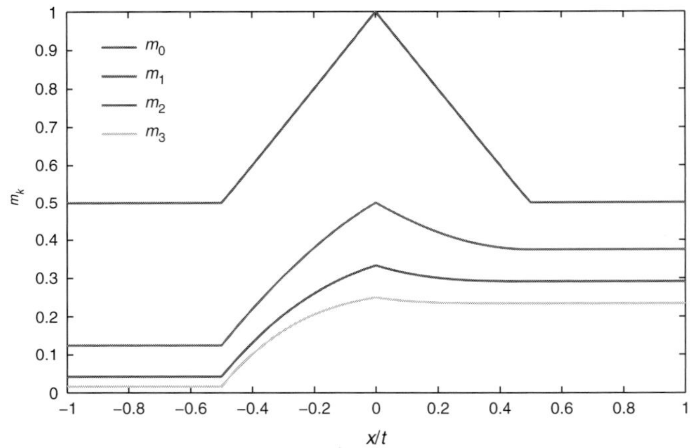

Figure C.3. Examples of moments found for mixing solution in Eq. (C.17) with $\xi^* = 1/2$.

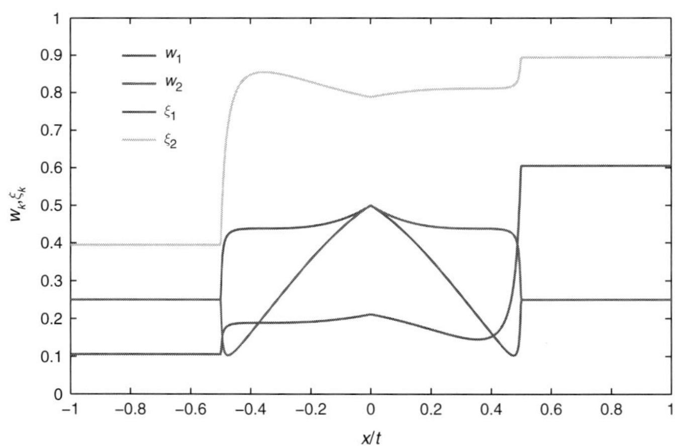

Figure C.4. Weights and abscissas found from the four moments in Figure C.3.

are continuous in x in Figure C.3, their spatial gradients are discontinuous at $x/t = \xi^* - 1$, $x = 0$, and $x/t = \xi^*$. The same is true for the weights and abscissas in Figure C.4.

Discrete NDF

For the discrete initial conditions, the moments are

$$m_k(t, x) = \begin{cases} (\xi^*/2)^k & \text{for } -\infty < x/t \leq -\xi^*/2, \\ (\xi^*/2)^k + (3\xi^*/2)^k & \text{for } -\xi^*/2 < x/t \leq \xi^*/2, \\ (3\xi^*/2)^k & \text{for } \xi^*/2 \leq x/t < \infty. \end{cases} \tag{C.18}$$

Reader can note that the upper and lower bounds on the central region correspond to the characteristics computed from Eq. (C.8) starting at $x(0) = 0$. The weights and abscissas for this case have a simple form:

$$\begin{array}{ll} w_1 = 1,\ \xi_1 = \xi^*/2 & \text{for } -\infty < x/t \leq -\xi^*/2, \\ w_1 = 1,\ \xi_1 = \xi^*/2,\ w_2 = 1,\ \xi_2 = 3\xi^*/2 & \text{for } -\xi^*/2 < x/t \leq \xi^*/2, \\ w_1 = 1,\ \xi_1 = 3\xi^*/2 & \text{for } \xi^*/2 \leq x/t < \infty. \end{array} \tag{C.19}$$

However, note that the number of abscissas changes at the points where the characteristics cross. As pointed out earlier, crossing characteristics lead to "mixing" of the initial NDF.

C.4 Application of QBMM

The application of QBMM to Eq. (C.1) will require a closure when $u(\xi)$ depends on ξ. Nevertheless, the resulting moment equations (used for the QMOM or the EQMOM) and transport equations for the weights and abscissas (used for the DQMOM) will still be hyperbolic. In terms of hyperbolic conservation laws, the moments are *conserved* variables (which result from a linear operation on f), while the weights and abscissas are *primitive* variables. Because conservation of moments is important to the stability of the moment-inversion algorithms, it is imperative that the numerical algorithm guarantee conservation. For hyperbolic systems, this is most easily accomplished using finite-volume methods (FVM) (or, more specifically, realizable FVM). The other important consideration is the accuracy of the moment closure used to close the function, as will be described below.

C.4.1 The moment-transport equation

The moment-transport equation found from Eq. (C.1) is

$$\frac{\partial m_k}{\partial t} + \frac{\partial F_k}{\partial t} = 0, \tag{C.20}$$

where the flux function is defined for the two cases considered above by

$$F_k = \int_0^\infty u(\xi)\xi^k f(\xi)\mathrm{d}\xi = \begin{cases} m_{k+1} - \xi^* m_k & \text{segregating solution,} \\ \xi^* m_k - m_{k+1} & \text{mixing solution.} \end{cases} \tag{C.21}$$

In order to solve the moment-transport equation, a closure must be provided for m_{k+1}. Using the QMOM, the first $2N$ moments can be used to find N weights w_α and N abscissas ξ_α. The flux function can then be closed using

$$m_{2N+1} = \sum_{\alpha=1}^{N} w_\alpha \xi_\alpha^{2N+1}. \quad \text{(C.22)}$$

Alternatively, we could use the EQMOM and transport $2N + 1$ moments, from which N weights w_α, N abscissas ξ_α, and σ could be found. In this case, the closure of m_{2N+2} would be found from the reconstructed NDF as described in Section 3.3. As is usually the case for hyperbolic systems, the reader can note that Eq. (C.20) is written in conservative form and can be solved numerically using KBFVM (see Section B.3 of Appendix B). For the discrete NDF, the QMOM closure will be exact with $N = 1$ (the segregating solution in Eq. (C.15)) or $N = 2$ (the mixing solution in Eq. (C.19)). Note that, for the mixing solution, the adaptive Wheeler algorithm will switch from $N = 1$ to $N = 2$ at $x/t = |\xi^*/2|$, which corresponds to crossing characteristics. Moreover, in a properly formulated kinetics-based finite-volume algorithm, the abscissas found from inverting the moments will remain at their original values, namely $\pm\xi*$, and the weights should be very near unity (differing only due to numerical diffusion). For the continuous NDF, neither the QMOM nor the EQMOM closure will be exact because of the error made in closing F_k. However, by increasing N it should be possible to approach the evolution of the moments shown in Figures C.1 and C.3. In all cases, the solution to the closed set of moment equations found with KBFVM will be well-behaved (e.g., the moments will be realizable), and there is no justification for adding diffusion terms on the right-hand side of Eq. (C.20).

C.4.2 Transport equations for weights and abscissas

In the DQMOM, the weights and abscissas are found by solving transport equations derived from Eq. (C.20) by defining the moments as

$$m_k = \sum_{\alpha=1}^{N} w_\alpha \xi_\alpha^k \quad \text{for } k = 0, \dots, 2N - 1. \quad \text{(C.23)}$$

For Eq. (C.1), this leads to

$$\frac{\partial w_\alpha}{\partial t} + \frac{\partial u(\xi_\alpha) w_\alpha}{\partial x} = 0 \quad \text{(C.24)}$$

and

$$\frac{\partial \xi_\alpha}{\partial t} + u(\xi_\alpha)\frac{\partial \xi_\alpha}{\partial x} = 0. \quad \text{(C.25)}$$

This system of hyperbolic equations has an exact solution:

$$w_\alpha(t, x) = w_\alpha(0, x - u(\xi_\alpha^*)t), \quad \xi_\alpha(t, x) = \xi_\alpha^*, \quad \text{(C.26)}$$

where ξ_α^* are constants, which corresponds to the discrete NDF in Eq. (C.15) and Eq. (C.19). As noted earlier, the solution in Eq. (C.26) is equivalent to the characteristics of the original hyperbolic equation, and for the mixing solutions the *characteristics cross*. This observation implies that the number of abscissas needed in order to describe the moments can change discontinuously in space and time. Furthermore, because Eq. (C.25) is of first order, only one integration constant is available in the exact solution. *Thus, setting the value of $\xi_\alpha(0, x)$ for a given α to be a function of x will lead to multi-valued* *solutions at characteristic crossings.* Instead, $\xi_\alpha(0,x)$ must be independent of x and the range of α should be made large enough to capture $f(0,x)$ for all x.[2] While adding diffusion terms to the right-hand sides of Eqs. (C.24) and (C.25) would make the DQMOM equations parabolic (and hence allow smoothing of the shocks generated at characteristic crossings), such a practice is unphysical and does not respect the hyperbolic nature of Eq. (C.20). Another difficulty associated with solving the DQMOM equations numerically is that it will be extremely difficult to guarantee that the moments are conserved. Only first-order advection schemes can guarantee conservation, but the numerical diffusion will be too large for one to obtain accurate solutions. Thus, in the authors' opinion, hyperbolic moment systems are best solved using Eq. (C.20) with appropriate QBMM closures for F_k (e.g. closures that are based on the EQMOM) combined with realizable high-order schemes (Vikas *et al.*, 2011a).

---

[2]This is precisely the approach which was taken in the original paper describing the DQMOM for hyperbolic systems (Fan *et al.*, 2004). When this approach is used, the weights $w_\alpha(0,x)$ with fixed abscissas ξ^*_α are used to approximate $f(0,x)$.

Appendix D

The direct quadrature method of moments fully conservative

In this appendix we discuss the problem of moment conservation with the DQMOM, and present a variation called DQMOM fully conservative (DQMOM-FC) that can be employed to partially overcome it. To illustrate the issue, we consider a one-dimensional spatially inhomogeneous PBE for the univariate NDF $n(t, x, \xi)$ with constant velocity and constant diffusion coefficient as described in Section 8.3. However, as will become clearer below, the extension to other, more complex, problems is straightforward.

D.1 Inhomogeneous PBE

Let us consider the simple spatially inhomogeneous PBE:

$$\frac{\partial n}{\partial t} + \frac{\partial}{\partial x}(un) = \frac{\partial}{\partial x}\left(\Gamma\frac{\partial n}{\partial x}\right) + S(t, x, \xi), \tag{D.1}$$

where the particle velocity u and diffusion coefficient Γ are constant (i.e. they do not depend on t, x, and ξ). This is not a significant limitation because with the methods presented in Chapter 8 the material presented in this appendix can be easily extended to more general cases. Moreover, let us assume that the initial NDF $n(0, x, \xi)$ as well as the source term $S(t, x, \xi)$ for phase-space processes such as nucleation, growth, aggregation, and breakage are known. When operator splitting is employed, the source term is generally treated separately from the spatial-transport terms as described in Chapter 8, and therefore could be taken as null. However, for completeness and consistency with our original DQMOM derivation (Marchisio & Fox, 2005), we will consider this term in the discussion.

D.2 Standard DQMOM

When standard DQMOM is applied, transport equations for weights w'_α and weighted abscissas $\varsigma'_\alpha = w'_\alpha \xi'_\alpha$ are solved:

$$\frac{\partial w'_\alpha}{\partial t} + \frac{\partial}{\partial x}(uw'_\alpha) = \frac{\partial}{\partial x}\left(\Gamma\frac{\partial w'_\alpha}{\partial x}\right) + a'_\alpha \tag{D.2}$$

and

$$\frac{\partial \varsigma'_\alpha}{\partial t} + \frac{\partial}{\partial x}(u\varsigma'_\alpha) = \frac{\partial}{\partial x}\left(\Gamma \frac{\partial \varsigma'_\alpha}{\partial x}\right) + b'_\alpha \tag{D.3}$$

with $\alpha = 1, \ldots, N$, where the prime $'$ is used, for reasons that will become clearer below, to indicate the solution obtained with standard DQMOM. An initial condition for the weights and weighted abscissas must also be defined, and this is generally done by calculating the moments associated with the initial NDF, $m_k(0, x) = \int \xi^k n(0, x, \xi) \mathrm{d}\xi$, and by subsequently applying one of the moment-inversion algorithms presented in Chapter 3. (More details can be found in Section 7.4.) The source terms a'_α and b'_α are calculated by choosing a set of $2N$ moments (for simplicity, here we use $2N$ integer moments) and by solving the following linear system (Marchisio & Fox, 2005):

$$(1-k)\sum_{\alpha=1}^{N}[\xi'_\alpha]^k a'_\alpha + k\sum_{\alpha=1}^{N}[\xi'_\alpha]^{k-1} b'_\alpha = \overline{S}_k + \overline{C}_k \tag{D.4}$$

with $k = 0, \ldots, 2N-1$, where $\overline{S}_k = \int \xi^k S \, \mathrm{d}\xi$ is the moment transform of order k of the source term and the correction term due to diffusion is

$$\overline{C}_k = k(k-1)\sum_{\alpha=1}^{N} w_\alpha [\xi'_\alpha]^{k-2} \Gamma \frac{\partial \xi'_\alpha}{\partial x}\frac{\partial \xi'_\alpha}{\partial x}. \tag{D.5}$$

It is important to remind the reader that this correction term is necessary in order to correctly transport moments of order greater than one (i.e. $k \geq 2$).

Finite-volume discretization

As discussed in Chapter 8, when the FVM with uniform grid spacing Δx is employed, the following volume-average quantities in the cell centered at x_i are defined:

$$w'_{\alpha,i}(t) \equiv \frac{1}{\Delta x}\int_{x_i-\Delta x/2}^{x_i+\Delta x/2} w'_\alpha(t, x)\mathrm{d}x \tag{D.6}$$

and

$$\varsigma'_{\alpha,i}(t) \equiv \frac{1}{\Delta x}\int_{x_i-\Delta x/2}^{x_i+\Delta x/2} \varsigma'_\alpha(t, x)\mathrm{d}x, \tag{D.7}$$

where $x_i - \Delta x/2$ and $x_i + \Delta x/2$ denote the locations of the left and right interfaces of the cell centered at x_i. The application of this volume-averaging procedure to Eqs. (D.2) and (D.3) results in the following expressions:

$$\frac{\mathrm{d}w'_{\alpha,i}}{\mathrm{d}t} = a'^{\mathrm{c}}_{\alpha,i} + a'^{\mathrm{d}}_{\alpha,i} + a'_{\alpha,i} \tag{D.8}$$

and

$$\frac{\mathrm{d}\varsigma'_{\alpha,i}}{\mathrm{d}t} = b'^{\mathrm{c}}_{\alpha,i} + b'^{\mathrm{d}}_{\alpha,i} + b'_{\alpha,i}, \tag{D.9}$$

where the first terms on the right-hand sides of Eqs. (D.8) and (D.9), representing the rates of change of weights and weighted abscissas in the cell centered at x_i due to advection in physical space, can be written as

$$a'^{\mathrm{c}}_{\alpha,i}(t) = -\frac{u}{\Delta x}\left[w'_{k,i+1/2}(t) - w'_{k,i-1/2}(t)\right]. \tag{D.10}$$

Here, $w'_{k,i\pm1/2}$ are the weights at the cell interfaces reconstructed from the volume-average weights $\{w'_{k,i-1}(t), w'_{k,i}(t), w'_{k,i+1}(t)\}$ and, consistently with the notation of Chapter 8, $w'_{k,i+1}$, $w'_{k,i}$, and $w'_{k,i-1}$ are the volume-average weights in cells $i+1$, i, and $i-1$. Analogously the source term for the volume-average weighted abscissas is

$$b'^{\mathrm{c}}_{\alpha,i}(t) = -\frac{u}{\Delta x}\left[\varsigma'_{k,i+1/2}(t) - \varsigma'_{k,i-1/2}(t)\right], \tag{D.11}$$

where $\varsigma'_{k,i\pm1/2}$ are the weighted abscissas at the cell interfaces reconstructed from the volume-average weighted abscissas $\{\varsigma'_{k,i-1}(t), \varsigma'_{k,i}(t), \varsigma'_{k,i+1}(t)\}$. The second terms on the right-hand sides of Eqs. (D.8) and (D.9) are due to diffusion and can be calculated in the same manner by using a realizable finite-volume scheme:

$$a'^{\mathrm{d}}_{\alpha,i}(t) = \frac{\Gamma}{\Delta x^2}\left[w'_{k,i+1}(t) - 2w'_{k,i}(t) + w'_{k,i-1}(t)\right] \tag{D.12}$$

and

$$b'^{\mathrm{d}}_{\alpha,i}(t) = \frac{\Gamma}{\Delta x^2}\left[\varsigma'_{k,i+1}(t) - 2\varsigma'_{k,i}(t) + \varsigma'_{k,i-1}(t)\right]. \tag{D.13}$$

More details on how reconstructed values are calculated from the volume-average values in neighboring cells in standard finite-volume schemes are reported in Chapter 8. The last terms appearing on the right-hand sides of Eqs. (D.8) and (D.9), $a'_{\alpha,i}$ and $b'_{\alpha,i}$, represent the rates of change of weights and weighted abscissas in the cell centered at x_i due to phase-space processes and are calculated from the linear system in Eq. (D.4), by using the volume-average source and correction terms, $\overline{S}_{k,i}$ and $\overline{C}_{k,i}$, in the cell centered at x_i. This can be easily done by evaluating $\overline{S}_k$ and $\overline{C}_k$ with volume-average weights $w'_{\alpha,i}$ and volume-average abscissas $\xi'_{\alpha,i} = \varsigma'_{\alpha,i}/w'_{\alpha,i}$. Time integration is typically carried out with explicit or implicit schemes, which are discussed below.

Using this procedure, the resulting numerical scheme conserves only the weights and weighted abscissas and their linear combinations. In fact, due to the approximations made (e.g. when evaluating the gradients in Eq. (D.5)) and due to discretization errors (e.g. numerical diffusion), the FVM applied to Eqs. (D.2) and (D.3) results in conservation errors for the moments.[1] It is immediately evident that only two moments of the NDF will be conserved, namely the moments of order zero and one (i.e. m_0 and m_1). By using the FVM and the quadrature approximation, these two moments can be written as $m'_{0,i} = \sum_{\alpha=1}^{N} w'_{\alpha_i}$ and $m'_{1,i} = \sum_{\alpha=1}^{N} \varsigma'_{\alpha_i}$. It is interesting to point out that the orders of the only two moments conserved (i.e. $k = 0, 1$) correspond to null correction terms, as can readily be confirmed by inspection of Eq. (D.5). This means that the correct transport (and therefore conservation) of moments for real-space diffusive terms (real or numerical) is always guaranteed for moments of order zero and one, but one needs a corrective term for higher-order moments. In fact, for all the other quantities involved that are not linear combinations of these primitive variables, notably moments of order greater than one, conservation cannot be guaranteed since the errors due to numerical diffusion do not cancel out. Moreover, as discussed in Appendix C, the hyperbolic system found with $\Gamma = 0$ (for which $\overline{C}_k = 0$) cannot be treated using the DQMOM in the presence of shocks and discontinuous weights and abscissas.

[1] In the context of FVM, conservative means that the sum of the transport terms over all grid cells i is zero for a closed system.

D.3 DQMOM-FC

For $\Gamma > 0$, a conservative scheme (for transport in physical space) can be developed by correctly treating the terms due to advection and diffusion in physical space. On rewriting the PBE in Eq. (D.1) as

$$\frac{\partial n}{\partial t} = \mathcal{H} \equiv -\frac{\partial}{\partial x}(un) + \frac{\partial}{\partial x}\left(\Gamma \frac{\partial n}{\partial x}\right) + S(t, x, \xi), \tag{D.14}$$

we can apply the moment transform to obtain

$$\frac{\partial m_k}{\partial t} = \overline{\mathcal{H}}_k = \overline{C}_k + \overline{\mathcal{D}}_k + \overline{S}_k, \tag{D.15}$$

where $m_k = \int \xi^k n \, \mathrm{d}\xi$ and $\overline{S}_k$ is the moment of order k of the source term corresponding to phase-space processes. The term corresponding to advection in physical space[2] is

$$\overline{C}_k = -\frac{\partial}{\partial x}\left(\int u \xi^k n \, \mathrm{d}\xi\right) = -\frac{\partial}{\partial x}(u m_k), \tag{D.16}$$

and the term for diffusion in physical space is

$$\overline{\mathcal{D}}_k = \frac{\partial^2}{\partial x^2}\left(\int \Gamma \xi^k n \, \mathrm{d}\xi\right) = \frac{\partial^2}{\partial x^2}(\Gamma m_k). \tag{D.17}$$

If we now apply the DQMOM methodology[3] to Eq. (D.15), the following equations are obtained:

$$\frac{\partial w''_\alpha}{\partial t} = a''^{\mathrm{c}}_\alpha + a''^{\mathrm{d}}_\alpha + a''_\alpha \tag{D.18}$$

and

$$\frac{\partial \varsigma''_\alpha}{\partial t} = b''^{\mathrm{c}}_\alpha + b''^{\mathrm{d}}_\alpha + b''_\alpha, \tag{D.19}$$

where the double prime $''$ indicates the solution obtained with DQMOM-FC. The source terms on the right-hand sides are calculated from the following linear systems:

$$(1-k)\sum_{\alpha=1}^{N} [\xi''_\alpha]^k a''^{\mathrm{c}}_\alpha + k \sum_{\alpha=1}^{N} [\xi''_\alpha]^{k-1} b''^{\mathrm{c}}_\alpha = \overline{C}_k, \tag{D.20}$$

$$(1-k)\sum_{\alpha=1}^{N} [\xi'']^k_\alpha a''^{\mathrm{d}}_\alpha + k \sum_{\alpha=1}^{N} [\xi''_\alpha]^{k-1} b''^{\mathrm{d}}_\alpha = \overline{\mathcal{D}}_k, \tag{D.21}$$

---

[2]It is straightforward to show that, if u is not constant but depends on the internal coordinate, the resulting closure problem can be overcome with the use of the quadrature approximation:

$$\int u(\xi)\xi^k n \, \mathrm{d}\xi \approx \sum_{\alpha=1}^{N} w_\alpha u(\xi_\alpha)\xi_\alpha^k.$$

A similar approximation can be used to treat a size-dependent diffusion coefficient $\Gamma(\xi)$ as described in Chapter 8.

[3]More precisely, we substitute the definition

$$m_k(t, x) = \sum_{\alpha=1}^{N} w_\alpha(t, x)[\xi_\alpha(t, x)]^k = \sum_{\alpha=1}^{N} [w_\alpha(t, x)]^{1-k}[\varsigma_\alpha(t, x)]^k$$

into the left-hand side of Eq. (D.15), and take the partial derivative with respect to t.

and

$$(1-k)\sum_{\alpha=1}^{N}[\xi''_\alpha]^k a''_\alpha + k\sum_{\alpha=1}^{N}[\xi''_\alpha]^{k-1} b''_\alpha = \overline{S}_k \tag{D.22}$$

with $k = 0, \ldots, 2N-1$. Notice that, since no spatial derivatives are used, no correction terms are needed to ensure conservation in space of the moments of order greater than one when $\Gamma > 0$. This observation suggests that a conservative finite-volume scheme can be derived by starting directly from Eq. (D.15).

Finite-volume discretization

We can derive the volume-average moment-evolution equations starting from Eq. (D.15):

$$\frac{\mathrm{d}m_{k,i}}{\mathrm{d}t} = \overline{\mathcal{H}}_{k,i} = \overline{\mathcal{C}}_{k,i} + \overline{\mathcal{D}}_{k,i} + \overline{S}_{k,i}, \tag{D.23}$$

where the volume-average moments and right-hand sides are defined by

$$m_{k,i}(t) \equiv \frac{1}{\Delta x}\int_{x_i-\Delta x/2}^{x_i+\Delta x/2} m_k(t,x)\mathrm{d}x \tag{D.24}$$

and

$$\overline{\mathcal{H}}_{k,i}(t) \equiv \frac{1}{\Delta x}\int_{x_i-\Delta x/2}^{x_i+\Delta x/2} \overline{\mathcal{H}}_k(t,x)\mathrm{d}x. \tag{D.25}$$

By definition, for a given time t, the weights and weighted abscissas for each i are related to the volume-average moments[4] by

$$m_{k,i}(t) \equiv \sum_{\alpha=1}^{N}[w''_{\alpha,i}(t)]^{1-k}[\varsigma''_{\alpha,i}(t)]^k \tag{D.26}$$

for $k = 0, \ldots, 2N-1$. Here, it is very important to note that $w''_{\alpha,i}$ and $\varsigma''_{\alpha,i}$ are not equal to the volume-average values found from $w_\alpha(t,x)$ and $\varsigma_\alpha(t,x)$ (i.e. they are not the variables defined in Eqs. (D.6) and (D.7)). By applying again the DQMOM methodology, the evolution equations for the weights and weighted abscissas are

$$\frac{\mathrm{d}w''_{\alpha,i}}{\mathrm{d}t} = a''^{\mathrm{c}}_{\alpha,i} + a''^{\mathrm{d}}_{\alpha,i} + a''_{\alpha,i} \tag{D.27}$$

and

$$\frac{\mathrm{d}\varsigma''_{\alpha,i}}{\mathrm{d}t} = b''^{\mathrm{c}}_{\alpha,i} + b''^{\mathrm{d}}_{\alpha,i} + b''_{\alpha,i}, \tag{D.28}$$

where $a''_{\alpha,i}$ and $b''_{\alpha,i}$ are the rates of change of the weights and weighted abscissas due to phase-space processes in the cell centered at x_i, which can readily be calculated from

[4] These quantities can be found from the moments $m_{k,i}(t)$ by, for example, using a moment-inversion algorithm described in Chapter 3.

Eq. (D.22) by replacing the source term $\overline{S}_k$ with its volume-average value, $\overline{S}_{k,i}$. The latter can easily be determined by evaluating the source term $\overline{S}_k$ with the weights $w''_{\alpha,i}$ and abscissas $\xi''_{\alpha,i} = \varsigma''_{\alpha,i}/w''_{\alpha,i}$. The advection terms $a''^{\mathrm{c}}_{\alpha,i}$ and $b''^{\mathrm{c}}_{\alpha,i}$ are calculated from Eq. (D.20) in the same manner by using a realizable finite-volume scheme to represent $\overline{C}_{k,i}$:

$$\overline{C}_{k,i}(t) = -\frac{u}{\Delta x}\left[m_{k,i+1/2}(t) - m_{k,i-1/2}(t)\right], \tag{D.29}$$

where $m_{k,i\pm 1/2}$ are the moments at the cell interfaces reconstructed from the volume-average moments $\{m_{k,i-1}(t), m_{k,i}(t), m_{k,i+1}(t)\}$. We refer the reader interested in the numerous implications of realizability of this reconstruction to Chapter 8. Analogously, the diffusive terms $a^{\mathrm{d}}_{\alpha,i}$ and $b^{\mathrm{d}}_{\alpha,i}$ are calculated in the same manner from Eq. (D.21) by using a realizable finite-volume scheme to represent $\overline{\mathcal{D}}_{k,i}$:

$$\overline{\mathcal{D}}_{k,i}(t) = \frac{\Gamma}{\Delta x^2}\left[m_{k,i+1}(t) - 2m_{k,i}(t) + m_{k,i-1}(t)\right], \tag{D.30}$$

where, consistently with the notation of Chapter 8, $m_{k,i+1}$, $m_{k,i}$, and $m_{k,i-1}$ are the volume-average moments in cells $i+1$, i, and $i-1$, respectively. Using DQMOM-FC, conservation of the moments is automatically respected for fixed t because the finite-volume schemes used to represent spatial transport are conservative.[5] However, the overall conservation properties of the numerical solutions for Eqs. (D.27) and (D.28) will depend on the time-integration scheme, as described next.

D.4 Time integration

Let us now discuss in detail the question of moment conservation during time integration. Consistently with Chapter 8, the source terms due to phase-space processes are set to zero so that only transport terms in real space are considered in this discussion. When Eq. (D.23) is integrated using an explicit Euler scheme, the volume-average moment of order k in the cell centered at x_i at time $(n+1)\Delta t$ is directly calculated from the volume-average moment of order k at time $n\,\Delta t$ from the following equation:

$$m^{n+1}_{k,i} = m^{n}_{k,i} + \Delta t\left(\overline{C}^{n}_{k,i} + \overline{\mathcal{D}}^{n}_{k,i}\right). \tag{D.31}$$

If a first-order upwind spatial-differencing scheme is used, the reconstructed moments $m_{k,i\pm 1/2}$ at the cell interfaces are equal to the upwind values and, in the case of positive u, the following equation is obtained:

$$m^{n+1}_{k,i} = m^{n}_{k,i-1}\left(\frac{\Delta t}{\Delta x}u + \frac{\Delta t}{\Delta x^2}\Gamma\right) + m^{n}_{k,i}\left(1 - \frac{\Delta t}{\Delta x}u - 2\frac{\Delta t}{\Delta x^2}\Gamma\right) + m^{n}_{k,i+1}\left(\frac{\Delta t}{\Delta x^2}\Gamma\right). \tag{D.32}$$

This method (i.e. the QMOM in the FVM) is always conservative, meaning that the following relations always hold (in the case of null source terms for phase-space processes):

$$\sum_i m^{n+1}_{k,i} = \sum_i m^{n}_{k,i} = \sum_i m^{n}_{k,i-1} = \sum_i m^{n}_{k,i+1},$$

where the two final equalities follow from the system being closed.

[5] In other words, the sum over all grid cells i of Eq. (D.23) yields zero contributions on the right-hand side due to advection (Eq. (D.29)) and diffusion (Eq. (D.30)) for a closed system.

With standard DQMOM, when the explicit Euler scheme in time and the first-order upwind differencing scheme for space are employed, the volume-average weights in the cell centered at x_i at time $(n+1)\Delta t$ are

$$w'^{n+1}_{\alpha,i} = w'^{n}_{k,i-1}\left(\frac{\Delta t}{\Delta x}u + \frac{\Delta t}{\Delta x^2}\Gamma\right) + w'^{n}_{\alpha,i}\left(1 - \frac{\Delta t}{\Delta x}u - 2\frac{\Delta t}{\Delta x^2}\Gamma\right) + w'^{n}_{\alpha,i+1}\left(\frac{\Delta t}{\Delta x^2}\Gamma\right) \tag{D.33}$$

with $\alpha = 1, \ldots, N$, whereas the volume-average weighted abscissas are

$$\varsigma'^{n+1}_{\alpha,i} = \varsigma'^{n}_{k,i-1}\left(\frac{\Delta t}{\Delta x}u + \frac{\Delta t}{\Delta x^2}\Gamma\right) + \varsigma'^{n}_{\alpha,i}\left(1 - \frac{\Delta t}{\Delta x}u - 2\frac{\Delta t}{\Delta x^2}\Gamma\right) + \varsigma'^{n}_{\alpha,i+1}\left(\frac{\Delta t}{\Delta x^2}\Gamma\right) \tag{D.34}$$

with $\alpha = 1, \ldots, N$. If, at time $n\,\Delta t$, the weights and weighted abscissas are exact, i.e. the relation

$$m^{n}_{k,i} = \sum_{\alpha=1}^{N} [w'^{n}_{\alpha,i}]^{1-k} [\varsigma'^{n}_{\alpha,i}]^{k} \tag{D.35}$$

is valid for $k = 0, \ldots, 2N-1$, then the moments calculated at time $(n+1)\Delta t$ by directly solving Eq. (D.32) (i.e. using the QMOM) might not be identical to the moments calculated from the weights and weighted abscissas determined from Eqs. (D.33) and (D.34) (i.e. using the DQMOM). In fact, it is easy to show that the moments calculated from Eq. (D.32), $m^{n+1}_{k,i}$, and those calculated from $w'^{n+1}_{\alpha,i}$ and $\varsigma'^{n+1}_{\alpha,i}$,

$$m'^{n+1}_{k,i} = \sum_{\alpha=1}^{N} [w'^{n+1}_{\alpha,i}]^{1-k} [\varsigma'^{n+1}_{\alpha,i}]^{k}, \tag{D.36}$$

are identical only for $k = 0$ and $k = 1$, whereas for $k \geq 2$ the two solutions will lead to different $m^{n+1}_{k,i}$ and $m'^{n+1}_{k,i}$ values. Moreover, on transporting moments with the QMOM in the FVM, time integration is always conservative. In contrast, on transporting weights and weighted abscissas with standard DQMOM in the FVM, time integration is conservative only for moments $k = 0$ and $k = 1$, or, in other words,

$$\sum_i m'^{n+1}_{k,i} = \sum_i m'^{n}_{k,i},$$

only for $k = 0$ and $k = 1$. It is interesting to estimate the conservation error committed at time $(n+1)\Delta t$ in calculating a moment with standard DQMOM in comparison with the QMOM, starting at time $n\,\Delta t$ with the same moment set. For example, in the case of no diffusion (i.e. $\Gamma = 0$), for $k = 2$ the error at time step n is given by

$$\mathrm{Err}^{n} \propto \sum_{\alpha=1}^{N} w'^{n}_{\alpha,i} w'^{n}_{\alpha,i-1} \left(\xi'^{n}_{\alpha,i} - \xi'^{n}_{\alpha,i-1}\right)^2 \left(\frac{u\,\Delta t}{\Delta x}\right)^2, \tag{D.37}$$

where $\xi'^{n}_{\alpha,i} = \varsigma'^{n}_{\alpha,i}/w'^{n}_{\alpha,i}$ and $\xi'^{n}_{\alpha,i-1} = \varsigma'^{n}_{\alpha,i-1}/w'^{n}_{\alpha,i-1}$. Equation (D.37) clearly shows that for a spatially inhomogeneous system the QMOM and standard DQMOM predictions are identical only if the abscissas are spatially uniform. If the abscissas are non-uniform (or, even worse, if the abscissas are discontinuous due, for example, to shocks) the error will be large. Note that the dimensionless quantity $u\,\Delta t/\Delta x$ is the CFL number, which is typically held constant during a simulation. Thus, reducing the time step will not lower the total error at the end of the simulation because more time steps will be required to complete the simulation and the total error is the sum $\sum_n \mathrm{Err}^n$.

When DQMOM-FC is used with an explicit Euler scheme for time, the equations

$$w''^{n+1}_{\alpha,i} = w''^{n}_{\alpha,i} + \Delta t \, A^{n}_{\alpha,i} \tag{D.38}$$

with $\alpha = 1, \ldots, N$ for the weights and

$$\varsigma''^{n+1}_{\alpha,i} = \varsigma''^{n}_{\alpha,i} + \Delta t \, B^{n}_{\alpha,i} \tag{D.39}$$

with $\alpha = 1, \ldots, N$ for the weighted abscissas are employed for time integration. For $k = 0, \ldots, 2N - 1$, the terms $A^{n}_{\alpha,i}$ and $B^{n}_{\alpha,i}$ are related to the moment source terms in Eq. (D.31) by the linear system

$$(1-k)\sum_{\alpha=1}^{N} [\xi''^{n}_{\alpha,i}]^{k} A^{n}_{\alpha,i} + k \sum_{\alpha=1}^{N} [\xi''^{n}_{\alpha,i}]^{k-1} B^{n}_{\alpha,i} = \overline{C}^{n}_{k,i} + \overline{\mathcal{D}}^{n}_{k,i}, \tag{D.40}$$

where $\xi''^{n}_{\alpha,i} = \varsigma''^{n}_{\alpha,i}/w''^{n}_{\alpha,i}$. On rewriting Eq. (D.38) to obtain $A^{n}_{\alpha,i}$ and Eq. (D.39) to obtain $B^{n}_{\alpha,i}$, substitution into Eq. (D.40) yields

$$\sum_{\alpha=1}^{N} w''^{n+1}_{\alpha,i} \left\{ (1-k)[\xi''^{n}_{\alpha,i}]^{k} + k[\xi''^{n}_{\alpha,i}]^{k-1} \xi''^{n+1}_{\alpha,i} \right\} = m''^{n}_{k,i} + \Delta t \left(\overline{C}^{n}_{k,i} + \overline{\mathcal{D}}^{n}_{k,i} \right). \tag{D.41}$$

If at time $n \, \Delta t$ the weights and weighted abscissas exactly represent the moment set,

$$m^{n}_{k,i} = m''^{n}_{k,i} = \sum_{\alpha=1}^{N} [w''^{n}_{\alpha,i}]^{1-k} [\varsigma''^{n}_{\alpha,i}]^{k}, \quad k = 0, \ldots, 2N-1, \tag{D.42}$$

i.e. the moment set for the QMOM is identical to that for DQMOM-FC, it is clear that Eq. (D.31) is consistent with Eq. (D.41) only for $k = 0$,

$$\sum_{\alpha=1}^{N} w''^{n+1}_{\alpha,i} = m''^{n+1}_{0,i} = m''^{n}_{0,i} + \Delta t \left(\overline{C}^{n}_{0,i} + \overline{\mathcal{D}}^{n}_{0,i} \right), \tag{D.43}$$

and $k = 1$,

$$\sum_{\alpha=1}^{N} w''^{n+1}_{\alpha,i} \xi''^{n+1}_{\alpha,i} = m''^{n+1}_{1,i} = m''^{n}_{1,i} + \Delta t \left(\overline{C}^{n}_{1,i} + \overline{\mathcal{D}}^{n}_{1,i} \right). \tag{D.44}$$

This means that calculating the evolution of the moments with Eq. (D.31) (i.e. using the QMOM) or through the weights and weighted abscissas with Eqs. (D.38) and (D.39) will result in the same moments only for $k = 0$ and for $k = 1$. In fact, after some tedious algebra and by expanding in powers of Δt, it is possible to show that

$$m''^{n+1}_{k,i} = m^{n+1}_{k,i} + k(k-1) \sum_{\alpha=1}^{N} w''^{n}_{\alpha,i} [\xi''^{n}_{\alpha,i}]^{k-2} \left(\frac{\xi''^{n}_{\alpha,i} A^{n}_{\alpha,i}}{w''^{n}_{\alpha,i}} - \frac{B^{n}_{\alpha,i}}{w''^{n}_{\alpha,i}} \right)^{2} \Delta t^{2} + O(\Delta t^{3}). \tag{D.45}$$

Thus, the explicit Euler scheme with DQMOM-FC has a conservation error of order Δt^{2} for moments of order greater than one when the weights and abscissas vary smoothly in time. Note that the coefficient for the error term has a form similar to the correction term in Eq. (D.5). Using the definitions of $A^{n}_{\alpha,i}$ and $B^{n}_{\alpha,i}$ in Eqs. (D.38) and (D.39), the coefficient in Eq. (D.45) can be rewritten as

$$m''^{n+1}_{k,i} = m^{n+1}_{k,i} + k(k-1) \sum_{\alpha=1}^{N} w''^{n+1}_{\alpha,i} \left(\frac{w''^{n+1}_{\alpha,i}}{w''^{n}_{\alpha,i}} \right) [\xi''^{n}_{\alpha,i}]^{k-2} \left(\xi''^{n+1}_{\alpha,i} - \xi''^{n}_{\alpha,i} \right)^{2} + O(\Delta t^{3}). \tag{D.46}$$

This form clearly shows that the moment-conservation error due to time integration in DQMOM-FC is not proportional (as in the DQMOM) to the abscissas' spatial gradients, but is instead proportional to $(\xi''^{n+1}_{\alpha,i} - \xi''^{n}_{\alpha,i})^2$, and thus that it will be proportional to Δt^2 only if the abscissas vary smoothly with time. For example, hyperbolic systems (for which the abscissas are discontinuous) and systems with small Γ (for which spatial gradients in the abscissas are very large) can have large moment-conservation errors due to advection. Another system that is problematic is one containing time-dependent void zones: i.e. $w^{n}_{\alpha,i} = 0$ but $w^{n+1}_{\alpha,i} > 0$. For all such systems, it is therefore preferable to work directly with the moment-transport equations given in Eq. (D.15).

References

AAMIR, E., NAGY, Z. K. & RIELLY, C. D. 2010 Optimal seed recipe design for crystal size distribution control for batch cooling crystallisation processes. *Chemical Engineering Science* **65**, 3602–3614.

AAMIR, E., NAGY, Z. K., RIELLY, C. D., KLEINERT, T. & JUDAT, B. 2009 Combined quadrature method of moments and method of characteristics approach for efficient solution of population balance models for dynamic modeling and crystal size distribution control of crystallization processes. *Industrial & Engineering Chemistry Research* **48**, 8575–8584.

AARONS, L., SUN, J. & SUNDARESAN, S. 2010 Unsteady shear of dense assemblies of cohesive granular materials under constant volume conditions. *Industrial & Engineering Chemistry Research* **49** (11), 5153–5165.

AIDUN, C. K. & CLAUSEN, J. R. 2010 Lattice-Boltzmann method for complex flows. *Annual Review of Fluid Mechanics* **42**, 439–472.

ALEXIADIS, A., VANNI, M. & GARDIN, P. 2004 Extension of the method of moments for population balances involving fractional moments and application to a typical agglomeration problem. *Journal of Colloid and Interface Science* **276**, 106–112.

ALEXOPOULOS, A. H. & KIPARISSIDES, C. 2007 Solution of the bivariate dynamic population balance equation in batch particulate systems: combined aggregation and breakage. *Chemical Engineering Science* **62**, 5048–5053.

ALOPAEUS, V., KOSKINEN, J., KESKINEN, K. I. & MAJANDER, J. 2002 Simulation of the population balances for liquid–liquid systems in a non ideal stirred tank. II. Parameter fitting and the use of the multiblock model for dense dispersions. *Chemical Engineering Science* **57**, 1815–1825.

ALOPAEUS, V., LAAKKONEN, M. & AITTAMAA, J. 2006 Numerical solution of moment-transformed population balance equation with fixed quadrature points. *Chemical Engineering Science* **61**, 4919–4929.

AMMAR, Y. & REEKS, M. 2009 Agglomeration of inertial particles in a random rotating symmetric straining flow. *International Journal of Multiphase Flow* **35**, 840–853.

ANDERSSON, R. & ANDERSSON, B. 2006 On the breakup of fluid particles in turbulent flows. *AIChE Journal* **52**, 2020–2030.

ANDRIES, P., AOKI, K. & PERTHAME, B. 2002a A consistent BGK-type model for gas mixtures. *Journal of Statistical Physics* **106**, 993–1018.

Andries, P., Bourgat, J.-F., le Tallec, P. & Perthame, B. 2002b Numerical comparison between the Boltzmann and ES-BGK models for rarefied gases. *Computer Methods in Applied Mechanics and Engineering* **191**, 3369–3390.

Arfken, J. B. 1985 *Mathematical Methods for Physicists*. New York (NY): Academic Press.

Aris, R. 1962 *Vectors, Tensors, and the Basic Equations of Fluid Mechanics*. Englewood Cliffs (NJ): Prentice-Hall.

Aristov, V. V. 2001 *Direct Methods for Solving the Boltzmann Equation and Study of Nonequilibrium Flows*. Dordrecht: Kluwer Academic Publishers.

Armenante, P. M. & Kirwan, D. J. 1989 Mass-transfer to microparticles in agitated systems. *Chemical Engineering Science* **44**, 2781–2796.

Asinari, P. 2010 Nonlinear Boltzmann equation for the homogeneous isotropic case: minimal deterministic Matlab program. *Computer Physics Communications* **181**, 1776–1788.

Attarakih, M. M., Drumm, C. & Bart, H.-J. 2009 Solution of the population balance equation using the sectional quadrature method of moments (SQMOM). *Chemical Engineering Science* **64**, 742–752.

Auton, T. R. 1987 The lift force on a spherical body in a rotational flow. *Journal of Fluid Mechanics* **183**, 199–218.

Auton, T. R., Hunt, J. C. R. & Prud'homme, M. 1988 The force exerted on a body in inviscid unsteady non-uniform rotational flow. *Journal of Fluid Mechanics* **197**, 241–257.

Ayazi Shamlou, P., Stravrinides, S., Titchener-Hooker, N. & Hoare, M. 1994 Growth-independent breakage frequency of protein precipitates in turbulently agitated bioreactors. *Chemical Engineering Science* **49**, 2647–2656.

Bakker, R. A. & van den Akker, H. E. A. 1994 A computational study of chemical reactors on the basis of micromixing models. *Chemical Engineering Research and Design* **72**, 733–738.

Balachandar, S. 2009 A scaling analysis for point-particle approaches to turbulent multiphase flows. *International Journal of Multiphase Flow* **35**, 801–810.

Balachandar, S. & Eaton, J. K. 2010 Turbulent dispersed multiphase flow. *Annual Review of Fluid Mechanics* **42**, 111–133.

Balmelli, M. & Steiner, L. 2000 Monte Carlo simulation of drop dispersion behaviour. *Chemical Engineering and Processing* **39**, 201–206.

Balthasar, M. & Kraft, M. 2003 A stochastic approach to calculate the particle size distribution function of soot particles in laminar premixed flames. *Combustion and Flame* **133**, 289–298.

Bardos, C., Golse, F. & Levermore, D. 1991 Fluid dynamic limits of kinetic equations. I. Formal derivations. *Journal of Statistical Physics* **63** (1–2), 323–344.

Barrett, J. C. & Jheeta, J. S. 1996 Improving the accuracy of the moments method for solving the aerosol dynamic equation. *Journal of Aerosol Science* **27**, 1135–1142.

Barton, E. 1995 Computation of particle tracks over a backward-facing step. *Journal of Aerosol Science* **26**, 887–901.

Basset, A. B. 1888 *Treatise on Hydrodynamics: With Numerous Examples*, vol. 2. Cambridge: Deighton, Bell and Company.

Batterham, R. J., Hall, J. S. & Barton, G. 1981 Pelletizing kinetics and simulation of full scale balling circuits, in *3rd International Symposium on Agglomeration*, Nuremberg, p. A136.

Beetstra, R., van der Hoef, M. & Kuipers, J. 2007 Drag force of intermediate Reynolds number flow past mono- and bidisperse arrays of spheres. *AIChE Journal* **53**, 489–501.

Bensaid, S., Marchisio, D. L. & Fino, D. 2010 Numerical simulation of soot filtration and combustion within diesel particulate filters. *Chemical Engineering Science* **65**, 357–363.

Bensaid, S., Marchisio, D. L., Fino, D., Saracco, G. & Specchia, V. 2009 Modeling of diesel particulate filtration in wall-flow traps. *Chemical Engineering Journal* **154**, 211–218.

Bensberg, A., Roth, P., Brink, R. & Lange, H. 1999 Modeling of particle evolution in aerosol reactors with coflowing gaseous reactants. *AIChE Journal* **45**, 2097–2107.

Bertola, F., Baldi, G., Marchisio, D. L. & Vanni, M. 2004 Momentum transfer in a swarm of bubbles: estimates from fluid-dynamic simulations. *Chemical Engineering Science* **59**, 5209–5215.

Bhatnagar, P. L., Gross, E. P. & Krook, M. 1954 A model for collisional processes in gases. I. Small amplitude processes in charged and neutral one-component systems. *Physical Reviews* **94**, 511–524.

Biesheuvel, A. & Spoelstra, S. 1989 The added mass coefficient of a dispersion of spherical gas bubbles in liquid. *International Journal of Multiphase Flow* **15**, 911–924.

Bird, G. A. 1963 Approach to translational equilibrium in a rigid sphere gas. *Physics of Fluids* **6**, 1518–1530.

Bird, G. A. 1994 *Molecular Gas Dynamics and the Direct Simulation of Gas Flows*. Oxford: Oxford University Press.

Bird, R. B., Stewart, W. E. & Lightfoot, E. N. 1960 *Transport Phenomena*. New York (NY): John Wiley and Sons.

Boltzmann, L. 1872 Weitere Studien über das Wärmegleichgewicht unter Gasmolekülen. *Sitzungsberichten der Akademie der Wissenschaften, Wien* **66**, 275–370.

Bouchut, F., Jin, S. & Li, X. T. 2003 Numerical approximations of pressureless gas and isothermal gas dynamics. *SIAM Journal of Numerical Analysis* **41**, 135–158.

Boussinesq, J. V. 1885 Sur la résistance qu'oppose un fluide indéfini au repos, sans pesanteur, au mouvement varié d'une sphère solide qu'il mouille sur toute sa surface, quand les vitesses restent bien continues et assez faibles pour que leurs carrés et produits soient négligeables. *Comptes Rendus de l'Académie des Sciences* **100**, 935–937.

Bove, S., Solberg, T. & Hjertager, B. H. 2005 A novel algorithm for solving population balance equations: the parallel parent and daughter classes. Derivation, analysis and testing. *Chemical Engineering Science* **60**, 1449–1464.

Braumann, A., Kraft, M. & Wagner, W. 2010 Numerical study of a stochastic particle algorithm solving a multidimensional population balance model for high shear granulation. *Journal of Computational Physics* **229**, 7672–7691.

Briesen, H. 2004 Simulation of habit transients by means of a reduced two-dimensional population balance model, in *2nd International Conference on Population Balance Modelling*, Valencia.

Buffo, A., Marchisio, D. L. & Vanni, M. 2012 Multidimensional population balance model for the simulation of turbulent gas–liquid systems in stirred tank reactors. *Chemical Engineering Science* **74**, 31–44.

Buffo, A., Vanni, M. & Marchisio, D. L. 2011 Comparison between different methods for the investigation of turbulent gas–liquid systems by using multivariate population balance models, in *8th International Conference on CFD in Oil and Gas, Metallurgical and Process Industries*, Trondheim.

Buffo, A., Vanni, M., Marchisio, D. L. & Fox, R. O. 2013 Multivariate quadrature-based moments methods for turbulent polydisperse gas–liquid systems. *International Journal of Multiphase Flow* (accepted for publication).

Buhrer-Skinner, M., Sundaresan, S., Menon, A. & Gordon, R. 2009 Drag law for bidisperse gas–solid suspensions containing equally sized spheres. *Industrial & Engineering Chemistry Research* **48** (1), 227–241.

Buyevich, Y. A. 1972 Statistical methods as applied to hydromechanics of concentrated disperse systems. *Zeitschrift für Angewandte Mathematik und Mechanik* **52** (10), T358–T363.

Cameron, I. T., Wang, F. Y., Immanuel, C. D. & Stepanek, F. 2005 Process systems modelling and applications in granulation: a review. *Chemical Engineering Science* **60**, 3723–3750.

Carnahan, N. F. & Starling, K. E. 1969 Equation of state for nonattracting rigid spheres. *Journal of Chemical Physics* **51**, 635–636.

Cartellier, A., Andreotti, M. & Sechet, P. 2009 Induced agitation in homogeneous bubbly flows at moderate particle Reynolds number. *Physical Review E* **80**, 065301(R).

Celnik, M., Patterson, R., Kraft, M. & Wagner, W. 2007 Coupling a stochastic soot population balance to gas-phase chemistry using operator splitting. *Combustion and Flame* **148**, 158–176.

Cercignani, C. 1975 *Theory and Application of the Boltzmann Equation*. London: Scottish Academic Press.

Cercignani, C. 1988 *The Boltzmann Equation and Its Applications*. New York (NY): Springer.

Cercignani, C. 2000 *Rarefied Gas Dynamics*. Cambridge: Cambridge University Press.

Cercignani, C., Illner, R. & Pulvirenti, M. 1994 *The Mathematical Theory of Dilute Gases*. Berlin: Springer.

Chakraborty, J. & Kumar, S. 2007 A new framework for solution of multidimensional population balance equations. *Chemical Engineering Science* **62**, 4112–4125.

Chalons, C., Fox, R. O. & Massot, M. 2010 A multi-Gaussian quadrature method of moments for gas–particle flows in a LES framework, in *Proceedings of the 2010 Summer Program*, Stanford (CA): Center for Turbulence Research, pp. 347–358.

Chalons, C., Kah, D. & Massot, M. 2012 Beyond pressureless gas dynamics: quadrature-based moment models. *Communications in Mathematical Sciences* (accepted for publication).

Chapman, S. 1916 On the law of distribution of molecular velocities, and on the theory of viscosity and thermal conduction, in a non-uniform simple monatomic gas. *Philosophical Transactions of the Royal Society A* **216**, 279–348.

Chapman, S. & Cowling, T. G. 1961 *The Mathematical Theory of Non-Uniform Gases*. Cambridge: Cambridge University Press.

Chen, G.-Q. & Liu, H. 2003 Formation of δ-shocks and vacuum states in the vanishing pressure limit of solutions to the Euler equations for isentropic fluids. *SIAM Journal of Mathematical Analysis* **34**, 925–938.

Chiney, A., Shivendra, S. C. & Kumar, S. 2010 Pivotwise expanding solution of bivariate population balance equations for composition dependent aggregative mixing, in *4th International Conference on Population Balance Modelling*, Berlin.

Chiu, T. & Christofides, P. D. 1999 Nonlinear control of particulate processes. *AIChE Journal* **45**, 1279–1297.

Clamen, A. & Gauvin, W. H. 1969 Effects of turbulence on the drag coefficients of spheres in a supercritical flow regime. *AIChE Journal* **15**, 184–189.

Clift, R., Grace, J. R. & Weber, M. E. 1978 *Bubbles, Drops, and Particles*. New York (NY): Academic Press.

Collins, S. B. & Knudsen, J. G. 1970 Drop size distribution produced by turbulent flow of immiscible liquids. *AIChE Journal* **16**, 1972–1980.

Cook, T. L. & Harlow, F. H. 1986 Vortices in bubbly two-phase flow. *International Journal of Multiphase Flow* **12**, 35–61.

Coulaloglou, C. A. & Tavlarides, L. L. 1977 Description of interaction processes in agitated liquid–liquid dispersions. *Chemical Engineering Science* **32**, 1289–1297.

Crowe, C., Sommerfeld, M. & Tsuji, Y. 1998 *Multiphase Flows with Droplets and Particles*. Boca Raton (FL): CRC Press.

Crowe, C. T. 2006 *Multiphase Flow Handbook*. Boca Raton (FL): CRC Press.

Crowe, C. T., Troutt, T. R. & Chung, J. N. 1996 Numerical models for two-phase turbulent flows. *Annual Review of Fluid Mechanics* **28**, 11–43.

Crowley, T., Meadows, E., Kostoulas, E. & Doyle III, F. J. 2000 Control of particle size distribution described by a population balance model of semibatch emulsion polymerization. *Journal of Process Control* **10**, 419–432.

Cunningham, E. 1910 On the velocity of steady fall of spherical particles through fluid medium. *Proceedings of the Royal Society* **83A**, 357–365.

Danckwerts, P. V. 1951 Significance of liquid-film coefficients in gas absorption. *Industrial & Engineering Chemistry Research* **43**, 1460–1467.

David, R., Villermaux, J., Marchal, P. & Klein, J. P. 1991 Crystallization and precipitation engineering – IV. Kinetic model for adipic acid crystallization. *Chemical Engineering Science* **46**, 1129–1136.

De Chaisemartin, S., Freret, L., Kah, D. *et al.* 2009 Eulerian models for turbulent spray combustion with polydispersity and droplet crossing. *Comptes Rendus Mécanique* **337**, 438–448.

Deen, N. G., Solberg, T. & Hjertager, B. H. 2001 Large eddy simulation of the gas–liquid flow in a square cross-sectioned bubble column. *Chemical Engineering Science* **56**, 6341–6349.

Delnoij, E., Lammers, F. A., Kuipers, J. A. M. & van Swaaij, W. P. M. 1997 Dynamic simulation of dispersed gas–liquid two-phase flow using a discrete bubble model. *Chemical Engineering Science* **52**, 1429–1458.

Derevich, I. V. & Zaichik, L. I. 1990 An equation for the probability density, velocity, and temperature of particles in a turbulent-flow modeled by a random Gaussian field. *PMM Journal of Applied Mathematics and Mechanics* **54** (5), 631–637.

Desjardins, O., Fox, R. O. & Villedieu, P. 2008 A quadrature-based moment method for dilute fluid-particle flows. *Journal of Computational Physics* **227**, 2524–2539.

Dette, H. & Studden, W. J. 1997 *The Theory of Canonical Moments with Applications in Statistics, Probability, and Analysis*. New York (NY): John Wiley and Sons.

Di Felice, R. 1994 The voidage function for fluid–particle interaction systems. *International Journal of Multiphase Flow* **20**, 153–159.

Diaz, M. E., Iranzo, A., Cuadra, D. *et al.* 2008 Numerical simulation of the gas–liquid flow in a laboratory scale bubble column. Influence of bubble size distribution and non-drag forces. *Chemical Engineering Journal* **139**, 363–379.

Diemer, R. B. & Ehrman, S. H. 2005 Pipeline agglomerator design as a model test case. *Powder Technology* **156**, 129–145.

Diemer, R. B. & Olson, J. H. 2002a A moment methodology for coagulation and breakage problems: part 1 – analytical solution of the steady-state population balance. *Chemical Engineering Science* **57**, 2193–2209.

DIEMER, R. B. & OLSON, J. H. 2002b A moment methodology for coagulation and breakage problems: part 2 – moment models and distribution reconstruction. *Chemical Engineering Science* **57**, 2211–2228.

DORGAN, A. J. & LOTH, E. 2007 Efficient calculation of the history force at finite Reynolds numbers. *International Journal of Multiphase Flow* **33**, 833–848.

DREW, D. A. 1971 Averaged equations for two-phase flow. *Studies in Applied Mathematics* **1**, 205–231.

DREW, D. A., CHENG, L. & LAHEY, R. T. 1979 The analysis of virtual mass effects in two-phase flow. *International Journal of Multiphase Flow* **5**, 233–242.

DREW, D. A. & LAHEY, R. T. 1993 *Particulate Two-Phase Flow*. Boston (MA): Butterworth-Heinemann.

DREW, D. A. & PASSMAN, S. L. 1999 *Theory of Multicomponent Fluids*. New York (NY): Springer.

DRUZHININ, O. A. 1987 On the two-way interaction in two-dimensional particle-laden flows: the accumulation of particles and flow modification. *Journal of Fluid Mechanics* **297**, 49–76.

ELIMELECH, M., GREGORY, J., JIA, X. & WILLIAMS, R. 1998 *Particle Deposition and Aggregation: Measurement, Modelling and Simulation*. Woburn (MA): Butterworth-Heinemann.

ENKSOG, D. 1921 Die numerische Berechnung der Vorgänge in mäßig verdünnten Gasen. *Arkiv för Matematik, Astronomi och Fysik* **16**, 1–52.

ETHAYARAJA, M., DUTTA, K. & BANDYOPADHYAYA, R. 2006 Mechanism of nanoparticle formation in self-assembled colloidal templates: population balance model and Monte Carlo simulation. *Journal of Physical Chemistry B* **110**, 16471–16481.

FAN, R., MARCHISIO, D. L. & FOX, R. O. 2004 Application of the direct quadrature method of moments to polydisperse gas–solid fluidized beds. *Powder Technology* **139**, 7–20.

FAVERO, J. L. & LAGE, P. L. C. 2012 The dual-quadrature method of generalized moments using automatic integration packages. *Computers & Chemical Engineering* **38**, 1–10.

FERRY, J. & BALACHANDAR, S. 2001 A fast Eulerian method for two-phase flow. *International Journal of Multiphase Flow* **27**, 1199–1226.

FERZIGER, J. H. & KAPER, G. H. 1972 *Mathematical Theory of Transport Processes in Gases*. Amsterdam: North Holland.

FILBET, F. & RUSSO, G. 2003 High order numerical methods for the space non-homogeneous Boltzmann equation. *Journal of Computational Physics* **186**, 457–480.

FOX, R. O. 2003 *Computational Models for Turbulent Reacting Flows*. Cambridge: Cambridge University Press.

FOX, R. O. 2007 Introduction and fundamentals of modeling approaches for polydisperse multiphase flows, in *Computational Models for Turbulent Multiphase Reacting Flows*, Vienna: Springer, pp. 1–40.

FOX, R. O. 2008 A quadrature-based third order moment method for dilute gas–particle flows. *Journal of Computational Physics* **227**, 6313–6350.

FOX, R. O. 2009a Higher-order quadrature-based moment methods for kinetic equations. *Journal of Computational Physics* **228**, 7771–7791.

FOX, R. O. 2009b Optimal moment sets for multivariate direct quadrature method of moments. *Industrial Engineering & Chemistry Research* **48**, 9686–9696.

FOX, R. O. 2012 Large-eddy-simulation tools for multiphase flows. *Annual Review of Fluid Mechanics* **44**, 47–76.

FOX, R. O., LAURENT, F. & MASSOT, M. 2008 Numerical simulation of spray coalescence in an Eulerian framework: direct quadrature method of moments and multi-fluid method. *Journal of Computational Physics* **227**, 3058–3088.

FOX, R. O. & VEDULA, P. 2010 Quadrature-based moment model for moderately dense polydisperse gas-particle flows. *Industrial & Engineering Chemistry Research* **49**, 5174–5187.

FRENKLACH, M. 2002 Method of moments with interpolative closure. *Chemical Engineering Science* **57**, 2229–2239.

FRENKLACH, M. & HARRIS, S. J. 1987 Aerosol dynamics modeling using the method of moments. *Journal of Colloid and Interface Science* **118**, 252–261.

FRERET, L., LAURENT, F., DE CHAISEMARTIN, S. *et al.* 2008 Turbulent combustion of polydisperse evaporating sprays with droplet crossing: Eulerian modeling of collisions at finite Knudsen and validation, in *Proceedings of the 2008 CTR Summer Program*, Stanford (CA): Center for Turbulence Research, pp. 277–288.

FRIEDLANDER, S. K. 2000 *Smoke, Dust, and Haze: Fundamentals of Aerosol Dynamics*. Oxford: Oxford University Press.

FRIEDLANDER, S. K. & WANG, C. S. 1966 The self-preserving particle size distribution for coagulation by Brownian motion. *Journal of Colloid and Interface Science* **22**, 126–132.

FRIESEN, W. I. & DABROS, T. 2003 Constant-number Monte Carlo simulation of aggregating and fragmenting particles. *Journal of Chemical Physics* **119**, 2825–2839.

FUCHS, N. 1964 *Mechanics of Aerosols*. New York (NY): Pergamon.

GANTT, J. A. & GATZKE, E. P. 2006 A stochastic technique for multidimensional granulation modeling. *AIChE Journal* **52**, 3067–3077.

GARDINER, C. W. 2004 *Handbook of Stochastic Methods for Physics, Chemistry and the Natural Sciences*, 3rd edn., Berlin: Springer.

GARSIDE, J. & AL-DIBOUNI, M. R. 1977 Velocity–voidage relationships for fluidization and sedimentation. *Industrial and Engineering Chemistry Process Design and Development* **16**, 206–214.

GARSIDE, J., JANSSEN VAN ROSMALEN, R. & BENNEMA, P. 1975 Verification of crystal-growth rate equations. *Journal of Crystal Growth* **29**, 353–366.

Garzó, V., Dufty, J. W. & Hrenya, C. M. 2007 Enskog theory for polydisperse granular mixtures. I. Navier–Stokes order transport. *Physical Review E* **76**, 031303.

Garzó, V., Santos, A. & Brey, J. J. 1989 A kinetic model for a multicomponent gas. *Physics of Fluids A* **1**, 380–383.

Gautschi, W. 1997 *Numerical Analysis: An Introduction*. Boston (MA): Birkhäuser.

Gautschi, W. 2004 *Orthogonal Polynomials: Computation and Approximation*. Oxford: Oxford University Press.

Gavi, E., Marchisio, D. L. & Barresi, A. A. 2008 On the importance of mixing for the production of nanoparticles. *Journal of Dispersion Science and Technology* **29**, 548–554.

Gavi, E., Marchisio, D. L., Barresi, A. A., Olsen, M. G. & Fox, R. O. 2010 Turbulent precipitation in micromixers: CFD simulation and flow field validation. *Chemical Engineering Research and Design* **99**, 1182–1193.

Gavi, E., Rivautella, L., Marchisio, D. L. *et al.* 2007 CFD modelling of nano-particle precipitation in confined impinging jet reactors. *Chemical Engineering Research and Design* **85**, 735–744.

Gelbard, F. & Seinfeld, J. H. 1978 Numerical solution of the dynamic equation for particulate systems. *Journal of Computational Physics* **28**, 357–375.

Gerber, A. G. & Mousavi, A. 2007 Application of quadrature method of moments to the polydispersed droplet spectrum in transonic steam flows with primary and secondary nucleation. *Applied Mathematical Modelling* **31**, 1518–1533.

Ghadiri, M. & Zhang, Z. 2002 Impact attrition of particulate solids: part 1. A theoretical model of chipping. *Chemical Engineering Science* **57**, 3659–3669.

Ghalehchian, J. S. 2002 Prediction of the hydrodynamics of rotating disc contactors based on a new Monte-Carlo simulation method for drop breakage. *Journal of Chemical Engineering of Japan* **35**, 604–612.

Gidaspow, D. 1994 *Multiphase Flow and Fluidization: Continuum and Kinetic Theory Descriptions with Applications*. San Diego (CA): Academic Press.

Gimbun, J., Nagy, Z. K. & Rielly, C. D. 2009 Simultaneous quadrature method of moments for the solution of population balance equations, using a differential algebraic equation framework. *Industrial & Engineering Chemistry Research* **48**, 7798–7812.

Girolami, M. W. & Rousseau, R. W. 1985 Size-dependent crystal growth: a manifestation of growth rate dispersion in the potassium alum–water system. *AIChE Journal* **31**, 1821–1828.

Godin, F. B., Cooper, D. G. & Rey, A. D. 1999 Numerical methods for a population balance model of a periodic fermentation process. *AIChE Journal* **45**, 1359–1364.

Goldhirsch, I. 2003 Rapid granular flows. *Annual Review of Fluid Mechanics* **35**, 267–293.

Goldman, E. & Sirovich, L. 1967 Equations for gas mixtures. *The Physics of Fluids* **10** (9), 1928–1940.

Golub, G. H. & Welsch, J. H. 1969 Calculation of Gauss quadrature rules. *Mathematics of Computation* **23**, 221–230.

Goodson, M. & Kraft, M. 2004 Simulation of coalescence and breakage: an assessment of two stochastic methods suitable for simulating liquid–liquid extraction. *Chemical Engineering Science* **59**, 3865–3881.

Gordon, R. G. 1968 Error bounds in equilibrium statistical mechanics. *Journal of Mathematical Physics* **9**, 655–663.

Gorsch, D. 1999 Discrete model collision operators of Boltzmann type. *Journal of Computational and Applied Mathematics* **104**, 145–162.

Grad, H. 1949a Note on n-dimensional Hermite polynomials. *Communications on Pure and Applied Mathematics* **2**, 325–330.

Grad, H. 1949b On the kinetic theory of rarefied gases. *Communications on Pure and Applied Mathematics* **2**, 331–407.

Grad, H. 1958 Principles of kinetic theory of gases, in *Handbuch der Physik XII: Thermodynamik der Gase*, ed. S. Flugge, Berlin: Springer.

Greenberg, J. B., Albagli, D. & Tambour, Y. 1986 An opposed jet quasi-monodisperse spray diffusion flame. *Combustion Science and Technology* **50**, 255–270.

Greenberg, J. B., Silverman, I. & Tambour, Y. 1993 On the origin of spray sectional conservation equations. *Combustion and Flame* **93**, 90–96.

Grosch, R., Briesen, H., Marquardt, W. & Wulkow, M. 2007 Generalization and numerical investigation of QMOM. *AIChE Journal* **53**, 207–227.

Gross, E. P. & Krook, M. 1956 Model for collision processes in gases: small-amplitude oscillations of charged two-component systems. *Physical Review* **102**, 593–604.

Gunawan, R., Fusman, I. & Braatz, R. 2004 High resolution algorithms for multidimensional population balance equations. *AIChE Journal* **50**, 2738–2749.

Hadamard, J. S. 1911 Mouvement permanent lent d'une sphère liquide et visqueuse dans un liquide visqueux. *Comptes Rendus de l'Académie des Sciences* **152**, 1735–1738.

Haseltine, E. L., Patience, D. B. & Rawlings, J. B. 2005 On the stochastic simulation of particulate systems. *Chemical Engineering Science* **60**, 2627–2641.

Heineken, W., Flockerzi, D., Voigt, A. & Sundmacher, K. 2011 Dimension reduction of bivariate population balances using the quadrature method of moments. *Computers and Chemical Engineering* **35**, 50–62.

Higbie, R. 1935 The rate of absorption of a pure gas into a still liquid during short periods of exposure. *Transactions of the American Institute of Chemical Engineering* **31**, 364–389.

Hill, P. J. & Ng, K. M. 1996 Statistics of multiple particle breakage. *AIChE Journal* **42**, 1600–1611.

van Hinsberg, M. A. T., ten Thije Boonkkamp, J. H. M. & Clercx, H. J. H. 2011 An efficient, second order method for the approximation of the Basset history force. *Journal of Computational Physics* **230**, 1465–1478.

van der Hoef, M. A., van Sint Annaland, M., Deen, N. G. & Kuipers, J. A. M. 2008 Numerical simulation of dense gas–solid fluidized beds: a multiscale modeling strategy. *Annual Review of Fluid Mechanics* **40**, 47–70.

Holloway, W., Yin, X. & Sundaresan, S. 2010 Fluid–particle drag in inertial polydisperse gas–solid suspensions. *AIChE Journal* **56** (8), 1995–2004.

Holway, L. H. 1966 New statistical models for kinetic theory. Methods of construction. *Physics of Fluids* **9** (9), 1658–1673.

Holway, L. H. 1973 Construction of linearized kinetic models for gaseous mixtures and molecular gases. *Physics of Fluids* **16**, 193–215.

Homsy, G. M., El-Kaissy, M. M. & Didwania, A. 1980 Instability waves and the origin of bubbles in fluidized beds – II. Comparison with theory. *International Journal of Multiphase Flow* **6**, 305–318.

Hounslow, M. J., Ryall, R. L. & Marshall, V. R. 1988 A discretized population balance for nucleation, growth, and aggregation. *AIChE Journal* **34**, 1821–1832.

Hsia, M. A. & Tavlarides, L. L. 1983 Simulation analysis of drop breakage coalescence and micromixing in liquid–liquid stirred tanks. *Chemical Engineering Journal and the Biochemical Engineering Journal* **26**, 189–199.

Hsu, J. P. & Tsao, H. K. 1992 Dynamic simulation of floc breakage with a random breakage distribution. *Colloids and Surfaces* **62**, 23–30.

Hulburt, H. M. & Katz, S. 1964 Some problems in particle technology. A statistical mechanical formulation. *Chemical Engineering Science* **19**, 555–574.

Icardi, M., Asinari, P., Marchisio, D. L., Izquierdo, S. & Fox, R. O. 2012 Quadrature-based moment closures for non-equilibrium flows: hard-sphere collisions and approach to equilibrium. *Journal of Computational Physics* **231**, 7431–7449.

Immanuel, C. D., Cordeiro, C. F., Sundaram, S. S. *et al.* 2002 Modeling of particle size distribution in emulsion co-polymerization: comparison with experimental data and parametric sensitivity studies. *Computers and Chemical Engineering* **26**, 1133–1152.

Immanuel, C. D. & Doyle, F. J. 2003 Computationally efficient solution of population balance models incorporating nucleation, growth and coagulation: application to emulsion polymerization. *Chemical Engineering Science* **58**, 3681–3698.

Immanuel, C. D. & Doyle, F. J. 2005 Solution technique for a multi-dimensional population balance model describing granulation processes. *Powder Technology* **156**, 213–225.

Irizarry, R. 2008a Fast Monte Carlo methodology for multivariate particulate systems I: point ensemble Monte Carlo. *Chemical Engineering Science* **63**, 95–110.

Irizarry, R. 2008b Fast Monte Carlo methodology for multivariate particulate systems – II: tau-PEMC. *Chemical Engineering Science* **63**, 111–121.

Ishii, M. 1975 *Thermo-Fluid Model for Two-Phase Flow*. Paris: Eyrolles.

Israelachvili, J. 1992 *Intermolecular and Surface Forces*. London: Academic Press.

Jabin, P.-E. 2002 Various levels of models for aerosols. *Mathematical Models and Methods in Applied Science* **12**, 903–919.

Janssen Van Rosmalen, R., Bennema, P. & Garside, J. 1975 Influence of volume diffusion on crystal-growth. *Journal of Crystal Growth* **29**, 342–352.

Jenkins, J. T. & Mancini, F. 1989 Kinetic theory for binary mixtures of smooth, nearly elastic spheres. *Physics of Fluids A* **1**, 2050–2057.

Jenkins, J. T. & Richman, M. W. 1985 Grad's 13-moment system for a dense gas of inelastic spheres. *Archives of Rational Mechanics and Analysis* **87**, 355–377.

Jiang, X., Siamas, G. A., Jagus, K. & Karayiannis, T. G. 2010 Physical modelling and advanced simulations of gas–liquid two-phase jet flows in atomization and sprays. *Progress in Energy and Combustion Science* **36** (2), 131–167.

Kah, D., Laurent, F., Freret, L. *et al.* 2010 Eulerian quadrature-based moment models for dilute polydisperse evaporating sprays. *Flow, Turbulence, and Combustion* **85**, 649–676.

Kah, D., Laurent, F., Massot, M. & Jay, S. 2012 A high order moment method simulating evaporation and advection of a polydisperse liquid. *Journal of Computational Physics* **231** (2), 394–422.

Kalani, A. & Christofides, P. D. 2002 Simulation, estimation and control of size distribution in aerosol processes with simultaneous reaction, nucleation, condensation and coagulation. *Computers and Chemical Engineering* **26**, 1153–1169.

Kazakov, A. & Frenklach, M. 1998 Dynamic modelling of soot particle coagulation and aggregation: implementation with the method of moments and application to high-pressure laminar premixed flames. *Combustion and Flame* **114**, 484–501.

Kendal, D. G. 1950 An artificial realization of a simple birth-and-death process. *Journal of the Royal Statistical Society: Series B* **12**, 116–119.

Khalili, S., Lin, Y., Armaou, A. & Matsoukas, T. 2010 Constant number Monte Carlo simulation of population balances with multiple growth mechanisms. *AIChE Journal* **56**, 3137–3145.

Knudsen, M. & Weber, S. 1911 Luftwiderstand gegen die langsame Bewegung kleiner Kugeln. *Annals of Physics* **36**, 981–994.

Konno, M., Aoki, M. & Saito, S. 1983 Scale effect on breakup process in liquid–liquid agitated tanks. *Journal of Chemical Engineering of Japan* **16**, 312–319.

Kostoglou, M. & Karabelas, A. J. 2004 Optimal low order methods of moments for solving the fragmentation equation. *Powder Technology* **143–144**, 280–290.

Kosuge, S. 2009 Model Boltzmann equation for gas mixtures: construction and numerical comparison. *European Journal of Mechanics B/Fluids* **28**, 170–184.

Kramer, T. A. & Clark, M. M. 1999 Incorporation of aggregate breakup in the simulation of orthokinetic coagulation. *Journal of Colloid and Interface Science* **216**, 116–126.

Kruis, F. E., Kusters, K. A., Pratsinis, S. E. & Scarlett, B. 1993 A simple model for the evolution of the characteristics of aggregate particles undergoing coagulation and sintering. *Aerosol Science and Technology* **19**, 514–526.

Kruis, F. E., Maisels, A. & Fissan, H. 2000 Direct simulation Monte Carlo method for particle coagulation and aggregation. *AIChE Journal* **46**, 1735–1742.

Kumar, J., Peglow, M., Warnecke, G. & Heinrich, S. 2007 An efficient numerical technique for solving population balance equation involving aggregation, breakage, growth and nucleation. *Powder Technology* **179**, 205–228.

Kumar, J., Peglow, M., Warnecke, G. & Heinrich, S. 2008a The cell average technique for solving multi-dimensional aggregation population balance equations. *Computers and Chemical Engineering* **32**, 1810–1830.

Kumar, J., Peglow, M., Warnecke, G. & Heinrich, S. 2008b An efficient numerical technique for solving population balance equation involving aggregation, breakage, growth and nucleation. *Powder Technology* **182**, 81–104.

Kumar, J., Peglow, M., Warnecke, G., Heinrich, S. & Morl, J. 2006 Improved accuracy and convergence of discretized population balance for aggregation: the cell average technique. *Chemical Engineering Science* **61**, 3327–3342.

Kumar, J., Warnecke, G., Peglow, M. & Heinrich, S. 2009 Comparison of numerical methods for solving population balance equations incorporating aggregation and breakage. *Powder Technology* **189**, 218–229.

Kumar, S. & Ramkrishna, D. 1996a On the solution of population balance equations by discretization – I. A fixed pivot technique. *Chemical Engineering Science* **51**, 1311–1332.

Kumar, S. & Ramkrishna, D. 1996b On the solution of population balance equations by discretization – II. A moving pivot technique. *Chemical Engineering Science* **51**, 1333–1342.

Kumar, S. & Ramkrishna, D. 1997 On the solution of population balance equations by discretization – III. Nucleation, growth and aggregation of particles. *Chemical Engineering Science* **52**, 4659–4679.

Kuo, K. K. 1986 *Principles of Combustion*. Hoboken (NJ): John Wiley and Sons.

Laakkonen, M., Alopaeus, V. & Aittamaa, J. 2006 Validation of bubble breakage, coalescence and mass transfer models for gas–liquid dispersion in agitated vessel. *Chemical Engineering Science* **61**, 218–228.

Laakkonen, M., Moilanen, P., Alopaeus, V. & Aittamaa, J. 2007 Modelling local bubble size distribution in agitated vessels. *Chemical Engineering Science* **62**, 721–740.

Ladd, A. C. J. 1993 Dynamical simulations of sedimenting spheres. *Physics of Fluids A* **5**, 299–310.

LAGE, P. L. C. 2011 On the representation of QMOM as a weighted-residual method – the dual-quadrature method of generalized moments. *Computers & Chemical Engineering* **35** (11), 2186–2203.

LAIN, S., BROEDER, D., SOMMERFELD, M. & GOEZ, M. F. 2002 Modelling hydrodynamics and turbulence in a bubble column using the Euler–Lagrange procedure. *International Journal of Multiphase Flow* **28**, 1381–1407.

LALLEMAND, P. & LUO, L.-S. 2003 Theory of the lattice Boltzmann method: dispersion, dissipation, isotropy, Galilean invariance, and stability. *Physical Review E* **68**, 036706.

LAMONT, J. C. & SCOTT, D. S. 1970 An eddy cell model of mass transfer into the surface of a turbulent liquid. *AIChE Journal* **16**, 513–519.

LARSON, M. A., WHITE, E. T., RAMANARAYANAN, K. A. & BERGLUND, K. A. 1985 Growth rate dispersion in MSMPR crystallizers. *AIChE Journal* **31**, 90–94.

LASHERAS, J. C., EASTWOOD, C., MARTINEZ-BAZÁN, C. & MONTANES, G. L. 2002 A review of statistical models for the break-up of an immiscible fluid immersed into a fully developed turbulent flow. *International Journal of Multiphase Flow* **28**, 247–278.

LATTUADA, M., WU, H., SANDKUHLER, P., SEFCIK, J. & MORBIDELLI, M. 2004 Modelling of aggregation kinetics of colloidal systems and its validation by light scattering measurements. *Chemical Engineering Science* **59**, 1783–1798.

LAURENT, F. 2006 Numerical analysis of Eulerian multi-fluid models in the context of kinetic formulations for dilute evaporating sprays. *Mathematical Modelling and Numerical Analysis* **40**, 431–468.

LAURENT, F. & MASSOT, M. 2001 Multi-fluid modelling of laminar polydisperse spray flames: origin, assumptions and comparison of sectional and sampling methods. *Combustion Theory and Modelling* **5**, 537–572.

LAURENT, F., MASSOT, M. & VILLEDIEU, P. 2004 Eulerian multi-fluid modeling for the numerical simulation of coalescence in polydisperse dense liquid sprays. *Journal of Computational Physics* **194**, 505–543.

LAURENZI, I. J., BARTELS, J. D. & DIAMOND, S. L. 2002 A general algorithm for exact simulation of multicomponent aggregation process. *Journal of Computational Physics* **177**, 418–449.

LAURENZI, I. J. & DIAMOND, S. L. 1999 Monte Carlo simulation of the heterotypic aggregation kinetics of platelets and neutrophils. *Biophysical Journal* **77**, 1733–1746.

LEE, C. H., ERICKSON, L. E. & GLASGOW, L. A. 1987 Bubble breakup and coalescence in turbulent gas–liquid dispersions. *Chemical Engineering Communications* **59**, 65–84.

LEE, K., KIM, T., RAJNIAK, P. & MATSOUKAS, T. 2008 Compositional distributions in multicomponent aggregation. *Chemical Engineering Science* **63**, 1293–1303.

LEE, K. & MATSOUKAS, T. 2000 Simultaneous coagulation and break-up using constant-N Monte Carlo. *Powder Technology* **110**, 82–89.

Lee, K., Thring, M. W. & Beer, J. 1962 On the rate of combustion of soot in a laminar soot flame. *Combustion and Flame* **6**, 137–145.

Lee, K. W. 1983 Change of particle size distribution during Brownian coagulation. *Journal of Colloid and Interface Science* **92**, 315–325.

Lee, W. H. 1979 A pressure iteration scheme for two-phase modeling, Los Alamos (NM): Los Alamos Scientific Laboratory, Technical Report LA-UR 79-975.

Leveque, R. 2002 *Finite Volume Methods for Hyperbolic Problems*. Cambridge: Cambridge University Press.

Li, T. S., Rohl, A. L. & Ilievski, D. 2000 Modelling non-stationary precipitation systems: sources of error and their propagation. *Chemical Engineering Science* **55**, 6037–6047.

Li-Shi, L. 2000 Some recent results on discrete velocity models and ramifications for lattice Boltzmann equation. *Computer Physics Communications* **129**, 63–74.

Liffman, K. 1992 A direct simulation Monte-Carlo method for cluster coagulation. *Journal of Computational Physics* **100**, 116–127.

Lin, Y., Lee, K. & Matsoukas, T. 2002 Solution of the population balance equation using constant-number Monte Carlo. *Chemical Engineering Science* **57**, 2241–2252.

Lince, F., Marchisio, D. L. & Barresi, A. A. 2008 Strategies to control the particle size distribution of poly-e-caprolactone nanoparticles for pharmaceutical applications. *Journal of Colloid and Interface Science* **332**, 505–515.

Litster, J., Smit, D. J. & Hounslow, M. J. 1995 Adjustable discretized population balance for growth and aggregation. *AIChE Journal* **41**, 591–602.

Lochiel, A. C. & Calderbank, P. H. 1964 Mass transfer in the continuous phase around axisymmetric bodies of revolution. *Chemical Engineering Science* **19**, 471–484.

Luo, H. & Svendsen, F. 1996 Theoretical model for drop and bubble break-up in turbulent dispersions. *AIChE Journal* **42**, 1225–1233.

Ma, C. Y., Wang, X. Z. & Roberts, K. J. 2007 Multi-dimensional population balance modeling of the growth of rod-like L-glutamic acid crystals using growth rates estimated from in-process imaging. *Advanced Powder Technology* **18**, 707–723.

Ma, H., Pedel, J., Fife, P. & Johnson, W. P. 2009 Hemispheres-in-cell geometry to predict colloid deposition in porous media. *Environmental Science and Technology* **43**, 8573–8579.

Mahoney, A. W. & Ramkrishna, D. 2002 Efficient solution of population balance equations with discontinuities by finite elements. *Chemical Engineering Science* **57**, 1107–1119.

Maisels, A., Kruis, F. E. & Fissan, H. 2004a Coagulation in bipolar aerosol chargers. *Journal of Aerosol Science* **35**, 1333–1345.

Maisels, A., Kruis, F. E. & Fissan, H. 2004b Direct simulation Monte Carlo for simultaneous nucleation, coagulation, and surface growth in dispersed systems. *Chemical Engineering Science* **59**, 2231–2239.

Marchal, P., David, R., Klein, J. P. & Villermaux, J. 1988 Crystallization and precipitation engineering – I. An efficient method for solving population balances in crystallization with agglomeration. *Chemical Engineering Science* **43**, 59–67.

Marchisio, D. L. 2009 On the use of bi-variate population balance equations for modelling barium titanate nanoparticle precipitation. *Chemical Engineering Science* **64**, 697–708.

Marchisio, D. L. & Barresi, A. A. 2009 Investigation of soot formation in turbulent flames with a pseudo-bivariate population balance model. *Chemical Engineering Science* **64**, 294–303.

Marchisio, D. L., Barresi, A. A. & Fox, R. O. 2001a Simulation of turbulent precipitation in a semi-batch Taylor–Couette reactor using CFD. *AIChE Journal* **47**, 664–676.

Marchisio, D. L., Barresi, A. A. & Garbero, M. 2002 Nucleation, growth, and agglomeration in barium sulfate turbulent precipitation. *AIChE Journal* **48**, 2039–2050.

Marchisio, D. L. & Fox, R. O. 2005 Solution of population balance equations using the direct quadrature method of moments. *Journal of Aerosol Science* **36**, 43–73.

Marchisio, D. L. & Fox, R. O. 2007 *Multiphase Reacting Flows: Modelling and Simulation*. Udine: Springer.

Marchisio, D. L., Fox, R. O. & Barresi, A. A. 2001b On the simulation of turbulent precipitation in a tubular reactor via computational fluid dynamics. *Transactions of the Institution of Chemical Engineers* **79**, 998–1004.

Marchisio, D. L., Fox, R. O., Barresi, A. A. & Baldi, G. 2001c On the comparison between presumed and full PDF methods for turbulent precipitation. *Industrial & Engineering Chemistry Research* **40**, 5132–5139.

Marchisio, D. L., Fox, R. O., Cheng, T.-C., Stroefer, W. & Woelfert, A. 2003a CFD study of fine particle nucleation, growth, and agglomeration, in *11th European Conference on Mixing*, Bamberg, pp. 371–378.

Marchisio, D. L., Pikturna, J. T., Fox, R. O., Vigil, R. D. & Barresi, A. A. 2003b Quadrature method of moments for population-balance equations. *AIChE Journal* **49**, 1266–1276.

Marchisio, D. L., Soos, M., Sefcik, J. & Morbidelli, M. 2006 Role of turbulent shear rate distribution in aggregation and breakage processes. *AIChE Journal* **52**, 158–173.

Marchisio, D. L., Vigil, R. D. & Fox, R. O. 2003c Quadrature method of moments for aggregation–breakage processes. *Journal of Colloid and Interface Science* **258**, 322–334.

Marshall, C. L., Rajniak, P. & Matsoukas, T. 2011 Numerical simulations of two-component granulation: comparison of three methods. *Chemical Engineering Research and Design* **89**, 545–552.

Martinez-Bazán, C., Montanes, J. L. & Lasheras, J. C. 1999 On the breakup of an air bubble injected into a fully developed turbulent flow: part 2. Size PDF of the resulting daughter bubbles. *Journal of Fluid Mechanics* **401**, 183–207.

Mashayek, F. & Pandya, R. V. R. 2003 Analytical description of particle/droplet-laden turbulent flows. *Progress in Energy and Combustion Science* **29** (4), 329–378.

Massot, M. 2007 Eulerian multi-fluid models for polydisperse evaporating sprays, in *Computational Models for Turbulent Multiphase Reacting Flows*, Vienna: Springer, pp. 79–123.

Massot, M., Laurent, F., Kah, D. & de Chaisemartin, S. 2010 A robust moment method for evaluation of the disappearance rate of evaporating sprays. *SIAM Journal of Applied Mathematics* **70** (8), 3203–3234.

Matsoukas, T., Kim, T. & Lee, K. 2009 Bicomponent aggregation with composition-dependent rates and the approach to well-mixed state. *Chemical Engineering Science* **64**, 787–799.

Matsoukas, T., Lee, K. & Kim, T. 2006 Mixing of components in two-component aggregation. *AIChE Journal* **52**, 3088–3099.

Maxey, M. R. 1987 The gravitational settling of aerosol particles in homogeneous turbulence and random flow fields. *Journal of Fluid Mechanics* **174**, 441–465.

Maxey, R. & Riley, J. 1983 Equation of motion of a small rigid sphere in a nonuniform flow. *Physics of Fluids* **26**, 883–889.

Maxwell, J. C. 1867 On the dynamical theory of gases. *Philosophical Transactions of the Royal Society London* **157**, 49–88.

Maxwell, J. C. 1879 On stresses in rarefied gases arising from inequalities of temperature. *Philosophical Transactions of the Royal Society London* **170**, 231–256.

Mazzei, L. 2011 Limitations of quadrature-based moment methods for modeling inhomogeneous polydisperse fluidized powders. *Chemical Engineering Science* **66**, 3628–3640.

Mazzei, L. & Lettieri, P. 2007 A drag force closure for uniformly dispersed fluidized suspensions. *Chemical Engineering Science* **62**, 6129–6142.

McCormack, F. J. 1973 Construction of linearized kinetic models for gaseous mixtures and molecular gases. *The Physics of Fluids* **16**, 2095–2105.

McGraw, R. 1997 Description of aerosol dynamics by the quadrature method of moments. *Aerosol Science and Technology* **27**, 255–265.

McGraw, R. 2006 Correcting moment sequences for errors associated with advective transport. http://www.ecd.bnl.gov/pubs/momentcorrection_mcgraw2006.pdf.

McGraw, R., Nemesure, S. & Schwartz, S. E. 1998 Properties and evolution of aerosols with size distribution having identical moments. *Journal of Aerosol Science* **29**, 761–772.

McGraw, R. & Wright, D. L. 2003 Chemically resolved aerosol dynamics for internal mixtures by the quadrature method of moments. *Journal of Aerosol Science* **34**, 189–209.

Mead, L. R. & Papanicolaou, N. 1984 Maximum entropy in the problem of moments. *Journal of Mathematical Physics* **25**, 2404–2417.

Megaridis, C. M. & Dobbins, R. A. 1990 A bimodal integral solution of the dynamic equation for an aerosol undergoing simultaneous particle inception and coagulation. *Aerosol Science and Technology* **12**, 240–255.

Meimaroglou, D. & Kiparissides, C. 2007 Monte Carlo simulation for the solution of the bi-variate dynamic population balance equation in batch particulate systems. *Chemical Engineering Science* **62**, 5295–5299.

Mendelson, H. D. 1967 The prediction of bubble terminal velocities from wave theory. *AIChE Journal* **13**, 250–253.

Michaelides, E. E. & Feng, Z.-G. 1994 Heat transfer from a rigid sphere in a nonuniform flow and temperature field. *International Journal of Heat and Mass Transfer* **37**, 2069–2076.

Millikan, R. A. 1910 The isolation of an ion, a precision measurement of its charge, and the correction of Stokes's law. *Science* **32**, 436–448.

Minier, J.-P. & Peirano, E. 2001 The PDF approach to turbulent polydispersed two-phase flows. *Physics Reports* **352**, 1–214.

Mishra, B. K. 2000 Monte Carlo simulation of particle breakage process during grinding. *Powder Technology* **110**, 246–252.

Monaco, R. & Preziosi, L. 1990 *Fluid Dynamic Applications of the Discrete Boltzmann Equation*. Singapore: World Scientific Publishing Co.

Monahan, S. M. & Fox, R. O. 2007 Effect of model formulation on flow-regime predictions for bubble columns. *AIChE Journal* **53**, 9–18.

Monahan, S. M., Vitankar, V. S. & Fox, R. O. 2005 CFD predictions of flow-regime transitions in bubble columns. *AIChE Journal* **51**, 1897–1923.

Montante, G., Paglianti, A. & Magelli, F. 2007 Experimental analysis and computational modelling of gas–liquid stirred vessels. *Chemical Engineering Research and Design* **85**, 647–653.

Moore, D. W. 1965 The velocity of rise of distorted gas bubbles in a liquid of small viscosity. *Journal of Fluid Mechanics* **23**, 749–766.

Moreno-Atanasio, R. & Ghadiri, M. 2006 Mechanistic analysis and computer simulation of impact breakage of agglomerates: effect of surface energy. *Chemical Engineering Science* **61**, 2476–2481.

Morsi, S. A. & Alexander, A. J. 1972 An investigation of particle trajectories in two-phase flow systems. *Journal of Fluid Mechanics* **55**, 193–208.

Mudde, R. F., Harteveld, W. K. & van den Akker, H. E. A. 2009 Uniform flow in bubble columns. *Industrial & Engineering Chemistry Research* **48** (1), 148–158.

Mydlarz, J. & Briedis, D. 1992 Growth rate dispersion vs size-dependent growth rate for MSMPR crystallizer data. *Computers and Chemical Engineering* **16**, 917–922.

Myerson, A. S. 2002 *Handbook of Industrial Crystallization*. Boston (MA): Butterworth-Heinemann.

Nagy, Z. K. 2009 Model based robust control approach for batch crystallization product design. *Computer and Chemical Engineering* **33**, 1685–1691.

Nandanwar, M. & Kumar, S. 2008 A new discretization of space for the solution of multi-dimensional population balance equations. *Chemical Engineering Science* **63**, 2198–2210.

Narsimhan, G., Gupta, J. P. & Ramkrishna, D. 1979 A model for transitional breakage probability of droplets in agitated lean liquid–liquid dispersion. *Chemical Engineering Science* **34**, 257–265.

Nyvlt, J., Matuchove, M. & Broul, M. 1985 *The Kinetics of Industrial Crystallization*. Amsterdam: Elsevier.

Ohara, M. & Reid, R. C. 1973 *Modeling Crystal Growth Rates from Solution*. Englewood Cliffs (NJ): Prentice-Hall.

Oles, V. 1991 Shear-induced aggregation and breakup of polystyrene latex particles. *Journal of Colloid and Interface Science* **154**, 351–362.

Oullion, M., Reynolds, G. K. & Hounslow, M. J. 2009 Simulating the early stage of high-shear granulation using a two-dimensional Monte-Carlo approach. *Chemical Engineering Science* **64**, 673–685.

Pandya, J. D. & Spielman, L. A. 1982 Particle breakage: theory and processing strategy. *Journal of Colloid and Interface Science* **90**, 517–531.

Pandya, R. V. R. & Mashayek, F. 2003a Kinetic equation for particle transport and heat transfer in nonisothermal turbulent flows. *AIAA Journal* **41** (5), 841–847.

Pandya, R. V. R. & Mashayek, F. 2003b Non-isothermal dispersed phase of particles in turbulent flow. *Journal of Fluid Mechanics* **475**, 205–245.

Passalacqua, A. & Fox, R. O. 2011 Advanced continuum modelling of gas–particle flows beyond the hydrodynamic limit. *Applied Mathematical Modelling* **35**, 1616–1627.

Passalacqua, A., Fox, R. O., Garg, R. & Subramaniam, S. 2010 A fully coupled quadrature-based moment method for dilute to moderately dilute fluid–particle flows. *Chemical Engineering Science* **65**, 2267–2283.

Passalacqua, A., Galvin, J. E., Vedula, P., Hrenya, C. M. & Fox, R. O. 2011 A quadrature-based kinetic model for dilute non-isothermal granular flows. *Communications in Computational Physics* **10**, 216–252.

Patruno, L. E., Dorao, C. A., Svendsen, H. F. & Jakobsen, H. A. 2009 Analysis of breakage kernels for population balance modelling. *Chemical Engineering Science* **64**, 501–508.

Peirano, E. & Minier, J.-P. 2002 Probabilistic formalism and hierarchy of models for polydispersed turbulent two-phase flows. *Physical Review E* **65**, 046301.

Peng, S. J. & Williams, R. A. 1994 Direct measurement of floc breakage in flowing suspension. *Journal of Colloid and Interface Science* **166**, 321–332.

Perthame, B. 2003 *Kinetic Formulation of Conservation Laws*. Oxford: Oxford University Press.

Petitti, M., Caramellino, M., Marchisio, D. L. & Vanni, M. 2007 Two-scale simulation of mass transfer in an agitated gas–liquid tank, in *6th International Conference on Multiphase Flow*, Leipzig, p. S6 Tue A 16.

Petitti, M., Marchisio, D. L., Vanni, M. *et al.* 2009 Effect of drag modeling on the prediction of critical regime transitions in agitated gas–liquid reactors with bubble size distribution modeling. *Multiphase Science and Technology* **21**, 95–106.

Petitti, M., Nasuti, A., Marchisio, D. L. *et al.* 2010 Bubble size distribution modeling in stirred gas–liquid reactors with QMOM augmented by a new correction algorithm. *AIChE Journal* **56**, 36–53.

Pinto, M. A., Immanuel, C. D. & Doyle III, F. J. 2007 A feasible solution technique for higher-dimensional population balance models. *Computers and Chemical Engineering* **31**, 1242–1256.

Pope, S. B. 2000 *Turbulent Flows*. Cambridge: Cambridge University Press.

Portela, L. M. & Oliemans, R. V. A. 2006 Possibilities and limitations of computer simulations of industrial turbulent dispersed multiphase flows. *Flow, Turbulence and Combustion* **77**, 381–403.

Prat, O. P. & Ducoste, J. J. 2006 Modeling spatial distribution of floc size in turbulent processes using the quadrature method of moment and computational fluid dynamics. *Chemical Engineering Science* **61**, 75–86.

Press, W. H., Teukolsky, S. A., Vetterling, W. T. & Flannery, B. P. 1992 *Numerical Recipes in Fortran 77: The Art of Scientific Computing*, 2nd edn., Cambridge: Cambridge University Press.

Prosperetti, A. & Tryggvason, G. 2009 *Computational Methods for Multiphase Flows*. Cambridge: Cambridge University Press.

Puel, F., Fevotte, G. & Klein, J. P. 2003a Simulation and analysis of industrial crystallization processes through multidimensional population balance equations. Part 1: a resolution algorithm based on the method of classes. *Chemical Engineering Science* **58**, 3715–3727.

Puel, F., Fevotte, G. & Klein, J. P. 2003b Simulation and analysis of industrial crystallization processes through multidimensional population balance equations. Part 2: a study of semi-batch crystalization. *Chemical Engineering Science* **58**, 3729–3740.

Qamar, S., Mukhtar, S., Ali, Q. & Seidel-Morgenstern, A. 2011a A Gaussian quadrature method for solving batch crystallization models. *AIChE Journal* **57**, 149–159.

Qamar, S., Noora, S., Rehman, M. & Seidel-Morgenstern, A. 2011b Numerical solution of a multi-dimensional batch crystallization model with fines dissolution. *Computers and Chemical Engineering* **35**, 412–422.

Qamar, S. & Seidel-Morgenstern, A. 2009 An efficient numerical technique for solving multi-dimensional batch crystallization models with size independent growth rates. *Computers and Chemical Engineering* **33**, 1221–1226.

Qamar, S. & Warnecke, G. 2007 Solving population balance equations for two-component aggregation by a finite volume scheme. *Chemical Engineering Science* **62**, 679–693.

Rajniak, P., Dhanasekharan, K., Sinka, C., MacPhail, N. & Chern, R. 2008 Modeling and measurement of granule attrition during pneumatic conveying in a laboratory scale system. *Powder Technology* **185**, 202–210.

Rajniak, P., Mancinelli, C., Chern, R. T. *et al.* 2007 Experimental study of wet granulation in fluidized bed: impact of the binder properties on the granule morphology. *International Journal of Pharmaceutics* **334**, 92–102.

Ramachandran, R., Immanuel, C. D., Stepanek, F., Litster, J. D. & Doyle III, F. J. 2009 A mechanistic model for breakage in population balances of granulation: theoretical kernel development and experimental validation. *Chemical Engineering Research and Design* **87**, 598–614.

Ramkrishna, D. 1981 Analysis of population balance 4. The precise connection between Monte-Carlo simulation and population balances. *Chemical Engineering Science* **36**, 1203–1209.

Ramkrishna, D. 2000 *Population Balances*. San Diego (CA): Academic Press.

Randolph, A. D. & Larson, L. A. 1971 *Theory of Particulate Processes: Analysis and Techniques of Continuous Crystallization*. London: Academic Press.

Ranz, W. E. & Marshall, W. R. 1952 Evaporation from drops. *Chemical Engineering Progress* **48**, 141–146.

Reeks, M. W. 1977 Dispersion of small particles suspended in an isotropic turbulent fluid. *Journal of Fluid Mechanics* **83**, 529–546.

Reeks, M. W. 1983 The transport of discrete particles in inhomogeneous turbulence. *Journal of Aerosol Science* **14** (6), 729–739.

Reeks, M. W. 1991 On a kinetic-equation for the transport of particles in turbulent flows. *Physics of Fluids A* **3** (3), 446–456.

Reeks, M. W. 1992 On the continuum equations for dispersed particles in nonuniform flows. *Physics of Fluids A* **4** (6), 1290–1303.

Reeks, M. W. 1993 On the constitutive relations for dispersed particles in nonuniform flows. 1. Dispersion in a simple shear-flow. *Physics of Fluids A* **5** (3), 750–761.

Reeks, M. W. 2005a On model equations for particle dispersion in inhomogeneous turbulence. *International Journal of Multiphase Flow* **31**, 93–114.

Reeks, M. W. 2005b On probability density function equations for particle dispersion in a uniform shear flow. *Journal of Fluid Mechanics* **522**, 263–302.

Reerink, H. & Overbeek, J. T. G. 1954 The rate of coagulation as a measure of the stability of silver iodide sols. *Discussions of the Faraday Society* **18**, 74–84.

Renksizbulut, M. & Yuen, M. 1983 Experimental study of droplet evaporation in a high-temperature air stream. *Journal of Heat Transfer* **105**, 384–388.

Riber, E., Moureau, V., Garcia, M., Poinsot, T. & Simonin, O. 2009 Evaluation of numerical strategies for large eddy simulation of particulate two-phase recirculating flows. *Journal of Computational Physics* **228** (2), 539–564.

Riboux, G., Risso, F. & Legendre, D. 2010 Experimental characterization of the agitation generated by bubbles rising at high Reynolds number. *Journal of Fluid Mechanics* **643**, 509–539.

Richardson, J. F. & Zaki, W. N. 1954 Sedimentation and fluidization. *Transactions of the Institute of Chemical Engineering* **32**, 35–53.

Rigopoulos, S. 2010 Population balance modelling of polydispersed particles in reactive flows. *Progress in Energy and Combustion Science* **36**, 412–443.

Rosner, D. & Yu, S. 2001 MC simulation of aerosol aggregation and simultaneous spheroidization. *AIChE Journal* **47**, 545–561.

Rosner, D. E., McGraw, R. & Tandon, P. 2003 Multivariate population balances via moment and Monte Carlo simulation methods: an important sol reaction engineering bivariate example and mixed moments for the estimation of deposition, scavenging, and optical properties for populations of nonspherical suspended particles. *Industrial & Engineering Chemistry Research* **42**, 2699–2711.

Sachdev, J. S., Groth, C. P. T. & Gottlieb, J. J. 2007 Numerical solution scheme for inert, disperse, and dilute gas–particle flows. *International Journal of Multiphase Flow* **33**, 282–299.

Sack, R. A. & Donovan, A. F. 1971 An algorithm for Gaussian quadrature given modified moments. *Numerische Mathematik* **18**, 465–478.

Saffman, P. G. 1965 The lift on a small sphere in a slow shear flow. *Journal of Fluid Mechanics* **22**, 385–400.

Saffman, P. G. & Turner, J. S. 1956 On the collision of drops in turbulent clouds. *Journal of Fluid Mechanics* **1**, 16–30.

Salazar, J. P. L. C. & Collins, L. R. 2009 Two-particle dispersion in isotropic turbulent flows. *Annual Review of Fluid Mechanics* **41**, 405–432.

Saliakas, V., Kotoulas, C., Meimaroglou, D. & Kiparissides, C. 2008 Dynamic evolution of the particle size distribution in suspension polymerization reactors: a comparative study on Monte Carlo and sectional grid methods. *Canadian Journal of Chemical Engineering* **86**, 924–936.

Sankaranarayanan, K., Shan, X., Kevrekidis, I. G. & Sundaresan, S. 2002 Analysis of drag and virtual mass forces in bubbly suspensions using an implicit formulation of the lattice Boltzmann method. *Journal of Fluid Mechanics* **452**, 61–96.

Sankaranarayanan, K. & Sundaresan, S. 2002 Lift force in bubbly suspensions. *Chemical Engineering Science* **57**, 3521–3542.

Sanyal, J., Marchisio, D. L., Fox, R. O. & Dhanasekharan, K. 2005 On the comparison between population balance models for CFD simulation of bubble columns. *Industrial & Engineering Chemistry Research* **44**, 5063–5072.

Sathyagal, A. N. & Ramkrishna, D. 1996 Droplet breakage in stirred dispersions. Breakage functions from experimental drop-size distributions. *Chemical Engineering Science* **51**, 1377–1391.

Sato, Y. & Sekoguchi, K. 1975 Liquid velocity distribution in two-phase bubble flow. *International Journal of Multiphase Flow* **2**, 79–95.

Saurel, R., Daniel, E. & Loraud, J. C. 1994 Two-phase flows: second-order schemes and boundary conditions. *AAIA Journal* **32**, 1214–1221.

Sazhin, S. S. 2006 Advanced models of fuel droplet heating and evaporation. *Progress in Energy and Combustion Science* **32**, 162–214.

Schiller, L. & Naumann, Z. 1935 Über die grundlegenden Berechnung bei der Schwerkraftaufbereitung. *Zeitschrift des Vereines deutscher Ingenieure* **77**, 318–320.

Serra, T. & Casamitjana, X. 1998 Effect of the shear and volume fraction on the aggregation and breakup of particles. *AIChE Journal* **44**, 1724–1730.

Shah, B. H., Borwanker, J. D. & Ramkrishna, D. 1976 Monte-Carlo simulation of microbial-population growth. *Mathematical Biosciences* **31**, 1–23.

Shah, B. H., Ramkrishna, D. & Borwanker, J. D. 1977 Simulation of particulate systems using concept of interval of quiescence. *AIChE Journal* **23**, 897–904.

Shohat, J. A. & Tamarkin, J. D. 1943 *The Problem of Moments*. New York (NY): American Mathematical Society.

Silva, L. F. L. R., Rodrigues, R. C., Mitre, J. F. & Lage, P. L. C. 2010 Comparison of the accuracy and performance of quadrature-based methods for population balance problems with simultaneous breakage and aggregation. *Computer and Chemical Engineering* **34**, 286–297.

Simonin, O., Deutsch, E. & Minier, J. P. 1993 Eulerian prediction of the fluid/particle correlated motion in turbulent two-phase flows. *Applied Scientific Research* **51**, 275–283.

Sirovich, L. 1962 Kinetic modeling of gas mixtures. *Physics of Fluids* **5**, 908–918.

Smith, M. & Matsoukas, T. 1998 Constant-number Monte Carlo simulation of population balances. *Chemical Engineering Science* **53**, 1777–1786.

von Smoluchowski, M. 1917 Versuch einer mathematischen Theorie der Koagulationskinetik kolloidaler Lösungen. *Zeitschrift für Physikalische Chemie* **92**, 129–168.

Sommerfeld, M. & Ho, C. A. 2003 Numerical calculation of particle transport in turbulent wall bounded flows. *Powder Technology* **131**, 1–6.

Song, M. & Qiu, X.-J. 1999 An alternative to the concept of the interval of quiescence (IQ) in the Monte Carlo simulation of population balances. *Chemical Engineering Science* **54**, 5711–5715.

Spicer, P. T. & Pratsinis, S. E. 1996 Coagulation and fragmentation: universal steady-state particle-size distribution. *AIChE Journal* **42**, 1612–1620.

Spielman, L. A. & Levenspiel, O. 1965 A Monte-Carlo treatment for reacting and coalescing dispersed phase systems. *Chemical Engineering Science* **20**, 247–254.

Stokes, G. G. 1851 On the effect of the internal friction of fluids on the motion of pendulums. *Transactions of the Cambridge Philosophical Society* **9**, 8–106.

STRUCHTRUP, H. 2005 *Macroscopic Transport Equations for Rarefied Gas Flows*. Berlin: Springer.

STRUMENDO, M. & ARASTOOPOUR, H. 2008 Solution of PBE by MOM in finite size domains. *Chemical Engineering Science* **63** (10), 2624–2640.

STRUMENDO, M. & ARASTOOPOUR, H. 2009 Solution of bivariate population balance equations using the finite size domain complete set of trial functions method of moments (FCMOM). *Industrial & Engineering Chemistry Research* **48**, 262–273.

STRUMENDO, M. & ARASTOOPOUR, H. 2010 Solution of population balance equations by the finite size domain complete set of trial functions method of moments (FCMOM) for inhomogeneous systems. *Industrial & Engineering Chemistry Research* **49**, 5222–5230.

SU, J., GU, Z., LI, Y., FENG, S. & XU, X. Y. 2007 Solution of population balance equation using quadrature method of moments with an adjustable factor. *Chemical Engineering Science* **62**, 5897–5911.

SU, J., GU, Z., LI, Y., FENG, S. & XU, X. Y. 2008 An adaptive direct quadrature method of moment for population balance equations. *AIChE Journal* **54**, 2872–2887.

SUBRAMANIAM, S. 2000 Statistical representation of a spray as a point process. *Physics of Fluids* **12**, 2413–2431.

SWIFT, P. T. & FRIEDLANDER, S. K. 1964 The coagulation of hydrosols by Brownian motion and laminar shear flow. *Journal of Colloid and Interface Science* **19**, 621–647.

SYAMLAL, M. & O'BRIEN, T. J. 1989 Computer simulation of bubbles in a fluidized bed. *AIChE Symposium Series* **62**, 22–31.

TAGLIANI, A. 1999 Hausdorff moment problem and maximum entropy: a unified approach. *Applied Mathematics and Computation* **105**, 291–305.

TALBOT, L. 1981 Thermophoresis – a review, in *12th International Symposium on Rarefied Gas Dynamics*, New York (NY): American Institute of Aeronautics and Astronautics, pp. 467–488.

TANASAWA, I. 1991 Advances in condensation heat transfer. *Advances in Heat Transfer* **21**, 55–139.

TANDON, P. & ROSNER, D. E. 1999 Monte Carlo simulation of particle aggregation and simultaneous restructuring. *Journal of Colloid and Interface Science* **213**, 273–286.

TENNETI, S., FOX, R. O. & SUBRAMANIAM, S. 2012 Stochastic acceleration model for gas–solid flows at moderate Reynolds numbers. *Journal of Fluid Mechanics* (submitted).

TENNETI, S., GARG, R., HRENYA, C. M., FOX, R. O. & SUBRAMANIAM, S. 2010 Direct numerical simulation of gas–solid suspensions at moderate Reynolds number: quantifying the coupling between hydrodynamic forces and particle velocity fluctuations. *Powder Technology* **203**, 57–69.

TERRAZAS-VELARDE, K., PEGLOW, M. & TSOTSAS, E. 2009 Stochastic simulation of agglomerate formation in fluidized bed spray drying: a micro-scale approach. *Chemical Engineering Science* **64**, 2631–2643.

Tomiyama, A. 2004 Drag, lift and virtual mass forces acting on a single bubble, in *3rd International Symposium on Two-Phase Flow: Modeling and Experiments*, Pisa.

Tomiyama, A., Celata, G. P., Hosokawa, S. & Yoshida, S. 2002a Terminal velocity of single bubbles in surface tension force dominant regime. *International Journal of Multiphase Flow* **28**, 1497–1519.

Tomiyama, A., Kataoka, I., Zun, I. & Sakaguchi, T. 1998 Drag coefficient of single bubbles under normal and microgravity conditions. *The J.S.M.E. International Journal* **41**, 472–479.

Tomiyama, A., Tamai, H., Zun, I. & S., H. 2002b Transverse migration of single bubbles in simple shear flows. *Chemical Engineering Science* **57**, 1849–1858.

Tong, C. 2001 Measurements of conserved scalar filtered density function in a turbulent jet. *Physics of Fluids* **13**, 2923–2937.

Torobin, L. B. & Gauvin, W. H. 1960 Fundamental aspects of solids–gas flow. Part V: the effects of fluid turbulence on the particle drag coefficient. *Canadian Journal of Chemical Engineering* **38**, 189–200.

Truesdell, C. & Muncaster, R. G. 1980 *Fundamentals of Maxwell's Kinetic Theory of a Simple Monoatomic Gas*. New York (NY): Academic Press.

Upadhyay, R. R. & Ezekoye, O. A. 2003 Evaluation of the 1-point quadrature approximation in QMOM for combined aerosol growth laws. *Journal of Aerosol Science* **34**, 1665–1683.

Upadhyay, R. R. & Ezekoye, O. A. 2005 Smoke buildup and light scattering in a cylindrical cavity above a uniform flow. *Journal of Aerosol Science* **36**, 471–493.

Vale, M. & McKenna, F. 2005 Solution of the population balance equation for two-component aggregation by an extended fixed pivot technique. *Industrial & Engineering Chemistry Research* **44**, 7885–7891.

Valentas, K. J., Bilous, O. & Amundson, N. R. 1966 Analysis of breakage in dispersed phase systems. *Industrial and Engineering Chemistry Fundamentals* **5**, 271–279.

Van Peborgh Gooch, J. R. & Hounslow, M. J. 1996 Monte Carlo simulation of size-enlargement mechanisms in crystallization. *AIChE Journal* **42**, 1864–1874.

Vanni, M. 1999 Discretized procedure for the breakage equation. *AIChE Journal* **45**, 916–919.

Vanni, M. 2000a Approximate population balance equation for aggregation–breakage processes. *Journal of Colloid and Interface Science* **221**, 143–160.

Vanni, M. 2000b Creeping flow over spherical permeable aggregates. *Chemical Engineering Science* **55**, 685–698.

Vanni, M. & Baldi, G. 2002 Coagulation efficiency of colloidal particles in shear flow. *Advances in Colloid and Interface Science* **97**, 151–177.

Venables, D. S. & Schmuttenmaer, C. A. 2000 Spectroscopy and dynamics of mixtures of water with acetone, acetonitrile, and methanol. *Journal of Chemical Physics* **113**, 11222.

Venneker, B. C. H., Derksen, J. J. & Van den Akker, H. E. A. 2002 Population balance modelling of aerated stirred vessels based on CFD. *AIChE Journal* **48**, 673–685.

Vikas, V., Wang, Z. J. & Fox, R. O. 2012 Realizable high-order finite-volume schemes for quadrature-based moment methods applied to diffusion population balance equations. *Journal of Computational Physics* (submitted).

Vikas, V., Wang, Z. J., Fox, R. O. & Passalacqua, A. 2009 High-order realizable finite-volume schemes for quadrature-based moment methods, in *Proceedings of the 48th AIAA Aerospace Sciences Meeting*, pp. 1–2.

Vikas, V., Wang, Z. J., Passalacqua, A. & Fox, R. O. 2011a Realizable high-order finite-volume schemes for quadrature-based moment methods. *Journal of Computational Physics* **230**, 5328–5352.

Vikas, V., Yuan, C., Wang, Z. J. & Fox, R. O. 2011b Modeling of bubble-column flows using quadrature-based moment methods. *Chemical Engineering Science* **66**, 3058–3070.

Vikhansky, A. & Kraft, M. 2004 A Monte Carlo methods for identification and sensitivity analysis of coagulation processes. *Journal of Computational Physics* **200**, 50–59.

Walpole, R. E. & Myers, R. H. 1985 *Probability and Statistics for Engineers and Scientists*. New York (NY): Macmillan Publishing Company.

Wang, L. & Fox, R. O. 2003 Application of in-situ adaptive tabulation to CFD simulation of nanoparticle formation by reactive precipitation. *Chemical Engineering Science* **58**, 4387–4401.

Wang, L., Marchisio, D. L., Vigil, R. D. & Fox, R. O. 2005a CFD simulation of shear-induced aggregation and breakage in laminar Taylor–Couette flow. *Journal of Colloid and Interface Science* **282**, 380–396.

Wang, L., Vigil, R. D. & Fox, R. O. 2005b CFD simulation of shear-induced aggregation and breakage in turbulent Taylor–Couette flow. *Journal of Colloid and Interface Science* **285**, 167–178.

Wang, T. F., Wang, J. F. & Jin, Y. 2003 A novel theoretical breakup kernel function for bubbles/droplets in a turbulent flow. *Chemical Engineering Science* **58**, 4629–4637.

Wang, X. Z., Roberts, K. J. & Ma, C. 2008 Crystal growth measurement using 2D and 3D imaging and the perspectives for shape control. *Chemical Engineering Science* **63**, 1173–1184.

Welander, P. 1954 On the temperature jump in a rarefied gas. *Arkiv för Fysik* **7**, 507–553.

Wellek, R. M., Arawal, A. K. & Skelland, A. H. P. 1966 Shapes of liquid drops moving in liquid media. *AIChE Journal* **12**, 854–862.

Wen, C. Y. & Yu, Y. H. 1966 Mechanics of fluidization. *Chemical Engineering Progress Symposium Series* **62**, 100–111.

Weroński, P. & Elimelech, M. 2008 Novel numerical method for calculating initial flux of colloid particle adsorption through an energy barrier. *Journal of Colloid and Interface Science* **319**, 406–415.

Wheeler, J. C. 1974 Modified moments and Gaussian quadratures. *Rocky Mountain Journal of Mathematics* **4**, 287–296.

Wildeboer, W. J., Litster, J. D. & Cameron, I. T. 2005 Modelling nucleation in wet granulation. *Chemical Engineering Science* **60**, 3751–3761.

Wilf, H. S. 1962 *Mathematics for the Physical Sciences*. New York (NY): John Wiley and Sons.

Williams, F. A. 1985 *Combustion Theory*, 2nd edn., Redwood City (CA): Addison-Wesley.

Wojcik, J. A. & Jones, A. G. 1998 Particle disruption of precipitated $CaCO_3$ crystal agglomerates in turbulently agitated suspensions. *Chemical Engineering Science* **53**, 1097–1101.

Wright, D. L. 2007 Numerical advection of moments of the particle size distribution in Eulerian models. *Journal of Aerosol Science* **38**, 352–369.

Wright, D. L., Kasibhatla, P. S., McGraw, R. & Schwartz, S. E. 2001a Description and evaluation of a six-moment aerosol microphysical module for use in atmospheric chemical transport models. *Journal of Geophysical Research – Atmospheres* **106**, 20275–20291.

Wright, D. L., McGraw, R., Benkovitz, C. M. & Schwartz, S. E. 2000 Six-moment representation of multiple aerosol populations in a sub-hemispheric chemical transformation model. *Geophysical Research Letters* **27**, 967–970.

Wright, D. L., McGraw, R. & Rosner, D. E. 2001b Bivariate extension of the quadrature method of moments for modeling simultaneous coagulation and sintering particle populations. *Journal of Colloid and Interface Science* **236**, 242–251.

Wu, J. J., Nguyen, H. V., Flagan, R. C., Okuyama, K. & Kousaka, Y. 1988 Evolution and control of particle properties in aerosol reactors. *AIChE Journal* **34**, 1249–1256.

Wulkow, M., Gerstlauer, A. & Nieken, U. 2001 Modeling and simulation of crystallization processes using Parsival. *Chemical Engineering Science* **56**, 2575–2588.

Wynn, E. J. W. 2004 Simulating aggregation and reaction: new Hounslow DPB and four-parameter summary. *AIChE Journal* **50**, 578–588.

Xiong, Y. & Pratsinis, S. E. 1991 Gas phase production of particles in reactive turbulent flows. *Journal of Aerosol Science* **22**, 637–655.

Xiong, Y. & Pratsinis, S. E. 1993 Formation of agglomerate particles by coagulation and sintering – part I. A two-dimensional solution of the population balance equation. *Journal of Aerosol Science* **24**, 283–300.

Xu, G. W. & Kato, K. 1999 Hydrodynamic equivalent diameter for clusters in heterogeneous gas–solid flow. *Chemical Engineering Science* **54**, 1837–1847.

Yearling, P. R. & Gould, R. D. S. 1993 Experimental determination of gas-phase turbulence effects on vaporizing water and alcohol liquid droplets, in *Proceedings of the 1993 ASME Winter Annual Meeting, New Orleans, LA*, New York (NY): ASME, pp. 19–24.

Yin, X. & Sundaresan, S. 2009 Fluid–particle drag in low-Reynolds-number polydisperse gas–solid suspensions. *AIChE Journal* **55** (6), 1352–1368.

Yoon, C. & McGraw, R. 2004a Representation of generally mixed multivariate aerosols by the quadrature method of moments: I. Statistical foundation. *Journal of Aerosol Science* **35**, 561–576.

Yoon, C. & McGraw, R. 2004b Representation of generally mixed multivariate aerosols by the quadrature method of moments: II. Aerosol dynamics. *Journal of Aerosol Science* **35**, 577–598.

Yuan, C. & Fox, R. O. 2011 Conditional quadrature method of moments for kinetic equations. *Journal of Computational Physics* **230**, 8216–8246.

Yuan, C., Laurent, F. & Fox, R. O. 2012 An extended quadrature method of moments for population balance equations. *Journal of Aerosol Science* **51**, 1–23.

Zaccone, A., Gäbler, A., Maass, S., Marchisio, D. & Kraume, M. 2007 Drop breakage in liquid–liquid stirred dispersions: modelling of single drop breakage. *Chemical Engineering Science* **62**, 6297–6307.

Zaichik, L., Oesterle, B. & Alipchenkov, V. 2004 On the probability density function model for the transport of particles in anisotropic turbulent flow. *Physics of Fluids* **16** (6), 1956–1964.

Zaichik, L. I. 1997 Modelling of the motion of particles in non-uniform turbulent flow using the equation for the probability density function. *PMM Journal of Applied Mathematics and Mechanics* **61** (1), 127–133.

Zaichik, L. I. 1999 A statistical model of particle transport and heat transfer in turbulent shear flows. *Physics of Fluids* **11**, 1521–1534.

Zaichik, L. I. & Alipchenkov, V. M. 1998 Kinetic equation for the probability density function of velocity and temperature of particles in an inhomogeneous turbulent flow: analysis of flow in a shear layer. *High Temperature* **36** (4), 572–582.

Zaichik, L. I., Alipchenkov, V. M. & Sinaiski, E. G. 2008 *Particles in Turbulent Flows*. Weinheim: Wiley-VCH.

Zaichik, L. I., Simonin, O. & Alipchenkov, V. M. 2003 Two statistical models for predicting collision rates of inertial particles in homogeneous isotropic turbulence. *Physics of Fluids* **15** (10), 2995–3005.

Zauner, R. & Jones, A. G. 2000 Determination of nucleation, growth, agglomeration and disruption kinetics from experimental precipitation data: the calcium oxalate system. *Chemical Engineering Science* **55**, 4219–4232.

Zeitlin, M. A. & Tavlarides, L. L. 1972a Dispersed phase reactor model for predicting conversion and mixing effects. *AIChE Journal* **16**, 1268–1271.

Zeitlin, M. A. & Tavlarides, L. L. 1972b Fluid–fluid interactions and hydrodynamics in agitated dispersions – simulation model. *Canadian Journal of Chemical Engineering* **50**, 207–215.

Zhang, Y. & Doherty, M. F. 2004 Simultaneous prediction of crystal shape and size for solution crystallization. *AIChE Journal* **50**, 2101–2112.

Zhao, H., Kruis, F. E. & Zheng, C. 2009 Reducing statistical noise and extending the size spectrum by applying weighted simulation particles in Monte Carlo simulation of coagulation. *Aerosol Science and Technology* **43**, 781–793.

Zhao, H., Kruis, F. E. & Zheng, C. 2010 A differentially weighted Monte Carlo method for two-component coagulation. *Journal of Computational Physics* **229**, 6931–6945.

Zhao, H., Maisels, A., Matsoukas, T. & Zheng, C. 2007 Analysis of four Monte Carlo methods for the solution of population balances in dispersed systems. *Powder Technology* **173**, 38–50.

Zhao, H. & Zheng, C. 2008 A stochastic simulation for the collection process of fly ashes in single-stage electrostatic precipitators. *Fuel* **87**, 2082–2089.

Zhao, H. & Zheng, C. 2009a Correcting the multi-Monte Carlo method for particle coagulation. *Powder Technology* **193**, 120–123.

Zhao, H. & Zheng, C. 2009b A new event-driven constant-volume method for solution of the time evolution of particle size distribution. *Journal of Computational Physics* **228**, 1412–1428.

Zhao, H. B., Zheng, C. G. & Xu, M. H. 2005a Multi-Monte Carlo approach for general dynamic equation considering simultaneous particle coagulation and breakage. *Powder Technology* **154**, 164–178.

Zhao, H. B., Zheng, C. G. & Xu, M. H. 2005b Multi-Monte Carlo method for particle coagulation: description and validation. *Applied Mathematics and Computation* **167**, 1383–1399.

Zou, Y., Kavousanakis, M. E., Kevrekidis, I. G. & Fox, R. O. 2010 Coarse-grained computation for particle coagulation and sintering processes by linking quadrature method of moments with Monte-Carlo. *Journal of Computational Physics* **229**, 5299–5314.

Zucca, A. 2005 *Modelling of Turbulence–Chemistry Interaction and Soot Formation in Turbulent Flames* (PhD Thesis), Turin: Politecnico di Torino.

Zucca, A., Marchisio, D. L., Barresi, A. A. & Fox, R. O. 2006 Implementation of the population balance equation in CFD codes for modelling soot formation in turbulent flames. *Chemical Engineering Science* **61**, 87–95.

Zucca, A., Marchisio, D. L., Vanni, M. & Barresi, A. A. 2007 Validation of bivariate DQMOM for nanoparticle processes simulation. *AIChE Journal* **53**, 918–931.

Zumstein, R. C. & Rousseau, R. W. 1987 Growth rate dispersion by initial growth rate distributions and growth rate fluctuations. *AIChE Journal* **33**, 121–129.

Index